PROBLEMS &
SOLUTIONS IN
GROUP THEORY
FOR PHYSICISTS

PROBLEMS &
SOLUTIONS IN
GROUP THEORY
FOR PHYSICISTS

Zhong-Qi Ma
Xiao-Yan Gu

Institute of High Energy Physics
China

World Scientific

NEW JERSEY • LONDON • SINGAPORE • BEIJING • SHANGHAI • HONG KONG • TAIPEI • CHENNAI

Published by

World Scientific Publishing Co. Pte. Ltd.

5 Toh Tuck Link, Singapore 596224

USA office: 27 Warren Street, Suite 401-402, Hackensack, NJ 07601

UK office: 57 Shelton Street, Covent Garden, London WC2H 9HE

Library of Congress Cataloging-in-Publication Data
Ma, Zhongqi, 1940–
   Problems and solutions in group theory for physicists / by Z.Q. Ma and X.Y. Gu.
      p. cm.
   Includes bibliographical references and index.
   ISBN-13 978-981-238-832-2 (alk. paper) -- ISBN-10 981-238-832-X (alk. paper)
   ISBN-13 978-981-238-833-9 (pbk.: alk. paper) -- ISBN-10 981-238-833-8 (pbk.: alk. paper)
   1. Group theory. 2. Mathematical physics. I. Gu, X.Y. (Xiao-Yan). II. Title.
   QC20.7.G76 M3 2004
   530.15'22-dc22   2004041980

British Library Cataloguing-in-Publication Data
A catalogue record for this book is available from the British Library.

Preface

Group theory is a powerful tool for studying the symmetry of a physical system, especially the symmetry of a quantum system. Since the exact solution of the dynamic equation in the quantum theory is generally difficult to obtain, one has to find other methods to analyze the property of the system. Group theory provides an effective method by analyzing symmetry of the system to obtain some precise information of the system verifiable with observations. Now, Group Theory is a required course for graduate students major in physics and theoretical chemistry.

The course of Group Theory for the students major in physics is very different from the same course for those major in mathematics. A graduate student in physics needs to know the theoretical framework of group theory and more importantly to master the techniques in application of group theory to various fields of physics, which is actually his main objective for taking the class. However, no course or textbook on group theory can be expected to include all explicit solutions to every problem of group theory in his research field of physics. A student of physics has to know the fundamental theory of group theory, otherwise he may not be able to apply the techniques creatively. On the other hand, physics students are not expected to completely grasp all the mathematics behind group theory due to the breadth of the knowledge required.

One of the authors (Ma) first taught the group theory course in 1962. Since 1986, he has been teaching Group Theory to graduate students mainly major in physics at the Graduate School of Chinese Academy of Sciences. In addition, most of his research work has been related to applications of group theory to physics. In 1996 the Chinese Academy of Sciences decided to publish a series of textbooks for graduate students. He was invited to write a textbook on group theory for the series. In his book, based on his experience in teaching and research work, he explained the fundamental concepts and techniques of group theory progressively and systematically using the language familiar to physicists, and also emphasized the ways with which group theory is applied to physics. The textbook (in Chinese) has been widely used for Group Theory classes in China since it was published by Science Press in Beijing six years ago. He is honored and flattered by the tremendous reception the book has received.

By the request of the readers, an exercise book on group theory by the same author was published in 2002 by Science Press to form a complete set of textbooks on group theory. In order to make the exercise book self-contained, a brief review of the main concepts and techniques is given before the problems in each section. The reviews can be used as a concise textbook on group theory. The present book is the new edition of that book. A great deal of new materials drawn from teaching and research experiences since the publication of the previous edition are included. The reviews of each chapter has been extensively revised. Last four chapters are essentially new.

This book consists of ten chapters. Chapter 1 is a short review on linear algebra. The reader is required not only to be familiar with its basic concepts but also to master its applications, especially the similarity transformation method. In Chapter 2, the concepts of a group and its subsets are studied through examples of some finite groups, where the importance of the multiplication table of a finite group is emphasized. Readers should pay special attention to Problem 17 of Chapter 2 and Problem 14 of Chapter 3 which demonstrate a systematic method for analyzing a finite group. The theory of representations of a group is studied in Chapter 3. The transformation operator P_R for the scalar functions bridges the gap between the representation theory and the physical application. The subduced and induced representations of groups are used to construct the character tables of finite groups in Chapter 3, and to study the outer product of representations of the permutation groups in Chapter 6. The symmetry groups **T** for a tetrahedron, **O** for a cube and **I** for an icosahedron are studied in Chapters 3 and 4. The Clebsch-Gordan series and coefficients are introduced in Chapter 3 and are calculated for various situations in the subsequent chapters. The calculated results of the Clebsch-Gordan coefficients and the Clebsch-Gordan series for the group **I** and for the permutation groups listed in Problem 27 of Chapter 3 and Problem 31 of Chapter 6, due to its complexity, are only for reference.

The classification and representations of semisimple Lie algebras are introduced in Chapter 7 and partly in Chapter 4 by the language familiar to physicists. The methods of block weight diagrams and dominant weight diagrams are recommended for calculating the representation matrices of the generators and the Clebsch-Gordan series in a simple Lie algebra. The readers who are interested in the strict mathematical definitions and proofs in the theory of semisimple Lie algebras are recommended to read the more mathematically oriented books, e.g. [Bourbaki (1989)].

The remaining part of the book is devoted to the properties of some important symmetry groups of physical systems. In Chapter 4 the symmetry group SO(3) of a spherically symmetric system in three dimensions is studied. The unitary representations with infinite dimensions of the non-compact group are discussed with the simplest example SO(2,1) in Problem 25 of Chapter 4. In Chapter 5 we introduce the symmetry of the crystals. More attention should be paid to the analysis method for the symmetry of a crystal from its International Notation (Problem 6). The commonly used matrix groups are studied in the last three chapters, while the Lorentz group is briefly discussed in Chapter 9.

The systematic examination of Young operators is an important characteristic of this book. We calculate the characters, the representation matrices, and the outer product of the irreducible representations of the permutation groups using the Young operators. The method of block weight diagrams can only symbolically give the basis states in the representation space of a Lie algebra. However, for the matrix groups $SU(N)$, $SO(N)$ and $Sp(2\ell)$, which are related to four classical Lie algebras, the basis states can be explicitly calculated using the Young operators. The relationship between two methods for the irreducible representations of the classical Lie algebras is demonstrated in the last three chapters in detail. The dimensions of the irreducible representations of the permutation groups, the $SU(N)$ groups, the $SO(N)$ groups, and the $Sp(2\ell)$ groups are all calculated with the hook rule, a method based on the Young diagram.

In summary, this book is written mainly for physics students and young physicists. Great care has been taken to make the book as self-contained as possible. However, this book reflects mainly the experiences of the authors. We sincerely welcome any suggestions and comments from the readers. This book was supported by the National Natural Science Foundation of China.

Institute of High Energy Physics
Beijing, China
December, 2003

Zhong-Qi Ma
Xiao-Yan Gu

Contents

Chapter 1

REVIEW ON LINEAR ALGEBRAS

1.1 Eigenvalues and Eigenvectors of a Matrix

★ The eigenequation of a matrix R is

$$R\mathbf{a} = \lambda \mathbf{a}, \tag{1.1}$$

where λ is the eigenvalue and $\mathbf{a}$ is the eigenvector for the eigenvalue. An eigenvector is determined up to a constant factor. A null vector is a trivial eigenvector of any matrix. We only discuss nontrivial eigenvectors.

Equation (1.1) is a set of linear homogeneous equations with respect to the components a_μ. The necessary and sufficient condition for the existence of a nontrivial solution to Eq. (1.1) is the vanishing of the coefficient determinant:

$$\det(R - \lambda \mathbf{1}) = \begin{vmatrix} (R_{11} - \lambda) & R_{12} & \ldots & R_{1m} \\ R_{21} & (R_{22} - \lambda) & \ldots & R_{2m} \\ \ldots & \ldots & \ldots & \ldots \\ R_{m1} & R_{m2} & \ldots & (R_{mm} - \lambda) \end{vmatrix} \tag{1.2}$$

$$= (-\lambda)^m + (-\lambda)^{m-1} \mathrm{Tr}R + \ldots + \det R = 0,$$

where m is the dimension of the matrix R, $\mathrm{Tr}R$ is its trace (the sum of the diagonal elements), and $\det R$ denotes its determinant. Equation (1.2) is called the secular equation of R. Evidently, this is an algebraic equation of order m with respect to λ, and there are m complex roots including multiple roots. Each root is an eigenvalue of the matrix. For a given eigenvalue λ, there is at least one eigenvector $\mathbf{a}$ obtained from solving Eq. (1.1). However, for a root λ with multiplicity n, it is not certain to obtain n linearly independent eigenvectors from Eq. (1.1).

The rank of a matrix R is said to be r if only r vectors among m row-vectors (or column-vectors) of R are linearly independent. If the rank of $(R - \lambda \mathbf{1})$ is r, then $(m - r)$ linearly independent eigenvectors can be obtained from Eq. (1.1). Any linear combination of eigenvectors for a given eigenvalue is still an eigenvector for this eigenvalue.

For a lower-dimensional matrix or for a sparse matrix (with many zero matrix entries), its eigenvalues and eigenvectors can be obtained by guess from experience or from some known results. So long as the eigenvalue and the eigenvector satisfy the eigenequation (1.1), they are the correct results. On the other hand, even if the results are calculated, they should also be checked whether Eq. (1.1) is satisfied.

★ A matrix R is said to be hermitian if $R^\dagger = R$. R is said to be unitary if $R^\dagger = R^{-1}$. A real and hermitian matrix is a real symmetric matrix. A real and unitary matrix is a real orthogonal matrix. R is said to be positive definite if its eigenvalues are all positive. R is said to be positive semi-definite if its eigenvalues are non-negative. A negative definite or negative semidefinite matrix can be defined similarly.

1. Prove that the sum of the eigenvalues of a matrix is equal to the trace of the matrix, and the product of eigenvalues is equal to the determinant of the matrix.

Solution. The secular equation of an m-dimensional matrix R is an algebraic equation of order m. Its m complex roots, including multiple roots, are the eigenvalues of the matrix R. So, the secular equation can also be expressed in the following way:

$$
\begin{aligned}
\det(R - \lambda \mathbf{1}) &= \prod_{j=1}^{m} (\lambda_j - \lambda) \\
&= (-\lambda)^m + (-\lambda)^{m-1} \sum_{j=1}^{m} \lambda_j + \ \cdots \ + \prod_{j=1}^{m} \lambda_j \\
&= 0.
\end{aligned}
$$

Comparing it with Eq. (1.2), we obtain that the sum of the eigenvalues is the trace of the matrix and the product of the eigenvalues is equal to the determinant of the matrix.

2. Calculate the eigenvalues and eigenvectors of the Pauli matrices σ_1 and σ_2:

$$\sigma_1 = \begin{pmatrix} 0 & 1 \\ 1 & 0 \end{pmatrix}, \quad \sigma_2 = \begin{pmatrix} 0 & -i \\ i & 0 \end{pmatrix}.$$

Solution. The action of σ_1 on a vector is to interchange two components of the vector. A vector is an eigenvector of σ_1 for the eigenvalue 1 if its two components are equal. A vector is an eigenvector of σ_1 for the eigenvalue -1 if its two components are different by sign.

$$\sigma_1 \begin{pmatrix} 1 \\ 1 \end{pmatrix} = \begin{pmatrix} 1 \\ 1 \end{pmatrix}, \quad \sigma_1 \begin{pmatrix} 1 \\ -1 \end{pmatrix} = - \begin{pmatrix} 1 \\ -1 \end{pmatrix}.$$

Similarly, if two components of a vector are different by a factor $\pm i$, then it is an eigenvector of σ_2:

$$\sigma_2 \begin{pmatrix} 1 \\ i \end{pmatrix} = \begin{pmatrix} 1 \\ i \end{pmatrix}, \quad \sigma_2 \begin{pmatrix} 1 \\ -i \end{pmatrix} = - \begin{pmatrix} 1 \\ -i \end{pmatrix}.$$

Sometimes a matrix R may contain a submatrix σ_1 or σ_2 in the form of direct sum or direct product. Thus, some eigenvalues and eigenvectors of R can be obtained in terms of the above results. A matrix R is said to contain a two-dimensional submatrix in the form of direct sum if for given a and b, $R_{ac} = R_{ca} = R_{bc} = R_{cb} = 0$, where c is not equal to a and b.

3. Calculate the eigenvalues and eigenvectors of the matrix R

$$R = \begin{pmatrix} 0 & 0 & 0 & 1 \\ 0 & 0 & 1 & 0 \\ 0 & 1 & 0 & 0 \\ 1 & 0 & 0 & 0 \end{pmatrix}.$$

Solution. R can be regarded as the direct sum of two submatrices σ_1, one lies in the first and fourth rows (columns), the other in the second and third rows (columns). From the result of Problem 2, two eigenvalues of R are 1, the remaining two are -1. The relative eigenvectors are as follows.

$$1: \quad \begin{pmatrix} 1 \\ 0 \\ 0 \\ 1 \end{pmatrix}, \begin{pmatrix} 0 \\ 1 \\ 1 \\ 0 \end{pmatrix}, \quad -1: \quad \begin{pmatrix} 1 \\ 0 \\ 0 \\ -1 \end{pmatrix}, \begin{pmatrix} 0 \\ 1 \\ -1 \\ 0 \end{pmatrix}.$$

4. Calculate the eigenvalues and eigenvectors of the matrix R

$$R = \begin{pmatrix} 0 & 0 & 1 \\ 1 & 0 & 0 \\ 0 & 1 & 0 \end{pmatrix}.$$

Solution. This is the generalization of σ_1. The action of R on a vector is to make a cycle on the three components of the vector in order. $R^3 = \mathbf{1}$, so the eigenvalues of the matrix R is 1, ω and ω^2, where $\omega = \exp\{-i2\pi/3\}$. The ratio of two adjacent components is the eigenvalue. The eigenvector for each eigenvalue is:

$$1: \quad \begin{pmatrix} 1 \\ 1 \\ 1 \end{pmatrix}, \qquad \omega: \quad \begin{pmatrix} 1 \\ \omega^2 \\ \omega \end{pmatrix}, \qquad \omega^2: \quad \begin{pmatrix} 1 \\ \omega \\ \omega^2 \end{pmatrix}.$$

5. If $\det R \neq 0$, prove that both $R^\dagger R$ and $RR^\dagger$ are positive definite hermitian matrices.

Solution. It is obvious that both $R^\dagger R$ and $RR^\dagger$ are hermitian. Letting λ and $\underline{a}$ be the eigenvalue and the eigenvector of $R^\dagger R$, where the eigenvector is written in the form of a column matrix:

$$(R^\dagger R)\underline{a} = \lambda \underline{a}, \qquad \underline{a}^\dagger \underline{a} > 0,$$

we have

$$\lambda \{\underline{a}^\dagger \underline{a}\} = \underline{a}^\dagger (R^\dagger R)\underline{a} = (R\underline{a})^\dagger (R\underline{a}) \geq 0,$$

namely, $\lambda \geq 0$. Since $\det R \neq 0$, $\det(R^\dagger R) \neq 0$ and $\lambda \neq 0$. Thus, $(R^\dagger R)$ is positive definite. Since $\det R^\dagger \neq 0$, one can similarly prove that $RR^\dagger$ is positive definite.

1.2   Some Special Matrices

★ The inner product of two column matrices $\underline{a}$ and $\underline{b}$ is defined as

$$\underline{a}^\dagger \underline{b} = \sum_\mu a_\mu^* b_\mu = \left(\underline{b}^\dagger \underline{a}\right)^*. \tag{1.3}$$

Two column matrices are called orthogonal to each other if their inner product is vanishing. $\underline{a}^\dagger \underline{a}$ must be a non-negative real number, called the square of the module of $\underline{a}$. $\underline{a}^\dagger \underline{a} = 0$ only when $\underline{a} = 0$.

★ R is called a unitary matrix if $R^\dagger R = \mathbf{1}$. The column (or row) matrices of R are normalized and orthogonal to each other:

$$\sum_\rho \left(R^\dagger\right)_{\mu\rho} R_{\rho\nu} = \sum_\rho R^*_{\rho\mu} R_{\rho\nu} = \delta_{\mu\nu},$$
$$\sum_\rho R_{\mu\rho} \left(R^\dagger\right)_{\rho\nu} = \sum_\rho R_{\mu\rho} R^*_{\nu\rho} = \delta_{\mu\nu}. \tag{1.4}$$

A unitary transformation R does not change the inner product of any two vectors:

$$(R\underline{a})^\dagger (R\underline{b}) = \underline{a}^\dagger R^\dagger R \underline{b} = \underline{a}^\dagger \underline{b}. \tag{1.5}$$

The module of any eigenvalue of a unitary matrix is one, and its two eigenvectors for different eigenvalues are orthogonal to each other. The module of the determinant of a unitary matrix is one, $|\det R| = 1$. Letting $R\underline{a} = \lambda \underline{a}$ and $R\underline{b} = \tau \underline{b}$, we have

$$0 \neq \underline{a}^\dagger \underline{a} = (R\underline{a})^\dagger (R\underline{a}) = |\lambda|^2 \underline{a}^\dagger \underline{a}, \quad |\lambda|^2 = 1,$$
$$\underline{a}^\dagger \underline{b} = (R\underline{a})^\dagger (R\underline{b}) = \lambda^* \tau \underline{a}^\dagger \underline{b} = 0, \quad \text{if } \lambda \neq \tau. \tag{1.6}$$

The product of two unitary matrices is a unitary matrix. There are m linearly independent orthonormal eigenvectors for an m-dimensional unitary matrix.

A real unitary matrix R is called a real orthogonal matrix. The product of two real orthogonal matrices is a real orthogonal matrix. The complex conjugate vector $\underline{a}^*$ of any eigenvector $\underline{a}$ of a real orthogonal matrix R for the eigenvalue λ is an eigenvector of R for the eigenvalue λ^*. The module of the determinant of an orthogonal matrix is $\det R = \pm 1$.

★ R is called a hermitian matrix if $R^\dagger = R$, or $R^*_{\mu\nu} = R_{\nu\mu}$. A hermitian matrix R satisfies

$$(R\underline{a})^\dagger \underline{b} = (\underline{a})^\dagger (R\underline{b}). \tag{1.7}$$

Any eigenvalue of a hermitian matrix R is real, and two eigenvectors of R for different eigenvalues are orthogonal to each other. The determinant of a hermitian matrix is real. Letting $R\underline{a} = \lambda \underline{a}$ and $R\underline{b} = \tau \underline{b}$, we have

$$\lambda^* \left(\underline{a}^\dagger \underline{a}\right) = (R\underline{a})^\dagger \underline{a} = \underline{a}^\dagger (R\underline{a}) = \lambda \left(\underline{a}^\dagger \underline{a}\right), \quad \lambda^* = \lambda,$$
$$(R\underline{a})^\dagger \underline{b} - \underline{a}^\dagger (R\underline{b}) = (\lambda - \tau) \underline{a}^\dagger \underline{b} = 0, \quad \underline{a}^\dagger \underline{b} = 0, \text{ if } \lambda \neq \tau. \tag{1.8}$$

The sum of two hermitian matrices is a hermitian matrix. There are m linearly independent orthonormal eigenvectors for an m-dimensional hermitian matrix.

A real hermitian matrix R is called the real symmetric matrix. The sum of two real symmetric matrices is a real symmetric matrix. There are m linearly independent orthonormal real eigenvectors for an m-dimensional real symmetric matrix.

6. Prove: (1) if $R^\dagger R = \mathbf{1}$, then $RR^\dagger = \mathbf{1}$;

   (2) if $R^{-1} R = \mathbf{1}$, then $RR^{-1} = \mathbf{1}$;

   (3) if $R^T R = \mathbf{1}$, then $RR^T = \mathbf{1}$.

Solution. The method for proving these three equalities are similar. We prove the first equality as example.

Since $R^\dagger R = \mathbf{1}$, $R^\dagger$ are nonsingular. Let S be the inverse matrix of $R^\dagger$, $SR^\dagger = \mathbf{1}$. So $RR^\dagger = (SR^\dagger) RR^\dagger = S(R^\dagger R) R^\dagger = SR^\dagger = \mathbf{1}$. Replacing $R^\dagger$ with R^{-1} or R^T, we can prove the remaining two equalities.

7. Find the independent real parameters in a 2×2 unitary matrix, a real orthogonal matrix and a hermitian matrix, and give their general expressions.

Solution. A 2×2 complex matrix contains four complex parameters, i.e., eight real parameters. For a unitary matrix, the normalization conditions of two columns give two real constraints for the parameters, and the orthogonal condition of two column matrices gives a complex constraint, i.e., two real constraints. Altogether, there are four real constraints. A 2×2 unitary matrix contains four independent real parameters. For a hermitian matrix, its diagonal elements are real, that is, their imaginary parts are zero. It leads to two real constraints. Two non-diagonal matrix entries are complex conjugate to each other. It leads to a complex constraint. Thus, a 2×2 hermitian matrix also contains four independent real parameters.

A 2×2 real matrix contains four real parameters. For a real orthogonal matrix, the normalization conditions of two columns give two real constraints for the parameters. The orthogonal condition of two column matrices gives a real constraint. There are three real constraints altogether. Thus, a 2×2 real orthogonal matrix contains only one independent real parameter.

Letting u be an arbitrary 2×2 unitary matrix, we have $\det u = \exp\{i\varphi\}$. Attracting the factor $\exp\{i\varphi/2\}$, we obtain

$$u = e^{i\varphi/2} \begin{pmatrix} a & c \\ b & d \end{pmatrix}.$$

From the definition,

$$aa^* + bb^* = cc^* + dd^* = 1, \qquad ac^* + bd^* = 0, \qquad ad - bc = 1,$$

we have

$$a = a\,(cc^* + dd^*) = d^*\,(-bc + ad) = d^*,$$
$$b = b\,(cc^* + dd^*) = c^*\,(bc - ad) = -c^*.$$

So, the general expression for a 2×2 unitary matrix is

$$u = e^{i\varphi/2} \begin{pmatrix} a & -b^* \\ b & a^* \end{pmatrix} \qquad aa^* + bb^* = 1,$$

where two complex parameters a and b with a constraint contain three real parameters. Together with φ, there are four independent real parameters.

A real orthogonal matrix is also a unitary matrix. Its determinant may be ± 1. When its determinant is 1, $\varphi = 0$ and both a and b are real numbers, satisfying $a^2 + b^2 = 1$. Usually, we take $a = \cos\theta$ and $b = \sin\theta$. When the determinant is -1, one may change the sign of the matrix elements in the first row. Thus, we obtain the general expression of a 2×2 real orthogonal matrix

$$R = \begin{pmatrix} \cos\theta & -\sin\theta \\ \sin\theta & \cos\theta \end{pmatrix}, \quad \text{or} \quad R' = \begin{pmatrix} -\cos\theta & \sin\theta \\ \sin\theta & \cos\theta \end{pmatrix}.$$

The general expression of a 2×2 hermitian matrix is

$$R = \begin{pmatrix} a & c + id \\ c - id & b \end{pmatrix},$$

where a, b, c and d all are independent real parameters.

1.3 Similarity Transformation

★ In an m-dimensional space $\mathcal{L}$, any vector $\mathbf{a}$ can be expanded in the chosen bases $\mathbf{e}_\mu$,

$$\mathbf{a} = \sum_{\mu=1}^{m} \mathbf{e}_\mu a_\mu. \tag{1.9}$$

a_μ is called the component of the vector **a** in the basis $\mathbf{e}_\mu$. Arrange these components in one column matrix $\underline{a}$ with m rows, which is called the column matrix form of a vector **a** with respect to the bases $\mathbf{e}_\mu$.

★ An m-dimensional space $\mathcal{L}$ is said to be invariant for an operator R if any vector in $\mathcal{L}$ is transformed by R to a vector belonging to $\mathcal{L}$. Specially, the basis $\mathbf{e}_\mu$ is transformed by R to a linear combination of the bases,

$$R\mathbf{e}_\mu = \sum_{\nu=1}^{m} \mathbf{e}_\nu D_{\nu\mu}(R). \tag{1.10}$$

One may arrange the combination coefficients $D_{\nu\mu}(R)$ as an $m \times m$ matrix $D(R)$ which is called the matrix form of an operator R in the bases $\mathbf{e}_\mu$.

★ Choose a new set of the bases $\mathbf{e}_\nu'$. Their column matrix forms $S_{\cdot\nu}$ in the original bases $\mathbf{e}_\mu$ can be arranged as a matrix S:

$$\mathbf{e}_\nu' = \sum_{\mu=1}^{m} \mathbf{e}_\mu S_{\mu\nu}. \tag{1.11}$$

This matrix S is called the transformation matrix between two sets of bases. S is an m-dimensional nonsingular matrix. In the new bases $\mathbf{e}_\nu'$, the matrix forms of a vector **a** and an operator R are

$$\mathbf{a} = \sum_{\mu=1}^{m} \mathbf{e}_\mu' a_\mu', \qquad R\mathbf{e}_\nu' = \sum_{\rho=1}^{m} \mathbf{e}_\rho' \overline{D}_{\rho\nu}(R). \tag{1.12}$$

The relations between two matrix forms in two sets of bases are

$$\underline{a}' = S^{-1}\underline{a}, \qquad \overline{D}(R) = S^{-1}D(R)S. \tag{1.13}$$

The matrix $D(R)$ is transformed into the matrix $\overline{D}(R)$ through the similarity transformation S. The matrices related by a similarity transformation have the same eigenvalues. They are the matrix forms of one operator in two sets of bases. It is also said in literature that $\overline{D}(R)$ is equivalent to $D(R)$.

Write the similarity transformation in the form of matrix entries:

$$\sum_{\rho=1}^{m} D_{\mu\rho}(R)S_{\rho\nu} = \sum_{\rho=1}^{m} S_{\mu\rho}\overline{D}_{\rho\nu}(R), \qquad D(R)S_{\cdot\nu} = \sum_{\rho=1}^{m} S_{\cdot\rho}\overline{D}_{\rho\nu}(R). \tag{1.14}$$

$S_{\cdot\nu}$ is the column matrix form of the new basis $\mathbf{e}_\nu'$ in the original bases, while $\overline{D}_{\rho\nu}(R)$ in Eq. (1.14) plays the role of the combination coefficients.

So Eq. (1.14) is the matrix form of Eq. (1.12) in the original bases $\mathbf{e}_\mu$. This form is very useful in later calculation.

★ If the first column of S is an eigenvector for λ_1 of the matrix $D(R)$, then all the components in the first column of $\overline{D}(R)$ are vanishing except for the first one, which is equal to λ_1. If all columns of S are the linearly independent eigenvectors of $D(R)$, then $\overline{D}(R)$ is a diagonal matrix where the diagonal elements are the eigenvalues of the corresponding eigenvectors. This is the essence of diagonalization of a matrix through a similarity transformation.

The necessary and sufficient condition for diagonalization of an m-dimensional matrix $D(R)$ is that there must exist m linearly independent eigenvectors for $D(R)$. Each column of the similarity transformation matrix S which diagonalizes $D(R)$ is the eigenvector of $D(R)$. Such a matrix S is not unique because it may be right-multiplied by a matrix X which commutes with the diagonal matrix $\overline{D}(R)$. If the m eigenvalues of $D(R)$ are different from each other, then X is a diagonal matrix. This means that each eigenvector may include an arbitrary constant factor. If the multiplicity of an eigenvalue λ_j of $D(R)$ is n_j, $\sum_j n_j = m$, then X is a block matrix, containing $\sum_j n_j^2$ parameters. These parameters reflect the arbitrary linear combination between eigenvectors for the same eigenvalue.

★ A unitary matrix or a hermitian matrix can be diagonalized by a unitary similarity transformation. A real symmetric matrix can be diagonalized by a real orthogonal similarity transformation. A real orthogonal matrix can be diagonalized by a unitary similarity transformation, but generally not by a real orthogonal similarity transformation. A unitary matrix remains unitary in any unitary similarity transformation. A hermitian matrix remains hermitian in any unitary similarity transformation. A real symmetric matrix remains real symmetric in any real orthogonal similarity transformation.

8. Find the similarity transformation to diagonalize the following matrices:

$$(1) \ \begin{pmatrix} 1 & -\sqrt{2} & 1 \\ \sqrt{2} & 0 & -\sqrt{2} \\ 1 & \sqrt{2} & 1 \end{pmatrix}; \quad (2) \ \begin{pmatrix} \cos\theta & -\sin\theta \\ \sin\theta & \cos\theta \end{pmatrix}.$$

Solution. We will denote the given matrix by $D(R)$.

(1)   The secular equation of $D(R)$ is

$$-\lambda^3 + 2\lambda^2 - 4\lambda + 8 = (2 - \lambda)(\lambda^2 + 4) = 0.$$

We obtain the eigenvalues 2, $2i$ and $-2i$. Noticing that the matrix $D(R)/2$ is a real orthogonal matrix, so its eigenvectors are orthogonal to each other. If the second component of an eigenvector is zero, we see from the second row of $D(R)$ that the first component of the eigenvector has to be equal to its third component. Then, from the first or third row of $D(R)$ we know that $(1, 0, 1)^T$ is the eigenvector for the eigenvalue 2. Due to orthogonality, the remaining eigenvectors take the form of $(1, a, -1)^T$. Substituting it into the eigenequation

$$\begin{pmatrix} 1 & -\sqrt{2} & 1 \\ \sqrt{2} & 0 & -\sqrt{2} \\ 1 & \sqrt{2} & 1 \end{pmatrix} \begin{pmatrix} 1 \\ a \\ -1 \end{pmatrix} = \pm 2i \begin{pmatrix} 1 \\ a \\ -1 \end{pmatrix},$$

we obtain $a = \mp i\sqrt{2}$. After normalization, we have

$$X^{-1}D(R)X = \begin{pmatrix} 2i & 0 & 0 \\ 0 & 2 & 0 \\ 0 & 0 & -2i \end{pmatrix}, \qquad X = \frac{1}{2}\begin{pmatrix} 1 & \sqrt{2} & 1 \\ -i\sqrt{2} & 0 & i\sqrt{2} \\ -1 & \sqrt{2} & -1 \end{pmatrix}.$$

(2)   The secular equation of $D(R)$ is

$$\lambda^2 - 2(\cos\theta)\lambda + 1 = (\lambda - e^{i\theta})(\lambda - e^{-i\theta}) = 0.$$

The eigenvalues are $\exp(\pm i\theta)$. Substituting them into the eigenequation,

$$\begin{pmatrix} \cos\theta & -\sin\theta \\ \sin\theta & \cos\theta \end{pmatrix} \begin{pmatrix} 1 \\ a \end{pmatrix} = e^{\pm i\theta} \begin{pmatrix} 1 \\ a \end{pmatrix},$$

we obtain $a = \mp i$. After normalization, we have

$$X^{-1}D(R)X = \begin{pmatrix} e^{i\theta} & 0 \\ 0 & e^{-i\theta} \end{pmatrix}, \qquad X = \frac{1}{\sqrt{2}}\begin{pmatrix} 1 & 1 \\ -i & i \end{pmatrix}.$$

9. Find a similarity transformation matrix M which satisfies

$$M^{-1}\begin{pmatrix} 0 & -\cos\theta & \sin\theta\sin\varphi \\ \cos\theta & 0 & -\sin\theta\cos\varphi \\ -\sin\theta\sin\varphi & \sin\theta\cos\varphi & 0 \end{pmatrix} M = \begin{pmatrix} 0 & -1 & 0 \\ 1 & 0 & 0 \\ 0 & 0 & 0 \end{pmatrix}.$$

Solution. Denote by $D(R)$ and $\overline{D}(R)$ the matrices before and after the similarity transformation M, respectively. $\overline{D}(R)$ is the direct sum of $-i\sigma_2$ and 0, so it is commutable with matrix X:

$$X = \begin{pmatrix} \alpha & -\beta & 0 \\ \beta & \alpha & 0 \\ 0 & 0 & \gamma \end{pmatrix}.$$

If M satisfies $M^{-1}D(R)M = \overline{D}(R)$, MX also satisfies this equation. Namely, the similarity transformation M is determined up to the three parameters.

From $\overline{D}(R)$ we see that the eigenvalues of $D(R)$ are 0 and $\pm i$. The third column of M is the eigenvector for the zero eigenvalue, and the first two columns are the linear combinations of eigenvectors for the eigenvalues $\pm i$. We first calculate the eigenvector for the zero eigenvalue. Take the third component of the eigenvector to be $\cos\theta$, which is a choice for the parameter γ. From the eigenequation, the first two components of the eigenvectors are calculated to be $\sin\theta\cos\varphi$ and $\sin\theta\sin\varphi$, respectively. Let

$$M = \begin{pmatrix} a_1 & a_2 & \sin\theta\cos\varphi \\ b_1 & b_2 & \sin\theta\sin\varphi \\ c_1 & c_2 & \cos\theta \end{pmatrix}.$$

Substituting M into

$$D(R)M = M\begin{pmatrix} 0 & -1 & 0 \\ 1 & 0 & 0 \\ 0 & 0 & 0 \end{pmatrix} = \begin{pmatrix} a_2 & -a_1 & 0 \\ b_2 & -b_1 & 0 \\ c_2 & -c_1 & 0 \end{pmatrix},$$

we obtain

$$\begin{aligned}
a_2 &= -b_1\cos\theta + c_1\sin\theta\sin\varphi, \\
b_2 &= a_1\cos\theta - c_1\sin\theta\cos\varphi, \\
c_2 &= -a_1\sin\theta\sin\varphi + b_1\sin\theta\cos\varphi, \\
-a_1 &= -b_2\cos\theta + c_2\sin\theta\sin\varphi, \\
-b_1 &= a_2\cos\theta - c_2\sin\theta\cos\varphi, \\
-c_1 &= -a_2\sin\theta\sin\varphi + b_2\sin\theta\cos\varphi.
\end{aligned}$$

Substituting the first three equations into the last three, or vice versa, we have

$$\begin{aligned}
a_1\sin\theta\cos\varphi + b_1\sin\theta\sin\varphi + c_1\cos\theta &= 0, \\
a_2\sin\theta\cos\varphi + b_2\sin\theta\sin\varphi + c_2\cos\theta &= 0.
\end{aligned}$$

These two relations reflect the fact that the first two column vectors of M are orthogonal to the third one. The reason is that $D(R)$ is antisymmetric. In fact, denoting by $\underline{b}$ the eigenvector for the zero eigenvalue and $D(R)\underline{a} = \lambda\underline{a} \neq 0$, we have $\underline{b}^T D(R) = 0$, and

$$0 = \underline{b}^T D(R)\underline{a} = \lambda\underline{b}^T\underline{a}.$$

Now, choosing $c_1 = -\sin\theta$, we obtain a solution from the orthogonality: $a_1 = \cos\theta\cos\varphi$ and $b_1 = \cos\theta\sin\varphi$. It is a choice for the parameters α and β. Substituting the results into the preceding formulas, we have $a_2 = -\sin\varphi$, $b_2 = \cos\varphi$ and $c_2 = 0$. Finally, we obtain the real orthogonal matrix M:

$$M = \begin{pmatrix} \cos\theta\cos\varphi & -\sin\varphi & \sin\theta\cos\varphi \\ \cos\theta\sin\varphi & \cos\varphi & \sin\theta\sin\varphi \\ -\sin\theta & 0 & \cos\theta \end{pmatrix}.$$

10. Find a similarity transformation matrix M which satisfies the following three equations simultaneously

$$M^{-1}\begin{pmatrix} 0 & -i & 0 \\ i & 0 & 0 \\ 0 & 0 & 0 \end{pmatrix}M = \begin{pmatrix} 1 & 0 & 0 \\ 0 & 0 & 0 \\ 0 & 0 & -1 \end{pmatrix},$$

$$M^{-1}\begin{pmatrix} 0 & 0 & 0 \\ 0 & 0 & -i \\ 0 & i & 0 \end{pmatrix}M = \frac{1}{\sqrt{2}}\begin{pmatrix} 0 & 1 & 0 \\ 1 & 0 & 1 \\ 0 & 1 & 0 \end{pmatrix},$$

$$M^{-1}\begin{pmatrix} 0 & 0 & i \\ 0 & 0 & 0 \\ -i & 0 & 0 \end{pmatrix}M = \frac{i}{\sqrt{2}}\begin{pmatrix} 0 & -1 & 0 \\ 1 & 0 & -1 \\ 0 & 1 & 0 \end{pmatrix}.$$

Solution. The first equation is the diagonalization of a matrix by a similarity transformation M, whose three column matrices are eigenvectors of the matrix. In order to satisfy the last two equations, we have to keep the multiplied factors in the eigenvectors temporarily, i.e.,

$$M = \begin{pmatrix} a & 0 & c \\ ia & 0 & -ic \\ 0 & b & 0 \end{pmatrix}.$$

Choose the common factor of M such that $b = 1$. Substituting M into the second equation, where M^{-1} is moved to the right-hand side of the equation to avoid the calculation of M^{-1}, we obtain

$$\begin{pmatrix} 0 & 0 & 0 \\ 0 & -i & 0 \\ -a & 0 & c \end{pmatrix} = \frac{1}{\sqrt{2}} \begin{pmatrix} 0 & a+c & 0 \\ 0 & i(a-c) & 0 \\ 1 & 0 & 1 \end{pmatrix}.$$

Thus, $a = -c = -\sqrt{1/2}$,

$$M = \frac{1}{\sqrt{2}} \begin{pmatrix} -1 & 0 & 1 \\ -i & 0 & -i \\ 0 & \sqrt{2} & 0 \end{pmatrix}.$$

After checking, it does satisfy the third equation.

11. Let

$$R = \begin{pmatrix} 1 & 0 \\ 0 & -1 \end{pmatrix}, \quad S = \frac{1}{2} \begin{pmatrix} -1 & -\sqrt{3} \\ \sqrt{3} & -1 \end{pmatrix}.$$

Find the common similarity transformation matrix X satisfying

$$X^{-1} (R \times R) X = \begin{pmatrix} 1 & 0 & 0 & 0 \\ 0 & -1 & 0 & 0 \\ 0 & 0 & 1 & 0 \\ 0 & 0 & 0 & -1 \end{pmatrix},$$

$$X^{-1} (S \times S) X = \frac{1}{2} \begin{pmatrix} 2 & 0 & 0 & 0 \\ 0 & 2 & 0 & 0 \\ 0 & 0 & -1 & -\sqrt{3} \\ 0 & 0 & \sqrt{3} & -1 \end{pmatrix}.$$

Solution. According to the definition of the direct product of matrices,

$$R \times R = \begin{pmatrix} 1 & 0 & 0 & 0 \\ 0 & -1 & 0 & 0 \\ 0 & 0 & -1 & 0 \\ 0 & 0 & 0 & 1 \end{pmatrix}, \quad S \times S = \frac{1}{4} \begin{pmatrix} 1 & \sqrt{3} & \sqrt{3} & 3 \\ -\sqrt{3} & 1 & -3 & \sqrt{3} \\ -\sqrt{3} & -3 & 1 & \sqrt{3} \\ 3 & -\sqrt{3} & -\sqrt{3} & 1 \end{pmatrix}.$$

Since the matrices before and after the similarity transformation X are both real orthogonal, we can also choose X to be real orthogonal. $R \times R$ is a diagonal matrix. It is easy to see that its eigenvectors for the eigenvalues

1 and -1 respectively are

$$
+1: \begin{pmatrix} a \\ 0 \\ 0 \\ b \end{pmatrix}, \qquad -1: \begin{pmatrix} 0 \\ c \\ d \\ 0 \end{pmatrix}.
$$

Thus, X is in the form of

$$
X = \begin{pmatrix} a & 0 & a' & 0 \\ 0 & c & 0 & c' \\ 0 & d & 0 & d' \\ b & 0 & b' & 0 \end{pmatrix}.
$$

Since the first two columns of X have to be the eigenvectors of $S \times S$ for the eigenvalue 1, we obtain $a = b$ and $c = -d$. Because X is a real orthogonal matrix, we have $a' = -b'$ and $c' = d'$. Choosing $a = c = c'$ and calculating the application of $S \times S$ to the fourth column of X, we have

$$
(S \times S) \begin{pmatrix} 0 \\ a \\ a \\ 0 \end{pmatrix} = -\frac{\sqrt{3}}{2} \begin{pmatrix} -a \\ 0 \\ 0 \\ a \end{pmatrix} - \frac{1}{2} \begin{pmatrix} 0 \\ a \\ a \\ 0 \end{pmatrix}.
$$

Thus, $a' = -a$. After normalization we obtain

$$
X = \frac{1}{\sqrt{2}} \begin{pmatrix} 1 & 0 & -1 & 0 \\ 0 & 1 & 0 & 1 \\ 0 & -1 & 0 & 1 \\ 1 & 0 & 1 & 0 \end{pmatrix}.
$$

12. Find the similarity transformation matrix X to diagonalize the following three matrices simultaneously,

$$
\begin{pmatrix} 0&0&0&1&0&0 \\ 0&0&0&0&1&0 \\ 0&0&0&0&0&1 \\ 1&0&0&0&0&0 \\ 0&1&0&0&0&0 \\ 0&0&1&0&0&0 \end{pmatrix}, \quad
\begin{pmatrix} 0&0&0&1&0&0 \\ 0&0&0&0&0&1 \\ 0&0&0&0&1&0 \\ 1&0&0&0&0&0 \\ 0&0&1&0&0&0 \\ 0&1&0&0&0&0 \end{pmatrix}, \quad
\begin{pmatrix} 0&0&0&1&1&1 \\ 0&0&0&1&1&1 \\ 0&0&0&1&1&1 \\ 1&1&1&0&0&0 \\ 1&1&1&0&0&0 \\ 1&1&1&0&0&0 \end{pmatrix}.
$$

Solution. Both the first two matrices are the direct sum of three σ_1, but different in the rows (columns). In the first matrix, the submatrices relate to the rows (columns) $(1,4)$, $(2,5)$ and $(3,6)$, while in the second matrix they relate to $(1,4)$, $(2,6)$ and $(3,5)$. So, three eigenvalues for each of two matrices are 1, and the other three are -1. The corresponding eigenvectors are as follows:

$$
\begin{pmatrix} a \\ b \\ c \\ a \\ b \\ c \\ 1 \end{pmatrix}, \quad
\begin{pmatrix} a' \\ b' \\ c' \\ -a' \\ -b' \\ -c' \\ -1 \end{pmatrix}; \quad
\begin{pmatrix} d \\ e \\ f \\ d \\ f \\ e \\ 1 \end{pmatrix}, \quad
\begin{pmatrix} d' \\ e' \\ f' \\ -d' \\ -f' \\ -e' \\ -1 \end{pmatrix},
$$

The eigenequations for the third matrix are

$$
\begin{pmatrix} 0 & 0 & 0 & 1 & 1 & 1 \\ 0 & 0 & 0 & 1 & 1 & 1 \\ 0 & 0 & 0 & 1 & 1 & 1 \\ 1 & 1 & 1 & 0 & 0 & 0 \\ 1 & 1 & 1 & 0 & 0 & 0 \\ 1 & 1 & 1 & 0 & 0 & 0 \end{pmatrix}
\begin{pmatrix} p \\ q \\ r \\ s \\ t \\ u \end{pmatrix}
=
\begin{pmatrix} s+t+u \\ s+t+u \\ s+t+u \\ p+q+r \\ p+q+r \\ p+q+r \end{pmatrix}
= \lambda
\begin{pmatrix} p \\ q \\ r \\ s \\ t \\ u \end{pmatrix}.
$$

Except for $\lambda = 0$, the first three components must be equal to each other, so do the last three components. Namely, the eigenvalues are ± 3, and the eigenvectors are

$$
+3: \quad \frac{1}{\sqrt{6}} \begin{pmatrix} 1 \\ 1 \\ 1 \\ 1 \\ 1 \\ 1 \end{pmatrix}, \quad
-3: \quad \frac{1}{\sqrt{6}} \begin{pmatrix} 1 \\ 1 \\ 1 \\ -1 \\ -1 \\ -1 \end{pmatrix}.
$$

These two vectors are also the eigenvectors of the first two matrices for the eigenvalues 1 and -1, respectively. In the remaining subspace, all the eigenvalues of the third matrix are zero, but the eigenvalues of the first two matrices are ± 1, respectively. Arranging the six eigenvectors, we obtain the similarity transformation matrix X

$$X = \frac{1}{2\sqrt{3}} \begin{pmatrix} \sqrt{2} & \sqrt{2} & 2 & 0 & 0 & 2 \\ \sqrt{2} & \sqrt{2} & -1 & \sqrt{3} & -\sqrt{3} & -1 \\ \sqrt{2} & \sqrt{2} & -1 & -\sqrt{3} & \sqrt{3} & -1 \\ \sqrt{2} & -\sqrt{2} & 2 & 0 & 0 & -2 \\ \sqrt{2} & -\sqrt{2} & -1 & \sqrt{3} & \sqrt{3} & 1 \\ \sqrt{2} & -\sqrt{2} & -1 & -\sqrt{3} & -\sqrt{3} & 1 \end{pmatrix},$$

which diagonalize all three matrices simultaneously:

$$\text{diag}\{1, \ -1, \ 1, \ 1, \ -1, \ -1\},$$
$$\text{diag}\{1, \ -1, \ 1, \ -1, \ 1, \ -1\},$$
$$\text{diag}\{3, \ -3, \ 0, \ 0, \ 0, \ 0\}.$$

13. Show the general form of an $m \times m$ matrix, both unitary and hermitian.

Solution. Both a unitary matrix and a hermitian matrix can be diagonalized by a unitary similarity transformation. After diagonalization, the module of each diagonal element of the unitary matrix is 1, while each diagonal element of the hermitian matrix is real. So, each diagonal element in the diagonalized matrix, both unitary and hermitian, must be 1 or -1. We denoted by Γ_n the diagonalized matrix where the first n diagonal elements are 1, and the last $m - n$ diagonal elements are -1. In general, a matrix both unitary and hermitian is in the form of $U\Gamma_n U^{-1}$, where U is a unitary matrix with determinant one.

14. Prove that any unitary matrix R can be diagonalized by a unitary similarity transformation, and any hermitian matrix R can be diagonalized by a unitary similarity transformation.

Solution. We prove this proposition by induction. Obviously, the proposition holds when the dimension of R is one. Assuming that the proposition holds when the dimension of R is $(m - 1)$, we are going to prove that it holds when the dimension of R is m. Let λ_1 be an eigenvalue of a unitary matrix R with dimension m and $S^{(1)}_{\cdot 1}$ be the normalized eigenvector of it. Arbitrarily find a complete set of orthonormal column matrices $S^{(1)}_{\cdot \mu}$, where $S^{(1)}_{\cdot 1}$ is the first one in the set. Thus, we obtain a unitary matrix $S^{(1)}$. After the similarity transformation $S^{(1)}$, $\left(S^{(1)}\right)^{-1} R S^{(1)}$ keeps unitary, where all its matrix entries in the first column are vanishing except for the first one which is λ_1. Since $|\lambda_1|^2 = 1$ and $\left(S^{(1)}\right)^{-1} R S^{(1)}$ is a unitary matrix, all its matrix entries in the first row vanish except the first one.

$$\left(S^{(1)}\right)^{-1}RS^{(1)} = \begin{pmatrix} \lambda_1 & 0 & \cdots & 0 \\ 0 & & & \\ \vdots & & R^{(1)} & \\ 0 & & & \end{pmatrix},$$

where the submatrix $R^{(1)}$ is unitary and $(m-1)$-dimensional. Therefore, $R^{(1)}$ as well as R can be diagonalized by a unitary similarity transformation. It means that there exist m orthonormal eigenvectors for an m-dimensional unitary matrix.

If R is a hermitian matrix, we introduce the similarity transformation $S^{(1)}$ in the same way. After the similarity transformation $S^{(1)}$, $\left(S^{(1)}\right)^{-1}RS^{(1)}$ keeps hermitian, where all its matrix entries in the first column are vanishing except for the first one which is λ_1. Since $\left(S^{(1)}\right)^{-1}RS^{(1)}$ is hermitian, all its matrix entries in the first row vanish except the first one. The remaining proof is the same as that for a unitary matrix. It means that there exist m orthonormal eigenvectors for an m-dimensional hermitian matrix.

15. Prove that R and $R^\dagger$ can be diagonalized by a common unitary similarity transformation if $R^\dagger$ is commutable with R. Further prove that the necessary and sufficient condition for a matrix R which can be diagonalized by a unitary similarity transformation is that $R^\dagger$ is commutable with R.

Solution. The first part of this problem is similar to the proof given in the preceding problem. We prove the first part of the present problem by induction. Obviously, it holds when the dimension of R is one. Assuming that it holds when the dimension of R is $(m-1)$, we are going to prove that it holds when the dimension of R is m. We introduce the similarity transformation $S^{(1)}$ in the same way. After the similarity transformation $S^{(1)}$, $\left(S^{(1)}\right)^{-1}R^\dagger\left(S^{(1)}\right)$ and $\left(S^{(1)}\right)^{-1}RS^{(1)}$ are still conjugate to each other and commutable with each other. Because

$$\left(S^{(1)}\right)^{-1}RS^{(1)} = \begin{pmatrix} \lambda_1 & r_2 & \cdots & r_m \\ 0 & & & \\ \vdots & & R^{(1)} & \\ 0 & & & \end{pmatrix} = \left\{\left(S^{(1)}\right)^{-1}R^\dagger S^{(1)}\right\}^\dagger,$$

we have

$$|\lambda_1|^2 = \left[\left(S^{(1)}\right)^{-1} R^\dagger R S^{(1)}\right]_{11} = \left[\left(S^{(1)}\right)^{-1} R R^\dagger S^{(1)}\right]_{11} = |\lambda_1|^2 + \sum_{n=2}^{m} |r_n|^2.$$

Thus, all r_n vanish. The submatrix $R^{(1)}$ is still commutable with $\left(R^{(1)}\right)^\dagger$ and is $(m-1)$-dimensional. Therefore, R and $R^\dagger$ can be diagonalized by a unitary similarity transformation simultaneously.

Conversely, if R can be diagonalized by a unitary similarity transformation S, then $S^{-1} R^\dagger S = \left(S^{-1} R S\right)^\dagger$ is also diagonal. Since two diagonal matrices are commutable, R and $R^\dagger$ are also commutable.

16. Prove that any matrix can be transformed into a direct sum of the standard Jordan forms, each of which is in the form

$$R_{a,b} = \begin{cases} \lambda & \text{when} \quad a = b \\ 0 \ \text{ or } \ 1 & \text{when} \quad a + 1 = b \\ 0 & \text{the remaining cases.} \end{cases}$$

Solution. As is well known, for each eigenvalue, there exists at least one corresponding eigenvector. But when the eigenvalue is multiple, it is not certain if the same number of linearly independent eigenvectors as the multiplicity of the eigenvalue can be found. If there are m linear independent eigenvectors of an m-dimensional matrix R, the similarity transformation matrix X by arranging the eigenvectors as its column matrices can diagonalize R. Now, we only need to deal with the case where the number of linearly independent eigenvectors is less than, at least for one eigenvalue, the multiplicity of the eigenvalue. We will divide this problem into three propositions. First, any matrix R can be transformed to an upper triangular matrix, where the diagonal elements are the eigenvalues of R and the same eigenvalues are arranged together. Second, by a further similarity transformation, R can be transformed into a block matrix, where each submatrix relates to only one eigenvalue. In other words, any nondiagonal matrix entry related to two different eigenvalues is vanishing. Third, each submatrix related to a given eigenvalue can be transformed into the standard Jordan form by a similarity transformation.

We will prove the first proposition by induction. The proposition obviously holds for a one-dimensional matrix. If the proposition holds for any n-dimensional matrix where $n < m$. We are going to show that the proposition also holds for any m-dimensional matrix R.

Let R have d different eigenvalues λ_j with multiplicity n_j, respectively. For λ_1, say, we can find one eigenvector $S_{\cdot 1}$. Arbitrary choose a complete set of vector bases $S_{\cdot \mu}$ where $S_{\cdot 1}$ is the first one in the set. Arrange them into an m-dimensional non-singular matrix S. After the similarity transformation S, R becomes $R' = S^{-1}RS$, where all matrix entries in the first column are vanishing except for the first one which is λ_1. Removing the first row and the first column in the matrix R', we obtain an $(m-1)$-dimensional submatrix A of R'. In comparison with the set of eigenvalues of R, the multiplicity of λ_1 in the set of eigenvalues of A decreases by one. Being an $(m-1)$-dimensional matrix, as we supposed, A can be transformed into an upper triangular matrix $A' = X^{-1}AX$ by an $(m-1)$-dimensional similarity transformation X. On the diagonal line the matrix entries of A' are arranged in the order: $\lambda_1, \ldots, \lambda_1, \lambda_2, \ldots, \lambda_2, \ldots, \lambda_d$. Let an m-dimensional matrix T be the direct sum of the digit one and the $(m-1)$-dimensional matrix X. Then, $R'' = T^{-1}R'T = (ST)^{-1}R(ST)$ satisfies the first proposition: R'' is an upper triangular matrix, where the diagonal elements are the eigenvalues of R and the same eigenvalues are arranged together.

For convenience, we remove the double primes on R''. Its matrix entry is denoted by $R_{ja,kb}$, where the row (column) subscript is designated by two indices (ja), $1 \le j \le d$, $1 \le a \le n_j$:

$$
R_{ja,kb} = \begin{cases}
0 & \text{when } j > k \\
0 & \text{when } j = k \text{ and } a > b \\
\lambda_j & \text{when } j = k \text{ and } a = b \\
R_{ja,kb} & \text{the remaining cases.}
\end{cases}
$$

The second proposition says that there exists a similarity transformation X such that after the transformation all $R'_{ja,kb}$ with $j \ne k$ become vanishing. Let $S^{(ab)}$ be a matrix, where all the diagonal matrix entries are one, and the nondiagonal matrix entries are vanishing except for $S^{(ab)}_{ja,kb} = \omega$, where $j < k$. Denote by $T^{(ab)}$ the inverse of $S^{(ab)}$ for convenience. In fact, the difference between $S^{(ab)}$ and $T^{(ab)}$ is that ω is replaced with $-\omega$. By this similarity transformation, $R' = T^{(ab)}RS^{(ab)}$, only the following matrix entries are changed:

$$
R'_{ja,kb} = R_{ja,kb} + R_{ja,ja}S^{(ab)}_{ja,kb} + T^{(ab)}_{ja,kb}R_{kb,kb} = R_{ja,kb} + \omega\left(\lambda_j - \lambda_k\right),
$$

$$R'_{ja,k'b'} = R_{ja,k'b'} + T^{(ab)}_{ja,kb}R_{kb,k'b'} = R_{ja,kb'} - \omega R_{kb,kb'},$$
$$\text{when } k < k' \text{ or when } k = k' \text{ and } b < b',$$

$$R'_{j'a',kb} = R_{j'a',kb} + R_{j'a',ja}S^{(ab)}_{ja,kb} = R_{j'a',kb} + \omega R_{j'a',ja},$$
$$\text{when } j' < j \text{ or when } j' = j \text{ and } a' < a.$$

Choosing

$$\omega = \frac{-R_{ja,kb}}{\lambda_j - \lambda_k},$$

we have $R'_{ja,kb} = 0$. Now, we annihilate the nondiagonal matrix entries $R_{ja,kb}$ with $j < k$ by a series of the above similarity transformations $S^{(ab)}$ one by one in the following order. First, beginning with $j = 1$, $k = 2$ and $b = 1$, we annihilate $R_{1a,21}$ one by one in the decreasing order of a from n_1 to 1. Second, increasing b by one from 1 to n_k, we repeat the above process to annihilate $R_{1a,2b}$ in the decreasing order of a. Third, increasing k by one from 2 to d, for a given k we decrease j by one from $k - 1$ to 1, and for the given j and k we repeat the above two processes to annihilate $R_{ja,kb}$.

Thus, after a series of similarity transformations, R is transformed into a block matrix, where each submatrix is an upper triangular matrix, namely,

$$R_{ja,kb} = \begin{cases} 0 & \text{when } j \neq k \\ 0 & \text{when } j = k \text{ and } a > b \\ \lambda_j & \text{when } j = k \text{ and } a = b \\ R_{ja,jb} & \text{when } j = k \text{ and } a < b. \end{cases}$$

The third proposition says that each submatrix related to a given eigenvalue can be transformed into the standard Jordan form by a similarity transformation. Let a submatrix, denoted still by R for convenience, be n-dimensional and have the same eigenvalue λ. Assume that the rank of the matrix $(R - \lambda\mathbf{1})$ is $(n - t)$, namely, there are t linearly independent eigenvectors of R for the eigenvalue λ. According to the first proposition, we can find a similarity transformation matrix X, where the first t columns are the eigenvectors of R, such that after the similarity transformation X, R becomes an upper triangle matrix, where the diagonal matrix entries are λ and the nondiagonal matrix entries in the first t columns are vanishing. Thus, the problem becomes how to transform an n-dimensional echelon matrix R into the standard Jordan form:

$$R = \begin{pmatrix} \Lambda & T \\ 0 & S \end{pmatrix},$$

where Λ is a $t \times t$ constant matrix, $\Lambda = \lambda\mathbf{1}$, S is an $(n - t) \times (n - t)$ upper triangular matrix with the diagonal elements λ, and T is a $t \times (n-t)$ matrix. We denote the indices of R by $\alpha = 1, 2, \ldots, n$. When α takes the first t values, we denote α by the first lowercase Latin letter, say, $a = 1, 2, \ldots, t$. When α takes the last $(n - t)$ values, we denote α by the middle lowercase Latin letter, say, $j = (t + 1), (t + 2), \ldots, n$.

$$\Lambda_{a,b} = \lambda\delta_{ab}, \qquad T_{a,j} \text{ no limit}, \qquad S_{j,k} = \begin{cases} \lambda & \text{when } j = k \\ 0 & \text{when } j > k \\ S_{j,k} & \text{when } j < k. \end{cases} \quad (1.15)$$

In the following we will introduce a series of similarity transformations, each of which keeps the form (1.15) invariant, but simplifies the nondiagonal matrix entries of R. Our aim is to transform the nondiagonal matrix entries of R such that in each row and in each column of R there is at most only one nonzero nondiagonal matrix entry, which is equal to 1. Then, by a simple similarity transformation, which only changes the order of the rows and the columns, R becomes the standard Jordan form.

Define a similarity transformation X, where the similarity transformation matrix $X(M, b)$ is the direct sum of a b-dimensional unit matrix, a $(t - b)$-dimensional non-singular matrix M^{-1} and an $(n - t)$-dimensional unit matrix, whose action is to transform each column from the $(b + 1)$th row to the tth row of T by M, but to keep Λ and S invariant. We can choose M to annihilate all the matrix entries in one column from the $(b+1)$th row to the tth row of T except for one, which is equal to 1.

We study the nondiagonal matrix entries in the $(t + 1)$ column of T. They are not all vanishing, otherwise R have $(t + 1)$ linearly independent eigenvectors. We choose M in the similarity transformation $X(M, 0)$ to annihilate all matrix entries in the $(t+1)$th column of T except for $T_{1,(t+1)} = 1$.

Define a similarity transformation Y, where the similarity transformation matrix $Y_\omega(\beta_1, \alpha_1; \beta_2, \alpha_2; \ldots)$ is a unit matrix plus a few nonzero nondiagonal matrix entries. Those matrix entries are all equal to ω and are located in the α_1 column of the β_1 row, the α_2 column of the β_2 row, and so on. All α_μ and β_ν are different from each other. Its inverse matrix can be obtained from it just by replacing ω with $-\omega$.

We can make a similarity transformation $Y_\omega(t + 1, k)$ with $\omega = -T_{1,k}$ to annihilate $T_{1,k}$ where $k > t+1$. From the viewpoint of basis transformation, the original basis is the unit vector $\mathbf{e}_\alpha$ in the rectangular coordinate system and the application of R on the basis is

$$Re_{t+1} = \lambda e_{t+1} + e_1, \qquad Re_k = \lambda e_k + T_{1,k} e_1 + \ldots .$$

The similarity transformation $Y_\omega(t+1, k)$ will transform the kth basis e_k into a new basis $e_k' = e_k - T_{1,k} e_{t+1}$, which just cancels the term related to e_1 in Re_k:

$$Re_k' = R(e_k - T_{1,k} e_{t+1}) = \lambda e_k + T_{1,k} e_1 + \ldots - T_{1,k}(\lambda e_{t+1} + e_1)$$
$$= \lambda(e_k - T_{1,k} e_{t+1}) + \ldots .$$

After a series of similarity transformations $Y_\omega(t+1, k)$, $\omega = -T_{1,k}$, where k runs from $t+2$ to n one by one, the matrix entries in the first row of T will all be annihilated except for $T_{1,(t+1)} = 1$. This method for simplifying the matrix entries in the $(t+1)$th column and in the first row of T is called the method of the first kind.

Now, we investigate the $(t+2)$th column of R. There are two cases. The first case is $S_{(t+1),(t+2)} = 0$. In this case, the method of the first kind can be used to simplify the matrix entries in the $(t+2)$th column and in the second row of T. We choose M in the similarity transformation $X(M, 1)$ to annihilate all matrix entries in the $(t+2)$th column of T except for $T_{2,(t+2)} = 1$. In the transformation the matrix entries in the $(t+1)$th column and in the first row of T are kept invariant. Then, by a series of similarity transformations $Y_\omega(t+2, k)$, where $\omega = -T_{2,k}$, $t+3 \leq k \leq n$, we can annihilate all matrix entries $T_{2,k}$ except for $T_{2,(t+2)} = 1$.

The second case is $S_{(t+1),(t+2)} \neq 0$. Define a similarity transformation Z, where the similarity transformation matrix $Z_\zeta(\alpha)$ is a diagonal matrix with all diagonal matrix entries to be 1 except for the αth entry to be ζ. This similarity transformation enlarges the matrix entries in the αth column of R by ζ times and reduces the matrix entries in the αth row by ζ times, but keeps the diagonal matrix entries invariant. We first choose the similarity transformation $Z_\zeta(t+2)$ with $\zeta = \{S_{(t+1),(t+2)}\}^{-1}$ to change $S_{(t+1),(t+2)}$ to be 1. Second, we make a series of similarity transformations $Y_\omega(a, t+1)$ with $\omega = T_{a,(t+2)}$, $2 \leq a \leq t$, to annihilate the matrix entries in the $(t+2)$th column of T, where the matrix entries in the $(t+1)$th column of T are kept invariant. From the viewpoint of base transformation, the application of R on the original basis is

$$Re_{t+2} = \lambda e_{t+2} + \left(e_{t+1} + \sum_{a=2}^{t} T_{a,(t+2)} e_a \right).$$

A series of similarity transformations $Y_\omega(a, t+1)$ with $\omega = T_{a,(t+2)}$, $2 \leq a \leq t$, define the vector in the bracket of the above formula as a new basis $\mathbf{e}'_{t+1}$ to annihilate all $T_{a,(t+2)}$.

Third, we annihilate the matrix entries $S_{(t+1),k}$, $k > t+2$, in the $(t+1)$th row of S by a series of similarity transformations $Y_\omega(t+2,k)$ with $\omega = -S_{(t+1),k}$, where k runs from $t+3$ to n one by one. The action of this series of similarity transformations is the same as that to annihilate $T_{1,k}$ in the method of the first kind. This method for simplifying the matrix entries in the $(t+2)$th column of T and in the $(t+1)$th row of S in the second case is called the method of the second kind. In the later application some revision of this method will be made.

Based on the above simplification, we are going to prove by induction that R can be transformed into the standard Jordan form by a series of similarity transformations. We define the d-simplified R-matrix as the R matrix in the form (1.15) satisfying that in each column of R from the $(t+1)$th column to the $(t+d)$th column there is only one nonzero nondiagonal matrix entry which is 1 and those nonzero nondiagonal matrix entries are all located in different rows. We will prove that the d-simplified R-matrix can be transformed into the $(d+1)$-simplified R-matrix by a series of similarity transformations. We have proved that the 1-simplified R-matrix can be transformed into the 2-simplified R-matrix by a series of similarity transformations.

For definiteness, we assume that from the $(t+1)$th column to the $(t+d)$th column of a d-simplified R-matrix, there are g nonzero nondiagonal matrix entries located in the first g rows of T and $(d-g)$ nonzero nondiagonal matrix entries located in S. There are two cases for the positions of the nonzero nondiagonal matrix entries in the $(t+d+1)$th column of R. In the first case, all nonzero nondiagonal matrix entries in the $(t+d+1)$th column of R appear in the submatrix T. In the second case, at least one nonzero nondiagonal matrix entry in the $(t+d+1)$th column of R appears in the submatrix S. We will use the method of the first kind to simplify R in the first case, but the method of the second kind in the second case.

In the first case, we choose M in the similarity transformation $X(M, d)$ to annihilate all matrix entries in the $(t+d+1)$th column of T except for $T_{(g+1),(t+d+1)} = 1$, where the matrix entries in the first g rows of T are kept invariant. Then, we annihilate all $T_{(g+1)k}$, where k runs from $t+d+2$ to n one by one, by the similarity transformations $Y_\omega(t+d+1,k)$ with $\omega = -T_{(g+1)k}$. The result is nothing but the $(d+1)$-simplified R-matrix.

In the second case, there is at least one nonzero nondiagonal matrix entry in the $(t+d+1)$th column of S. We have known that there is only one nonzero nondiagonal matrix entry (which is 1) in each column from the $(t+1)$th column to the $(t+d)$th column of the d-simplified R-matrix. For a nonzero nondiagonal matrix entry $S_{j_1,(t+d+1)}$ in the $(t+d+1)$th column of S, if the nonzero nondiagonal matrix entry in the j_1th column of R lies in the α_2th row, the nonzero nondiagonal matrix entry in the α_2th column lies in the α_3th row, and so on, till $\alpha_u \leq t$, then we define the degree of $S_{j_1,(t+d+1)}$ to be u and the degree indices of $S_{j_1,(t+d+1)}$ to be $(j_1, \alpha_2, \alpha_3, \ldots, \alpha_u)$. Recall that in the α_uth column of R, all nondiagonal matrix entries are vanishing. Among nonzero nondiagonal matrix entries in the $(t+d+1)$th column of S, we choose the one with the highest degree, say $S_{j_1,(t+d+1)}$ with the degree u and the degree indices $(j_1, \alpha_2, \alpha_3, \ldots, \alpha_u)$. We first choose a similarity transformation $Z_\zeta(t+d+1)$ with $\zeta = \left(S_{j_1,(t+d+1)}\right)^{-1}$ to make $S_{j_1,(t+d+1)} = 1$. This similarity transformation does not change the positions of the nonzero nondiagonal matrix entries in the $(t+d+1)$th column of S.

The degree v of any other nonzero nondiagonal matrix entry in the $(t+d+1)$th column of S is not larger than u, say $S_{j_2,(t+d+1)} = \sigma$ with the degree v and the degree indices $(j_2, \beta_2, \ldots, \beta_v)$, $v \leq u$. In the following we will show that the similarity transformation $Y_\omega(j_2, j_1; \beta_2, \alpha_2; \ldots; \beta_v, \alpha_v)$ with $\omega = \sigma$ can annihilate the matrix entry $S_{j_2,(t+d+1)}$. The reason for this can be understood by the viewpoint of the basis transformation. The new basis induced by this similarity transformation is

$$\mathbf{e}'_\tau = \begin{cases} \mathbf{e}_{j_1} + \sigma \mathbf{e}_{j_2} & \text{when } \tau = j_1, \\ \mathbf{e}_{\alpha_2} + \sigma \mathbf{e}_{\beta_2} & \text{when } \tau = \alpha_2, \\ \ldots & \ldots \\ \mathbf{e}_{\alpha_v} + \sigma \mathbf{e}_{\beta_v} & \text{when } \tau = \alpha_v, \\ \mathbf{e}_\tau & \text{the remaining cases.} \end{cases}$$

Before this transformation we have

$$R\mathbf{e}_{t+d+1} = \lambda \mathbf{e}_{t+d+1} + (\mathbf{e}_{j_1} + \sigma \mathbf{e}_{j_2}) + \ldots,$$

$$R\mathbf{e}_{j_1} = \lambda \mathbf{e}_{j_1} + \mathbf{e}_{\alpha_2}, \quad R\mathbf{e}_{\alpha_2} = \lambda \mathbf{e}_{\alpha_2} + \mathbf{e}_{\alpha_3}, \quad \ldots,$$

$$R\mathbf{e}_{\alpha_v} = \begin{cases} \lambda \mathbf{e}_{\alpha_v} + \mathbf{e}_{\alpha_{v+1}} & \text{when } v < u \\ \lambda \mathbf{e}_{\alpha_v} & \text{when } v = u, \end{cases}$$

$$R\mathbf{e}_{j_2} = \lambda \mathbf{e}_{j_2} + \mathbf{e}_{\beta_2}, \quad R\mathbf{e}_{\beta_2} = \lambda \mathbf{e}_{\beta_2} + \mathbf{e}_{\beta_3}, \quad \ldots, \quad R\mathbf{e}_{\beta_v} = \lambda \mathbf{e}_{\beta_v}.$$

After the similarity transformation we have

$$R\mathbf{e}'_{t+d+1} = R\mathbf{e}_{t+d+1} = \lambda\mathbf{e}'_{t+d+1} + \mathbf{e}'_{j_1} + \cdots,$$
$$R\mathbf{e}'_{j_1} = \lambda\left(\mathbf{e}_{j_1} + \sigma\mathbf{e}_{j_2}\right) + \mathbf{e}_{\alpha_2} + \sigma\mathbf{e}_{\beta_2} = \lambda\mathbf{e}'_{j_1} + \mathbf{e}'_{\alpha_2},$$
$$R\mathbf{e}'_{\alpha_2} = \lambda\left(\mathbf{e}_{\alpha_2} + \sigma\mathbf{e}_{\beta_2}\right) + \mathbf{e}_{\alpha_3} + \sigma\mathbf{e}_{\beta_3} = \lambda\mathbf{e}'_{\alpha_2} + \mathbf{e}'_{\alpha_3},$$

$$\cdots$$

$$R\mathbf{e}'_{\alpha_v} = \lambda\left(\mathbf{e}_{\alpha_v} + \sigma\mathbf{e}_{\beta_v}\right) + \mathbf{e}_{\alpha_{v+1}} = \lambda\mathbf{e}'_{\alpha_v} + \mathbf{e}'_{\alpha_{v+1}}.$$

When $v = u$, the terms of $\mathbf{e}_{\alpha_{v+1}}$ and $\mathbf{e}'_{\alpha_{v+1}}$ in the last formula disappear. Thus, the action of R on the new basis $\mathbf{e}'_{\alpha_\mu}$ is the same as that on the original basis $\mathbf{e}_{\alpha_\mu}$ except for that on the basis $\mathbf{e}'_{t+d+1}$, where the term of $\sigma\mathbf{e}_{j_2}$ disappears. The action of R on the bases $\mathbf{e}'_k$, $k > t + d + 1$, may also be changed, namely, the nondiagonal matrix entries in the kth column of R may be changed. But we do not care about them. Through a series of similarity transformations, the matrix entries in the $(t + d + 1)$th column of S become vanishing except for $S_{j_1,(t+d+1)} = 1$.

If a matrix entry in the $(t + d + 1)$th column of T is nonzero, say $T_{a(t+d+1)} \neq 0$, we can use the similarity transformation $Y_\omega(a, j_1)$ with $\omega = T_{a(t+d+1)}$ to annihilate $T_{a(t+d+1)}$. Finally, we choose a series of similarity transformations $Y_\omega(t + d + 1, k)$ with $\omega = -S_{j_1,k}$, $t + d + 2 \le k \le n$, to annihilate all $S_{j_1,k}$ in the j_1th row of S. Thus, we have found a series of similarity transformations to change a d-simplified R-matrix into a $(d + 1)$-simplified R-matrix, namely, we have proved the third proposition. Therefore, we have proved that any matrix R can be transformed into a direct sum of the standard Jordan forms.

We will give a typical and instructive example to show how to transform a d-simplified R-matrix into a $(d + 1)$-simplified R-matrix by a series of similarity transformations. Assume a 7-simplified R-matrix with $t = 6$, $n = 15$, $g = 4$, and $d = 7$,

$$T = \begin{pmatrix} 1 & 0 & 0 & 0 & 0 & 0 & 0 & 0 & 0 \\ 0 & 1 & 0 & 0 & 0 & 0 & 0 & 0 & 0 \\ 0 & 0 & 0 & 0 & 1 & 0 & 0 & 0 & 0 \\ 0 & 0 & 0 & 0 & 0 & 1 & 0 & 0 & 0 \\ 0 & 0 & 0 & 0 & 0 & 0 & 0 & T_{5,14} & T_{5,15} \\ 0 & 0 & 0 & 0 & 0 & 0 & 0 & T_{6,14} & T_{6,15} \end{pmatrix},$$

$$S = \begin{pmatrix} \lambda & 0 & 1 & 0 & 0 & 0 & 0 & 0 & 0 \\ 0 & \lambda & 0 & 0 & 0 & 0 & 0 & S_{8,14} & S_{8,15} \\ 0 & 0 & \lambda & 1 & 0 & 0 & 0 & 0 & 0 \\ 0 & 0 & 0 & \lambda & 0 & 0 & 0 & S_{10,14} & S_{10,15} \\ 0 & 0 & 0 & 0 & \lambda & 0 & 0 & S_{11,14} & S_{11,15} \\ 0 & 0 & 0 & 0 & 0 & \lambda & 1 & 0 & 0 \\ 0 & 0 & 0 & 0 & 0 & 0 & \lambda & S_{13,14} & S_{13,15} \\ 0 & 0 & 0 & 0 & 0 & 0 & 0 & \lambda & S_{14,15} \\ 0 & 0 & 0 & 0 & 0 & 0 & 0 & 0 & \lambda \end{pmatrix} .$$

According to the matrix entries in the 14th column (in general, the $(t + d + 1)$th column) of S, there are two cases. One is the case where the nondiagonal matrix entries in the 14th column of S are all zero. In this case the method of first kind can be used to simplify R. We choose a two-dimensional matrix M in the similarity transformation $X(M, 4)$ to make $T_{5,14} = 1$ and $T_{6,14} = 0$, but to keep the first four rows of T invariant. Then, we use the similarity transformation $Y_\omega(14, 15)$ with $\omega = -T_{5,15}$ to make $T_{5,15} = 0$. This has been the 8-simplified R-matrix. The other is the case where the nondiagonal matrix entries in the $(t + d + 1)$th column of S are not all zero. In the example we assume that the nondiagonal matrix entries given in S are all nonzero. First to determine the degrees and the degree indices for these matrix entries. $S_{8,14}$ has the degree two and the degree indices $(8, 2)$. $S_{10,14}$ has the highest degree four and the degree indices $(10, 9, 7, 1)$. $S_{11,14}$ has the degree two and the degree indices $(11, 3)$. $S_{13,14}$ has the degree three and the degree indices $(13, 12, 4)$. We use the similarity transformation $Z_\zeta(14)$, with $\zeta = (S_{10,14})^{-1}$ to make $S_{10,14} = 1$. The remaining nondiagonal matrix entries in the 14th column of S have changed their values, but are still denoted by the original symbols. Now, we use the following similarity transformations one by one to obtain the 8-simplified R-matrix:

$Y_\omega(8, 10; 2, 9)$	with $\omega = S_{8,14}$	transforms $S_{8,14}$ to be zero,
$Y_\omega(11, 10; 3, 9)$	with $\omega = S_{11,14}$	transforms $S_{11,14}$ to be zero,
$Y_\omega(13, 10; 12, 9; 4, 7)$	with $\omega = S_{13,14}$	transforms $S_{13,14}$ to be zero,
$Y_\omega(5, 10)$	with $\omega = T_{5,14}$	transforms $T_{5,14}$ to be zero,
$Y_\omega(6, 10)$	with $\omega = T_{6,14}$	transforms $T_{6,14}$ to be zero,
$Y_\omega(14, 15)$	with $\omega = -S_{10,15}$	transforms $S_{10,15}$ to be zero.

Chapter 2

GROUP AND ITS SUBSETS

2.1 Definition of a Group

★ A group is a set G of elements R satisfying four axioms with respect to the given multiplication rule of elements. The axioms are: a) The set is closed to this multiplication; b) The multiplication between elements satisfy the associative law; c) There is an identity $E \in G$ satisfying $ER = R$; d) The set contains the inverse R^{-1} of any element $R \in G$ satisfying $R^{-1}R = E$. The multiplication rule of elements completely describes the structure and property of a group. A group G is called a finite group if it contains finite number g of elements, and g is called the order of G. Otherwise, the group is called an infinite group. For a finite group, the multiplication rule can be given by the multiplication table, or called the group table. A group is called the Abelian group if the product of its elements is commutable. A few elements in G are called the generators of G if any element in G can be expressed as their product. The rearrangement theorem says $RG = GR = G$, namely, there are no duplicate elements in each row and in each column of the multiplication table.

★ Two groups are called isomorphic, $G \approx G'$, if there is a one-to-one correspondence between elements of two groups in such a way products correspond to products. From the viewpoint of group theory, two isomorphic groups are the same as each other.

1. Let E be the identity of a group G, R and S be any two elements in the group G, R^{-1} and S^{-1} be the inverses of R and S, respectively. Try to show from the definition of a group: (a) $RR^{-1} = E$; (b) $RE = R$; (c) if $TR = R$, then $T = E$; (d) if $TR = E$, then $T = R^{-1}$; (e) The inverse of (RS) is $S^{-1}R^{-1}$.

Solution. The key to the proof is that each element in a group has its inverse. Recall that only the definition of a group and the proved conclusion can be used in the later proof.

(a) We prove the conclusion by two methods. In the first method, since R^{-1} is an element in the group, there exists its inverse in the group, denoted by S, $SR^{-1} = E$. So, from the definition of a group we have

$$RR^{-1} = ERR^{-1} = \left(SR^{-1}\right) RR^{-1} = S\left(R^{-1}R\right) R^{-1}$$
$$= SER^{-1} = SR^{-1} = E.$$

Another method is to denote by W^{-1} the inverse of $RR^{-1} = W$, $W^{-1}W = E$. Then, we have:

$$WW = R\left(R^{-1}R\right) R^{-1} = RER^{-1} = W,$$
$$RR^{-1} = W = EW = \left(W^{-1}W\right) W = W^{-1}\left(WW\right) = W^{-1}W = E.$$

(b) $RE = R\left(R^{-1}R\right) = \left(RR^{-1}\right) R = ER = R.$

(c) If $TR = R$, then

$$T = T\left(RR^{-1}\right) = (TR) R^{-1} = RR^{-1} = E.$$

The conclusion says that the identity in a group is unique.

(d) If $TR = E$, then

$$T = TE = T\left(RR^{-1}\right) = (TR) R^{-1} = ER^{-1} = R^{-1}.$$

The conclusion says that the inverse of any element in a group is unique.

(e) Since $\left(S^{-1}R^{-1}\right) (RS) = S^{-1}\left(R^{-1}R\right) S = S^{-1}ES = S^{-1}S = E$ and the conclusion (d), $S^{-1}R^{-1}$ is the inverse of RS.

2. Show that a group composed of all positive real numbers where the multiplication rule of elements is defined with the product of digits, is isomorphic onto a group composed of all real numbers where the multiplication rule of elements is defined with the addition of digits.

Solution. Denote by H the set of all positive real numbers R. The multiplication of elements is defined to be the product of digits. The set is closed for the multiplication. The product of digits satisfies the associative law. The positive real number 1 is the identity in the set. The reciprocal digit $1/R$ is still a positive real number, which is the inverse of R. Therefore, the set of H is a group, called the multiplication group of positive real numbers.

Denote by G the set of all real numbers S. The multiplication of elements is defined to be the addition of digits. The set is closed for the multiplication. The addition of digits satisfies the associative law. The real number 0 is the identity in the set. $-S$ is still a real number, which is the inverse of S. Therefore, the set of G is a group, called the addition group of real numbers.

The exponential relation $R = e^S$ gives a one-to-one correspondence between the positive real number R in H and the real number S in G, and it is invariant for the multiplication of elements:

$$R = e^S, \qquad R' = e^{S'}, \qquad RR' = e^S e^{S'} = e^{S+S'}.$$

Therefore, the group H is isomorphic onto the group G.

2.2 Subsets in a Group

★ A subset of a group is called a subgroup of the group if in the multiplication rule of the group elements the subset satisfies four axioms in the group definition. The set composed of the powers of any element in a group forms a cyclic subgroup, also called the period of the element. For a finite group, a subset of a group is its subgroup if and only if the subset is closed with respect to the multiplication rule because the subset contains the period of any element in it. In the same reasoning, a finite set of elements is a group if and only if the set is closed with respect to an associative multiplication rule. It is easy to judge by the multiplication table of the group whether or not a subset is a subgroup. The order of the cyclic subgroup generated by the element is called the order of the element. Only the order of the identity is 1. The identity and the whole group are two trivial subgroups. We will only consider the nontrivial subgroups.

★ A subset in a group G obtained by left-multiplying (or right-multiplying) an element $R \in G$ to a subgroup H of G, where R does not belong to H, is called a left coset RH (or a right coset HR) of the subgroup H. For a finite group, the number of the elements in a coset is equal to the order of the subgroup. Since a coset does not contain any element in the subgroup, any coset must not be a subgroup due to lack of the identity. Different left cosets (or right cosets) do not contain any common element. The necessary and sufficient condition for two elements R and S belonging to the same left coset is $R^{-1}S \in H$, and that for them belonging to the same right coset is $RS^{-1} \in H$. Hence, the order h of a subgroup H must be a whole number divisor of the order g of the group G. The integer $n = g/h$ is called the index of the subgroup H in G. The number of the different left cosets (or right cosets) of a subgroup is $(n - 1)$. In the columns of the multiplication table, related to the elements of a subgroup, the set of elements in each row is the subgroup itself or the left coset of the subgroup. In the rows of the multiplication table, related to the elements of a subgroup, the set of elements in each column is the subgroup itself or the right coset of the subgroup.

★ The element $R' = SRS^{-1}$ is said to be an element conjugate to R in a group G, where R, R' and S all belong to G. The conjugate relation of two elements is mutual. If two elements are both conjugate to a third element, they are also conjugate to each other. The subset of all mutually conjugate elements in a group G is called a class C_α in G. Any element in G belongs and only belongs to one class. No class forms a subgroup of G except for the class C_1 composed of the identity. The set of all the inverse elements R^{-1} of $R \in C_\alpha$ is also a class, denoted by C_α^{-1}. These two classes are called the reciprocal classes mutually. It is called the self-reciprocal class if $C_\alpha = C_\alpha^{-1}$. RS is conjugate to SR. Conversely, two conjugate elements can be expressed as two products RS and SR where R and S are two elements in G. For a finite group, the elements in one class have the same order. However, two elements with the same order are not necessary conjugate to each other. For a multiplication table of a finite group where the arrangement for the rows is the same as that for the columns, two conjugate elements appear and must appear at least once in two symmetrical positions with respect to the diagonal line of the multiplication table. This is the main method to check by the multiplication table whether two elements with the same order are conjugate to each other.

★ A subgroup H is called an invariant subgroup (or normal subgroup) of the group G if for any element $R \in G$ the left coset is equal to the right coset,

$$RH = HR, \qquad RHR^{-1} = H.$$

A subgroup with index 2 must be an invariant subgroup. An invariant subgroup is composed of a few whole classes and satisfies the criterion for a subgroup. It is the main method to find an invariant subgroup of a finite group G that one first gathers a few classes including the identity to see whether the number of elements in the set is a whole number divisor of the order of G and whether the set contains the whole period of any element in the set, then, to check whether the set is closed for the multiplication rule of the group elements.

★ The multiplication of two subsets is defined to be a set composed of all the products of any two elements respectively belonging to two subsets. Note that in a set only one element is taken for the duplicated element. Denote the invariant subgroup H of G and its cosets uniformly by R_jH, where $R_1 = E$. The aggregate of the subsets R_jH satisfies the four axioms with respect to the multiplication rule of subsets. In fact,

$$(R_jH)(R_kH) = R_jR_kHH = (R_jR_k)H.$$

The multiplication rule satisfies the associate law. $R_1H = H$ is the identity in the aggregate. $R_j^{-1}H$ is a coset of H and is the inverse of R_jH in the aggregate. Therefore, the aggregate of the subsets R_jH forms a group, called the quotient group G/H of the invariant subgroup H.

3. If H_1 and H_2 are two subgroups of a group G, prove that the common elements in H_1 and H_2 also form a subgroup of G.

Solution. Denote the set of the common elements of the subgroups H_1 and H_2 by $H_3 = \{R_1, R_2, \cdots\}$, where $R_j \in H_1$ and $R_j \in H_2$. The set of H_3 is also a subset of G, and the product of elements in H_3 satisfies the multiplication rule of G. So the associative law is satisfied for H_3. Since H_1 and H_2 are two subgroups of G, the product R_iR_j of any two elements in H_3 must belong to both subgroups H_1 and H_2 such that it belongs to the set H_3. For the same reason, the identity E in G and the inverse R_j^{-1} of any element R_j in H_3 belong to both subgroups H_1 and H_2, so that they also belong to the set H_3. Therefore, the set H_3 is a subgroup of G.

4. Prove that a group whose order g is a prime number must be a cyclic group C_g.

Solution. The order of an element is nothing but the order of the cyclic subgroup generated by the element. As we have known, the order h of the subgroup $H \subset G$ is a whole number divisor of the order g of the group G. When g is a prime number, except for the identity, the order of any element in G must be the order g of the group, so the cyclic subgroup is the group G itself. From it we come to the conclusion that any two groups whose orders are the same prime number must be isomorphic onto each other.

5. Show that up to isomorphism, there are only two different fourth-order groups: The cyclic group C_4 and the fourth-order inversion group V_4.

Solution. Since the order of any element in a fourth-order group, except for the identity, has to be 2 or 4. If there is at least one fourth-order element R, the group is the cyclic group $C_4 = \{E, R, R^2, R^3\}$. If the order of any element in the group, except for the identity, is 2, denote the group by $V_4 = \{e, \sigma, \tau, \rho\}$, where e is the identity and $\sigma^2 = \tau^2 = \rho^2 = e$. From Problem 1, the product of any two elements, say $\sigma\tau$, cannot be equal to e, σ and τ, so $\sigma\tau = \rho$ and the group is the fourth-order inversion group. This completes the proof.

6. Show that up to isomorphism, there are only two different sixth-order groups: The cyclic group C_6 and the symmetry group D_3 for the regular triangle.

Solution. Since the period of any element is a cyclic subgroup, in a sixth-order group, except for the identity, the order of any element has to be 2, 3 or 6. There are three cases. In the first case where there is at least one sixth-order element, the group is a cyclic group C_6. In the second case the group contains no sixth-order element, but contains at least one third-order element, denoted by R. The cyclic subgroup $H = \{E, R, R^2\}$ generated by R has the index two so that it is an invariant subgroup of the group. Denote its coset by $HS_0 = \{S_0, S_1, S_2\}$, where $R^m S_0 = S_m$, $R^m S_j = S_{j+m}$, and $S_{j+3} = S_j$. S_j^2 cannot be equal to R or R^2, otherwise the order of S_j is 6. From the rearrangement theorem, S_j^2 cannot be equal to S_k. Thus, $S_j^2 = E$. It can be deduced that $R^m = S_{j+m}S_j$ and $S_j R^m = S_{j-m}$. This group is nothing but D_3. In the third case, the order of any element in the group is 2 except for the identity. Arbitrarily take two elements R and S in the group, and let $RS = T$. From Problem 1 we conclude that T is not equal to E, R or S. The subset composed of E, R, S and T forms a subgroup. It is isomorphic onto the inversion group V_4 of order 4. Since $6/4$ is not an integer, the third case cannot exist. This completes the proof.

7. Show that a group must be an Abelian group if the order of any element in the group, except for the identity, is 2.

Solution. What we need to show is whether $RS = SR$ for any two different elements R and S in the group G. Letting $RS = T$, we have $R^2 = S^2 = T^2 = E$. From Problem 1, $TS = RS^2 = R$, $TR = T^2S = S$, and $SR = TR^2 = T = RS$. Therefore, the group is an Abelian group.

2.3 Homomorphism of Groups

★ A group G is said to be homomorphic onto another group G', $G' \sim G$, if one and only one element of G' corresponds to any element of G, if at least one element of G corresponds to any element of G', and if the correspondence is such that the product of two elements of G is in the same way to map onto the product of two corresponding elements of G'. In other words, if $G' \sim G$, there is a one-to-many correspondence between elements of G' and G and the correspondence is invariant in the multiplication of elements. The set H of elements in G which corresponds to the identity in G' forms an invariant subgroup of G, which is called the kernel of homomorphism. The quotient group G/H is isomorphic onto G'. G' describes only part of properties of G, namely, G' describes the property of the quotient group G/H, but does not describe the difference among elements in the kernel of homomorphism.

8. The Pauli matrices σ_a are defined as follows:

$$\sigma_1 = \begin{pmatrix} 0 & 1 \\ 1 & 0 \end{pmatrix}, \quad \sigma_2 = \begin{pmatrix} 0 & -i \\ i & 0 \end{pmatrix}, \quad \sigma_3 = \begin{pmatrix} 1 & 0 \\ 0 & -1 \end{pmatrix},$$

$$\sigma_a \sigma_b = \delta_{ab} \mathbf{1} + i \sum_{d=1}^{3} \epsilon_{abd} \sigma_d, \quad \text{e.g.,} \quad \sigma_a^2 = \mathbf{1}, \quad \sigma_1 \sigma_2 = i\sigma_3,$$

where ϵ_{abd} is the totally antisymmetrical tensor of rank 3. Show that all possible products generated by σ_1 and σ_2 form a group. List the multiplication table of this group. Point out the order of this group, the order of each element, the classes, the invariant subgroups and their quotient groups. Prove that this group is isomorphic onto the symmetry group D_4 for the square.

Solution. According to the multiplication rule of the Pauli matrices, there are 8 matrices generated by σ_1 and σ_2. The multiplication table is given in the Table. From the multiplication table, the set of these 8 elements is closed for the multiplication of elements. The multiplication of matrices satisfies the associative law. $\mathbf{1}$ is the identity E in the set. Any element in the set is self-inverse except for $\pm i\sigma_3$. $-i\sigma_3$ is the inverse of $i\sigma_3$. Therefore, this set forms a group G with order 8. The order of the identity E is 1. The orders of $-\mathbf{1}$, $\pm\sigma_1$ and $\pm\sigma_2$ all are 2. The orders of $\pm i\sigma_3$ are both 4. $\mathbf{1}$ and $-\mathbf{1}$ form two classes. $\{\pm\sigma_1\}$, $\{\pm\sigma_2\}$ and $\{\pm i\sigma_3\}$ form the classes, respectively. There are five classes altogether. The invariant subgroups in the group are $\{1, -1\}$, $\{1, -1, \sigma_1, -\sigma_1\}$, $\{1, -1, \sigma_2, -\sigma_2\}$, and $\{1, -1, i\sigma_3, -i\sigma_3\}$. The cosets of $\{1, -1\}$ are $\{\pm\sigma_1\}$, $\{\pm\sigma_2\}$ and $\{\pm i\sigma_3\}$. The square of each coset is the invariant subgroup $\{\pm1\}$, so the quotient group is isomorphic onto the fourth-order inversion group V_4. The quotient groups of the next three invariant subgroups are all isomorphic onto the second-order inversion group V_2. G is homomorphic onto V_4 with a $2 : 1$ correspondence and onto V_2 with the $4 : 1$ correspondence where there are three different kernels of homomorphism.

	1	σ_1	σ_2	$i\sigma_3$	-1	$-\sigma_1$	$-\sigma_2$	$-i\sigma_3$
1	1	σ_1	σ_2	$i\sigma_3$	-1	$-\sigma_1$	$-\sigma_2$	$-i\sigma_3$
σ_1	σ_1	1	$i\sigma_3$	σ_2	$-\sigma_1$	-1	$-i\sigma_3$	$-\sigma_2$
σ_2	σ_2	$-i\sigma_3$	1	$-\sigma_1$	$-\sigma_2$	$i\sigma_3$	-1	σ_1
$i\sigma_3$	$i\sigma_3$	$-\sigma_2$	σ_1	-1	$-i\sigma_3$	σ_2	$-\sigma_1$	1
-1	-1	$-\sigma_1$	$-\sigma_2$	$-i\sigma_3$	1	σ_1	σ_2	$i\sigma_3$
$-\sigma_1$	$-\sigma_1$	-1	$-i\sigma_3$	$-\sigma_2$	σ_1	1	$i\sigma_3$	σ_2
$-\sigma_2$	$-\sigma_2$	$i\sigma_3$	-1	σ_1	σ_2	$-i\sigma_3$	1	$-\sigma_1$
$-i\sigma_3$	$-i\sigma_3$	σ_2	$-\sigma_1$	1	$i\sigma_3$	$-\sigma_2$	σ_1	-1

Let the fourfold rotation C_4 around the Z axis in the group D_4 correspond to $i\sigma_3$, and the twofold rotation C_2' around the X axis correspond to σ_1. The correspondence between the remaining elements can be obtained by the products of elements:

$$C_4^2 \longleftrightarrow -1, \qquad C_4^3 \longleftrightarrow -i\sigma_3, \qquad C_4 C_2' C_4^3 \longleftrightarrow -\sigma_1,$$
$$C_4 C_2' \longleftrightarrow -\sigma_2, \qquad C_2' C_4 \longleftrightarrow \sigma_2, \qquad E \longleftrightarrow 1.$$

Thus, the multiplications of elements also satisfy the one-to-one correspondence, so two groups are isomorphic.

9. Show that the set of all possible products generated by $i\sigma_1$ and $i\sigma_2$ forms a group. List the multiplication table of this group. Point out the order of this group, the order of each element, the classes, the invariant subgroups and their quotient groups in the group, respectively. Show that this group is not isomorphic onto the group D_4.

Solution. There are 8 matrices generated by the products of $i\sigma_1$ and $i\sigma_2$. The multiplication table is as follows.

	1	$i\sigma_1$	$i\sigma_2$	$i\sigma_3$	-1	$-i\sigma_1$	$-i\sigma_2$	$-i\sigma_3$
1	1	$i\sigma_1$	$i\sigma_2$	$i\sigma_3$	-1	$-i\sigma_1$	$-i\sigma_2$	$-i\sigma_3$
$i\sigma_1$	$i\sigma_1$	-1	$-i\sigma_3$	$i\sigma_2$	$-i\sigma_1$	1	$i\sigma_3$	$-i\sigma_2$
$i\sigma_2$	$i\sigma_2$	$i\sigma_3$	-1	$-i\sigma_1$	$-i\sigma_2$	$-i\sigma_3$	1	$i\sigma_1$
$i\sigma_3$	$i\sigma_3$	$-i\sigma_2$	$i\sigma_1$	-1	$-i\sigma_3$	$i\sigma_2$	$-i\sigma_1$	1
-1	-1	$-i\sigma_1$	$-i\sigma_2$	$-i\sigma_3$	1	$i\sigma_1$	$i\sigma_2$	$i\sigma_3$
$-i\sigma_1$	$-i\sigma_1$	1	$i\sigma_3$	$-i\sigma_2$	$i\sigma_1$	-1	$-i\sigma_3$	$i\sigma_2$
$-i\sigma_2$	$-i\sigma_2$	$-i\sigma_3$	1	$i\sigma_1$	$i\sigma_2$	$i\sigma_3$	-1	$-i\sigma_1$
$-i\sigma_3$	$-i\sigma_3$	$i\sigma_2$	$-i\sigma_1$	1	$i\sigma_3$	$-i\sigma_2$	$i\sigma_1$	-1

From the multiplication table, the set of these 8 elements is closed for the multiplication of elements. The multiplication of matrices satisfies the associative law. Therefore, this set forms a group G with order 8. The order of the identity 1 is 1. The order of -1 is 2. The orders of $\pm i\sigma_a$ all are 4. 1 and -1 form two classes. $\{\pm i\sigma_1\}$, $\{\pm i\sigma_2\}$ and $\{\pm i\sigma_3\}$ form the classes, respectively. There are five classes altogether. The invariant subgroups in the group are $\{1, -1\}$, $\{1, -1, i\sigma_1, -i\sigma_1\}$, $\{1, -1, i\sigma_2, -i\sigma_2\}$, and $\{1, -1, i\sigma_3, -i\sigma_3\}$. The cosets of $\{\pm 1\}$ are $\{\pm i\sigma_1\}$, $\{\pm i\sigma_2\}$ and $\{\pm i\sigma_3\}$. The square of each coset is the invariant subgroup $\{\pm 1\}$, so the quotient group is isomorphic onto the fourth-order inversion group V_4. The quotient groups of the next three invariant subgroups all are isomorphic onto the second-order inversion group V_2. G is homomorphic onto V_4 with a $2:1$ correspondence and onto V_2 with the $4:1$ correspondences where there are three different kernels of homomorphism. Because this group contains six fourth-order elements, this group is isomorphic neither onto the group D_4, nor onto the group C_{4h}. In literature, this group is usually called the quaternion group Q_8.

10. Up to isomorphism, prove that there are only five different eighth-order group: The cyclic group C_8, $C_{4h} = C_4 \times V_2$, the symmetry group D_4 of a square, the quaternion group Q_8 and $D_{2h} = D_2 \times V_2$.

Solution. The order of an element in an eighth-order group G, except for the identity, has to be 2, 4 or 8. If there is at least one eighth-order element, G is the cyclic group C_8. If the orders of all elements in G, except for the identity E, are 2, we obtain from Problem 7 that G is an Abelian group, which is isomorphic onto $D_{2h} = D_2 \times V_2$.

If there is no eighth-order element G, but there is at least one fourth-order element, denoted by R, the cyclic subgroup $H = \{E, R, R^2, R^3\}$ generated by R is an invariant subgroup of G because its index is 2. Denote its coset by $HS_0 = \{S_0, S_1, S_2, S_3\}$, where $R^m S_0 = S_m$, $R^m S_j = S_{j+m}$, and

$S_{j+4} = S_j$. S_j cannot be equal to R or R^3, otherwise the order of S_j is 8. From the rearrangement theorem, S_j^2 cannot be equal to S_k. If $S_j^2 = R^2$, then S_j is a fourth-order element. Thus, $S_j^{-1} = S_j^3 = R^2 S_j = S_{j+2}$, and $S_{j+2}^2 = S_j^{-2} = R^2$. Now, there are three cases. The first case is that all four S_j satisfy $S_j^2 = R^2$. Then, $S_1 S_0 = R S_0^2 = R^3$, $S_0 S_1 = R^3 S_1^2 = R$, and this group is isomorphic onto the quaternion group Q_8. The correspondence is

$$R \longleftrightarrow i\sigma_3, \qquad S_0 \longleftrightarrow i\sigma_2, \qquad S_1 \longleftrightarrow i\sigma_1.$$

The second case is that only a pair of S_j satisfy $S_j^2 = R^2$, say $S_1^2 = S_3^2 = R^2$. Namely, S_1 and S_3 are the fourth-order elements, and S_0 and S_2 are the second-order elements, $S_0^2 = S_2^2 = E$. Thus, $S_1 S_0 = R S_0^2 = R$, $S_0 S_1 = R^3 S_1^2 = R$, $S_0 R = S_0^2 S_1 = S_1 = R S_0$, $S_1 R = S_1^2 S_0 = R^2 S_0 = S_2 = R S_1$, and this group is isomorphic onto the Abelian group $C_{4h} = C_4 \times V_2$. The correspondence is

$$R \longleftrightarrow C_4, \qquad S_0 \longleftrightarrow \sigma, \qquad S_1 \longleftrightarrow \sigma C_4,$$

where σ is the space inversion. The third case is that the orders of all S_j are 2, $S_j^2 = E$, then from $R^m S_j = S_{j+m}$ we have $R^m = S_{j+m} S_j$ and $S_j R^m = S_{j-m}$. Thus, the group is isomorphic onto D_4. The correspondence is

$$R \longleftrightarrow i\sigma_3, \qquad S_0 \longleftrightarrow \sigma_2, \qquad S_1 \longleftrightarrow \sigma_1.$$

The orders of elements and the classes in these five groups are listed as follows.

Group	No. of elements				No. of classes
	1st-order	2nd-order	4th-order	8th-order	
C_8	1	1	2	4	8
Q_8	1	1	6	0	5
C_{4h}	1	3	4	0	6
D_4	1	5	2	0	5
D_{2h}	1	7	0	0	8

11. Investigate all ninth-order group which are not isomorphic onto each other.

Solution. The order of any element in a ninth-order group G, except for the identity, has to be 3 or 9. If there is at least one ninth-order element in the group, then this group is a cyclic group C_9.

If there is no ninth-order element, all elements in G, except for the identity, are the third-order elements. Arbitrary take a third-order element, denoted by A. The cyclic subgroup generated by A is $H = \{E, A, A^2\}$. Denote one right coset by $HB = \{B, C, D\}$ where $AB = C$, $AC = D$, and $AD = B$. Since B, C and D are all third-order elements, their square cannot be equal to E, A or A^2. From the rearrangement theorem, their square cannot be equal to B, C or D. They have to be different from each other due to the uniqueness of the inverse element. Therefore, the remaining three elements in the group are B^2, C^2 and D^2, which form another right coset of H. From the rearrangement theorem, $AB^2 = CB$ cannot be equal to C^2 and B^2, so it has to be equal to D^2. Similarly, we can calculate the remaining products and give the following multiplication table of the group. From the multiplication table, we know that this group is an Abelian group with nine classes. There are only two different ninth-order groups.

	E	A	A^2	B	C	D	B^2	C^2	D^2
E	E	A	A^2	B	C	D	B^2	C^2	D^2
A	A	A^2	E	C	D	B	D^2	B^2	C^2
A^2	A^2	E	A	D	B	C	C^2	D^2	B^2
B	B	C	D	B^2	D^2	C^2	E	A^2	A
C	C	D	B	D^2	C^2	B^2	A	E	A^2
D	D	B	C	C^2	B^2	D^2	A^2	A	E
B^2	B^2	D^2	C^2	E	A	A^2	B	D	C
C^2	C^2	B^2	D^2	A^2	E	A	D	C	B
D^2	D^2	C^2	B^2	A	A^2	E	C	B	D

12. Investigate all tenth-order groups which are not isomorphic.

Solution. The order of any element in a tenth-order group G, except for the identity, has to be 2, 5 and 10. If there is at least one tenth-order element in the group, then this group is a cyclic group C_{10}.

If all elements in a tenth-order group, except for the identity, are second-order elements, arbitrary taking two different elements R and S, we have $R^2 = S^2 = E$, $RS = T$. T has to be a new element as we have known in Problem 1. Thus, we obtain a fourth-order subgroup $\{E, R, S, T\}$ isomorphic onto the fourth-order inversion group V_4. Since the order of the subgroup is not a whole number divisor of ten, a contradiction.

If there is no tenth-order element in G, but there is at least one fifth-order element, denoted as R, then the cyclic subgroup $\{E, R, R^2, R^3, R^4\}$ generated by R is an invariant subgroup because its index is 2. Denote its coset by $\{S_0, S_1, S_2, S_3, S_4\}$, where $R^m S_0 = S_m$, $R^m S_j = S_{j+m}$, and $S_{j+5} = S_j$. From the rearrangement theorem, S_j^2 cannot be equal to S_k.

If it is equal to R^j, where j is not a multiple of 5, then S_j is a tenth-order element, thus a contradiction. So $S_j^2 = E$, S_j all are second-order elements. Thus, we have $R^m = S_{j+m}S_j$ and $S_j R^m = S_{j-m}$. This group is the group D_5. Up to isomorphism, there are only two tenth-order group, C_{10} and D_5. From the investigation on the sixth-order group (Problem 6) and the tenth-order group, we conclude that when the order g of a group G is the double of a prime number $n > 1$, the group has to be a cyclic group C_{2n} or the symmetry group D_n for a regular n-sided polygon.

13. Give an example to show that the invariant subgroup of an invariant subgroup of a group G is not necessary an invariant subgroup of the group G. Conversely, show that if an invariant subgroup of a group G completely belongs to a subgroup H of G, then it is also an invariant subgroup of H.

Solution. Let H be a subgroup of G, and let H_1 be a subgroup of H. For an arbitrary element R in H_1, an element SRS^{-1} conjugate to R for G may not be an element conjugate to R for H, because $S \in G$ may not belong to H. Conversely, an element SRS^{-1} conjugate to R for H must be an element conjugate to R for G, because $S \in H$ must belong to G. Therefore, the invariant subgroup of an invariant subgroup H of a group G is not necessary an invariant subgroup of G, but the invariant subgroup H_1 of a group G must be an invariant subgroup of the subgroup $H \subset G$ if $H_1 \subset H$.

An example is as follows. The group **T** contains three twofold axes and four threefold axes. **T** contains an invariant subgroup D_2 composed of the identity E and three twofold rotations. However, three second-order subgroups in D_2 are all the invariant subgroups of D_2, but not the invariant subgroup of T.

14. In a finite group G of order g, let $C_\alpha = \{S_1, S_2, \ldots, S_{n(\alpha)}\}$ be a class of G containing $n(\alpha)$ elements. For any two elements S_i and S_j in the class C_α (may be different or the same), show that the number $m(\alpha)$ of elements $P \in G$ satisfying $S_i = PS_jP^{-1}$ is $g/n(\alpha)$.

Solution. For a given element $S_j \in C_\alpha$, let the number of elements $R \in G$, commutable with S_j, be $m(\alpha)$. It is easy to show that $m(\alpha)$ does not depend on S_j. In fact, if $S_i = PS_jP^{-1}$, PRP^{-1} commutes with S_i.

Under the multiplication rule of G, the set H of $R \in G$, commutable with S_j, forms a subgroup of the finite group G, because if both R and

R' can commute with S_j, RR' also commutes with S_j. The order of H is $m(\alpha)$. Note that H is not necessary an invariant subgroup of G. Any element $T \in G$, which does not belong to the subgroup H, cannot commute with S_j. Let $TS_jT^{-1} = S_i \in C_\alpha$ and $S_j \neq S_i$. Any element TR in the left coset TH of the subgroup H satisfies $TRS_jR^{-1}T^{-1} = S_i$. The number of elements in the left coset TH is still $m(\alpha)$. On the other hand, for any element P which satisfies $PS_jP^{-1} = S_i$, we have

$$\left(P^{-1}T\right) S_j \left(P^{-1}T\right)^{-1} = P^{-1}S_iP = S_j.$$

Thus, $P^{-1}T \in H$, and P belongs to the left coset TH. Therefore, we obtain a one-to-one correspondence between the element $S_i \in C_\alpha$ and the left coset TH by the relation

$$TRS_jR^{-1}T^{-1} = S_i.$$

When $S_i = S_j$, $T \in H$ and $TH = H$. The index $g/m(\alpha)$ of the subgroup H is equal to the number $n(\alpha)$ of the elements S_j in the class C_α. Namely, $g = m(\alpha)n(\alpha)$. Both the order $m(\alpha)$ of a subgroup H and the number $n(\alpha)$ of elements in a class are the divisors of the order g of the group G.

15. Prove that being the product of two subsets, the product of two classes in a group G must be a sum aggregate of a few whole classes. Namely, the sum aggregate contains all elements conjugate to any product of two elements belonging to the two classes, respectively.

Solution. Let C_α and C_β be two classes in a group G. For given $R \in C_\alpha$ and $S \in C_\beta$, we have $TRT^{-1} \in C_\alpha$ and $TST^{-1} \in C_\beta$, where T is an arbitrary element in G. Therefore,

$$T(RS)T^{-1} = TRT^{-1}TST^{-1} \in C_\alpha C_\beta.$$

The conclusion is proved. It is easy to prove that the multiplicity of RS in the product space of $C_\alpha C_\beta$ is the same as that of the element conjugate to RS. Therefore, considering the multiplicity $f(\alpha, \beta, \gamma)$, we have

$$C_\alpha C_\beta = \sum_\gamma f(\alpha, \beta, \gamma)C_\gamma.$$

16. Calculate the multiplication table of the group **T** by extending the multiplication table of the subgroup $C_3 = \{E, \ R_1, \ R_1^2\}$ of **T**.

Solution. The symmetry group $\mathbf{T}$ for a tetrahedron contains three twofold axes orthogonal to each other and four threefold axes. Denote the twofold rotations by T_x^2, T_y^2 and T_z^2, respectively. $T_z^2 = T_x^2 T_y^2$. The threefold axes in T are respectively along the directions $\mathbf{e}_x \pm \mathbf{e}_y \pm \mathbf{e}_z$, and $-\mathbf{e}_x \mp \mathbf{e}_y \pm \mathbf{e}_z$. Denoting by R_1 the rotation through $2\pi/3$ around the direction $\mathbf{e}_x + \mathbf{e}_y + \mathbf{e}_z$, we have

$$R_1^2 T_x^2 R_1 = T_z^2, \qquad R_1^2 T_y^2 R_1 = T_x^2, \qquad R_1^2 T_z^2 R_1 = T_y^2,$$
$$R_1 T_x^2 R_1^2 = T_y^2, \qquad R_1 T_y^2 R_1^2 = T_z^2, \qquad R_1 T_z^2 R_1^2 = T_x^2.$$

The remaining threefold rotations can be defined as

$$T_x^2 R_1 = R_4, \qquad T_y^2 R_1 = R_3, \qquad T_z^2 R_1 = R_2.$$

Substituting them into the above formulas, we obtain

$$R_1 T_x^2 = R_3, \qquad R_1 T_y^2 = R_2, \qquad R_1 T_z^2 = R_4.$$

Taking the inverse of the formulas, we have

$$R_1^2 T_x^2 = R_4^2, \qquad R_1^2 T_y^2 = R_3^2, \qquad R_1^2 T_z^2 = R_2^2,$$
$$T_x^2 R_1^2 = R_3^2, \qquad T_y^2 R_1^2 = R_2^2, \qquad T_z^2 R_1^2 = R_4^2.$$

In terms of $T_x^2 T_y^2 = T_z^2$ we obtain

$$
\begin{array}{lll}
T_x^2 R_2 = R_3, & T_x^2 R_3 = R_2, & T_x^2 R_4 = R_1, \\
T_y^2 R_2 = R_4, & T_y^2 R_3 = R_1, & T_y^2 R_4 = R_2, \\
T_z^2 R_2 = R_1, & T_z^2 R_3 = R_4, & T_z^2 R_4 = R_3, \\
T_x^2 R_2^2 = R_4^2, & T_x^2 R_3^2 = R_1^2, & T_x^2 R_4^2 = R_2^2, \\
T_y^2 R_2^2 = R_1^2, & T_y^2 R_3^2 = R_4^2, & T_y^2 R_4^2 = R_3^2, \\
T_z^2 R_2^2 = R_3^2, & T_z^2 R_3^2 = R_2^2, & T_z^2 R_4^2 = R_1^2.
\end{array}
$$

The above equations give the formulas for the right cosets of the subgroup C_3 and the formulas for left-multiplying T_x^2, T_y^2 and T_z^2 on the elements in $\mathbf{T}$. From them we can obtain the multiplication table of $\mathbf{T}$.

Arrange the rows in the multiplication table (the first column of the table) in the order of the subgroup C_3 and its left cosets $T_x^2 C_3$, $T_y^2 C_3$ and $T_z^2 C_3$, and the columns (the first row of the table) in the order of C_3 and its right cosets $C_3 T_x^2$, $C_3 T_y^2$ and $C_3 T_z^2$. Thus, the multiplication table is divided into 16 subtables. Each subtable contains 3×3 multiplication elements. The first subtable in the first row is just the multiplication table of the subgroup C_3. The remaining three subtables are obtained by right-multiplying the first subtable with T_x^2, T_y^2 and T_z^2, respectively. The next three rows are obtained by left-multiplying the first row with T_x^2, T_y^2 and T_z^2, respectively. The relative arrangement of elements in each subtable is the same as the multiplication table of C_3, where the only difference is the contained elements, which are given in the above formulas. Finally, we obtain multiplication table of the symmetry group **T** for the regular tetrahedral.

	E	R_1	R_1^2	T_x^2	R_3	R_4^2	T_y^2	R_2	R_3^2	T_z^2	R_4	R_2^2
E	E	R_1	R_1^2	T_x^2	R_3	R_4^2	T_y^2	R_2	R_3^2	T_z^2	R_4	R_2^2
R_1	R_1	R_1^2	E	R_3	R_4^2	T_x^2	R_2	R_3^2	T_y^2	R_4	R_2^2	T_z^2
R_1^2	R_1^2	E	R_1	R_4^2	T_x^2	R_3	R_3^2	T_y^2	R_2	R_2^2	T_z^2	R_4
T_x^2	T_x^2	R_4	R_3^2	E	R_2	R_2^2	T_z^2	R_3	R_1^2	T_y^2	R_1	R_4^2
R_4	R_4	R_3^2	T_x^2	R_2	R_2^2	E	R_3	R_1^2	T_z^2	R_1	R_4^2	T_y^2
R_3^2	R_3^2	T_x^2	R_4	R_2^2	E	R_2	R_1^2	T_z^2	R_3	R_4^2	T_y^2	R_1
T_y^2	T_y^2	R_3	R_2^2	T_z^2	R_1	R_3^2	E	R_4	R_4^2	T_x^2	R_2	R_1^2
R_3	R_3	R_2^2	T_y^2	R_1	R_3^2	T_z^2	R_4	R_4^2	E	R_2	R_1^2	T_x^2
R_2^2	R_2^2	T_y^2	R_3	R_3^2	T_z^2	R_1	R_4^2	E	R_4	R_1^2	T_x^2	R_2
T_z^2	T_z^2	R_2	R_4^2	T_x^2	R_4	R_1^2	T_y^2	R_1	R_2^2	E	R_3	R_3^2
R_2	R_2	R_4^2	T_z^2	R_4	R_1^2	T_x^2	R_1	R_2^2	T_y^2	R_3	R_3^2	E
R_4^2	R_4^2	T_z^2	R_2	R_1^2	T_x^2	R_4	R_2^2	T_y^2	R_1	R_3^2	E	R_3

17. The multiplication table of the finite group G is as follows.

	E	A	B	C	D	F	I	J	K	L	M	N
E	E	A	B	C	D	F	I	J	K	L	M	N
A	A	E	F	I	J	B	C	D	M	N	K	L
B	B	F	A	K	L	E	M	N	I	J	C	D
C	C	I	L	A	K	N	E	M	J	F	D	B
D	D	J	K	L	A	M	N	E	F	I	B	C
F	F	B	E	M	N	A	K	L	C	D	I	J
I	I	C	N	E	M	L	A	K	D	B	J	F
J	J	D	M	N	E	K	L	A	B	C	F	I
K	K	M	J	F	I	D	B	C	N	E	L	A
L	L	N	I	J	F	C	D	B	E	M	A	K
M	M	K	D	B	C	J	F	I	L	A	N	E
N	N	L	C	D	B	I	J	F	A	K	E	M

a) Find the inverse of each element in G;

b) Point out which elements can commute with any element in G;

c) List the period and order of each element;

d) Find the elements in each class of G;

e) Find all invariant subgroups in G. For each invariant subgroup, list its cosets and point out onto which group its quotient group is isomorphic;

f) Make a judgment whether G is isomorphic onto the tetrahedral symmetric group **T**, or isomorphic onto the regular six-sided polygon symmetric group D_6.

Solution. a) The pairs of elements which are inverse to each other are given as follows:

$$E^{-1} = E, \quad A^{-1} = A, \quad B^{-1} = F, \quad C^{-1} = I$$
$$D^{-1} = J, \quad K^{-1} = L, \quad M^{-1} = N.$$

b) E and A can commute with any element in the group.

c) The element whose order is 1 is the identity E.

The element of order 2 is A. Its period is $\{E, A\}$.

The elements of order 3 are M and N. They belong to a common period $\{E, M, N\}$.

The elements of order 4 are B, C, D, F, I and J. They belong to the periods $\{E, B, A, F\}$, $\{E, C, A, I\}$, and $\{E, D, A, J\}$, respectively.

The sixth-order elements are K and L. They belong to a common period $\{E, K, N, A, M, L\}$.

d) E and A form two classes, $\{E\}$ and $\{A\}$. The elements of order 3 and 6 respectively form two classes, $\{M, N\}$ and $\{K, L\}$. The elements of order 4 are divided into two classes, $\{B, C, D\}$ and $\{F, I, J\}$.

e) The invariant subgroup consists of several whole classes, and contains the period of each element. Its order has to be a divisor of 12, the order of G, namely, it has to be 2, 3, 4 or 6.

$\{E, A\}$ is an invariant subgroup of G. Its cosets are $\{B, F\}$, $\{C, I\}$, $\{D, J\}$, $\{K, M\}$, and $\{L, N\}$. Because $B^2 = C^2 = D^2 = A$, the squares of the first three cosets all are the invariant subgroup. Thus, its quotient group is isomorphic onto the regular triangle symmetric group D_3.

$\{E, M, N\}$ is an invariant subgroup of G. Its cosets are $\{A, K, L\}$, $\{B, C, D\}$ and $\{F, I, J\}$. Due to $B^2 = F^2 = A$, the squares of the last two cosets are not the invariant subgroup. Thus, its quotient group is isomorphic onto fourth-order cyclic group C_4.

$\{E, K, N, A, M, L\}$ is an invariant subgroup with index 2. Its coset is $\{B, I, D, F, C, J\}$. Its quotient group is isomorphic onto two order inversion group V_2.

f) Since the group **T** does not contain any sixth-order element, the group G is not isomorphic onto **T**. Since D_6 does not contain any fourth-order element, the group G is not isomorphic onto D_6.

Chapter 3

THEORY OF REPRESENTATIONS

3.1 Transformation Operators for a Scalar Function

★ An m-dimensional matrix group $D(G)$ is called a representation of the given group G, if G is isomorphic or homomorphic onto $D(G)$. The element $D(R)$ in $D(G)$, which is nonsingular, is called the representation matrix of the group element $R \in G$ in the representation $D(G)$. The trace $\mathrm{Tr}D(R) \equiv \chi(R)$ is called the character of R in the representation $D(G)$. The representation matrix $D(E)$ of the identity E is a unit matrix, and the representation matrices of two elements R^{-1} and R are mutually inverse matrices, $D(R^{-1}) = D(R)^{-1}$. The representation $D(G)$ is said to be faithful if G is isomorphic onto $D(G)$. If all matrices $D(R)$ are unitary, $D(G)$ is called a unitary representation. If all matrices $D(R)$ are real orthogonal, $D(G)$ is called a real orthogonal representation. $D(G)^*$ composed of the complex conjugate $D(R)^*$ of the representation matrices $D(R)$ is the conjugate representation of G with respect to $D(G)$. The representation is called a self-conjugate representation if all characters in the representation are real. The representation $D(G)$ is called the identical representation if $D(R) = 1$ for any element R in G. The dimension m of the representation is assumed to be finite in this book if without special notification.

★ Denote simply by x all coordinates of degrees of freedom in a quantum system, and by $\psi(x)$ the scalar wave function. Under a transformation R, x is transformed into $x' = Rx$, the wave function $\psi(x)$ is transformed into $\psi'(x) \equiv P_R\psi(x)$. Being a scalar wave function, the value of the transformed wave function $P_R\psi$ at the point Rx must be equal to that of the wave function ψ before transformation at the point x, namely

$$P_R\psi(Rx) = \psi(x), \qquad P_R\psi(x) = \psi(R^{-1}x). \qquad (3.1)$$

The transformed wave function $P_R\psi(x)$ can be obtained from the original wave function $\psi(x)$ according to Eq. (3.1), namely, first to replace x with $R^{-1}x$ in the wave function $\psi(x)$, then to regard it as the function of x, which is nothing but the transformed function $P_R\psi(x)$. P_R is a linear operator and has a one-to-one correspondence with the transformation R. This correspondence is invariant in the product of transformations, namely, $P_R P_S$ corresponds to RS in the same rule. If the set of transformations R forms a group G, then the set P_G of P_R also forms a group and is isomorphic onto G. A linear operator $L(x)$ for the scalar wave functions is transformed into $L'(x)$ in the transformation R:

$$L(x) \xrightarrow{R} L'(x) = P_R L(x) P_R^{-1}.$$

★ Assume that a quantum system with the hamiltonian $H(x)$ is described by a scalar wave function. R is said to be a symmetry transformation if P_R can commute with $H(x)$. The set of symmetry transformations R forms the symmetry group G of the system. If the energy level E is m degeneracy, we denote by $\psi_\mu(x)$ the basis in the eigenfunction space. Then, the m-dimensional eigenfunction space for the energy E keeps invariant under the application of P_R. The matrix form of P_R in this space with respect to the basis $\psi_\mu(x)$ is $D(R)$:

$$P_R\psi_\mu(x) = \sum_\nu \psi_\nu(x) D_{\nu\mu}(R). \qquad (3.2)$$

The set of $D(R)$ forms a representation of the symmetry group.

1. Let G be a non-Abelian group, $D(G)$ be a faithful representation of the group G, and $D(R)$ be the representation matrix of element R. If the element R in G corresponds to the following matrix in the set, please decide whether the following set forms a representation of the group G. For example, in a), if $R \longleftrightarrow D(R)^\dagger$, please decide whether the set $D(G)^\dagger$ composed of $D(R)^\dagger$ forms a representation of G.
   a) $D(R)^\dagger$; b) $D(R)^T$; c) $D(R^{-1})$; d) $D(R)^*$; e) $D(R^{-1})^\dagger$;
   f) $\det D(R)$; g) $\text{Tr } D(R)$.

Solution. Each subproblem gives a one-to-one correspondence between the element in G and the matrix in the given set of matrices. We should judge whether their products satisfy the same one-to-one correspondence in the given rule. If yes, it is a representation of G. Otherwise, it is not. Notice that $D(RS) = D(R)D(S)$.

a) Since $D(R)^\dagger D(S)^\dagger \neq D(RS)^\dagger$, the set of $D(R)^\dagger$ is not a representation of G.

b) Since $D(R)^T D(S)^T \neq D(RS)^T$, the set of $D(R)^T$ is not a representation of G.

c) Since $D(R^{-1})D(S^{-1}) \neq D\left[(RS)^{-1}\right]$, the set of $D(R^{-1})$ is not a representation of G.

d) Since $D(R)^* D(S)^* = D(RS)^*$, the set of $D(R)^*$ is a representation of G, which is called the conjugate representation of $D(G)$.

e) Since $D(R^{-1})^\dagger D(S^{-1})^\dagger = D\left[(RS)^{-1}\right]^\dagger$, the set of $D(R^{-1})^\dagger$ is a representation of G.

f) Since $\det D(R) \det D(S) = \det D(RS)$, the set of $\det D(R)$ is a one-dimensional representation of G.

g) Since $\mathrm{Tr}D(R)\,\mathrm{Tr}D(S) \neq \mathrm{Tr}D(RS)$, the set of $\mathrm{Tr}D(R)$ is not a representation of G.

2. The homogeneous function space of degree 2 spanned by the basis functions $\psi_1(x,y) = x^2$, $\psi_2(x,y) = xy$, and $\psi_3(x,y) = y^2$ is invariant in the following rotations R in the two-dimensional coordinate space. Calculate the matrix form $D(R)$ of the corresponding transformation operator P_R for the scalar function in the three-dimensional function space:

$$\begin{pmatrix} x' \\ y' \end{pmatrix} = R \begin{pmatrix} x \\ y \end{pmatrix},$$

$$a) \quad R = \begin{pmatrix} 1 & 0 \\ 0 & -1 \end{pmatrix},$$

$$b) \quad R = \frac{1}{2}\begin{pmatrix} -1 & -\sqrt{3} \\ \sqrt{3} & -1 \end{pmatrix}, \qquad c) \quad R = \begin{pmatrix} \cos\alpha & -\sin\alpha \\ \sin\alpha & \cos\alpha \end{pmatrix}.$$

If replacing the basis function $\psi_2(x,y) = xy$ with $\sqrt{2}xy$, how will the matrix form of the operator P_R change?

Solution. Three transformation matrices in the problem are all real orthogonal. The inverse matrix of R is equal to its transpose.

$$a) \quad \begin{aligned} P_R\psi_1(x,y) &= x^2 = \psi_1(x,y), \\ P_R\psi_2(x,y) &= -xy = -\psi_2(x,y), \\ P_R\psi_3(x,y) &= y^2 = \psi_3(x,y), \end{aligned} \qquad D(R) = \begin{pmatrix} 1 & 0 & 0 \\ 0 & -1 & 0 \\ 0 & 0 & 1 \end{pmatrix}.$$

b) $\quad P_R\psi_1(x,y) = (1/4)\left(-x+\sqrt{3}y\right)^2$

$\qquad = (1/4)\left\{\psi_1(x,y) - 2\sqrt{3}\psi_2(x,y) + 3\psi_3(x,y)\right\},$

$\quad P_R\psi_2(x,y) = (1/4)\left(-x+\sqrt{3}y\right)\left(-\sqrt{3}x-y\right)$

$\qquad = (1/4)\left\{\sqrt{3}\psi_1(x,y) - 2\psi_2(x,y) - \sqrt{3}\psi_3(x,y)\right\},$

$\quad P_R\psi_3(x,y) = (1/4)\left(-\sqrt{3}x-y\right)^2$

$\qquad = (1/4)\left\{3\psi_1(x,y) + 2\sqrt{3}\psi_2(x,y) + \psi_3(x,y)\right\},$

$$D(R) = \frac{1}{4}\begin{pmatrix} 1 & \sqrt{3} & 3 \\ -2\sqrt{3} & -2 & 2\sqrt{3} \\ 3 & -\sqrt{3} & 1 \end{pmatrix}.$$

c) $\quad P_R\psi_1(x,y) = (x\cos\alpha + y\sin\alpha)^2$

$\qquad = \cos^2\alpha\,\psi_1(x,y) + \sin(2\alpha)\psi_2(x,y) + \sin^2\alpha\,\psi_3(x,y),$

$\quad P_R\psi_2(x,y) = (x\cos\alpha + y\sin\alpha)(-x\sin\alpha + y\cos\alpha)$

$\qquad = -\sin\alpha\cos\alpha\,\psi_1(x,y) + \cos(2\alpha)\psi_2(x,y) + \sin\alpha\cos\alpha\,\psi_3(x,y),$

$\quad P_R\psi_3(x,y) = (-x\sin\alpha + y\cos\alpha)^2$

$\qquad = \sin^2\alpha\,\psi_1(x,y) - \sin(2\alpha)\psi_2(x,y) + \cos^2\alpha\,\psi_3(x,y),$

$$D(R) = \frac{1}{2}\begin{pmatrix} 2\cos^2\alpha & -\sin(2\alpha) & 2\sin^2\alpha \\ 2\sin(2\alpha) & 2\cos(2\alpha) & -2\sin(2\alpha) \\ 2\sin^2\alpha & \sin(2\alpha) & 2\cos^2\alpha \end{pmatrix}.$$

Multiplying the basis function $\psi_2(x,y)$ by $\sqrt{2}$, we obtain the matrix form of P_R by a diagonal similarity transformation M.

$$M = \text{diag}\,\{1,\ \sqrt{2},\ 1\}, \quad \overline{D}(R) = M^{-1}D(R)M.$$

a) $\qquad \overline{D}(R) = D(R),$

b) $\qquad \overline{D}(R) = \frac{1}{4}\begin{pmatrix} 1 & \sqrt{6} & 3 \\ -\sqrt{6} & -2 & \sqrt{6} \\ 3 & -\sqrt{6} & 1 \end{pmatrix},$

c) $\qquad \overline{D}(R) = \frac{1}{\sqrt{2}}\begin{pmatrix} \sqrt{2}\cos^2\alpha & -\sin(2\alpha) & \sqrt{2}\sin^2\alpha \\ \sin(2\alpha) & \sqrt{2}\cos(2\alpha) & -\sin(2\alpha) \\ \sqrt{2}\sin^2\alpha & \sin(2\alpha) & \sqrt{2}\cos^2\alpha \end{pmatrix}.$

Each $\overline{D}(R)$ becomes a real orthogonal matrix.

3.2 Inequivalent and Irreducible Representations

★ Two representations $D(G)$ and $\overline{D}(G)$ with the same dimension are called equivalent to each other if there is a similarity transformation X relating two representation matrices for each element R in the group G, $\overline{D}(R) = X^{-1}D(R)X$. The characters of each element R in two equivalent representations must be the same. For a finite group and a compact Lie group, any representation is equivalent to a unitary representation, two equivalent unitary representations can be related by a unitary similarity transformation, and two representations are equivalent if and only if the corresponding characters of each element in two representations are equal to each other.

★ A representation $D(G)$ of the group G is said to be reducible if the representation matrix $D(R)$ of any element R of G in $D(G)$ can be transformed into the same form of the echelon matrix by a common similarity transformation X,

$$X^{-1}D(R)X = \begin{pmatrix} D^{(1)}(R) & T(R) \\ 0 & D^{(2)}(R) \end{pmatrix} . \tag{3.3}$$

Otherwise, it is called an irreducible representation. The necessary and sufficient condition for a reducible representation is that there is a nontrivial invariant subspace with respect to $D(G)$ in its representation space. If both complementary subspaces are invariant with respect to $D(G)$, the reducible representation is called the completely reducible one where there exists a common similarity transformation X such that the $T(R)$ in Eq. (3.3) for each element R of G is vanishing:

$$X^{-1}D(R)X = \begin{pmatrix} D^{(1)}(R) & 0 \\ 0 & D^{(2)}(R) \end{pmatrix} . \tag{3.4}$$

This form of reducible representation, where the representation matrix $D(R)$ of each element R of G is a direct sum of two submatrices $D^{(1)}(R)$ and $D^{(2)}(R)$, is called the reduced representation. The representation $D(G)$ is said to be the direct sum of two representations $D^{(1)}(G)$ and $D^{(2)}(G)$. For a finite group and a compact Lie group, any reducible representation is complete reducible, and can be transformed into the direct sum of two representations by a similarity transformation.

★ The Schur theorem says that X must be a null matrix if $D^{(1)}(R)X = XD^{(2)}(R)$ holds for each element R of G where $D^{(1)}(G)$ and $D^{(2)}(G)$ are two inequivalent and irreducible representations of G, and X must be a constant matrix if $D(R)X = XD(R)$ holds for each element R of G where $D(G)$ is an irreducible representations of G.

★ The necessary and sufficient condition for two inequivalent and irreducible representations $D^i(G)$ and $D^j(G)$ of a finite group G is

$$\frac{1}{g} \sum_{R \in G} \chi^i(R)^* \chi^j(R) = \frac{1}{g} \sum_{\alpha} n(\alpha) \left(\chi_\alpha^i\right)^* \chi_\alpha^j = \delta_{ij}, \qquad (3.5)$$

where $n(\alpha)$ denotes the number of elements in the class C_α, and g is the order of G. Equation (3.5) says that the characters of two inequivalent and irreducible representations are orthogonal to each other in the class space with the weight $n(\alpha)$. When $i = j$, Eq. (3.5) is the necessary and sufficient condition for an irreducible representation of G. The characters in all inequivalent and irreducible representations of G form a complete bases in the class space, namely the characters satisfy

$$\sum_j \chi_\alpha^j \chi_\beta^{j\,*} = \frac{g}{n(\alpha)} \delta_{\alpha\beta}. \qquad (3.6)$$

The number of all inequivalent and irreducible representations of a finite group G is equal to the number g_c of the classes in G. The sum of dimension squares m_j^2 of all inequivalent and irreducible representations is equal to the order g of G,

$$\sum_j 1 = g_c, \qquad \sum_j m_j^2 = g. \qquad (3.7)$$

Equations (3.5-7) are the necessary conditions satisfied by the characters in any finite group G. However, they are not enough to determine all characters of G. Considering some other methods, such as the representations of the quotient group of the invariant subgroups of G, the subduced and induced representations, the representations of the direct product of two subgroups etc., we can determine the characters in all inequivalent and irreducible representations of G, and obtain the character table of G. Based on the character table, we can find out the convenient forms of the unitary representation matrices of G. Usually, we choose as many diagonal representation matrices of generators as possible. For the complicated groups, the special method are needed.

★ For a given quantum system with the Hamiltonian $H(x)$, we first find its symmetry group G, where the transformation operator P_R for each element

R in G can commute with the Hamiltonian $H(x)$,

$$P_R H(x) = H(x)P_R. \tag{3.8}$$

P_R is called the symmetry operator of the system.

Second, we study the symmetry group G of the system. We want to find out its character table and the convenient forms of the representation matrices of all generators R in G. The representation matrices of other elements can be calculated from those of the generators.

Third, if the energy level E is m degeneracy, we can find arbitrarily m linearly independent eigenfunctions $\psi_\mu(x)$ for the energy E:

$$H(x)\psi_\mu(x) = E\psi_\mu(x), \qquad \mu = 1, 2, \ldots, m.$$

Due to Eq. (3.8), $P_R\psi_\mu(x)$ must be an eigenfunction of $H(x)$ with the same energy E, namely, the m-dimensional space spanned by $\psi_\mu(x)$ is invariant in the action of the symmetry operator P_R. We can calculate the matrix form $D(R)$ of P_R in the basis function $\psi_\mu(x)$:

$$P_R\psi_\mu(x) = \psi_\mu(R^{-1}x) = \sum_{\nu=1}^{m} \psi_\nu(x)D_{\nu\mu}(R). \tag{3.9}$$

The set of $D(R)$ forms a representation $D(G)$ of the symmetry group G of the system, called the representation corresponding to the energy level E. The character $\chi(R) = \text{Tr}D(R)$ of R is easy to calculate. Generally speaking, the representation $D(G)$ is reducible and not in the convenient form. In terms of the following method, the representation $D(G)$ can be reduced into the direct sum of the irreducible representations and the eigenfunctions $\psi_\mu(x)$ can be combined into the basis functions transformed according to the irreducible representation.

Make a similarity transformation X,

$$X^{-1}D(R)X = \bigoplus_j a_j D^j(R), \qquad \chi_\alpha = \sum_j a_j \chi_\alpha^j. \tag{3.10}$$

The multiplicity a_j of the irreducible representation $D^j(G)$ in the reducible representation $D(G)$ can be calculated from the orthogonal relation (3.5):

$$a_j = \frac{1}{g}\sum_{R \in G} \chi^j(R)^*\chi(R) = \frac{1}{g}\sum_\alpha n(\alpha)\chi_\alpha^{j\ *}\chi_\alpha. \tag{3.11}$$

Substituting it into Eq. (3.10), one is able to calculate the similarity transformation matrix X. Let R be the generator A whose representation matrix

$D^j(A)$ is diagonal, where the problem becomes that of finding a similarity transformation X to diagonalize a matrix $D(A)$. An important point is to keep all the undetermined parameters in X waiting for the subsequent calculation. Substituting the matrix X into Eq. (3.10) for the remaining generator B whose representation matrix $D^j(B)$ may not be diagonal. In the calculation, X^{-1} in Eq. (3.10) should be moved to the right-hand side of the equation for avoiding the calculation for X^{-1}. After the calculation of Eq. (3.10) for all generators of G, if there are still some parameters undetermined, they should be chosen by any feasible way. Do not leave them as the undetermined parameters. Make sure that $\det X$ should not be vanishing. The row index μ of X is the same as that of $D(R)$, and the column index of X is a combined set of the index j for the irreducible representation and its row index ρ. When $a_j > 1$, an additional index r is needed to distinguish different D^j in the reduction (3.10). The new eigenfunctions are the combinations of ψ_μ by X:

$$
\begin{aligned}
\Phi^j_{\rho r}(x) &= \sum_\mu \psi_\mu(x) X_{\mu,j\rho r}, \\
P_R \Phi^j_{\rho r}(x) &= \sum_\lambda \Phi^j_{\lambda r}(x) D^j_{\lambda\rho}(R).
\end{aligned}
\tag{3.12}
$$

$\Phi^j_{\rho r}(x)$ is called a function belonging to the ρth row of the irreducible representation D^j of G. When $a_j > 1$, any linear combination of the functions $\Phi^j_{\rho r}(x)$ with the same j and ρ, where the combination coefficients are independent of j and ρ, is the eigenfunction belonging to the ρth row of D^j. This combination reflects the arbitrary choice of the undetermined parameters. The number of the undetermined parameters is $\sum_j a_j^2$.

★ For the inner product of two wave functions used in quantum mechanics, the symmetry operator P_R is usually unitary (there is exceptional case). Under this condition, the functions belonging to two inequivalent and irreducible representations of the symmetry group G are orthogonal to each other.

$$
\begin{aligned}
P_R \phi^i_\nu(x) &= \sum_\lambda \phi^i_\lambda(x) D^i_{\lambda\nu}(R), \\
P_R \psi^j_\mu(x) &= \sum_\rho \psi^j_\rho(x) D^j_{\rho\mu}(R), \\
\langle \phi^i_\nu(x),\ \psi^j_\mu(x) \rangle &= \delta_{ij} \delta_{\nu\mu} \langle \phi^i || \Psi^j \rangle,
\end{aligned}
\tag{3.13}
$$

where $\langle \phi^i || \Psi^j \rangle$ is called the reduced matrix elements, which is a parameter independent of the subscripts ν and μ. Equation (3.13) is called the Wigner-

Eckart theorem.

★ A group G is called the direct product of two subgroups, $G = H_1 \otimes H_2$, if each element in G can be expressed as a product of $R \in H_1$ and $S \in H_2$, $RS = SR$, and there is no common element in two subgroups H_1 and H_2 except for the identity E. A pure rotation in the three-dimensional space is called a proper rotation. A rotation together with a space inversion is called an improper rotation. A proper point group consists of the proper rotations, and improper point group consists of the proper rotations and the improper rotations. The subset of the proper rotations in an improper point group G forms an invariant subgroup H in G with index 2. An improper point group G is said to be I-type if $G = H \otimes V_2$, where $V_2 = \{e, \sigma\}$ is the two-order inversion group. An improper point group G is said to be P-type if G does not contain the space inversion σ. Multiplying σ on each improper rotation in a P-type improper point group G, we obtain a proper point group G'. G' is isomorphic onto G and contains an invariant subgroup with index 2.

3. Prove that the module of any representation matrix in a one-dimensional representation of a finite group is equal to 1.

Solution. In fact, the representation matrix in a one-dimensional representation is a complex number. For a finite group, any representation is equivalent to the unitary one, while a one-dimensional representation is invariant under any similarity transformation. Therefore, it is a unitary one, where the module of any representation matrix is 1.

There are many other methods to show this conclusion. In a finite group G, any element R in G with the order n satisfies $R^n = E$, and the representation matrix of the identity E in any representation is the unit matrix. Therefore, in a one-dimensional representation $D(G)$ of G, $D(R)^n = D(E) = 1$, namely, its module is 1.

4. Prove that any irreducible representation of an infinite Abelian group is one-dimensional.

Solution. In an Abelian group G, the product of elements are commutable, so the representation matrix $D(R)$ of any element R in an irreducible representation is commutable with the representation matrix of any other element in G. From the Schur theorem, $D(R)$ has to be a constant matrix. Since R is an arbitrary element in G, and the representation is irreducible, the dimension of the representation has to be one.

5. Prove that the similarity transformation matrix between two equivalent irreducible unitary representations of a finite group, if restricting its determinant to be 1, has to be unitary.

Solution. Let $D(G)$ and $\overline{D}(G)$ be two equivalent irreducible unitary representations of a finite group. There exists a unitary similarity transformation M, $M^\dagger M = \mathbf{1}$ and $\overline{D}(R) = M^{-1}D(R)M$. If they can be related by another similarity transformation X, $\overline{D}(R) = X^{-1}D(R)X$, then

$$D(R) = \left(XM^{-1}\right)^{-1} D(R) \left(XM^{-1}\right).$$

From the Schur theorem, $XM^{-1} = c\mathbf{1}$ and $X = cM$, where c is a constant. Since $\det X = 1$, $|c| = 1$. Thus, X is a unitary matrix.

6. Show that for a finite group G, the sum of the characters of all elements in any irreducible representation of G, except for the identical representation, is equal to zero.

Solution. For a finite group G, the characters of two inequivalent and irreducible representation satisfy Eq. (3.5). If the representation $D^i(R)$ is the identical representation, $D^i(R) = \chi^i(R) = 1$, Eq. (3.5) shows that the sum of characters $\chi^j(R)$ of any irreducible representation $D^j(G)$ of G, except for the identical representation, is equal to zero.

7. The linear space spanned by the elements of a finite group G is called the group space, where the group element is the basis and the vector is a linear combination of the group elements. The group space is invariant for left- or right-multiplication by any element of G. The set of the matrix forms $D(S)$ of left-multiplying an element S in the group space with respect to the basis R forms the regular representation $D(G)$ of G. The set of the matrix forms $\overline{D}(S)$ of right-multiplication forms an equivalent representation $\overline{D}(G)$ of G.

$$SR = \sum_{T \in G} T D_{TR}(S), \qquad RS = \sum_{T \in G} \overline{D}_{RT}(S)T,$$

$$D_{TR}(S) = \begin{cases} 1 & \text{when } SR = T \\ 0 & \text{when } SR \neq T, \end{cases}$$

$$\overline{D}_{RT}(S) = \begin{cases} 1 & \text{when } RS = T \\ 0 & \text{when } RS \neq T, \end{cases}$$

$$\chi(S) = \overline{\chi}(S) = \begin{cases} g & \text{when } S = E \\ 0 & \text{when } S \neq E. \end{cases}$$

Calculate the similarity transformation matrix X between two equivalent regular representations in the group space of the D_3 group. How to generalize the result to any other finite group?

Solution. Let X be the similarity transformation relating two equivalent regular representations:

$$\sum_{P \in G} \overline{D}_{TP}(S) X_{PR} = \sum_{P \in G} X_{TP} D_{PR}(S).$$

Substituting the values of the representation matrix entries into the equation, we have $X_{(TS)R} = X_{T(SR)}$. Noticing that the rows and the columns are designated by the group elements. From the associative law which is satisfied by the multiplication of elements, the matrix entries of X are equal if the products of their row index and the column index as the group elements are equal. Namely, we may choose the matrix entries of X such that the entries are one when the product of the row index and the column index as the group elements is equal to a given element, say E, and the remaining entries are zero. In fact, the similarity transformation matrix X can be calculated from the multiplication table of the group, where the arrangement of rows and columns is the same as those in X, in the following way: The matrix entry in X is one if its position is the same as that of a given element, say E, in the multiplication table, otherwise it is zero. For the group D_3, we may choose X by the positions of the identity E in the multiplication table. The reader is encouraged to write the matrix form of the generators of D_3 in two equivalent regular representations according to the multiplication table of D_3, and to check whether they are related by the similarity transformation X.

The multiplication table of D_3

	E	D	F	A	B	C
E	E	D	F	A	B	C
D	D	F	E	B	C	A
F	F	E	D	C	A	B
A	A	C	B	E	F	D
B	B	A	C	D	E	F
C	C	B	A	F	D	E

$$X = \begin{pmatrix} 1 & 0 & 0 & 0 & 0 & 0 \\ 0 & 0 & 1 & 0 & 0 & 0 \\ 0 & 1 & 0 & 0 & 0 & 0 \\ 0 & 0 & 0 & 1 & 0 & 0 \\ 0 & 0 & 0 & 0 & 1 & 0 \\ 0 & 0 & 0 & 0 & 0 & 1 \end{pmatrix}.$$

For the different choice of the given element, we can find g linearly independent similarity transformation matrices X, where g is the order of the group. In fact, the reduced form of the regular representation is the direct sum of all irreducible representations, where the multiplicity of each irreducible representation is equal to its dimension m_j. The matrix, which can commute with all representation matrices in the reduced form of the regular representation, contains $\sum_j m_j^2 = g$ parameters.

8. C_α is a class in a finite group G, and W is a vector in the group space composed of the sum of all elements in C_α. Prove that the representation matrix of W in an irreducible representation is a constant matrix, and calculate this constant.

Solution. Let $C_\alpha = \{R_1, R_2, \ldots, R_{n(\alpha)}\}$ be a class in G, where $n(\alpha)$ is the number of elements in the class. Denote by χ_α^j the character of R_k in the m_j-dimensional irreducible representation $D^j(R)$. For any element S in G, we have $SR_kS^{-1} \in C_\alpha$. Obviously, if $R_k \neq R_i$, then $SR_kS^{-1} \neq SR_iS^{-1}$. Therefore, the set of SR_kS^{-1} is the same as the class C_α, so that $SWS^{-1} = W$.

$$W = \sum_{R_k \in C_\alpha} R_k, \quad D^j(W) = \sum_{R_k \in C_\alpha} D^j(R_k), \quad \mathrm{Tr}D^j(W) = n(\alpha)\chi_\alpha^j.$$

Since $D^j(S)D^j(W)D^j(S)^{-1} = D^j(W)$, according to the Schur theorem, $D^j(W) = c\mathbf{1}$. Taking its trace, we obtain $n(\alpha)\chi_\alpha^j = m_j c$. Then,

$$D^j(W) = \left\{ \frac{n(\alpha)\chi_\alpha^j}{m_j} \right\} \mathbf{1}.$$

9. Prove that the number of the self-reciprocal classes in a finite group G is equal to the number of the inequivalent and irreducible self-conjugate representations of G. In other words, the number of pairs of the reciprocal classes is equal to the number of pairs of the inequivalent and irreducible non-self-conjugate representations.

Solution. Let n be the number of the self-reciprocal classes C_i in G, and m be the number of pairs of the reciprocal classes C_α and C_α^{-1}. The number of classes in G is $g_c = n + 2m$. In any representation, the character χ_i of a self-reciprocal class is real, and the characters of two reciprocal classes are complex conjugate to each other, $\chi_\alpha^* = \chi_{\alpha^{-1}}$. On the other hand, any character in a self-conjugate representation is real, but two characters of a class in the pair of two non-self-conjugate irreducible representations are complex conjugate to each other. The complex characters only occur for the non-self-reciprocal class in a non-self-conjugate representation. The sum of two characters of a class in two irreducible representations which are complex conjugate to each other is real, and the difference of two characters is pure imaginary for a non-self-reciprocal class and is zero for a self-reciprocal class.

Being a function of the classes, the characters in the inequivalent and irreducible representations of G are linearly independent, and form a complete set of bases in the class space. Any class function can be expressed as a linear combination of characters of the inequivalent and irreducible representations of G. On the one hand, since the differences of two characters in the different pairs of non-self-conjugate irreducible representations, being a function of the classes, is linear independent, the number of pairs of non-self-conjugate irreducible representations cannot be greater than the number m of pairs of the reciprocal classes C_α and C_α^{-1}. On the other hand, define m functions of the classes F_β,

$$F_\beta(C_i) = 0, \qquad F_\beta(C_\alpha) = -F_\beta(C_\alpha^{-1}) = \delta_{\alpha\beta}.$$

Being a function of the classes, F_β can be expressed as a linear combination of characters of all inequivalent irreducible representations of G. Due to the explicit form of F_β, the characters of the self-conjugate representations will not appear in the combination, and the characters of the non-self-conjugate representations will appear only in the form of difference of two characters in a pair of representations. Therefore, the numbers of pairs of non-self-conjugate irreducible representations cannot be smaller than m. In summary, the number of pairs of non-self-conjugate irreducible representations of a finite group G is equal to the number m of pairs of the reciprocal classes in G. As a result, the number of the self-conjugate irreducible representations of G is equal to the number n of the self-reciprocal classes in G. This completes the proof.

10. If the group G is a direct product $H_1 \otimes H_2$ of two subgroups, show that the direct product of two irreducible representations of two subgroups is an irreducible representation of G.

Solution. Suppose that $R \in H_1$, $S \in H_2$, $RS = SR \in G = H_1 \otimes H_2$, $D^j(H_1)$ is an irreducible representation of H_1 with the representation space $\mathcal{L}_1$, and $D^k(H_2)$ is an irreducible representation of H_2 with the representation space $\mathcal{L}_2$. $\mathcal{L}_1$ and $\mathcal{L}_2$ have no non-trivial invariant subspace for the groups H_1 and H_2, respectively.

First, we will show that the direct product of two irreducible representations of subgroups is a representation of the direct product group. In fact, if there is a correspondence from the elements RS and $R'S'$ in G to the matrices $D(RS) = D^j(R) \times D^k(S)$ and $D(R'S') = D^j(R') \times D^k(S')$, respectively, we have the correspondence from $RSR'S' = RR'SS'$ to

$$
\begin{aligned}
D(RS)D(R'S') &= \left[D^j(R) \times D^k(S)\right]\left[D^j(R') \times D^k(S')\right] \\
&= \left[D^j(R)D^j(R')\right] \times \left[D^k(S)D^k(S')\right] \\
&= D^j(RR') \times D^k(SS') = D(RR'SS').
\end{aligned}
$$

Therefore, the set of $D(RS)$ forms a representation of the group G, denoted by $D(G) = D^j(H_1) \times D^k(H_2)$. Its representation space, denoted by $\mathcal{L}$, is the direct product of the spaces $\mathcal{L}_1$ and $\mathcal{L}_2$. Let $\mathbf{a}$ and $\mathbf{b}$ respectively be the arbitrary vectors in the spaces $\mathcal{L}_1$ and $\mathcal{L}_2$. Then, their direct product $\mathbf{a} \times \mathbf{b}$ is a vector in the space $\mathcal{L}$.

Second, we will demonstrate that this representation is irreducible. According to the definition of irreducible representation, we are going to show by reduction to absurdity that there is no non-trivial invariant subspace in its representation space. If there is a nonzero invariant subspace $\mathcal{L}'$ in the space $\mathcal{L}$ with respect to $D(G)$, where $\mathbf{a} \times \mathbf{b} \in \mathcal{L}'$, $\mathcal{L}'$ must contain all vectors in the subspace $\mathcal{L}_1 \times \mathbf{b}$ because $D^i(H_1) \times D^j(E) \subset D(G)$. Further, $\mathcal{L}'$ must contain all vectors in the space $\mathcal{L}_1 \times \mathcal{L}_2 = \mathcal{L}$ because $D^i(E) \times D^j(H_2) \subset D(G)$. This leads to $\mathcal{L} = \mathcal{L}'$.

If $D^j(R)$ or $D^k(S)$ is replaced with another inequivalent irreducible representation, it is easy to prove by reduction to absurdity that the representation by the direct product is also changed into another inequivalent irreducible representation. If G is a finite group, the number of the inequivalent irreducible representations $D^j(H_1)$ is equal to the number of classes in the subgroup H_1, and the number of the inequivalent irreducible representations $D^k(H_2)$ is equal to the number of classes in the subgroup H_2. While the number of classes in the direct product group G is equal to the product of the numbers of classes in two subgroups, which is just equal to the number of the inequivalent irreducible representations of G obtained from the direct product of representations. Therefore, each irreducible representation of the direct product group can all be expressed as the direct product of the irreducible representations of two subgroups.

11. Let each element in D_3 be the coordinate transformation in two-dimensional space:

$$
\begin{pmatrix} x' \\ y' \end{pmatrix} = R \begin{pmatrix} x \\ y \end{pmatrix}, \qquad R \in D_3,
$$

where R is equal to the representation matrix in the two-dimensional representation $D^E(\text{D}_3)$. For the generators D and A in D_3, we have

$$D = D^E(D) = \frac{1}{2}\begin{pmatrix} -1 & -\sqrt{3} \\ \sqrt{3} & -1 \end{pmatrix}, \qquad A = D^E(A) = \begin{pmatrix} 1 & 0 \\ 0 & -1 \end{pmatrix}.$$

The four-dimensional function space spanned by the following basis functions is invariant in the group D_3:

$$\psi_1(x,y) = x^3, \quad \psi_2(x,y) = x^2y, \quad \psi_3(x,y) = xy^2, \quad \psi_4(x,y) = y^3.$$

Calculate the representation matrices of the generators D and A of D_3 in this representation. Then, reduce this representation into the direct sum of the irreducible representations of D_3, and construct new basis function which is the linear combination of the original basis functions and belongs to the irreducible representation.

Solution. First, according to the formula

$$P_R\psi_\mu(x) = \psi_\mu(R^{-1}x) = \sum_\nu \psi_\nu(x)D_{\nu\mu}(R),$$

one is able to calculate the representation matrices of the generators D and A of D_3 in the four-dimensional space spanned by the given basis functions, where $R^{-1}x$ means

$$\begin{pmatrix} x'' \\ y'' \end{pmatrix} = R^{-1}\begin{pmatrix} x \\ y \end{pmatrix}.$$

For the element D, we have

$$x'' = \left(-x + \sqrt{3}y\right)/2, \qquad y'' = \left(-\sqrt{3}x - y\right)/2,$$

$$\begin{aligned}
P_D\psi_1(x,y) &= \psi_1(x'',y'') = \left\{-x^3 + 3\sqrt{3}x^2y - 9xy^2 + 3\sqrt{3}y^3\right\}/8 \\
&= \left\{-\psi_1 + 3\sqrt{3}\psi_2 - 9\psi_3 + 3\sqrt{3}\psi_4\right\}/8, \\
P_D\psi_2(x,y) &= \psi_2(x'',y'') = \left\{-\sqrt{3}x^3 + 5x^2y - \sqrt{3}xy^2 - 3y^3\right\}/8 \\
&= \left\{-\sqrt{3}\psi_1 + 5\psi_2 - \sqrt{3}\psi_3 - 3\psi_4\right\}/8, \\
P_D\psi_3(x,y) &= \psi_3(x'',y'') = \left\{-3x^3 + \sqrt{3}x^2y + 5xy^2 + \sqrt{3}y^3\right\}/8 \\
&= \left\{-3\psi_1 + \sqrt{3}\psi_2 + 5\psi_3 + \sqrt{3}\psi_4\right\}/8, \\
P_D\psi_4(x,y) &= \psi_4(x'',y'') = \left\{-3\sqrt{3}x^3 - 9x^2y - 3\sqrt{3}xy^2 - y^3\right\}/8 \\
&= \left\{-3\sqrt{3}\psi_1 - 9\psi_2 - 3\sqrt{3}\psi_3 - \psi_4\right\}/8.
\end{aligned}$$

Hence,

$$D(D) = \frac{1}{8} \begin{pmatrix} -1 & -\sqrt{3} & -3 & -3\sqrt{3} \\ 3\sqrt{3} & 5 & \sqrt{3} & -9 \\ -9 & -\sqrt{3} & 5 & -3\sqrt{3} \\ 3\sqrt{3} & -3 & \sqrt{3} & -1 \end{pmatrix}.$$

For the element A, we have

$$x'' = x, \qquad y'' = -y,$$

$$P_A\psi_1(x,y) = \psi_1(x'',y'') = x^3 = \psi_1,$$

$$P_A\psi_2(x,y) = \psi_2(x'',y'') = -x^2y = -\psi_2,$$

$$P_A\psi_3(x,y) = \psi_3(x'',y'') = xy^2 = \psi_3,$$

$$P_A\psi_4(x,y) = \psi_4(x'',y'') = -y^3 = -\psi_4.$$

Hence,

$$D(A) = \begin{pmatrix} 1 & 0 & 0 & 0 \\ 0 & -1 & 0 & 0 \\ 0 & 0 & 1 & 0 \\ 0 & 0 & 0 & -1 \end{pmatrix}.$$

The characters are

$$\chi(D) = 1 = \chi^{A_1}(D) + \chi^{A_2}(D) + \chi^E(D),$$

$$\chi(A) = 0 = \chi^{A_1}(A) + \chi^{A_2}(A) + \chi^E(A).$$

The similarity transformation matrix X satisfies

$$D(D)X = \frac{X}{2} \begin{pmatrix} 2 & 0 & 0 & 0 \\ 0 & 2 & 0 & 0 \\ 0 & 0 & -1 & -\sqrt{3} \\ 0 & 0 & \sqrt{3} & -1 \end{pmatrix}, \qquad D(A)X = X \begin{pmatrix} 1 & 0 & 0 & 0 \\ 0 & -1 & 0 & 0 \\ 0 & 0 & 1 & 0 \\ 0 & 0 & 0 & -1 \end{pmatrix}.$$

The second equality gives

$$X = \begin{pmatrix} a_1 & 0 & c_1 & 0 \\ 0 & b_1 & 0 & d_1 \\ a_2 & 0 & c_2 & 0 \\ 0 & b_2 & 0 & d_2 \end{pmatrix}.$$

Substituting it into the first equality, we have

$$
\begin{pmatrix}
-a_1 - 3a_2 & -\sqrt{3}\,(b_1 + 3b_2) & -c_1 - 3c_2 & -\sqrt{3}\,(d_1 + 3d_2) \\
\sqrt{3}\,(3a_1 + a_2) & 5b_1 - 9b_2 & \sqrt{3}\,(3c_1 + c_2) & 5d_1 - 9d_2 \\
-9a_1 + 5a_2 & -\sqrt{3}\,(b_1 + 3b_2) & -9c_1 + 5c_2 & -\sqrt{3}\,(d_1 + 3d_2) \\
\sqrt{3}\,(3a_1 + a_2) & -3b_1 - b_2 & \sqrt{3}\,(3c_1 + c_2) & -3d_1 - d_2
\end{pmatrix}
$$

$$
= 4
\begin{pmatrix}
2a_1 & 0 & -c_1 & -\sqrt{3}c_1 \\
0 & 2b_1 & \sqrt{3}d_1 & -d_1 \\
2a_2 & 0 & -c_2 & -\sqrt{3}c_2 \\
0 & 2b_2 & \sqrt{3}d_2 & -d_2
\end{pmatrix}.
$$

The solutions are $a_2 = -3a_1$, $b_1 = -3b_2$ and $c_1 = c_2 = d_1 = d_2$. Arbitrary choose three constants, we obtain

$$
X =
\begin{pmatrix}
1 & 0 & 1 & 0 \\
0 & 3 & 0 & 1 \\
-3 & 0 & 1 & 0 \\
0 & -1 & 0 & 1
\end{pmatrix}.
$$

The new basis ϕ_ρ^Γ, which belongs to the ρth row of the irreducible representation Γ, can be calculated by X,

$$
\phi^{A_1}(x, y) = \psi_1(x, y) - 3\psi_3(x, y) = x(x^2 - 3y^2),
$$

$$
\phi^{A_2}(x, y) = 3\psi_2(x, y) - \psi_4(x, y) = y(3x^2 - y^2),
$$

$$
\phi_1^E(x, y) = \psi_1(x, y) + \psi_3(x, y) = x(x^2 + y^2),
$$

$$
\phi_2^E(x, y) = \psi_2(x, y) + \psi_4(x, y) = y(x^2 + y^2).
$$

12. Calculate the characters and the representation matrices of the proper symmetry group **O** of a cube with the method of the quotient group and the method of coordinate transformations.

Solution. The group **O** is the proper symmetry group of a cube, composed of three fourfold axes, four threefold axes and six twofold axes. Its order is $g = 24$. The generators of three fourfold axes are the rotations around the directions of three coordinate axes through π angle, denoted by T_x, T_y and T_z, respectively. Four threefold axes are along the following four directions with the generators R_j:

$$
\begin{array}{ll}
R_1: \ (\mathbf{e}_x + \mathbf{e}_y + \mathbf{e}_z)/\sqrt{3}, & R_2: \ (\mathbf{e}_x - \mathbf{e}_y - \mathbf{e}_z)/\sqrt{3}, \\
R_3: \ (-\mathbf{e}_x - \mathbf{e}_y + \mathbf{e}_z)/\sqrt{3}, & R_4: \ (-\mathbf{e}_x + \mathbf{e}_y - \mathbf{e}_z)/\sqrt{3}.
\end{array}
$$

The twofold axes are along the following six directions with the generators S_k:

$$S_1: \ (\mathbf{e}_x + \mathbf{e}_y)/\sqrt{2}, \qquad S_2: \ (\mathbf{e}_x - \mathbf{e}_y)/\sqrt{2},$$
$$S_3: \ (\mathbf{e}_y + \mathbf{e}_z)/\sqrt{2}, \qquad S_4: \ (\mathbf{e}_y - \mathbf{e}_z)/\sqrt{2},$$
$$S_5: \ (\mathbf{e}_x + \mathbf{e}_z)/\sqrt{2}, \qquad S_6: \ (\mathbf{e}_x - \mathbf{e}_z)/\sqrt{2}.$$

$\mathbf{T}$ is an invariant subgroup of $\mathbf{O}$ with the index 2. Through the direct calculation we obtain the coset formulas of T:

	E	T_x^2	T_y^2	T_z^2	R_1	R_2	R_3	R_4	R_1^2	R_2^2	R_3^2	R_4^2
right-multiply by S_1	S_1	T_z^3	T_z	S_2	T_y^3	S_5	T_y	S_6	T_x	S_4	T_x^3	S_3
left-multiply by S_1	S_1	T_z	T_z^3	S_2	T_x^3	S_4	T_x	S_3	T_y	S_5	T_y^3	S_6

Due to $S_1^2 = E$, the following formulas are just the opposite to the second set of formulas:

	S_1	T_z	T_z^3	S_2	T_x^3	S_4	T_x	S_3	T_y	S_5	T_y^3	S_6
left-multiply by S_1	E	T_x^2	T_y^2	T_z^2	R_1	R_2	R_3	R_4	R_1^2	R_2^2	R_3^2	R_4^2

Now, right-multiplying S_1 on the multiplication table of $\mathbf{T}$ given in Problem 16 of Chapter 2, we obtain the right-upper part of the multiplication table of $\mathbf{O}$. Then, left-multiplying S_1 on the upper part of the multiplication table of $\mathbf{O}$, we obtain its lower part. In the calculation, we only make the replacement of elements according to the above coset formulas of $\mathbf{T}$.

The right-upper part of multiplication table of O

	S_1	T_y^3	T_x	T_z^3	T_y	S_3	T_z	S_5	T_x^3	S_2	S_6	S_4
E	S_1	T_y^3	T_x	T_z^3	T_y	S_3	T_z	S_5	T_x^3	S_2	S_6	S_4
R_1	T_y^3	T_x	S_1	T_y	S_3	T_z^3	S_5	T_x^3	T_z	S_6	S_4	S_2
R_1^2	T_x	S_1	T_y^3	S_3	T_z^3	T_y	T_x^3	T_z	S_5	S_4	S_2	S_6
T_x^2	T_z^3	S_6	T_x^3	S_1	S_5	S_4	S_2	T_y	T_x	T_z	T_y^3	S_3
R_4	S_6	T_x^3	T_z^3	S_5	S_4	S_1	T_y	T_x	S_2	T_y^3	S_3	T_z
R_3^2	T_x^3	T_z^3	S_6	S_4	S_1	S_5	T_x	S_2	T_y	S_3	T_z	T_y^3
T_y^2	T_z	T_y	S_4	S_2	T_y^3	T_x^3	S_1	S_6	S_3	T_z^3	S_5	T_x
R_3	T_y	S_4	T_z	T_y^3	T_x^3	S_2	S_6	S_3	S_1	S_5	T_x	T_z^3
R_2^2	S_4	T_z	T_y	T_x^3	S_2	T_y^3	S_3	S_1	S_6	T_x	T_z^3	S_5
T_z^2	S_2	S_5	S_3	T_z	S_6	T_x	T_y^3	T_x^3	S_4	S_1	T_y	T_z^3
R_2	S_5	S_3	S_2	S_6	T_x	T_z	T_y^3	S_4	T_z^3	T_y	T_x^3	S_1
R_4^2	S_3	S_2	S_5	T_x	T_z	S_6	S_4	T_z^3	T_y^3	T_x^3	S_1	T_y

The left-lower part of multiplication table of O

	E	R_1	R_1^2	T_x^2	R_3	R_4^2	T_y^2	R_2	R_3^2	T_z^2	R_4	R_2^2
S_1	S_1	T_x^3	T_y	T_z	T_x	S_6	T_z^3	S_4	T_y^3	S_2	S_3	S_5
T_x^3	T_x^3	T_y	S_1	T_x	S_6	T_z	S_4	T_y^3	T_z^3	S_3	S_5	S_2
T_y	T_y	S_1	T_x^3	S_6	T_z	T_x	T_y^3	T_z^3	S_4	S_5	S_2	S_3
T_z	T_z	S_3	T_y^3	S_1	S_4	S_5	S_2	T_x	T_y	T_z^3	T_x^3	S_6
S_3	S_3	T_y^3	T_z	S_4	S_5	S_1	T_x	T_y	S_2	T_z^3	S_6	T_z^3
T_y^3	T_y^3	T_z	S_3	S_5	S_1	S_4	T_y	S_2	T_x	S_6	T_z^3	T_z^3
T_z^3	T_z^3	T_x	S_5	S_2	T_x^3	T_y^3	S_1	S_3	S_6	T_z	S_4	T_y
T_x	T_x	S_5	T_z^3	T_x^3	T_y^3	S_2	S_3	S_6	S_1	S_4	T_y	T_z
S_5	S_5	T_z^3	T_x	T_y^3	S_2	T_x^3	S_6	S_1	S_3	T_y	T_z	S_4
S_2	S_2	S_4	S_6	T_z^3	S_3	T_y	T_z	T_x^3	S_5	S_1	T_x	T_y^3
S_4	S_4	S_6	S_2	S_3	T_y	T_z^3	T_x^3	S_5	T_z	T_x	T_y^3	S_1
S_6	S_6	S_2	S_4	T_y	T_z^3	S_3	S_5	T_z	T_x^3	T_y^3	S_1	T_x

The right-lower part of multiplication table of O

	S_1	T_y^3	T_x	T_z^3	T_y	S_3	T_z	S_5	T_x^3	S_2	S_6	S_4
S_1	E	R_3^2	R_3	T_y^2	R_1^2	R_4	T_x^2	R_2^2	R_1	T_z^2	R_4^2	R_2
T_x^3	R_3^2	R_3	E	R_1^2	R_4	T_y^2	R_2^2	R_1	T_x^2	R_4^2	R_2	T_z^2
T_y	R_3	E	R_3^2	R_4	T_y^2	R_1^2	R_1	T_x^2	R_2^2	R_2	T_z^2	R_4^2
T_z	T_y^2	R_4^2	R_1	E	R_2^2	R_2	T_z^2	R_1^2	R_3	T_x^2	R_3^2	R_4
S_3	R_4^2	R_1	T_y^2	R_2^2	R_2	E	R_1^2	R_3	T_z^2	R_3^2	R_4	T_x^2
T_y^3	R_1	T_y^2	R_4^2	R_2	E	R_2^2	R_3	T_z^2	R_1^2	R_4	T_x^2	R_3^2
T_z^3	T_x^2	R_1^2	R_2	T_z^2	R_3^2	R_1	E	R_4^2	R_4	T_y^2	R_2^2	R_3
T_x	R_1^2	R_2	T_x^2	R_3^2	R_1	T_z^2	R_4^2	R_4	E	R_2^2	R_3	T_y^2
S_5	R_2	T_x^2	R_1^2	R_1	T_z^2	R_3^2	R_4	E	R_4^2	R_3	T_y^2	R_2^2
S_2	T_z^2	R_2^2	R_4	T_x^2	R_4^2	R_3	T_y^2	R_3^2	R_2	E	R_1^2	R_1
S_4	R_2^2	R_4	T_z^2	R_4^2	R_3	T_y^2	R_3^2	R_2	T_x^2	R_1^2	R_1	E
S_6	R_4	T_z^2	R_2^2	R_3	T_x^2	R_4	R_2	T_y^2	R_3^2	R_1	E	R_1^2

The group O has five classes: $C_1 = \{E\}$, $C_2 = \{T_x^2, T_y^2, T_z^2\}$, $C_3 = \{T_x, T_x^3, T_y, T_y^3, T_z, T_z^3\}$, $C_4 = \{R_j, R_j^2, 1 \leq j \leq 4\}$, and $C_5 = \{S_k, 1 \leq k \leq 6\}$. From $1^2 + 1^2 + 2^2 + 3^2 + 3^2 = 24$, we know that the group O has five inequivalent irreducible representations, denoted by A, B, E, T_1, and T_2, with dimensions 1, 1, 2, 3, and 3, respectively. The quotient group O/T is isomorphic onto V_2. From the quotient group, we obtain two one-dimensional representations: The identical representation $D^A = 1$ and the antisymmetric representation D^B with the characters:

$$\chi^B(C_1) = \chi^B(C_2) = \chi^B(C_4) = 1, \qquad \chi^B(C_3) = \chi^B(C_5) = -1.$$

For a one-dimensional representation, the character is the same as the representation matrix.

The group $\mathbf{O}$ contains another invariant subgroup D_2 of order 6, composed of four elements E, T_x^2, T_y^2 and T_z^2. Its cosets are R_1D_2, $R_1^2D_2$, S_1D_2, S_3D_2 and S_5D_2. As we know that two cosets RH and SH are different if $R^{-1}S$ does not belong to H. Now, it is easy to check that these five cosets all are different. The quotient group contains two three-order elements $(R_1D_2$ and $R_1^2D_2)$ and three two-order elements $(S_1D_2$, S_3D_2 and $S_5D_2)$ so that it is isomorphic onto the group D_3. Let R_1D_2, $R_1^2D_2$ and S_1D_2 respectively correspond to D, F and A in D_3. S_5D_2 has to correspond to B in D_3, because $R_1S_1 = T_y^3 = S_5T_x^2 \in S_5D_3$. Then, S_3D_2 corresponds to C in D_3. In constructing the multiplication table of D_3, the only independent multiplication formula is $DA = B$. The remaining formulas can be derived from it and the orders of elements. Therefore, the correspondence we just established is invariant to the product of elements.

The group D_3 has a two-dimensional irreducible representation, which is a non-faithful representation of $\mathbf{O}$. The elements in $\mathbf{O}$ belonging to the invariant subgroup D_3 correspond to the same representation matrix in this representation, so do those belonging to its coset. Note that $R_1D_2 = \{R_1, R_2, R_3, R_4\}$, $R_1^2D_2 = \{R_1^2, R_2^2, R_3^2, R_4^2\}$, $S_1D_2 = \{S_1, S_2, T_z, T_z^3\}$, $S_3D_2 = \{S_3, S_4, T_x, T_x^3\}$, and $S_5D_2 = \{S_5, S_6, T_y, T_y^3\}$.

$$\chi^E(E) = 2, \quad \chi^E(R_1) = \chi^E(R_1^2) = -1,$$
$$\chi^E(S_1) = \chi^E(S_3) = \chi^E(S_5) = 0,$$

$$D^E(E) = \begin{pmatrix} 1 & 0 \\ 0 & 1 \end{pmatrix}, \qquad D^E(R_1) = \frac{1}{2}\begin{pmatrix} -1 & -\sqrt{3} \\ \sqrt{3} & -1 \end{pmatrix},$$

$$D^E(R_1^2) = \frac{1}{2}\begin{pmatrix} -1 & \sqrt{3} \\ -\sqrt{3} & -1 \end{pmatrix}, \qquad D^E(S_1) = \begin{pmatrix} 1 & 0 \\ 0 & -1 \end{pmatrix},$$

$$D^E(S_5) = \frac{1}{2}\begin{pmatrix} -1 & \sqrt{3} \\ \sqrt{3} & 1 \end{pmatrix}, \qquad D^E(S_3) = \frac{1}{2}\begin{pmatrix} -1 & -\sqrt{3} \\ -\sqrt{3} & 1 \end{pmatrix}.$$

We can establish one three-dimensional representation of $\mathbf{O}$ by the coordinate transformation matrices. In fact, being a rotation in the three-dimensional space, the representation matrices of some elements in $\mathbf{O}$ are easy to obtain:

$$D^{T_1}(T_z) = \begin{pmatrix} 0 & -1 & 0 \\ 1 & 0 & 0 \\ 0 & 0 & 1 \end{pmatrix}, \qquad D^{T_1}(T_z^2) = \begin{pmatrix} -1 & 0 & 0 \\ 0 & -1 & 0 \\ 0 & 0 & 1 \end{pmatrix},$$

$$D^{T_1}(R_1) = \begin{pmatrix} 0 & 0 & 1 \\ 1 & 0 & 0 \\ 0 & 1 & 0 \end{pmatrix}, \qquad D^{T_1}(S_1) = \begin{pmatrix} 0 & 1 & 0 \\ 1 & 0 & 0 \\ 0 & 0 & -1 \end{pmatrix}.$$

Thus, the characters of classes in **O** are

$$\chi^{T_1}(C_1) = 3, \quad \chi^{T_1}(C_2) = -1, \quad \chi^{T_1}(C_3) = 1,$$
$$\chi^{T_1}(C_4) = 0, \quad \chi^{T_1}(C_5) = -1.$$

Another three-dimensional irreducible representation $D^{T_2}(\mathbf{O})$ can be obtained by the direct product of the antisymmetric representation $D^B(\mathbf{O})$ and the three-dimensional representation $D^{T_1}(\mathbf{O})$:

$$\chi^{T_2}(C_1) = 3, \quad \chi^{T_2}(C_2) = -1, \quad \chi^{T_2}(C_3) = -1,$$
$$\chi^{T_2}(C_4) = 0, \quad \chi^{T_2}(C_5) = 1,$$

$$D^{T_2}(T_z) = \begin{pmatrix} 0 & 1 & 0 \\ -1 & 0 & 0 \\ 0 & 0 & -1 \end{pmatrix}, \qquad D^{T_2}(R_1) = \begin{pmatrix} 0 & 0 & 1 \\ 1 & 0 & 0 \\ 0 & 1 & 0 \end{pmatrix}.$$

13. The multiplication table for a group G of order 12 is as follows.

	E	A	B	C	D	F	I	J	K	L	M	N
E	E	A	B	C	D	F	I	J	K	L	M	N
A	A	B	E	I	L	K	N	D	M	J	F	C
B	B	E	A	N	J	M	C	L	F	D	K	I
C	C	K	L	D	E	B	M	I	N	F	J	A
D	D	N	F	E	C	L	J	M	A	B	I	K
F	F	D	N	K	M	I	E	B	L	C	A	J
I	I	M	J	L	A	E	F	N	C	K	D	B
J	J	I	M	B	N	D	L	K	E	A	C	F
K	K	L	C	M	F	N	A	E	J	I	B	D
L	L	C	K	A	I	J	D	F	B	E	N	M
M	M	J	I	F	K	C	B	A	D	N	E	L
N	N	F	D	J	B	A	K	C	I	M	L	E

a) Find out the inverse of each element in G;

b) Point out which elements can commute with any element in the group;

c) List the period and the order of each element;

d) Find out the elements in each class of G;

e) Find out all invariant subgroups in G. For each invariant subgroup, list its cosets and point out onto which group its quotient group is isomorphic;

f) Establish the character table of G;

g) Decide whether G is isomorphic onto the symmetry group **T** of the tetrahedron, or the symmetry group D_6 of the regular six-sided polygon.

Solution. a) $E^{-1} = E$, $A^{-1} = B$, $C^{-1} = D$, $F^{-1} = I$, $J^{-1} = K$, $L^{-1} = L$, $M^{-1} = M$, and $N^{-1} = N$.

b) Only the identity E can commute with any element in G.

c) The order of the identity E is one. The orders of L, M and N are two, and the orders of A, B, C, D, F, I, J and K all are three.

d) The group G contains four classes, $\{E\}$, $\{L, M, N\}$, $\{A, C, F, J\}$, and $\{B, D, I, K\}$. The first two are self-reciprocal classes, but the last two are the mutual reciprocal classes.

e) G contains only one non-trivial invariant subgroup $\{E, L, M, N\}$ with index 3. Its cosets are $\{A, C, F, J\}$ and $\{B, D, I, K\}$, and its quotient group is isomorphic onto the cyclic group C_3.

f) From the number of the classes in G, we know there are four inequivalent irreducible representations in G. Since $1^2 + 1^2 + 1^2 + 3^2 = 12$, there are three one-dimensional representations and one three-dimensional irreducible representation in G. Three one-dimensional representations can be calculated from the quotient group of the invariant subgroup. The characters in the three-dimensional representation of G can be determined by the orthogonality.

	E	(L, M, N)	(A, C, F, J)	(B, D, I, K)
A	1	1	1	1
E	1	1	$e^{-i2\pi/3}$	$e^{i2\pi/3}$
E'	1	1	$e^{i2\pi/3}$	$e^{-i2\pi/3}$
T	3	-1	0	0

g) From the orders of elements and the classes in G, we know that G is isomorphic onto the group **T**, but not isomorphic onto the group D_6.

14. Calculate the character table of the group G given in Problem 17 of Chapter 2.

Solution. The group G in Problem 17 of Chapter 2 has 12 elements, divided into 6 classes. Since $1^2 + 1^2 + 1^2 + 1^2 + 2^2 + 2^2 = 12$, there are four one-dimensional and two two-dimensional inequivalent irreducible representations in G. There are three invariant subgroups in G. The invariant subgroup $\{E, K, N, A, M, L\}$ has index 2. Its coset is $\{B, I, D, F, C, J\}$. Its quotient group is isomorphic onto the two-order inversion group V_2. From the irreducible representations of the quotient group we obtain two one-dimensional representations, including the identity representation $D^{(1)}(R) = 1$. The invariant subgroup $\{E, N, M\}$ has index 4. Its cosets are $\{A, K, L\}$, $\{B, C, D\}$ and $\{F, I, J\}$. The quotient group is isomorphic onto the cyclic group C_4, where $\{A, K, L\}$ is a two-order element. From this quotient group we obtain four one-dimensional representations, including two which are known. The invariant subgroup $\{E, A\}$ has index 6. Its cosets are $\{B, F\}$, $\{C, I\}$, $\{D, J\}$, $\{K, M\}$ and $\{L, N\}$. The quotient group is isomorphic onto the group D_3, where $\{K, M\}$ and $\{L, N\}$ are three-order elements. From the two-dimensional irreducible representation of D_3 we obtain the irreducible representation D^5 of G. The representation matrices for the generators K and B, which correspond respectively to D and A in D_3, are

$$D^5(B) = \begin{pmatrix} 1 & 0 \\ 0 & -1 \end{pmatrix}, \qquad D^5(K) = \frac{1}{2} \begin{pmatrix} -1 & -\sqrt{3} \\ \sqrt{3} & -1 \end{pmatrix}.$$

Another two-dimensional irreducible representation D^6 of G is the direct product $D^3 \times D^5$,

$$D^6(B) = iD^5(B), \qquad D^6(K) = -D^5(K).$$

Thus, we obtain the character tables of G.

	E	A	BCD	FIJ	KL	MN
χ^1	1	1	1	1	1	1
χ^2	1	1	-1	-1	1	1
χ^3	1	-1	i	$-i$	-1	1
χ^4	1	-1	$-i$	i	-1	1
χ^5	2	2	0	0	-1	-1
χ^6	2	-2	0	0	1	-1

3.3 Subduced and Induced Representations

★ Let the order of a finite group G be g, and $H = \{T_1 = E, T_2, \ldots, T_h\}$ be a subgroup of G with the order h and the index $n = g/h$. Denote its left-cosets by $R_r H$, $2 \le r \le n$. For unification, the subgroup is denoted by $R_1 H$ with $R_1 = E$. Although R_j are not determined uniquely, we make choice of them arbitrarily. Then, any element in the group G can be expressed as $R_r T_t$. Let $D^j(G)$ be an m_j-dimensional irreducible representation of G. The set of those matrices $D^j(T_t)$ corresponding to the elements in H

constitutes a representation of the subgroup H, denoted by $D^j(H)$, which is called the subduced representation from an irreducible representation $D^j(G)$ of the group G with respect to the subgroup H. Generally, the subduced representation is reducible, and can be reduced with respect to the irreducible representations $\overline{D}^k(H)$ of the subgroup H

$$X^{-1}D^j(T_t)X = \bigoplus_k a_{jk}\overline{D}^k(T_t), \qquad m_j = \sum_k a_{jk}\overline{m}_k,$$
$$a_{jk} = \frac{1}{h}\sum_{T_t \in H} \overline{\chi}^k(T_t)^*\chi^j(T_t) = \frac{1}{h}\sum_\beta \overline{n}(\beta)\left(\overline{\chi}_\beta^k\right)^*\chi_\beta^j, \qquad (3.14)$$

where $\overline{m}_k$ is the dimension of the representation $\overline{D}^k(H)$ of H, and $\overline{n}(\beta)$ is the number of elements contained in the class $\overline{C}_\beta$ of H.

★ Denote by ψ_μ the $\overline{m}_k$ bases in the representation space of $\overline{D}^k(H)$

$$P_{T_t}\psi_\mu = \sum_\nu \psi_\nu \overline{D}_{\nu\mu}^k(T_t).$$

We define an extended space of dimension $n\overline{m}_k$ with the bases $\psi_{r\mu} = P_{R_r}\psi_\mu$, where $\psi_{1\mu} = \psi_\mu$. The extended space is invariant to the group G. We can calculate an $n\overline{m}_k$-dimensional representation $\Delta^k(G)$ of G in the following way. For any given element S in G, we calculate SR_r for each R_r, which can be expressed in the form of $R_u T_t$, where u and t are completely determined by S and r. Because

$$P_S\psi_{r\mu} = P_{SR_r}\psi_\mu = P_{R_u}P_{T_t}\psi_\mu = \sum_\nu \psi_{u\nu}\overline{D}_{\nu\mu}^k(T_t),$$

we obtain

$$\Delta_{u\nu,r\mu}^k(S) = \overline{D}_{\nu\mu}^k(T_t), \qquad \chi^k(S) = \sum_{r\mu} \Delta_{r\mu,r\mu}^k(S). \qquad (3.15)$$

This representation $\Delta^k(G)$ is called the induced representation from the irreducible representation $\overline{D}^k(H)$ of the subgroup H with respect to the group G. In general, the induced representation is reducible and can be reduced with respect to the irreducible representation $D^j(G)$ of G:

$$Y^{-1}\Delta^k(S)Y = \bigoplus_j b_{jk}D^j(S), \qquad (g/h)\overline{m}_k = \sum_j b_{jk}m_j,$$

$$b_{jk} = \frac{1}{g} \sum_{S \in G} \chi^j(S)^* \chi^k(S) = \frac{1}{g} \sum_{\alpha} n(\alpha) \left(\chi_\alpha^j\right)^* \chi_\alpha^k, \qquad (3.16)$$

where $n(\alpha)$ is the number of the class C_α of G.

★ Calculate the character $\chi^k(S)$ of the element S in the induced representation $\Delta^k(G)$. Let S belong to the class C_α in G. In general, some elements in C_α belong to the subgroup H, and constitute a few classes of the subgroup H, denoted by $\overline{C}_\beta$. It is possible that no element in C_α belongs to the subgroup H. For this case we say that no β exists. From Eq. (3.15), the diagonal element of $\Delta^k(S)$ may appear only when $r = u$, i.e., $SR_r = R_r T_t$. Namely, $\chi^k(S)$ is non-vanishing only when the class C_α contains a few elements belonging to the subgroup H. Denoting by κ_β the number of the different R_r satisfying $SR_r = R_r T_t$, where T_t belongs to the class $\overline{C}_\beta$ with the character $\overline{\chi}_\beta^k$ in the irreducible representation $\overline{D}^k(H)$ of H, we have $\chi_\alpha^k = \sum_\beta \kappa_\beta \overline{\chi}_\beta^k$.

From Problem 14 in Chapter 2, the number of elements R in G satisfying $SR = RT_t$ is $m(\alpha) = g/n(\alpha)$. Such elements R, in general, can be expressed in the form of $R_r T_x$, so $SR_r = R_r \left(T_x T_t T_x^{-1}\right)$, while $\overline{\chi}^k\left(T_x T_t T_x^{-1}\right) = \overline{\chi}^k(T_t)$. On the other hand, the number of T_y in the subgroup H which satisfy $T_y T_t T_y^{-1} = T_t$ is $\overline{m}(\beta) = h/\overline{n}(\beta)$. If $R_r T_x$ satisfy $S(R_r T_x) = (R_r T_x)T_t$, then $R_r T_x T_y$ also satisfy this formula. However, the latter does not make new contribution to the characters $\chi^k(S)$, so $\kappa_\beta = m(\alpha)/\overline{m}(\beta)$,

$$\chi_\alpha^k = \frac{g}{hn(\alpha)} \sum_\beta \overline{n}(\beta)\overline{\chi}_\beta^k. \qquad (3.17)$$

By the way, we do not need to consider the case of $SR' = R'T_{t'}$ with another $T_{t'} \in \overline{C}_\beta$, because its contribution to the characters $\chi^k(S)$ has been calculated. In fact, letting $T_{t'} = T_z T_t T_z^{-1}$, we have $S(R'T_z) = (R'T_z)T_t$.

Note that the elements T_t in the class $\overline{C}_\beta$ of H all belong to the class C_α of G. Obviously, the different classes C_α correspond to the different classes $\overline{C}_\beta$. Therefore, the sum over the class α in Eq. (3.16) is equivalent to the sum over all classes $\overline{C}_\beta$ in H. Due to $\chi^j(S) = \chi^j(R^{-1}SR)$, it is easy to show from Eq. (3.17) that two multiplicities a_{jk} in Eq. (3.14) and b_{jk} in Eq. (3.16) are equal:

$$b_{jk} = \frac{1}{g} \sum_\alpha n(\alpha) \left(\chi_\alpha^j\right)^* \chi_\alpha^k = \frac{1}{h} \sum_\beta \overline{n}(\beta) \left(\chi_\beta^j\right)^* \overline{\chi}_\beta^k = a_{jk}. \qquad (3.18)$$

The formula (3.18) is called the Frobenius theorem.

15. Calculate all inequivalent irreducible representations of the group D_{2n+1} with the method of induced representation.

Solution. The group D_{2n+1} contains a $(2n+1)$-fold axis, called the principal axis, and $(2n+1)$ equivalent twofold axes, located in the plane orthogonal to the principal axis. Denote by C_{2n+1} the generator of the $(2n+1)$-fold axis and by $C_{2'}$ the generator of one twofold axis. C_{2n+1} and $C_{2'}$ are two generators of D_{2n+1}. Due to the orthogonality, $C_{2n+1}C_{2'} = C_{2'}C_{2n+1}^{-1}$. The order of D_{2n+1} is $g = 4n + 2$. The number of the classes in D_{2n+1} is $g_c = n+2$. There are two one-dimensional and n two-dimensional inequivalent representations of D_{2n+1}.

The group D_{2n+1} contains an invariant subgroup C_{2n+1} with index two. Its coset consists of all twofold rotations whose axes are orthogonal to the principal axis. The quotient group is isomorphic onto the two-order inversion group V_2, which gives two one-dimensional inequivalent irreducible representations of D_{2n+1}, respectively denoted by A and B:

$$D^A(C_{2n+1}) = D^B(C_{2n+1}) = D^A(C_{2'}) = 1, \qquad D^B(C_{2'}) = -1.$$

The invariant subgroup C_{2n+1} is a cyclic group. It has $(2n + 1)$ one-dimensional inequivalent representations. Denote the basis in each one-dimensional representation space by ψ^j,

$$C_{2n+1}\psi^j = e^{-i2j\pi/(2n+1)}\psi^j, \qquad 0 \le j \le 2n.$$

Extending the one-dimensional space by defining another basis $\phi^j = C_{2'}\psi^j$, we have

$$C_{2'}\psi^j = \phi^j, \qquad C_{2'}\phi^j = \psi^j, \qquad C_{2n+1}\phi^j = e^{i2j\pi/(2n+1)}\phi^j,$$

where the formula $C_{2n+1}C_{2'} = C_{2'}C_{2n+1}^{-1}$ is used. Thus, we obtain n two-dimensional inequivalent irreducible representations of the group D_{2n+1}, denoted by E_j, $1 \le j \le n$,

$$D^{E_j}(C_{2n+1}) = \begin{pmatrix} e^{-i2j\pi/(2n+1)} & 0 \\ 0 & e^{i2j\pi/(2n+1)} \end{pmatrix}, \qquad D^{E_j}(C_{2'}) = \begin{pmatrix} 0 & 1 \\ 1 & 0 \end{pmatrix}.$$

The remaining representations with the different j are either equivalent to one of the above representations or reducible.

16. Calculate all inequivalent irreducible representations of the group D_{2n} with the method of induced representation.

Solution. The group D_{2n} contains a $(2n)$-fold axis, called the principal axis, and $(2n)$ twofold axes, located in the plane orthogonal to the principal axis. The twofold axes are divided into two sets, each of which contains n equivalent twofold axes. They form two classes, respectively. Denote by C_{2n} the generator of the $(2n)$-fold axis and by $C_{2'}$ the generator of one twofold axis belonging to one class. Define $C_{2''} = C_{2n}C_{2'} = C_{2'}C_{2n}^{-1}$, belonging to another class. The angle between two twofold axes for $C_{2'}$ and $C_{2''}$ is π/n. C_{2n} and $C_{2'}$ are two generators of D_{2n}. The order of D_{2n} is $g = 4n$. The number of the classes in D_{2n} is $g_c = n + 3$. There are four one-dimensional and $n - 1$ two-dimensional inequivalent irreducible representations of D_{2n}.

The group D_{2n} contains four invariant subgroups. One subgroup is C_{2n} composed of all rotations around the principal axis. Its index is two, and its quotient group is isomorphic onto the two-order inversion group V_2, which gives two one-dimensional inequivalent representations of D_{2n}, denoted by A_1 and A_2:

$$D^{A_1}(C_{2n}) = D^{A_2}(C_{2n}) = D^{A_1}(C_{2'}) = 1, \qquad D^{A_2}(C_{2'}) = -1.$$

The subgroup C_n of C_{2n} composed of the even powers of C_{2n} is also an invariant subgroup of D_{2n}. Its index is four. One of its coset consists of the odd powers of C_{2n}. The remaining cosets are two classes of the twofold rotations, respectively. The quotient group is isomorphic onto the four-order inversion group V_4, which gives four one-dimensional inequivalent representations of D_{2n} including two known. The new representations are denoted by B_1 and B_2:

$$D^{B_1}(C_{2n}) = D^{B_2}(C_{2n}) = D^{B_2}(C_{2'}) = -1, \qquad D^{B_1}(C_{2'}) = 1.$$

The set composed of C_n and one class of the twofold rotations also forms an invariant subgroup of D_{2n}. These two subgroups are both the two-order groups. The representations of D_{2n} provided by their quotient groups are nothing but A_1, B_1 and A_1, B_2, respectively.

The invariant subgroup C_{2n} is a cyclic group. It has $(2n)$ one-dimensional inequivalent representation. Denote the basis in each one-dimensional representation space by ψ^j,

$$C_{2n}\psi^j = e^{-ij\pi/n}\psi^j \qquad 0 \le j \le 2n - 1.$$

Extending the one-dimensional space by defining another basis $\phi^j = C_{2'}\psi^j$, we have

$$C_{2'}\psi^j = \phi^j, \qquad C_{2'}\phi^j = \psi^j, \qquad C_{2n}\phi^j = e^{ij\pi/n}\phi^j.$$

Thus, we obtain $(n-1)$ two-dimensional inequivalent irreducible representations of the group D_{2n}, denoted by E_j, $1 \le j \le n-1$,

$$D^{E_j}(C_{2n}) = \begin{pmatrix} e^{-ij\pi/n} & 0 \\ 0 & e^{ij\pi/n} \end{pmatrix}, \qquad D^{E_j}(C_{2'}) = \begin{pmatrix} 0 & 1 \\ 1 & 0 \end{pmatrix}.$$

The remaining representations with the different j are either equivalent to one of the above representations or reducible.

17. Calculate all inequivalent irreducible representations of the symmetry group **O** of a cube with the method of induced representation.

Solution. In Problem 12 we have calculated all inequivalent irreducible representations of the group **O** with the method of the quotient groups and the method of coordinate transformations. In the present problem we will calculate them with the method of induced representation. In Problem 12 we have given the multiplication table of the group **O**. Now, we use the same notation as that used there.

The group **O** contains a subgroup D_4 of order 8, composed of one fourfold axis along the z axis and four twofold axes in the xy plane. The generator of the fourfold axis is denoted by T_z, and four twofold rotations are denoted by T_x^2, S_1, T_y^2, and S_2. The angle of two neighboring twofold axes is $\pi/4$. The index of D_4 in **O** is 3. The characters table of the subgroup D_4 is as follows.

<table>
<tr><td colspan="6">The characters table of O</td></tr>
<tr><td></td><td>E</td><td>$3T_z^2$</td><td>$6T_z$</td><td>$8R_1$</td><td>$6S_1$</td></tr>
<tr><td>A</td><td>1</td><td>1</td><td>1</td><td>1</td><td>1</td></tr>
<tr><td>B</td><td>1</td><td>1</td><td>−1</td><td>1</td><td>−1</td></tr>
<tr><td>E</td><td>2</td><td>2</td><td>0</td><td>−1</td><td>0</td></tr>
<tr><td>T_1</td><td>3</td><td>−1</td><td>1</td><td>0</td><td>−1</td></tr>
<tr><td>T_2</td><td>3</td><td>−1</td><td>−1</td><td>0</td><td>1</td></tr>
</table>

<table>
<tr><td colspan="6">The characters table of D4</td></tr>
<tr><td></td><td>E</td><td>$2T_z$</td><td>T_z^2</td><td>$2T_x^2$</td><td>$2S_1$</td></tr>
<tr><td>A_1</td><td>1</td><td>1</td><td>1</td><td>1</td><td>1</td></tr>
<tr><td>A_2</td><td>1</td><td>1</td><td>1</td><td>−1</td><td>−1</td></tr>
<tr><td>B_1</td><td>1</td><td>−1</td><td>1</td><td>1</td><td>−1</td></tr>
<tr><td>B_2</td><td>1</td><td>−1</td><td>1</td><td>−1</td><td>1</td></tr>
<tr><td>E</td><td>2</td><td>0</td><td>−2</td><td>0</td><td>0</td></tr>
</table>

We calculate the character of the induced representation from B_1 of D_4 with respect to **O** by Eq. (3.17),

$$\chi^{B_1}(E) = \frac{24 \times 1}{8} = 3, \quad \chi^{B_1}(T_z) = \frac{24 \times 2 \times (-1)}{8 \times 6} = -1,$$

$$\chi^{B_1}(T_z^2) = \frac{24}{8 \times 3}\{1 \times 1 + 2 \times 1\} = 3,$$

$$\chi^{B_1}(S_1) = \frac{24 \times 2 \times (-1)}{8 \times 6} = -1, \qquad \chi^{B_1}(R_1) = 0,$$

$$1 \times 3^2 + 6 \times (-1)^2 + 3 \times (3)^2 + 6 \times (-1)^2 + 0 = 48.$$

Because the characters is orthogonal to that of the identical representation A, this representation is the direct sum of a one-dimensional representation B and a two-dimensional representation E of the group $\mathbf{O}$. Denote by ϕ_1 the basis of the representation B_1 of D_4. Extend the space by defining new bases $\phi_2 = R_1\phi_1$ and $\phi_3 = R_1^2\phi_1$. By making use of the multiplication table of $\mathbf{O}$ given in Problem 12, we have $T_z R_1 = S_3 = R_1^2 T_z^3$ and $T_z R_1^2 = T_y^3 = R_1 S_1$. Thus,

$$T_z\phi_1 = -\phi_1, \quad T_z\phi_2 = -\phi_3, \quad T_z\phi_3 = -\phi_2,$$
$$R_1\phi_1 = \phi_2, \quad R_1\phi_2 = \phi_3, \quad R_1\phi_3 = \phi_1.$$

$$D(T_z) = \begin{pmatrix} -1 & 0 & 0 \\ 0 & 0 & -1 \\ 0 & -1 & 0 \end{pmatrix}, \qquad D(R_1) = \begin{pmatrix} 0 & 0 & 1 \\ 1 & 0 & 0 \\ 0 & 1 & 0 \end{pmatrix}.$$

We obtain the reduced form of the direct sum of B and E:

$$X = \begin{pmatrix} 1 & 0 & 2 \\ 1 & -\sqrt{3} & -1 \\ 1 & \sqrt{3} & -1 \end{pmatrix}, \quad X^{-1}D(T_z)X = \begin{pmatrix} -1 & 0 & 0 \\ 0 & 1 & 0 \\ 0 & 0 & -1 \end{pmatrix},$$

$$X^{-1}D(R_1)X = \frac{1}{2}\begin{pmatrix} 2 & 0 & 0 \\ 0 & -1 & -\sqrt{3} \\ 0 & \sqrt{3} & -1 \end{pmatrix}.$$

We calculate the characters for the induced representation from A_2 of D_4 with respect to $\mathbf{O}$ by Eq. (3.17)

$$\chi^{A_2}(E) = \frac{24 \times 1}{8} = 3, \quad \chi^{A_2}(T_z) = \frac{24 \times 2 \times 1}{8 \times 6} = 1,$$
$$\chi^{A_2}(T_z^2) = \frac{24}{8 \times 3}\{1 \times 1 + 2 \times (-1)\} = -1,$$
$$\chi^{A_2}(S_1) = \frac{24 \times 2 \times (-1)}{8 \times 6} = -1, \quad \chi^{A_2}(R_1) = 0,$$
$$1 \times 3^2 + 6 \times 1^2 + 3 \times (-1)^2 + 6 \times (-1)^2 + 0 = 24.$$

This is the irreducible representation T_1 of $\mathbf{O}$. Denote by ψ_1 the basis of the representation A_2 of D_4. Extending the space by defining new bases $\psi_2 = R_1\psi_1$, $\psi_3 = R_1^2\psi_1$, we have

$$R_1\psi_1 = \psi_2, \quad R_1\psi_2 = \psi_3, \quad R_1\psi_3 = \psi_1,$$
$$T_z\psi_1 = \psi_1, \quad T_z\psi_2 = \psi_3, \quad T_z\psi_3 = -\psi_2.$$

$$DT_1(T_z) = \begin{pmatrix} 1 & 0 & 0 \\ 0 & 0 & -1 \\ 0 & 1 & 0 \end{pmatrix}, \quad D^{T_1}(R_1) = \begin{pmatrix} 0 & 0 & 1 \\ 1 & 0 & 0 \\ 0 & 1 & 0 \end{pmatrix}.$$

By a simple similarity transformation (the cyclic of 1,2,3), this form of representation coincides with the form given in Problem 12.

We calculate the characters for the induced representation from B_2 of D_4 with respect to **O** by Eq. (3.17)

$$\chi^{B_2}(E) = \frac{24 \times 1}{8} = 3, \quad \chi^{B_2}(T_z) = \frac{24 \times 2 \times (-1)}{8 \times 6} = -1,$$

$$\chi^{B_2}(T_z^2) = \frac{24}{8 \times 3}\{1 \times 1 + 2 \times (-1)\} = -1,$$

$$\chi^{B_2}(S_1) = \frac{24 \times 2 \times 1}{8 \times 6} = 1, \quad \chi^{B_2}(R_1) = 0,$$

$$1 \times 3^2 + 6 \times (-1)^2 + 3 \times (-1)^2 + 6 \times 1^2 + 0 = 24.$$

This is the irreducible representation T_2 of **O**. The characters are the same as those in T_1 except for the signs of the characters of T_z and S_1. For the group **O**, $T_2 = T_1 \times B$. The calculation is similar.

18. The regular icosahedron is shown in Fig. 3.1. The opposite vertices are denoted by A_j and B_j, $0 \le j \le 5$. Choose the coordinate frame such that the origin is at the center O of the regular icosahedron, and the z axis is in the direction from the vertex B_0 to the vertex A_0. A_j are located above the xy plane.

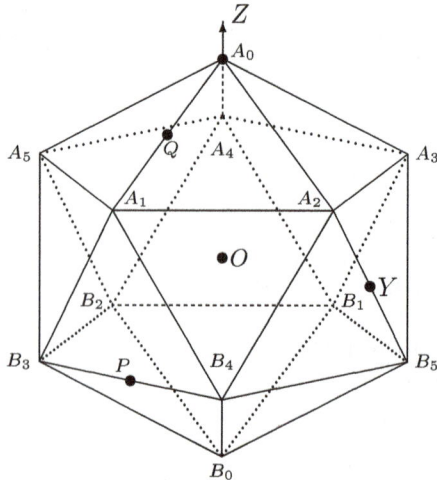

Fig. 3.1   The regular icosahedron.

In the regular icosahedron, there are six fivefold axes around the directions from B_j to A_j with the generators T_j, $0 \le j \le 5$. Except for one fivefold axis ($j = 0$) which is along the positive z axis, the polar angles of the remaining fivefold axes all are θ_1, and their azimuthal angles are $2(j-1)\pi/5$, respectively. There are 10 threefold axes along the lines connecting the centers of two opposite triangles with the generators R_j, $1 \le j \le 10$. The polar angles of the threefold axes are θ_2 when $1 \le j \le 5$, and θ_3 when $6 \le j \le 10$. Their azimuthal angles respectively are $(2j-1)\pi/5$. There are 15 twofold axes along the lines connecting the central points of two opposite edges with the generators S_j, $1 \le j \le 15$. The polar angles of the twofold axes are θ_4 when $1 \le j \le 5$, θ_5 when $6 \le j \le 10$, and $\pi/2$ when $11 \le j \le 15$. Their azimuthal angles are $2(j-1)\pi/5$ when $1 \le j \le 5$, $(2j-1)\pi/5$ when $6 \le j \le 10$, and $(4j-3)\pi/10$ when $11 \le j \le 15$, respectively.

$$\tan\theta_1 = 2, \qquad \tan\theta_2 = 3 - \sqrt{5}, \qquad \tan\theta_3 = 3 + \sqrt{5},$$
$$\tan\theta_4 = (\sqrt{5}-1)/2, \qquad \tan\theta_5 = (\sqrt{5}+1)/2. \tag{3.19}$$

All axes are the non-polar axes, and any two axes with the same fold are equivalent to each other.

The proper symmetry group of an icosahedron is denoted by $\mathbf{I}$, which contains 60 elements and five classes. The classes, denoted by E, C_5, C_5^2, C_3 and C_2, contain 1, 12, 12, 20, and 15 elements, respectively. Calculate the character table of the proper symmetric group $\mathbf{I}$ of an icosahedron with the method of induced representation.

Solution. Since $1^2 + 3^2 + 3^2 + 4^2 + 5^2 = 60$, the dimensions of five inequivalent irreducible representations respectively are 1, 3, 3, 4 and 5, denoted by A, T_1, T_2, G and H. The representation A is the identical representation.

The character table of T

	E	$3C_2$	$4C_3'$	$4C_3'^2$
A	1	1	1	1
E	1	1	ω	ω^2
E'	1	1	ω^2	ω
T	3	-1	0	0

The character table of D_5

	E	$2C_5$	$2C_5^2$	$5C_2'$
A_1	1	1	1	1
A_2	1	1	1	-1
E_1	2	p	$-p^{-1}$	0
E_2	2	$-p^{-1}$	p	0

$$\omega = \exp\{-i2\pi/3\}, \qquad \eta = \exp\{-i2\pi/5\}, \qquad p = \eta + \eta^{-1} = (\sqrt{5}-1)/2.$$

The group $\mathbf{I}$ does not contain any non-trivial invariant subgroup. There are two bigger subgroups $\mathbf{T}$ and $\mathbf{D}_5$. The subgroup $\mathbf{T}$ consists of three twofold axes (S_8, S_{12} and S_1) and four threefold axes (R_6, R_2^{-1}, R_4, and R_{10}^{-1}), and the subgroup $\mathbf{D}_5$ contains one fivefold (T_0) and five twofold axes (S_j, $11 \le j \le 15$). Their character tables are given in the Table.

For the induced representation from the identical representation A of $\mathbf{T}$ with respect to $\mathbf{I}$, we have from Eq. (3.17)

$$\chi^A(E) = 5, \quad \chi^A(C_5) = \chi^A(C_5^2) = 0, \quad \chi^A(C_2) = \frac{60 \times 3 \times 1}{12 \times 15} = 1,$$

$$\chi^A(C_3) = \frac{60}{12 \times 20}(4 \times 1 + 4 \times 1) = 2,$$

$$1 \times 5^2 + 0 + 0 + 15 \times 1^2 + 20 \times 2^2 = 120.$$

Because there is no negative character, this representation must be equivalent to the direct sum of an identity representation A and a 4-dimensional representation G. Subtracting the characters of A, we obtain the characters of the irreducible representation G of $\mathbf{I}$:

$$\chi^A(E) = \chi^A(C_5) = \chi^A(C_5^2) = \chi^A(C_2) = \chi^A(C_3) = 1,$$

$$\chi^G(E) = 4, \quad \chi^G(C_5) = \chi^G(C_5^2) = -1, \quad \chi^G(C_2) = 0, \quad \chi^G(C_3) = 1.$$

For the induced representation from the representation E of T with respect to $\mathbf{I}$, we have

$$\chi^E(E) = 5, \quad \chi^E(C_5) = \chi^E(C_5^2) = 0, \quad \chi^E(C_2) = \frac{60 \times 3}{12 \times 15} = 1,$$

$$\chi^E(C_3) = \frac{60}{12 \times 20}\left(4 \times \omega + 4 \times \omega^2\right) = -1,$$

$$1 \times 5^2 + 0 + 0 + 15 \times 1^2 + 20 \times (-1)^2 = 60.$$

It is the five-dimensional irreducible representation H of $\mathbf{I}$:

$$\chi^H(E) = 5, \quad \chi^H(C_5) = \chi^H(C_5^2) = 0, \quad \chi^H(C_2) = 1, \quad \chi^H(C_3) = -1.$$

By the induced representations from the irreducible representation of $\mathbf{T}$, it is impossible to distinguish two classes C_5 and C_5^2.

For the induced representation from E_1 of $\mathbf{D}_5$ with respect to $\mathbf{I}$, we have

$$\chi^{E_1}(E) = \frac{60 \times 2}{10 \times 1} = 12, \quad \chi^{E_1}(C_5) = \frac{60 \times 2 \times p}{10 \times 12} = p,$$

$$\chi^{E_1}(C_5^2) = \frac{60 \times 2 \times (-p^{-1})}{10 \times 12} = -p^{-1}, \quad \chi^{E_1}(C_2) = \chi^{E_1}(C_3) = 0,$$

$$p = (\sqrt{5} - 1)/2, \quad p^{-1} = (\sqrt{5} + 1)/2,$$

$$1 \times 12^2 + 12 \times p^2 + 12 \times \left(-p^{-1}\right)^2 + 0 + 0 = 180.$$

It is the direct sum of three irreducible representations. Because the characters are not orthogonal to those of the representations G and H, so this representation contains one representation G and one representation H. Subtracting the characters of G and H, we obtain the characters of the three-dimensional irreducible representation T_1 of **I**:

$$\chi^{T_1}(E) = 3, \quad \chi^{T_1}(C_5) = p^{-1}, \quad \chi^{T_1}(C_5^2) = -p,$$
$$\chi^{T_1}(C_2) = -1, \quad \chi^{T_1}(C_3) = 0.$$

By the induced representation from E_2 of D_5, we can calculate the characters of T_2 similarly. In comparison with the characters of T_1, only the characters of C_5 and C_5^2 are interchanged. Finally, the character table of the group **I** is listed in the table.

The characters table of the group I

	E	$12C_5$	$12C_5^2$	$15C_2$	$20C_3$
A	1	1	1	1	1
T_1	3	p^{-1}	$-p$	-1	0
T_2	3	$-p$	p^{-1}	-1	0
G	4	-1	-1	0	1
H	5	0	0	1	-1

19. Calculate the reduction of the subduced representation from each irreducible representation of the group **I** with respect to the subgroups C_5, D_5 and **T**.

Solution. The character tables of **I**, D_5 and **T** have been given in Problem 18. C_5 is a cyclic group, whose character table is easy to write. The important thing in calculation is to identify which class of **I** each element in the subgroup belongs to. Then, the character formula (3.5) can be used to calculate the reduction of the subduced representation from each irreducible representation of the group **I** with respect to the subgroups C_5, D_5 and **T**. Here we neglect the calculation process, but only list the results.

The reduction of the subduced representation from each irreducible representation of I with respect to the subgroup C_5

	E	C_5	C_5^2	C_5^3	C_5^4	Reduction of the subduced representation
A_1	1	1	1	1	1	
A_2	1	η	η^2	η^3	η^4	
A_3	1	η^2	η^4	η	η^3	
A_4	1	η^3	η	η^4	η^2	
A_5	1	η^4	η^3	η^2	η	
A	1	1	1	1	1	A_1
T_1	3	p^{-1}	$-p$	$-p$	p^{-1}	$A_1 \oplus A_2 \oplus A_5$
T_2	3	$-p$	p^{-1}	p^{-1}	$-p$	$A_1 \oplus A_3 \oplus A_4$
G	4	-1	-1	-1	-1	$A_2 \oplus A_3 \oplus A_4 \oplus A_5$
H	5	0	0	0	0	$A_1 \oplus A_2 \oplus A_3 \oplus A_4 \oplus A_5$

The reduction of the subduced representation from each irreducible representation of I with respect to the subgroup D_5

	E	$2C_5$	$2C_5^2$	$5C_2'$	Reduction of the subduced representation
A_1	1	1	1	1	
A_2	1	1	1	-1	
E_1	2	p	$-p^{-1}$	0	
E_2	2	$-p^{-1}$	p	0	
A	1	1	1	1	A_1
T_1	3	p^{-1}	$-p$	-1	$A_2 \oplus E_1$
T_2	3	$-p$	p^{-1}	-1	$A_2 \oplus E_2$
G	4	-1	-1	0	$E_1 \oplus E_2$
H	5	0	0	1	$A_1 \oplus E_1 \oplus E_2$

The reduction of the subduced representation from each irreducible representation of I with respect to the subgroup T

	E	$3C_2$	$4C_3'$	$4C_3'^2$	Reduction of the subduced representation
A_1	1	1	1	1	
E	1	1	ω	ω^2	
E'	1	1	ω^2	ω	
T	3	-1	0	0	
A	1	1	1	1	A_1
T_1	3	-1	0	0	T
T_2	3	-1	0	0	T
G	4	0	1	1	$A_1 \oplus T$
H	5	1	-1	-1	$E \oplus E' \oplus T$

$$\eta = \exp\{-i2\pi/5\}, \qquad \omega = \exp\{-i2\pi/3\},$$
$$p = \eta + \eta^4 = (\sqrt{5}-1)/2, \qquad p^{-1} = -\eta^2 - \eta^3 = (\sqrt{5}+1)/2.$$

20. Calculate the reduction of the subduced representation from the regular representation of the improper symmetry group I_h of an icosahedron with respect to the subgroup C_{5i}, D_{5d} and T_h.

Solution. The I_h group is an improper point group of I-type. It is the direct product of the group I and the two-order inversion group $V_2 = \{E, \sigma\}$. Any irreducible representation of I_h can be expressed as the direct product of two irreducible representations respectively belonging to I and V_2. Since the space inversion σ is commutable with all elements in I_h and its square is equal to the identity, the representation matrix of σ in an irreducible representation of I_h is a constant matrix ± 1. Based on $D(\sigma) = 1$ and -1, the irreducible representations of I_h are divided into Γ_g and Γ_u, where Γ denotes the irreducible representations of I. g and u are the abbreviation of the Germanic gerade and ungerade, respectively.

The subgroups C_{5i}, D_{5d} and T_h are also the improper point groups of

I-type, their irreducible representations are the direct product of an irreducible representation of V_2 and an irreducible representation of respectively the proper point groups C_5, D_5 or $\mathbf{T}$. Similarly, based on $D(\sigma) = \mathbf{1}$ and $-\mathbf{1}$, the irreducible representations of the subgroups are also divided into γ_g and γ_u, where γ denotes the irreducible representations of C_5, D_5 or $\mathbf{T}$. Therefore, the reduction of the subduced representation from the regular representation of $\mathbf{I}_h$ with respect to the subgroup C_{5i}, D_{5d} and $\mathbf{T}_h$ is similar to the reduction of the subduced representation from the regular representation of $\mathbf{I}$ with respect to the subgroup C_5, D_5 and $\mathbf{T}$, where the only difference is that all the representations are duplicated by the subscripts g and u.

Remind that the character of each element in the regular representation is vanishing except for the character of the identity which is equal to the order of the group. In the reduction of the subduced representation from the regular representation of a group G of order g with respect to its subgroup H of order h, the multiplicity of an irreducible representation of H is equal to its dimension multiplied by $n = g/h$. The results of the reduction are as follows.

For group C_{5i},

$$D = 12 \{ A_{1g} \oplus A_{1u} \oplus A_{2g} \oplus A_{2u} \oplus A_{3g} \oplus A_{3u}$$
$$\oplus A_{4g} \oplus A_{4u} \oplus A_{5g} \oplus A_{5u} \}.$$

For group D_{5d},

$$D = 6 \{ A_{1g} \oplus A_{1u} \oplus A_{2g} \oplus A_{2u} \oplus 2E_{1g} \oplus 2E_{1u} \oplus 2E_{2g} \oplus 2E_{2u} \}.$$

For group $\mathbf{T}_h$,

$$D = 5 \left\{ A_g \oplus A_u \oplus E_g \oplus E_u \oplus E'_g \oplus E'_u \oplus 3T_g \oplus 3T_u \right\}.$$

In all formulas, the series in the bracket is just the reduction series of the regular representation of the subgroup.

21. Calculate the representation matrices of R_6 and S_2 of $\mathbf{I}$ in its irreducible representations by $R_6 = S_1 T_0^2 S_1 T_0^{-1}$ and $S_{12} = T_0 S_1 T_0^3 R_6$ [see Problem 4 in Chapter 4], where the representation matrices of the generators T_0 and S_1 are [see Problem 12 in Chapter 4]:

$$D^{T_1}(T_0) = \text{diag}\left\{\eta, 1, \eta^{-1}\right\}, \quad D^G(T_0) = \text{diag}\left\{\eta^2, \eta, \eta^{-1}, \eta^{-2}\right\},$$
$$D^{T_2}(T_0) = \text{diag}\left\{\eta^2, 1, \eta^{-2}\right\}, \quad D^H(T_0) = \text{diag}\left\{\eta^2, \eta, 1, \eta^{-1}, \eta^{-2}\right\},$$

$$D^{T_1}(S_1) = \frac{-1}{\sqrt{5}} \begin{pmatrix} p^{-1} & \sqrt{2} & p \\ \sqrt{2} & -1 & -\sqrt{2} \\ p & -\sqrt{2} & p^{-1} \end{pmatrix}, \quad D^{T_2}(S_1) = \frac{1}{\sqrt{5}} \begin{pmatrix} -p & \sqrt{2} & p^{-1} \\ \sqrt{2} & -1 & \sqrt{2} \\ p^{-1} & \sqrt{2} & -p \end{pmatrix},$$

$$D^{G}(S_1) = \frac{1}{\sqrt{5}} \begin{pmatrix} -1 & -p & -p^{-1} & 1 \\ -p & 1 & -1 & -p^{-1} \\ -p^{-1} & -1 & 1 & -p \\ 1 & -p^{-1} & -p & -1 \end{pmatrix},$$

$$D^{H}(S_1) = \frac{1}{5} \begin{pmatrix} p^{-2} & 2p^{-1} & \sqrt{6} & 2p & p^2 \\ 2p^{-1} & p^2 & -\sqrt{6} & -p^{-2} & -2p \\ \sqrt{6} & -\sqrt{6} & -1 & \sqrt{6} & \sqrt{6} \\ 2p & -p^{-2} & \sqrt{6} & p^2 & -2p^{-1} \\ p^2 & -2p & \sqrt{6} & -2p^{-1} & p^{-2} \end{pmatrix}.$$

Solution. Any element in **I** can be expressed by the product of two elements respectively belonging to the subgroup C_5 and **T**: $T_0^a \left(R_6^b S_1^c S_{12}^d\right)$, where T_0 and S_1 are two generators of **I** and R_6 and S_2 can be expressed as the product of the generators. In this problem we are going to calculate the representation matrices of R_6 and S_2 explicitly. In the calculation we express R_6 and S_{12} as a product of two elements and make use of the property that the representation matrix of T_0 is diagonal. Remind that $\eta = \exp\{-i2\pi/5\}$ satisfies the following useful formulas:

$$\sum_{a=0}^{4} \eta^a = 0, \quad \eta^{a+5n} = \eta^a, \quad 1 + 2\eta + 2\eta^4 = -1 - 2\eta^2 - 2\eta^3 = \sqrt{5}$$
$$p = \eta + \eta^4 = (\sqrt{5} - 1)/2, \quad p^{-1} = -\eta^2 - \eta^3 = (\sqrt{5} + 1)/2.$$

For the identical representation, $D^A(T_0) = D^A(S_1) = D^A(R_6) = D^A(S_{12}) = 1$. The results for other representations are as follows.

$$D^{T_1}(R_6) = \frac{1}{\sqrt{5}} \begin{pmatrix} -\eta^2 - \eta^4 & \sqrt{2}\eta^2 & \eta^3 + \eta^4 \\ \sqrt{2}\eta & -1 & -\sqrt{2}\eta^4 \\ \eta + \eta^2 & -\sqrt{2}\eta^3 & -\eta - \eta^3 \end{pmatrix},$$

$$D^{T_2}(R_6) = \frac{1}{\sqrt{5}} \begin{pmatrix} \eta^3 + \eta^4 & -\sqrt{2}\eta^4 & \eta + \eta^3 \\ -\sqrt{2}\eta^2 & 1 & -\sqrt{2}\eta^3 \\ \eta^2 + \eta^4 & -\sqrt{2}\eta & \eta + \eta^2 \end{pmatrix},$$

$$D^G(R_6) = \frac{1}{\sqrt{5}} \begin{pmatrix} \eta & \eta^2+\eta^3 & -\eta^2-\eta^4 & -\eta^2 \\ \eta+\eta^2 & -\eta^3 & \eta & -\eta-\eta^4 \\ -\eta-\eta^4 & \eta^4 & -\eta^2 & \eta^3+\eta^4 \\ -\eta^3 & -\eta-\eta^3 & \eta^2+\eta^3 & \eta^4 \end{pmatrix},$$

$$D^H(R_6)$$
$$= \frac{1}{5} \begin{pmatrix} -1+\eta-\eta^2 & -2\eta-2\eta^4 & \sqrt{6}\eta^4 & 2+2\eta & -1+\eta^2-\eta^4 \\ -2-2\eta^3 & -1-\eta+\eta^3 & -\sqrt{6}\eta^2 & 1-\eta+\eta^2 & -2\eta^2-2\eta^3 \\ \sqrt{6}\eta^2 & -\sqrt{6}\eta & -1 & \sqrt{6}\eta^4 & \sqrt{6}\eta^3 \\ 2\eta^2+2\eta^3 & 1+\eta^3-\eta^4 & \sqrt{6}\eta^3 & -1+\eta^2-\eta^4 & 2+2\eta^2 \\ -1-\eta+\eta^3 & -2-2\eta^4 & \sqrt{6}\eta & 2\eta+2\eta^4 & -1-\eta^3+\eta^4 \end{pmatrix},$$

$$D^{T_1}(S_{12}) = \begin{pmatrix} 0 & 0 & 1 \\ 0 & -1 & 0 \\ 1 & 0 & 0 \end{pmatrix}, \qquad D^{T_2}(S_{12}) = \begin{pmatrix} 0 & 0 & -1 \\ 0 & -1 & 0 \\ -1 & 0 & 0 \end{pmatrix},$$

$$D^G(S_{12}) = \begin{pmatrix} 0 & 0 & 0 & 1 \\ 0 & 0 & 1 & 0 \\ 0 & 1 & 0 & 0 \\ 1 & 0 & 0 & 0 \end{pmatrix}, \qquad D^H(S_{12}) = \begin{pmatrix} 0 & 0 & 0 & 0 & 1 \\ 0 & 0 & 0 & -1 & 0 \\ 0 & 0 & 1 & 0 & 0 \\ 0 & -1 & 0 & 0 & 0 \\ 1 & 0 & 0 & 0 & 0 \end{pmatrix}.$$

3.4 The Clebsch-Gordan Coefficients

★ If a quantum system consists of two subsystems, the wave function of the total system can be expressed as the combination of the products of the wave functions of two subsystems. Denote by G the common symmetry group of the system and two subsystems. We choose the wave function of each subsystem belonging to a given row of a given irreducible representation of G

$$P_R \psi^j_\mu(1) = \sum_\rho \psi^j_\rho(1) D^j_{\rho\mu}(R), \qquad P_R \psi^k_\nu(2) = \sum_\sigma \psi^k_\sigma(2) D^k_{\sigma\nu}(R). \tag{3.20}$$

Then, the wave function for the total system is expressed as their product, transforming according to the direct product representation:

$$\Psi^{jk}_{\mu\nu}(1,2) = \psi^j_\mu(1)\psi^k_\nu(2),$$
$$P_R \Psi^{jk}_{\mu\nu}(1,2) = \sum_{\rho\sigma} \psi^j_\rho(1)\psi^k_\sigma(2) D^j_{\rho\mu}(R) D^k_{\sigma\nu}(R), \tag{3.21}$$
$$\{D^j(R) \times D^k(R)\}_{\rho\sigma,\mu\nu} = D^j_{\rho\mu}(R) D^k_{\sigma\nu}(R).$$

The character in the direct product representation is equal to the product of the characters in two representations.

The direct product of two irreducible representations is reducible in general, and can be reduced by a similarity transformation [see Eqs. (3.10-12)] :

$$\left(C^{jk}\right)^{-1}\left\{D^j(R) \times D^k(R)\right\}C^{jk} = \bigoplus_J a_J D^J(R). \qquad (3.22)$$

For a finite group or a compact Lie group, all representations in Eq. (3.22) are chosen to be unitary, then C^{jk} can be chosen to be a unitary matrix with determinant one. The series in the right-hand side of Eq. (3.22) is called the Clebsch-Gordan (C-G) series, the matrix entries in the similarity transformation matrix C^{jk} are called the C-G coefficients. Rewrite Eq. (3.22) in the form of characters, we may calculate the multiplicities a_J by the orthogonal relation (3.5) of characters:

$$\chi^j(R)\chi^k(R) = \sum_J a_J \chi^J(R),$$
$$a_J = \frac{1}{g}\sum_{R \in G}\left[\chi^j(R)\chi^k(R)\right]^* \chi^J(R). \qquad (3.23)$$

Substituting Eq. (3.23) into Eq. (3.22), we are able to calculate the C-G coefficients $C^{jk}_{\mu\nu,JMr}$. When $a_J > 1$, the parameter r is needed to distinguish these a_J irreducible representation $D^J(G)$. The method has been demonstrated in section 2 in detail. The total wave function $\Phi^J_{Mr}(1,2)$ with the given symmetry is combined from the wave functions $\Psi^{jk}_{\mu\nu}(1,2)$ by the C-G coefficients $C^{jk}_{\mu\nu,JMr}$:

$$\Phi^J_{Mr}(1,2) = \sum_{\mu\nu} \psi^j_\mu(1)\psi^k_\nu(2)C^{jk}_{\mu\nu,JMr},$$
$$P_R\Phi^J_{Mr}(1,2) = \sum_{M'} \Phi^J_{M'r}(1,2)D^J_{M'M}(R). \qquad (3.24)$$

In physics, it is convenient to express Eq. (3.24) in the Dirac notations. $\psi^j_\mu(1)\psi^k_\nu(2)$ is denoted by $|j,\mu\rangle|k,\nu\rangle$, where the indices j and k are sometimes neglected. $\Phi^J_{Mr}(1,2)$ is denoted by $||J,(r),M\rangle$:

$$||J,(r),M\rangle = \sum_{\mu\nu} |j,\mu\rangle|k,\nu\rangle C^{jk}_{\mu\nu,JMr}. \qquad (3.25)$$

The action of the symmetric operator P_R on the wave function is

$$P_R||J, (r), M\rangle = \sum_{\mu\nu} (P_R|j, \mu)) (P_R|k, \nu)) C^{jk}_{\mu\nu, JMr}. \qquad (3.26)$$

First, we choose R to be the generator A whose representation matrices are diagonal such that Eq. (3.26) becomes an eigenequation, and gives some relations of the C-G coefficients. Sometimes, the method of projection operators is useful in this step. Then, the C-G coefficients can be further determined in terms of the remaining generators. Before completion of the calculation, some undetermined parameters should be chosen arbitrarily.

★ The inverse transformation of Eq. (3.24) is

$$\psi^j_\mu(1)\psi^k_\nu(2) = \sum_{JMr} \Phi^J_{Mr}(1,2) \left(C^{jk}_{\mu\nu, JMr}\right)^*.$$

$\psi^j_\mu(1)\psi^k_\nu(2)$ contains some functions $\Phi^J_{Mr}(1,2)$ belonging to different irreducible representation. In terms of the projection operator

$$P^J_M = \frac{m_J}{g} \sum_{S\in G} D^J_{MM}(S)^* P_S, \qquad (3.27)$$

one can pick up from $\psi^j_\mu(1)\psi^k_\nu(2)$ the part belonging to the Mth row of the irreducible representation $D^J(G)$. In fact,

$$\begin{aligned}
P^J_M \psi^j_\mu(1)\psi^k_\nu(2) &= \frac{m_J}{g} \sum_{S\in G} D^J_{MM}(S)^* \sum_{KNr} P_S \Phi^K_{Nr}(1,2) \left(C^{jk}_{\mu\nu, KNr}\right)^* \\
&= \frac{m_J}{g} \sum_{KNr\rho} \sum_{S\in G} D^J_{MM}(S)^* D^K_{\rho N}(S) \Phi^K_{\rho r}(1,2) \left(C^{jk}_{\mu\nu, KNr}\right)^* \\
&= \sum_r \Phi^J_{Mr}(1,2) \left(C^{jk}_{\mu\nu, JMr}\right)^*.
\end{aligned}$$

The projection operator is helpful in the calculation of the Clebsch-Gordan coefficients.

22. Please use the projection operators to reduce the regular representation of group **T**.

Solution. In the notation for the group **T** given in Problem 16 of Chapter 2, R_1 is a three-order element. Its eigenvalues are denoted by ω^μ, where $\omega = \exp\{-i2\pi/3\}$, $\omega^{\mu+3} = \omega^\mu$. The value of μ can be taken as module 3, namely $\mu + 3$ is equal to μ. Taking the representation matrix of R_1 to be diagonal, we can designate the rows and columns of each irreducible representation by the power μ of the eigenvalue ω^μ for R_1. By making use of the projection operator for the cyclic subgroup of R_1, we can calculate the common eigenfunctions $\Phi_{\mu\nu}^{(a)}$ for left- or right-multiplying of R_1:

$$\Phi_{\mu\nu}^{(a)} = c_a P_\mu R^{(a)} P_\nu, \qquad P_\mu = \{E + \omega^{-\mu} R_1 + \omega^\mu R_1^2\}/3,$$

$$R_1 \Phi_{\mu\nu}^{(a)} = \omega^\mu \Phi_{\mu\nu}^{(a)}, \qquad \Phi_{\mu\nu}^{(a)} R_1 = \omega^\nu \Phi_{\mu\nu}^{(a)},$$

where c_a is the normalization factor and can be chosen for convenience. Take $R^{(1)} = E$ and $c_1 = 9$,

$$\begin{aligned}
\Phi_{\mu\nu}^{(1)} &= 9P_\mu E P_\nu = (1 + \omega^{\nu-\mu} + \omega^{\mu-\nu})\left\{E + R_1\omega^{-\mu} + R_1^2\omega^\mu\right\} \\
&= 3\delta_{\mu\nu}\left\{E + R_1\omega^{-\mu} + R_1^2\omega^\mu\right\}.
\end{aligned}$$

Then, taking an element $R^{(2)}$ which does not appear in $\Phi_{\mu\nu}^{(1)}$, say $R^{(2)} = T_z^2$, we have

$$\begin{aligned}
\Phi_{\mu\nu}^{(2)} &= 9P_\mu T_z^2 P_\nu = T_z^2 + \omega^{-\nu} T_z^2 R_1 + \omega^\nu T_z^2 R_1^2 \\
&\quad + \omega^{-\mu+\nu} R_1 T_z^2 R_1^2 + \omega^{-\mu} R_1 T_z^2 + \omega^{-\mu-\nu} R_1 T_z^2 R_1 \\
&\quad + \omega^{\mu-\nu} R_1^2 T_z^2 R_1 + \omega^{\mu+\nu} R_1^2 T_z^2 R_1^2 + \omega^\mu R_1^2 T_z^2 \\
&= \left\{T_z^2 + \omega^{-\nu} R_2 + \omega^\nu R_4^2\right\} + \omega^{\nu-\mu}\left\{T_x^2 + \omega^{-\nu} R_4 + \omega^\nu R_3^2\right\} \\
&\quad + \omega^{\mu-\nu}\left\{T_y^2 + \omega^{-\nu} R_3 + \omega^\nu R_2^2\right\}.
\end{aligned}$$

Now, all elements in **T** appear in the functions $\Phi_{\mu\nu}^{(1)}$ and $\Phi_{\mu\nu}^{(2)}$. One will not obtain any other linearly independent function $\Phi_{\mu\nu}^{(a)}$ by choosing new $R^{(a)}$.

Generally, $\Phi_{\mu\nu}^{(a)}$ does not belong to a given irreducible representation, and can be combined to new basis functions belonging to the irreducible representation by the class operator $W = R_1 + R_2 + R_3 + R_4$, which take constant values $4\chi^j(R_1)/m_j$ (see Problem 8) in an irreducible representation. In the representation A, E, E' and T of the group **T**, the values of W are 4, 4ω, $4\omega^2$ and 0, respectively. In fact, for given subscripts μ and ν, we calculate the matrix form of W in the basis functions $\Phi_{\mu\nu}^{(a)}$, and then diagonalize it. Due to completeness of the function bases, we only need to list those terms of E and T_z^2 in the calculation

$$\begin{aligned}
W\Phi_{\mu\mu}^{(1)} &= 3\omega^\mu\left\{E + T_z^2\right\} + \cdots = \omega^\mu\left\{\Phi_{\mu\mu}^{(1)} + 3\Phi_{\mu\mu}^{(2)}\right\}, \\
W\Phi_{\mu\nu}^{(2)} &= \left\{E + T_z^2\right\}\omega^\nu\left\{1 + \omega^{\nu-\mu} + \omega^{\mu-\nu}\right\} + \cdots \\
&= \delta_{\mu\nu}\omega^\mu\left\{\Phi_{\mu\mu}^{(1)} + 3\Phi_{\mu\mu}^{(2)}\right\}.
\end{aligned}$$

When $\mu = \nu$, the matrix form of W in these basis functions is

$$\omega^\mu \begin{pmatrix} 1 & 1 \\ 3 & 3 \end{pmatrix}.$$

The eigenvalues are $4\omega^\mu$ and 0, and the transpose of the corresponding eigenvectors respectively are $(1/3, 1)$ and $(1, -1)$. Combining the basis functions according to the eigenvectors, we obtain the irreducible basis functions which respectively belong to representations A, E, E' and T.

$$\Psi^A = (1/3)\Phi_{00}^{(1)} + \Phi_{00}^{(2)} = E + T_x^2 + T_y^2 + T_z^2$$
$$+ R_1 + R_2 + R_3 + R_4 + R_1^2 + R_2^2 + R_3^2 + R_4^2,$$

$$\Psi^E = (1/3)\Phi_{11}^{(1)} + \Phi_{11}^{(2)} = E + T_x^2 + T_y^2 + T_z^2$$
$$+ \omega^2 (R_1 + R_2 + R_3 + R_4) + \omega (R_1^2 + R_2^2 + R_3^2 + R_4^2),$$

$$\Psi^{E'} = (1/3)\Phi_{22}^{(1)} + \Phi_{22}^{(2)} = E + T_x^2 + T_y^2 + T_z^2$$
$$+ \omega (R_1 + R_2 + R_3 + R_4) + \omega^2 (R_1^2 + R_2^2 + R_3^2 + R_4^2),$$

$$\Psi_{\mu\mu}^T = c_{\mu\mu} \left\{ \Phi_{\mu\mu}^{(1)} - \Phi_{\mu\mu}^{(2)} \right\}.$$

When $\mu \neq \nu$, there is only one basis function $\Phi_{\mu\nu}^{(2)}$. $\Phi_{\mu\nu}^{(2)}$ belongs to the representation T:

$$\Psi_{\mu\nu}^T = c_{\mu\nu} \Phi_{\mu\nu}^{(2)}.$$

In the calculation, being the eigenfunctions, each basis function can contain an arbitrary constant factor. For a one-dimensional representation, this factor is the normalization factor, which is irrelevant to the representation matrix and can be omitted. But for the three-dimensional representation T, the choice of the factor is related to the representation matrix. In order to make the representation unitary, we choose the basis functions to be normalized, $c_{00} = 1/6$ and $c_{10} = c_{20} = 1/3$. In calculating the representation matrix of another generator T_z^2 with respect to the basis function $\Psi_{\mu 0}^T$, we only need to list those terms of E, T_x^2, and T_z^2, and notice $T_z^2 T_y^2 = T_x^2$.

$$\Psi_{00}^T = \left\{ 3E + 3R_1 + 3R_1^2 - T_x^2 - T_y^2 - T_z^2 - R_4 - R_3 \right.$$
$$\left. - R_2 - R_3^2 - R_2^2 - R_4^2 \right\}/6,$$

$$\Psi_{10}^T = \left\{ T_z^2 + R_2 + R_4^2 + \omega^2 (T_x^2 + R_4 + R_3^2) \right.$$
$$\left. + \omega (T_y^2 + R_3 + R_2^2) \right\}/3,$$

$$\Psi_{20}^T = \{T_z^2 + R_2 + R_4^2 + \omega\left(T_x^2 + R_4 + R_3^2\right)$$
$$+ \omega^2\left(T_y^2 + R_3 + R_2^2\right)\}/3.$$

Thus,

$$T_z^2\Psi_{00}^T = (1/6)\left\{-E - T_x^2 + 3T_z^2 + \cdots\right\}$$
$$= \left\{-\Psi_{00}^T + 2\Psi_{10}^T + 2\Psi_{20}^T\right\}/3 = \sum_\nu \Psi_{\nu 0}^T \overline{D}_{\nu 0}^T(T_z^2),$$
$$T_z^2\Psi_{10}^T = (1/3)\left\{E + \omega T_x^2 + \cdots\right\}$$
$$= \left\{2\Psi_{00}^T - \Psi_{10}^T + 2\Psi_{20}^T\right\}/3 = \sum_\nu \Psi_{\nu 0}^T \overline{D}_{\nu 1}^T(T_z^2),$$
$$T_z^2\Psi_{20}^T = (1/3)\left(E + \omega^2 T_x^2 + \cdots\right\}$$
$$= \left\{2\Psi_{00}^T + 2\Psi_{10}^T - \Psi_{20}^T\right\}/3 = \sum_\nu \Psi_{\nu 0}^T \overline{D}_{\nu 2}^T(T_z^2).$$

The representation matrices of the generators are

$$\overline{D}^T(R_1) = \begin{pmatrix} 1 & 0 & 0 \\ 0 & \omega & 0 \\ 0 & 0 & \omega^2 \end{pmatrix}, \qquad \overline{D}^T(T_z^2) = \frac{1}{3}\begin{pmatrix} -1 & 2 & 2 \\ 2 & -1 & 2 \\ 2 & 2 & -1 \end{pmatrix}.$$

The factors $c_{\mu\nu}$ in other basis functions are determined by the representation matrices and the following formula:

$$\Psi_{\mu\nu}^T T_z^2 = \sum_\rho \overline{D}_{\nu\rho}^T(T_z^2)\Psi_{\mu\rho}^T.$$

$$\Psi_{00}^T T_z^2 = (1/6)\left\{-E - T_x^2 + 3T_z^2 + \cdots\right\} = (1/3)\left\{-\Psi_{00}^T + 2\Psi_{01}^T + 2\Psi_{02}^T\right\},$$
$$\Psi_{10}^T T_z^2 = (1/3)\left\{E + \omega T_x^2 + \cdots\right\} = (1/3)\left\{-\Psi_{10}^T + 2\Psi_{11}^T + 2\Psi_{12}^T\right\},$$
$$\Psi_{20}^T T_z^2 = (1/3)\left\{E + \omega^2 T_x^2 + \cdots\right\} = (1/3)\left\{-\Psi_{20}^T + 2\Psi_{21}^T + 2\Psi_{22}^T\right\}.$$

Namely, $c_{11} = c_{22} = 1/6$, $c_{01} = c_{02} = c_{12} = c_{21} = 1/3$,

$$\Psi_{01}^T = \{T_z^2 + R_4 + R_2^2 + \omega\left(T_x^2 + R_3 + R_4^2\right)$$
$$+ \omega^2\left(T_y^2 + R_2 + R_3^2\right)\}/3,$$
$$\Psi_{02}^T = \{T_z^2 + R_4 + R_2^2 + \omega\left(T_y^2 + R_2 + R_3^2\right)$$
$$+ \omega^2\left(T_x^2 + R_3 + R_4^2\right)\}/3,$$
$$\Psi_{11}^T = \{3E - T_x^2 - T_y^2 - T_z^2 + \omega\left(3R_1^2 - R_2^2 - R_3^2 - R_4^2\right)$$
$$+ \omega^2\left(3R_1 - R_2 - R_3 - R_4\right)\}/6,$$

$$\Psi_{12}^T = \{T_z^2 + R_3 + R_3^2 + w\left(T_x^2 + R_2 + R_2^2\right)$$
$$+ w^2\left(T_y^2 + R_4 + R_4^2\right)\}/3,$$

$$\Psi_{21}^T = \{T_z^2 + R_3 + R_3^2 + w\left(T_y^2 + R_4 + R_4^2\right)$$
$$+ w^2\left(T_x^2 + R_2 + R_2^2\right)\}/3,$$

$$\Psi_{22}^T = \{3E - T_x^2 - T_y^2 - T_z^2 + w\left(3R_1 - R_2 - R_3 - R_4\right)$$
$$+ w^2\left(3R_1^2 - R_2^2 - R_3^2 - R_4^2\right)\}/6.$$

Usually, the three-dimensional irreducible representation T of the group **T** is calculated by the coordinate transformation method in three-dimensional space. The results are as follows:

$$D^T(R_1) = \begin{pmatrix} 0 & 0 & 1 \\ 1 & 0 & 0 \\ 0 & 1 & 0 \end{pmatrix}, \qquad D^T(T_z^2) = \begin{pmatrix} -1 & 0 & 0 \\ 0 & -1 & 0 \\ 0 & 0 & 1 \end{pmatrix}.$$

We can find a similarity transformation X to relate two equivalent representation. For the generators R_1,

$$X^{-1}\begin{pmatrix} 0 & 0 & 1 \\ 1 & 0 & 0 \\ 0 & 1 & 0 \end{pmatrix} X = \begin{pmatrix} 1 & 0 & 0 \\ 0 & w & 0 \\ 0 & 0 & w^2 \end{pmatrix},$$

we have

$$X = \begin{pmatrix} a & b & c \\ a & bw^2 & cw \\ a & bw & cw^2 \end{pmatrix}.$$

Substituting it into $D^T(T_z^2)X = X\overline{D}^T(T_z^2)$,

$$\begin{pmatrix} -a & -b & -c \\ -a & -bw^2 & -cw \\ a & bw & cw^2 \end{pmatrix} = \frac{1}{3}\begin{pmatrix} -a+2b+2c & 2a-b+2c & 2a+2b-c \\ -a+2bw^2+2cw & 2a-bw^2+2cw & 2a+2bw^2-cw \\ -a+2bw+2cw^2 & 2a-bw+2cw^2 & 2a+2bw-cw^2 \end{pmatrix},$$

we obtain $a = wb = w^2c = \sqrt{1/3}$,

$$X = \frac{1}{\sqrt{3}}\begin{pmatrix} 1 & w^2 & w \\ 1 & w & w^2 \\ 1 & 1 & 1 \end{pmatrix}, \qquad X^{-1} = \frac{1}{\sqrt{3}}\begin{pmatrix} 1 & 1 & 1 \\ w & w^2 & 1 \\ w^2 & w & 1 \end{pmatrix}.$$

23. Please use the projection operators to reduce the regular representation of group **O**.

Solution. We follow the notations and formulas given in Problem 16 of Chapter 2 and Problem 12 of this Chapter. Just like the method used in the preceding problem, we first calculate the common eigenfunctions $\Phi_{\mu\nu}^{(a)} = c_a P_\mu R^{(a)} P_\nu$ of left- or right-multiplying R_1 in terms of the projection operators. Letting $R^{(1)} = E$ and $R^{(2)} = T_z^2$, we have

$$\Phi_{\mu\nu}^{(1)} = 9P_\mu E P_\nu = 3\delta_{\mu\nu} \left\{ E + R_1 \omega^{-\mu} + R_1^2 \omega^\mu \right\},$$

$$\Phi_{\mu\nu}^{(2)} = 9P_\mu T_z^2 P_\nu$$

$$= T_z^2 + \omega^{-\nu} R_2 + \omega^\nu R_4^2 + \omega^{\nu-\mu} \left\{ T_x^2 + \omega^{-\nu} R_4 + \omega^\nu R_3^2 \right\}$$

$$+ \omega^{\mu-\nu} \left\{ T_y^2 + \omega^{-\nu} R_3 + \omega^\nu R_2^2 \right\}.$$

There still exist some elements in **O** which do not appear in the functions $\Phi_{\mu\nu}^{(1)}$ and $\Phi_{\mu\nu}^{(2)}$. Letting $R^{(3)} = T_z$ and $R^{(4)} = S_2$, we have

$$\Phi_{\mu\nu}^{(3)} = 9P_\mu T_z P_\nu = T_z + \omega^{-\nu} T_z R_1 + \omega^\nu T_z R_1^2 + \omega^{-\mu+\nu} R_1 T_z R_1^2$$

$$+ \omega^{-\mu} R_1 T_z + \omega^{-\mu-\nu} R_1 T_z R_1 + \omega^{\mu-\nu} R_1^2 T_z R_1$$

$$+ \omega^{\mu+\nu} R_1^2 T_z R_1^2 + \omega^\mu R_1^2 T_z$$

$$= T_z + \omega^{-\nu} S_3 + \omega^\nu T_y^3 + \omega^{\nu-\mu} \left\{ T_x + \omega^{-\nu} S_5 + \omega^\nu T_z^3 \right\}$$

$$+ \omega^{\mu-\nu} \left\{ T_y + \omega^{-\nu} S_1 + \omega^\nu T_x^3 \right\},$$

$$\Phi_{\mu\nu}^{(4)} = 9P_\mu S_2 P_\nu$$

$$= S_2 + \omega^{-\nu} S_2 R_1 + \omega^\nu S_2 R_1^2$$

$$+ \omega^{-\mu+\nu} R_1 S_2 R_1^2 + \omega^{-\mu} R_1 S_2 + \omega^{-\mu-\nu} R_1 S_2 R_1$$

$$+ \omega^{\mu-\nu} R_1^2 S_2 R_1 + \omega^{\mu+\nu} R_1^2 S_2 R_1^2 + \omega^\mu R_1^2 S_2$$

$$= \left\{ S_2^2 + \omega^{-\nu} S_4 + \omega^\nu S_6 \right\} \left\{ 1 + \omega^{-\mu-\nu} + \omega^{\mu+\nu} \right\}$$

$$= 3\delta_{\mu(-\nu)} \left\{ S_2 + \omega^\mu S_4 + \omega^{-\mu} S_6 \right\},$$

where $\delta_{\mu(-\nu)}$ is equal to 1 when $(\mu + \nu)$ is a multiple of 3, otherwise it is zero. Thus, all elements in **O** appear in the four kinds of basis functions, so they are complete.

Now, we will combine the basis functions by making use of the class operator $W = T_x + T_x^3 + T_y + T_y^3 + T_z + T_z^3$ such that they belong to the irreducible representations. W takes a constant value $4\chi^j(R_1)/m_j$ (see Problem 8) in an irreducible representation. Through calculation, the constants in the irreducible representations A_1, A_2, E, T_1 and T_2 of **O** are

6, -6, 0, 2 and -2, respectively. In fact, for given subscripts μ and ν, we calculate the matrix form of W in the basis functions $\Phi_{\mu\nu}^{(a)}$, and then diagonalize it. Due to completeness of the basis functions, we only need to list those terms of E, T_z^2, T_z and S_2. In the calculation of left-multiplying W on the basis functions $\Phi_{\mu\nu}^{(a)}$, the following formulas obtained from the multiplication table will be useful:

$$T_x T_x^3 = T_x^3 T_x = T_y T_y^3 = T_y^3 T_y = T_z T_z^3 = T_z^3 T_z = E,$$

$$T_z E = T_z^3 T_z^2 = T_x R_2^2 = T_x^3 R_4^2 = T_y R_3 = T_y^3 R_1 = T_z,$$

$$T_z T_z = T_z^3 T_z^3 = T_x S_3 = T_x^3 S_4 = T_y S_6 = T_y^3 S_5 = T_z^2,$$

$$T_z T_y^2 = T_z^3 T_x^2 = T_x R_4^2 = T_x^3 R_2^2 = T_y R_4 = T_y^3 R_2 = S_2.$$

Hence,

$$W\Phi_{\mu\mu}^{(1)} = 3T_z \left\{ 1 + \omega^{-\mu} \right\} + \cdots = 3\omega^\mu \left(3\delta_{\mu 0} - 1 \right) \Phi_{\mu\mu}^{(3)},$$

$$\begin{aligned}
W\Phi_{\mu\nu}^{(2)} &= T_z \left\{ 1 + \omega^\mu + \omega^\nu + \omega^{\mu+\nu} \right\} \\
&\quad + S_2 \left\{ \omega^{\nu-\mu} + \omega^{\mu-\nu} + \omega^{-\mu} + \omega^{-\nu} + \omega^\mu + \omega^\nu \right\} + \cdots \\
&= \omega^{-\mu-\nu} \left(1 + 9\delta_{\mu 0}\delta_{\nu 0} - 3\delta_{\mu 0} - 3\delta_{\nu 0} \right) \Phi_{\mu\nu}^{(3)} \\
&\quad + \delta_{\mu(-\nu)} \left(3\delta_{\mu 0} - 1 \right) \Phi_{\mu\nu}^{(4)},
\end{aligned}$$

$$\begin{aligned}
W\Phi_{\mu\nu}^{(3)} &= E \left\{ \omega^\mu + \omega^{\nu-\mu} + \omega^\nu + \omega^{\mu-\nu} + \omega^{-\nu-\mu} + 1 \right\} \\
&\quad + T_z^2 \left\{ 1 + \omega^{-\nu-\mu} + \omega^{-\nu} + \omega^{-\mu} \right\} + \cdots \\
&= \delta_{\mu\nu} \left(3\delta_{\mu 0} - 1 \right) \omega^{-\mu} \Phi_{\mu\mu}^{(1)} \\
&\quad + \omega^{\mu+\nu} \left(1 + 9\delta_{\mu 0}\delta_{\nu 0} - 3\delta_{\mu 0} - 3\delta_{\nu 0} \right) \Phi_{\mu\nu}^{(2)},
\end{aligned}$$

$$W\Phi_{\mu\nu}^{(4)} = 3\delta_{\mu(-\nu)} T_z^2 \left\{ \omega^\mu + \omega^{-\mu} \right\} + \cdots = 3\delta_{\mu(-\nu)} \left(3\delta_{\mu 0} - 1 \right) \Phi_{\mu\nu}^{(2)}.$$

When $\mu = \nu = 0$, there are four bases, $\Phi_{00}^{(1)}$, $\Phi_{00}^{(2)}$, $\Phi_{00}^{(3)}$ and $\Phi_{00}^{(4)}$. The matrix form of W in the basis functions and its eigenvectors for the eigenvalues 6, -6, 2 and -2 are

$$\begin{pmatrix} 0 & 0 & 2 & 0 \\ 0 & 0 & 4 & 6 \\ 6 & 4 & 0 & 0 \\ 0 & 2 & 0 & 0 \end{pmatrix}, \quad \begin{pmatrix} 1 \\ 3 \\ 3 \\ 1 \end{pmatrix}, \quad \begin{pmatrix} 1 \\ 3 \\ -3 \\ -1 \end{pmatrix}, \quad \begin{pmatrix} 1 \\ -1 \\ 1 \\ -1 \end{pmatrix}, \quad \begin{pmatrix} 1 \\ -1 \\ -1 \\ 1 \end{pmatrix}.$$

The corresponding eigenfunctions respectively belong to the representations

A_1, A_2, T_1 and T_2:

$$
\begin{aligned}
\Psi_{00}^{A_1} &= \Phi_{00}^{(1)}/3 + \Phi_{00}^{(2)} + \Phi_{00}^{(3)} + \Phi_{00}^{(4)}/3 \\
&= E + R_1 + R_1^2 + T_x^2 + T_y^2 + T_z^2 + R_2 + R_2^2 \\
&\quad + R_3 + R_3^2 + R_4 + R_4^2 + T_x + T_x^3 + T_y + T_y^3 \\
&\quad + T_z + T_z^3 + S_1 + S_3 + S_5 + S_2 + S_4 + S_6, \\
\Psi_{00}^{A_2} &= \Phi_{00}^{(1)}/3 + \Phi_{00}^{(2)} - \Phi_{00}^{(3)} - \Phi_{00}^{(4)}/3 \\
&= E + R_1 + R_1^2 + T_x^2 + T_y^2 + T_z^2 + R_2 + R_2^2 \\
&\quad + R_3 + R_3^2 + R_4 + R_4^2 - T_x - T_x^3 - T_y - T_y^3 \\
&\quad - T_z - T_z^3 - S_1 - S_3 - S_5 - S_2 - S_4 - S_6, \\
\Psi_{00}^{T_1} &= b_{00}\left\{ \Phi_{00}^{(1)} - \Phi_{00}^{(2)} + \Phi_{00}^{(3)} - \Phi_{00}^{(4)} \right\} \\
&= b_{00}\left\{ 3E + 3R_1 + 3R_1^2 - T_x^2 - T_y^2 - T_z^2 - R_2 - R_2^2 \right. \\
&\quad - R_3 - R_3^2 - R_4 - R_4^2 + T_x + T_x^3 + T_y + T_y^3 \\
&\quad \left. + T_z + T_z^3 + S_1 + S_3 + S_5 - 3S_2 - 3S_4 - 3S_6 \right\}, \\
\Psi_{00}^{T_2} &= c_{00}\left\{ \Phi_{00}^{(1)} - \Phi_{00}^{(2)} - \Phi_{00}^{(3)} + \Phi_{00}^{(4)} \right\} \\
&= c_{00}\left\{ 3E + 3R_1 + 3R_1^2 - T_x^2 - T_y^2 - T_z^2 - R_2 - R_2^2 \right. \\
&\quad - R_3 - R_3^2 - R_4 - R_4^2 - T_x - T_x^3 - T_y - T_y^3 \\
&\quad \left. - T_z - T_z^3 - S_1 - S_3 - S_5 + 3S_2 + 3S_4 + 3S_6 \right\}.
\end{aligned}
$$

Each eigenfunction may contain a constant factor. For one-dimensional representation, this constant is the normalized factor, which is irrelevant to the representation matrix and may be omitted. But for two-dimensional or three-dimensional representations, the choices of factors are related to the explicit forms of the representation matrices.

When $\mu = \nu \neq 0$, there are three bases, $\Phi_{\mu\mu}^{(1)}$, $\Phi_{\mu\mu}^{(2)}$ and $\Phi_{\mu\mu}^{(3)}$. The matrix form of W in the basis functions and its eigenvectors for the eigenvalues 0, 2 and -2 are

$$
\begin{pmatrix} 0 & 0 & -\omega^{-\mu} \\ 0 & 0 & \omega^{-\mu} \\ -3\omega^{\mu} & \omega^{\mu} & 0 \end{pmatrix}, \quad
\begin{pmatrix} 1 \\ 3 \\ 0 \end{pmatrix}, \quad
\begin{pmatrix} 1 \\ -1 \\ -2\omega^{\mu} \end{pmatrix}, \quad
\begin{pmatrix} 1 \\ -1 \\ 2\omega^{\mu} \end{pmatrix}.
$$

The corresponding eigenfunctions respectively belong to the representation

E, T_1 and T_2:

$$\Psi^E_{\mu\mu} = a_{\mu\mu}\left\{\Phi^{(1)}_{\mu\mu}/3 + \Phi^{(2)}_{\mu\mu}\right\}$$
$$= a_{\mu\mu}\left\{E + T_x^2 + T_y^2 + T_z^2 + \omega^{-\mu}\left(R_1 + R_2 + R_3 + R_4\right)\right.$$
$$\left. + \omega^\mu\left(R_1^2 + R_2^2 + R_3^2 + R_4^2\right)\right\},$$

$$\Psi^{T_1}_{\mu\mu} = b_{\mu\mu}\left\{\Phi^{(1)}_{\mu\mu} - \Phi^{(2)}_{\mu\mu} - 2\omega^\mu\Phi^{(3)}_{\mu\mu}\right\}$$
$$= b_{\mu\mu}\left\{3E - T_x^2 - T_y^2 - T_z^2 - 2S_1 - 2S_3 - 2S_5\right.$$
$$+ \omega^{-\mu}\left(3R_1 - R_2 - R_3 - R_4 - 2T_x^3 - 2T_y^3 - 2T_z^3\right)$$
$$\left. + \omega^\mu\left(3R_1^2 - R_2^2 - R_3^2 - R_4^2 - 2T_x - 2T_y - 2T_z\right)\right\},$$

$$\Psi^{T_2}_{\mu\mu} = c_{\mu\mu}\left\{\Phi^{(1)}_{\mu\mu} - \Phi^{(2)}_{\mu\mu} + 2\omega^\mu\Phi^{(3)}_{\mu\mu}\right\}$$
$$= c_{\mu\mu}\left\{3E - T_x^2 - T_y^2 - T_z^2 + 2S_1 + 2S_3 + 2S_5\right.$$
$$+ \omega^{-\mu}\left(3R_1 - R_2 - R_3 - R_4 + 2T_x^3 + 2T_y^3 + 2T_z^3\right)$$
$$\left. + \omega^\mu\left(3R_1^2 - R_2^2 - R_3^2 - R_4^2 + 2T_x + 2T_y + 2T_z\right)\right\}.$$

When $\mu \neq \nu$ and $\mu + \nu = 3$, there are three bases, $\Phi^{(2)}_{\mu\nu}$, $\Phi^{(3)}_{\mu\nu}$ and $\Phi^{(4)}_{\mu\nu}$. The matrix form of W in the basis functions and its eigenvectors for the eigenvalues 0, 2 and -2 are

$$\begin{pmatrix} 0 & 1 & -3 \\ 1 & 0 & 0 \\ -1 & 0 & 0 \end{pmatrix}, \quad \begin{pmatrix} 0 \\ 3 \\ 1 \end{pmatrix}, \quad \begin{pmatrix} 2 \\ 1 \\ -1 \end{pmatrix}, \quad \begin{pmatrix} 2 \\ -1 \\ 1 \end{pmatrix}.$$

The corresponding eigenfunctions respectively belong to the representation E, T_1 and T_2:

$$\Psi^E_{\mu\nu} = a_{\mu\nu}\left\{\Phi^{(3)}_{\mu\nu} + \Phi^{(4)}_{\mu\nu}/3\right\}$$
$$= a_{\mu\nu}\left\{T_z + T_z^3 + S_1 + S_2 + \omega^{-\mu}\left(T_y + T_y^3 + S_5 + S_6\right)\right.$$
$$\left. + \omega^\mu\left(T_x + T_x^3 + S_3 + S_4\right)\right\},$$

$$\Psi^{T_1}_{\mu\nu} = b_{\mu\nu}\left\{2\Phi^{(2)}_{\mu\nu} + \Phi^{(3)}_{\mu\nu} - \Phi^{(4)}_{\mu\nu}\right\}$$
$$= b_{\mu\nu}\left\{2T_z^2 + 2R_3 + 2R_3^2 + T_z + T_z^3 + S_1 - 3S_2\right.$$
$$+ \omega^{-\mu}\left(2T_y^2 + 2R_4 + 2R_4^2 + T_y + T_y^3 + S_5 - 3S_6\right)$$
$$\left. + \omega^\mu\left(2T_x^2 + 2R_2 + 2R_2^2 + T_x + T_x^3 + S_3 - 3S_4\right)\right\},$$

$$\Psi_{\mu\nu}^{T_2} = c_{\mu\nu}\left\{2\Phi_{\mu\nu}^{(2)} - \Phi_{\mu\nu}^{(3)} + \Phi_{\mu\nu}^{(4)}\right\}$$

$$= c_{\mu\nu}\left\{2T_z^2 + 2R_3 + 2R_3^2 - T_z - T_z^3 - S_1 + 3S_2\right.$$

$$+ \omega^{-\mu}\left(2T_y^2 + 2R_4 + 2R_4^2 - T_y - T_y^3 - S_5 + 3S_6\right)$$

$$\left. + \omega^{\mu}\left(2T_x^2 + 2R_2 + 2R_2^2 - T_x - T_x^3 - S_3 + 3S_4\right)\right\}.$$

When $\mu = 0$ and $\nu \neq 0$, there are two bases, $\Phi_{0\nu}^{(2)}$ and $\Phi_{0\nu}^{(3)}$. The matrix form of W in the basis functions and its eigenvectors for the eigenvalues 2 and -2 are

$$\begin{pmatrix} 0 & -2\omega^\nu \\ -2\omega^{-\nu} & 0 \end{pmatrix}, \quad \begin{pmatrix} 1 \\ -\omega^{-\nu} \end{pmatrix}, \quad \begin{pmatrix} 1 \\ \omega^{-\nu} \end{pmatrix}.$$

The corresponding eigenfunctions respectively belong to the representation T_1 and T_2:

$$\Psi_{0\nu}^{T_1} = b_{0\nu}\left\{\Phi_{0\nu}^{(2)} - \omega^{-\nu}\Phi_{0\nu}^{(3)}\right\}$$

$$= b_{0\nu}\left\{T_z^2 + R_4 + R_2^2 - T_x - T_y^3 - S_1\right.$$

$$+ \omega^{-\nu}\left(T_y^2 + R_2 + R_3^2 - T_z - T_x^3 - S_5\right)$$

$$\left. + \omega^{\nu}\left(T_x^2 + R_3 + R_4^2 - T_y - T_z^3 - S_3\right)\right\},$$

$$\Psi_{0\nu}^{T_2} = c_{0\nu}\left\{\Phi_{0\nu}^{(2)} + \omega^{-\nu}\Phi_{0\nu}^{(3)}\right\}$$

$$= c_{0\nu}\left\{T_z^2 + R_4 + R_2^2 + T_x + T_y^3 + S_1\right.$$

$$+ \omega^{-\nu}\left(T_y^2 + R_2 + R_3^2 + T_z + T_x^3 + S_5\right)$$

$$\left. + \omega^{\nu}\left(T_x^2 + R_3 + R_4^2 + T_y + T_z^3 + S_3\right)\right\}.$$

When $\nu = 0$ and $\mu \neq 0$, there are two bases, $\Phi_{\mu 0}^{(2)}$ and $\Phi_{\mu 0}^{(3)}$. The matrix form of W in the basis functions and its eigenvectors for the eigenvalues 2 and -2 are

$$\begin{pmatrix} 0 & -2\omega^\mu \\ -2\omega^{-\mu} & 0 \end{pmatrix}, \quad \begin{pmatrix} 1 \\ -\omega^{-\mu} \end{pmatrix}, \quad \begin{pmatrix} 1 \\ \omega^{-\mu} \end{pmatrix}.$$

The corresponding eigenfunctions respectively belong to the representation T_1 and T_2:

$$\Psi_{\mu 0}^{T_1} = b_{\mu 0}\left\{\Phi_{\mu 0}^{(2)} - \omega^{-\mu}\Phi_{\mu 0}^{(3)}\right\}$$

$$= b_{\mu 0}\left\{T_z^2 + R_2 + R_4^2 - T_y - S_1 - T_x^3\right.$$

$$+ \omega^{-\mu} \left(T_x^2 + R_4 + R_3^2 - T_z - S_3 - T_y^3 \right)$$
$$+ \omega^{\mu} \left(T_y^2 + R_3 + R_2^2 - T_x - S_5 - T_z^3 \right) \},$$

$$\Psi_{\mu 0}^{T_2} = c_{\mu 0} \left\{ \Phi_{\mu 0}^{(2)} + \omega^{-\mu} \Phi_{\mu 0}^{(3)} \right\}$$
$$= c_{\mu 0} \{ T_z^2 + R_2 + R_4^2 + T_y + S_1 + T_x^3$$
$$+ \omega^{-\mu} \left(T_x^2 + R_4 + R_3^2 + T_z + S_3 + T_y^3 \right)$$
$$+ \omega^{\mu} \left(T_y^2 + R_3 + R_2^2 + T_x + S_5 + T_z^3 \right) \}.$$

Now, we calculate the representation matrix of the generator T_z. The representation matrix of the generator R_1 is known. First, for the representation E, we take the normalized bases, $a_{11} = a_{21} = \sqrt{1/12}$, to make the representation unitary,

$$\Psi_{11}^E = \sqrt{1/12} \{ E + T_x^2 + T_y^2 + T_z^2 + \omega^2 \left(R_1 + R_2 + R_3 + R_4 \right)$$
$$+ \omega \left(R_1^2 + R_2^2 + R_3^2 + R_4^2 \right) \},$$
$$\Psi_{21}^E = \sqrt{1/12} \{ T_z + T_z^3 + S_1 + S_2 + \omega \left(T_y + T_y^3 + S_5 + S_6 \right)$$
$$+ \omega^2 \left(T_x + T_x^3 + S_3 + S_4 \right) \}.$$

We only need to list the terms of E and T_z:

$$T_z \Psi_{11}^E = \sqrt{1/12} \{ T_z + \cdots \} = \Psi_{21}^E.$$
$$T_z \Psi_{21}^E = \sqrt{1/12} \{ E + \cdots \} = \Psi_{11}^E.$$

Hence, the representation matrices are

$$\overline{D}^E (R_1) = \begin{pmatrix} \omega & 0 \\ 0 & \omega^2 \end{pmatrix}, \qquad \overline{D}^E (T_z) = \begin{pmatrix} 0 & 1 \\ 1 & 0 \end{pmatrix}.$$

According to these representation matrices we can calculate the remaining factors $a_{12} = a_{22} = \sqrt{1/12}$:

$$\Psi_{\mu\nu}^T T_z = \sum_{\rho} \overline{D}_{\nu\rho}^E (T_z) \Psi_{\mu\rho}^E.$$

$$\Psi_{12}^E = \sqrt{1/12} \{ T_z + T_z^3 + S_1 + S_2 + \omega^2 \left(T_y + T_y^3 + S_5 + S_6 \right)$$
$$+ \omega \left(T_x + T_x^3 + S_3 + S_4 \right) \},$$
$$\Psi_{22}^E = \sqrt{1/12} \{ E + T_x^2 + T_y^2 + T_z^2 + \omega \left(R_1 + R_2 + R_3 + R_4 \right)$$
$$+ \omega^2 \left(R_1^2 + R_2^2 + R_3^2 + R_4^2 \right) \}.$$

In comparison with the representation matrices calculated in Problem 12 with the method of the quotient group,

$$D^E(R_1) = \frac{1}{2}\begin{pmatrix} -1 & -\sqrt{3} \\ \sqrt{3} & -1 \end{pmatrix}, \qquad D^E(T_z) = \begin{pmatrix} 1 & 0 \\ 0 & -1 \end{pmatrix},$$

the similarity transformation X satisfying $X^{-1}D^E(R)X = \overline{D}^E(R)$ is

$$X = \frac{1}{\sqrt{2}}\begin{pmatrix} 1 & 1 \\ i & -i \end{pmatrix}, \qquad X^{-1} = \frac{1}{\sqrt{2}}\begin{pmatrix} 1 & -i \\ 1 & i \end{pmatrix}.$$

Second, for the representation T_1, we take the normalized bases, $b_{00} = \sqrt{1/72}$ and $b_{10} = b_{20} = \sqrt{1/18}$,

$$\begin{aligned}
\Psi_{00}^{T_1} &= \sqrt{1/72}\,\{3E + 3R_1 + 3R_1^2 - T_x^2 - T_y^2 - T_z^2 - R_2 - R_2^2 \\
&\quad - R_3 - R_3^2 - R_4 - R_4^2 + T_x + T_x^3 + T_y + T_y^3 \\
&\quad + T_z + T_z^3 + S_1 + S_3 + S_5 - 3S_2 - 3S_4 - 3S_6\}\,, \\
\Psi_{10}^{T_1} &= \sqrt{1/18}\,\{T_z^2 + R_2 + R_4^2 - T_y - T_x^3 - S_1 \\
&\quad + \omega^2\,(T_x^2 + R_4 + R_3^2 - T_z - T_y^3 - S_3) \\
&\quad + \omega\,(T_y^2 + R_3 + R_2^2 - T_x - T_z^3 - S_5)\}\,, \\
\Psi_{20}^{T_1} &= \sqrt{1/18}\,\{T_z^2 + R_2 + R_4^2 - T_y - T_x^3 - S_1 \\
&\quad + \omega\,(T_x^2 + R_4 + R_3^2 - T_z - T_y^3 - S_3) \\
&\quad + \omega^2\,(T_y^2 + R_3 + R_2^2 - T_x - T_z^3 - S_5)\}\,.
\end{aligned}$$

We only need to list the terms of E, T_z^2 and T_x^2. Notice that $T_z S_2 = S_1 T_z = T_x^2$.

$$\begin{aligned}
T_z\Psi_{00}^{T_1} &= \sqrt{\frac{1}{72}}\,\{E + T_z^2 - 3T_x^2 + \cdots\} = \frac{1}{3}\left\{\Psi_{00}^{T_1} - 2\omega\Psi_{10}^{T_1} - 2\omega^2\Psi_{20}^{T_1}\right\}, \\
T_z\Psi_{10}^{T_1} &= \sqrt{\frac{1}{18}}\,\{-\omega E - \omega^2 T_z^2 + \cdots\} = \frac{1}{3}\left\{-2\omega\Psi_{00}^{T_1} - 2\omega^2\Psi_{10}^{T_1} + \Psi_{20}^{T_1}\right\}, \\
T_z\Psi_{20}^{T_1} &= \sqrt{\frac{1}{18}}\,\{-\omega^2 E - \omega T_z^2 + \cdots\} = \frac{1}{3}\left\{-2\omega^2\Psi_{00}^{T_1} + \Psi_{10}^{T_1} - 2\omega\Psi_{20}^{T_1}\right\}.
\end{aligned}$$

Hence, the representation matrices are

$$\overline{D}^{T_1}(R_1) = \begin{pmatrix} 1 & 0 & 0 \\ 0 & \omega & 0 \\ 0 & 0 & \omega^2 \end{pmatrix}, \qquad \overline{D}^{T_1}(T_z) = \frac{1}{3}\begin{pmatrix} 1 & -2\omega & -2\omega^2 \\ -2\omega & -2\omega^2 & 1 \\ -2\omega^2 & 1 & -2\omega \end{pmatrix}.$$

They are symmetric. The factors in the remaining basis functions can be calculated by the following formulas:

$$\Psi_{\mu\nu}^{T_1} T_z = \sum_{\rho} \overline{D}_{\nu\rho}^{T_1}(T_z)\Psi_{\mu\rho}^{T_1}.$$

$$\Psi_{00}^{T_1} T_z = \sqrt{\frac{1}{72}}\left\{E + T_z^2 + T_x^2 + \cdots\right\} = \frac{1}{3}\left\{\Psi_{00}^{T_1} - 2\omega\Psi_{01}^{T_1} - 2\omega^2\Psi_{02}^{T_1}\right\},$$

$$\Psi_{10}^{T_1} T_z = \sqrt{\frac{1}{18}}\left\{-\omega E - \omega^2 T_z^2 - T_x^2 + \cdots\right\} = \frac{1}{3}\left\{\Psi_{10}^{T_1} - 2\omega\Psi_{11}^{T_1} - 2\omega^2\Psi_{12}^{T_1}\right\},$$

$$\Psi_{20}^{T_1} T_z = \sqrt{\frac{1}{18}}\left\{-\omega^2 E - \omega T_z^2 - T_x^2 + \cdots\right\} = \frac{1}{3}\left\{\Psi_{20}^{T_1} - 2\omega\Psi_{21}^{T_1} - 2\omega^2\Psi_{22}^{T_1}\right\}.$$

Thus,

$$
\begin{aligned}
\Psi_{01}^{T_1} =\ & \sqrt{1/18}\left\{T_z^2 + R_4 + R_2^2 - T_x - T_y^3 - S_1\right.\\
& + \omega^2\left(T_y^2 + R_2 + R_3^2 - T_z - T_x^3 - S_5\right)\\
& \left. + \omega\left(T_x^2 + R_3 + R_4^2 - T_y - T_z^3 - S_3\right)\right\},
\end{aligned}
$$

$$
\begin{aligned}
\Psi_{02}^{T_1} =\ & \sqrt{1/18}\left\{T_z^2 + R_4 + R_2^2 - T_x - T_y^3 - S_1\right.\\
& + \omega\left(T_y^2 + R_2 + R_3^2 - T_z - T_x^3 - S_5\right)\\
& \left. + \omega^2\left(T_x^2 + R_3 + R_4^2 - T_y - T_z^3 - S_3\right)\right\},
\end{aligned}
$$

$$
\begin{aligned}
\Psi_{11}^{T_1} =\ & \sqrt{1/72}\left\{3E - T_x^2 - T_y^2 - T_z^2 - 2S_1 - 2S_3 - 2S_5\right.\\
& + \omega^2\left(3R_1 - R_2 - R_3 - R_4 - 2T_x^3 - 2T_y^3 - 2T_z^3\right)\\
& \left. + \omega\left(3R_1^2 - R_2^2 - R_3^2 - R_4^2 - 2T_x - 2T_y - 2T_z\right)\right\},
\end{aligned}
$$

$$
\begin{aligned}
\Psi_{12}^{T_1} =\ & \sqrt{1/72}\left\{2T_z^2 + 2R_3 + 2R_3^2 + T_z + T_z^3 + S_1 - 3S_2\right.\\
& + \omega^2\left(2T_y^2 + 2R_4 + 2R_4^2 + T_y + T_y^3 + S_5 - 3S_6\right)\\
& \left. + \omega\left(2T_x^2 + 2R_2 + 2R_2^2 + T_x + T_x^3 + S_3 - 3S_4\right)\right\},
\end{aligned}
$$

$$
\begin{aligned}
\Psi_{21}^{T_1} =\ & \sqrt{1/72}\left\{2T_z^2 + 2R_3 + 2R_3^2 + T_z + T_z^3 + S_1 - 3S_2\right.\\
& + \omega\left(2T_y^2 + 2R_4 + 2R_4^2 + T_y + T_y^3 + S_5 - 3S_6\right)\\
& \left. + \omega^2\left(2T_x^2 + 2R_2 + 2R_2^2 + T_x + T_x^3 + S_3 - 3S_4\right)\right\},
\end{aligned}
$$

$$
\begin{aligned}
\Psi_{22}^{T_1} =\ & \sqrt{1/72}\left\{3E - T_x^2 - T_y^2 - T_z^2 - 2S_1 - 2S_3 - 2S_5\right.\\
& + \omega\left(3R_1 - R_2 - R_3 - R_4 - 2T_x^3 - 2T_y^3 - 2T_z^3\right)\\
& \left. + \omega^2\left(3R_1^2 - R_2^2 - R_3^2 - R_4^2 - 2T_x - 2T_y - 2T_z\right)\right\}.
\end{aligned}
$$

In comparison with the representation matrices calculated in Problem 12 with the method of coordinate transformation in three-dimensional space,

$$D^{T_1}(R_1) = \begin{pmatrix} 0 & 0 & 1 \\ 1 & 0 & 0 \\ 0 & 1 & 0 \end{pmatrix}, \qquad D^{T_1}(T_z) = \begin{pmatrix} 0 & -1 & 0 \\ 1 & 0 & 0 \\ 0 & 0 & 1 \end{pmatrix}.$$

The similarity transformation Y, satisfying $Y^{-1}D^{T_1}(R)Y = \overline{D}^{T_1}(R)$ is

$$Y = \frac{1}{\sqrt{3}} \begin{pmatrix} 1 & \omega^2 & \omega \\ 1 & \omega & \omega^2 \\ 1 & 1 & 1 \end{pmatrix}, \qquad Y^{-1} = \frac{1}{\sqrt{3}} \begin{pmatrix} 1 & 1 & 1 \\ \omega & \omega^2 & 1 \\ \omega^2 & \omega & 1 \end{pmatrix}.$$

At last, for the representation T_2, if we choose $c_{\mu\nu} = b_{\mu\nu}$, the coefficients in the expansions of the basis functions $\Psi_{\mu\nu}^{T_2}$ are the same as that of $\Psi_{\mu\nu}^{T_1}$ except for the terms related to the elements in the coset of the subgroup **T**, which change signs. So, the representation matrix $D^{T_2}(R)$ is just the direct product of $D^{T_1}(R)$ and $D^{A_2}(R)$.

24. The multiplication table of the group G is given as follows.

left \ right	E	A	B	C	F	K	M	N
E	E	A	B	C	F	K	M	N
A	A	E	M	K	N	C	B	F
B	B	N	E	M	K	F	C	A
C	C	K	N	E	M	A	F	B
F	F	M	K	N	E	B	A	C
K	K	C	F	A	B	E	N	M
M	M	F	A	B	C	N	K	E
N	N	B	C	F	A	M	E	K

a) Write the character table for the group G.

b) If we know that the representation matrices of the generators A and B in a two-dimensional irreducible representation D are

$$D(A) = \begin{pmatrix} 1 & 0 \\ 0 & -1 \end{pmatrix}, \qquad D(B) = \begin{pmatrix} 0 & 1 \\ 1 & 0 \end{pmatrix},$$

and two sets of basis functions ψ_μ and ϕ_ν respectively transform according to this two-dimensional representation of G:

$$P_R\psi_\mu = \sum_{\mu'} \psi_{\mu'} D_{\mu'\mu}(R), \qquad P_R\phi_\nu = \sum_{\nu'} \phi_{\nu'} D_{\nu'\nu}(R),$$

combine the product function $\psi_\mu \phi_\nu$ such that the basis function belongs to the irreducible representation of G.

Solution. The group G consists of 8 elements, where the order of the identity E is one, the orders of A, B, C, F and K are two, and the orders of M and N are four. E and K are commutable with any element in G. G contains five classes, $\{E\}$, $\{K\}$, $\{A,\ C\}$, $\{B,\ F\}$ and $\{M,\ N\}$. They are all self-reciprocal classes. Since $1^1 + 1^2 + 1^2 + 1^2 + 2^2 = 8$, the group G has four one-dimensional and one two-dimensional inequivalent irreducible representations. These representations all are self-conjugate.

There are four invariant subgroups in G. The cosets of the invariant subgroup $\{E,\ K\}$ are $\{A,\ C\}$, $\{B,\ F\}$ and $\{M,\ N\}$. Its quotient group is isomorphic onto the group V_4. The indices of the remaining three invariant subgroups, $\{E,\ A,\ K,\ C\}$, $\{E,\ B,\ K,\ F\}$ and $\{E,\ M,\ K,\ N\}$ all are two. We can obtain four one-dimensional representations of G from these quotient groups. The characters in the two-dimensional representation can be obtained by Eqs. (3.5-6). The character table of G is as follows. The characters for the self-direct product of the two-dimensional representation are also listed in the table.

	E	K	(A,C)	(B,F)	(M,N)
χ^1	1	1	1	1	1
χ^2	1	1	1	-1	-1
χ^3	1	1	-1	1	-1
χ^4	1	1	-1	-1	1
χ^5	2	-2	0	0	0
χ	4	4	0	0	0

From Eqs. (3.10-11), one obtains:

$$X^{-1}\left[D^5(R) \times D^5(R)\right] X^{-1} = D^1(R) \oplus D^2(R) \oplus D^3(R) \oplus D^4(R).$$

Substituting the matrix forms of A and B into the above formula, one has

$$X^{-1}\begin{pmatrix} 1 & 0 & 0 & 0 \\ 0 & -1 & 0 & 0 \\ 0 & 0 & -1 & 0 \\ 0 & 0 & 0 & 1 \end{pmatrix} X = \begin{pmatrix} 1 & 0 & 0 & 0 \\ 0 & 1 & 0 & 0 \\ 0 & 0 & -1 & 0 \\ 0 & 0 & 0 & -1 \end{pmatrix},$$

$$X^{-1}\begin{pmatrix} 0 & 0 & 0 & 1 \\ 0 & 0 & 1 & 0 \\ 0 & 1 & 0 & 0 \\ 1 & 0 & 0 & 0 \end{pmatrix} X = \begin{pmatrix} 1 & 0 & 0 & 0 \\ 0 & -1 & 0 & 0 \\ 0 & 0 & 1 & 0 \\ 0 & 0 & 0 & -1 \end{pmatrix}.$$

The eigenvectors of the product matrices for the eigenvalues ± 1 are:

$$\text{eigenvalue} \quad 1 \qquad -1 \qquad\qquad 1 \qquad -1$$

$$A: \begin{pmatrix} a \\ 0 \\ 0 \\ b \end{pmatrix}, \begin{pmatrix} 0 \\ c \\ d \\ 0 \end{pmatrix}, \qquad B: \begin{pmatrix} r \\ s \\ s \\ r \end{pmatrix}, \begin{pmatrix} p \\ q \\ -q \\ -p \end{pmatrix}.$$

Thus, the similarity transformation matrix X is

$$X = \frac{1}{\sqrt{2}} \begin{pmatrix} 1 & 1 & 0 & 0 \\ 0 & 0 & 1 & 1 \\ 0 & 0 & 1 & -1 \\ 1 & -1 & 0 & 0 \end{pmatrix}.$$

The new basis functions belonging to the irreducible representations are calculated by X:

$$\Psi^{(1)} = (\psi_1 \phi_1 + \psi_2 \phi_2)/\sqrt{2}, \qquad \Psi^{(2)} = (\psi_1 \phi_1 - \psi_2 \phi_2)/\sqrt{2},$$
$$\Psi^{(3)} = (\psi_1 \phi_2 + \psi_2 \phi_1)/\sqrt{2}, \qquad \Psi^{(4)} = (\psi_1 \phi_2 - \psi_2 \phi_1)/\sqrt{2}.$$

25. Calculate the unitary similarity transformation matrix X for reducing the self-direct product of the three-dimensional irreducible unitary representation D^T of the group $\mathbf{T}$:

$$X^{-1} \left\{ D^T(R) \times D^T(R) \right\} X = \sum_j a_j D^j(R).$$

Solution. This problem will be solved by two methods. One is to calculate the similarity transformation matrix directly. Another is to calculate the expansions of states in the product space. These two methods are the same in principle. However, when the dimension of the representation is quite large, the first method may not be convenient. The aim of this problem is to show the relation between these two methods, and to demonstrate the merit of the second method.

To begin with, we solve this problem with the first method. Take the generators of the group $\mathbf{T}$ to be T_z^2 and R_1. Their representation matrices in the three-dimensional representation T are

$$D^T(T_z^2) = \begin{pmatrix} -1 & 0 & 0 \\ 0 & -1 & 0 \\ 0 & 0 & 1 \end{pmatrix}, \qquad D^T(R_1) = \begin{pmatrix} 0 & 0 & 1 \\ 1 & 0 & 0 \\ 0 & 1 & 0 \end{pmatrix}.$$

Denote by $D(R)$ the self-direct product representation of D^T,

$$X^{-1}D(R)X \equiv X^{-1}\left\{D^T(R) \times D^T(R)\right\}X = \sum_j a_j D^j(R).$$

The multiplicities a_j can be calculated with the character formula (3.11):

$$a_A = (12)^{-1}\left\{1 \times 9 \times 1 + 3 \times 1 \times 1\right\} = 1,$$
$$a_E = (12)^{-1}\left\{1 \times 9 \times 1 + 3 \times 1 \times 1\right\} = 1,$$
$$a_{E'} = (12)^{-1}\left\{1 \times 9 \times 1 + 3 \times 1 \times 1\right\} = 1,$$
$$a_T = (12)^{-1}\left\{1 \times 9 \times 3 + 3 \times 1 \times (-1)\right\} = 2.$$

Namely,

$$X^{-1}D(R)X = D^A(R) \oplus D^E(R) \oplus D^{E'}(R) \oplus D^T(R) \oplus D^T(R).$$

Let $R = T_z^2$ in the formula of the similarity transformation.

$$D(T_z^2) = \mathrm{diag}\left\{1, 1, -1, 1, 1, -1, -1, -1, 1\right\}$$
$$X^{-1}D(T_z^2)X = \mathrm{diag}\left\{1, 1, 1, -1, -1, 1, -1, -1, 1\right\}.$$

We conclude that for the 1st, 2nd, 3rd, 6th and 9th columns of X, the matrix entries in the 3rd, 6th, 7th and 8th rows are zero, while for the 4th, 5th, 7th and 8th columns, the matrix entries in the 1st, 2nd, 4th, 5th and 9th rows are zero. Then, let $R = R_1$. The action of $D^T(R_1)$ is to cycle the matrix entries in three rows as $1 \to 2 \to 3 \to 1$. In the direct product space, the rows (columns) can be designated by two indices ab, and the action of $D(R_1)$ is to change the 12-th component in the vector to the 23-th position, etc. If we change the designation of the rows (columns) by one index α running from 1 to 9, corresponding to the order of the indices ab as 11, 12, 13, 21, 22, 23, 31, 32, 33, then the action of $D(R_1)$ is

$$11 \longrightarrow 22 \longrightarrow 33 \longrightarrow 11, \quad 12 \longrightarrow 23 \longrightarrow 31 \longrightarrow 12,$$
$$1 \longrightarrow 5 \longrightarrow 9 \longrightarrow 1, \quad\quad 2 \longrightarrow 6 \longrightarrow 7 \longrightarrow 2,$$
$$13 \longrightarrow 21 \longrightarrow 32 \longrightarrow 13,$$
$$3 \longrightarrow 4 \longrightarrow 8 \longrightarrow 3.$$

The matrix after the similarity transformation is

$$X^{-1}D(R_1)X = \begin{pmatrix} 1 & 0 & 0 & 0 & 0 & 0 & 0 & 0 & 0 \\ 0 & \omega & 0 & 0 & 0 & 0 & 0 & 0 & 0 \\ 0 & 0 & \omega^2 & 0 & 0 & 0 & 0 & 0 & 0 \\ 0 & 0 & 0 & 0 & 0 & 1 & 0 & 0 & 0 \\ 0 & 0 & 0 & 1 & 0 & 0 & 0 & 0 & 0 \\ 0 & 0 & 0 & 0 & 1 & 0 & 0 & 0 & 0 \\ 0 & 0 & 0 & 0 & 0 & 0 & 0 & 0 & 1 \\ 0 & 0 & 0 & 0 & 0 & 1 & 0 & 0 & 0 \\ 0 & 0 & 0 & 0 & 0 & 0 & 1 & 0 \end{pmatrix},$$

where $\omega = \exp\{-i2\pi/3\}$. We have known that the third and the 6th matrix entries in each of the first three columns of X are zero. The three column matrices are also the eigenvectors of $D(R_1)$. From the above cyclic property, their matrix entries in the 2nd, 4th, 7th and 8th rows must also be zero, namely only the matrix entries in the 1st, 5th and 9th rows can be nonzero. The values of those matrix entries can be determined from the eigenequation:

$$X_{11} = X_{51} = X_{91} = X_{12} = X_{13} = \sqrt{1/3},$$
$$X_{92} = X_{53} = \omega/\sqrt{3}, \qquad X_{52} = X_{93} = \omega^2/\sqrt{3}.$$

From the unitary property of X, for the 6th and 9th columns, only the matrix entries in the 2nd and 4th rows may be nonvanishing. Since the multiplicity of the representation T in the reduction of the direct product representation is two, we can simply choose the matrix entries in the 6th and 9th columns to be zero except for $X_{26} = X_{49} = 1$. Substituting them into the formula for the similarity transformation on $D(R_1)$, one obtains

$$D(R_1)X_{\cdot 6} = X_{\cdot 4}, \qquad D(R_1)X_{\cdot 4} = X_{\cdot 5},$$
$$D(R_1)X_{\cdot 9} = X_{\cdot 7}, \qquad D(R_1)X_{\cdot 7} = X_{\cdot 8}.$$

The solutions are

$$X_{64} = X_{75} = X_{87} = X_{38} = 1,$$

and the remaining matrix entries all are zero. Finally, the matrix X is

$$X = \begin{pmatrix} \sqrt{1/3} & \sqrt{1/3} & \sqrt{1/3} & 0 & 0 & 0 & 0 & 0 & 0 \\ 0 & 0 & 0 & 0 & 0 & 1 & 0 & 0 & 0 \\ 0 & 0 & 0 & 0 & 0 & 0 & 0 & 1 & 0 \\ 0 & 0 & 0 & 0 & 0 & 0 & 0 & 0 & 1 \\ \sqrt{1/3} & \omega^2/\sqrt{3} & \omega/\sqrt{3} & 0 & 0 & 0 & 0 & 0 & 0 \\ 0 & 0 & 0 & 1 & 0 & 0 & 0 & 0 & 0 \\ 0 & 0 & 0 & 0 & 1 & 0 & 0 & 0 & 0 \\ 0 & 0 & 0 & 0 & 0 & 0 & 1 & 0 & 0 \\ \sqrt{1/3} & \omega/\sqrt{3} & \omega^2/\sqrt{3} & 0 & 0 & 0 & 0 & 0 & 0 \end{pmatrix}.$$

It can be checked that X does satisfy the similarity transformation formulas.

Now we turn to the second method. For convenience we use the Dirac notation to express the basis state in the representation space before and after the similarity transformation. The representation space before the transformation is the direct product of two subspaces, and the basis state is denoted by $|T,\mu\rangle|T,\nu\rangle$. Sometimes, two T in this basis state can be omitted. The actions of the generators T_z^2 and R_1 on the basis state are:

$$T_z^2|T,1\rangle = -|T,1\rangle, \quad T_z^2|T,2\rangle = -|T,2\rangle, \quad T_z^2|T,3\rangle = |T,3\rangle,$$
$$R_1|T,1\rangle = |T,2\rangle, \quad R_1|T,2\rangle = |T,3\rangle, \quad R_1|T,3\rangle = |T,1\rangle.$$

The representation space after the transformation is the direct sum of five subspaces, where the basis state is respectively denoted by $||A\rangle$, $||E\rangle$, $||E'\rangle$, $||T,(1),\rho\rangle$ and $||T,(2),\rho\rangle$, ρ is 1, 2 or 3, satisfying

$$T_z^2||A\rangle = R_1||A\rangle = ||A\rangle, \qquad T_z^2||E\rangle = \omega^2 R_1||E\rangle = ||E\rangle,$$
$$T_z^2||E'\rangle = \omega R_1||E'\rangle = ||E'\rangle, \quad T_z^2||T,(r),1\rangle = -||T,(r),1\rangle,$$
$$T_z^2||T,(r),2\rangle = -||T,(r),2\rangle, \quad T_z^2||T,(r),3\rangle = ||T,(r),3\rangle,$$
$$R_1||T,(r),1\rangle = ||T,(r),2\rangle, \qquad R_1||T,(r),3\rangle = ||T,(r),3\rangle,$$
$$R_1||T,(r),3\rangle = ||T,(r),1\rangle,$$

where the index $r = 1$ or 2 is used to identify two representations T. The basis state which is invariant in the action of T_z^2 must be in the following form:

$$c_1|T,1\rangle|T,1\rangle + c_2|T,1\rangle|T,2\rangle + c_3|T,2\rangle|T,1\rangle$$
$$+ c_4|T,2\rangle|T,2\rangle + c_5|T,3\rangle|T,3\rangle.$$

The basis state which changes its sign in the action of T_z^2 must be in the following form:

$$d_1|T,1\rangle|T,3\rangle + d_2|T,2\rangle|T,3\rangle + d_3|T,3\rangle|T,1\rangle + d_4|T,3\rangle|T,2\rangle.$$

Applying R_1 to the basis state $||A\rangle$, we have

$$
\begin{aligned}
R_1||A\rangle &= c_1^A|T,2\rangle|T,2\rangle + c_2^A|T,2\rangle|T,3\rangle + c_3^A|T,3\rangle|T,2\rangle \\
&\quad + c_4^A|T,3\rangle|T,3\rangle + c_5^A|T,1\rangle|T,1\rangle \\
&= ||A\rangle = c_1^A|T,1\rangle|T,1\rangle + c_2^A|T,1\rangle|T,2\rangle + c_3^A|T,2\rangle|T,1\rangle \\
&\quad + c_4^A|T,2\rangle|T,2\rangle + c_5^A|T,3\rangle|T,3\rangle.
\end{aligned}
$$

By comparison, we obtain $c_2^A = c_3^A = 0$, and $c_1^A = c_4^A = c_5^A$. After normal-ization, $c_1^A = \sqrt{1/3}$, and the normalized basis state is

$$||A\rangle = \sqrt{1/3}\{|T,1\rangle|T,1\rangle + |T,2\rangle|T,2\rangle + |T,3\rangle|T,3\rangle\}.$$

Applying R_1 to the basis state $||E\rangle$, we have

$$
\begin{aligned}
R_1||E\rangle &= c_1^E|T,2\rangle|T,2\rangle + c_2^E|T,2\rangle|T,3\rangle + c_3^E|T,3\rangle|T,2\rangle \\
&\quad + c_4^E|T,3\rangle|T,3\rangle + c_5^E|T,1\rangle|T,1\rangle \\
&= \omega||E\rangle = c_1^E\omega|T,1\rangle|T,1\rangle + c_2^E\omega|T,1\rangle|T,2\rangle + c_3^E\omega|T,2\rangle|T,1\rangle \\
&\quad + c_4^E\omega|T,2\rangle|T,2\rangle + c_5^E\omega|T,3\rangle|T,3\rangle.
\end{aligned}
$$

By comparison, we have $c_2^E = c_3^E = 0$, and $c_1^E = c_4^E\omega = c_5^E\omega^2$. Taking $c_1^E = \sqrt{1/3}$, we obtain the normalized basis state

$$||E\rangle = \sqrt{1/3}\{|T,1\rangle|T,1\rangle + \omega^2|T,2\rangle|T,2\rangle + \omega|T,3\rangle|T,3\rangle\}.$$

Similarly, we have

$$||E'\rangle = \sqrt{1/3}\{|T,1\rangle|T,1\rangle + \omega|T,2\rangle|T,2\rangle + \omega^2|T,3\rangle|T,3\rangle\}.$$

Applying R_1 to the basis state $||T,(r),3\rangle$, we have

$$
\begin{aligned}
R_1||T,(r),3\rangle &= c_1^{T(r)}|T,2\rangle|T,2\rangle + c_2^{T(r)}|T,2\rangle|T,3\rangle + c_3^{T(r)}|T,3\rangle|T,2\rangle \\
&\quad + c_4^{T(r)}|T,3\rangle|T,3\rangle + c_5^{T(r)}|T,1\rangle|T,1\rangle \\
&= ||T,(r),1\rangle = d_1^{T(r)}|T,1\rangle|T,3\rangle + d_2^{T(r)}|T,2\rangle|T,3\rangle \\
&\quad + d_3^{T(r)}|T,3\rangle|T,1\rangle + d_4^{T(r)}|T,3\rangle|T,2\rangle.
\end{aligned}
$$

By comparison, we have $c_1^{T(r)} = c_4^{T(r)} = c_5^{T(r)} = 0$. We may take $c_2^{T(1)} = c_3^{T(2)} = 1$, and $c_3^{T(1)} = c_2^{T(2)} = 0$,

$$||T, (1), 3\rangle = |T, 1\rangle |T, 2\rangle, \qquad ||T, (2), 3\rangle = |T, 2\rangle |T, 1\rangle.$$

Applying R_1 to them, we have

$$||T, (1), 1\rangle = R_1 ||T, (1), 3\rangle = |T, 2\rangle |T, 3\rangle,$$

$$||T, (1), 2\rangle = R_1 ||T, (1), 1\rangle = |T, 3\rangle |T, 1\rangle,$$

$$||T, (2), 1\rangle = R_1 ||T, (2), 3\rangle = |T, 3\rangle |T, 2\rangle,$$

$$||T, (2), 2\rangle = R_1 ||T, (2), 1\rangle = |T, 1\rangle |T, 3\rangle.$$

These results are consistent with that obtained by the first method.

26. Calculate the C-G series and the C-G coefficients for the direct product representation of each pair of two irreducible representations of the group **O**.

Solution. In Problem 17, we have calculated the representation matrices of the generators T_z and R_1 of the group **O** in five inequivalent irreducible representations A, B, E, T_1 and T_2. Now take a similarity transformation to diagonalize the representation matrices of R_1:

$$D^A(T_z) = D^A(R_1) = D^B(R_1) = 1, \qquad D^B(T_z) = -1,$$

$$D^E(T_z) = \begin{pmatrix} 0 & 1 \\ 1 & 0 \end{pmatrix}, \qquad D^E(R_1) = \begin{pmatrix} \omega & 0 \\ 0 & \omega^{-1} \end{pmatrix},$$

$$D^{T_1}(T_z) = -D^{T_2}(T_z) = \frac{1}{3}\begin{pmatrix} -2\omega^2 & 2\omega & -1 \\ 2\omega & 1 & -2\omega^2 \\ -1 & -2\omega^2 & -2\omega \end{pmatrix},$$

$$D^{T_1}(R_1) = D^{T_2}(R_1) = \begin{pmatrix} \omega & 0 & 0 \\ 0 & 1 & 0 \\ 0 & 0 & \omega^{-1} \end{pmatrix},$$

where $\omega = \exp\{-i2\pi/3\}$. Since the representation matrix of R_1 is diagonal, and the diagonal elements ω^m are different in each irreducible representation, this power m can be used to designate the row (column) in each irreducible representation. Notice that the value of the power m is module 3, $m + 3 = m$. The representations A and B are one-dimensional, so that the index 0 for the row (column) can be omitted. The index for the row (column) in the representation E takes the values 1 and -1. The index in the representations T_1 and T_2 takes the values 1, 0 and -1.

The direct product of the identical representation A and any representation is still equal to that representation. The direct product of the antisymmetric representation B and any representation is easy to calculate:

$$D^B \times D^B = D^A, \qquad D^B \times D^E = \sigma_3^{-1} D^E \sigma_3,$$
$$D^B \times D^{T_1} = D^{T_2}, \qquad D^B \times D^{T_2} = D^{T_1},$$

where σ_3 is the Pauli matrix. The Clebsch-Gordan series for the direct product representations is listed in the Table, and the Clebsch-Gordan coefficients are calculated in the following.

The representation of **O**	E	$3T_z^2$	$6T_z$	$8R_1$	$6S_1$	The reduction
A	1	1	1	1	1	
B	1	1	-1	1	-1	
E	2	2	0	-1	0	
T_1	3	-1	1	0	-1	
T_2	3	-1	-1	0	1	
$E \times E$	4	4	0	1	0	$A \oplus B \oplus E$
$E \times T_1 \simeq E \times T_2$	6	-2	0	0	0	$T_1 \oplus T_2$
$T_1 \times T_1 \simeq T_2 \times T_2$	9	1	1	0	1	$A \oplus E \oplus T_1 \oplus T_2$
$T_1 \times T_2$	9	1	-1	0	-1	$B \oplus E \oplus T_2 \oplus T_1$

(1) $E \times E \simeq A \oplus B \oplus E$.

Since the representation matrices of R_1 are all diagonal, the eigenvalue of R_1 for the product state is the product of the eigenvalues of two states, i.e., the sum of $m's$. Thus, we have:

$$\|A\rangle = a_1|E, 1\rangle|E, -1\rangle + a_2|E, -1\rangle|E, 1\rangle,$$
$$\|B\rangle = b_1|E, 1\rangle|E, -1\rangle + b_2|E, -1\rangle|E, 1\rangle,$$
$$\|E, 1\rangle = c_1|E, -1\rangle|E, -1\rangle, \qquad \|E, -1\rangle = c_2|E, 1\rangle|E, 1\rangle.$$

Applying T_z to them, we have

$$\|A\rangle = T_z\|A\rangle = a_1|E, -1\rangle|E, 1\rangle + a_2|E, 1\rangle|E, -1\rangle,$$
$$-\|B\rangle = T_z\|B\rangle = b_1|E, -1\rangle|E, 1\rangle + b_2|E, 1\rangle|E, -1\rangle,$$
$$\|E, -1\rangle = T_z\|E, 1\rangle = c_1|E, 1\rangle|E, 1\rangle.$$

So, $a_1 = a_2$, $b_1 = -b_2$, and $c_1 = c_2$. After normalization we have

$$\|A\rangle = 2^{-1/2}\{|E, 1\rangle|E, -1\rangle + |E, -1\rangle|E, 1\rangle\},$$
$$\|B\rangle = 2^{-1/2}\{|E, 1\rangle|E, -1\rangle - |E, -1\rangle|E, 1\rangle\},$$
$$\|E, 1\rangle = |E, -1\rangle|E, -1\rangle, \qquad \|E, -1\rangle = |E, 1\rangle|E, 1\rangle.$$

(2) $E \times T_1 \simeq E \oplus T_2$.

From the eigenvalues of R_1, we have:

$$||T_1, 1\rangle = a_1|E, 1\rangle|T_1, 0\rangle + a_2|E, -1\rangle|T_1, -1\rangle,$$
$$||T_1, 0\rangle = b_1|E, 1\rangle|T_1, -1\rangle + b_2|E, -1\rangle|T_1, 1\rangle,$$
$$||T_1, -1\rangle = c_1|E, 1\rangle|T_1, 1\rangle + c_2|E, -1\rangle|T_1, 0\rangle.$$

Applying T_z to them, we have

$$
\begin{aligned}
T_z||T_1, 1\rangle &= \frac{a_1}{3}|E, -1\rangle \left\{ 2\omega|T_1, 1\rangle + |T_1, 0\rangle - 2\omega^2|T_1, -1\rangle \right\} \\
&\quad + \frac{a_2}{3}|E, 1\rangle \left\{ -|T_1, 1\rangle - 2\omega^2|T_1, 0\rangle - 2\omega|T_1, -1\rangle \right\}, \\
&= \frac{1}{3} \left\{ -2\omega^2||T_1, 1\rangle + 2\omega||T_1, 0\rangle - ||T_1, -1\rangle \right\} \\
&= \frac{1}{3}|E, -1\rangle \left\{ 2\omega b_2|T_1, 1\rangle - c_2|T_1, 0\rangle - 2\omega^2 a_2|T_1, -1\rangle \right\} \\
&\quad + \frac{1}{3}|E, 1\rangle \left\{ -c_1|T_1, 1\rangle - 2\omega^2 a_1|T_1, 0\rangle + 2\omega b_1|T_1, -1\rangle \right\}.
\end{aligned}
$$

The solution is $a_1 = b_2 = -c_2 = a_2 = c_1 = -b_1$. After normalization we have

$$||T_1, 1\rangle = 2^{-1/2} \left\{ |E, 1\rangle|T_1, 0\rangle + |E, -1\rangle|T_1, -1\rangle \right\},$$
$$||T_1, 0\rangle = 2^{-1/2} \left\{ -|E, 1\rangle|T_1, -1\rangle + |E, -1\rangle|T_1, 1\rangle \right\},$$
$$||T_1, -1\rangle = 2^{-1/2} \left\{ |E, 1\rangle|T_1, 1\rangle - |E, -1\rangle|T_1, 0\rangle \right\}.$$

For the representation of T_2, we obtain $a_1 = -b_2 = c_2 = -a_2 = c_1 = -b_1$ by the similar calculation. After normalization we have

$$||T_2, 1\rangle = 2^{-1/2} \left\{ |E, 1\rangle|T_1, 0\rangle - |E, -1\rangle|T_1, -1\rangle \right\},$$
$$||T_2, 0\rangle = 2^{-1/2} \left\{ -|E, 1\rangle|T_1, -1\rangle - |E, -1\rangle|T_1, 1\rangle \right\},$$
$$||T_2, -1\rangle = 2^{-1/2} \left\{ |E, 1\rangle|T_1, 1\rangle + |E, -1\rangle|T_1, 0\rangle \right\}.$$

For the C-G coefficients in $E \times T_2$, one only needs to exchange T_1 and T_2 in the above results. Namely,

$$||T_2, 1\rangle = 2^{-1/2} \left\{ |E, 1\rangle|T_2, 0\rangle + |E, -1\rangle|T_2, -1\rangle \right\},$$
$$||T_2, 0\rangle = 2^{-1/2} \left\{ -|E, 1\rangle|T_2, -1\rangle + |E, -1\rangle|T_2, 1\rangle \right\},$$
$$||T_2, -1\rangle = 2^{-1/2} \left\{ |E, 1\rangle|T_2, 1\rangle - |E, -1\rangle|T_2, 0\rangle \right\},$$
$$||T_1, 1\rangle = 2^{-1/2} \left\{ |E, 1\rangle|T_2, 0\rangle - |E, -1\rangle|T_2, -1\rangle \right\},$$
$$||T_1, 0\rangle = 2^{-1/2} \left\{ -|E, 1\rangle|T_2, -1\rangle - |E, -1\rangle|T_2, 1\rangle \right\},$$
$$||T_1, -1\rangle = 2^{-1/2} \left\{ |E, 1\rangle|T_2, 1\rangle + |E, -1\rangle|T_2, 0\rangle \right\}.$$

(3) $T_1 \times T_1 \simeq T_2 \times T_2 \simeq A \oplus E \oplus T_1 \oplus T_2$.

From the eigenvalues of R_1, we have:

$$||A\rangle = a_1|T_1,1\rangle|T_1,-1\rangle + a_2|T_1,0\rangle|T_1,0\rangle + a_3|T_1,-1\rangle|T_1,1\rangle,$$

$$||E,1\rangle = b_1|T_1,1\rangle|T_1,0\rangle + b_2|T_1,0\rangle|T_1,1\rangle + b_3|T_1,-1\rangle|T_1,-1\rangle,$$

$$||E,-1\rangle = b_4|T_1,1\rangle|T_1,1\rangle + b_5|T_1,0\rangle|T_1,-1\rangle + b_6|T_1,-1\rangle|T_1,0\rangle.$$

Applying T_z to them, we have

$$
\begin{aligned}
||A\rangle = T_z||A\rangle = \frac{1}{9} &\left\{ \left(2\omega^2 a_1 + 4\omega^2 a_2 + 2\omega^2 a_3\right)|T_1,1\rangle|T_1,1\rangle \right. \\
&+ \left(4\omega a_1 + 2\omega a_2 - 2\omega a_3\right)|T_1,1\rangle|T_1,0\rangle \\
&\left.+ \left(4a_1 - 4a_2 + a_3\right)|T_1,1\rangle|T_1,-1\rangle\right\} + \ldots,
\end{aligned}
$$

$$
\begin{aligned}
||E,-1\rangle = T_z||E,1\rangle = \frac{1}{9} &\left\{(-4b_1 - 4b_2 + b_3)|T_1,1\rangle|T_1,1\rangle \right. \\
&+ \left(-2\omega^2 b_1 + 4\omega^2 b_2 + 2\omega^2 b_3\right)|T_1,1\rangle|T_1,0\rangle \\
&+ \left(4\omega b_1 - 2\omega b_2 + 2\omega b_3\right)|T_1,1\rangle|T_1,-1\rangle \\
&+ \left(-4b_1 - b_2 + 4b_3\right)|T_1,0\rangle|T_1,-1\rangle \\
&\left.+ \left(-b_1 - 4b_2 + 4b_3\right)|T_1,-1\rangle|T_1,0\rangle\right\} + \ldots.
\end{aligned}
$$

Thus, $a_1 = -a_2 = a_3$, $b_1 = b_2 = -b_3 = -b_4 = -b_5 = -b_6$. After normalization we have

$$||A\rangle = 3^{-1/2}\left\{|T_1,1\rangle|T_1,-1\rangle - |T_1,0\rangle|T_1,0\rangle + |T_1,-1\rangle|T_1,1\rangle\right\},$$

$$||E,1\rangle = 3^{-1/2}\left\{|T_1,1\rangle|T_1,0\rangle + |T_1,0\rangle|T_1,1\rangle - |T_1,-1\rangle|T_1,-1\rangle\right\},$$

$$||E,-1\rangle = 3^{-1/2}\left\{-|T_1,1\rangle|T_1,1\rangle - |T_1,0\rangle|T_1,-1\rangle - |T_1,-1\rangle|T_1,0\rangle\right\}.$$

The states belonging to the representation T_1 must be orthogonal to the above states,

$$||T_1,1\rangle = c_1|T_1,1\rangle|T_1,0\rangle + c_2|T_1,0\rangle|T_1,1\rangle + (c_1 + c_2)|T_1,-1\rangle|T_1,-1\rangle,$$

$$||T_1,0\rangle = c_3|T_1,1\rangle|T_1,-1\rangle + (c_3 + c_4)|T_1,0\rangle|T_1,0\rangle + c_4|T_1,-1\rangle|T_1,1\rangle,$$

$$||T_1,-1\rangle = c_5|T_1,1\rangle|T_1,1\rangle + c_6|T_1,0\rangle|T_1,-1\rangle - (c_5 + c_6)|T_1,-1\rangle|T_1,0\rangle.$$

Applying T_z to them, we have

$$
\begin{aligned}
T_z||T_1,1\rangle = \frac{1}{9}&\left\{[-4c_1 - 4c_2 + (c_1 + c_2)]|T_1,1\rangle|T_1,1\rangle \right. \\
&+ \left[-2\omega^2 c_1 + 4\omega^2 c_2 + 2\omega^2(c_1 + c_2)\right]|T_1,1\rangle|T_1,0\rangle
\end{aligned}
$$

$$+ \left[4\omega c_1 - 2\omega c_2 + 2\omega\left(c_1 + c_2\right)\right]|T_1,1\rangle|T_1,-1\rangle$$
$$+ \left[4\omega^2 c_1 - 2\omega^2 c_2 + 2\omega^2\left(c_1 + c_2\right)\right]|T_1,0\rangle|T_1,1\rangle$$
$$+ \left[2\omega c_1 + 2\omega c_2 + 4\omega\left(c_1 + c_2\right)\right]|T_1,0\rangle|T_1,0\rangle$$
$$+ \left[-4c_1 - c_2 + 4\left(c_1 + c_2\right)\right]|T_1,0\rangle|T_1,-1\rangle$$
$$+ \left[-2\omega c_1 + 4\omega c_2 + 2\omega\left(c_1 + c_2\right)\right]|T_1,-1\rangle|T_1,1\rangle$$
$$+ \left[-c_1 - 4c_2 + 4\left(c_1 + c_2\right)\right]|T_1,-1\rangle|T_1,0\rangle$$
$$+ \left[2\omega^2 c_1 + 2\omega^2 c_2 + 4\omega^2\left(c_1 + c_2\right)\right]|T_1,-1\rangle|T_1,-1\rangle\}$$
$$= \frac{1}{3}\left\{-2\omega^2\|T_1,1\rangle + 2\omega\|T_1,0\rangle - \|T_1,-1\rangle\right\}$$
$$= \frac{-2\omega^2}{3}\left\{c_1|T_1,1\rangle|T_1,0\rangle + c_2|T_1,0\rangle|T_1,1\rangle + \left(c_1 + c_2\right)|T_1,-1\rangle|T_1,-1\rangle\right\}$$
$$+ \frac{2\omega}{3}\left\{c_3|T_1,1\rangle|T_1,-1\rangle + \left(c_3 + c_4\right)|T_1,0\rangle|T_1,0\rangle + c_4|T_1,-1\rangle|T_1,1\rangle\right\}$$
$$- \frac{1}{3}\left\{c_5|T_1,1\rangle|T_1,1\rangle + c_6|T_1,0\rangle|T_1,-1\rangle - \left(c_5 + c_6\right)|T_1,-1\rangle|T_1,0\rangle\right\}.$$

Thus, $c_1 = -c_2 = c_3 = -c_4 = -c_6$, $c_5 = 0$. After normalization we have

$$\|T_1,1\rangle = 2^{-1/2}\left\{|T_1,1\rangle|T_1,0\rangle - |T_1,0\rangle|T_1,1\rangle\right\},$$
$$\|T_1,0\rangle = 2^{-1/2}\left\{|T_1,1\rangle|T_1,-1\rangle - |T_1,-1\rangle|T_1,1\rangle\right\},$$
$$\|T_1,-1\rangle = 2^{-1/2}\left\{-|T_1,0\rangle|T_1,-1\rangle + |T_1,-1\rangle|T_1,0\rangle\right\}.$$

From the orthogonality of states we have

$$\|T_2,1\rangle = d_1\left\{|T_1,1\rangle|T_1,0\rangle + |T_1,0\rangle|T_1,1\rangle + 2|T_1,-1\rangle|T_1,-1\rangle\right\},$$
$$\|T_2,0\rangle = d_2\left\{|T_1,1\rangle|T_1,-1\rangle + 2|T_1,0\rangle|T_1,0\rangle + |T_1,-1\rangle|T_1,1\rangle\right\},$$
$$\|T_2,-1\rangle = d_3\left\{2|T_1,1\rangle|T_1,1\rangle - |T_1,0\rangle|T_1,-1\rangle - |T_1,-1\rangle|T_1,0\rangle\right\}.$$

Applying T_z to them, we have

$$T_z\|T_2,1\rangle = \frac{d_1}{9}\left\{-6|T_1,1\rangle|T_1,1\rangle + 6\omega^2|T_1,1\rangle|T_1,0\rangle\right.$$
$$\left. + 6\omega|T_1,1\rangle|T_1,-1\rangle + \ldots\right\}$$
$$= \frac{1}{3}\left\{2\omega^2\|T_1,1\rangle - 2\omega\|T_1,0\rangle + \|T_1,-1\rangle\right\}.$$

Thus, $d_1 = -d_2 = -d_3$. After normalization we have

$$\|T_2,1\rangle = 6^{-1/2}\left\{-|T_1,1\rangle|T_1,0\rangle - |T_1,0\rangle|T_1,1\rangle - 2|T_1,-1\rangle|T_1,-1\rangle\right\},$$
$$\|T_2,0\rangle = 6^{-1/2}\left\{|T_1,1\rangle|T_1,-1\rangle + 2|T_1,0\rangle|T_1,0\rangle + |T_1,-1\rangle|T_1,1\rangle\right\},$$
$$\|T_2,-1\rangle = 6^{-1/2}\left\{2|T_1,1\rangle|T_1,1\rangle - |T_1,0\rangle|T_1,-1\rangle - |T_1,-1\rangle|T_1,0\rangle\right\}.$$

For the case of $T_2 \times T_2$, the term $|T_1, \mu\rangle|T_1, \nu\rangle$ in the expansion is replaced with $|T_2, \mu\rangle|T_2, \nu\rangle$, and the coefficients are exactly the same.

(4) $T_1 \times T_2 \simeq B \oplus E \oplus T_2 \oplus T_1$.

Multiplying the antisymmetric representation B on the formula in the case (3), we obtain the formula for the present case. Note that the similarity transformation σ_3 appearing in the direct product of B and E. The results are as follows:

$$\|B\rangle = 3^{-1/2} \left\{ |T_1, 1\rangle|T_2, -1\rangle - |T_1, 0\rangle|T_2, 0\rangle + |T_1, -1\rangle|T_2, 1\rangle \right\},$$
$$\|E, 1\rangle = 3^{-1/2} \left\{ |T_1, 1\rangle|T_2, 0\rangle + |T_1, 0\rangle|T_2, 1\rangle - |T_1, -1\rangle|T_2, -1\rangle \right\},$$
$$\|E, -1\rangle = 3^{-1/2} \left\{ |T_1, 1\rangle|T_2, 1\rangle + |T_1, 0\rangle|T_2, -1\rangle + |T_1, -1\rangle|T_2, 0\rangle \right\}.$$

$$\|T_2, 1\rangle = 2^{-1/2} \left\{ |T_1, 1\rangle|T_2, 0\rangle - |T_1, 0\rangle|T_2, 1\rangle \right\},$$
$$\|T_2, 0\rangle = 2^{-1/2} \left\{ |T_1, 1\rangle|T_2, -1\rangle - |T_1, -1\rangle|T_2, 1\rangle \right\},$$
$$\|T_2, -1\rangle = 2^{-1/2} \left\{ -|T_1, 0\rangle|T_2, -1\rangle + |T_1, -1\rangle|T_2, 0\rangle \right\}.$$

$$\|T_1, 1\rangle = 6^{-1/2} \left\{ -|T_1, 1\rangle|T_2, 0\rangle - |T_1, 0\rangle|T_2, 1\rangle - 2|T_1, -1\rangle|T_2, -1\rangle \right\},$$
$$\|T_1, 0\rangle = 6^{-1/2} \left\{ |T_1, 1\rangle|T_2, -1\rangle + 2|T_1, 0\rangle|T_2, 0\rangle + |T_1, -1\rangle|T_2, 1\rangle \right\},$$
$$\|T_1, -1\rangle = 6^{-1/2} \left\{ 2|T_1, 1\rangle|T_2, 1\rangle - |T_1, 0\rangle|T_2, -1\rangle - |T_1, -1\rangle|T_2, 0\rangle \right\}.$$

27. Calculate the C-G series and C-G coefficients in the reduction of direct product representation in terms of the character table of the group **I** given in Problem 18:

(1) $D^{T_1} \times D^{T_1}$; (2) $D^{T_1} \times D^{T_2}$; (3) $D^{T_2} \times D^{T_2}$; (4) $D^{T_1} \times D^{G}$;

(5) $D^{T_2} \times D^{G}$; (6) $D^{T_1} \times D^{H}$; (7) $D^{T_2} \times D^{H}$; (8) $D^{G} \times D^{G}$;

(9) $D^{G} \times D^{H}$; (10) $D^{H} \times D^{H}$.

Solution. We will only list the results for this problem. Notice that the character for the direct product representation is equal to the product of characters of two representations.

In Problem 23 the representation matrices of the generators T and S_1 in each irreducible representation of the group **I** were given, where the representation matrix of T is diagonal. Since the order of T is five, the diagonal element of $D(T)$ takes the power η^m where $\eta = \exp\{-i2\pi/5\}$. The power m is different in each irreducible representation of **I**, so this power m can be used to designate the row (column) in each irreducible representation. The possible values of the power in each representation are as follows: 0 for the representation A, 1, 0 and -1 for the representation T_1,

2, 0 and −2 for the representation T_2, 2, 1, −1 and −2 for the representation G, and 2, 1, 0, −1 and −2 for the representation H. Note that the power m is module 5, $m + 5 = m$.

The reductions of the direct product representations in I

	E	$12C_5$	$12C_5^2$	$15C_2$	$20C_3$	The reduction
A	1	1	1	1	1	
T_1	3	p^{-1}	$-p$	-1	0	
T_2	3	$-p$	p^{-1}	-1	0	
G	4	-1	-1	0	1	
H	5	0	0	1	-1	
$T_1 \times T_1$	9	p^{-2}	p^2	1	0	$A \oplus T_1 \oplus H$
$T_1 \times T_2$	9	-1	-1	1	0	$G \oplus H$
$T_2 \times T_2$	9	p^2	p^{-2}	1	0	$A \oplus T_2 \oplus H$
$T_1 \times G$	12	$-p^{-1}$	p	0	0	$T_2 \oplus G \oplus H$
$T_2 \times G$	12	p	$-p^{-1}$	0	0	$T_1 \oplus G \oplus H$
$T_1 \times H$	15	0	0	-1	0	$T_1 \oplus T_2 \oplus G \oplus H$
$T_2 \times H$	15	0	0	-1	0	$T_1 \oplus T_2 \oplus G \oplus H$
$G \times G$	16	1	1	0	1	$A \oplus T_1 \oplus T_2 \oplus G \oplus H$
$G \times H$	20	0	0	0	-1	$T_1 \oplus T_2 \oplus G \oplus 2H$
$H \times H$	25	0	0	1	1	$A \oplus T_1 \oplus T_2 \oplus 2G \oplus 2H$

(1) $D^{T_1} \times D^{T_1} \simeq D^A \oplus D^{T_1} \oplus D^H$.

$$\|A, 0\rangle = 3^{-1/2} \{|T_1, 1\rangle|T_1, -1\rangle - |T_1, 0\rangle|T_1, 0\rangle + |T_1, -1\rangle|T_1, 1\rangle\},$$
$$\|T_1, 1\rangle = 2^{-1/2} \{|T_1, 1\rangle|T_1, 0\rangle - |T_1, 0\rangle|T_1, 1\rangle\},$$
$$\|T_1, 0\rangle = 2^{-1/2} \{|T_1, 1\rangle|T_1, -1\rangle - |T_1, -1\rangle|T_1, 1\rangle\},$$
$$\|T_1, -1\rangle = 2^{-1/2} \{|T_1, 0\rangle|T_1, -1\rangle - |T_1, -1\rangle|T_1, 0\rangle\},$$
$$\|H, 2\rangle = |T_1, 1\rangle|T_1, 1\rangle,$$
$$\|H, 1\rangle = 2^{-1/2} \{|T_1, 1\rangle|T_1, 0\rangle + |T_1, 0\rangle|T_1, 1\rangle\},$$
$$\|H, 0\rangle = 6^{-1/2} \{|T_1, 1\rangle|T_1, -1\rangle + 2|T_1, 0\rangle|T_1, 0\rangle + |T_1, -1\rangle|T_1, 1\rangle\},$$
$$\|H, -1\rangle = 2^{-1/2} \{|T_1, 0\rangle|T_1, -1\rangle + |T_1, -1\rangle|T_1, 0\rangle\},$$
$$\|H, -2\rangle = |T_1, -1\rangle|T_1, -1\rangle,$$

(2) $D^{T_1} \times D^{T_2} \simeq D^G \oplus D^H$.

$$\|G, 2\rangle = 3^{-1/2} \{\sqrt{2}|T_1, 0\rangle|T_2, 2\rangle - |T_1, -1\rangle|T_2, -2\rangle\},$$
$$\|G, 1\rangle = 3^{-1/2} \{\sqrt{2}|T_1, 1\rangle|T_2, 0\rangle - |T_1, -1\rangle|T_2, 2\rangle\},$$
$$\|G, -1\rangle = 3^{-1/2} \{|T_1, 1\rangle|T_2, -2\rangle - \sqrt{2}|T_1, -1\rangle|T_2, 0\rangle\},$$
$$\|G, -2\rangle = 3^{-1/2} \{|T_1, 1\rangle|T_2, 2\rangle + \sqrt{2}|T_1, 0\rangle|T_2, -2\rangle\},$$

$$\|H,2\rangle = 3^{-1/2}\left\{|T_1,0\rangle|T_2,2\rangle + \sqrt{2}|T_1,-1\rangle|T_2,-2\rangle\right\},$$

$$\|H,1\rangle = 3^{-1/2}\left\{-|T_1,1\rangle|T_2,0\rangle - \sqrt{2}|T_1,-1\rangle|T_2,2\rangle\right\},$$

$$\|H,0\rangle = |T_1,0\rangle|T_2,0\rangle,$$

$$\|H,-1\rangle = 3^{-1/2}\left\{-\sqrt{2}|T_1,1\rangle|T_2,-2\rangle - |T_1,-1\rangle|T_2,0\rangle\right\},$$

$$\|H,-2\rangle = 3^{-1/2}\left\{-\sqrt{2}|T_1,1\rangle|T_2,2\rangle + |T_1,0\rangle|T_2,-2\rangle\right\},$$

(3) $D^{T_2} \times D^{T_2} \simeq D^A \oplus D^{T_2} \oplus D^H.$

$$\|A,0\rangle = 3^{-1/2}\left\{|T_2,2\rangle|T_2,-2\rangle + |T_2,0\rangle|T_2,0\rangle + |T_2,-2\rangle|T_2,2\rangle\right\},$$

$$\|T_2,2\rangle = 2^{-1/2}\left\{|T_2,2\rangle|T_2,0\rangle - |T_2,0\rangle|T_2,2\rangle\right\},$$

$$\|T_2,0\rangle = 2^{-1/2}\left\{-|T_2,2\rangle|T_2,-2\rangle + |T_2,-2\rangle|T_2,2\rangle\right\},$$

$$\|T_2,-2\rangle = 2^{-1/2}\left\{|T_2,0\rangle|T_2,-2\rangle - |T_2,-2\rangle|T_2,0\rangle\right\},$$

$$\|H,2\rangle = 2^{-1/2}\left\{|T_2,2\rangle|T_2,0\rangle + |T_2,0\rangle|T_2,2\rangle\right\},$$

$$\|H,1\rangle = |T_2,-2\rangle|T_2,-2\rangle,$$

$$\|H,0\rangle = 6^{-1/2}\left\{|T_2,2\rangle|T_2,-2\rangle - 2|T_2,0\rangle|T_2,0\rangle + |T_2,-2\rangle|T_2,2\rangle\right\},$$

$$\|H,-1\rangle = -|T_2,2\rangle|T_2,2\rangle,$$

$$\|H,-2\rangle = 2^{-1/2}\left\{|T_2,0\rangle|T_2,-2\rangle + |T_2,-2\rangle|T_2,0\rangle\right\},$$

(4) $D^{T_1} \times D^G \simeq D^{T_2} \oplus D^G \oplus D^H.$

$$\|T_2,2\rangle = 2^{-1}\left\{|T_1,1\rangle|G,1\rangle + \sqrt{2}|T_1,0\rangle|G,2\rangle - |T_1,-1\rangle|G,-2\rangle\right\},$$

$$\|T_2,0\rangle = 2^{-1/2}\left\{|T_1,1\rangle|G,-1\rangle - |T_1,-1\rangle|G,1\rangle\right\},$$

$$\|T_2,-2\rangle = 2^{-1}\left\{|T_1,1\rangle|G,2\rangle + \sqrt{2}|T_1,0\rangle|G,-2\rangle - |T_1,-1\rangle|G,-1\rangle\right\},$$

$$\|G,2\rangle = 3^{-1/2}\left\{\sqrt{2}|T_1,1\rangle|G,1\rangle - |T_1,0\rangle|G,2\rangle\right\},$$

$$\|G,1\rangle = 3^{-1/2}\left\{|T_1,0\rangle|G,1\rangle - \sqrt{2}|T_1,-1\rangle|G,2\rangle\right\},$$

$$\|G,-1\rangle = 3^{-1/2}\left\{-\sqrt{2}|T_1,1\rangle|G,-2\rangle - |T_1,0\rangle|G,-1\rangle\right\},$$

$$\|G,-2\rangle = 3^{-1/2}\left\{|T_1,0\rangle|G,-2\rangle + \sqrt{2}|T_1,-1\rangle|G,-1\rangle\right\},$$

$$\|H,2\rangle = 12^{-1/2}\left\{|T_1,1\rangle|G,1\rangle + \sqrt{2}|T_1,0\rangle|G,2\rangle + 3|T_1,-1\rangle|G,-2\rangle\right\},$$

$$\|H,1\rangle = 3^{-1/2}\left\{-\sqrt{2}|T_1,0\rangle|G,1\rangle - |T_1,-1\rangle|G,2\rangle\right\},$$

$$\|H,0\rangle = 2^{-1/2}\left\{|T_1,1\rangle|G,-1\rangle + |T_1,-1\rangle|G,1\rangle\right\},$$

$$\|H,-1\rangle = 3^{-1/2}\left\{|T_1,1\rangle|G,-2\rangle - \sqrt{2}|T_1,0\rangle|G,-1\rangle\right\},$$

$$\|H,-2\rangle = 12^{-1/2}\left\{3|T_1,1\rangle|G,2\rangle - \sqrt{2}|T_1,0\rangle|G,-2\rangle \right.$$
$$\left. + |T_1,-1\rangle|G,-1\rangle\right\},$$

(5) $D^{T_2} \times D^G \simeq D^{T_1} \oplus D^G \oplus D^H$.

$\||T_1, 1\rangle = 2^{-1} \left\{ |T_2, 2\rangle |G, -1\rangle + \sqrt{2}|T_2, 0\rangle |G, 1\rangle + |T_2, -2\rangle |G, -2\rangle \right\}$,

$\||T_1, 0\rangle = 2^{-1/2} \left\{ |T_2, 2\rangle |G, -2\rangle + |T_2, -2\rangle |G, 2\rangle \right\}$,

$\||T_1, -1\rangle = 2^{-1} \left\{ -|T_2, 2\rangle |G, 2\rangle - \sqrt{2}|T_2, 0\rangle |G, -1\rangle - |T_2, -2\rangle |G, 1\rangle \right\}$,

$\||G, 2\rangle = 3^{-1/2} \left\{ |T_2, 0\rangle |G, 2\rangle + \sqrt{2}|T_2, -2\rangle |G, -1\rangle \right\}$,

$\||G, 1\rangle = 3^{-1/2} \left\{ |T_2, 0\rangle |G, 1\rangle - \sqrt{2}|T_2, -2\rangle |G, -2\rangle \right\}$,

$\||G, -1\rangle = 3^{-1/2} \left\{ \sqrt{2}|T_2, 2\rangle |G, 2\rangle - |T_2, 0\rangle |G, -1\rangle \right\}$,

$\||G, -2\rangle = 3^{-1/2} \left\{ -\sqrt{2}|T_2, 2\rangle |G, 1\rangle - |T_2, 0\rangle |G, -2\rangle \right\}$,

$\||H, 2\rangle = 3^{-1/2} \left\{ \sqrt{2}|T_2, 0\rangle |G, 2\rangle - |T_2, -2\rangle |G, -1\rangle \right\}$,

$\||H, 1\rangle = 12^{-1/2} \left\{ -3|T_2, 2\rangle |G, -1\rangle + \sqrt{2}|T_2, 0\rangle |G, 1\rangle \right.$
$\left. + |T_2, -2\rangle |G, -2\rangle \right\}$,

$\||H, 0\rangle = 2^{-1/2} \left\{ |T_2, 2\rangle |G, -2\rangle - |T_2, -2\rangle |G, 2\rangle \right\}$,

$\||H, -1\rangle = 12^{-1/2} \left\{ |T_2, 2\rangle |G, 2\rangle + \sqrt{2}|T_2, 0\rangle |G, -1\rangle \right.$
$\left. - 3|T_2, -2\rangle |G, 1\rangle \right\}$,

$\||H, -2\rangle = 3^{-1/2} \left\{ |T_2, 2\rangle |G, 1\rangle - \sqrt{2}|T_2, 0\rangle |G, -2\rangle \right\}$,

(6) $D^{T_1} \times D^H \simeq D^{T_1} \oplus D^{T_2} \oplus D^G \oplus D^H$.

$\||T_1, 1\rangle = 10^{-1/2} \left\{ |T_1, 1\rangle |H, 0\rangle - \sqrt{3}|T_1, 0\rangle |H, 1\rangle \right.$
$\left. + \sqrt{6}|T_1, -1\rangle |H, 2\rangle \right\}$,

$\||T_1, 0\rangle = 10^{-1/2} \left\{ \sqrt{3}|T_1, 1\rangle |H, -1\rangle - 2|T_1, 0\rangle |H, 0\rangle \right.$
$\left. + \sqrt{3}|T_1, -1\rangle |H, 1\rangle \right\}$,

$\||T_1, -1\rangle = 10^{-1/2} \left\{ \sqrt{6}|T_1, 1\rangle |H, -2\rangle - \sqrt{3}|T_1, 0\rangle |H, -1\rangle \right.$
$\left. + |T_1, -1\rangle |H, 0\rangle \right\}$,

$\||T_2, 2\rangle = 5^{-1/2} \left\{ \sqrt{2}|T_1, 1\rangle |H, 1\rangle + |T_1, 0\rangle |H, 2\rangle \right.$
$\left. + \sqrt{2}|T_1, -1\rangle |H, -2\rangle \right\}$,

$\||T_2, 0\rangle = 5^{-1/2} \left\{ |T_1, 1\rangle |H, -1\rangle + \sqrt{3}|T_1, 0\rangle |H, 0\rangle + |T_1, -1\rangle |H, 1\rangle \right\}$,

$\||T_2, -2\rangle = 5^{-1/2} \left\{ -\sqrt{2}|T_1, 1\rangle |H, 2\rangle + |T_1, 0\rangle |H, -2\rangle \right.$
$\left. + \sqrt{2}|T_1, -1\rangle |H, -1\rangle \right\}$,

$\||G, 2\rangle = 15^{-1/2} \left\{ -2|T_1, 1\rangle |H, 1\rangle - \sqrt{2}|T_1, 0\rangle |H, 2\rangle \right.$
$\left. + 3|T_1, -1\rangle |H, -2\rangle \right\}$,

$$\|G, 1\rangle = 15^{-1/2} \left\{ \sqrt{6}|T_1, 1\rangle|H, 0\rangle + \sqrt{8}|T_1, 0\rangle|H, 1\rangle \right.$$
$$\left. + |T_1, -1\rangle|H, 2\rangle \right\},$$
$$\|G, -1\rangle = 15^{-1/2} \left\{ |T_1, 1\rangle|H, -2\rangle + \sqrt{8}|T_1, 0\rangle|H, -1\rangle \right.$$
$$\left. + \sqrt{6}|T_1, -1\rangle|H, 0\rangle \right\},$$
$$\|G, -2\rangle = 15^{-1/2} \left\{ 3|T_1, 1\rangle|H, 2\rangle + \sqrt{2}|T_1, 0\rangle|H, -2\rangle \right.$$
$$\left. + 2|T_1, -1\rangle|H, -1\rangle \right\},$$
$$\|H, 2\rangle = 3^{-1/2} \left\{ |T_1, 1\rangle|H, 1\rangle - \sqrt{2}|T_1, 0\rangle|H, 2\rangle \right\},$$
$$\|H, 1\rangle = 6^{-1/2} \left\{ \sqrt{3}|T_1, 1\rangle|H, 0\rangle - |T_1, 0\rangle|H, 1\rangle \right.$$
$$\left. - \sqrt{2}|T_1, -1\rangle|H, 2\rangle \right\},$$
$$\|H, 0\rangle = 2^{-1/2} \left\{ |T_1, 1\rangle|H, -1\rangle - |T_1, -1\rangle|H, 1\rangle \right\},$$
$$\|H, -1\rangle = 6^{-1/2} \left\{ \sqrt{2}|T_1, 1\rangle|H, -2\rangle + |T_1, 0\rangle|H, -1\rangle \right.$$
$$\left. - \sqrt{3}|T_1, -1\rangle|H, 0\rangle \right\},$$
$$\|H, -2\rangle = 3^{-1/2} \left\{ \sqrt{2}|T_1, 0\rangle|H, -2\rangle - |T_1, -1\rangle|H, -1\rangle \right\},$$

(7) $D^{T_2} \times D^H \simeq D^{T_1} \oplus D^{T_2} \oplus D^G \oplus D^H.$

$$\|T_1, 1\rangle = 5^{-1/2} \left\{ -\sqrt{2}|T_2, 2\rangle|H, -1\rangle - |T_2, 0\rangle|H, 1\rangle \right.$$
$$\left. - \sqrt{2}|T_2, -2\rangle|H, -2\rangle \right\},$$
$$\|T_1, 0\rangle = 5^{-1/2} \left\{ |T_2, 2\rangle|H, -2\rangle + \sqrt{3}|T_2, 0\rangle|H, 0\rangle + |T_2, -2\rangle|H, 2\rangle \right\},$$
$$\|T_1, -1\rangle = 5^{-1/2} \left\{ \sqrt{2}|T_2, 2\rangle|H, 2\rangle - |T_2, 0\rangle|H, -1\rangle \right.$$
$$\left. - \sqrt{2}|T_2, -2\rangle|H, 1\rangle \right\},$$
$$\|T_2, 2\rangle = 10^{-1/2} \left\{ |T_2, 2\rangle|H, 0\rangle + \sqrt{3}|T_2, 0\rangle|H, 2\rangle \right.$$
$$\left. - \sqrt{6}|T_2, -2\rangle|H, -1\rangle \right\},$$
$$\|T_2, 0\rangle = 10^{-1/2} \left\{ \sqrt{3}|T_2, 2\rangle|H, -2\rangle - 2|T_2, 0\rangle|H, 0\rangle \right.$$
$$\left. + \sqrt{3}|T_2, -2\rangle|H, 2\rangle \right\},$$
$$\|T_2, -2\rangle = 10^{-1/2} \left\{ \sqrt{6}|T_2, 2\rangle|H, 1\rangle + \sqrt{3}|T_2, 0\rangle|H, -2\rangle \right.$$
$$\left. + |T_2, -2\rangle|H, 0\rangle \right\},$$
$$\|G, 2\rangle = 15^{-1/2} \left\{ \sqrt{6}|T_2, 2\rangle|H, 0\rangle - \sqrt{8}|T_2, 0\rangle|H, 2\rangle \right.$$
$$\left. - |T_2, -2\rangle|H, -1\rangle \right\},$$
$$\|G, 1\rangle = 15^{-1/2} \left\{ 3|T_2, 2\rangle|H, -1\rangle - \sqrt{2}|T_2, 0\rangle|H, 1\rangle \right.$$
$$\left. - 2|T_2, -2\rangle|H, -2\rangle \right\},$$
$$\|G, -1\rangle = 15^{-1/2} \left\{ 2|T_2, 2\rangle|H, 2\rangle - \sqrt{2}|T_2, 0\rangle|H, -1\rangle \right.$$
$$\left. + 3|T_2, -2\rangle|H, 1\rangle \right\},$$

$$\|G,-2\rangle = 15^{-1/2}\left\{-|T_2,2\rangle|H,1\rangle + \sqrt{8}|T_2,0\rangle|H,-2\rangle\right.$$
$$\left. - \sqrt{6}|T_2,-2\rangle|H,0\rangle\right\},$$
$$\|H,2\rangle = 6^{-1/2}\left\{\sqrt{3}|T_2,2\rangle|H,0\rangle + |T_2,0\rangle|H,2\rangle\right.$$
$$\left. + \sqrt{2}|T_2,-2\rangle|H,-1\rangle\right\},$$
$$\|H,1\rangle = 3^{-1/2}\left\{-\sqrt{2}|T_2,0\rangle|H,1\rangle + |T_2,-2\rangle|H,-2\rangle\right\},$$
$$\|H,0\rangle = 2^{-1/2}\left\{-|T_2,2\rangle|H,-2\rangle + |T_2,-2\rangle|H,2\rangle\right\},$$
$$\|H,-1\rangle = 3^{-1/2}\left\{|T_2,2\rangle|H,2\rangle + \sqrt{2}|T_2,0\rangle|H,-1\rangle\right\},$$
$$\|H,-2\rangle = 6^{-1/2}\left\{\sqrt{2}|T_2,2\rangle|H,1\rangle - |T_2,0\rangle|H,-2\rangle\right.$$
$$\left. - \sqrt{3}|T_2,-2\rangle|H,0\rangle\right\},$$

(8) $D^G \times D^G \simeq D^A \oplus D^{T_1} \oplus D^{T_2} \oplus D^G \oplus D^H.$

$$\|A,0\rangle = 2^{-1}\left\{|G,2\rangle|G,-2\rangle + |G,1\rangle|G,-1\rangle\right.$$
$$\left. + |G,-1\rangle|G,1\rangle + |G,-2\rangle|G,2\rangle\right\},$$
$$\|T_1,1\rangle = 2^{-1/2}\left\{|G,2\rangle|G,-1\rangle - |G,-1\rangle|G,2\rangle\right\},$$
$$\|T_1,0\rangle = 2^{-1}\left\{-|G,2\rangle|G,-2\rangle + |G,1\rangle|G,-1\rangle\right.$$
$$\left. - |G,-1\rangle|G,1\rangle + |G,-2\rangle|G,2\rangle\right\},$$
$$\|T_1,-1\rangle = 2^{-1/2}\left\{-|G,1\rangle|G,-2\rangle + |G,-2\rangle|G,1\rangle\right\},$$
$$\|T_2,2\rangle = 2^{-1/2}\left\{-|G,-1\rangle|G,-2\rangle + |G,-2\rangle|G,-1\rangle\right\},$$
$$\|T_2,0\rangle = 2^{-1}\left\{-|G,2\rangle|G,-2\rangle - |G,1\rangle|G,-1\rangle\right.$$
$$\left. + |G,-1\rangle|G,1\rangle + |G,-2\rangle|G,2\rangle\right\},$$
$$\|T_2,-2\rangle = 2^{-1/2}\left\{-|G,2\rangle|G,1\rangle + |G,1\rangle|G,2\rangle\right\},$$
$$\|G,2\rangle = 3^{-1/2}\left\{|G,1\rangle|G,1\rangle - |G,-1\rangle|G,-2\rangle - |G,-2\rangle|G,-1\rangle\right\},$$
$$\|G,1\rangle = 3^{-1/2}\left\{|G,2\rangle|G,-1\rangle + |G,-1\rangle|G,2\rangle - |G,-2\rangle|G,-2\rangle\right\},$$
$$\|G,-1\rangle = 3^{-1/2}\left\{-|G,2\rangle|G,2\rangle + |G,1\rangle|G,-2\rangle + |G,-2\rangle|G,1\rangle\right\},$$
$$\|G,-2\rangle = 3^{-1/2}\left\{-|G,2\rangle|G,1\rangle - |G,1\rangle|G,2\rangle + |G,-1\rangle|G,-1\rangle\right\},$$
$$\|H,2\rangle = 6^{-1/2}\left\{2|G,1\rangle|G,1\rangle + |G,-1\rangle|G,-2\rangle + |G,-2\rangle|G,-1\rangle\right\},$$
$$\|H,1\rangle = 6^{-1/2}\left\{|G,2\rangle|G,-1\rangle + |G,-1\rangle|G,2\rangle + 2|G,-2\rangle|G,-2\rangle\right\},$$
$$\|H,0\rangle = 2^{-1}\left\{|G,2\rangle|G,-2\rangle - |G,1\rangle|G,-1\rangle\right.$$
$$\left. - |G,-1\rangle|G,1\rangle + |G,-2\rangle|G,2\rangle\right\},$$
$$\|H,-1\rangle = 6^{-1/2}\left\{-2|G,2\rangle|G,2\rangle - |G,1\rangle|G,-2\rangle - |G,-2\rangle|G,1\rangle\right\},$$
$$\|H,-2\rangle = 6^{-1/2}\left\{|G,2\rangle|G,1\rangle + |G,1\rangle|G,2\rangle + 2|G,-1\rangle|G,-1\rangle\right\},$$

(9) $D^G \times D^H \simeq D^{T_1} \oplus D^{T_2} \oplus D^G \oplus 2D^H$.

$$\|T_1, 1\rangle = 20^{-1/2} \left\{ 2|G, 2\rangle|H, -1\rangle + \sqrt{6}|G, 1\rangle|H, 0\rangle \right.$$
$$\left. + |G, -1\rangle|H, 2\rangle + 3|G, -2\rangle|H, -2\rangle \right\},$$

$$\|T_1, 0\rangle = 10^{-1/2} \left\{ -|G, 2\rangle|H, -2\rangle - 2|G, 1\rangle|H, -1\rangle \right.$$
$$\left. - 2|G, -1\rangle|H, 1\rangle + |G, -2\rangle|H, 2\rangle \right\},$$

$$\|T_1, -1\rangle = 20^{-1/2} \left\{ 3|G, 2\rangle|H, 2\rangle + |G, 1\rangle|H, -2\rangle \right.$$
$$\left. + \sqrt{6}|G, -1\rangle|H, 0\rangle - 2|G, -2\rangle|H, 1\rangle \right\},$$

$$\|T_2, 2\rangle = 20^{-1/2} \left\{ \sqrt{6}|G, 2\rangle|H, 0\rangle - 3|G, 1\rangle|H, 1\rangle \right.$$
$$\left. + 2|G, -1\rangle|H, -2\rangle + |G, -2\rangle|H, -1\rangle \right\},$$

$$\|T_2, 0\rangle = 10^{-1/2} \left\{ -2|G, 2\rangle|H, -2\rangle + |G, 1\rangle|H, -1\rangle \right.$$
$$\left. + |G, -1\rangle|H, 1\rangle + 2|G, -2\rangle|H, 2\rangle \right\},$$

$$\|T_2, -2\rangle = 20^{-1/2} \left\{ |G, 2\rangle|H, 1\rangle - 2|G, 1\rangle|H, 2\rangle \right.$$
$$\left. - 3|G, -1\rangle|H, -1\rangle - \sqrt{6}|G, -2\rangle|H, 0\rangle \right\},$$

$$\|G, 2\rangle = 15^{-1/2} \left\{ \sqrt{3}|G, 2\rangle|H, 0\rangle + \sqrt{2}|G, 1\rangle|H, 1\rangle \right.$$
$$\left. + \sqrt{2}|G, -1\rangle|H, -2\rangle - \sqrt{8}|G, -2\rangle|H, -1\rangle \right\},$$

$$\|G, 1\rangle = 15^{-1/2} \left\{ -\sqrt{2}|G, 2\rangle|H, -1\rangle - \sqrt{3}|G, 1\rangle|H, 0\rangle \right.$$
$$\left. + \sqrt{8}|G, -1\rangle|H, 2\rangle + \sqrt{2}|G, -2\rangle|H, -2\rangle \right\},$$

$$\|G, -1\rangle = 15^{-1/2} \left\{ \sqrt{2}|G, 2\rangle|H, 2\rangle + \sqrt{8}|G, 1\rangle|H, -2\rangle \right.$$
$$\left. - \sqrt{3}|G, -1\rangle|H, 0\rangle + \sqrt{2}|G, -2\rangle|H, 1\rangle \right\},$$

$$\|G, -2\rangle = 15^{-1/2} \left\{ \sqrt{8}|G, 2\rangle|H, 1\rangle + \sqrt{2}|G, 1\rangle|H, 2\rangle \right.$$
$$\left. - \sqrt{2}|G, -1\rangle|H, -1\rangle + \sqrt{3}|G, -2\rangle|H, 0\rangle \right\},$$

$$\|H, (1), 2\rangle = 3^{-1/2} \left\{ -|G, 1\rangle|H, 1\rangle - |G, -1\rangle|H, -2\rangle \right.$$
$$\left. - |G, -2\rangle|H, -1\rangle \right\},$$

$$\|H, (1), 1\rangle = 12^{-1/2} \left\{ -2|G, 2\rangle|H, -1\rangle + \sqrt{6}|G, 1\rangle|H, 0\rangle \right.$$
$$\left. + |G, -1\rangle|H, 2\rangle - |G, -2\rangle|H, -2\rangle \right\},$$

$$\|H, (1), 0\rangle = 2^{-1/2} \left\{ |G, 2\rangle|H, -2\rangle + |G, -2\rangle|H, 2\rangle \right\},$$

$$\|H, (1), -1\rangle = 12^{-1/2} \left\{ |G, 2\rangle|H, 2\rangle - |G, 1\rangle|H, -2\rangle \right.$$
$$\left. - \sqrt{6}|G, -1\rangle|H, 0\rangle - 2|G, -2\rangle|H, 1\rangle \right\},$$

$$\|H, (1), -2\rangle = 3^{-1/2} \left\{ |G, 2\rangle|H, 1\rangle - |G, 1\rangle|H, 2\rangle + |G, -1\rangle|H, -1\rangle \right\},$$

$$\|H, (2), 2\rangle = 12^{-1/2} \left\{ -\sqrt{6}|G, 2\rangle|H, 0\rangle - |G, 1\rangle|H, 1\rangle \right.$$
$$\left. + 2|G, -1\rangle|H, -2\rangle - |G, -2\rangle|H, -1\rangle \right\},$$

$$||H,(2),1\rangle = 3^{-1/2} \{|G,2\rangle|H,-1\rangle + |G,-1\rangle|H,2\rangle - |G,-2\rangle|H,-2\rangle\},$$
$$||H,(2),0\rangle = 2^{-1/2} \{|G,1\rangle|H,-1\rangle - |G,-1\rangle|H,1\rangle\},$$
$$||H,(2),-1\rangle = 3^{-1/2} \{|G,2\rangle|H,2\rangle - |G,1\rangle|H,-2\rangle + |G,-2\rangle|H,1\rangle\},$$
$$||H,(2),-2\rangle = 12^{-1/2} \{|G,2\rangle|H,1\rangle + 2|G,1\rangle|H,2\rangle$$
$$+ |G,-1\rangle|H,-1\rangle - \sqrt{6}|G,-2\rangle|H,0\rangle\},$$

(10) $D^H \times D^H \simeq D^A \oplus D^{T_1} \oplus D^{T_2} \oplus 2D^G \oplus 2D^H.$

$$||A,0\rangle = 5^{-1/2} \{|H,2\rangle|H,-2\rangle - |H,1\rangle|H,-1\rangle + |H,0\rangle|H,0\rangle$$
$$- |H,-1\rangle|H,1\rangle + |H,-2\rangle|H,2\rangle\},$$
$$||T_1,1\rangle = 10^{-1/2} \{\sqrt{2}|H,2\rangle|H,-1\rangle - \sqrt{3}|H,1\rangle|H,0\rangle$$
$$+ \sqrt{3}|H,0\rangle|H,1\rangle - \sqrt{2}|H,-1\rangle|H,2\rangle\},$$
$$||T_1,0\rangle = 10^{-1/2} \{2|H,2\rangle|H,-2\rangle - |H,1\rangle|H,-1\rangle$$
$$+ |H,-1\rangle|H,1\rangle - 2|H,-2\rangle|H,2\rangle\},$$
$$||T_1,-1\rangle = 10^{-1/2} \{\sqrt{2}|H,1\rangle|H,-2\rangle - \sqrt{3}|H,0\rangle|H,-1\rangle$$
$$+ \sqrt{3}|H,-1\rangle|H,0\rangle - \sqrt{2}|H,-2\rangle|H,1\rangle\},$$
$$||T_2,2\rangle = 20^{-1/2} \{\sqrt{6}|H,2\rangle|H,0\rangle - \sqrt{6}|H,0\rangle|H,2\rangle$$
$$+ 2|H,-1\rangle|H,-2\rangle - 2|H,-2\rangle|H,-1\rangle\},$$
$$||T_2,0\rangle = 10^{-1/2} \{|H,2\rangle|H,-2\rangle + 2|H,1\rangle|H,-1\rangle$$
$$- 2|H,-1\rangle|H,1\rangle - |H,-2\rangle|H,2\rangle\},$$
$$||T_2,-2\rangle = 20^{-1/2} \{-2|H,2\rangle|H,1\rangle + 2|H,1\rangle|H,2\rangle$$
$$+ \sqrt{6}|H,0\rangle|H,-2\rangle - \sqrt{6}|H,-2\rangle|H,0\rangle\},$$
$$||G,(1),2\rangle = 10^{-1/2} \{-\sqrt{2}|H,2\rangle|H,0\rangle + \sqrt{2}|H,0\rangle|H,2\rangle$$
$$+ \sqrt{3}|H,-1\rangle|H,-2\rangle - \sqrt{3}|H,-2\rangle|H,-1\rangle\},$$
$$||G,(1),1\rangle = 10^{-1/2} \{\sqrt{3}|H,2\rangle|H,-1\rangle + \sqrt{2}|H,1\rangle|H,0\rangle$$
$$- \sqrt{2}|H,0\rangle|H,1\rangle - \sqrt{3}|H,-1\rangle|H,2\rangle\},$$
$$||G,(1),-1\rangle = 10^{-1/2} \{\sqrt{3}|H,1\rangle|H,-2\rangle + \sqrt{2}|H,0\rangle|H,-1\rangle$$
$$- \sqrt{2}|H,-1\rangle|H,0\rangle - \sqrt{3}|H,-2\rangle|H,1\rangle\},$$
$$||G,(1),-2\rangle = 10^{-1/2} \{\sqrt{3}|H,2\rangle|H,1\rangle - \sqrt{3}|H,1\rangle|H,2\rangle$$
$$+ \sqrt{2}|H,0\rangle|H,-2\rangle - \sqrt{2}|H,-2\rangle|H,0\rangle\},$$
$$||G,(2),2\rangle = 30^{-1/2} \{\sqrt{6}|H,2\rangle|H,0\rangle + 4|H,1\rangle|H,1\rangle$$
$$+ \sqrt{6}|H,0\rangle|H,2\rangle - |H,-1\rangle|H,-2\rangle - |H,-2\rangle|H,-1\rangle\},$$

$$||G,(2),1\rangle = 30^{-1/2}\left\{|H,2\rangle|H,-1\rangle + \sqrt{6}|H,1\rangle|H,0\rangle + \sqrt{6}|H,0\rangle|H,1\rangle \right.$$
$$\left. + |H,-1\rangle|H,2\rangle - 4|H,-2\rangle|H,-2\rangle\right\},$$

$$||G,(2),-1\rangle = 30^{-1/2}\left\{-4|H,2\rangle|H,2\rangle - |H,1\rangle|H,-2\rangle \right.$$
$$\left. - \sqrt{6}|H,0\rangle|H,-1\rangle - \sqrt{6}|H,-1\rangle|H,0\rangle - |H,-2\rangle|H,1\rangle\right\},$$

$$||G,(2),-2\rangle = 30^{-1/2}\left\{|H,2\rangle|H,1\rangle + |H,1\rangle|H,2\rangle \right.$$
$$\left. + \sqrt{6}|H,0\rangle|H,-2\rangle + 4|H,-1\rangle|H,-1\rangle + \sqrt{6}|H,-2\rangle|H,0\rangle\right\},$$

$$||H,(1),2\rangle = 7^{-1/2}\left\{\sqrt{2}|H,2\rangle|H,0\rangle - \sqrt{3}|H,1\rangle|H,1\rangle \right.$$
$$\left. + \sqrt{2}|H,0\rangle|H,2\rangle\right\},$$

$$||H,(1),1\rangle = 14^{-1/2}\left\{\sqrt{6}|H,2\rangle|H,-1\rangle - |H,1\rangle|H,0\rangle \right.$$
$$\left. - |H,0\rangle|H,1\rangle + \sqrt{6}|H,-1\rangle|H,2\rangle\right\},$$

$$||H,(1),0\rangle = 14^{-1/2}\left\{2|H,2\rangle|H,-2\rangle + |H,1\rangle|H,-1\rangle - 2|H,0\rangle|H,0\rangle \right.$$
$$\left. + |H,-1\rangle|H,1\rangle + 2|H,-2\rangle|H,2\rangle\right\},$$

$$||H,(1),-1\rangle = 14^{-1/2}\left\{\sqrt{6}|H,1\rangle|H,-2\rangle - |H,0\rangle|H,-1\rangle \right.$$
$$\left. - |H,-1\rangle|H,0\rangle + \sqrt{6}|H,-2\rangle|H,1\rangle\right\},$$

$$||H,(1),-2\rangle = 7^{-1/2}\left\{\sqrt{2}|H,0\rangle|H,-2\rangle - \sqrt{3}|H,-1\rangle|H,-1\rangle \right.$$
$$\left. + \sqrt{2}|H,-2\rangle|H,0\rangle\right\},$$

$$||H,(2),2\rangle = 210^{-1/2}\left\{\sqrt{3}|H,2\rangle|H,0\rangle + \sqrt{8}|H,1\rangle|H,1\rangle \right.$$
$$\left. + \sqrt{3}|H,0\rangle|H,2\rangle + \sqrt{98}|H,-1\rangle|H,-2\rangle + \sqrt{98}|H,-2\rangle|H,-1\rangle\right\},$$

$$||H,(2),1\rangle = 105^{-1/2}\left\{-2|H,2\rangle|H,-1\rangle - \sqrt{24}|H,1\rangle|H,0\rangle \right.$$
$$\left. - \sqrt{24}|H,0\rangle|H,1\rangle - 2|H,-1\rangle|H,2\rangle - 7|H,-2\rangle|H,-2\rangle\right\},$$

$$||H,(2),0\rangle = 70^{-1/2}\left\{|H,2\rangle|H,-2\rangle + 4|H,1\rangle|H,-1\rangle \right.$$
$$\left. + 6|H,0\rangle|H,0\rangle + 4|H,-1\rangle|H,1\rangle + |H,-2\rangle|H,2\rangle\right\},$$

$$||H,(2),-1\rangle = 105^{-1/2}\left\{7|H,2\rangle|H,2\rangle - 2|H,1\rangle|H,-2\rangle \right.$$
$$\left. - \sqrt{24}|H,0\rangle|H,-1\rangle - \sqrt{24}|H,-1\rangle|H,0\rangle - 2|H,-2\rangle|H,1\rangle\right\},$$

$$||H,(2),-2\rangle = 210^{-1/2}\left\{-\sqrt{98}|H,2\rangle|H,1\rangle - \sqrt{98}|H,1\rangle|H,2\rangle \right.$$
$$\left. + \sqrt{3}|H,0\rangle|H,-2\rangle + \sqrt{8}|H,-1\rangle|H,-1\rangle \right.$$
$$\left. + \sqrt{3}|H,-2\rangle|H,0\rangle\right\}.$$

Chapter 4

THREE-DIMENSIONAL
ROTATION GROUP

4.1 SO(3) Group and Its Covering Group SU(2)

★ In the real three-dimensional space, a spatial rotation which keeps both the position of the origin and the distance of any two points invariant:

$$\begin{pmatrix} x_1' \\ x_2' \\ x_3' \end{pmatrix} = \begin{pmatrix} R_{11} & R_{12} & R_{13} \\ R_{21} & R_{22} & R_{23} \\ R_{31} & R_{32} & R_{33} \end{pmatrix} \begin{pmatrix} x_1 \\ x_2 \\ x_3 \end{pmatrix}, \qquad \underline{x}' = R\underline{x},$$

$$(\underline{x}')^T \underline{x}' = \underline{x}^T \underline{x}, \qquad R^T R = \mathbf{1}, \qquad R^* = R,$$

is described by a real orthogonal matrix R. R is a proper rotation if $\det R = 1$, and an improper rotation if $\det R = -1$. The set of all three-dimensional real orthogonal matrices with $\det R = 1$, in the multiplication rule of matrices, satisfies the four axioms and forms the simplest non-Abelian compact Lie group SO(3). It is a very important Lie group from the viewpoints of both physics and mathematics. The element $R(\hat{\mathbf{n}}, \omega)$ in SO(3) is a rotation around the direction $\hat{\mathbf{n}}$ through the angle ω, satisfying

$$R(\hat{\mathbf{n}}, \omega + 2\pi) = R(\hat{\mathbf{n}}, \omega) = R(-\hat{\mathbf{n}}, 2\pi - \omega),$$

$$R(\hat{\mathbf{n}}, \pi) = R(-\hat{\mathbf{n}}, \pi). \tag{4.1}$$

When $\hat{\mathbf{n}}$ is along the coordinate axis, the rotation matrix can be written in the exponential form of matrix:

$$R(\mathbf{e}_1, \omega) = \begin{pmatrix} 1 & 0 & 0 \\ 0 & \cos\omega & -\sin\omega \\ 0 & \sin\omega & \cos\omega \end{pmatrix} = \exp\{-i\omega T_1\},$$

$$R(\mathbf{e}_2, \omega) = \begin{pmatrix} \cos\omega & 0 & \sin\omega \\ 0 & 1 & 0 \\ -\sin\omega & 0 & \cos\omega \end{pmatrix} = \exp\{-i\omega T_2\},$$

$$R(\mathbf{e}_3, \omega) = \begin{pmatrix} \cos\omega & -\sin\omega & 0 \\ \sin\omega & \cos\omega & 0 \\ 0 & 0 & 1 \end{pmatrix} = \exp\{-i\omega T_3\},$$

where

$$T_1 = \begin{pmatrix} 0 & 0 & 0 \\ 0 & 0 & -i \\ 0 & i & 0 \end{pmatrix}, \quad T_2 = \begin{pmatrix} 0 & 0 & i \\ 0 & 0 & 0 \\ -i & 0 & 0 \end{pmatrix}, \quad T_3 = \begin{pmatrix} 0 & -i & 0 \\ i & 0 & 0 \\ 0 & 0 & 0 \end{pmatrix}, \tag{4.2}$$

$$(T_a)_{bd} = -i\epsilon_{abd}.$$

They are the generators in the self-representation of SO(3), satisfying the typical commutation relations for the angular momentum operators

$$[T_a, T_b] = i \sum_{d=1}^{3} \epsilon_{abd} T_d. \tag{4.3}$$

★ The rotation $S(\varphi, \theta)$

$$S(\varphi, \theta) = R(\mathbf{e}_3, \varphi) R(\mathbf{e}_2, \theta) \tag{4.4}$$

transforms $\mathbf{e}_3$ to the direction $\hat{\mathbf{n}}(\theta, \varphi)$. Since

$$ST_3 S^{-1} = \hat{\mathbf{n}} \cdot \mathbf{T}, \qquad \mathbf{T} = \mathbf{e}_1 T_1 + \mathbf{e}_2 T_2 + \mathbf{e}_3 T_3.$$

The rotation $R(\hat{\mathbf{n}}, \omega)$ around $\hat{\mathbf{n}}$ through ω can be expressed as

$$R(\hat{\mathbf{n}}, \omega) = S(\varphi, \theta) R(\mathbf{e}_3, \omega) S(\varphi, \theta)^{-1} = \exp\{-i\omega\hat{\mathbf{n}} \cdot \mathbf{T}\}. \tag{4.5}$$

Thus, the rotations with the same rotational angle ω are conjugate to each other. The class of SO(3) is described by the rotational angle ω. The SO(3) group is a simple Lie group because it does not contain any nontrivial invariant Lie subgroup.

Define a vector $\vec{\omega}$, which is along the direction $\hat{\mathbf{n}}$ with the length ω. Let the polar angle and the azimuthal angle of $\hat{\mathbf{n}}$ be θ and φ, respectively. $R(\hat{\mathbf{n}}, \omega)$ can be described by the spherical coordinates $(\omega, \theta, \varphi)$ of $\vec{\omega}$, or its rectangular coordinates $(\omega_1, \omega_2, \omega_3)$, which are the group parameters of SO(3). The varied region of the group parameters is called the group space. The group space of SO(3) is a spheroid with radius π, where in the sphere, two end points of a diameter describe the same rotation. Namely, the group space of SO(3) is a doubly-connected closed region. It is the reason why SO(3) has the double-valued representations.

★ The set of all two-dimensional unitary matrices u with det $u = 1$, in the multiplication rule of matrices, form the Lie group SU(2). Its element can be generally expressed by

$$
\begin{aligned}
u(\hat{\mathbf{n}}, \omega) &= \mathbf{1} \cos(\omega/2) - i(\vec{\sigma} \cdot \hat{\mathbf{n}}) \sin(\omega/2) \\
&= \begin{pmatrix} \cos(\omega/2) - i \sin(\omega/2) \cos\theta & -i \sin(\omega/2) \sin\theta e^{-i\varphi} \\ -i \sin(\omega/2) \sin\theta e^{i\varphi} & \cos(\omega/2) + i \sin(\omega/2) \cos\theta \end{pmatrix},
\end{aligned} \tag{4.6}
$$

satisfying

$$
\begin{aligned}
u(\hat{\mathbf{n}}, \omega_1) u(\hat{\mathbf{n}}, \omega_2) &= u(\hat{\mathbf{n}}, \omega_1 + \omega_2), \\
u(\hat{\mathbf{n}}, 4\pi) &= \mathbf{1}, \quad u(\hat{\mathbf{n}}, 2\pi) = -\mathbf{1}, \\
u(\hat{\mathbf{n}}, \omega) &= u(-\hat{\mathbf{n}}, 4\pi - \omega) = -u(-\hat{\mathbf{n}}, 2\pi - \omega).
\end{aligned} \tag{4.7}
$$

Choosing the group parameters of SU(2) to be the vector $\vec{\omega}$ along the direction $\hat{\mathbf{n}}(\theta, \varphi)$ with the length ω, we obtain the group space of SU(2) to be a spheroid with radius 2π, where all points on the sphere describe the same element $-\mathbf{1}$. The group space for SU(2) is a simply-connected closed region, and the SU(2) group is a compact simple Lie group. The generators in the self-representation of SU(2) are $\sigma_a/2$. There is a two-to-one correspondence between $\pm u(\hat{\mathbf{n}}, \omega) \in$ SU(2) and $R(\hat{\mathbf{n}}, \omega) \in$ SO(3):

$$
u(\hat{\mathbf{n}}, \omega) \sigma_a u(\hat{\mathbf{n}}, \omega)^{-1} = \sum_{b=1}^{3} \sigma_b R_{ba}(\hat{\mathbf{n}}, \omega). \tag{4.8}
$$

So, the SU(2) group is the covering group of the SO(3) group. The subset of $u(\hat{\mathbf{n}}, \omega)$ with the same ω forms a class of SU(2).

1. Prove the preliminary formula by induction:

$$
\begin{aligned}
e^{\alpha} \beta e^{-\alpha} &= \beta + \frac{1}{1!} [\alpha, \beta] + +\frac{1}{2!} [\alpha, [\alpha, \beta]] + \cdots \\
&= \sum_{n=0}^{\infty} \frac{1}{n!} \overbrace{[\alpha, [\alpha, \cdots [\alpha, \beta] \cdots]]}^{n},
\end{aligned}
$$

where α and β are two matrices with the same dimension. Then, show Eq. (4.8) and prove that SU(2) is homomorphic onto SO(3).

Solution.

$$e^{\alpha}\beta e^{-\alpha} = \left\{\sum_{m=0}^{\infty} \frac{1}{m!}\alpha^m\right\}\beta\left\{\sum_{r=0}^{\infty} \frac{(-1)^r}{r!}\alpha^r\right\}$$

$$= \beta + \sum_{n=1}^{\infty} \frac{1}{n!}\left\{\sum_{m=0}^{n} \frac{(-1)^{n-m}n!}{m!(n-m)!}\alpha^m\beta\alpha^{n-m}\right\}.$$

We are going to show by induction that the expression in the brackets is equal the following commutator with n-multiplicity

$$\sum_{m=0}^{n} \frac{(-1)^{n-m}n!}{m!(n-m)!}\alpha^m\beta\alpha^{n-m} = \overbrace{[\alpha, [\alpha, \cdots [\alpha, \beta] \cdots]]}^{n}.$$

The above formula is true for $n = 1$ obviously. Now, if it is true for $n - 1$, we will show that it is also true for n.

$$\overbrace{[\alpha, [\alpha, \cdots [\alpha, \beta] \cdots]]}^{n}$$

$$= \alpha\overbrace{[\alpha, [\alpha, \cdots [\alpha, \beta] \cdots]]}^{n-1} - \overbrace{[\alpha, [\alpha, \cdots [\alpha, \beta] \cdots]]}^{n-1}\alpha$$

$$= \alpha \sum_{m=1}^{n} \frac{(-1)^{n-m}(n-1)!}{(m-1)!(n-m)!}\alpha^{m-1}\beta\alpha^{n-m}$$

$$- \sum_{m=0}^{n-1} \frac{(-1)^{n-m-1}(n-1)!}{(m)!(n-m-1)!}\alpha^m\beta\alpha^{n-m-1}\alpha$$

$$= \alpha^n\beta + \sum_{m=1}^{n-1} \frac{(-1)^{n-m}(n-1)!}{(m)!(n-m)!}[m+(n-m)]\alpha^m\beta\alpha^{n-m} + (-1)^n\beta\alpha^n$$

$$= \sum_{m=0}^{n} \frac{(-1)^{n-m}n!}{m!(n-m)!}\alpha^m\beta\alpha^{n-m}.$$

This completes the proof. Further, due to $(T_c)_{ba} = i\epsilon_{cab}$,

$$2^{-1}[\vec{\sigma}\cdot\hat{\mathbf{n}}, \sigma_a] = i\sum_{bc} n_c\epsilon_{cab}\sigma_b = \sum_{b=1}^{3} \sigma_b\,(\hat{\mathbf{n}}\cdot\mathbf{T})_{ba}\,,$$

$$2^{-2}[\vec{\sigma}\cdot\hat{\mathbf{n}}, [\vec{\sigma}\cdot\hat{\mathbf{n}}, \sigma_a]] = \sum_{b=1}^{3} \sigma_b\left[(\hat{\mathbf{n}}\cdot\mathbf{T})^2\right]_{ba}.$$

Repeating the commutation relation by n times, we have

$$2^{-n}\overbrace{[\vec{\sigma}\cdot\hat{\mathbf{n}}, [\vec{\sigma}\cdot\hat{\mathbf{n}}, \cdots [\vec{\sigma}\cdot\hat{\mathbf{n}}, \sigma_a] \cdots]]}^{n} = \sum_{b=1}^{3} \sigma_b\,[(\hat{\mathbf{n}}\cdot\mathbf{T})^n]_{ba}.$$

By making use of the preliminary formula and

$$u(\hat{\mathbf{n}}, \omega) = \exp\{-i\omega\vec{\sigma} \cdot \hat{\mathbf{n}}/2\}, \qquad R(\hat{\mathbf{n}}, \omega) = \exp\left(-i\omega\hat{\mathbf{n}} \cdot \mathbf{T}\right),$$

we obtain

$$\begin{aligned}
u(\hat{\mathbf{n}}, \omega)\sigma_a u(\hat{\mathbf{n}}, \omega)^{-1} &= \sigma_a + \sum_{n=1}^{\infty} \frac{(-i\omega)^n}{n!} \sum_{b=1}^{3} \sigma_b \left[(\hat{\mathbf{n}} \cdot \mathbf{T})^n\right]_{ba} \\
&= \sum_{b=1}^{3} \sigma_b R_{ba}(\hat{\mathbf{n}}, \omega).
\end{aligned}$$

Now, Eq. (4.8) is proved. Any $u(\hat{\mathbf{n}}, \omega) \in SU(2)$ can uniquely determine a matrix $R(\hat{\mathbf{n}}, \omega) \in SO(3)$ by Eq. (4.8). If both $u_1 \in SU(2)$ and $u_2 \in SU(2)$ determine the same $R \in SO(3)$ by Eq. (4.8), $u_2^{-1}u_1$ can commute with three Pauli matrices σ_a, so it has to be a constant matrix. Due to their determinants, we have $u_1 = \pm u_2$. Therefore, Eq. (4.8) gives a two-to-one correspondence between $\pm u(\hat{\mathbf{n}}, \omega) \in SU(2)$ and $R(\hat{\mathbf{n}}, \omega) \in SO(3)$, which is obviously invariant in the product of elements. It means that SU(2) is homomorphic onto SO(3).

2. Expand $R(\hat{\mathbf{n}}, \omega) = \exp\left(-i\omega\hat{\mathbf{n}} \cdot \mathbf{T}\right)$ as a sum of matrices with the finite terms.

Hint: $(\hat{\mathbf{n}} \cdot \mathbf{T})^3 = \hat{\mathbf{n}} \cdot \mathbf{T}$.

Solution. From

$$\hat{\mathbf{n}} \cdot \mathbf{T} = \begin{pmatrix} 0 & -in_3 & in_2 \\ in_3 & 0 & -in_1 \\ -in_2 & in_1 & 0 \end{pmatrix},$$

$$(\hat{\mathbf{n}} \cdot \mathbf{T})^2 = \begin{pmatrix} 1 - n_1 n_1 & -n_1 n_2 & -n_1 n_3 \\ -n_2 n_1 & 1 - n_2 n_2 & -n_2 n_3 \\ -n_3 n_1 & -n_3 n_2 & 1 - n_3 n_3 \end{pmatrix},$$

$$(\hat{\mathbf{n}} \cdot \mathbf{T})^3 = \hat{\mathbf{n}} \cdot \mathbf{T},$$

one has

$$R(\hat{\mathbf{n}}, \omega) = \exp\left(-i\omega\hat{\mathbf{n}} \cdot \mathbf{T}\right)$$

$$= 1 - i\omega\hat{\mathbf{n}}\cdot\mathbf{T} - \frac{1}{2!}\omega^2(\hat{\mathbf{n}}\cdot\mathbf{T})^2 + i\omega^3(\hat{\mathbf{n}}\cdot\mathbf{T})^3 + \cdots$$

$$= 1 - (\hat{\mathbf{n}}\cdot\mathbf{T})^2 + (\hat{\mathbf{n}}\cdot\mathbf{T})^2\left\{1 - \frac{1}{2!}\omega^2 + \frac{1}{4!}\omega^4 + \cdots\right\}$$

$$- i(\hat{\mathbf{n}}\cdot\mathbf{T})\left\{\omega - \frac{1}{3!}\omega^3 + \frac{1}{5!}\omega^5 + \cdots\right\}$$

$$= 1 - (\hat{\mathbf{n}}\cdot\mathbf{T})^2 + (\hat{\mathbf{n}}\cdot\mathbf{T})^2\cos\omega - i(\hat{\mathbf{n}}\cdot\mathbf{T})\sin\omega.$$

This problem can also be solved with the method of similarity transformation:

$$\hat{\mathbf{n}}\cdot\mathbf{T} = ST_3S^{-1}, \qquad S = R(\mathbf{e}_3,\varphi)R(\mathbf{e}_2,\theta),$$

where θ and φ are the polar angle and azimuthal angle of the direction $\hat{\mathbf{n}}(\theta,\varphi)$, and S is the rotation transforming the X_3 axis to the direction $\hat{\mathbf{n}}$. Since $R(\mathbf{e}_3,\omega)$ can be expanded in the following form

$$R(\mathbf{e}_3,\omega) = \exp(-i\omega T_3) = 1 - T_3^2 + T_3^2\cos\omega - iT_3\sin\omega,$$

we have

$$R(\hat{\mathbf{n}},\omega) = SR(\mathbf{e}_3,\omega)S^{-1}$$
$$= 1 - ST_3^2S^{-1} + ST_3^2S^{-1}\cos\omega - iST_3S^{-1}\sin\omega$$
$$= 1 - (\hat{\mathbf{n}}\cdot\mathbf{T})^2 + (\hat{\mathbf{n}}\cdot\mathbf{T})^2\cos\omega - i(\hat{\mathbf{n}}\cdot\mathbf{T})\sin\omega.$$

3. Check the following formulas in the group **O** [see the notation given in Problem 12 of Chapter 3] in terms of the homomorphism of SU(2) onto SO(3):

$$T_z R_1 = S_3, \qquad T_z T_x = R_1, \qquad R_1 R_2 = R_3^2.$$

Solution. Since the elements in SU(2) are two-dimensional matrices, the homomorphism of SU(2) onto SO(3) can be used to simplify the calculation for the multiplication of elements in SO(3). The elements T_z, T_x, R_1, R_2, R_3^2 and S_3 in the group **O** correspond to those elements in the group SU(2), where a sign in $u(R)$ is irrelevant to us:

$$u(T_z) = u(\mathbf{e}_3,\pi/2) = \sqrt{1/2}\,\{1 - i\sigma_3\},$$
$$u(T_x) = u(\mathbf{e}_1,\pi/2) = \sqrt{1/2}\,\{1 - i\sigma_1\},$$
$$u(R_1) = u\left[(\mathbf{e}_1 + \mathbf{e}_2 + \mathbf{e}_3)/\sqrt{3}, 2\pi/3\right] = (1/2)\,\{1 - i(\sigma_1 + \sigma_2 + \sigma_3)\},$$
$$u(R_2) = u\left[(\mathbf{e}_1 - \mathbf{e}_2 - \mathbf{e}_3)/\sqrt{3}, 2\pi/3\right] = (1/2)\,\{1 - i(\sigma_1 - \sigma_2 - \sigma_3)\},$$

$$u(R_3^2) = u\left[(-\mathbf{e}_1 - \mathbf{e}_2 + \mathbf{e}_3)/\sqrt{3}, 4\pi/3\right] = -(1/2)\left\{1 - i\left(\sigma_1 + \sigma_2 - \sigma_3\right)\right\},$$
$$u(S_3) = u\left[(\mathbf{e}_2 + \mathbf{e}_3)/\sqrt{2}, \pi\right] = -i\sqrt{1/2}\left(\sigma_2 + \sigma_3\right).$$

Now, replacing the elements R of $\mathbf{O}$ in the multiplication formulas with the corresponding elements $u(R)$ of $SU(2)$, we are able to check whether the multiplication formulas are correct up to a sign.

$$u(T_z)u(R_1) = \sqrt{1/2}\left\{1 - i\sigma_3\right\}(1/2)\left\{1 - i\left(\sigma_1 + \sigma_2 + \sigma_3\right)\right\}$$
$$= \sqrt{1/2}\left\{-i\sigma_2 - i\sigma_3\right\} = u(S_3),$$
$$u(T_z)u(T_x) = \sqrt{1/2}\left\{1 - i\sigma_3\right\}\sqrt{1/2}\left\{1 - i\sigma_1\right\}$$
$$= (1/2)\left\{1 - i\left(\sigma_1 + \sigma_2 + \sigma_3\right)\right\} = u(R_1),$$
$$u(R_1)u(R_2) = (1/2)\left\{1 - i\left(\sigma_1 + \sigma_2 + \sigma_3\right)\right\}(1/2)\left\{1 - i\left(\sigma_1 - \sigma_2 - \sigma_3\right)\right\}$$
$$= (1/2)\left\{1 - i\left(\sigma_1 + \sigma_2 - \sigma_3\right)\right\} = -u(R_3^2).$$

4. Prove the following formulas for the group $\mathbf{I}$ in terms of the homomorphism of $SU(2)$ onto $SO(3)$

$$R_6 = S_1 T_0^2 S_1 T_0^{-1}, \qquad S_{12} = T_0 S_1 T_0^3 R_6,$$

which were used in Problem 21 of Chapter 3.

Solution. According to the notations given in Problem 18 of Chapter 3, we are able to calculate the matrices $u(R)$ in $SU(2)$ which corresponds to elements of $\mathbf{I}$, where a sign in $u(R)$ is irrelevant to us.

$$u(S_{12}) = -i\sigma_2, \qquad u(T_0) = \mathbf{1}\cos(\pi/5) - i\sigma_3\sin(\pi/5),$$
$$u(S_1) = -i\left(\hat{\mathbf{m}} \cdot \vec{\sigma}\right), \qquad u(R_6) = \mathbf{1}\cos(\pi/3) - i\left(\hat{\mathbf{n}} \cdot \vec{\sigma}\right)\sin(\pi/3),$$

where the polar angle and azimuthal angle of $\hat{\mathbf{n}}$ are θ_3 and $\pi/5$, and the polar angle and azimuthal angle of $\hat{\mathbf{m}}$ are θ_4 and zero, $\tan\theta_3 = 3 + \sqrt{5}$, and $\tan\theta_4 = (\sqrt{5} - 1)/2$. Thus,

$$u(S_{12}) = \begin{pmatrix} 0 & -1 \\ 1 & 0 \end{pmatrix}, \qquad u(T_0) = \begin{pmatrix} e^{-i\pi/5} & 0 \\ 0 & e^{i\pi/5} \end{pmatrix},$$

$$u(S_1) = \frac{-i}{2\sqrt{5}}\begin{pmatrix} \sqrt{10 + 2\sqrt{5}} & \sqrt{10 - 2\sqrt{5}} \\ \sqrt{10 - 2\sqrt{5}} & -\sqrt{10 + 2\sqrt{5}} \end{pmatrix},$$

$$u(R_6) = \frac{1}{2}\begin{pmatrix} 1 - i\sqrt{1 - 2/\sqrt{5}} & -i\sqrt{2 + 2/\sqrt{5}}e^{-i\pi/5} \\ -i\sqrt{2 + 2/\sqrt{5}}e^{i\pi/5} & 1 + i\sqrt{1 - 2/\sqrt{5}} \end{pmatrix}.$$

Let $A = S_1 \left(T_0^2 S_1 T_0^{-1} \right)$ and $B = \left(T_0^2 S_1 T_0^{-1} \right) S_{12}$. We are going to calculate the matrix entries of $u(A)$ and $u(B)$ to show that they are both equal to $-u(R_6)$. Because

$$u(T_0^2 S_1 T_0^{-1}) = \frac{-i}{2\sqrt{5}} \begin{pmatrix} \sqrt{10 + 2\sqrt{5}}e^{-i\pi/5} & \sqrt{10 - 2\sqrt{5}}e^{-i3\pi/5} \\ \sqrt{10 - 2\sqrt{5}}e^{i3\pi/5} & -\sqrt{10 + 2\sqrt{5}}e^{i\pi/5} \end{pmatrix},$$

$$\cos(\pi/5) = \frac{\sqrt{5} + 1}{4}, \qquad \sin(\pi/5) = \sqrt{\frac{5 - \sqrt{5}}{8}},$$

$$\cos(3\pi/5) = -\frac{\sqrt{5} - 1}{4}, \qquad \sin(2\pi/5) = \sin(3\pi/5) = \sqrt{\frac{5 + \sqrt{5}}{8}},$$

we have

$$A_{11} = A_{44}^* = -\frac{1}{20} \left[(10 + 2\sqrt{5})e^{-i\pi/5} + (10 - 2\sqrt{5})e^{i3\pi/5} \right]$$

$$= -\frac{1}{20} \left[(10 + 2\sqrt{5})\frac{\sqrt{5} + 1}{4} - (10 - 2\sqrt{5})\frac{\sqrt{5} - 1}{4} \right.$$

$$\left. - i(10 + 2\sqrt{5})\sqrt{\frac{5 - \sqrt{5}}{8}} + i(10 - 2\sqrt{5})\sqrt{\frac{5 + \sqrt{5}}{8}} \right]$$

$$= -\frac{1}{2} + \frac{i}{40\sqrt{10 - 2\sqrt{5}}} \left[(10 + 2\sqrt{5})(5 - \sqrt{5}) - (10 - 2\sqrt{5})2\sqrt{5} \right]$$

$$= -\frac{1}{2} + i\frac{(3 - \sqrt{5})\sqrt{5 - 2\sqrt{5}}}{2\sqrt{10 - 2\sqrt{5}}\sqrt{5 - 2\sqrt{5}}}$$

$$= -\frac{1}{2} + i\frac{\sqrt{5 - 2\sqrt{5}}}{2\sqrt{5}} = -u(R_6)_{11}$$

$$A_{12} = -A_{21}^* = \frac{-1}{\sqrt{5}} \left[e^{-i3\pi/5} - e^{i\pi/5} \right] = \frac{ie^{-i\pi/5}}{\sqrt{5}} \sqrt{\frac{5 + \sqrt{5}}{2}} = -u(R_6)_{12}$$

$$B_{11} = B_{44}^* = -i\sqrt{\frac{10 - 2\sqrt{5}}{20}} e^{-i3\pi/5}$$

$$= -i\sqrt{\frac{10 - 2\sqrt{5}}{20}} \left(-\frac{\sqrt{5} - 1}{4} - i\sqrt{\frac{5 + \sqrt{5}}{8}} \right) = -\frac{1}{2} + i\sqrt{\frac{5 - 2\sqrt{5}}{20}}$$

$$= -u(R_6)_{11}$$

$$B_{12} = -B_{21}^* = i\sqrt{\frac{10 + 2\sqrt{5}}{20}} e^{-i\pi/5} = -u(R_6)_{12},$$

where $\sqrt{10 - 2\sqrt{5}}\sqrt{5 - 2\sqrt{5}} = 3\sqrt{5} - 5$ is used.

4.2 Inequivalent and Irreducible Representations

★ In the three-dimensional space, any rotation can be expressed as a product of three rotations around the coordinate axes

$$R(\alpha, \beta, \gamma) = R(\mathbf{e}_3, \alpha)R(\mathbf{e}_2, \beta)R(\mathbf{e}_3, \gamma)$$

$$= \begin{pmatrix} c_\alpha c_\beta c_\gamma - s_\alpha s_\gamma & -c_\alpha c_\beta s_\gamma - s_\alpha c_\gamma & c_\alpha s_\beta \\ s_\alpha c_\beta c_\gamma + c_\alpha s_\gamma & -s_\alpha c_\beta s_\gamma + c_\alpha c_\gamma & s_\alpha s_\beta \\ -s_\beta c_\gamma & s_\beta s_\gamma & c_\beta \end{pmatrix}, \tag{4.9}$$

where $c_\alpha = \cos\alpha$, $s_\alpha = \sin\alpha$, etc. Three angles α, β and γ are called the Euler angles. There are two methods for calculating the Euler angles. One is based on the matrix form of R. The third column of R is a unit vector, where its polar angle is β and its azimuthal angle is α. The third row of R is also a unit vector, where its polar angle is β and its azimuthal angle is $(\pi - \gamma)$. The other is based on the relative position of the coordinate frames before and after the rotation R. If R transforms the coordinate frame K to K', then in the K coordinate frame, the polar angle and the azimuthal angle of the Z' axis of K' are β and α, respectively. In the K' coordinate frame, the polar angle and the azimuthal angle of the Z axis of K are β and $(\pi - \gamma)$. The domain of definition for the Euler angles in SO(3) are

$$-\pi \le \alpha \le \pi, \qquad 0 \le \beta \le \pi, \qquad -\pi \le \gamma \le \pi. \tag{4.10}$$

For the SU(2) group, $-2\pi \le \gamma \le 2\pi$. When $\beta = 0$ or π, only one parameter between α and γ is independent.

★ For the SU(2) group, its inequivalent irreducible representation can be designated by a half-integer $j = 0, 1/2, 1, 3/2, \ldots$,

$$D^j_{\nu\mu}(\alpha, \beta, \gamma) = \left\{ D^j(\mathbf{e}_3, \alpha)D^j(\mathbf{e}_2, \beta)D^j(\mathbf{e}_3, \gamma) \right\}_{\nu\mu} = e^{-i\nu\alpha} d^j_{\nu\mu}(\beta) e^{-i\mu\gamma},$$

$$d^j_{\nu\mu}(\omega) \equiv D^j_{\nu\mu}(\mathbf{e}_2, \omega)$$

$$= \sum_n \frac{(-1)^n \{(j+\nu)!(j-\nu)!(j+\mu)!(j-\mu)!\}^{1/2}}{(j+\nu-n)!(j-\mu-n)!n!(n-\nu+\mu)!}$$

$$\cdot \{\cos(\omega/2)\}^{2j+\nu-\mu-2n} \{\sin(\omega/2)\}^{2n-\nu+\mu},$$

$$n : \quad \max\begin{pmatrix} 0 \\ \nu - \mu \end{pmatrix}, \quad \cdots, \quad \min\begin{pmatrix} j+\nu \\ j-\mu \end{pmatrix}.$$

$$\tag{4.11}$$

The matrix $d^j(\omega)$ satisfies the following symmetric relations:

$$(-1)^{\mu-\nu}d^j_{\nu\mu}(\omega) = d^j_{\nu\mu}(-\omega) = d^j_{\mu\nu}(\omega) = d^j_{-\nu-\mu}(\omega),$$

$$d^j_{\nu\mu}(\pi) = (-1)^{j-\mu}\delta_{\nu(-\mu)}, \quad d^j_{\nu\mu}(2\pi) = (-1)^{2j}\delta_{\nu\mu}, \qquad (4.12)$$

$$d^j_{\nu\mu}(\pi - \omega) = (-1)^{j-\mu}d^j_{-\nu\mu}(\omega) = (-1)^{j+\nu}d^j_{\nu-\mu}(\omega).$$

The character of the element $u(\hat{\mathbf{n}}, \omega)$ in the representation D^j is

$$\chi^j(\omega) = \sum_{\mu=-j}^{j} e^{-i\mu\omega} = \sum_{\mu=-j}^{j} e^{i\mu\omega} = \frac{\sin\{(j + 1/2)\omega\}}{\sin(\omega/2)}. \qquad (4.13)$$

★ D^j is a unitary representation of SU(2) with dimension $(2j + 1)$. When j is an integer, D^j is a single-valued representation of SO(3) and a non-faithful representation of SU(2). It is equivalent to a real representation by a similarity transformation. When j is a half-odd-integer, D^j is a double-valued representation of SO(3) and a faithful representation of SU(2). It is a self-conjugate representation, but cannot be transformed into a real representation by a similarity transformation. $D^0(u) = 1$ is the identical representation, $D^{1/2}(u) = u$ is the self-representation of the SU(2) group. $D^1(u)$ is equivalent to the self-representation of the SO(3) group:

$$M^{-1}R(\alpha, \beta, \gamma)M = D^1(\alpha, \beta, \gamma), \quad M = \frac{1}{\sqrt{2}} \begin{pmatrix} -1 & 0 & 1 \\ -i & 0 & -i \\ 0 & \sqrt{2} & 0 \end{pmatrix}. \qquad (4.14)$$

The generators in the representation D^j are

$$\left(I^j_+\right)_{\nu\mu} = \left(I^j_-\right)_{\mu\nu} = \delta_{(\nu-1)\mu}\Gamma^j_\nu = \delta_{\nu(\mu+1)}\Gamma^j_{-\mu},$$

$$\left(I^j_3\right)_{\nu\mu} = \mu\delta_{\nu\mu}, \quad I^j_\pm = I^j_1 \pm iI^j_2, \qquad (4.15)$$

$$\Gamma^j_\nu = \Gamma^j_{-\nu+1} = \{(j + \nu)(j - \nu + 1)\}^{1/2}.$$

Letting $c = \cos(\beta/2)$ and $s = \sin(\beta/2)$, we have

$$d^0(\beta) = 1, \quad d^{1/2}(\beta) = \begin{pmatrix} c & -s \\ s & c \end{pmatrix},$$

$$d^1(\beta) = \begin{pmatrix} c^2 & -\sqrt{2}cs & s^2 \\ \sqrt{2}cs & c^2 - s^2 & -\sqrt{2}cs \\ s^2 & \sqrt{2}cs & c^2 \end{pmatrix},$$

$$d^{3/2}(\beta) = \begin{pmatrix} c^3 & -\sqrt{3}c^2 s & \sqrt{3}cs^2 & -s^3 \\ \sqrt{3}c^2 s & c^3 - 2cs^2 & -2c^2 s + s^3 & \sqrt{3}cs^2 \\ \sqrt{3}cs^2 & 2c^2 s - s^3 & c^3 - 2cs^2 & -\sqrt{3}c^2 s \\ s^3 & \sqrt{3}cs^2 & \sqrt{3}c^2 s & c^3 \end{pmatrix},$$

$$d^j_{\mu j}(\beta) = d^j_{-j-\mu}(\beta) = (-1)^{j-\mu} d^j_{j\mu}(\beta) = (-1)^{j-\mu} d^j_{-\mu-j}(\beta)$$

$$= \left\{ \frac{(2j)!}{(j+\mu)!(j-\mu)!} \right\}^{1/2} c^{j+\mu} s^{j-\mu},$$

$$d^\ell_{00}(\beta) = \sum_{n=0}^{\ell} (-1)^n \left\{ \frac{\ell! c^{\ell-n} s^n}{n!(\ell-n)!} \right\}^2. \tag{4.16}$$

5. Calculate the Euler angles for the following transformation matrices R, S and T, and write their representation matrices in D^j of SO(3):

a) $R(\alpha, \beta, \gamma) = \dfrac{1}{4} \begin{pmatrix} -\sqrt{3} - 2 & \sqrt{3} - 2 & -\sqrt{2} \\ \sqrt{3} - 2 & -\sqrt{3} - 2 & \sqrt{2} \\ -\sqrt{2} & \sqrt{2} & 2\sqrt{3} \end{pmatrix},$

b) $S(\alpha, \beta, \gamma) = \dfrac{1}{8} \begin{pmatrix} \sqrt{6} + 2\sqrt{3} & 3\sqrt{2} - 2 & 2\sqrt{6} \\ \sqrt{2} - 6 & \sqrt{6} + 2\sqrt{3} & 2\sqrt{2} \\ -2\sqrt{2} & -2\sqrt{6} & 4\sqrt{2} \end{pmatrix},$

c) $T(\alpha, \beta, \gamma) = \dfrac{1}{2} \begin{pmatrix} \sqrt{3} & -1 & 0 \\ -1 & -\sqrt{3} & 0 \\ 0 & 0 & -2 \end{pmatrix}.$

Solution. a) From the third column of R, we have

$$\cos \beta = \sqrt{3}/2, \quad \sin \beta = 1/2, \quad \beta = \pi/6,$$
$$\cos \alpha = -(\sqrt{2}/4) \cdot 2 = -\sqrt{1/2},$$
$$\sin \alpha = (\sqrt{2}/4) \cdot 2 = \sqrt{1/2}, \qquad \alpha = 3\pi/4.$$

From the third row of R, we have

$$\cos(\pi - \gamma) = -(\sqrt{2}/4) \cdot 2 = -\sqrt{1/2},$$
$$\sin(\pi - \gamma) = (\sqrt{2}/4) \cdot 2 = \sqrt{1/2},$$
$$\pi - \gamma = 3\pi/4, \qquad \gamma = \pi/4,$$
$$R(\alpha, \beta, \gamma) = R(3\pi/4, \pi/6, \pi/4).$$

Hence,

$$D^j_{\nu\mu}(R) = e^{-i3\nu\pi/4}d^j_{\nu\mu}(\pi/6)e^{-i\mu\pi/4}.$$

b) From the third column of S, we have

$$\cos\beta = \sqrt{2}/2, \quad \sin\beta = \sqrt{2}/2, \quad \beta = \pi/4,$$
$$\cos\alpha = (\sqrt{6}/4)\cdot\sqrt{2} = \sqrt{3}/2,$$
$$\sin\alpha = (\sqrt{2}/4)\cdot\sqrt{2} = 1/2, \quad\quad \alpha = \pi/6.$$

From the third row of S, we have

$$\cos(\pi - \gamma) = -(\sqrt{2}/4)\cdot\sqrt{2} = -1/2,$$
$$\sin(\pi - \gamma) = -(\sqrt{6}/4)\cdot\sqrt{2} = -\sqrt{3}/2,$$
$$\pi - \gamma = 4\pi/3, \quad\quad \gamma = -\pi/3,$$
$$S(\alpha,\beta,\gamma) = S(\pi/6, \pi/4, -\pi/3).$$

Hence,

$$D^j_{\nu\mu}(S) = e^{-i\nu\pi/6}d^j_{\nu\mu}(\pi/4)e^{i\mu\pi/3}.$$

c) From the third column or the third row of T, we have $\beta = \pi$, but α and β are undetermined. In fact, when $\beta = \pi$, only $(\alpha - \gamma)$ is independent. For definiteness, we may choose $\alpha = 0$ and calculate $R(\mathbf{e}_2, \pi)^{-1}T$

$$R(\mathbf{e}_2, \pi)^{-1}T = \begin{pmatrix} -1 & 0 & 0 \\ 0 & 1 & 0 \\ 0 & 0 & -1 \end{pmatrix}\frac{1}{2}\begin{pmatrix} \sqrt{3} & -1 & 0 \\ -1 & -\sqrt{3} & 0 \\ 0 & 0 & -2 \end{pmatrix}$$

$$= \frac{1}{2}\begin{pmatrix} -\sqrt{3} & 1 & 0 \\ -1 & -\sqrt{3} & 0 \\ 0 & 0 & 2 \end{pmatrix} = R(\mathbf{e}_3, -5\pi/6).$$

Hence,

$$T(\alpha,\beta,\gamma) = T(0, \pi, -5\pi/6)$$
$$D^j_{\nu\mu}(T) = d^j_{\nu\mu}(\pi)e^{i5\mu\pi/6} = (-1)^{j-\mu}\delta_{\nu(-\mu)}e^{i5\mu\pi/6}.$$

6. Calculate the Euler angles for the following rotations R, S and T, and write their representation matrices in D^j of SO(3).

a) R is a rotation around the direction $\hat{\mathbf{n}} = \mathbf{e}_1 \sin\theta + \mathbf{e}_3 \cos\theta$ through an acute angle θ;

b) S is a rotation around the direction $\hat{\mathbf{n}} = (\mathbf{e}_1 + \mathbf{e}_2 + \mathbf{e}_3)/\sqrt{3}$ through $2\pi/3$;

c) T is a rotation around the direction $\hat{\mathbf{n}} = (\mathbf{e}_1 + \mathbf{e}_2)/\sqrt{2}$ through π.

Solution. If a rotation R transforms the coordinate frame K to the frame K', the polar angle and the azimuthal angle of the Z' axis of K' in the coordinate frame K are β and α, respectively. In K', the polar angle and the azimuthal angle of the Z axis of K are β and $(\pi - \gamma)$, respectively.

a) The X' axis and the Z' axis of K' are both in the XZ plane of K, and the angle between two Z axes is 2θ, so $\beta = 2\theta$ and $\alpha = \pi - \gamma = 0$,

$$D^j_{\nu\mu}(R) = D^j_{\nu\mu}(0, 2\theta, \pi) = d^j_{\nu\mu}(2\theta)e^{-i\mu\pi}.$$

b) The X', Y' and Z' axes of K' are located in the Y, Z and X axes of K, respectively, so $\beta = \pi/2$, $\alpha = 0$ and $\pi - \gamma = \pi/2$,

$$D^j_{\nu\mu}(S) = D^j_{\nu\mu}(0, \pi/2, \pi/2) = d^j_{\nu\mu}(\pi/2)e^{-i\mu\pi/2}.$$

c) The Z' axes of K' is along the negative Z axis of K, so that $\beta = \pi$ and only $(\alpha - \gamma)$ is independent. Since the Y' axis of K' coincides with the X axis of K, we can choose $\gamma = 0$ and obtain $\alpha = -\pi/2$.

$$D^j_{\nu\mu}(T) = D^j_{\nu\mu}(-\pi/2, \pi, 0) = e^{i\nu\pi/2}d^j_{\nu\mu}(\pi) = (-1)^{j+\nu}\delta_{(-\nu)\mu}e^{i\nu\pi/2}.$$

7. Calculate the Euler angles for the rotations T_0, T_2, R_1, R_2, R_6, S_1, S_2, S_6, S_{11}, and S_{12} in the icosahedron group **I**, where the notations for the elements are given in Problem 18 of Chapter 3.

Solution. The key for solving this problem is to determine the directions of the coordinate axes after rotation.

The Euler angles of T_0 are obviously $\beta = 0$ and $\alpha + \gamma = 2\pi/5$. One may choose $\alpha = 0$ for definiteness.

After the rotation T_2, the Z axis is rotated to the direction OA_1, so we have $\alpha = 0$ and $\beta = \theta_1$. The Y axis is rotated to the direction pointing from O to the central point D_1 of the edge A_2A_3. The angle between two planes A_1OA_0 and A_1OD_1 is $3\pi/10$. So, $\pi - \gamma = \pi/2 + 3\pi/10$, and $\gamma = \pi/5$.

After the rotation R_1, the Z axis is rotated to the direction of OA_1, so we have $\alpha = 0$ and $\beta = \theta_1$. The Y axis is rotated to the direction pointing from the origin O to the central point D_2 of the edge A_0A_4. The angle between two planes A_1OA_0 and A_1OD_2 is $\pi/10$. So, $\pi - \gamma = \pi/2 - \pi/10$, and $\gamma = 3\pi/5$.

After the rotation R_2, the Z axis is rotated to the direction of OA_2, so $\alpha = 2\pi/5$ and $\beta = \theta_1$. The Y axis is rotated to the direction pointing from O to the central point D_3 of the edge A_3A_4. The angle between two planes A_2OA_0 and A_2OD_3 is $3\pi/10$. So, $\pi - \gamma = \pi/2 + 3\pi/10$, and $\gamma = \pi/5$.

After the rotation R_6, the Z axis is rotated to the direction OB_3, so we have $\alpha = -\pi/5$ and $\beta = 2\theta_5$. The Y axis is rotated to the direction pointing from O to the central point Q of the edge A_0A_1. The angle between two planes B_3OA_0 and B_3OQ is $\pi/10$. So, $\pi - \gamma = \pi/2 + \pi/10$, and $\gamma = 2\pi/5$.

From the result of the preceding problem, the Euler angles for S_1 are $\beta = 2\theta_4$, $\alpha = 0$ and $\gamma = \pi$.

After the rotation S_2, the Z axis is rotated to the direction OA_2, so we have $\alpha = 2\pi/5$ and $\beta = \theta_1$. The Y axis is rotated to the direction pointing from O to the central point D_4 of the edge A_0A_5. The angle between two planes A_2OA_0 and A_2OD_4 is $\pi/10$. So, $\pi - \gamma = \pi/2 - \pi/10$, and $\gamma = 3\pi/5$.

After the rotation S_6, the Z axis is rotated to the direction OB_4, so we have $\alpha = \pi/5$ and $\beta = 2\theta_5$. The Y axis is rotated to the direction pointing from O to the central point D_5 of the edge A_1A_5. The angle between two planes B_4OA_0 and B_4OD_5 is $3\pi/10$. So, $\pi - \gamma = \pi/2 - 3\pi/10$, and $\gamma = 4\pi/5$.

After the rotation S_{11}, the Z axis is rotated to the direction of the negative Z axis, so we have $\beta = \pi$ and only $\alpha - \gamma$ is independent. We may choose $\gamma = 0$. The Y axis is rotated to the direction pointing from O to the central point D_6 of the edge B_3A_5. The angle between OD_6 and the Y axis in the XY plane is $4\pi/5$. So, $\alpha = -4\pi/5$.

S_{12} is a rotation around the Y axis through π, so its Euler angles are $\beta = \pi$ and $\alpha = \gamma = 0$.

8. Express the representation matrix $D^j(\hat{\mathbf{n}}, \omega)$ of a rotation around the direction $\hat{\mathbf{n}}(\theta, \varphi)$ through ω in terms of $D^j(\mathbf{e}_3, \alpha)$ and $d^j(\beta)$.

Solution. Letting S be a rotation with the Euler angles $\alpha = \varphi$, $\beta = \theta$ and $\gamma = 0$, we have $SR(\mathbf{e}_3, \omega)S^{-1} = R(\hat{\mathbf{n}}, \omega)$,

$$D^j(\hat{\mathbf{n}}, \omega) = D^j(S)D^j(\mathbf{e}_3, \omega)D^j(S)^{-1}$$

$$= D^j(\mathbf{e}_3, \varphi)d^j(\theta)D^j(\mathbf{e}_3, \omega)d^j(\theta)^{-1}D^j(\mathbf{e}_3, -\varphi),$$

$$D^j_{\nu\mu}(\hat{\mathbf{n}}, \omega) = \sum_{\rho} e^{-i\nu\varphi}d^j_{\nu\rho}(\theta)e^{-i\rho\omega}d^j_{\mu\rho}(\theta)e^{i\mu\varphi}.$$

9. Calculate all the matrix entries $d^j_{\nu\mu}(\omega)$ by Eq. (4.11) where $j = 1/2, 1$, 3/2, 2, 5/2 and 3.

Solution. Due to the symmetric relation (4.12) of $d^j(\omega)$, we only need to calculate the matrix entries $d^j_{\nu\mu}(\omega)$ with $0 \le \nu \le \mu$. The remaining matrix entries can be calculated by the following formulas

$$d^j_{(-\nu)(-\mu)}(\omega) = d^j_{\mu\nu}(\omega) = (-1)^{\nu-\mu} d^j_{\nu\mu}(\omega),$$

$$d^j_{(-\nu)\mu}(\omega) = (-1)^{\mu+\nu} d^j_{\nu(-\mu)}(\omega) = (-1)^{j-\mu} d^j_{\nu\mu}(\pi - \omega).$$

It is useful to obtain the formulas for $d^j_{\nu\mu}(\omega)$ with $\mu = j,\ j-1,\ j-2$, and 0 from Eq. (4.16):

$$d^j_{\nu j}(\omega) = \left\{ \frac{(2j)!}{(j+\nu)!(j-\nu)!} \right\}^{1/2} c^{j+\nu} s^{j-\nu},$$

$$d^j_{\nu(j-1)}(\omega) = \left\{ \frac{(2j-1)!(j-\nu)}{(j+\nu)!(j-\nu-1)!} \right\}^{1/2} c^{j+\nu+1} s^{j-\nu-1}$$
$$- \left\{ \frac{(2j-1)!(j+\nu)}{(j+\nu-1)!(j-\nu)!} \right\}^{1/2} c^{j+\nu-1} s^{j-\nu+1},$$

$$d^j_{\nu(j-2)}(\omega) = \left\{ \frac{(2j-2)!(j-\nu)(j-\nu-1)}{2(j+\nu)!(j-\nu-2)!} \right\}^{1/2} c^{j+\nu+2} s^{j-\nu-2}$$
$$- \left\{ \frac{(2j-2)!2(j+\nu)(j-\nu)}{(j+\nu-1)!(j-\nu-1)!} \right\}^{1/2} c^{j+\nu} s^{j-\nu}$$
$$+ \left\{ \frac{(2j-2)!(j+\nu)(j+\nu-1)}{2(j+\nu-2)!(j-\nu)!} \right\}^{1/2} c^{j+\nu-2} s^{j-\nu+2},$$

$$d^\ell_{00}(\beta) = \sum_{n=0}^{\ell} (-1)^n \left\{ \frac{\ell! c^{\ell-n} s^n}{n!(\ell-n)!} \right\}^2.$$

For convenience, we use the brief notations $c = \cos(\omega/2)$ and $s = \sin(\omega/2)$. $d^j_{\nu\mu}(\pi - \omega)$ can be calculated from $d^j_{\nu\mu}(\omega)$ by interchanging c and s.

$$d^{1/2}_{(1/2)(1/2)}(\omega) = c, \qquad\qquad d^1_{11}(\omega) = c^2,$$
$$d^1_{01}(\omega) = \sqrt{2}cs, \qquad\qquad d^1_{00}(\omega) = c^2 - s^2,$$
$$d^{3/2}_{(3/2)(3/2)}(\omega) = c^3, \qquad\qquad d^{3/2}_{(1/2)(3/2)}(\omega) = \sqrt{3}c^2 s,$$
$$d^{3/2}_{(1/2)(1/2)}(\omega) = c^3 - 2cs^2, \qquad d^2_{22}(\omega) = c^4,$$
$$d^2_{12}(\omega) = 2c^3 s, \qquad\qquad d^2_{02}(\omega) = \sqrt{6}c^2 s^2,$$
$$d^2_{11}(\omega) = c^4 - 3c^2 s^2, \qquad\quad d^2_{01}(\omega) = \sqrt{6}cs\left(c^2 - s^2\right),$$
$$d^2_{00}(\omega) = c^4 - 4c^2 s^2 + s^4, \qquad d^{5/2}_{(5/2)(5/2)}(\omega) = c^5,$$

$$d^{5/2}_{(3/2)(5/2)}(\omega) = \sqrt{5}c^4 s,$$

$$d^{5/2}_{(1/2)(5/2)}(\omega) = \sqrt{10}c^3 s^2,$$

$$d^{5/2}_{(3/2)(3/2)}(\omega) = c^5 - 4c^3 s^2,$$

$$d^{5/2}_{(1/2)(3/2)}(\omega) = \sqrt{2}\left(2c^4 s - 3c^2 s^3\right),$$

$$d^{5/2}_{(1/2)(1/2)}(\omega) = c^5 - 6c^3 s^2 + 3cs^4, \quad d^3_{33}(\omega) = c^6,$$

$$d^3_{23}(\omega) = \sqrt{6}c^5 s, \qquad\qquad d^3_{13}(\omega) = \sqrt{15}c^4 s^2,$$

$$d^3_{03}(\omega) = 2\sqrt{5}c^3 s^3, \qquad\qquad d^3_{22}(\omega) = c^6 - 5c^4 s^2,$$

$$d^3_{12}(\omega) = \sqrt{10}\left(c^5 s - 2c^3 s^3\right), \qquad d^3_{02}(\omega) = \sqrt{30}\left(c^4 s^2 - c^2 s^4\right),$$

$$d^3_{11}(\omega) = c^6 - 8c^4 s^2 + 6c^2 s^4,$$

$$d^3_{01}(\omega) = 2\sqrt{3}\left(c^5 s - 3c^3 s^3 + cs^5\right),$$

$$d^3_{00}(\omega) = c^6 - 9c^4 s^2 + 9c^2 s^4 - s^6.$$

10. Reduce the subduced representation from the irreducible representation D^3 of SO(3) with respect to the subgroup D_3 and find the similarity transformation matrix.

Solution. In terms of the formula $\chi^3(\omega) = \sin(7\omega/2)/\sin(\omega/2)$ for the representation D^3 of SO(3), the characters of the elements with the rotational angles 0 (the identity E), $2\pi/3$ and π are calculated respectively to be $\chi^3(E) = 7$, $\chi^3(2\pi/3) = 1$ and $\chi^3(\pi) = -1$. The subduced representation from D^3 with respect to the subgroup D_3 is reducible, and the multiplicities a_j of the irreducible representations A_1, A_2 and E in it can be calculated by Eq. (3.11):

$$a_{A_1} = (1/6)\left\{1 \times 7 \times 1 + 2 \times 1 \times 1 + 3 \times (-1) \times 1\right\} = 1,$$

$$a_{A_2} = (1/6)\left\{1 \times 7 \times 1 + 2 \times 1 \times 1 + 3 \times (-1) \times (-1)\right\} = 2,$$

$$a_E = (1/6)\left\{1 \times 7 \times 2 + 2 \times 1 \times (-1) + 3 \times (-1) \times 0\right\} = 2.$$

Namely,

$$D^3(D_3) \simeq D^{A_1}(D_3) \oplus 2D^{A_2}(D_3) \oplus 2D^E(D_3).$$

The dimensions of the representations on two sides of the equation are both 7.

In order to obtain the similarity transformation matrix for reduction, we need the matrix forms of the generators D and A of the subgroup D_3 before and after the similarity transformation. The forms after the transformation have been given in Problem 11 of Chapter 3. The forms before the transformation depend upon the Euler angles of two generators:

$$D^3_{\nu\mu}(A) = D^3_{\nu\mu}(-\pi/2, \pi, \pi/2) = D^3_{\nu\mu}(0, \pi, \pi) = -\delta_{\nu(-\mu)},$$

$$D^3(D) = D^3(\hat{e}_3, 2\pi/3) = \mathrm{diag}\{1, \omega^2, \omega, 1, \omega^2, \omega, 1\},$$

$$\omega = \exp\{-i2\pi/3\} = \left(-1 - i\sqrt{3}\right)/2, \quad \omega^2 = \left(-1 + i\sqrt{3}\right)/2.$$

$$X^{-1}D^3(A)X = \mathrm{diag}\{1, -1, -1, 1, -1, 1, -1\},$$

$$X^{-1}D^3(D)X = \frac{1}{2}\begin{pmatrix} 2 & 0 & 0 & 0 & 0 & 0 & 0 \\ 0 & 2 & 0 & 0 & 0 & 0 & 0 \\ 0 & 0 & 2 & 0 & 0 & 0 & 0 \\ 0 & 0 & 0 & -1 & -\sqrt{3} & 0 & 0 \\ 0 & 0 & 0 & \sqrt{3} & -1 & 0 & 0 \\ 0 & 0 & 0 & 0 & 0 & -1 & -\sqrt{3} \\ 0 & 0 & 0 & 0 & 0 & \sqrt{3} & -1 \end{pmatrix}.$$

The row index of X is the same as that of D^3, which runs over $3, 2, \cdots, (-3)$. The column index of X is the same as that of the reduced representation, denoted by the representations and their rows after reduction, which runs over A_1, $A_2(1)$, $A_2(2)$, $[E(1)1]$, $[E(1)2]$, $[E(2)1]$ and $[E(2)2]$, where the digits in the bracket are used to distinguish the multiple representations. $D^3(D)$ and $X^{-1}D^3(A)X$ are diagonal, and $D^3(A)$ is a block matrix containing three $-\sigma_1$ and one digit -1. The first three column-matrices of X [designated by A_1, $A_2(1)$ and $A_2(2)$] are the eigenvectors of $D^3(D)$ for the eigenvalue 1 and the eigenvectors of $D^3(A)$ for the eigenvalues 1, -1 and -1, respectively, so that only the 1st, 4th and 7th components (designated by 3, 0, and -3) are not vanishing:

$$X_{3,A_1} = -X_{-3,A_1} = X_{3,A_2(1)} = X_{-3,A_2(1)} = \sqrt{1/2}, \qquad X_{0,A_2(2)} = 1.$$

The 4th and 6th column-matrices of X (designated by $[E(1)1]$ and $[E(2)1]$) are the eigenvectors of $D^3(A)$ for the eigenvalue 1. Due to the orthogonality, their 1st, 4th and 7th components are vanishing and the remaining components can be chosen as

$$X_{2,E(1)1} = -X_{-2,E(1)1} = X_{1,E(2)1} = -X_{-1,E(2)1} = \sqrt{1/2}.$$

Substituting them into the similarity transformation formula on $D^3(D)$, we can calculate the remaining colomns in X. In the calculation we are only interested in the rows where the matrix entries are not vanishing. For the 4th column of X (designated by $[E(1)1]$), we are only interested in the 2nd and 6th rows (designated by 2 and −2):

$$D^3(D)X_{\cdot,E(1)1} = \frac{1}{\sqrt{2}}\begin{pmatrix} \omega^2 \\ -\omega \end{pmatrix}$$

$$= -\frac{1}{2}X_{\cdot,E(1)1} + \frac{\sqrt{3}}{2}X_{\cdot,E(1)2} = -\frac{1}{2}\begin{pmatrix} \sqrt{1/2} \\ -\sqrt{1/2} \end{pmatrix} + \frac{\sqrt{3}}{2}\begin{pmatrix} i\sqrt{1/2} \\ i\sqrt{1/2} \end{pmatrix}.$$

For the 6th column of X (designated by $[E(2)1]$), we are only interested in the 3rd and 5th rows (designated by 1 and −1):

$$D^3(D)X_{\cdot,E(2)1} = \frac{1}{\sqrt{2}}\begin{pmatrix} \omega \\ -\omega^2 \end{pmatrix}$$

$$= -\frac{1}{2}X_{\cdot,E(2)1} + \frac{\sqrt{3}}{2}X_{\cdot,E(2)2} = -\frac{1}{2}\begin{pmatrix} \sqrt{1/2} \\ -\sqrt{1/2} \end{pmatrix} + \frac{\sqrt{3}}{2}\begin{pmatrix} -i\sqrt{1/2} \\ -i\sqrt{1/2} \end{pmatrix}.$$

Finally, we obtain the similarity transformation matrix X:

$$X = \frac{1}{\sqrt{2}}\begin{pmatrix} 1 & 1 & 0 & 0 & 0 & 0 & 0 \\ 0 & 0 & 0 & 1 & i & 0 & 0 \\ 0 & 0 & 0 & 0 & 0 & 1 & -i \\ 0 & 0 & \sqrt{2} & 0 & 0 & 0 & 0 \\ 0 & 0 & 0 & 0 & 0 & -1 & -i \\ 0 & 0 & 0 & -1 & i & 0 & 0 \\ -1 & 1 & 0 & 0 & 0 & 0 & 0 \end{pmatrix}.$$

$X_{\cdot,E(1)2}$ and $X_{\cdot,E(2)2}$ are the eigenvectors of $D^3(A)$ for the eigenvalue −1.

11. Reduce the subduced representations from the irreducible representations D^{20} and D^{18} of SO(3) with respect to the subgroup **I** (the proper symmetry group of the icosahedron), respectively.

Solution. From the character formula for $D^j(SO(3))$

$$\chi^j(\omega) = \sin[(j + 1/2)\omega]/\sin(\omega/2),$$

we calculate the characters for the elements with the rotational angles 0 (the identity E), $2\pi/5$, $4\pi/5$, $2\pi/3$, and π, respectively

$$\chi^{20}(E) = 41, \quad \chi^{20}(2\pi/5) = \frac{\sin(41\pi/5)}{\sin(\pi/5)} = 1,$$

$$\chi^{20}(4\pi/5) = \frac{\sin(82\pi/5)}{\sin(2\pi/5)} = 1, \quad \chi^{20}(2\pi/3) = \frac{\sin(41\pi/3)}{\sin(\pi/3)} = -1$$

$$\chi^{20}(\pi) = \frac{\sin(41\pi/2)}{\sin(\pi/2)} = 1, \quad \chi^{18}(E) = 37,$$

$$\chi^{18}(2\pi/5) = \frac{\sin(37\pi/5)}{\sin(\pi/5)} = \frac{-\sin(2\pi/5)}{\sin(\pi/5)} = -2\cos(\pi/5) = -p^{-1},$$

$$\chi^{18}(4\pi/5) = \frac{\sin(74\pi/5)}{\sin(2\pi/5)} = \frac{\sin(4\pi/5)}{\sin(2\pi/5)} = 2\cos(2\pi/5) = p,$$

$$\chi^{18}(2\pi/3) = \frac{\sin(37\pi/3)}{\sin(\pi/3)} = \frac{\sin(\pi/3)}{\sin(\pi/3)} = 1,$$

$$\chi^{18}(\pi) = \frac{\sin(37\pi/2)}{\sin(\pi/2)} = \frac{\sin(\pi/2)}{\sin(\pi/2)} = 1.$$

The multiplicity a_Γ^j of each irreducible representation Γ of $\mathbf{I}$ can be calculated with the character formula (3.11) and the character table of $\mathbf{I}$ (see Problem 18 in Chapter 3). For the representation D^{20}, we have

$$a_A^{20} = (1/60)\{1 \times 41 \times 1 + 12 \times 1 \times 1 + 12 \times 1 \times 1$$
$$+ 20 \times (-1) \times 1 + 15 \times 1 \times 1\} = 1,$$

$$a_{T_1}^{20} = (1/60)\{1 \times 41 \times 3 + 12 \times 1 \times p^{-1} + 12 \times 1 \times (-p)$$
$$+ 20 \times (-1) \times 0 + 15 \times 1 \times (-1)\} = 2,$$

$$a_{T_2}^{20} = (1/60)\{1 \times 41 \times 3 + 12 \times 1 \times (-p) + 12 \times 1 \times p^{-1}$$
$$+ 20 \times (-1) \times 0 + 15 \times 1 \times (-1)\} = 2,$$

$$a_G^{20} = (1/60)\{1 \times 41 \times 4 + 12 \times 1 \times (-1) + 12 \times 1 \times (-1)$$
$$+ 20 \times (-1) \times 1 + 15 \times 1 \times 0\} = 2,$$

$$a_H^{20} = (1/60)\{1 \times 41 \times 5 + 12 \times 1 \times 0 + 12 \times 1 \times 0$$
$$+ 20 \times (-1) \times (-1) + 15 \times 1 \times 1\} = 4.$$

For the representation D^{18}, we have

$$a_A^{18} = (1/60)\{1 \times 37 \times 1 + 12 \times (-p)^{-1} \times 1 + 12 \times p \times 1$$
$$+ 20 \times 1 \times 1 + 15 \times 1 \times 1\} = 1,$$

$$a_{T_1}^{18} = (1/60)\{1 \times 37 \times 3 + 12 \times (-p)^{-1} \times p^{-1} + 12 \times p \times (-p)$$
$$+ 20 \times 1 \times 0 + 15 \times 1 \times (-1)\} = 1,$$

$$a_{T_2}^{18} = (1/60)\{1 \times 37 \times 3 + 12 \times (-p)^{-1} \times (-p) + 12 \times p \times p^{-1}$$
$$+ 20 \times 1 \times 0 + 15 \times 1 \times (-1)\} = 2,$$

$$a_G^{18} = (1/60) \left\{ 1 \times 37 \times 4 + 12 \times (-p)^{-1} \times (-1) + 12 \times p \times (-1) \right.$$
$$\left. + 20 \times 1 \times 1 + 15 \times 1 \times 0 \right\} = 3,$$
$$a_H^{18} = (1/60) \left\{ 1 \times 37 \times 5 + 12 \times (-p)^{-1} \times 0 + 12 \times p \times 0 \right.$$
$$\left. + 20 \times 1 \times (-1) + 15 \times 1 \times 1 \right\} = 3.$$

Thus,

$$D^{20}(I) \simeq D^A(I) \oplus 2D^{T_1}(I) \oplus 2D^{T_2}(I) \oplus 2D^G(I) \oplus 4D^H(I),$$
$$D^{18}(I) \simeq D^A(I) \oplus D^{T_1}(I) \oplus 2D^{T_2}(I) \oplus 3D^G(I) \oplus 3D^H(I).$$

The dimensions of the representations on both sides of each equation are the same.

12. Reduce the subduced representations from the irreducible representations D^1, D^2 and D^3 of SO(3) with respect to the subgroup **I**, respectively, and calculate the similarity transformation matrices. From the results, calculate the polar angles of the axes in the icosahedron.

Solution. The representation matrices of the generators of **I** in each irreducible representation have been given in Problem 21 of Chapter 3, and the polar angles for all twofold axes, threefold axes and fivefold axes in **I** have been listed in Problem 18 of Chapter 3. In the present Problem we are going to show how to obtain these results. We need to calculate the irreducible bases in the group space of **I** by the projection operator method, just like we have done for the groups **T** and **O** in Problems 22 and 23 of Chapter 3. However, in the calculation we have to use the multiplication table of **I**, which is neglected in this book due to limited space. In the following we will only explain the calculation method by using the results for the multiplication of elements. The reader can find the multiplication table in the textbook *Group Theory for Physicists*, Chap. 3 [Ma]. However, it is suggested to pay more attention to the calculation method introduced here by accepting the multiplication results.

Since the representation matrix of T_0 we choose is diagonal and $T_0^5 = E$, each diagonal matrix entry of T_0 must be a power of $\eta = \exp\{-i2\pi/5\}$. From the character of T_0 in each representation, we have

$$D^A(T_0) = 1, \quad D^{T_1}(T_0) = \mathrm{diag}\left\{\eta, 1, \eta^{-1}\right\}, \quad D^{T_2}(T_0) = \mathrm{diag}\left\{\eta^2, 1, \eta^{-2}\right\},$$
$$D^G(T_0) = \mathrm{diag}\left\{\eta^2, \eta, \eta^{-1}, \eta^{-2}\right\}, \quad D^H(T_0) = \mathrm{diag}\left\{\eta^2, \eta, 1, \eta^{-1}, \eta^{-2}\right\}.$$

This power μ can be used to designate the row and column. Note that μ is module 5, $\mu + 5 = \mu$.

We first calculate the vectors $\Phi_{\mu\nu}^{(a)}$ in the group space of **I** by the projection operator P_μ:

$$P_\mu = \frac{1}{5} \sum_{\rho=-2}^{2} \eta^{-\mu\rho} T_0^\rho, \qquad \Phi_{\mu\nu}^{(a)} = c P_\mu R^{(a)} P_\nu,$$

where c is the normalization factor. $\Phi_{\mu\nu}^{(a)}$ satisfies

$$T_0 \Phi_{\mu\nu}^{(a)} = \eta^\mu \Phi_{\mu\nu}^{(a)}, \qquad \Phi_{\mu\nu}^{(a)} T_0 = \eta^\nu \Phi_{\mu\nu}^{(a)}.$$

Letting $R^{(1)} = E$, we obtain $\Phi_{\mu\mu}^{(1)}$, which is vanishing if $\mu \neq \nu$. Then, letting $R^{(2)}$ be an element in **I** which does not appear in $\Phi_{\mu\mu}^{(1)}$, say S_{11}, we calculate $\Phi_{\mu\nu}^{(2)}$, which is vanishing if $\mu + \nu \neq 0$. In this way, we calculate $\Phi_{\mu\nu}^{(3)}$ and $\Phi_{\mu\nu}^{(4)}$ by $R^{(3)} = S_5$ and $R^{(4)} = S_{10}$.

$$\Phi_{\mu\mu}^{(1)} = \left(E + \eta^{-\mu} T_0 + \eta^{-2\mu} T_0^2 + \eta^{2\mu} T_0^3 + \eta^\mu T_0^4 \right) / \sqrt{5},$$

$$\Phi_{\mu(-\mu)}^{(2)} = \left(S_{11} + \eta^{-\mu} S_{14} + \eta^{-2\mu} S_{12} + \eta^{2\mu} S_{15} + \eta^\mu S_{13} \right) / \sqrt{5},$$

$$\begin{aligned}
\Phi_{\mu\nu}^{(3)} = \{ & \left(S_5 + \eta^{-\mu} R_5^2 + \eta^{-2\mu} T_1^4 + \eta^{2\mu} T_4 + \eta^\mu R_4 \right) \\
& + \eta^{(\mu-\nu)} \left(S_4 + \eta^{-\mu} R_4^2 + \eta^{-2\mu} T_5^4 + \eta^{2\mu} T_3 + \eta^\mu R_3 \right) \\
& + \eta^{2(\mu-\nu)} \left(S_3 + \eta^{-\mu} R_3^2 + \eta^{-2\mu} T_4^4 + \eta^{2\mu} T_2 + \eta^\mu R_2 \right) \\
& + \eta^{-2(\mu-\nu)} \left(S_2 + \eta^{-\mu} R_2^2 + \eta^{-2\mu} T_3^4 + \eta^{2\mu} T_1 + \eta^\mu R_1 \right) \\
& + \eta^{-(\mu-\nu)} \left(S_1 + \eta^{-\mu} R_1^2 + \eta^{-2\mu} T_2^4 + \eta^{2\mu} T_5 + \eta^\mu R_5 \right) \} / 5,
\end{aligned}$$

$$\begin{aligned}
\Phi_{\mu\nu}^{(4)} = \{ & \left(S_{10} + \eta^{-\mu} T_1^3 + \eta^{-2\mu} R_6^2 + \eta^{2\mu} R_9 + \eta^\mu T_5^2 \right) \\
& + \eta^{(\mu-\nu)} \left(S_9 + \eta^{-\mu} T_5^3 + \eta^{-2\mu} R_{10}^2 + \eta^{2\mu} R_8 + \eta^\mu T_4^2 \right) \\
& + \eta^{2(\mu-\nu)} \left(S_8 + \eta^{-\mu} T_4^3 + \eta^{-2\mu} R_9^2 + \eta^{2\mu} R_7 + \eta^\mu T_3^2 \right) \\
& + \eta^{-2(\mu-\nu)} \left(S_7 + \eta^{-\mu} T_3^3 + \eta^{-2\mu} R_8^2 + \eta^{2\mu} R_6 + \eta^\mu T_2^2 \right) \\
& + \eta^{-(\mu-\nu)} \left(S_6 + \eta^{-\mu} T_2^3 + \eta^{-2\mu} R_7^2 + \eta^{2\mu} R_{10} + \eta^\mu T_1^2 \right) \} / 5.
\end{aligned}$$

Now, we consider the case of $\mu = \nu = 1$, where there are three independent bases

$$\Phi_{11}^{(1)} = \left\{ E + \eta^{-1} T_0 + \eta^{-2} T_0^2 + \eta^2 T_0^3 + \eta T_0^4 \right\} / \sqrt{5},$$

$$\Phi_{11}^{(3)} = \sum_{n=1}^{5} \left\{ S_n + \eta^{-1} R_n^2 + \eta^{-2} T_n^4 + \eta^2 T_n + \eta R_n \right\} / 5,$$

$$\Phi_{11}^{(4)} = \sum_{n=1}^{5} \left\{ S_{n+5} + \eta^{-1} T_n^3 + \eta^{-2} R_{n+5}^2 + \eta^2 R_{n+5} + \eta T_n^2 \right\} / 5.$$

Following the method used in Problems 22 and 23 of Chapter 3, we introduce W, which is the sum of elements in the class C_5, to pick up the linear combinations of the vectors $\Phi_{\mu\nu}^{(a)}$ such that the new bases belong to the irreducible representations:

$$W = \sum_{j=0}^{5} \left(T_j + T_j^4\right), \qquad D^\Gamma(W) = \alpha^\Gamma \mathbf{1},$$

$$\alpha^A = 12, \qquad \alpha^{T_1} = 4p^{-1}, \qquad \alpha^{T_2} = -4p, \qquad \alpha^G = -3, \qquad \alpha^H = 0,$$

$$p = (\sqrt{5}-1)/2 = \eta + \eta^{-1}, \qquad p^{-1} = (\sqrt{5}+1)/2 = -\eta^2 - \eta^{-2}.$$

In order to calculate the matrix form of W in the bases $\Phi_{11}^{(a)}$, we only need to calculate the coefficients of three terms of E, S_1 and S_6 in the action of W,

$$W\Phi_{11}^{(1)} = \left\{pE - p^{-1}S_1 + \ldots\right\}/\sqrt{5},$$

$$W\Phi_{11}^{(3)} = \left\{-5p^{-1}E + 2p^{-1}S_1 + p^{-2}S_6 + \ldots\right\}/5,$$

$$W\Phi_{11}^{(4)} = \left\{p^{-2}S_1 - p^2 S_6 + \ldots\right\}/5.$$

The matrix form of W in the bases $\Phi_{11}^{(a)}$ is

$$\begin{pmatrix} p & -\sqrt{5}p^{-1} & 0 \\ -\sqrt{5}p^{-1} & 2p^{-1} & p^{-2} \\ 0 & p^{-2} & -p^2 \end{pmatrix}.$$

The eigenvectors corresponding to the eigenvalues $4p^{-1}$, -3 and 0 respectively are

$$\frac{1}{2}\begin{pmatrix} 1 \\ -p^{-1} \\ -p \end{pmatrix}, \qquad \frac{1}{\sqrt{3}}\begin{pmatrix} 1 \\ 1 \\ -1 \end{pmatrix}, \qquad \frac{1}{\sqrt{12}}\begin{pmatrix} \sqrt{5} \\ p^2 \\ p^{-2} \end{pmatrix}.$$

Thus, we obtain the new basis belonging to the representation T_1, where we are only interested in the terms of E and S_1:

$$\Psi_{11}^{T_1} = \left\{\Phi_{11}^{(1)} - p^{-1}\Phi_{11}^{(3)} - p\Phi_{11}^{(4)}\right\}/2 = E/(2\sqrt{5}) - p^{-1}S_1/10 + \ldots$$

$$S_1\Psi_{11}^{T_1} = -p^{-1}E/10 + \ldots = -p^{-1}\Psi_{11}^{T_1}/\sqrt{5} + \ldots.$$

Solving it, we have $D_{11}^{T_1}(S_1) = -p^{-1}/\sqrt{5}$.

On the other hand, following the method used in the preceding Problem, we reduce the subduced representations from the irreducible representations D^1, D^2 and D^3 of SO(3) with respect to the subgroup I, respectively.

In fact, the subduced representations from D^1 and D^2 with respect to the subgroup $\mathbf{I}$ are the irreducible representations, and that from D^3 is equivalent to the direct sum of $D^{T_2}(I)$ and $D^G(I)$. Due to the forms of $D^\Gamma(T_0)$ we have chosen, we are able to define

$$D^1(R) = D^{T_1}(R), \quad D^2(R) = D^H(R),$$

$$X^{-1}D^3(R)X = D^{T_2}(R) \oplus D^G(R), \quad R \in I.$$

From the explicit form of $d^1(\beta)$ given in Problem 9, we have

$$D^1_{11}(S_1) = D^1_{11}(0, 2\theta_4, \pi) = -c^2 = D^{T_1}_{11}(S_1) = -p^{-1}/\sqrt{5},$$

where $c = \cos\theta_4$, $s = \sin\theta_4$ and θ_4 is an acute angle. Thus, we obtain

$$c^2 = p^{-1}/\sqrt{5}, \quad s^2 = p/\sqrt{5}, \quad cs = c^2 - s^2 = 1/\sqrt{5},$$

$$\tan\theta_4 = s/c = p = \left(\sqrt{5} - 1\right)/2, \quad \theta_4 = 31.72°.$$

Then,

$$D^1(S_1) = \begin{pmatrix} -c^2 & -\sqrt{2}cs & -s^2 \\ -\sqrt{2}cs & c^2 - s^2 & \sqrt{2}cs \\ -s^2 & \sqrt{2}cs & -c^2 \end{pmatrix}$$

$$= \frac{1}{\sqrt{5}}\begin{pmatrix} -p^{-1} & -\sqrt{2} & -p \\ -\sqrt{2} & 1 & \sqrt{2} \\ -p & \sqrt{2} & -p^{-1} \end{pmatrix} = D^{T_1}(S_1),$$

$$D^2(S_1)$$

$$= \begin{pmatrix} c^4 & 2c^3s & \sqrt{6}c^2s^2 & 2cs^3 & s^4 \\ 2c^3s & -c^4 + 3c^2s^2 & -\sqrt{6}cs(c^2 - s^2) & -3c^2s^2 + s^4 & -2cs^3 \\ \sqrt{6}c^2s^2 & -\sqrt{6}cs(c^2 - s^2) & c^4 - 4c^2s^2 + s^4 & \sqrt{6}cs(c^2 - s^2) & \sqrt{6}c^2s^2 \\ 2cs^3 & -3c^2s^2 + s^4 & \sqrt{6}cs(c^2 - s^2) & -c^4 + 3c^2s^2 & -2c^3s \\ s^4 & -2cs^3 & \sqrt{6}c^2s^2 & -2c^3s & c^4 \end{pmatrix}$$

$$= \frac{1}{\sqrt{5}}\begin{pmatrix} p^{-2} & 2p^{-1} & \sqrt{6} & 2p & p^2 \\ 2p^{-1} & p^2 & -\sqrt{6} & -p^{-2} & -2p \\ \sqrt{6} & -\sqrt{6} & -1 & \sqrt{6} & \sqrt{6} \\ 2p & -p^{-2} & \sqrt{6} & p^2 & -2p^{-1} \\ p^2 & -2p & \sqrt{6} & -2p^{-1} & p^{-2} \end{pmatrix} = D^H(S_1).$$

By the unitary similarity transformation X,

$$X^{-1}D^3(T_0)X = D^{T_2}(T_0) \oplus D^G(T_0),$$

we obtain

$$X = \begin{pmatrix} 0 & 0 & c & 0 & 0 & 0 & -d^* \\ a & 0 & 0 & -b^* & 0 & 0 & 0 \\ 0 & 0 & 0 & 0 & 1 & 0 & 0 \\ 0 & 1 & 0 & 0 & 0 & 0 & 0 \\ 0 & 0 & 0 & 0 & 0 & 1 & 0 \\ 0 & 0 & d & 0 & 0 & 0 & c^* \\ b & 0 & 0 & a^* & 0 & 0 & 0 \end{pmatrix} = (X^{-1})^\dagger,$$

where $|a|^2 + |b|^2 = |c|^2 + |d|^2 = 1$. The row index of X runs from 3 to -3, and the column index runs over $(T_2, 1)$, $(T_2, 2)$, $(T_2, 3)$, $(G, 1)$, $(G, 2)$, $(G, 3)$, and $(G, 4)$. We calculate the first column and the third column in the fifth row of the formula $X^{-1} D^3(S_1) X = D^{T_2}(S_1) \oplus D^G(S_1)$:

$$D^3_{12}(S_1)a + D^3_{1(-3)}(S_1)b = 0, \qquad D^3_{13}(S_1)c + D^3_{1(-2)}(S_1)d = 0.$$

From Problems 7 and 9, we have $D^3_{\nu\mu}(S_1) = d^3_{\nu\mu}(2\theta_4)\exp\{-i\mu\pi\}$, and

$$D^3_{12}(S_1) = \sqrt{10}c^3 s(c^2 - 2s^2) = \sqrt{2/25}p,$$

$$D^3_{1(-3)}(S_1) = -\sqrt{15}c^2 s^4 = -\sqrt{3/25}p,$$

$$D^3_{13}(S_1) = -\sqrt{15}c^4 s^2 = -\sqrt{3/25}p^{-1},$$

$$D^3_{1(-2)}(S_1) = -\sqrt{10}cs^3(2c^2 - s^2) = -\sqrt{2/25}p^{-1}.$$

The solution is $a/b = \sqrt{3/2}$ and $c/d = -\sqrt{2/3}$. Due to normalization, we have $a = d = \sqrt{3/5}$, $b = -c = \sqrt{2/5}$, and

$$D^{T_2}(S_1) = \frac{1}{\sqrt{5}} \begin{pmatrix} -p & \sqrt{2} & p^{-1} \\ \sqrt{2} & -1 & \sqrt{2} \\ p^{-1} & \sqrt{2} & -p \end{pmatrix},$$

$$D^G(S_1) = \frac{1}{\sqrt{5}} \begin{pmatrix} -1 & -p & -p^{-1} & 1 \\ -p & 1 & -1 & -p^{-1} \\ -p^{-1} & -1 & 1 & -p \\ 1 & -p^{-1} & -p & -1 \end{pmatrix}.$$

Now, we are going to calculate the directions for the symmetric axes in a regular icosahedron. The azimuthal angles can be determined directly from Fig. 3.1. Here, we only calculate their polar angles. Following the notations used in Problem 18 of Chapter 3, we denote by θ_1 the polar angle for the fivefold axes except for T_0, by θ_2 and θ_3 the polar angles of two sets of the threefold axes, and by θ_4 and θ_5 the polar angles of two sets of the twofold axes, respectively. Another set of the twofold axes lies in the XY plane with the polar angle $\pi/2$. We have calculate θ_4 from $D^{T_1}(S_1)$. The remaining polar angles can be calculated by the common geometry knowledge. Let the length of the edge of the regular icosahedron be 1. Denote by R the radius of the circumcircle, and by r the radius of the incircle. The face of the regular icosahedral is a regular triangle. The distance from the center of the triangle to the vertex is $d = \sqrt{1/3}$. From Fig. 3.1 we see that $\sin\theta_4 = (2R)^{-1}$, $\tan\theta_2 = d/r$, and $R^2 = d^2 + r^2$. Thus,

$$R = (2\sin\theta_4)^{-1} = \frac{1}{2}\left(1 + \cot^2\theta_4\right)^{1/2} = \left(\frac{5 + \sqrt{5}}{8}\right)^{1/2} = 0.9511,$$

$$r = \left(R^2 - d^2\right)^{1/2} = \left(\frac{14 + 6\sqrt{5}}{48}\right)^{1/2} = \frac{3 + \sqrt{5}}{4\sqrt{3}} = 0.7558,$$

$$\tan\theta_2 = \frac{d}{r} = \frac{4}{3 + \sqrt{5}} = 3 - \sqrt{5} = 2p^2, \qquad \theta_2 = 37.38°,$$

$$\theta_1 = 2\theta_4 = 63.43°, \qquad \theta_5 = \frac{\pi - \theta_1}{2} = \frac{\pi}{2} - \theta_4 = 58.28°,$$

$$\tan\theta_1 = \frac{2\tan\theta_4}{1 - \tan^2\theta_4} = 2, \qquad \tan\theta_5 = \cot\theta_4 = p^{-1} = \frac{\sqrt{5} + 1}{2}$$

$$\theta_2 + \theta_3 = 2\theta_5 = \pi - \theta_1, \qquad \theta_3 = \pi - \theta_1 - \theta_2 = 79.19°$$

$$\tan\theta_3 = -\frac{\tan\theta_1 + \tan\theta_2}{1 - \tan\theta_1\tan\theta_2} = 3 + \sqrt{5} = 2p^{-2}.$$

13. Prove that the elements $u(\hat{\mathbf{n}}, \omega)$ with the same ω form a class of the SU(2) group.

Solution. It has been proven that the SU(2) group is homomorphic onto the SO(3) group through a two-to-one correspondence:

$$u(\hat{\mathbf{n}}, \omega)\sigma_a u(\hat{\mathbf{n}}, \omega)^{-1} = \sum_{b=1}^{3} \sigma_b R_{ba}(\hat{\mathbf{n}}, \omega), \qquad \pm u(\hat{\mathbf{n}}, \omega) \longrightarrow R(\hat{\mathbf{n}}, \omega).$$

Denote by θ and φ the polar angle and azimuthal angle of the direction $\hat{\mathbf{n}}$, respectively. We define

$$S = R(\mathbf{e}_3, \varphi)R(\mathbf{e}_2, \theta) \in SO(3), \qquad v = u(\mathbf{e}_3, \varphi)u(\mathbf{e}_2, \theta) \in SU(2).$$

Thus, $S_{b3} = n_b$,

$$v\sigma_3 v^{-1} = \sum_{b=1}^{3} \sigma_b S_{b3} = \vec{\sigma} \cdot \hat{\mathbf{n}}.$$

Since $u(\hat{\mathbf{n}}, \omega) = \mathbf{1}\cos(\omega/2) - i\,(\vec{\sigma} \cdot \hat{\mathbf{n}})\sin(\omega/2)$, we have

$$v^{-1}u(\hat{\mathbf{n}}, \omega)v = \mathbf{1}\cos(\omega/2) - i\sigma_3\sin(\omega/2) = u(\mathbf{e}_3, \omega).$$

Namely, all elements $u(\hat{\mathbf{n}}, \omega)$ with a given ω are conjugate to $u(\mathbf{e}_3, \omega)$. They belong to the same class. Since $\operatorname{Tr} u(\hat{\mathbf{n}}, \omega) = 2\cos(\omega/2)$, $0 \le \omega \le 2\pi$, those $u(\hat{\mathbf{n}}, \omega)$ with different ω must not be conjugate to each other. They belong to the different classes. This completes the proof.

4.3   Lie Groups and Lie Theorems

★ A group G is called a continuous group if its element R can be described by g independent and continuous parameters r_μ, $1 \le \mu \le g$, varying in a g-dimensional region. g is called the order of a continuous group G, and the region is called the group space of G. It is required that the set of parameters r_μ is one-to-one correspondence onto the group element R at least in the region where the measure is not vanishing. The parameters t_μ of the product $T = RS$ are the functions $t_\mu = f_\mu(r_1 \ldots, r_g; s_1 \ldots, s_g) = f_\mu(r; s)$ of the parameters of the factors R and S. $f_\mu(r; s)$ are called the composition functions. The domain of the definition of $f_\mu(r; s)$ is square of the group space, and the domain of function is the group space. The composition functions have to satisfy some conditions such as

$$f_\mu(f(r; s); t) = f_\mu(r; f(s; t)), \qquad f_\mu(e; r) = r_\mu, \qquad f_\mu(\bar{r}; r) = e_\mu,$$

where $\bar{r}_\mu$ are the parameters of the inverse R^{-1}, and e_μ are the parameters of the identity E. e_μ are usually taken to be zero for convenience. The continuous group is called the Lie group if the composition functions are analytic, or at least piecewise continuously differentiable. A Lie group is said to be a mixed Lie group if its group space falls into several disjoint regions. A mixed Lie group contains an invariant Lie subgroup G whose group space is a connected region in which the identity element of the group lies. The set of elements related to the other connected region is the coset of the invariant Lie subgroup. If the group space of a Lie group G is multiply-connected, there exists another Lie group G' with simply-connected group space such that G' is homomorphic onto G. G' is called the covering group of G. The group space of SO(3) is doubly-connected, and its covering group is SU(2). The Lie group is said to be compact if its group space is compact. If the group space of G is a closed Euclidean region, G is compact.

★ The group elements are said to be adjacent if their parameters differ only slightly from one another. An element is said to be infinitesimal if it is adjacent with the identity E. The infinitesimal elements describe the local property of the Lie group. The representation matrix of an infinitesimal element $A(\alpha)$ is determined by g generators I_μ

$$D(A) = 1 - i \sum_{\mu=1}^{g} \alpha_\mu I_\mu, \qquad I_\mu = i \left. \frac{\partial D(A)}{\partial \alpha_\mu} \right|_{\alpha=0}. \qquad (4.17)$$

I_μ are linearly independent of one another if the representation $D(G)$ is faithful.

★ **The First Lie Theorem**: The representation of a Lie group G with a connected group space is completely determined by its generators. Namely, the representation matrix of any element R in G can be solved from the following differential equation and the boundary condition:

$$\frac{\partial D(R)}{\partial r_\nu} = -i \left\{ \sum_\mu I_\mu S_{\mu\nu}(r) \right\} D(R), \qquad D(R)|_{R=E} = 1, \qquad (4.18)$$

where $S_{\mu\nu}(r)$ is independent of the representation. $S_{\mu\nu}(r)$ depends upon the choice of the group parameters and the composition functions of the Lie group:

$$S_{\mu\nu}(r) = \left. \frac{\partial f_\mu(r; s)}{\partial r_\nu} \right|_{s=\bar{r}}.$$

★ **The Second Lie Theorem**: The generators of any representation of a Lie group satisfy the common commutation relations

$$I_\mu I_\nu - I_\nu I_\mu = i \sum_\ell C_{\mu\nu}{}^\tau I_\tau, \qquad (4.19)$$

where $C_{\mu\nu}{}^\tau$ are called the structure constants, independent of the representation:

$$C_{\mu\nu}{}^\tau = \left. \left\{ \frac{\partial S_{\tau\nu}(r)}{\partial r_\mu} - \frac{\partial S_{\tau\mu}(r)}{\partial r_\nu} \right\} \right|_{r=0}.$$

Usually, they are calculated by the commutation relations of the generators in a known representation. The structure constants depend upon the choice of the group parameters.

★ **The Third Lie Theorem**: The necessary and sufficient condition for a set of parameters $C_{\mu\nu}{}^{\tau}$ to be the structure constants of a Lie group is the the parameters satisfy

$$C_{\mu\nu}{}^{\tau} = -C_{\nu\mu}{}^{\tau},$$

$$\sum_{\rho} \left\{ C_{\mu\nu}{}^{\rho} C_{\rho\tau}{}^{\sigma} + C_{\nu\tau}{}^{\rho} C_{\rho\mu}{}^{\sigma} + C_{\tau\mu}{}^{\rho} C_{\rho\nu}{}^{\sigma} \right\} = 0. \tag{4.20}$$

★ A representation $D(G)$ and its generators I_{μ} of a Lie group G satisfy

$$D(R)I_{\mu}D(R)^{-1} = \sum_{\nu} I_{\nu} D_{\nu\mu}^{\text{adj}}(R), \tag{4.21}$$

where $D^{\text{adj}}(G)$ is called the adjoint representation of the Lie group. The generators in the adjoint representation is related to the structure constants:

$$\left(I_{\tau}^{\text{adj}} \right)_{\nu\mu} = iC_{\tau\mu}{}^{\nu}. \tag{4.22}$$

The self-representation of SO(3) is the adjoint representation of both the SU(2) group and the SO(3) group:

$$D^j(R)I_a^j D^j(R)^{-1} = \sum_{\nu} I_b^j R_{ba},$$

$$P_R L_a P_R^{-1} = \sum_{k} L_b R_{ba}, \tag{4.23}$$

$$O_R J_a O_R^{-1} = \sum_{k} J_b R_{ba},$$

where P_R is the transformation operators for the scalar functions, and $O_R = Q_R P_R$ is the transformation operators for the spinor functions of rank s:

$$O_R \Psi^{(s)}(x) = D^s(R)\Psi^{(s)}(R^{-1}x),$$

$$Q_R \Psi^{(s)}(x) = D^s(R)\Psi^{(s)}(x), \tag{4.24}$$

$$P_R \Psi^{(s)}(x) = \Psi^{(s)}(R^{-1}x).$$

L_a, S_a, and J_a are the orbital, the spinor, and the total angular momentum operators, respectively.

If the spinor function $\Psi_{\mu}^j(x)$ of rank s belongs to the μth row of the irreducible representation $D^j(\text{SO}(3))$,

$$O_R \Psi_{\mu}^j(x) = D^s(R)\Psi_{\mu}^j(R^{-1}x) = \sum_{\nu} \Psi_{\nu}^j(x) D_{\nu\mu}^j(R), \tag{4.25}$$

$\Psi_\mu^j(x)$ is the simultaneous eigenfunctions of J^2 and J_3 for the eigenvalues $j(j+1)$ and μ. Similar formulas hold for the orbital space and the spinor space.

14. Calculate the representation matrices of the generators in an irreducible representation of SU(2) by the second Lie theorem.

Solution. This is a fundamental problem both in the matrix mechanics theory of quantum mechanics and in the representation theory of the Lie groups. The angular momentum operators are the generators in the transformation operators of SU(2). From the second Lie theorem, the generators in an irreducible representation of SU(2) satisfy the common commutation relations like that of the angular momentum operators:

$$J^2 = J_3^2 + J_3 + J_-J_+ = J_3^2 - J_3 + J_+J_-, \qquad J_\pm = J_1 \pm iJ_2,$$

$$[J_3,\ J_\pm] = \pm J_\pm, \qquad [J_+,\ J_-] = 2J_3, \qquad [J^2,\ J_a] = 0.$$

Since J^2 is commutable with all generators J_a, it takes a constant in an irreducible representation due to the Schur theorem. In terms of the Dirac notation, we denote by $|j\rangle$ the eigenfunction of J_3 with the highest eigenvalue j in the representation space:

$$J_3|j\rangle = j|j\rangle, \qquad J_+|j\rangle = 0, \qquad J^2|j\rangle = j(j+1)|j\rangle.$$

If the dimension of the representation is finite, there must exist a positive integer n such that

$$(J_-)^{n+1}|j\rangle = 0, \qquad (J_-)^\mu|j\rangle \neq 0, \qquad 0 \leq \mu \leq n.$$

Thus,

$$J_3\{J_-|j\rangle\} = (J_-J_3 - J_-)|j\rangle = (j-1)\{J_-|j\rangle\},$$

$$J_3(J_-)^n|j\rangle = (j-n)(J_-)^n|j\rangle,$$

$$J^2(J_-)^n|j\rangle = (j-n)(j-n-1)(J_-)^n|j\rangle = j(j+1)(J_-)^n|j\rangle.$$

The positive root for n is $2j$. Therefore, for an irreducible representation with a finite dimension, j must be an integer or a half-odd-integer. Let

$$J_-|\mu\rangle = A_\mu|\mu-1\rangle, \qquad 0 \leq j - \mu \leq n = 2j,$$

where, except for $A_{-j} = 0$, A_μ is a nonzero constant to be determined. We are going to show by induction that this $2j + 1$-dimensional space spanned by $|\mu\rangle$ is closed for the angular momentum operators J_a, namely,

$$J_+|\mu\rangle = B_{\mu+1}|\mu+1\rangle,$$

where $B_{\mu+1}$ is a constant to be determined. It is true when $\mu = j$ because of $B_{j+1} = 0$. If it is true when $\nu \le \mu \le j$, we want to show that it is also true for $\mu = \nu - 1 \ge -j$,

$$2\nu|\nu\rangle = 2J_3|\nu\rangle = [J_+, \ J_-]|\nu\rangle = A_\nu J_+|\nu-1\rangle - A_{\nu+1}B_{\nu+1}|\nu\rangle,$$
$$J_+|\nu-1\rangle = (A_\nu)^{-1}(2\nu + A_{\nu+1}B_{\nu+1})|\nu\rangle = B_\nu|\nu\rangle.$$

This completes the proof. The above equation gives a recursive formula:

$$
\begin{aligned}
A_\nu B_\nu &= A_{\nu+1}B_{\nu+1} + 2\nu = A_{\nu+2}B_{\nu+2} + 2\nu + 2(\nu+1) \\
&= \cdots \\
&= A_{\nu+(j-\nu+1)}B_{j+1} + (j+\nu)(j-\nu+1) \\
&= (j+\nu)(j-\nu+1) = \left(\Gamma_\nu^j\right)^2.
\end{aligned}
$$

Choosing the suitable factor in the basis, we are able to make $A_\mu = B_\mu = \Gamma_\mu^j$ to be a positive real number,

$$\Gamma_j^j = [2j\cdot 1]^{1/2}, \quad \Gamma_{j-1}^j = [(2j-1)2]^{1/2}, \quad \Gamma_{j-n}^j = [(2j-n)(n+1)]^{1/2},$$
$$\Gamma_\mu^j = [(j+\mu)(j-\mu+1)]^{1/2}, \quad \Gamma_{-\mu}^j = \Gamma_{\mu+1}^j,$$
$$J_3|\mu\rangle = \mu|\mu\rangle, \quad J^2|\mu\rangle = j(j+1)|\mu\rangle, \quad J_\pm|\mu\rangle = \Gamma_{\mp\mu}^j|\mu\pm 1\rangle.$$

15. For any one-order Lie group with the composition function $f(r; s)$, please try to find a new parameter r' such that the new composition function is the additive function, $f'(r'; s') = r' + s'$. The Lorentz transformation $A(v)$ for the boost along the Z axis with the relative velocity v is taken in the following form. The set of them forms a one-order Lie group:

$$
A(v) = \begin{pmatrix} 1 & 0 & 0 & 0 \\ 0 & 1 & 0 & 0 \\ 0 & 0 & \gamma & -i\gamma v/c \\ 0 & 0 & i\gamma v/c & \gamma \end{pmatrix}, \qquad
\begin{aligned}
&\gamma = \left(1 - v^2/c^2\right)^{-1/2}, \\[2mm]
&f(v_1; v_2) = \frac{v_1 + v_2}{1 + v_1 v_2/c^2}.
\end{aligned}
$$

Find the new parameter with the additive composition function.

Solution. For any given one-order Lie group with the composition function

$f(r; s)$,

$$S(r) = \left(\frac{\partial f(r; s)}{\partial r} \right)_{s=\bar{r}}.$$

From the first Lie theorem,

$$\frac{dD(r)}{dr} = -iIS(r)D(r), \qquad D(r)|_{r=e} = 1,$$

where I is the generator of the representation $D(r)$, e is the parameter of the identity, taken to be zero. Solving the differential equation with the boundary condition, we obtain

$$D(r) = \exp\left\{ -iI \int_0^r S(t)dt \right\}.$$

Define the new parameter

$$\omega(r) = \int_0^r S(t)dt.$$

The new parameter $\omega(0)$ of the identity is still zero. Thus,

$$D(r) = \exp\left\{ -iI\omega(r) \right\}, \qquad D(s) = \exp\left\{ -iI\omega(s) \right\},$$

$$\exp\left\{ -iI[\omega(r) + \omega(s)] \right\} = D(r)D(s) = D(rs)$$

$$= \exp\left\{ -iI\omega(rs) \right\},$$

$$\omega(rs) = \omega(r) + \omega(s).$$

For the multiplication of elements, the new parameter is additive.

As an example, for the Lorentz transformation with the relative velocity v along the Z axis, the composition function for the parameter v is

$$f(v_1; v_2) = \frac{v_1 + v_2}{1 + v_1 v_2/c^2}, \qquad S(v) = \frac{\partial f(v; v')}{\partial v}\bigg|_{v'=-v} = \frac{1}{1 - v^2/c^2}.$$

Defining a new parameter ω

$$\omega = \frac{1}{c} \int_0^v S(t)dt = \operatorname{arctanh}\left(\frac{t}{c} \right)\bigg|_0^v = \operatorname{arctanh}\left(\frac{v}{c} \right),$$

we have

$$\tanh \omega = v/c, \qquad \cosh \omega = \left(1 - v^2/c^2 \right)^{-1/2} = \gamma, \qquad \sinh \omega = \gamma v/c,$$

$$A(\omega) = \begin{pmatrix} 1 & 0 & 0 & 0 \\ 0 & 1 & 0 & 0 \\ 0 & 0 & \cosh\omega & -i\sinh\omega \\ 0 & 0 & i\sinh\omega & \cosh\omega \end{pmatrix}.$$

In the multiplication of two Lorentz transformations, the parameter ω is additive.

4.4 Irreducible Tensor Operators

★ The set of $(2k+1)$ operators $L_\rho^k(x)$, $-k \le \rho \le k$, is called the irreducible tensor operators of rank k if those operators transform according to the irreducible representation $D^k(R)$ in the rotation R of SO(3):

$$O_R L_\rho^k(x) O_R^{-1} = \sum_{\lambda=-k}^{k} L_\lambda^k(x) D_{\lambda\rho}^k(R). \tag{4.26}$$

For the infinitesimal rotation, we have

$$\begin{aligned} & [J_3,\ L_\rho^k(x)] = \rho L_\rho^k(x), \\ & [J_\pm,\ L_\rho^k(x)] = \{(k \mp \rho)(k \pm \rho + 1)\}^{1/2}\, L_{\rho\pm1}^k(x), \\ & \sum_{a=1}^{3} [J_a, [J_a, L_\rho^k(x)]] = k(k+1) L_\rho^k(x). \end{aligned} \tag{4.27}$$

Equation (4.26) is equivalent to Eq. (4.27). They are both regarded as the definition of the irreducible tensor operators of rank k. $L_\rho^k(x)$ are called the irreducible tensor operators of rank k with respect to the orbital space or the spinor space, if replacing O_R with P_R or Q_R, respectively.

The irreducible tensor operator of rank one is called the vector operator. Since the representation D^1 is equivalent to the self-representation R of SO(3) [see Eq. (4.14)], there are two kinds of components for the vector operators:

$$\begin{aligned} & O_R L_\rho^1(x) O_R^{-1} = \sum_{\lambda=-1}^{1} L_\lambda^1(x) D_{\lambda\rho}^1(R), \\ & O_R V_a(x) O_R^{-1} = \sum_{b=1}^{3} V_b(x) R_{ba}, \end{aligned} \tag{4.28}$$

$$L_1^1 = -\sqrt{1/2}\,(V_1 + iV_2), \qquad V_1 = \sqrt{1/2}\,(L_{-1}^1 - L_1^1),$$
$$L_0^1 = V_3, \qquad\qquad\qquad V_2 = i\sqrt{1/2}\,(L_{-1}^1 + L_1^1),$$
$$L_{-1}^1 = \sqrt{1/2}\,(V_1 - iV_2), \qquad V_3 = L_0^1.$$

$L_\rho^1(x)$ are called the spherical components of the vector operators, and $V_a(x)$ are called the rectangular components of the vector operators. The examples for the vector operators in common use are the electric dipole operators x_a or $Y_\rho^1(\hat{\mathbf{n}})$, the momentum operators p_a, the angular momentum operators J_a, L_a and S_a etc.

★ Being a vector operator, J_a satisfies

$$O_R J_a(x) O_R^{-1} = \sum_{b=1}^{3} J_b(x) R_{ba},$$

where J_a is the generator for the transformation O_R, and R_{ba} is the representation matrix in the adjoint representation of SO(3) [see Eq. (4.23)]. Letting $a = 3$, we have

$$O_R J_3(x) O_R^{-1} = \mathbf{J} \cdot \hat{\mathbf{n}}(\theta, \varphi), \qquad R = R(\varphi, \theta, \gamma). \tag{4.29}$$

γ is often taken to be zero. If $\Psi_\mu^j(x)$ is the function belonging to the μth row of the irreducible representation D^j [see Eq. (4.25)], $\Psi_\mu^j(x)$ is the simultaneous eigenfunctions of J^2 and J_3 for the eigenvalues $j(j+1)$ and μ. Now, from Eq. (4.29), $O_R \Psi_\mu^j(x)$ is the eigenfunction for $\mathbf{J} \cdot \hat{\mathbf{n}}(\theta, \varphi)$ with the eigenvalue μ.

★ For a spherical symmetrical system, the eigenfunction of the energy can be chosen to belong to a certain row of an irreducible unitary representation:

$$O_R \Psi_\mu^j(x) = \sum_\nu \Psi_\nu^j(x) D_{\nu\mu}^j(R),$$
$$O_R \Phi_{\mu'}^{j'}(x) = \sum_{\nu'} \Phi_{\nu'}^{j'}(x) D_{\nu'\mu'}^{j'}(R).$$

The Wigner-Eckart theorem says that if O_R is a unitary operator for the inner product of functions, then

$$\langle \Phi_{\mu'}^{j'}(x) | \Psi_\mu^j(x) \rangle = \delta_{j'j} \delta_{\mu'\mu} \langle \Phi^{j'} || \Psi^j \rangle, \tag{4.30}$$

where $\langle \Phi^{j'} || \Psi^j \rangle$, called the reduced matrix entry, is independent of μ. Namely, the functions belonging to two inequivalent irreducible representations are orthogonal to each other, the functions belonging to different rows of an irreducible unitary representation are orthogonal to each other, and the inner product of two functions belonging to the same row of an irreducible unitary representation is independent of the row. The reduced matrix entry depends upon j as well as the explicit forms of Φ and Ψ. The theorem can be generalized to the matrix entries of the irreducible tensor operators in such basis functions. Because

$$O_R \left\{ L_\rho^k(x) \Psi_\mu^j(x) \right\} = \left\{ O_R L_\rho^k(x) O_R^{-1} \right\} \left\{ O_R \Psi_\mu^j(x) \right\}$$
$$= \sum_{\lambda\nu} \left\{ L_\lambda^k(x) \Psi_\nu^j(x) \right\} \left\{ D_{\lambda\rho}^k(R) D_{\nu\mu}^j(R) \right\},$$

$L_\rho^k(x) \Psi_\mu^j(x)$ is transformed according to the direct product representation, which can be reduced by the unitary similarity transformation C^{jk}:

$$\left(C^{jk} \right)^{-1} \left(D^j(R) \times D^k(R) \right) C^{jk} = \bigoplus_{J=|j-k|}^{j+k} D^J(R).$$

The matrix entries of C^{jk} are called the Clebsch-Gordan (C-G) coefficients. According to the phase convention, we make the C-G coefficients real and satisfying

$$C_{j\nu J(j+\nu)}^{jk} \quad \text{and} \quad C_{\mu(-k)J(\mu-k)}^{jk} \quad \text{are positive,}$$

$$C_{\mu\nu JM}^{jk} \neq 0 \text{ only when } M = \mu + \nu \text{ and } |j-k| \leq J \leq j+k$$

$$C_{\mu\nu JM}^{jk} = (-1)^{j+k-J} \, C_{\nu\mu JM}^{kj} = (-1)^{j+k-J} \, C_{(-\mu)(-\nu)J(-M)}^{jk}$$

$$= (-1)^{k-J-\mu} \left(\frac{2J+1}{2k+1} \right)^{1/2} C_{(-M)\mu k(-\nu)}^{Jj} \tag{4.31}$$

$$= (-1)^{j-J+\nu} \left(\frac{2J+1}{2j+1} \right)^{1/2} C_{\nu(-M)j(-\mu)}^{kJ}.$$

The Clebsch-Gordan (C-G) coefficients can be used to combine $L_\rho^k(x) \Psi_\mu^j(x)$ into the functions $F_M^J(x)$ belonging to the irreducible representation:

$$F_M^J(x) = \sum_\rho L_\rho^k(x) \Psi_{M-\rho}^j(x) C_{\rho(M-\rho)JM}^{kj},$$

$$L_\rho^k(x) \Psi_\mu^j(x) = \sum_J F_{\rho+\mu}^J(x) C_{\rho\mu J(\rho+\mu)}^{kj},$$

$$O_R F_M^J(x) = \sum_{M'} F_{M'}^J(x) D_{M'M}^J(R).$$

From the Wigner-Eckart theorem, we have

$$\langle \Phi_{\mu'}^{j'}(x)|F_M^J(x)\rangle = c\,\delta_{j'J}\delta_{\mu'M}\;,$$

where the constant c is independent of the subscripts μ' and M. c depends on the indices j', k and j as well as the explicit forms of Φ, L and Ψ. c is still called the reduced matrix entry, denoted by $\langle \Phi^{j'}||L^k||\Psi^j\rangle$. Thus,

$$\begin{aligned}
\langle \Phi_{\mu'}^{j'}(x)|L_\rho^k(x)|\Psi_\mu^j(x)\rangle &= \sum_J C_{\rho\mu J(\rho+\mu)}^{kj}\langle \Phi_{\mu'}^{j'}(x)|F_{\rho+\mu}^J(x)\rangle \\
&= C_{\rho\mu j'\mu'}^{kj}\langle \Phi^{j'}||L^k||\Psi^j\rangle\;.
\end{aligned} \tag{4.32}$$

There are $(2j'+1)(2k+1)(2j+1)$ matrix entries in the form of $\langle \Phi_{\mu'}^{j'}(x)|L_\rho^k(x)|\Psi_\mu^j(x)\rangle$. Due to their property in rotation, they are related by the C-G coefficients. The Wigner-Eckart theorem simplifies the calculation of $(2j'+1)(2k+1)(2j+1)$ matrix entries to the calculation of only one parameter $\langle \Phi^{j'}||L^k||\Psi^j\rangle$. If there is one matrix entry with given subscripts μ', ρ and μ which can be calculated, the remaining matrix entries all are calculable. In most cases, even one matrix entry is hard to be calculated. For example, the explicit forms of the wave functions $\Phi_\mu^j(x)$ and $\Phi_{\mu'}^{j'}(x)$ are unknown. However, Eq. (4.32) gives some relations among the matrix entries, which sometimes can be compared with experiments.

16. Directly calculate the C-G coefficients for the direct product representation of two irreducible representations of SU(2) in terms of the raising and lowering operators $J_\pm$: a) $D^{1/2} \times D^{1/2}$, b) $D^{1/2} \times D^1$, c) $D^1 \times D^1$, d) $D^1 \times D^{3/2}$.

Solution. In the representation space of $D^j \times D^k$, the basis state $|j,\mu\rangle|k,\nu\rangle$ is the eigenfunction of J_3 for the eigenvalue $\mu+\nu$. The state with the highest eigenvalue of J_3 is $|j,j\rangle|k,k\rangle$. It means that in the reduction of $D^j \times D^k$ there is an irreducible representation D^{j+k}. $|j,j\rangle|k,k\rangle$ is also the state with the highest eigenvalue of J_3 in the representation space of D^{j+k},

$$||j+k,j+k\rangle = |j,j\rangle|k,k\rangle.$$

Applying the lowering operator J_- to it, we can calculate all remaining basis states in the representation D^{j+k}. In the subspace which is orthogonal to the basis states belonging to D^{j+k}, we try to find the state with the highest eigenvalues of J_3, which belongs to another irreducible representation D^J in the reduction of $D^j \times D^k$. Then, calculate the basis states in D^J by the lowering operator. In the calculation the following formulas are useful:

$$J_-||J,M\rangle = [(J+M)(J-M+1)]^{1/2}\,||J,M-1\rangle,$$

$$= J_- \left\{ \sum_{\mu=-j}^{j} |j,\mu\rangle|k,M-\mu\rangle C^{jk}_{\mu(M-\mu)JM} \right\}$$

$$= \sum_{\mu=-j}^{j} \left\{ |j,\mu\rangle\,(J_-|k,M-\mu\rangle) + (J_-|j,\mu\rangle)\,|k,M-\mu\rangle \right\}$$
$$\times\, C^{jk}_{\mu(M-\mu)JM},$$

$$J_+||J,J\rangle = 0.$$

a) In the direct product representation space, the state with the highest eigenvalue of J_3 is $|1/2,1/2\rangle|1/2,1/2\rangle$, which belongs to the irreducible representation D^1.

$$||1,1\rangle = |1/2,1/2\rangle|1/2,1/2\rangle,$$
$$||1,0\rangle = 2^{-1/2} J_-||1,1\rangle$$
$$= 2^{-1/2}\left\{|1/2,1/2\rangle|1/2,-1/2\rangle + |1/2,-1/2\rangle|1/2,1/2\rangle\right\},$$
$$||1,-1\rangle = 2^{-1/2} J_-||1,0\rangle$$
$$= 2^{-1}\left\{|1/2,-1/2\rangle|1/2,-1/2\rangle + |1/2,-1/2\rangle|1/2,-1/2\rangle\right\}$$
$$= |1/2,-1/2\rangle|1/2,-1/2\rangle.$$

The remaining state which is orthogonal to $||1,0\rangle$ is

$$||0,0\rangle = 2^{-1/2}\left\{|1/2,1/2\rangle|1/2,-1/2\rangle - |1/2,-1/2\rangle|1/2,1/2\rangle\right\}.$$

It belongs to the irreducible representation D^0. From the phase convention (4.31) for the C-G coefficients, we choose the coefficient in the term $|1/2,1/2\rangle|1/2,-1/2\rangle$ to be positive. Finally, we obtain the Clebsch-Gordan series for the direct product representation

$$D^{1/2} \times D^{1/2} \simeq D^1 \oplus D^0.$$

b) In the direct product representation space, the state with the highest eigenvalue of J_3 is $|1/2,1/2\rangle|1,1\rangle$, which belongs to the irreducible representation $D^{3/2}$.

$$||3/2, 3/2\rangle = |1/2, 1/2\rangle |1, 1\rangle,$$

$$||3/2, 1/2\rangle = 3^{-1/2} J_- ||3/2, 3/2\rangle$$
$$= 3^{-1/2} \left\{ \sqrt{2}|1/2, 1/2\rangle |1, 0\rangle + |1/2, -1/2\rangle |1, 1\rangle \right\},$$

$$||3/2, -1/2\rangle = 2^{-1} J_- ||3/2, 1/2\rangle$$
$$= (12)^{-1/2} \left\{ 2|1/2, 1/2\rangle |1, -1\rangle + \sqrt{2}|1/2, -1/2\rangle |1, 0\rangle \right.$$
$$\left. + \sqrt{2}|1/2, -1/2\rangle |1, 0\rangle \right\},$$
$$= 3^{-1/2} \left\{ |1/2, 1/2\rangle |1, -1\rangle + \sqrt{2}|1/2, -1/2\rangle |1, 0\rangle \right\},$$

$$||3/2, -3/2\rangle = 3^{-1/2} J_- ||3/2, -1/2\rangle$$
$$= 3^{-1} \left\{ |1/2, -1/2\rangle |1, -1\rangle + 2|1/2, -1/2\rangle |1, -1\rangle \right\}$$
$$= |1/2, -1/2\rangle |1, -1\rangle.$$

The basis state which is orthogonal to $||3/2, 1/2\rangle$ belongs to the irreducible representation $D^{1/2}$. From the phase convention for the C-G coefficients, we take the coefficient in the term $|1/2, 1/2\rangle |1, 0\rangle$ to be positive:

$$||1/2, 1/2\rangle = 3^{-1/2} \left\{ |1/2, 1/2\rangle |1, 0\rangle - \sqrt{2}|1/2, -1/2\rangle |1, 1\rangle \right\},$$

$$||1/2, -1/2\rangle = J_- ||1/2, 1/2\rangle$$
$$= 3^{-1/2} \left\{ \sqrt{2}|1/2, 1/2\rangle |1, -1\rangle + |1/2, -1/2\rangle |1, 0\rangle \right.$$
$$\left. - 2|1/2, -1/2\rangle |1, 0\rangle \right\},$$
$$= 3^{-1/2} \left\{ \sqrt{2}|1/2, 1/2\rangle |1, -1\rangle - |1/2, -1/2\rangle |1, 0\rangle \right\}.$$

Finally, we obtain the Clebsch-Gordan series for the direct product representation

$$D^{1/2} \times D^1 \simeq D^{3/2} \oplus D^{1/2}.$$

c) In the direct product representation space, the state with the highest eigenvalue of J_3 is $|1, 1\rangle |1, 1\rangle$, which belongs to the irreducible representation D^2.

$$||2, 2\rangle = |1, 1\rangle |1, 1\rangle,$$

$$||2, 1\rangle = 2^{-1} J_- ||2, 2\rangle = 2^{-1/2} \left\{ |1, 1\rangle |1, 0\rangle + |1, 0\rangle |1, 1\rangle \right\},$$

$$||2, 0\rangle = 6^{-1/2} J_- ||2, 1\rangle$$
$$= 6^{-1/2} \left\{ |1, 1\rangle |1, -1\rangle + |1, 0\rangle |1, 0\rangle + |1, 0\rangle |1, 0\rangle + |1, -1\rangle |1, 1\rangle \right\},$$
$$= 6^{-1/2} \left\{ |1, 1\rangle |1, -1\rangle + 2|1, 0\rangle |1, 0\rangle + |1, -1\rangle |1, 1\rangle \right\}.$$

The remaining basis states can be calculated similarly. They can also be calculated by the symmetry (4.31) of the C-G coefficients:

$$C^{jk}_{\mu\nu JM} = (-1)^{j+k-J} C^{jk}_{(-\mu)(-\nu)J(-M)} = C^{kj}_{(-\nu)(-\mu)J(-M)}$$

$$\|2,-1\rangle = 2^{-1/2}\left\{|1,0\rangle|1,-1\rangle + |1,-1\rangle|1,0\rangle\right\},$$

$$\|2,-2\rangle = |1,-1\rangle|1,-1\rangle.$$

The basis state which is orthogonal to $\|2,1\rangle$ belongs to the irreducible representation D^1. From the phase convention for the C-G coefficients, we take the coefficient in the term $|1,1\rangle|1,0\rangle$ to be positive:

$$\|1,1\rangle = 2^{-1/2}\left\{|1,1\rangle|1,0\rangle - |1,0\rangle|1,1\rangle\right\},$$

$$\begin{aligned}\|1,0\rangle &= 2^{-1/2}J_-\|1,1\rangle\\ &= 2^{-1/2}\left\{|1,1\rangle|1,-1\rangle + |1,0\rangle|1,0\rangle - |1,0\rangle|1,0\rangle - |1,-1\rangle|1,1\rangle\right\},\\ &= 2^{-1/2}\left\{|1,1\rangle|1,-1\rangle - |1,-1\rangle|1,1\rangle\right\},\end{aligned}$$

$$\|1,-1\rangle = 2^{-1/2}J_-\|1,0\rangle = 2^{-1/2}\left\{|1,0\rangle|1,-1\rangle - |1,-1\rangle|1,0\rangle\right\}.$$

In the direct product representation space, there are three basis states with the zero eigenvalue of J_3. We have obtained two orthogonal basis states $\|2,0\rangle$ and $\|1,0\rangle$. The third state $\|0,0\rangle$ can be calculated from the orthogonal condition. But here we prefer to use another method. This method is based on the condition for the state with the highest eigenvalue of J_3 in an irreducible representation of SO(3) that it is annihilated by the raising operator J_+. Let

$$\|0,0\rangle = a|1,1\rangle|1,-1\rangle + b|1,0\rangle|1,0\rangle + c|1,-1\rangle|1,1\rangle.$$

J_+ annihilates this state:

$$\begin{aligned}J_+\|0,0\rangle = \;&\sqrt{2}\left\{a|1,1\rangle|1,0\rangle + b|1,1\rangle|1,0\rangle\right.\\ &\left. + b|1,0\rangle|1,1\rangle + c|1,0\rangle|1,1\rangle\right\} = 0.\end{aligned}$$

The solution is $a = -b = c$. After normalization we obtain

$$\|0,0\rangle = 3^{-1/2}\left\{|1,1\rangle|1,-1\rangle - |1,0\rangle|1,0\rangle + |1,-1\rangle|1,1\rangle\right\}.$$

Finally, we obtain the Clebsch-Gordan series for the direct product representation

$$D^1 \times D^1 \simeq D^2 \oplus D^1 \oplus D^0.$$

d) In the direct product representation space, the state with the highest eigenvalue of J_3 is $|1,1\rangle|3/2,3/2\rangle$, which belongs to the irreducible representation $D^{5/2}$.

$$\|5/2,5/2\rangle = |1,1\rangle|3/2,3/2\rangle,$$

$$\begin{aligned}\|5/2,3/2\rangle &= 5^{-1/2}J_-\|5/2,5/2\rangle \\ &= 5^{-1/2}\left\{\sqrt{3}|1,1\rangle|3/2,1/2\rangle + \sqrt{2}|1,0\rangle|3/2,3/2\rangle\right\},\end{aligned}$$

$$\begin{aligned}\|5/2,1/2\rangle &= 8^{-1/2}J_-\|5/2,3/2\rangle \\ &= (40)^{-1/2}\left\{2\sqrt{3}|1,1\rangle|3/2,-1/2\rangle + \sqrt{6}|1,0\rangle|3/2,1/2\rangle \right. \\ &\quad\left. + \sqrt{6}|1,0\rangle|3/2,1/2\rangle + 2|1,-1\rangle|3/2,3/2\rangle\right\}, \\ &= (10)^{-1/2}\left\{\sqrt{3}|1,1\rangle|3/2,-1/2\rangle + \sqrt{6}|1,0\rangle|3/2,1/2\rangle \right. \\ &\quad\left. + |1,-1\rangle|3/2,3/2\rangle\right\}.\end{aligned}$$

The basis state which is orthogonal to $\|5/2,3/2\rangle$ belongs to the irreducible representation $D^{3/2}$. According to the phase convention of the C-G coefficients, we take the coefficient in the term $|1,1\rangle|3/2,1/2\rangle$ to be positive.

$$\|3/2,3/2\rangle = 5^{-1/2}\left\{\sqrt{2}|1,1\rangle|3/2,1/2\rangle - \sqrt{3}|1,0\rangle|3/2,3/2\rangle\right\},$$

$$\begin{aligned}\|3/2,1/2\rangle &= 3^{-1/2}J_-\|3/2,3/2\rangle \\ &= (15)^{-1/2}\left\{2\sqrt{2}|1,1\rangle|3/2,-1/2\rangle + 2|1,0\rangle|3/2,1/2\rangle \right. \\ &\quad\left. - 3|1,0\rangle|3/2,1/2\rangle - \sqrt{6}|1,-1\rangle|3/2,3/2\rangle\right\}, \\ &= (15)^{-1/2}\left\{2\sqrt{2}|1,1\rangle|3/2,-1/2\rangle - |1,0\rangle|3/2,1/2\rangle \right. \\ &\quad\left. - \sqrt{6}|1,-1\rangle|3/2,3/2\rangle\right\}.\end{aligned}$$

In the direct product representation space, there are three basis states with eigenvalue $1/2$ of J_3. We have obtained two orthogonal basis states $\|5/2,1/2\rangle$ and $\|3/2,1/2\rangle$. The third state belongs to the representation $D^{1/2}$, and can be calculated by the orthogonal condition or by the annihilation condition of J_+. Now, we use the latter method. Let

$$\|1/2,1/2\rangle = a|1,1\rangle|3/2,-1/2\rangle + b|1,0\rangle|3/2,1/2\rangle + c|1,-1\rangle|3/2,3/2\rangle.$$

J_+ annihilates this state:

$$\begin{aligned}J_+\|1/2,1/2\rangle &= 2a|1,1\rangle|3/2,1/2\rangle + b\sqrt{2}|1,1\rangle|3/2,1/2\rangle \\ &\quad + b\sqrt{3}|1,0\rangle|3/2,3/2\rangle + c\sqrt{2}|1,0\rangle|3/2,3/2\rangle = 0.\end{aligned}$$

The solution is $a\sqrt{6} = -b\sqrt{3} = c\sqrt{2}$. After normalization we obtain

$$\|1/2, 1/2\rangle = 6^{-1/2} \left\{ |1, 1\rangle |3/2, -1/2\rangle - \sqrt{2}|1, 0\rangle |3/2, 1/2\rangle \right.$$
$$\left. + \sqrt{3}|1, -1\rangle |3/2, 3/2\rangle \right\}.$$

The expansions for the remaining basis states can be obtained by the symmetry (4.31) of the C-G coefficients:

$$\|5/2, -5/2\rangle = |1, -1\rangle |3/2, -3/2\rangle,$$
$$\|5/2, -3/2\rangle = 5^{-1/2} \left\{ \sqrt{3}|1, -1\rangle |3/2, -1/2\rangle + \sqrt{2}|1, 0\rangle |3/2, -3/2\rangle \right\},$$
$$\|5/2, -1/2\rangle = (10)^{-1/2} \left\{ \sqrt{3}|1, -1\rangle |3/2, 1/2\rangle + \sqrt{6}|1, 0\rangle |3/2, -1/2\rangle \right.$$
$$\left. + |1, 1\rangle |3/2, -3/2\rangle \right\},$$
$$\|3/2, -3/2\rangle = 5^{-1/2} \left\{ -\sqrt{2}|1, -1\rangle |3/2, -1/2\rangle + \sqrt{3}|1, 0\rangle |3/2, -3/2\rangle \right\},$$
$$\|3/2, -1/2\rangle = (15)^{-1/2} \left\{ -2\sqrt{2}|1, -1\rangle |3/2, 1/2\rangle + |1, 0\rangle |3/2, -1/2\rangle \right.$$
$$\left. + \sqrt{6}|1, 1\rangle |3/2, -3/2\rangle \right\},$$
$$\|1/2, -1/2\rangle = 6^{-1/2} \left\{ |1, -1\rangle |3/2, 1/2\rangle - \sqrt{2}|1, 0\rangle |3/2, -1/2\rangle \right.$$
$$\left. + \sqrt{3}|1, 1\rangle |3/2, -3/2\rangle \right\}.$$

Finally, we obtain the Clebsch-Gordan series for the direct product representation

$$D^1 \times D^{3/2} \simeq D^{5/2} \oplus D^{3/2} \oplus D^{1/2}.$$

17. Prove two sets of formulas for the C-G coefficients in terms of the raising and lowering operators $J_{\pm}$:
a) In the reduction of $D^{1/2} \times D^j$,

$$\|(j + 1/2), M\rangle = \left(\frac{j + M + 1/2}{2j + 1} \right)^{1/2} |1/2, 1/2\rangle |j, M - 1/2\rangle$$
$$+ \left(\frac{j - M + 1/2}{2j + 1} \right)^{1/2} |1/2, -1/2\rangle |j, M + 1/2\rangle,$$
$$\|(j - 1/2), M\rangle = \left(\frac{j - M + 1/2}{2j + 1} \right)^{1/2} |1/2, 1/2\rangle |j, M - 1/2\rangle$$
$$- \left(\frac{j + M + 1/2}{2j + 1} \right)^{1/2} |1/2, -1/2\rangle |j, M + 1/2\rangle.$$

b) In the reduction of $D^1 \times D^j$,

$$\|(j+1), M\rangle = \left\{\frac{(j+M)(j+M+1)}{2(2j+1)(j+1)}\right\}^{1/2} |1,1\rangle|j, M-1\rangle$$

$$+ \left\{\frac{(j-M+1)(j+M+1)}{(2j+1)(j+1)}\right\}^{1/2} |1,0\rangle|j, M\rangle$$

$$+ \left\{\frac{(j-M)(j-M+1)}{2(2j+1)(j+1)}\right\}^{1/2} |1,-1\rangle|j, M+1\rangle,$$

$$\|j, M\rangle = \left\{\frac{(j+M)(j-M+1)}{2j(j+1)}\right\}^{1/2} |1,1\rangle|j, M-1\rangle$$

$$- \frac{M}{[j(j+1)]^{1/2}}|1,0\rangle|j, M\rangle$$

$$- \left\{\frac{(j-M)(j+M+1)}{2j(j+1)}\right\}^{1/2} |1,-1\rangle|j, M+1\rangle,$$

$$\|(j-1), M\rangle = \left\{\frac{(j-M)(j-M+1)}{2j(2j+1)}\right\}^{1/2} |1,1\rangle|j, M-1\rangle$$

$$- \left\{\frac{(j-M)(j+M)}{j(2j+1)}\right\}^{1/2} |1,0\rangle|j, M\rangle$$

$$+ \left\{\frac{(j+M)(j+M+1)}{2j(2j+1)}\right\}^{1/2} |1,-1\rangle|j, M+1\rangle.$$

Solution. We proof them by induction.

a) We only prove the first equation. The second equation can be obtained by orthogonality of the basis states and the phase convention. In the direct product representation space, the state with the highest eigenvalue of J_3 is $|1/2, 1/2\rangle|j, j\rangle$, which belongs to the irreducible representation $D^{j+1/2}$,

$$\|(j+1/2), (j+1/2)\rangle = |1/2, 1/2\rangle|j, j\rangle.$$

It satisfies the first equation with $M = j + 1/2$. Suppose that the first equation holds for $M = m$. We are going to show it also holds for $M = m - 1$.

$$\|(j+1/2), m-1\rangle = \frac{J_- \|(j+1/2), m\rangle}{[(j+m+1/2)(j-m+3/2)]^{1/2}}$$

$$= \left(\frac{j+m-1/2}{2j+1}\right)^{1/2} |1/2, 1/2\rangle|j, m-3/2\rangle$$

$$+ \frac{1}{[(2j+1)(j-m+3/2)]^{1/2}} |1/2, -1/2\rangle |j, m-1/2\rangle$$

$$+ \frac{(j-m+1/2)}{[(2j+1)(j-m+3/2)]^{1/2}} |1/2, -1/2\rangle |j, m-1/2\rangle,$$

$$= \left(\frac{j+m-1/2}{2j+1}\right)^{1/2} |1/2, 1/2\rangle |j, m-3/2\rangle$$

$$+ \left(\frac{j-m+3/2}{2j+1}\right)^{1/2} |1/2, -1/2\rangle |j, m-1/2\rangle.$$

The first equation is proved.

b) In the direct product representation space, the state with the highest eigenvalue of J_3 is $|1, 1\rangle |j, j\rangle$, which belongs to the irreducible representation D^{j+1},

$$\|j+1, j+1\rangle = |1, 1\rangle |j, j\rangle.$$

It satisfies the first equation with $M = j+1$. Suppose that the first equation holds for $M = m$. We are going to show it also holds for $M = m - 1$.

$$\|j+1, m-1\rangle = \frac{J_- \|j+1, m\rangle}{[(j+m+1)(j-m+2)]^{1/2}}$$

$$= \left\{\frac{(j+m)(j+m-1)}{2(2j+1)(j+1)}\right\}^{1/2} |1, 1\rangle |j, m-2\rangle$$

$$+ \left\{\frac{(j+m)}{(2j+1)(j+1)(j-m+2)}\right\}^{1/2} |1, 0\rangle |j, m-1\rangle$$

$$+ \left\{\frac{(j-m+1)^2(j+m)}{(2j+1)(j+1)(j-m+2)}\right\}^{1/2} |1, 0\rangle |j, m-1\rangle$$

$$+ \left\{\frac{2(j-m+1)}{(2j+1)(j+1)(j-m+2)}\right\}^{1/2} |1, -1\rangle |j, m\rangle$$

$$+ \left\{\frac{(j-m)^2(j-m+1)}{2(2j+1)(j+1)(j-m+2)}\right\}^{1/2} |1, -1\rangle |j, m\rangle$$

$$= \left\{\frac{(j+m)(j+m-1)}{2(2j+1)(j+1)}\right\}^{1/2} |1, 1\rangle |j, m-2\rangle$$

$$+ \left\{\frac{(j+m)(j-m+2)}{(2j+1)(j+1)}\right\}^{1/2} |1, 0\rangle |j, m-1\rangle$$

$$+ \left\{\frac{(j-m+1)(j-m+2)}{2(2j+1)(j+1)}\right\}^{1/2} |1, -1\rangle |j, m\rangle.$$

The first equation is proved.

We substitute $M = j$ in the second equation:

$$||j,j\rangle = \left(\frac{1}{j+1}\right)^{1/2}|1,1\rangle|j,j-1\rangle - \left(\frac{j}{j+1}\right)^{1/2}|1,0\rangle|j,j\rangle.$$

This equation is correct because $J_+||j,j\rangle = 0$. Suppose that the second equation holds for $M = m$. We are going to show it also holds for $M = m - 1$.

$$\begin{aligned}
||j,m-1\rangle &= \frac{J_-||j,m\rangle}{[(j+m)(j-m+1)]^{1/2}} \\
&= \left\{\frac{(j+m-1)(j-m+2)}{2j(j+1)}\right\}^{1/2}|1,1\rangle|j,m-2\rangle \\
&\quad + \left\{\frac{1}{[j(j+1)]^{1/2}} - \frac{m}{[j(j+1)]^{1/2}}\right\}|1,0\rangle|j,m-1\rangle \\
&\quad - \left\{\frac{2m+(j-m)(j+m+1)}{[2j(j+1)(j+m)(j-m+1)]^{1/2}}\right\}|1,-1\rangle|j,m\rangle \\
&= \left\{\frac{(j+m-1)(j-m+2)}{2j(j+1)}\right\}^{1/2}|1,1\rangle|j,m-2\rangle \\
&\quad - \frac{m-1}{[j(j+1)]^{1/2}}|1,0\rangle|j,m-1\rangle \\
&\quad - \left\{\frac{(j+m)(j-m+1)}{2j(j+1)}\right\}^{1/2}|1,-1\rangle|j,m\rangle.
\end{aligned}$$

The second equation is proved. The third equation can be proved with the same method. However, it can also be proved by orthogonality of the basis states and the phase convention.

18. Calculate the eigenfunctions of the total spinor angular momentum in a three-electron system.

Solution. We first calculate the eigenfunctions of the spinor angular momentum in a two-electron system. The spin of each electron is 1/2. The spinor state for a two-electron system is the product of the spinor states of two electrons, $|\mu\rangle|\nu\rangle$, where μ and ν are taken to be $\pm 1/2$, denoted by $\pm$. The eigenfunction of the total spinor angular momentum in a two-electron system is denoted by $|S_{12}, M_{12}\rangle$, where S_{12} takes 0 or 1, $|M_{12}| \leq S_{12}$. From the result given in Problem 16 we have

$$|1,1\rangle = |+\rangle|+\rangle, \qquad |1,0\rangle = \sqrt{1/2}\,(|+\rangle|-\rangle + |-\rangle|+\rangle),$$
$$|1,-1\rangle = |-\rangle|-\rangle, \qquad |0,0\rangle = \sqrt{1/2}\,(|+\rangle|-\rangle - |-\rangle|+\rangle).$$

One may regard the three-electron system as a compound system composed of a two-electron subsystem and a one-electron subsystem. Its state is the product of the states of two subsystems, $|S_{12}, M_{12}\rangle|\rho\rangle$. Let $||S, S_{12}, M\rangle$ be the state in the three-electron system with the spin S and M, calculated in terms of the results given in Problem 16:

$$||3/2, 1, 3/2\rangle = |1,1\rangle|+\rangle = |+\rangle|+\rangle|+\rangle,$$

$$\begin{aligned}||3/2, 1, 1/2\rangle &= \sqrt{1/3}|1,1\rangle|-\rangle + \sqrt{2/3}|1,0\rangle|+\rangle \\ &= \sqrt{1/3}\,\{|+\rangle|+\rangle|-\rangle + |+\rangle|-\rangle|+\rangle + |-\rangle|+\rangle|+\rangle\},\end{aligned}$$

$$\begin{aligned}||3/2, 1, -1/2\rangle &= \sqrt{2/3}|1,0\rangle|-\rangle + \sqrt{1/3}|1,-1\rangle|+\rangle \\ &= \sqrt{1/3}\,\{|+\rangle|-\rangle|-\rangle + |-\rangle|+\rangle|-\rangle + |-\rangle|-\rangle|+\rangle\},\end{aligned}$$

$$||3/2, 1, -3/2\rangle = |1,-1\rangle|-\rangle = |-\rangle|-\rangle|-\rangle.$$

$$\begin{aligned}||1/2, 1, 1/2\rangle &= \sqrt{2/3}|1,1\rangle|-\rangle - \sqrt{1/3}|1,0\rangle|+\rangle \\ &= \sqrt{1/6}\,\{2|+\rangle|+\rangle|-\rangle - |+\rangle|-\rangle|+\rangle - |-\rangle|+\rangle|+\rangle\},\end{aligned}$$

$$\begin{aligned}||1/2, 1, -1/2\rangle &= \sqrt{1/3}|1,0\rangle|-\rangle - \sqrt{2/3}|1,-1\rangle|+\rangle \\ &= \sqrt{1/6}\,\{|+\rangle|-\rangle|-\rangle + |-\rangle|+\rangle|-\rangle - 2|-\rangle|-\rangle|+\rangle\}.\end{aligned}$$

$$||1/2, 0, 1/2\rangle = |0,0\rangle|+\rangle = \sqrt{1/2}\,\{|+\rangle|-\rangle|+\rangle - |-\rangle|+\rangle|+\rangle\},$$
$$||1/2, 0, -1/2\rangle = |0,0\rangle|-\rangle = \sqrt{1/2}\,\{|+\rangle|-\rangle|-\rangle - |-\rangle|+\rangle|-\rangle\}.$$

If the two-electron subsystem consists of the last two electrons, we calculate the eigenstate $|S, S_{23}, M\rangle$ in the three-electron system with the spin S and M in terms of the results given in Problem 16:

$$|3/2, 1, 3/2\rangle = |+\rangle|1,1\rangle = |+\rangle|+\rangle|+\rangle = ||3/2, 1, 3/2\rangle,$$

$$\begin{aligned}|3/2, 1, 1/2\rangle &= \sqrt{2/3}|+\rangle|1,0\rangle + \sqrt{1/3}|-\rangle|1,1\rangle \\ &= \sqrt{1/3}\,\{|+\rangle|+\rangle|-\rangle + |+\rangle|-\rangle|+\rangle + |-\rangle|+\rangle|+\rangle\} \\ &= ||3/2, 1, 1/2\rangle,\end{aligned}$$

$$|3/2, 1, -1/2\rangle = \sqrt{1/3}|+\rangle|1, -1\rangle + \sqrt{2/3}|-\rangle|1, 0\rangle$$
$$= \sqrt{1/3}\{|+\rangle|-\rangle|-\rangle + |-\rangle|+\rangle|-\rangle + |-\rangle|-\rangle|+\rangle\}$$
$$= ||3/2, 1, -1/2\rangle,$$

$$|3/2, 1, -3/2\rangle = |-\rangle|1, -1\rangle = |-\rangle|-\rangle|-\rangle = ||3/2, 1, -3/2\rangle,$$

$$|1/2, 1, 1/2\rangle = \sqrt{1/3}|+\rangle|1, 0\rangle - \sqrt{2/3}|-\rangle|1, 1\rangle$$
$$= \sqrt{1/6}\{|+\rangle|+\rangle|-\rangle + |+\rangle|-\rangle|+\rangle - 2|-\rangle|+\rangle|+\rangle\}$$
$$= (1/2)||1/2, 1, 1/2\rangle + (\sqrt{3}/2)||1/2, 0, 1/2\rangle,$$

$$|1/2, 1, -1/2\rangle = \sqrt{2/3}|+\rangle|1, -1\rangle - \sqrt{1/3}|-\rangle|1, 0\rangle$$
$$= \sqrt{1/6}\{2|+\rangle|-\rangle|-\rangle - |-\rangle|+\rangle|-\rangle - |-\rangle|-\rangle|+\rangle\}$$
$$= (1/2)||1/2, 1, -1/2\rangle + (\sqrt{3}/2)||1/2, 0, -1/2\rangle,$$

$$|1/2, 0, 1/2\rangle = |+\rangle|0, 0\rangle = \sqrt{1/2}\{|+\rangle|+\rangle|-\rangle - |+\rangle|-\rangle|+\rangle\}$$
$$= (\sqrt{3}/2)||1/2, 1, 1/2\rangle - (1/2)||1/2, 0, 1/2\rangle,$$

$$|1/2, 0, -1/2\rangle = |-\rangle|0, 0\rangle = \sqrt{1/2}\{|-\rangle|+\rangle|-\rangle - |-\rangle|-\rangle|+\rangle\}$$
$$= (\sqrt{3}/2)||1/2, 1, -1/2\rangle - (1/2)||1/2, 0, -1/2\rangle.$$

Comparing two sets of results, we find that the expressions for the states with total spin 3/2 are the same, but the expressions for the states with total spin 1/2 are mixed together. Actually, two sets of solutions are related by the Racah coefficients.

19. The spherical harmonic function $Y_m^\ell(\hat{n})$ belongs to the mth row of the representation D^ℓ of SO(3) so that it is the eigenfunction of the orbital angular momentum operator L_3 for the eigenvalue m. Calculate the eigenfunction for the eigenvalue m of the orbital angular momentum operator $\mathbf{L} \cdot \hat{\mathbf{a}}$ along the direction $\hat{\mathbf{a}} = (\mathbf{e}_1 - \mathbf{e}_2)/\sqrt{2}$ in terms of combining $Y_m^\ell(\hat{n})$ linearly.

Solution. The polar angle and the azimuthal angle of the direction $\hat{\mathbf{a}} = (\mathbf{e}_1 - \mathbf{e}_2)/2$ are $\theta = \pi/2$ and $\varphi = -\pi/4$, respectively. Letting $S = R(-\pi/4, \pi/2, 0)$, we have

$$P_S L_3 P_S^{-1} = \sum_{b=1}^{3} L_b S_{b3} = \mathbf{L} \cdot \hat{\mathbf{a}}.$$

The eigenfunction of $\mathbf{L} \cdot \hat{\mathbf{a}}$ for the eigenvalue m is

$$P_S Y_m^\ell(\hat{\mathbf{n}}) = \sum_{m'} Y_{m'}^\ell(\hat{\mathbf{n}}) D_{m'm}^\ell(-\pi/4, \pi/2, 0)$$
$$= \sum_{m'} Y_{m'}^\ell(\hat{\mathbf{n}}) e^{im'\pi/4} d_{m'm}^\ell(\pi/2),$$

because

$$\mathbf{L} \cdot \hat{\mathbf{a}} \left\{ P_S Y_m^\ell(\hat{\mathbf{n}}) \right\} = P_S L_3 Y_m^\ell(\hat{\mathbf{n}}) = m \left\{ P_S Y_m^\ell(\hat{\mathbf{n}}) \right\}.$$

20. Let the function $\psi_m^\ell(x)$ belong to the mth row of the irreducible representation D^ℓ of SO(3). Calculate the eigenfunction for the eigenvalue m of the orbital angular momentum operator $\mathbf{L} \cdot \hat{\mathbf{b}}$ along the direction $\hat{\mathbf{b}} = \left(\sqrt{3}\mathbf{e}_2 + \mathbf{e}_3 \right)/2$ in terms of combining $\psi_m^\ell(x)^*$ linearly.

Hint: Use the similarity transformation between the representation D^j of SO(3) and its complex conjugate representation.

Solution. The representation $D^j(R)^*$ is equivalent to the representation $D^j(R)$,

$$D^j(R) = d^j(\pi) D^j(R)^* d^j(\pi)^{-1},$$
$$d_{\mu\nu}^j(\pi)^{-1} = d_{\mu\nu}^j(-\pi) = \delta_{\mu(-\nu)}(-1)^{j+\nu}.$$

Combine the functions $\psi_m^\ell(x)^*$ into the function $\phi_m^\ell(x)$ belonging to the irreducible representation D^ℓ:

$$P_R \psi_m^\ell(x)^* = \sum_{m'} \psi_{m'}^\ell(x)^* D_{m'm}^\ell(R)^*,$$
$$\phi_m^\ell(x) = \sum_{m'} \psi_{m'}^\ell(x)^* d_{m'm}^\ell(-\pi) = (-1)^{\ell+m} \psi_{-m}^\ell(x)^*,$$
$$P_R \phi_m^\ell(x) = \sum_{m'} \phi_{m'}^\ell(x) D_{m'm}^\ell(R).$$

Thus, $\phi_m^\ell(x)$ is the eigenfunction of the orbital angular momentum L_3 along the Z axis for the eigenvalue m. The polar angle and azimuthal angle of $\hat{\mathbf{b}}$ are $\pi/3$ and $\pi/2$, respectively. By the method used in Problem 19, we obtain the eigenfunction of the orbital angular momentum $\mathbf{L} \cdot \hat{\mathbf{b}}$ along the direction $\hat{\mathbf{b}}$ as follows, where $S = R(\pi/2, \pi/3, 0)$:

$$P_S \phi_m^\ell(x) = \sum_{m'} \phi_{m'}^\ell(x) D_{m'm}^\ell(S)$$
$$= (-1)^\ell \sum_{m'} \psi_{-m'}^\ell(x)^* e^{im'\pi/2} d_{m'm}^\ell(\pi/3).$$

Another method of calculation is based on the property that $L_3\psi^\ell_{-m}(x)^* = m\psi^\ell_m(x)^*$. Thus, $P_S\psi^\ell_{-m}(x)^*$ is the eigenfunction of $P_S L_3 P_S^{-1}$ for the eigenvalue m. Letting $S = R(\pi/2, \pi/3, 0)$, we obtain the eigenfunction of the orbital angular momentum $\mathbf{L} \cdot \hat{\mathbf{b}}$ along the direction $\hat{\mathbf{b}}$ for the eigenvalue m:

$$
\begin{aligned}
P_S\psi^\ell_{-m}(x)^* &= \sum_{m'} \psi^\ell_{m'}(x)^* D^\ell_{m'(-m)}(S)^* \\
&= (-1)^m \sum_{m'} \psi^\ell_{m'}(x)^* e^{-im'\pi/2} d^\ell_{(-m')m}(\pi/3).
\end{aligned}
$$

Two results are the same up to the phase factor. Note that the summing index m' can be replaced with $-m'$.

21. Q_R is the rotational transformation operator in the spinor space. In the rotation Q_R, the spinor basis $e^{(s)}(\rho)$ belongs to the ρth row of the irreducible representation D^s, so that it is the common eigenfunction of the spinor angular momentum S^2 and S_3 for the eigenvalues $s(s+1)$ and ρ, respectively. Based on this property, calculate the eigenfunction of the spinor angular momentum $\mathbf{S} \cdot \hat{\mathbf{r}}$ along the radial direction, where $\hat{\mathbf{r}}$ is the unit vector in the radial direction.

Solution. The operator $\mathbf{S} \cdot \hat{\mathbf{r}}$ is a scalar operator in the whole spatial rotation, but a vector operator in the rotation of the spinor space. Letting the polar angle and azimuthal angle of $\mathbf{r}$ be θ and φ, respectively, we have

$$
\mathbf{S} \cdot \hat{\mathbf{r}} = Q_R S_3 Q_R^{-1}, \qquad R = R(\varphi, \theta, 0).
$$

Its eigenfunction for the eigenvalue ρ is

$$
Q_R e^{(s)}(\rho) = \sum_\lambda e^{(s)}(\lambda) D^s_{\lambda\rho}(R) = \sum_\lambda e^{(s)}(\lambda) e^{-i\lambda\varphi} d^s_{\lambda\rho}(\theta),
$$
$$
(\mathbf{S} \cdot \hat{\mathbf{r}}) Q_R e^{(s)}(\rho) = \rho Q_R e^{(s)}(\rho).
$$

22. There are three sets of the mutual commutable angular momentum operators. One set consists of L^2, L_3, S^2, and S_3. The other set consists of J^2, J_3, L^2 and S^2, and the third set consists of J^2, J_3, S^2 and $\mathbf{S} \cdot \hat{\mathbf{r}}$. Calculate the common eigenfunctions of three sets of operators, respectively.

Solution. The common eigenfunction of L^2, L_3, S^2, and S_3 is the product of the spherical harmonic function $Y^\ell_m(\hat{\mathbf{n}})$ and the spinor basis $e^{(s)}(\rho)$. Combining the products $Y^\ell_m(\hat{\mathbf{n}})e^{(s)}(\rho)$ by the C-G coefficients, we obtain the common eigenfunction of J^2, J_3, L^2 and S^2, which is called the spherical spinor function $Y_\mu^{j\ell s}(\hat{\mathbf{n}})$:

$$Y_\mu^{j\ell s}(\hat{\mathbf{n}}) = \sum_\rho C_{(\mu-\rho)\rho j\mu}^{\ell s} Y_{\mu-\rho}^{\ell}(\hat{\mathbf{n}}) e^{(s)}(\rho),$$

where the spinor basis $e^{(s)}(\rho)$ is a spinor of rank s. The above formula is a spinor equation, namely, a $(2s+1) \times 1$ matrix equation. Since the sum runs over the azimuthal quantum numbers, the spherical spinor function $Y_\mu^{j\ell s}(\hat{\mathbf{n}})$ is also the eigenfunction of L^2 and S^2:

$$J^2 Y_\mu^{j\ell s}(\hat{\mathbf{n}}) = j(j+1) Y_\mu^{j\ell s}(\hat{\mathbf{n}}) \ ,$$
$$J_3 Y_\mu^{j\ell s}(\hat{\mathbf{n}}) = \mu Y_\mu^{j\ell s}(\hat{\mathbf{n}}) \ ,$$
$$J_\pm Y_\mu^{j\ell s}(\hat{\mathbf{n}}) = \Gamma_{\mp\mu}^j Y_{\mu\pm1}^{j\ell s}(\hat{\mathbf{n}}) \ ,$$
$$L^2 Y_\mu^{j\ell s}(\hat{\mathbf{n}}) = \ell(\ell+1) Y_\mu^{j\ell s}(\hat{\mathbf{n}}) \ ,$$
$$S^2 Y_\mu^{j\ell s}(\hat{\mathbf{n}}) = s(s+1) Y_\mu^{j\ell s}(\hat{\mathbf{n}}).$$

By making use of the Clebsch-Gordan coefficients calculated in Problem 17, we obtain the explicit forms of spherical spinor function $Y_\mu^{j\ell s}(\hat{\mathbf{n}})$, where $s = 1/2$ and $\ell = j \mp 1/2$:

$$Y_\mu^{j(j-1/2)(1/2)}(\hat{\mathbf{n}}) = \begin{pmatrix} \left(\dfrac{j+\mu}{2j}\right)^{1/2} Y_{\mu-1/2}^{j-1/2}(\hat{\mathbf{n}}) \\ \left(\dfrac{j-\mu}{2j}\right)^{1/2} Y_{\mu+1/2}^{j-1/2}(\hat{\mathbf{n}}) \end{pmatrix},$$

$$Y_\mu^{j(j+1/2)(1/2)}(\hat{\mathbf{n}}) = \begin{pmatrix} -\left(\dfrac{j-\mu+1}{2j+2}\right)^{1/2} Y_{\mu-1/2}^{j+1/2}(\hat{\mathbf{n}}) \\ \left(\dfrac{j+\mu+1}{2j+2}\right)^{1/2} Y_{\mu+1/2}^{j+1/2}(\hat{\mathbf{n}}) \end{pmatrix}.$$

The common eigenfunction of J^2, J_3 and S^2 is a spinor of rank s and belongs to the irreducible representation D^j of SO(3) in the whole spatial rotation,

$$O_R \Psi_\mu^j(\mathbf{r}) = D^s(R) \Psi_\mu^j(R^{-1}\mathbf{r}) = \sum_\nu \Psi_\nu^j(\mathbf{r}) D_{\nu\mu}^j(R).$$

Let the spherical coordinates of the position vector $\mathbf{r}$ be (r, θ, φ). $T = R(\varphi, \theta, \gamma)$ is a rotation transforming the Z axis to the radial direction $\mathbf{r}$. $T\mathbf{r}_0 = \mathbf{r}$, where $\mathbf{r}_0 = (r, 0, 0)$ is the position vector $\mathbf{r}$ along the Z axis.

Replacing $\mathbf{r}$ with $\mathbf{r}_0$ and R with T^{-1} in the above formula, we have

$$\Psi_\mu^j(\mathbf{r}) = \Psi_\mu^j(T\mathbf{r}_0) = D^s(T)O_{T^{-1}}\Psi_\mu^j(\mathbf{r}_0)$$
$$= \sum_\nu D^s(T)\Psi_\nu^j(\mathbf{r}_0)D_{\nu\mu}^j(T^{-1}).$$

Write it in the component equation,

$$\Psi_\mu^j(\mathbf{r})_\rho = \sum_{\nu\sigma} D_{\rho\sigma}^s(\varphi,\theta,\gamma)\Psi_\nu^j(\mathbf{r}_0)_\sigma D_{\mu\nu}^j(\varphi,\theta,\gamma)^*$$
$$= \sum_{\nu\sigma} e^{i\varphi(\mu-\rho)}d_{\rho\sigma}^s(\theta)e^{i\gamma(\nu-\sigma)}\Psi_\nu^j(\mathbf{r}_0)_\sigma d_{\mu\nu}^j(\theta).$$

Since the left-hand side of the equation is independent of γ, its right-hand side has to be independent of γ, too. Namely, $\Psi_\nu^j(\mathbf{r}_0)_\sigma$ has to be zero except for the case $\nu = \sigma$,

$$\Psi_\nu^j(\mathbf{r}_0)_\sigma = \delta_{\nu\sigma}\phi_\nu^j(r).$$

Substituting it into $\Psi_\mu^j(\mathbf{r})_\rho$, we have

$$\Psi_\mu^j(\mathbf{r}) = \sum_\rho \Psi_\mu^j(\mathbf{r})_\rho e^{(s)}(\rho)$$
$$= \sum_{\rho\nu} e^{(s)}(\rho)D_{\rho\nu}^s(\varphi,\theta,0)D_{\mu\nu}^j(\varphi,\theta,0)^*\phi_\nu^j(r)$$
$$= \sum_\nu \left[Q_{T_0}e^{(s)}(\nu)\right]D_{\mu\nu}^j(T_0)^*\phi_\nu^j(r),$$

where $T_0 = R(\varphi,\theta,0)$. Remind that the quantity in the square bracket is nothing but the eigenfunction of $\mathbf{S}\cdot\hat{\mathbf{r}}$ for the eigenvalue ν [see Problem 21]. $\Psi_\mu^j(\mathbf{r})$ is the eigenfunction of $\mathbf{S}\cdot\hat{\mathbf{r}}$ if and only if the sum contains only one term with a fixed ν. Namely, $\phi_\nu^j(r)$ is nonvanishing only when ν takes a given value. Therefore, the common eigenfunction of J^2, J_3, S^2 and $\mathbf{S}\cdot\hat{\mathbf{r}}$ respectively for the eigenvalues $j(j+1)$, μ, $s(s+1)$ and ν is

$$\left[Q_{T_0}e^{(s)}(\nu)\right]D_{\mu\nu}^j(T_0)^*\phi_\nu^j(r) = \sum_\rho e^{(s)}(\rho)D_{\rho\nu}^s(\varphi,\theta,0)D_{\mu\nu}^j(\varphi,\theta,0)^*\phi_\nu^j(r),$$

where $\phi_\nu^j(r)$ plays the role of the radial function, or simply the role of the normalization factor.

23. Calculate $\left\{d^\ell(\theta)\left(I_3^\ell\right)^2 d^\ell(\theta)^{-1}\right\}_{mm}$, where $d^\ell(\theta)$ is the representation matrix of $R(\mathbf{e}_2,\theta)$ in D^ℓ of $SO(3)$ and I_3^ℓ is the third generator in the representation.

Hint: Use the property of the adjoint representation.

Solution. The self-representation of SO(3) is just its adjoint representation. From the definition (4.21) for the adjoint representation, we have

$$d^\ell(\theta)I_3^\ell d^\ell(\theta)^{-1} = \sum_{b=1}^{3} I_b^\ell R_{b3}(\mathbf{e}_2,\theta) = I_1^\ell \sin\theta + I_3^\ell \cos\theta,$$

$$d^\ell(\theta)\left(I_3^\ell\right)^2 d^\ell(\theta)^{-1} = \left\{d^\ell(\theta)I_3^\ell d^\ell(\theta)^{-1}\right\}^2$$

$$= \left(I_1^\ell\right)^2 \sin^2\theta + \left(I_3^\ell\right)^2 \cos^2\theta + \left(I_1^\ell I_3^\ell + I_3^\ell I_1^\ell\right)\sin\theta\cos\theta.$$

Since I_3^ℓ is diagonal, and $I_1^\ell = \left(I_+^\ell + I_-^\ell\right)/2$ does not contain any nonvanishing diagonal matrix entry, so

$$\left\{d^\ell(\theta)\left(I_3^\ell\right)^2 d^\ell(\theta)^{-1}\right\}_{mm}$$

$$= \frac{\sin^2\theta}{4}\left\{\left(I_+^\ell\right)^2 + \left(I_-^\ell\right)^2 + I_+^\ell I_-^\ell + I_-^\ell I_+^\ell\right\}_{mm} + m^2 \cos^2\theta$$

$$= \frac{\sin^2\theta}{4}\left\{(\ell+m)(\ell-m+1) + (\ell+m+1)(\ell-m)\right\} + m^2 \cos^2\theta$$

$$= \frac{\sin^2\theta}{2}\left\{\ell^2 + \ell - m^2\right\} + m^2 \cos^2\theta.$$

24. Establish the differential equation satisfied by the matrix entries $D_{\nu\mu}^j(\alpha,\beta,\gamma)$ of the representation D^j of SO(3).

Solution. $D_{\nu\mu}^j(\alpha,\beta,\gamma)$ is the matrix entry of the rotational operator $O_{R(\alpha,\beta,\gamma)}$ between the eigenstates of the angular momentum. Using the Dirac notation, we have:

$$D_{\nu\mu}^j(\alpha,\beta,\gamma) = \langle j,\nu|O_{R(\alpha,\beta,\gamma)}|j,\mu\rangle.$$

Define

$$R = R(\alpha,\beta,\gamma) = R_1 R_2 R_3,$$

$$R_1 = R(\mathbf{e}_3,\alpha), \qquad R_2 = R(\mathbf{e}_2,\beta), \qquad R_3 = R(\mathbf{e}_3,\gamma).$$

Thus, $O_{R(\alpha,\beta,\gamma)}$ can be expressed by the angular momentum operators J_a,

$$O_{R(\alpha,\beta,\gamma)} = O_R = O_{R_1}O_{R_2}O_{R_3} = e^{-iJ_3\alpha}e^{-iJ_2\beta}e^{-iJ_3\gamma},$$

$$O_{R_1} = e^{-iJ_3\alpha}, \qquad O_{R_2} = e^{-iJ_2\beta}, \qquad O_{R_3} = e^{-iJ_3\gamma}.$$

J_a is a vector operator, satisfying:

$$O_R J_a O_R^{-1} = \sum_{b=1}^{3} J_b R_{ba}.$$

Letting $s_\alpha = \sin\alpha$, $c_\alpha = \cos\alpha$ etc., we have

$$i\frac{\partial}{\partial\alpha}O_R = J_3 O_R,$$

$$i\frac{\partial}{\partial\beta}O_R = O_{R_1} J_2 O_{R_1}^{-1} O_R = \{-s_\alpha J_1 + c_\alpha J_2\} O_R,$$

$$i\frac{\partial}{\partial\gamma}O_R = O_{R_1} O_{R_2} J_3 O_{R_2}^{-1} O_{R_1}^{-1} O_R$$

$$= \{s_\beta c_\alpha J_1 + s_\beta s_\alpha J_2 + c_\beta J_3\} O_R.$$

Hence,

$$J_1 O_R = i\left\{-s_\alpha\frac{\partial}{\partial\beta} + \frac{c_\alpha}{s_\beta}\frac{\partial}{\partial\gamma} - \frac{c_\alpha c_\beta}{s_\beta}\frac{\partial}{\partial\alpha}\right\}O_R,$$

$$J_2 O_R = i\left\{c_\alpha\frac{\partial}{\partial\beta} + \frac{s_\alpha}{s_\beta}\frac{\partial}{\partial\gamma} - \frac{s_\alpha c_\beta}{s_\beta}\frac{\partial}{\partial\alpha}\right\}O_R,$$

$$J_3 O_R = i\frac{\partial}{\partial\alpha}O_R,$$

$$J_\pm O_R = (J_1 \pm iJ_2) O_R = ie^{\pm i\alpha}\left\{\pm i\frac{\partial}{\partial\beta} + \frac{1}{s_\beta}\frac{\partial}{\partial\gamma} - \frac{c_\beta}{s_\beta}\frac{\partial}{\partial\alpha}\right\}O_R.$$

Noticing the order of J_+ and J_-, we have

$$J_- J_+ O_R = ie^{i\alpha}\left\{i\frac{\partial}{\partial\beta} + \frac{1}{s_\beta}\frac{\partial}{\partial\gamma} - \frac{c_\beta}{s_\beta}\frac{\partial}{\partial\alpha}\right\}J_- O_R$$

$$= \left\{-\frac{\partial^2}{\partial\beta^2} - \frac{c_\beta}{s_\beta}\frac{\partial}{\partial\beta} - \frac{1}{s_\beta^2}\frac{\partial^2}{\partial\gamma^2} + \frac{2c_\beta}{s_\beta^2}\frac{\partial^2}{\partial\alpha\partial\gamma} - \frac{c_\beta^2}{s_\beta^2}\frac{\partial^2}{\partial\alpha^2} - i\frac{\partial}{\partial\alpha}\right\}O_R.$$

Finally, we obtain

$$J^2 O_R = \left(J_3^2 + J_3 + J_- J_+\right)O_R$$

$$= \left\{-\frac{1}{s_\beta}\frac{\partial}{\partial\beta}s_\beta\frac{\partial}{\partial\beta} - \frac{1}{s_\beta^2}\left(\frac{\partial^2}{\partial\gamma^2} + \frac{\partial^2}{\partial\alpha^2}\right) + \frac{2c_\beta}{s_\beta^2}\frac{\partial^2}{\partial\alpha\partial\gamma}\right\}O_R.$$

Calculating the matrix entry between two eigenstates of the angular momentum, $\langle j,\nu|$ and $|j,\mu\rangle$, we obtain the differential equation satisfied by

$D^j_{\nu\mu}(\alpha, \beta, \gamma)$:

$$\left\{-\frac{1}{s_\beta}\frac{\partial}{\partial\beta}s_\beta\frac{\partial}{\partial\beta} - \frac{1}{s_\beta^2}\left(\frac{\partial^2}{\partial\gamma^2} + \frac{\partial^2}{\partial\alpha^2}\right) + \frac{2c_\beta}{s_\beta^2}\frac{\partial^2}{\partial\alpha\partial\gamma}\right\} D^j_{\nu\mu}(\alpha, \beta, \gamma)$$
$$= j(j+1)D^j_{\nu\mu}(\alpha, \beta, \gamma),$$

$$\left\{\frac{1}{s_\beta}\frac{\partial}{\partial\beta}s_\beta\frac{\partial}{\partial\beta} - \frac{\nu^2 + \mu^2 - 2\nu\mu c_\beta}{s_\beta^2}\right\} d^j_{\nu\mu}(\beta) = -j(j+1)d^j_{\nu\mu}(\beta).$$

4.5 Unitary Representations with Infinite Dimensions

★ SO(3) group is a compact Lie group. It has irreducible unitary representations D^j of finite dimensions. However, SO(2,1) group is not a compact Lie group. Except for the identical representation, any irreducible representation of SO(2,1) with a finite dimension is not unitary, and the dimension of any irreducible unitary representation of SO(2,1) is infinite. In this section we will study the irreducible unitary representations of SO(2,1) in some detail as an example of the Lie groups which are not compact.

The set of all three-dimensional real matrices R satisfying

$$R^T J R = J, \quad \det R = 1, \quad J = \text{diag}\{1, 1, \sigma\}, \tag{4.33}$$

in the multiplication rule of matrices, forms the SO(3) group if $\sigma = 1$ and the SO(2,1) group if $\sigma = -1$. We will denote these two groups by G uniformly. The following inner product is invariant for the group G:

$$\langle \mathbf{x}, \mathbf{y} \rangle = \sum_{a,b=1}^{3} x_a J_{ab} y_b. \tag{4.34}$$

It can be seen from Eq. (4.33) that $R_{33}^2 = 1 - \sigma\left(R_{13}^2 + R_{23}^2\right)$. $|R_{33}| \leq 1$ for SO(3), but $|R_{33}| \geq 1$ for SO(2,1). Therefore, SO(3) group is a compact Lie group, while SO(2,1) group is not compact. The group space of SO(2,1) falls into two disjoint regions. The region with $R_{33} \geq 1$ corresponds to the invariant subgroup $SO_+(2,1)$. For convenience we will still denote this subgroup $SO_+(2,1)$ by the same symbol SO(2,1) if no confusion arises.

Discuss the infinitesimal elements and the generators of G:

$$R = \mathbf{1} - i\alpha X,$$

$$J = R^T J R = J - i\alpha \left\{ X^T J + JX \right\},$$

$$1 = \det R = 1 - i\alpha \operatorname{Tr} X.$$

Thus,

$$\operatorname{Tr} X = 0, \qquad X = -JX^T J = \sum_{a=1}^{3} \omega_a T_a.$$

Denoted by L_a three generators, and by T_a their representation matrices in the self-representation:

$$T_1 = \begin{pmatrix} 0 & 0 & 0 \\ 0 & 0 & -i\sigma \\ 0 & i & 0 \end{pmatrix}, \qquad T_2 = \begin{pmatrix} 0 & 0 & i\sigma \\ 0 & 0 & 0 \\ -i & 0 & 0 \end{pmatrix}, \qquad T_3 = \begin{pmatrix} 0 & -i & 0 \\ i & 0 & 0 \\ 0 & 0 & 0 \end{pmatrix}.$$

They satisfy the following commutation relations

$$[L_1 , L_2] = i\sigma L_3, \qquad [L_2 , L_3] = iL_1, \qquad [L_3 , L_1] = iL_2.$$

Define the raising and lowering operators

$$L_\pm = L_1 \pm iL_2, \qquad [L_3 , L_\pm] = \pm L_\pm, \qquad [L_+ , L_-] = 2\sigma L_3. \qquad (4.35)$$

The Casimir operators are

$$L^2 = \sigma \left(L_1^2 + L_2^2 \right) + L_3^2 = \sigma L_+ L_- + L_3 \left(L_3 - 1 \right)$$

$$= \sigma L_- L_+ + L_3 \left(L_3 + 1 \right), \qquad (4.36)$$

$$\left[L^2 , L_a \right] = 0, \qquad a = 1, \, 2, \, 3.$$

★ As we have known, the covering group of SO(3) is SU(2). Similarly, we will show that the covering group of SO(2,1) is SL(2, R), the set of two-dimensional real matrices with determinant $+1$ where the multiplication rule of elements is the matrix product.

The space composed by two-dimensional traceless real matrices is a real space with respect to three real bases

$$\tau_1 = \sigma_3, \qquad \tau_2 = \sigma_1, \qquad \tau_3 = i\,\sigma_2. \qquad (4.37)$$

Any two-dimensional traceless real matrix X maps onto the position vector $\mathbf{x} = (x_1, x_2, x_3)$ in the real three-dimensional space by a one-to-one correspondence:

$$X = x_1\,\tau_1 + x_2\,\tau_2 + x_3\,\tau_3, \qquad \det X = -\left(x_1^2 + x_2^2 - x_3^2\right),$$

$$x_1 = \frac{1}{2}\mathrm{Tr}\,(X\tau_1), \qquad x_2 = \frac{1}{2}\mathrm{Tr}\,(X\tau_2), \qquad x_3 = \frac{-1}{2}\mathrm{Tr}\,(X\tau_3). \tag{4.38}$$

After a similarity transformation $uXu^{-1} = X'$, $u \in SL(2, R)$, X' is still a traceless real matrix with the same determinant. Thus, the position vector $\mathbf{x}'$ corresponding to X' is related to the position vector $\mathbf{x}$ corresponding to X by a transformation $R \in SO(2,1)$.

$$u\tau_a u^{-1} = \sum_{b=1}^{3} \tau_b R_{ba}, \qquad u \in SL(2, R), \qquad R \in SO(2, 1).$$

Similar to the proof for the covering group of $SO(3)$, we can show the two-to-one correspondence between two matrices $\pm u \in SL(2, R)$ and $R \in SO(2,1)$. This correspondence is invariant in the product of elements. Therefore, $SL(2, R)$ group is the covering group of $SO(2,1)$. The representations for the $SO(3)$ group and the $SO(2,1)$ group including the double-valued representations are actually the representations of the $SU(2)$ group and the $SL(2, R)$ group, respectively.

25. Discuss all inequivalent and irreducible unitary representations of the $SO(3)$ group and the $SO(2,1)$ group.

Solution. Choose the basis states $|Q, m\rangle$ in the irreducible representation space where the representation matrices of L^2 and L_3 are diagonal,

$$L^2|Q, m\rangle = Q|Q, m\rangle, \qquad L_3|Q, m\rangle = m|Q, m\rangle. \tag{4.39}$$

In a unitary representation, the representation matrices of L_3 and L^2 are both hermitian, so Q and m are both real numbers. Let

$$Q = j(j + 1). \tag{4.40}$$

There are two solutions for j

$$j = -(1/2) \pm \sqrt{Q + 1/4}.$$

Two solutions are related with each other by the following transformation

$$j \longleftrightarrow -j - 1. \tag{4.41}$$

Recall that the irreducible representation is described by the real parameter Q. Two j's related by the transformation (4.41) describe the same representation. j is real when $Q \geq -1/4$, and j is complex when $Q < -1/4$.

From Eqs. (4.35) and (4.36), we have

$$L_3 (L_\pm|Q,m\rangle) = (m \pm 1)(L_\pm|Q,m\rangle),$$

$$L^2 (L_\pm|Q,m\rangle) = Q (L_\pm|Q,m\rangle).$$

Thus, beginning with a given basis state $|Q,m_0\rangle$, by the raising and lowering operators $L_\pm$, we can obtain a series of basis states $|Q,m\rangle$ where the neighboring m are different by ± 1:

$$
\begin{aligned}
L_+|Q,m-1\rangle = A_{Qm}|Q,m\rangle, &\qquad m > m_0, \\
L_-|Q,m\rangle = B_{Qm}|Q,m-1\rangle, &\qquad m \leq m_0,
\end{aligned}
\tag{4.42}
$$

until A_{Qm} or B_{Qm} equals to zero. Equation (4.42) is the definition for these basis states. They are all common eigenstates of L^2 and L_3, satisfying Eq. (4.39). We need to show that the set of basis states is closed for $L_\pm$, namely

$$
\begin{aligned}
L_-|Q,m\rangle = B_{Qm}|Q,m-1\rangle, &\qquad m > m_0, \\
L_+|Q,m-1\rangle = A_{Qm}|Q,m\rangle, &\qquad m \leq m_0.
\end{aligned}
\tag{4.43}
$$

The proofs for two equations are similar. We will prove the first equation in Eq. (4.43) as example. When $m > m_0$, we have

$$
\begin{aligned}
Q|Q,m-1\rangle &= L^2|Q,m-1\rangle \\
&= \{\sigma L_- L_+ + L_3 (L_3 + 1)\}|Q,m-1\rangle \\
&= \sigma A_{Qm} (L_-|Q,m\rangle) + (m^2 - m)|Q,m-1\rangle.
\end{aligned}
\tag{4.44}
$$

From Eq. (4.42), A_{Qm} is nonvanishing if $|Q,m\rangle$ exists, so $L_-|Q,m\rangle$ is proportional to $|Q,m-1\rangle$. Thus, Eq. (4.43) is proved. So, these basis states form a complete set, spanning an irreducible representation space of G. Note that the definition (4.42) for the basis state $|Q,m\rangle$ allows a multiplied factor in front of it. Namely, the values of A_{Qm} and B_{Qm} are undetermined. On the other hand, the basis state $|Q,m\rangle$ is also the eigenstate of the operators L_+L_- and L_-L_+ for the eigenvalue $A_{Qm}B_{Qm}$ and $A_{Q(m+1)}B_{Q(m+1)}$, respectively. In a unitary representation, the representation matrices for L_+ and L_- are conjugate to each other, so the representation matrices of L_+L_- and L_-L_+ are both positive semi-definite. From Eq. (4.44), when

$m > m_0$, we have

$$A_{Qm}B_{Qm} = \sigma\left(Q - m^2 + m\right) = \sigma(j + m)(j - m + 1) \geq 0. \qquad (4.45)$$

Similarly, it is also true for $m \leq m_0$. We may choose the factor in the basis state $|Q, m\rangle$ such that

$$A_{Qm} = B_{Qm} = [\sigma(j + m)(j - m + 1)]^{1/2} \geq 0. \qquad (4.46)$$

According to whether the state chain is truncated, i.e., whether A_{Qm} (B_{Qm}) is vanishing at some m, the irreducible unitary representations of the SO(3) group and the SO(2,1) group are classified into the unbounded spectrum $D(Q, m_0)$, the spectrum $D^+(j)$ bounded below, the spectrum $D^-(j)$ bounded above, and the bounded spectrum $D(j)$ [Adams et al. (1987)].

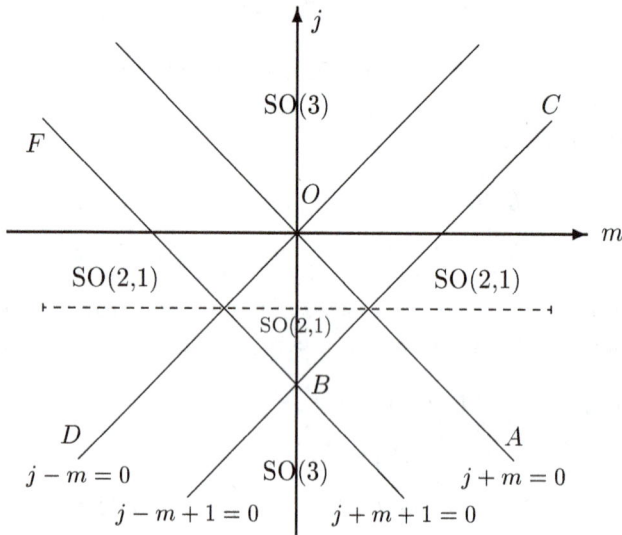

Fig. 4.1 The varied region for the parameters j and m in the irreducible unitary representation of SO(3) and SO(2,1).

Recall that the values of σ are different for two groups: $\sigma = 1$ for the SO(3) group and $\sigma = -1$ for the SO(2,1) group. From Eq. (4.45), the varied regions of the parameters j and m for two groups are different due to different σ. For SO(3), $Q \geq m^2 - m \geq -1/4$, and j is real. The possible values of j and m are restricted in the intersection of two regions: the region above or below two crossed line $j + m = 0$ and $j - m + 1 = 0$, and the region above or below two crossed lines $j - m = 0$ and $j + m + 1 = 0$. The intersection is divided into two infinite regions indicated by SO(3) in Fig. 4.1 [Adams et al. (1987)]. Two regions are equivalent because they are related by the transformation (4.41). In the above region of the intersection, $-j \leq m \leq j$, and $j \geq 0$, namely, the values of m are finite. Letting the upper boundary of m be M, from

$$A_{Q(M+1)} = [(j + M + 1)(j - M)]^{1/2} = 0,$$

we obtain $M = j \geq 0$. Here and the remaining part in this Problem we take k to be the non-negative integer. Letting the lower boundary of m be $j - k$, from

$$B_{Q(j-k)} = [(2j - k)(k + 1)]^{1/2} = 0,$$

we obtain $j = k/2 = 0$, $1/2$, 1, $3/2$, $\cdots$. Therefore, the bounded spectra $D(j)$, where j is a non-negative half integer and $m = j$, $j - 1$, $\cdots$, $-j$, are the only unitary representations for the SO(3) group. Obviously, they are the commonly used finite dimensional representations $D^j(SO(3))$.

For the SO(2,1) group, j may be real or complex. When j is a complex number, $j = -1/2 + i\beta$, $\beta > 0$, one may have

$$Q = -1/4 - \beta^2, \qquad m = m_0 \pm k, \qquad -1/2 < m_0 \leq 1/2. \qquad (4.47)$$

This is called the principal series of representations $D_p(\beta, m_0)$ for the unbounded spectrum.

When j is a real number, the possible values of j and m are restricted in the intersection of two regions: the region in the left or right side of two crossed lines $j + m = 0$ and $j - m + 1 = 0$ and the region in the left or right side of two crossed lines $j - m = 0$ and $j + m + 1 = 0$. The intersection is divided into two infinite areas and a finite area indicated by SO(2,1) in Fig. 4.1. Considering the equivalent transformation (4.41), we obtain a few kinds of representations of SO(2,1). First, if A_{Qm} (B_{Qm}) is always nonvanishing for any m in the representation, there is no truncated in the series of m. In this case, $-1/4 \leq Q < 0$, and it is the representation $D_s(Q, m_0)$ in the supplementary series for the unbounded spectrum :

$$-1/2 \leq j < 0, \qquad m = m_0 \pm k, \qquad |m_0| < -j = 1/2 - (Q + 1/4)^{1/2}. \qquad (4.48)$$

Second, if $Q = j = m = 0$, this is the identical representation. Third, if there is a lower boundary m_0 for m in the representation, the lower boundary lies only on the line of AO (or the equivalent line CB). It is the representation $D^+(j)$ for the spectrum bounded below where $j < 0$, $m > 0$, and $m_0 = -j$:

$$j < 0, \qquad m = -j + k.$$

Finally, if there is an upper boundary m_0 for m in the representation, the upper boundary lies only on the line DO (or the equivalent line FB). It is the representation $D^-(j)$ for the spectrum bounded above, where $j < 0$, $m < 0$, and $m_0 = j$:

$$j < 0, \qquad m = j - k.$$

Therefore, the dimension of any irreducible unitary representation of SO(2,1), except for the identical representation, is infinite.

Chapter 5

SYMMETRY OF CRYSTALS

5.1 Symmetric Operations and Space Groups

★ The fundamental character of a crystal is the spatial periodic array of the atoms composing the crystal, called the crystal lattice. By the periodic boundary condition, the crystal is invariant in the following translation:

$$\mathbf{r} \longrightarrow T(\vec{\ell})\mathbf{r} = \mathbf{r} + \vec{\ell}, \tag{5.1}$$

where $\vec{\ell}$ is called the vector of the crystal lattice. Three fundamental periods $\mathbf{a}_j$ of the crystal lattice, which are not coplanar, are taken to be the basis vectors of the crystal lattice, or briefly called the lattice bases. The lattice bases are said to be primitive if any vector of the crystal lattice is an integral combination of the lattice bases. For simplicity, we only use the primitive lattice bases if without special indication.

★ The multiplication of two translations is defined to be a translation where two translation vectors are added. The set of all translations $T(\vec{\ell})$ which leave the crystal invariant forms the Abelian translation group $\mathcal{T}$ of the crystal. Usually, in addition to the translation symmetry, a crystal has some other symmetric operations which leave the crystal invariant. A general symmetric operation may be a combinative operation composed of the spatial inversion, the rotation and the translation. Denoted by $g(R, \vec{\alpha})$ a general symmetric operation, where R is a proper or improper rotation, $R \in O(3)$, and $\vec{\alpha}$ is a translation vector, not necessary an integral combination of the lattice bases $\mathbf{a}_j$:

$$g(R, \vec{\alpha})\mathbf{r} = R\mathbf{r} + \vec{\alpha}. \tag{5.2}$$

Moving out the vector of the crystal lattice $\vec{\ell}$ from $\vec{\alpha}$, we express the general symmetric operation as

$$g(R, \vec{\alpha}) = T(\vec{\ell})g(R, \mathbf{t}), \qquad \vec{\alpha} = \vec{\ell} + \mathbf{t}$$
$$\mathbf{t} = \sum_{j=1}^{3} \mathbf{a}_j t_j, \qquad 0 \le t_j < 1. \tag{5.3}$$

When $R = E$, $\vec{\alpha}$ has to be a vector of the crystal lattice $\vec{\ell}$ and $g(E, \vec{\ell}) = T(\vec{\ell})$.

From the definition (5.2), the multiplication of two symmetric operations satisfies

$$g(R, \vec{\alpha})g(R', \vec{\beta}) = g(RR', \vec{\alpha} + R\vec{\beta}). \tag{5.4}$$

Thus, the inverse $g(R, \vec{\alpha})^{-1}$ and the identity E are

$$g(R, \vec{\alpha})^{-1} = g(R^{-1}, -R^{-1}\vec{\alpha}), \qquad E = g(E, \mathbf{0}). \tag{5.5}$$

The set of all symmetric operations $g(R, \vec{\alpha})$ for a crystal with the multiplication rule (5.4) forms a group S, called the space group of the crystal. In the multiplication (5.4) of two symmetric operations, the multiplication RR' of two rotations does not matter with whether there are the translations $\vec{\alpha}$ and $\vec{\beta}$ or not. Therefore, the set of the rotational parts R in $g(R, \vec{\alpha})$ forms a group G, called the crystallographic point group. For a given crystal with the space group S, it is easy to show by reduction to absurdity that $\mathbf{t}$ in the symmetric operation $g(R, \mathbf{t})$ depends upon R uniquely. Generally, R is not an element of S, and G is not a subgroup of S. The space group S is called the symmorphic space group if G is the subgroup of S. In a symmorphic space group, any element can be expressed as $g(R, \vec{\ell}) = T(\ell)R$.

From Eq. (5.4) we have

$$g(R, \vec{\alpha})T(\vec{\ell})g(R, \vec{\alpha})^{-1} = T(R\vec{\ell}). \tag{5.6}$$

Namely, the translation group $\mathcal{T}$ is the invariant subgroup of the space group S. Due to Eq. (5.3), the coset of $\mathcal{T}$ is completely determined by the rotation R. Thus, the quotient group of $\mathcal{T}$ with respect to S is the crystallographic point group, $G = \mathcal{T}/S$. From Eq. (5.6), the action of any element R of G on any vector of the crystal lattice $\vec{\ell}$ has to be a vector of the crystal lattice $\vec{\ell'}$

$$R\vec{\ell} = \vec{\ell'}. \tag{5.7}$$

This is the fundamental constraint for the possible crystallographic point groups, the crystal systems, and the Bravais lattices.

★ In the crystal theory, the lattice bases $\mathbf{a}_j$ are usually chosen as the basis vectors in the real three dimensional space. The merit for this choice is that, due to Eq. (5.7), any matrix entry R_{ij} of R in the bases is an integer:

$$R\mathbf{a}_j = \sum_{i=1}^{3} \mathbf{a}_i R_{ij}.$$

The shortcoming is that $\mathbf{a}_j$ are generally not orthonormal. Define a set of basis vectors $\mathbf{b}_j$ satisfying

$$\mathbf{b}_i \cdot \mathbf{a}_j = \delta_{ij}. \tag{5.8}$$

$\mathbf{b}_j$ is called the basis vector of the reciprocal lattice, or briefly called the reciprocal lattice basis. It is convenient in the crystal theory to express a rotation R in a double-vector form:

$$\vec{R} = \sum_{ij} R_{ij}\mathbf{a}_i\mathbf{b}_j, \qquad \vec{R} \cdot \mathbf{a}_j = \sum_{i} R_{ij}\mathbf{a}_i. \tag{5.9}$$

1. In the rectangular coordinate frame, write the double-vector forms and the matrix forms of the proper and improper six-fold rotations around the z axis.

Solution. An improper six-fold rotation S_6 around the z axis is equal to the corresponding proper one C_6 multiplying by a spatial inversion. Therefore, their double-vector forms and the matrix forms are different only by a sign.

$$\vec{C}_6 = -\vec{S}_6 = \left\{ \frac{1}{2}\mathbf{e}_1 + \frac{\sqrt{3}}{2}\mathbf{e}_2 \right\} \mathbf{e}_1 + \left\{ -\frac{\sqrt{3}}{2}\mathbf{e}_1 + \frac{1}{2}\mathbf{e}_2 \right\} \mathbf{e}_2 + \mathbf{e}_3\mathbf{e}_3,$$

$$C_6 = -S_6 = \frac{1}{2} \begin{pmatrix} 1 & -\sqrt{3} & 0 \\ \sqrt{3} & 1 & 0 \\ 0 & 0 & 2 \end{pmatrix}.$$

2. Express the directions of the proper and improper rotational axes in the groups $\mathbf{T}_d$ and $\mathbf{O}_h$ by the basis vectors $\mathbf{e}_j$ of the rectangular coordinate frame. Express the double-vector forms of the generators of the proper rotational axes in $\mathbf{O}_h$ by $\mathbf{e}_j$.

Solution. The group $\mathbf{T}_d$ is a subgroup of the group $\mathbf{O}_h$. $\mathbf{T}_d$ contains four proper three-fold axes, three improper four-fold axes, and six improper two-fold axes. For $\mathbf{O}_h$, all those axes are both the proper and improper ones.

The directions and the double-vector forms of the generators of the proper three-fold axes are

Direction : $\sqrt{1/3}\,\{e_1 + e_2 + e_3\}$, $\quad \vec{R}_1 = e_2e_1 + e_3e_2 + e_1e_3$,

Direction : $\sqrt{1/3}\,\{e_1 - e_2 - e_3\}$, $\quad \vec{R}_2 = -e_2e_1 + e_3e_2 - e_1e_3$,

Direction : $\sqrt{1/3}\,\{-e_1 - e_2 + e_3\}$, $\quad \vec{R}_3 = e_2e_1 - e_3e_2 - e_1e_3$,

Direction : $\sqrt{1/3}\,\{-e_1 + e_2 - e_3\}$, $\quad \vec{R}_4 = -e_2e_1 - e_3e_2 + e_1e_3$.

The directions and the double-vector forms of the generators of the proper four-fold axes are

Direction : e_1, $\quad \vec{T}_1 = e_1e_1 + e_3e_2 - e_2e_3$,

Direction : e_2, $\quad \vec{T}_2 = -e_3e_1 + e_2e_2 + e_1e_3$,

Direction : e_3, $\quad \vec{T}_3 = e_2e_1 - e_1e_2 + e_3e_3$.

The directions and the double-vector forms of the generators of the proper two-fold axes are

Direction : $\sqrt{1/2}\,\{e_1 + e_2\}$, $\quad \vec{S}_1 = e_2e_1 + e_1e_2 - e_3e_3$,

Direction : $\sqrt{1/2}\,\{e_1 - e_2\}$, $\quad \vec{S}_2 = -e_2e_1 - e_1e_2 - e_3e_3$,

Direction : $\sqrt{1/2}\,\{e_2 + e_3\}$, $\quad \vec{S}_3 = -e_1e_1 + e_3e_2 + e_2e_3$,

Direction : $\sqrt{1/2}\,\{e_2 - e_3\}$, $\quad \vec{S}_4 = -e_1e_1 - e_3e_2 - e_2e_3$,

Direction : $\sqrt{1/2}\,\{e_1 + e_3\}$, $\quad \vec{S}_5 = e_3e_1 - e_2e_2 + e_1e_3$,

Direction : $\sqrt{1/2}\,\{e_1 - e_3\}$, $\quad \vec{S}_6 = -e_3e_1 - e_2e_2 - e_1e_3$.

3. Write the double-vector form of a rotation $R(\hat{n}, \omega)$ around the direction $\hat{n}$ through the angle ω by the unit vector $\vec{I}$ and the unit vector $\hat{n}$.

Solution. Let $\hat{m}$ be a unit vector orthogonal to $\hat{n}$. Then, $\hat{n} \times \hat{m}$ is a unit vector orthogonal to both $\hat{n}$ and $\hat{m}$. Applying $R(\hat{n}, \omega)$ to those three unit vectors, we have

$$\vec{R}(\hat{n},\omega) \cdot \hat{n} = \hat{n},$$
$$\vec{R}(\hat{n},\omega) \cdot \hat{m} = \hat{m}\cos\omega + \hat{n} \times \hat{m}\sin\omega,$$
$$\vec{R}(\hat{n},\omega) \cdot (\hat{n} \times \hat{m}) = -\hat{m}\sin\omega + \hat{n} \times \hat{m}\cos\omega.$$

In terms of the project operators $\hat{n}\hat{n}$ and $\vec{1} - \hat{n}\hat{n}$ onto $\hat{n}$ and the plane orthogonal to $\hat{n}$, respectively, we have

$$\vec{R}(\hat{n},\omega) = \hat{n}\hat{n} + \left(\vec{1} - \hat{n}\hat{n}\right)\cos\omega + \left(\vec{1} \times \hat{n}\right)\sin\omega,$$

where the following formulas are used:

$$\left(\vec{1} \times \hat{n}\right) \cdot \hat{m} = \hat{n} \times \hat{m},$$
$$\left(\vec{1} \times \hat{n}\right) \cdot (\hat{n} \times \hat{m}) = \hat{n} \times (\hat{n} \times \hat{m}) = -\hat{m}.$$

5.2 Symmetric Elements

★ A point is called the symmetric center of a symmetric operation $g(R,\mathbf{t})$ if the point is invariant in $g(R,\mathbf{t})$. A straight line is called the symmetric straight line of a symmetric operation $g(R,\mathbf{t})$ if the straight line is invariant in $g(R,\mathbf{t})$. A plane is called the symmetric plane of a symmetric operation $g(R,\mathbf{t})$ if the plane is invariant in $g(R,\mathbf{t})$.

★ A symmetric operation $g(R,\mathbf{t})$ is called closed if its power may be equal to the identity, $g(R,\mathbf{t})^m = E$, otherwise, it is called an open operation. There are only two kinds of open operations. One is $g(C_N,\mathbf{t})$, $N \neq 1$, where C_N is a proper N-fold rotation, and the nonvanishing component $\mathbf{t}_\parallel$ of $\mathbf{t}$ along the direction $\hat{n}$ of the rotational axis is a multiple of the smallest vector of the crystal lattice in the direction $\hat{n}$ divided by N. This axis is called the screw axis, which is parallel to $\hat{n}$ through the point $\mathbf{r}_0$, where $\mathbf{r}_0$ satisfies

$$(E - C_N)\mathbf{r}_0 = \mathbf{t} - \mathbf{t}_\parallel = \mathbf{t}_\perp.$$

$\mathbf{t}_\parallel \neq 0$ is the gliding vector. A screw axis is a symmetric straight line of $g(C_N,\mathbf{t})$. The other is $g(S_2,\mathbf{t})$, where S_2 is an improper twofold rotation (reflection), and the component $\mathbf{t}_\perp$ of $\mathbf{t}$ along the reflection plane is half of the smallest vector of the crystal lattice in the direction. This reflection plane is called the gliding plane, which is orthogonal to the rotational axis $\hat{\mathbf{n}}$ of the improper rotation S_2 through the point $\mathbf{r}_0$, satisfying

$$(E - S_2)\mathbf{r}_0 = \mathbf{t} - \mathbf{t}_\perp = \mathbf{t}_\|.$$

$\mathbf{t}_\perp$ is the gliding vector on the plane. A gliding plane is a symmetric plane of $g(S_2, \mathbf{t})$. The closed transformation has a symmetric center $\mathbf{r}_0$, satisfying

$$(E - R)\mathbf{r}_0 = \mathbf{t}.$$

When R is a proper rotation C_N ($N \neq 1$), $\mathbf{t}$ has to be orthogonal to the rotational axis. The straight line which is through the point $\mathbf{r}_0$ and parallel to $\hat{\mathbf{n}}$ is a symmetric line. Each point on the symmetric line is the symmetric center for the closed operation. When R is the reflection S_2, $\mathbf{t}$ has to be parallel to the rotational axis, i.e. orthogonal to the reflection plane. The plane which is through the point $\mathbf{r}_0$ and orthogonal to the direction $\hat{\mathbf{n}}$ of the rotational axis is a symmetric plane. Each point on the symmetric plane is the symmetric center for the closed operation. When R is an improper rotation S_N, $N \neq 2$, the closed transformation only has a symmetric center.

★ A symbol for a space group of a crystal in the international notations contains all information on the symmetry of the crystal. In the international notations, the first character indicates the Bravais lattice, $\pm$ denotes the two order inversion group C_i, and the digit N or $\overline{N}$ denotes the proper or improper axis. Except for the rhombohedral system, the axis denoted by a digit without prime indicates the principal axis, along which the lattice basis $\mathbf{a}_3$ directs, the axis denoted by the digit 2 with a prime indicates a twofold axis orthogonal to the principal axis, along which the lattice basis $\mathbf{a}_1$ directs, and the axis denoted by a digit $N > 2$ with a prime or denoted by the digit 2 with double primes indicates the rotational axis along other direction.

In analysis of the symmetry of a crystal from its international symbol, one has to know the rule for choosing the lattice bases $\mathbf{a}_j$ in the given crystal system. Whether $\mathbf{a}_j$ are primitive or not depends upon the Bravais lattice. Denote by a_j the length of $\mathbf{a}_j$, and by α_3 the angle between $\mathbf{a}_1$ and $\mathbf{a}_2$, and so on. Let $\mathbf{a}_1$, $\mathbf{a}_2$ and $\mathbf{a}_3$ construct the right-handed system. The rule for choosing the lattice bases $\mathbf{a}_j$ is as follows.

(1). Triclinic system contains $P1$ and $P\overline{1}$.

The lattice bases $\mathbf{a}_j$ are taken to be primitive. There is no restriction on the lengths and the angles of the lattice bases.

(2). Monoclinic system contains P2, $P\overline{2}$, $P \pm 2$, A2, $A\overline{2}$, and $A \pm 2$.

Define the shortest vector of the crystal lattice along the twofold axis to be $\mathbf{a}_3$, and take two non-collinear smallest vectors of crystal lattice in the plane orthogonal to $\mathbf{a}_3$ to be $\mathbf{a}_1$ and $\mathbf{a}_2$, respectively. Two non-collinear vectors of the crystal lattice are called the smallest in a plane if any vector of the crystal lattice in the plane is their integral combination. The restriction for the lattice bases is $\alpha_1 = \alpha_2 = \pi/2$.

(3). Orthorhombic system contains $P22'$, $P2\overline{2}'$, $P \pm 22'$, $C22'$, $C2\overline{2}'$, $A2\overline{2}'$, $C \pm 22'$, $I22'$, $I2\overline{2}'$, $I \pm 22'$, $F22'$, $F2\overline{2}'$, $F \pm 22'$.

Define the shortest vectors of the crystal lattice along three twofold axes to be $\mathbf{a}_1$, $\mathbf{a}_2$ and $\mathbf{a}_3$, respectively. For C_{2v} $(2\overline{2}')$, $\mathbf{a}_3$ is along the proper twofold axis. The restriction for the lattice bases is $\alpha_1 = \alpha_2 = \alpha_3 = \pi/2$.

(4). Tetragonal system contains $P4$, $P\overline{4}$, $P \pm 4$, $P42'$, $P4\overline{2}'$, $P\overline{4}2'$, $P\overline{4}2''$, $P \pm 42'$, $I4$, $I\overline{4}$, $I \pm 4$, $I42'$, $I4\overline{2}'$, $I\overline{4}2'$, $I\overline{4}2''$, and $I \pm 42'$.

Define the shortest vector of the crystal lattice along the fourfold axis to be $\mathbf{a}_3$, and take one shortest vectors of the crystal lattice in the plane orthogonal to $\mathbf{a}_3$ to be $\mathbf{a}_1$. Define $\mathbf{a}_2 = C_4\mathbf{a}_1$. The restriction for the lattice bases is $a_1 = a_2$ and $\alpha_1 = \alpha_2 = \alpha_3 = \pi/2$. For the point group D_{2d}, there are two different symbols $\overline{4}2'$ and $\overline{4}2''$, according to whether $\mathbf{a}_1$ directs along proper or improper twofold axis.

(5). Cubic system contains $P3'22'$, $P\overline{3}'22'$, $P3'42''$, $P3'\overline{4}2''$, $P\overline{3}'42''$, $I3'22'$, $I\overline{3}'22'$, $I3'42''$, $I3'\overline{4}2''$, $I\overline{3}'42''$, $F3'22'$, $F\overline{3}'22'$, $F3'42''$, $F3'\overline{4}2''$, and $F\overline{3}'42''$.

Define three shortest vectors of the crystal lattice along three orthogonal twofold axes [for $\mathbf{T}$ $(3'22')$ and $\mathbf{T}_h$ $(\overline{3}'22')$] or fourfold axes [for $\mathbf{O}$ $(3'42'')$, $\mathbf{T}_d$ $(3'\overline{4}2'')$ and $\mathbf{O}_h$ $(\overline{3}'42'')$] to be $\mathbf{a}_1$, $\mathbf{a}_2$ and $\mathbf{a}_3$, respectively. The restriction for the lattice bases is $a_1 = a_2 = a_3$ and $\alpha_1 = \alpha_2 = \alpha_3 = \pi/2$.

(6). Hexagonal system contains $P3$, $P\overline{3}$, $P32'$, $P32''$, $P3\overline{2}'$, $P3\overline{2}''$, $P\overline{3}2'$, $P\overline{3}2''$, $P6$, $P\overline{6}$, $P \pm 6$, $P62'$, $P6\overline{2}'$, $P\overline{6}2'$, $P\overline{6}2''$ and $P \pm 62'$.

Define the shortest vector of the crystal lattice along a threefold axis or a sixfold axis to be $\mathbf{a}_3$, and take one shortest vectors of the crystal lattice in the plane orthogonal to $\mathbf{a}_3$ to be $\mathbf{a}_1$. Define $\mathbf{a}_2 = C_3\mathbf{a}_1$. The restriction for the lattice bases is $a_1 = a_2$, $\alpha_1 = \alpha_2 = \pi/2$ and $\alpha_3 = 2\pi/3$. For D_3, C_{3v} and D_{3d}, there are two kinds of symbols $32'$ or $32''$, $3\overline{2}'$ or $3\overline{2}''$, and $\overline{3}2'$ or $\overline{3}2''$, where $2'$ (or $\overline{2}'$) means that $\mathbf{a}_1$ directs along a twofold axis, and $2''$ (or $\overline{2}''$) means that $\mathbf{a}_1$ directs along the angular bisector of two twofold axes. For D_6 $(62')$, C_{6v} $(6\overline{2}')$ and D_{6h} $(\pm 62')$, $\mathbf{a}_1$ directs along a twofold axis. For D_{3h}, there are two kinds of symbols $\overline{6}2'$ and $\overline{6}2''$, according to whether $\mathbf{a}_1$ directs along a proper or improper twofold axis.

(7). Rhombohedral system contains $R3$, $R\overline{3}$, $R32'$, $R3\overline{2}'$ and $R\overline{3}2'$.

Three lattice bases are distributed symmetrically around the threefold axis with the same acute angle. The sum of three lattice bases is equal to the shortest vector of the crystal lattice along the threefold axis. The restriction for the lattice bases is $a_1 = a_2 = a_3$ and $\alpha_1 = \alpha_2 = \alpha_3$. For D_3 $(32')$, C_{3v} $(3\overline{2}')$ and D_{3d} $(\overline{3}2')$, the map of any lattice basis onto the plane orthogonal to the threefold axis directs along the angular bisection of two twofold axes.

For different Bravais lattices, in addition to the integral combinations $\vec{\ell}$, the vectors of the crystal lattice are allowed to be some fractional combinations $\mathbf{f}$:

$$
\begin{aligned}
&P \text{ type of Bravais lattice}: && \mathbf{f} = 0, \\
&R \text{ type of Bravais lattice}: && \mathbf{f} = 0, \\
&A \text{ type of Bravais lattice}: && \mathbf{f} = (\mathbf{a}_2 + \mathbf{a}_3)/2, \\
&B \text{ type of Bravais lattice}: && \mathbf{f} = (\mathbf{a}_1 + \mathbf{a}_3)/2, \\
&C \text{ type of Bravais lattice}: && \mathbf{f} = (\mathbf{a}_1 + \mathbf{a}_2)/2, \\
&I \text{ type of Bravais lattice}: && \mathbf{f} = (\mathbf{a}_1 + \mathbf{a}_2 + \mathbf{a}_3)/2, \\
&F \text{ type of Bravais lattice}: && \mathbf{f} = (\mathbf{a}_2 + \mathbf{a}_3)/2, \quad (\mathbf{a}_3 + \mathbf{a}_1)/2, \\
& && \text{and} \quad (\mathbf{a}_1 + \mathbf{a}_2)/2.
\end{aligned}
\tag{5.10}
$$

4. Let R be a rotation around the direction $\hat{n} = (\mathbf{e}_1 + \mathbf{e}_2)/\sqrt{2}$ through $2\pi/3$. Find the symmetric straight line for $g(R, \mathbf{t})$ where (a) $\mathbf{t} = \mathbf{e}_3$, (b) $\mathbf{t} = \mathbf{e}_1 + \mathbf{e}_3$. If the symmetric straight line is a screw line, find its gliding vector. Simplify $g(R, \mathbf{t})$ by moving the origin to the symmetric straight line for checking your result.

Solution. Write the double-vector form of $R = R(\hat{n}, 2\pi/3)$ with $\hat{n} = (\mathbf{e}_1 + \mathbf{e}_2)/\sqrt{2}$ in the rectangular coordinate frame:

$$
\begin{aligned}
\overleftrightarrow{R} &= \frac{1}{2}(\mathbf{e}_1 + \mathbf{e}_2)\{\mathbf{e}_1 + \mathbf{e}_2\} + \frac{1}{4}\left(-\mathbf{e}_1 + \mathbf{e}_2 - \sqrt{6}\mathbf{e}_3\right)\{\mathbf{e}_1 - \mathbf{e}_2\} \\
&\quad + \frac{1}{4}\left\{\sqrt{6}(\mathbf{e}_1 - \mathbf{e}_2) - 2\mathbf{e}_3\right\}\mathbf{e}_3 \\
&= \frac{1}{4}\mathbf{e}_1\mathbf{e}_1 + \frac{3}{4}\mathbf{e}_2\mathbf{e}_1 - \frac{\sqrt{6}}{4}\mathbf{e}_3\mathbf{e}_1 + \frac{3}{4}\mathbf{e}_1\mathbf{e}_2 + \frac{1}{4}\mathbf{e}_2\mathbf{e}_2 + \frac{\sqrt{6}}{4}\mathbf{e}_3\mathbf{e}_2 \\
&\quad + \frac{\sqrt{6}}{4}\mathbf{e}_1\mathbf{e}_3 - \frac{\sqrt{6}}{4}\mathbf{e}_2\mathbf{e}_3 - \frac{1}{2}\mathbf{e}_3\mathbf{e}_3.
\end{aligned}
$$

(a) $\mathbf{t} = \mathbf{t}_\perp = \mathbf{e}_3$.

$g(R, \mathbf{t})$ has a symmetric straight line through the point $\mathbf{r}_0$ and parallel to $\hat{\mathbf{n}}$, where $\mathbf{r}_0$ satisfies

$$g(R, \mathbf{t})\mathbf{r}_0 = R\mathbf{r}_0 + \mathbf{t} = \mathbf{r}_0.$$

Write the equation in the rectangular coordinate frame:

$$(1/4)r_{01} + (3/4)r_{02} + (\sqrt{6}/4)r_{03} = r_{01},$$
$$(3/4)r_{01} + (1/4)r_{02} - (\sqrt{6}/4)r_{03} = r_{02},$$
$$-(\sqrt{6}/4)r_{01} + (\sqrt{6}/4)r_{02} - (1/2)r_{03} + 1 = r_{03}.$$

Taking $r_{02} = 0$, we have $r_{01} = \sqrt{1/6}$ and $r_{03} = 1/2$. Each point on the symmetric straight line is the symmetric center. Moving the origin to the point $\mathbf{r}_0$, we transform $g(R, \mathbf{t})$ to g'

$$g' = T(-\mathbf{r}_0)g(R, \mathbf{t})T(\mathbf{r}_0) = g(R, -\mathbf{r}_0 + \mathbf{t} + R\mathbf{r}_0) = R.$$

(b) $\mathbf{t} = \mathbf{e}_1 + \mathbf{e}_3$.

$$\mathbf{t}_\perp = (\mathbf{e}_1 - \mathbf{e}_2)/2 + \mathbf{e}_3, \qquad \mathbf{t}_\| = (\mathbf{e}_1 + \mathbf{e}_2)/2,$$

$g(R, \mathbf{t})$ has a screw straight line which is through the point $\mathbf{r}_0$ and parallel to $\hat{\mathbf{n}}$ with the gliding vector $\mathbf{t}_\|$, where $\mathbf{r}_0$ satisfies

$$g(R, \mathbf{t}_\perp)\mathbf{r}_0 = R\mathbf{r}_0 + \mathbf{t}_\perp = \mathbf{r}_0.$$

Write the equation in the rectangular coordinate frame:

$$(1/4)r_{01} + (3/4)r_{02} + (\sqrt{6}/4)r_{03} + 1/2 = r_{01},$$
$$(3/4)r_{01} + (1/4)r_{02} - (\sqrt{6}/4)r_{03} - 1/2 = r_{02},$$
$$-(\sqrt{6}/4)r_{01} + (\sqrt{6}/4)r_{02} - (1/2)r_{03} + 1 = r_{03}.$$

Taking $r_{02} = 0$, we have $r_{01} = 1/2 + \sqrt{1/6}$ and $r_{03} = (1 - \sqrt{1/6})/2$. Moving the origin to the point $\mathbf{r}_0$, we transform $g(R, \mathbf{t})$ to g'

$$g' = T(-\mathbf{r}_0)g(R, \mathbf{t})T(\mathbf{r}_0) = g(R, -\mathbf{r}_0 + \mathbf{t} + R\mathbf{r}_0) = g(R, \mathbf{t}_\|).$$

5. Let R be a reflection with respect to the xy plane. Find the symmetric plane of $g(R, \mathbf{t})$ where (a) $\mathbf{t} = \mathbf{e}_3$, (b) $\mathbf{t} = \mathbf{e}_1 + \mathbf{e}_3$. If the symmetric plane is a gliding plane, find its gliding vector. Simplify $g(R, \mathbf{t})$ by moving the origin to the symmetric plane for checking your result.

Solution. $R = \sigma R(\mathbf{e}_3, \pi)$, where σ is the spatial inversion. The double-vector form of R in the rectangular coordinate frame is

$$\vec{\vec{R}} = \mathbf{e}_1\mathbf{e}_1 + \mathbf{e}_2\mathbf{e}_2 - \mathbf{e}_3\mathbf{e}_3.$$

(a) $\mathbf{t} = \mathbf{t}_{\parallel} = \mathbf{e}_3$.

$g(R, \mathbf{t})$ has a symmetric plane through the point $\mathbf{r}_0$ and orthogonal to the z axis, where $\mathbf{r}_0$ satisfies

$$g(R, \mathbf{t})\mathbf{r}_0 = R\mathbf{r}_0 + \mathbf{t} = \mathbf{r}_0.$$

Letting $\mathbf{r}_0$ be on the z axis, we have $\mathbf{r}_0 = \mathbf{e}_3/2$. Each point on the symmetric plane is the symmetric center. Moving the origin to the point $\mathbf{r}_0$, we transform $g(R, \mathbf{t})$ to g'

$$g' = T(-\mathbf{r}_0)g(R, \mathbf{t})T(\mathbf{r}_0) = g(R, \mathbf{t} + (R - E)\mathbf{r}_0) = R.$$

(b) $\mathbf{t} = \mathbf{e}_1 + \mathbf{e}_3$.

$\mathbf{t}_{\parallel} = \mathbf{e}_3$ and $\mathbf{t}_{\perp} = \mathbf{e}_1$. $g(R, \mathbf{t})$ has a gliding plane which is through the point $\mathbf{r}_0$ and orthogonal to the z axis with the gliding vector $t_{\perp} = \mathbf{e}_1$, where $\mathbf{r}_0$ satisfies

$$g(R, \mathbf{t}_{\parallel})\mathbf{r}_0 = R\mathbf{r}_0 + \mathbf{t}_{\parallel} = \mathbf{r}_0.$$

Letting $\mathbf{r}_0$ be on the z axis, we have $\mathbf{r}_0 = \mathbf{e}_3/2$. Moving the origin to the point $\mathbf{r}_0$, we transform $g(R, \mathbf{t})$ to g'

$$g' = T(-\mathbf{r}_0)g(R, \mathbf{t})T(\mathbf{r}_0) = g(R, \mathbf{t} + (R - E)\mathbf{r}_0) = g(R, \mathbf{e}_1).$$

6. Analyze the symmetry property of the crystal with the following space group: No. 52 [D_{2h}^6, $P \pm 2(\frac{1}{2}\frac{1}{2}0)2'(\frac{1}{2}\frac{1}{2}\frac{1}{2})$], No. 161 [$C_{3v}^6$, $R3\overline{2}'(\frac{1}{2}\frac{1}{2}\frac{1}{2})$], and No. 199 [$T^5$, $I3'2(\frac{1}{2}0\frac{1}{2})2'(\frac{1}{2}\frac{1}{2}0)$]. Point (a) The general form of the symmetric operation; (b) The relations of directions and lengths among three lattice bases; (c) The symmetric straight line, symmetric plane and the gliding vector of the generator in each cyclic subgroup of the space group, if they exist; (d) The equivalent point to an arbitrary point $\mathbf{r} = \mathbf{a}_1 x_1 + \mathbf{a}_2 x_2 + \mathbf{a}_3 x_3$ in the crystal cell.

Solution. Let us analyze the property of the space groups, respectively.
(1) The space group $D_{2h}^6 = P \pm 2_{1/2,1/2,0}2'_{1/2,1/2,1/2}$ belongs to P type of Bravais lattice of the tetragonal system. Its crystallographic point group is D_{2h}, which contains three orthogonal twofold axes, both proper and

improper. Three primitive lattice bases are along three twofold axes, respectively. $\alpha_1 = \alpha_2 = \alpha_3 = \pi/2$ but the lengths of the lattice bases may not be equal to each other. The double-vector forms of the generators are

$$\vec{S}_1 = -\mathbf{a}_1\mathbf{b}_1 - \mathbf{a}_2\mathbf{b}_2 - \mathbf{a}_3\mathbf{b}_3,$$

$$\vec{C}_2 = -\mathbf{a}_1\mathbf{b}_1 - \mathbf{a}_2\mathbf{b}_2 + \mathbf{a}_3\mathbf{b}_3,$$

$$\vec{C}_2' = \mathbf{a}_1\mathbf{b}_1 - \mathbf{a}_2\mathbf{b}_2 - \mathbf{a}_3\mathbf{b}_3.$$

The arbitrary element in the space group is expressed as

$$T(\vec{\ell})S_1^{n_1}g(C_2, (\mathbf{a}_1 + \mathbf{a}_2)/2)^{n_2}g(C_2', (\mathbf{a}_1 + \mathbf{a}_2 + \mathbf{a}_3)/2)^{n_3},$$

where S_1 denotes the spatial inversion, and n_1, n_2 and n_3 are taken to be 0 or 1, respectively. The origin is the symmetric center of the spatial inversion.

$g(C_2, \mathbf{q})$ with $\mathbf{q} = \mathbf{q}_\perp = (\mathbf{a}_1 + \mathbf{a}_2)/2$ has a symmetric straight line through the point $\mathbf{r}_0$ and parallel to $\mathbf{a}_3$. Each point on the symmetric straight line is the symmetric center. Let the components of $\mathbf{r}_0$ along the lattice bases be r_{01}, r_{02}, and r_{03}. Take $r_{03} = 0$. Thus,

$$0 = (C_2 - E)\mathbf{r}_0 + \mathbf{q} = \{-2r_{01} + 1/2\}\mathbf{a}_1 + \{-2r_{02} + 1/2\}\mathbf{a}_2.$$

We have $r_{01} = r_{02} = 1/4$. From another viewpoint, since the spatial inversion S_1 is a symmetric operation, there is also an improper twofold axis around the direction $\mathbf{a}_3$. $S_1g(C_2, \mathbf{q})$ has a gliding plane, which is through the origin and orthogonal to $\mathbf{a}_3$ with the gliding vector $\mathbf{q}$.

$g(C_2', \mathbf{p})$ with $\mathbf{p} = \mathbf{p}_\| + \mathbf{p}_\perp$,

$$\mathbf{p}_\| = \mathbf{a}_1/2, \qquad \mathbf{p}_\perp = (\mathbf{a}_2 + \mathbf{a}_3)/2,$$

has a screw axis which is through the point $\mathbf{r}_0'$ and parallel to $\mathbf{a}_1$ with the gliding vector $\mathbf{p}_\|$. The component of $\mathbf{r}_0'$ along the direction $\mathbf{a}_1$ can be taken to be zero, and the components along $\mathbf{a}_2$ and $\mathbf{a}_3$ are denoted by r_{02}' and r_{03}', satisfying:

$$0 = (C_2' - E)\mathbf{r}_0' + \mathbf{p}_\perp = \{-2r_{02}' + 1/2\}\mathbf{a}_2 + \{-2r_{03}' + 1/2\}\mathbf{a}_3.$$

Thus, $r_{02}' = r_{03}' = 1/4$. Similarly, there is also an improper twofold axis along the direction $\mathbf{a}_1$. $S_1g(C_2', \mathbf{p})$ has the gliding plane which is through the point $\mathbf{p}_\|/2 = \mathbf{a}_1/4$ and orthogonal to $\mathbf{a}_1$ with the gliding vector $\mathbf{p}_\perp$.

From an arbitrary point $\mathbf{r} = \mathbf{a}_1 x_1 + \mathbf{a}_2 x_2 + \mathbf{a}_3 x_3$ in the crystal cell, we can obtain eight equivalent points by the symmetric operations. Due to the spatial inversion, the equivalent points are paired by different signs:

$$\pm \{\mathbf{a}_1 x_1 + \mathbf{a}_2 x_2 + \mathbf{a}_3 x_3\},$$

$$\pm \{\mathbf{a}_1 (1/2 - x_1) + \mathbf{a}_2 (1/2 - x_2) + \mathbf{a}_3 x_3\},$$

$$\pm \{\mathbf{a}_1 (1/2 + x_1) + \mathbf{a}_2 (1/2 - x_2) + \mathbf{a}_3 (1/2 - x_3)\},$$

$$\pm \{-\mathbf{a}_1 x_1 + \mathbf{a}_2 x_2 + \mathbf{a}_3 (1/2 - x_3)\}.$$

(2) The space group $C_{3v}^6 = R3\bar{2}'_{1/2,1/2,1/2}$ belongs to the R type of Bravais lattice in the Rhombohedral system. The crystallographic point group C_{3v} contains a proper threefold axis, which is the principal axis, and three improper twofold axes, distributed symmetrically on the plane orthogonal to the principal axis. Recall that those improper twofold rotations are reflections with respect to the planes containing the principal axis. Three lattice bases have the same length and distributed around the principal axis symmetrically:

$$a_1 = a_2 = a_3, \qquad \alpha_1 = \alpha_2 = \alpha_3.$$

The double-vector forms for the generators are

$$\vec{C}_3 = \mathbf{a}_2 \mathbf{b}_1 + \mathbf{a}_3 \mathbf{b}_2 + \mathbf{a}_1 \mathbf{b}_3,$$

$$\vec{S}'_2 = \mathbf{a}_2 \mathbf{b}_1 + \mathbf{a}_1 \mathbf{b}_2 + \mathbf{a}_3 \mathbf{b}_3.$$

The general element in the space group is

$$T(\vec{\ell}) (C_3)^m g(S'_2, (\mathbf{a}_1 + \mathbf{a}_2 + \mathbf{a}_3)/2)^n,$$

where m is taken to be 0, 1 or 2, n is 0 or 1. The proper threefold axis is the principal axis. The origin is taken on the principal axis. Each point on the principal axis is the symmetric center for the threefold rotation.

$g(S'_2, (\mathbf{a}_1 + \mathbf{a}_2 + \mathbf{a}_3)/2)$ with $\mathbf{p} = \mathbf{p}_\perp = (\mathbf{a}_1 + \mathbf{a}_2 + \mathbf{a}_3)/2$ has a gliding plane which is through the origin and orthogonal to the direction $\mathbf{a}_1 - \mathbf{a}_2$ with the gliding vector $\mathbf{p}_\perp$.

From an arbitrary point $\mathbf{r} = \mathbf{a}_1 x_1 + \mathbf{a}_2 x_2 + \mathbf{a}_3 x_3$ in the crystal cell, we obtain six equivalent points by the symmetric operations:

$$\mathbf{a}_1 x_1 + \mathbf{a}_2 x_2 + \mathbf{a}_3 x_3,$$

$$\mathbf{a}_1 \left(1/2 + x_2 \right) + \mathbf{a}_2 \left(1/2 + x_1 \right) + \mathbf{a}_3 \left(1/2 + x_3 \right),$$

$$\mathbf{a}_1 x_3 + \mathbf{a}_2 x_1 + \mathbf{a}_3 x_2,$$

$$\mathbf{a}_1 \left(1/2 + x_1 \right) + \mathbf{a}_2 \left(1/2 + x_3 \right) + \mathbf{a}_3 \left(1/2 + x_2 \right),$$

$$\mathbf{a}_1 x_2 + \mathbf{a}_2 x_3 + \mathbf{a}_3 x_1,$$

$$\mathbf{a}_1 \left(1/2 + x_3 \right) + \mathbf{a}_2 \left(1/2 + x_2 \right) + \mathbf{a}_3 \left(1/2 + x_1 \right).$$

(3) The space group $\mathbf{T}^5 = I3'2_{1/2,0,1/2}2'_{1/2,1/2,0}$ belongs to the I type of Bravais lattice in the cubic system. The crystallographic point group $\mathbf{T}$ contains three orthogonal proper twofold axes and four proper threefold axes distributed symmetrically. Three lattice bases with the same length are along three twofold axes, respectively. The lattice bases are not primitive. The following fractional combination is allowed to be a vector of the crystal lattice:

$$\mathbf{f} = \left(\mathbf{a}_1 + \mathbf{a}_2 + \mathbf{a}_3 \right)/2.$$

The double-vector forms of generators are

$$\vec{\vec{C_3'}} = \mathbf{a}_2 \mathbf{b}_1 + \mathbf{a}_3 \mathbf{b}_2 + \mathbf{a}_1 \mathbf{b}_3,$$

$$\vec{\vec{C_2}} = -\mathbf{a}_1 \mathbf{b}_1 - \mathbf{a}_2 \mathbf{b}_2 + \mathbf{a}_3 \mathbf{b}_3,$$

$$\vec{\vec{C_2'}} = \mathbf{a}_1 \mathbf{b}_1 - \mathbf{a}_2 \mathbf{b}_2 - \mathbf{a}_3 \mathbf{b}_3.$$

The general element of the space group is

$$T(\vec{\ell})T(\mathbf{f})^{n_1} \left(C_3' \right)^m g(C_2, (\mathbf{a}_1 + \mathbf{a}_3)/2)^{n_2} g(C_2', (\mathbf{a}_1 + \mathbf{a}_2)/2)^{n_3},$$

where n_1, n_2 and n_3 are taken to be 0 or 1, m is 0, 1 or 2. The origin is on the threefold axis. C_3' is a pure rotation. $g(C_2, \mathbf{q})$ with $\mathbf{q} = \mathbf{q}_{\parallel} + \mathbf{q}_{\perp}$

$$\mathbf{q}_{\parallel} = \mathbf{a}_3/2, \qquad \mathbf{q}_{\perp} = \mathbf{a}_1/2,$$

has a screw axis which is through the point $\mathbf{r}_0$ and parallel to the direction $\mathbf{a}_3$ with the gliding vector $\mathbf{q}_{\parallel}$. The components of $\mathbf{r}_0$ along the lattice bases are denoted by r_{01}, r_{02} and r_{03}. We may choose $r_{03} = 0$. r_{01} and r_{02} satisfy:

$$0 = \left(C_2 - E \right) \mathbf{r}_0 + \mathbf{q}_{\perp} = \left\{ -2r_{01} + 1/2 \right\} \mathbf{a}_1 - 2r_{02}\mathbf{a}_2.$$

Thus, $r_{01} = 1/4$ and $r_{02} = r_{03} = 0$. $g(C_2', \mathbf{p})$ with $\mathbf{p} = \mathbf{p}_\| + \mathbf{p}_\perp$,

$$\mathbf{p}_\| = \mathbf{a}_1/2, \qquad \mathbf{p}_\perp = \mathbf{a}_2/2,$$

has a screw axis which is through the point $\mathbf{r}_0'$ and parallel to the direction $\mathbf{a}_1$ with the gliding vector $\mathbf{p}_\|$. The components of $\mathbf{r}_0'$ along the lattice bases are denoted by r_{01}', r_{02}' and r_{03}'. We may choose $r_{01}' = 0$. r_{02}' and r_{03}' satisfy:

$$0 = (C_2' - E)\,\mathbf{r}_0' + \mathbf{p}_\perp = \{-2r_{02}' + 1/2\}\,\mathbf{a}_2 - 2r_{03}'\mathbf{a}_3.$$

Thus, $r_{02}' = 1/4$ and $r_{01}' = r_{03}' = 0$.

From an arbitrary point $\mathbf{r} = \mathbf{a}_1 x_1 + \mathbf{a}_2 x_2 + \mathbf{a}_3 x_3$ in the crystal cell we obtain 24 equivalent points by the symmetric operations, where half points can be calculated by a translation $\mathbf{f}$. The remaining 12 points are:

$$\mathbf{a}_1 x_1 + \mathbf{a}_2 x_2 + \mathbf{a}_3 x_3,$$
$$\mathbf{a}_1 (1/2 - x_1) - \mathbf{a}_2 x_2 + \mathbf{a}_3 (1/2 + x_3),$$
$$\mathbf{a}_1 (1/2 + x_1) + \mathbf{a}_2 (1/2 - x_2) - \mathbf{a}_3 x_3,$$
$$-\mathbf{a}_1 x_1 + \mathbf{a}_2 (1/2 + x_2) + \mathbf{a}_3 (1/2 - x_3),$$
$$\mathbf{a}_1 x_3 + \mathbf{a}_2 x_1 + \mathbf{a}_3 x_2,$$
$$\mathbf{a}_1 (1/2 - x_3) - \mathbf{a}_2 x_1 + \mathbf{a}_3 (1/2 + x_2),$$
$$\mathbf{a}_1 (1/2 + x_3) + \mathbf{a}_2 (1/2 - x_1) - \mathbf{a}_3 x_2,$$
$$-\mathbf{a}_1 x_3 + \mathbf{a}_2 (1/2 + x_1) + \mathbf{a}_3 (1/2 - x_2),$$
$$\mathbf{a}_1 x_2 + \mathbf{a}_2 x_3 + \mathbf{a}_3 x_1,$$
$$\mathbf{a}_1 (1/2 - x_2) - \mathbf{a}_2 x_3 + \mathbf{a}_3 (1/2 + x_1),$$
$$\mathbf{a}_1 (1/2 + x_2) + \mathbf{a}_2 (1/2 - x_3) - \mathbf{a}_3 x_1,$$
$$-\mathbf{a}_1 x_2 + \mathbf{a}_2 (1/2 + x_3) + \mathbf{a}_3 (1/2 - x_1).$$

5.3 International Notations for Space Groups

★ Attaching as the subscripts the suitable translation vectors to the international symbol for the symmorphic space group, we obtain the international symbol for the general space group. The general element of a space group can be expressed as

$$T(\mathbf{L})\left\{g(R,\mathbf{t})\right\}^{n},$$

$$T(\mathbf{L})\left\{g(R,\mathbf{t})\right\}^{n}\left\{g(R_1,\mathbf{p})\right\}^{n_1}, \tag{5.11}$$

$$T(\mathbf{L})\left\{g(R,\mathbf{t})\right\}^{n}\left\{g(R_2,\mathbf{q})\right\}^{n_2}\left\{g(R_1,\mathbf{p})\right\}^{n_1},$$

where n, n_1 and n_2 all are integers, $\mathbf{L} = \vec{\ell} + \mathbf{f}$ is a vector of crystal lattice, $\mathbf{f}$ depends upon the Bravais lattice, and the translation vectors $\mathbf{t}$, $\mathbf{p}$ and $\mathbf{q}$ are the fractional combinations of the lattice bases $\mathbf{a}_j$:

$$\mathbf{t} = \sum_{j=1}^{3} \mathbf{a}_j t_j, \quad \mathbf{p} = \sum_{j=1}^{3} \mathbf{a}_j p_j, \quad \mathbf{q} = \sum_{j=1}^{3} \mathbf{a}_j q_j,$$

$$0 \le t_j < 1, \qquad 0 \le p_j < 1, \qquad 0 \le q_j < 1.$$

The translation vectors $\mathbf{t}$, $\mathbf{p}$ and $\mathbf{q}$ should satisfy the group property, i.e., the multiplication of two elements of the space group has to be expressed in the same form (5.11). First, due to the property of a cyclic subgroup, the translation vector $\mathbf{t}$ in $g(R,\mathbf{t})$ has to satisfy

$$\mathbf{t} = 0 \qquad\qquad \text{when } R = C_1 = E,$$

$$\left\{\hat{\mathbf{n}}\left(\hat{\mathbf{n}} \cdot \mathbf{t}\right)\right\} = \mathbf{t}_\parallel = m\mathbf{a}_\parallel/N \qquad \text{when } R = C_N, \ N \neq 1,$$

$$\left\{\mathbf{t} - \hat{\mathbf{n}}\left(\hat{\mathbf{n}} \cdot \mathbf{t}\right)\right\} = \mathbf{t}_\perp = m\mathbf{a}_\perp/2 \quad \text{when } R = S_2, \tag{5.12}$$

$$\mathbf{t} \ \text{ no restriction} \qquad\qquad \text{when } R = S_N, \ N \neq 2.$$

Second, there are constraints resulted from the multiplication of R, R_1 and R_2. For example, due to $R_1 = RR_1R$, we have

$$g(R_1,\mathbf{p} + \mathbf{L}) = g(R,\mathbf{t})g(R_1,\mathbf{p})g(R,\mathbf{t}) = g(RR_1R,\mathbf{t} + R\mathbf{p} + RR_1\mathbf{t})$$

$$(R - E)\mathbf{p} = \mathbf{L} - (E + RR_1)\mathbf{t}. \tag{5.13}$$

At last, we have to avoid the equivalent symbols related to the different choice of the origin. Let the vector from the origin O to the new origin O' be $\mathbf{r}_0$. If the symmetric operation is denoted by $g(R,\mathbf{t})$ for the origin O, it is changed to g' in the new origin O'

$$g' = T(-\mathbf{r}_0)g(R,\mathbf{t})T(\mathbf{r}_0) = g(R,\mathbf{t} + (R - E)\mathbf{r}_0). \tag{5.14}$$

The different choice of the origin does not change the space group, but changes its symbol. Those two symbols are called the equivalent symbols. We have to remove the equivalent symbols for the same space group.

★ Usually, we choose the origin such that the translation vector $\mathbf{t}$ in Eq. (5.11) becomes as simple as possible. In other words, we choose the origin such that the part of $\mathbf{t}$, which is not restricted in Eq. (5.12), vanishing. $\mathbf{t}$ can be more simplified if the remaining $\mathbf{t}$ is a component of $\mathbf{L}$. Then, we may choose the origin again to simplify the translation vectors $\mathbf{p}$ and $\mathbf{q}$ under the condition that $\mathbf{t}$ keeps invariant. This is the general method to determine the symbol of a space group.

Since the choice of the origin first depends upon the simplification of $\mathbf{t}$ in $g(R, \mathbf{t})$, The space groups are classified as type A ($R = S_N$, $N \neq 2$), type B ($R = C_N$) and type C ($R = S_2$). There are only two space groups in type C: $P\overline{2}_{\frac{1}{2}00}$ and $A\overline{2}_{\frac{1}{2}00}$.

7. Calculate the 12 space groups related to the crystallographic point group D_{2d}.

Solution. The point group D_{2d} contains an improper fourfold axis (the principal axis), two proper twofold axes and two improper twofold axes. Four twofold axes are distributed symmetrically on the plane orthogonal to the principal axis. The lattice basis $\mathbf{a}_3$ is along the principal axis. According to whether the lattice basis $\mathbf{a}_1$ is along a proper or an improper twofold axis, the international symbol of D_{2d} is denoted by $\overline{4}2'$ or $\overline{4}2''$, namely, denoted by the product of two subgroups $S_4 C_2'$ or $S_4 C_2''$. The space groups belong to the tetragonal system, where there are two Bravais lattices P and I. The restriction for the lattice bases are $\alpha_1 = \alpha_2 = \alpha_3$, and $a_1 = a_2$. The double-vector form of the fourfold rotation is

$$\vec{\overleftrightarrow{S_4}} = -\mathbf{a}_2 \mathbf{b}_1 + \mathbf{a}_1 \mathbf{b}_2 - \mathbf{a}_3 \mathbf{b}_3.$$

The space groups with the crystallographic point group D_{2d} belong to type A. Choosing the origin on the principal axis, we express the general element in the space group as

$$T(\mathbf{L}) (S_4)^m g(C_2', \mathbf{p})^n,$$

where m is taken 0, 1, 2 or 3, n is 0 or 1. C_2' in the formula may be changed to C_2'', depending on the discussed space groups. We may translate the origin by $\mathbf{r}_0$, $\mathbf{r}_0 = \mathbf{a}_1 r_{01} + \mathbf{a}_2 r_{02} + \mathbf{a}_3 r_{03}$, satisfying

$$(S_4 - E) \mathbf{r}_0 = \mathbf{a}_1 (-r_{01} + r_{02}) + \mathbf{a}_2 (-r_{01} - r_{02}) - 2\mathbf{a}_3 r_{03} = \mathbf{L}.$$

Thus, $\mathbf{r}_0$ may be taken the following values, in addition to $\mathbf{L}$

$$\mathbf{r}_0 = \begin{cases} (\mathbf{a}_1 + \mathbf{a}_2)/2, \quad \mathbf{a}_3/2 & \text{Bravais lattice } P \text{ or } I \\ (2\mathbf{a}_1 + \mathbf{a}_3)/4, \quad (2\mathbf{a}_2 + \mathbf{a}_3)/4, & \text{Bravais lattice } I. \end{cases}$$

In the following we study two kinds of space groups with the crystallo-graphic point groups $\bar{4}2'$ and $\bar{4}2''$, respectively.

(1) For the crystallographic point group $\bar{4}2'$, $\mathbf{a}_1$ is along the direction of a proper twofold axis. The double-vector form of the twofold rotation is

$$\vec{\vec{C}}_2' = \mathbf{a}_1\mathbf{b}_1 - \mathbf{a}_2\mathbf{b}_2 - \mathbf{a}_3\mathbf{b}_3.$$

Since $S_4 C_2' S_4 = C_2'$, we have

$$S_4 g(C_2', \mathbf{p}) S_4 = g(C_2', S_4\mathbf{p}) = g(C_2', \mathbf{p} + \mathbf{L}).$$

Letting $\mathbf{p} = \mathbf{a}_1 p_1 + \mathbf{a}_2 p_2 + \mathbf{a}_3 p_3$, we have:

$$(S_4 - E)\,\mathbf{p} = \mathbf{a}_1\left(-p_1 + p_2\right) + \mathbf{a}_2\left(-p_1 - p_2\right) - \mathbf{a}_3 2p_3 = \mathbf{L}.$$

For the P type of Bravais lattice, we have solutions $p_1 = p_2 = 0$ or $1/2$, and $p_3 = 0$ or $1/2$. For the I type of Bravais lattice, because $\mathbf{f} = (\mathbf{a}_1 + \mathbf{a}_2 + \mathbf{a}_3)/2$ is a vector of crystal lattice, the solution $\mathbf{p} = \mathbf{f}$ is ruled out, and two solutions $\mathbf{p} = (\mathbf{a}_1 + \mathbf{a}_2)/2$ and $\mathbf{p} = \mathbf{a}_3/2$ become equivalent to each other. In addition, there are new solutions $\mathbf{p} = (2\mathbf{a}_1 + \mathbf{a}_3)/4$ and $\mathbf{p} = (2\mathbf{a}_2 + \mathbf{a}_3)/4$. In all those solutions, the parallel component $\mathbf{p}_\parallel$ is equal to multiple of half vector of crystal lattice along the rotational axis.

Moving the origin by $\mathbf{r}_0$, $\mathbf{p}$ may be changed by:

$$(C_2' - E)\,\mathbf{r}_0 = -\mathbf{a}_2 2r_{02} - \mathbf{a}_3 2r_{03}.$$

For the P type of Bravais lattice, $\mathbf{p}$ is not simplified. But for the I type of Bravais lattice, the solution $\mathbf{p} = \mathbf{a}_3/2$ is ruled out, and two solutions $\mathbf{p} = (2\mathbf{a}_1 + \mathbf{a}_3)/4$ and $\mathbf{p} = (2\mathbf{a}_2 + \mathbf{a}_3)/4$ become equivalent. Therefore, we have the following space groups:

$$P\bar{4}2', \quad P\bar{4}2'_{00\frac{1}{2}}, \quad P\bar{4}2'_{\frac{1}{2}\frac{1}{2}0}, \quad P\bar{4}2'_{\frac{1}{2}\frac{1}{2}\frac{1}{2}}, \quad I\bar{4}2', \quad I\bar{4}2'_{0\frac{1}{2}\frac{1}{4}},$$

which are listed in the table of space groups with Nos. 111, 112, 113, 114, 121 and 122.

(2) For the crystallographic point group $\bar{4}2''$, $\mathbf{a}_1$ is along the direction of an improper twofold axis. The double-vector form of the twofold rotation is

$$\vec{\vec{C}}_2'' = -\mathbf{a}_2\mathbf{b}_1 - \mathbf{a}_1\mathbf{b}_2 - \mathbf{a}_3\mathbf{b}_3.$$

Since $S_4 C_2'' S_4 = C_2''$, we have

$$S_4 g(C_2'', \mathbf{p}) S_4 = g(C_2'', \mathbf{p} + \mathbf{L}),$$

$$(S_4 - E)\, \mathbf{p} = \mathbf{a}_1 \left(-p_1 + p_2\right) + \mathbf{a}_2 \left(-p_1 - p_2\right) - \mathbf{a}_3 2 p_3 = \mathbf{L}.$$

Although this equation is the same as the former case (1), the solution may be different from the former case because of the different lattice bases. For the P type of Bravais lattice, we have the same solutions as the former case: $p_1 = p_2 = 0$ or $1/2$, and $p_3 = 0$ or $1/2$. For the I type of Bravais lattice, we only have solutions $\mathbf{p} = \mathbf{a}_3/2$ or 0. The solutions $\mathbf{p} = (2\mathbf{a}_1 + \mathbf{a}_3)/4$ and $\mathbf{p} = (2\mathbf{a}_2 + \mathbf{a}_3)/4$ are ruled out because their parallel component $\mathbf{p}_\| = (\mathbf{a}_1 - \mathbf{a}_2)/4$ is less than half of the vector of crystal lattice along the rotational axis.

Moving the origin by $\mathbf{r}_0$, $\mathbf{p}$ may be changed by:

$$(C_2'' - E)\, \mathbf{r}_0 = -\left(\mathbf{a}_1 + \mathbf{a}_2\right) \left(r_{01} + r_{02}\right) - \mathbf{a}_3 2 r_{03}.$$

Both for the P type and the I type of Bravais lattice, $\mathbf{p}$ is not simplified. Therefore, we have the following space groups:

$$P\bar{4}2'', \quad P\bar{4}2''_{00\frac{1}{2}}, \quad P\bar{4}2''_{\frac{1}{2}\frac{1}{2}0}, \quad P\bar{4}2''_{\frac{1}{2}\frac{1}{2}\frac{1}{2}}, \quad I\bar{4}2'', \quad I\bar{4}2''_{00\frac{1}{2}},$$

which are listed in the table of space groups with Nos. 115, 116, 117, 118, 119 and 120.

8. Please calculate the 9 space groups related to the crystallographic point group D_2.

Solution. The international symbol of the crystallographic point group D_2 is $22'$. It contains three orthogonal proper twofold axes and can be expressed as the product of two subgroups: $C_2 C_2'$. The space groups with the crystallographic point group D_2 belong to the orthorhombic system with P, A, F and I types of Bravais lattices. Three lattice bases are orthogonal to each other, but have different lengths. The double-vector forms for twofold rotations are

$$\vec{\vec{C}}_2 = -\mathbf{a}_1 \mathbf{b}_1 - \mathbf{a}_2 \mathbf{b}_2 + \mathbf{a}_3 \mathbf{b}_3, \qquad \vec{\vec{C}}_2' = \mathbf{a}_1 \mathbf{b}_1 - \mathbf{a}_2 \mathbf{b}_2 - \mathbf{a}_3 \mathbf{b}_3.$$

The space groups with the crystallographic point group D_2 belong to type B. Choosing the origin on one of the twofold axes, we express the general element in the space group as

$$T(\mathbf{L}) g(C_2, \mathbf{t})^m g(C_2', \mathbf{p})^n,$$

where both m and n are taken to be 0 or 1. $\mathbf{t} = t\mathbf{a}_3$ with $t = 0$ or $1/2$. For the A type, F type and I type of Bravais lattices, $\mathbf{L}$ contains the component $\mathbf{a}_3/2$ so that t can be removed by a translation $\mathbf{r}_0$ of the origin

$$\mathbf{r}_0 = \mathbf{a}_1 r_{01} + \mathbf{a}_2 r_{02} + \mathbf{a}_3 r_{03},$$

$$(C_2 - E)\mathbf{r}_0 = -\mathbf{a}_1 2r_{01} - \mathbf{a}_2 2r_{02} = \mathbf{L} + \mathbf{t}.$$

Under the condition that the simplified $\mathbf{t}$ keeps invariant, the origin can be made a further translation $\mathbf{r}_0$

$$\mathbf{r}_0 = \begin{cases} \mathbf{a}_1/2, \quad \mathbf{a}_2/2, \quad r_{03}\mathbf{a}_3 & \text{Bravais lattices } P, A, F \text{ and } I \\ (\mathbf{a}_1 + \mathbf{a}_2)/4, & \text{Bravais lattice } F, \end{cases}$$

where r_{03} is arbitrary.

Since $C_2 C_2' C_2 = C_2'$ and $C_2 C_2' \mathbf{t} = -\mathbf{t}$, we have

$$g(C_2, \mathbf{t})g(C_2', \mathbf{p})g(C_2, \mathbf{t}) = g(C_2', C_2\mathbf{p}) = g(C_2', \mathbf{p} + \mathbf{L}).$$

Letting $\mathbf{p} = \mathbf{a}_1 p_1 + \mathbf{a}_2 p_2 + \mathbf{a}_3 p_3$, we have:

$$(C_2 - E)\mathbf{p} = -\mathbf{a}_1 2p_1 - \mathbf{a}_2 2p_2 = \mathbf{L}.$$

For P, A, F and I types of Bravais lattices, we obtain that p_1 and p_2 are respectively taken to be 0 or $1/2$, but p_3 is arbitrary. There is an additional solution for the F type of Bravais lattice: $\mathbf{p} = (\mathbf{a}_1 + \mathbf{a}_2)/4$. However, the latter is ruled out by the constraint that $\mathbf{p}_\parallel$ has to be a multiple of half vector of crystal lattice along the rotational axis.

In translating the origin by $\mathbf{r}_0$, $\mathbf{p}$ is made the following transformation:

$$(C_2' - E)\mathbf{r}_0 = -\mathbf{a}_2 2r_{02} - \mathbf{a}_3 2r_{03}.$$

Thus, p_3 is removed. Now, we study the space groups for different Bravais lattices.

For the P type of Bravais lattice, t, p_1 and p_2 are allowed to be 0 or $1/2$ independently. However, some space groups may be equivalent to each other. Obviously, two space groups with $t = p_1 = p_2 = 0$ and $1/2$ are inequivalent. They are denoted by $P22'$ and $P2_{00\frac{1}{2}}2'_{\frac{1}{2}\frac{1}{2}0}$, listed in the table of space groups with Nos. 16 and 19. If one of t, p_1 and p_2 is $1/2$ and the remaining are vanishing, we obtain three equivalent space groups because three twofold axes are distributed symmetrically. In other words, three solutions $t = 1/2$ $(p_1 = p_2 = 0)$, $p_1 = 1/2$ $(t = p_2 = 0)$ and $p_2 = 1/2$ $(t = p_1 = 0)$ are equivalent because one can be obtained from the other by interchanging the lattice bases. Those three equivalent space groups are denoted by $P22'_{0\frac{1}{2}0}$, listed in the table of space groups with No. 17. If two of t, p_1 and p_2 are $1/2$ and the third one is vanishing, we also obtain three equivalent space groups. In other words, the solution with $t = p_2 = 1/2$ can be obtained from the solution $t = p_1 = 1/2$ by changing $\mathbf{a}_1$ to the direction of $\mathbf{a}_2$. For the solution with $p_1 = p_2 = 1/2$ and $t = 0$, two twofold rotations are expressed as

$$C_2, \qquad g(C'_2, (\mathbf{a}_1 + \mathbf{a}_2)/2).$$

After moving the origin by $\mathbf{r}_0 = \mathbf{a}_2/4$, they become

$$g(C_2, 0) \longrightarrow g(C_2, (C_2 - E)\mathbf{a}_2/4) = g(C_2, -\mathbf{a}_2/2),$$
$$g(C'_2, (\mathbf{a}_1 + \mathbf{a}_2)/2) \longrightarrow g(C'_2, (\mathbf{a}_1 + \mathbf{a}_2)/2 + (C'_2 - E)\mathbf{a}_2/4)$$
$$= g(C'_2, \mathbf{a}_1/2).$$

Then, interchanging $\mathbf{a}_1$ and $\mathbf{a}_3$, the solution becomes that with $t = p_2 = 1/2$ and $p_1 = 0$. Those three equivalent space groups are denoted by $P22'_{\frac{1}{2}\frac{1}{2}0}$, listed in the table of space groups with No. 18.

For the A type of Bravais lattice, $t = 0$. Since $\mathbf{f} = (\mathbf{a}_2 + \mathbf{a}_3)/2$ is a vector of crystal lattice, the solution with $p_2 = 1/2$ can be removed. Thus, there are only two space groups for the A type of Bravais lattice $A22'$ and $A22'_{\frac{1}{2}00}$, listed in the table of space groups with Nos. 21 and 20.

For the F type of Bravais lattice, $t = 0$. Since $\mathbf{f} = (\mathbf{a}_1 + \mathbf{a}_3)/2$ and $(\mathbf{a}_2 + \mathbf{a}_3)/2$ are the vectors of crystal lattice, the solutions with $p_1 = 1/2$ and (or) $p_2 = 1/2$ can be removed. Thus, there is only one space group for the F type of Bravais lattice, $F22'$, listed in the table of space groups with No. 22.

For the I type of Bravais lattice, $t = 0$. Since $\mathbf{f} = (\mathbf{a}_1 + \mathbf{a}_2 + \mathbf{a}_3)/2$ is a vector of crystal lattice, the solution with $p_1 = p_2 = 1/2$ can be removed. The solution with $p_1 = 0$ and $p_2 = 1/2$ can be obtained from that with $p_1 = 1/2$ and $p_2 = 0$ by changing $\mathbf{a}_1$ to the direction of $\mathbf{a}_2$. Thus, there are only two space groups $I22'$ and $I22'_{\frac{1}{2}00}$ for the I type of Bravais lattice, listed in the table of the space groups with Nos. 23 and 24. Usually, the space group with No. 24 in the table of the space groups are denoted by $I2_{00\frac{1}{2}}2'_{\frac{1}{2}\frac{1}{2}0}$, which is equivalent to $I22'_{\frac{1}{2}00}$. One will convince himself by subtracting $\mathbf{f}$ from $\mathbf{p}$, removing p_3 through moving the origin, and interchanging $\mathbf{a}_1$ and $\mathbf{a}_3$.

In summary, there are nine inequivalent space groups related to the crystallographic point group D_2, listed in the table of space groups with Nos. 16-24.

Chapter 6

PERMUTATION GROUPS

6.1 Multiplication of Permutations

★ A rearrangement of n objects is called a permutation. Denote a permutation R which transforms the object in the position j into the position r_j by a $2 \times n$ matrix

$$R = \begin{pmatrix} 1 & 2 & \ldots & n \\ r_1 & r_2 & \ldots & r_n \end{pmatrix}.$$

In the matrix for R the order of columns does not matter, but the corresponding relation between two digits in each column is essential. The multiplication rule for two permutations is defined as successive applications of the two transformations.

$$\begin{pmatrix} r_1 & r_2 & \ldots & r_n \\ s_1 & s_2 & \ldots & s_n \end{pmatrix} \begin{pmatrix} 1 & 2 & \ldots & n \\ r_1 & r_2 & \ldots & r_n \end{pmatrix} = \begin{pmatrix} 1 & 2 & \ldots & n \\ s_1 & s_2 & \ldots & s_n \end{pmatrix}.$$

There are $n!$ permutations for the system of n objects. The set of those $n!$ permutations forms a group, called the permutation group S_n of n objects, where the identity and the inverse of R are

$$E = \begin{pmatrix} 1 & 2 & \ldots & n \\ 1 & 2 & \ldots & n \end{pmatrix}, \qquad R^{-1} = \begin{pmatrix} r_1 & r_2 & \ldots & r_n \\ 1 & 2 & \ldots & n \end{pmatrix}.$$

If a permutation S keeps $(n - \ell)$ objects invariant and changes the remaining ℓ objects in order, S is called a cycle with length ℓ, described by a one-row matrix:

$$\begin{aligned} S &= \begin{pmatrix} a_1 & a_2 & \ldots & a_{\ell-1} & a_\ell & b_1 & \ldots & b_{n-\ell} \\ a_2 & a_3 & \ldots & a_\ell & a_1 & b_1 & \ldots & b_{n-\ell} \end{pmatrix} \\ &= (a_1 \ a_2 \ \ldots \ a_{\ell-1} \ a_\ell) = (a_2 \ a_3 \ \ldots \ a_\ell \ a_1). \end{aligned}$$

In the one-row matrix for a cycle, the order of digits is essential, while the transformation of digits in sequence is permitted. The order of a cycle S with length ℓ is ℓ, namely, ℓ is the smallest power with $S^\ell = E$. A cycle with length 2 is called a transposition.

★ Any permutation can be decomposed into a product of cycles which contain no common object. The product of two cycles without any common object is commutable. Up to the order, the decomposition of a permutation into a product of cycles without any common object is unique. The set of lengths of those cycles is called the cycle structure of the permutation.

A cycle is equal to a product of two cycles by cutting it in any digit and repeating the digit:

$$(a \ \cdots \ b \ c \ d \ \cdots \ f) = (a \ \cdots \ b \ c)(c \ d \ \cdots \ f). \tag{6.1}$$

Conversely, two cycles which contain one common object may be connected to be one cycle through Eq. (6.1). The product of two cycles which contain more common objects can be simplified as a product of several cycles without any common object by using Eq. (6.1) repeatedly. For example,

$$
\begin{aligned}
&(a_1 \ \ldots \ a_i \ c \ a_{i+1} \ \ldots \ a_j \ d)(d \ b_1 \ \ldots \ b_r \ c \ b_{r+1} \ \ldots \ b_s) \\
&= (a_1 \ \ldots \ a_i \ c)(c \ a_{i+1} \ \ldots \ a_j \ d)(d \ b_1 \ \ldots \ b_r \ c)(c \ b_{r+1} \ \ldots \ b_s) \\
&= (a_1 \ \ldots \ a_i \ c)(a_{i+1} \ \ldots \ a_j \ d)(d \ c)(c \ d)(d \ b_1 \ \ldots \ b_r)(c \ b_{r+1} \ \ldots \ b_s) \\
&= (a_1 \ \ldots \ a_i \ c)(c \ b_{r+1} \ \ldots \ b_s)(a_{i+1} \ \ldots \ a_j \ d)(d \ b_1 \ \ldots \ b_r) \\
&= (a_1 \ \ldots \ a_i \ c \ b_{r+1} \ \ldots \ b_s)(a_{i+1} \ \ldots \ a_j \ d \ b_1 \ \ldots \ b_r).
\end{aligned}
$$
$$\tag{6.2}$$

A cycle with length ℓ may be decomposed into a product of $(\ell-1)$ transposition by Eq. (6.1). Any permutation may be decomposed into the product of transpositions by different ways, where the number of transpositions in the product is not unique, but it is determined whether the number is even or odd. A permutation is said to be even (or odd) if its decomposition contains even (or odd) number of transpositions. The permutation parity $\delta(R)$ of an even permutation is 1, and that of an odd permutation is -1. The permutation parity of a cycle with length ℓ is $(\ell - 1)$.

★ When a permutation S is moved from the left-side of a permutation R to its right-side, the digits (objects) in R are made the permutation S. Conversely, when a permutation S is moved from the right-side of R to its left-side, the digits in R are made the inverse permutation S^{-1}. In other words, if the permutation R is a cycle (or the product of cycles) containing some digits r_j, and the permutation S transforms r_j into t_j, then, $T =$

SRS^{-1} is the same cycle as R except that the containing digits r_j are changed to t_j, respectively. This property in the product of permutations is called the interchange rule. From the rule, two permutations with the same cycle structure are conjugate to each other. The class in a permutation group is described by the cycle structure of the permutations in the class.

★ Denote by $P_a = (a \ a+1)$ the transposition of two neighbored objects a and $(a+1)$. It is easy to know from the interchange rule in the product of permutations that P_a satisfy

$$P_a^2 = E, \qquad P_a P_{a+1} P_a = P_{a+1} P_a P_{a+1},$$
$$P_a P_b = P_b P_a, \qquad |a - b| \geq 2. \tag{6.3}$$

1. Simplify the following permutations into the product of cycles without any common object:

(1): $(1\ 2)(2\ 3)(1\ 2)$,   (2): $(1\ 2\ 3)(1\ 3\ 4)(3\ 2\ 1)$,

(3): $(1\ 2\ 3\ 4)^{-1}$,   (4): $(1\ 2\ 4\ 5)(4\ 3\ 2\ 6)$,

(5): $(1\ 2\ 3)(4\ 2\ 6)(3\ 4\ 5\ 6)$.

Solution: The first two problems can be solved by the interchange rule, and the last three problems can be solved by Eq. (6.1).

(1): $(1\ 3)$,   (2): $(2\ 1\ 4)$,   (3): $(4\ 3\ 2\ 1)$,

(4): $(5\ 1\ 2\ 6)(4\ 3)$,   (5): $(1\ 2\ 6)(4\ 5)$.

2. Decompose an arbitrary transposition $(b\ d)$ as a product of the transpositions P_a of the neighbored objects.

Solution. Without loss of generality, we suppose $b > d$ and obtain

$$(b\ d) = P_{b-1} P_{b-2} \cdots P_{d+1} P_d P_{d+1} \cdots P_{b-2} P_{b-1}$$
$$= P_d P_{d+1} \cdots P_{b-2} P_{b-1} P_{b-2} \cdots P_{d+1} P_d.$$

3. Prove that the order of a cycle R with length ℓ is ℓ, namely, $R^\ell = E$.

Solution. Let $R = (a_1\ a_2\ \cdots\ a_\ell)$ and b be an object different from any a_m, $1 \leq m \leq \ell$. Then, R transforms a_m into a_{m+1}, and a_ℓ into a_1. Thus, $R^{\ell-m}$ transforms a_m into a_ℓ, R transforms a_ℓ into a_1, and R^{m-1} transforms a_1 into a_m. Altogether, R^ℓ transforms an arbitrary a_m into itself, while keeping b invariant, namely, R^ℓ is the identity E.

4. Prove that $P_1 = (1\ 2)$ and $W = (1\ 2\ 3\ \ldots\ n)$ are two generators of the permutation group S_n.

Solution. From Problem 2, any permutation can be decomposed as a product of the transpositions P_a of the neighbored objects. From the interchange rule, $WP_a = P_{a+1}W$. From Problem 3 we have $W^{-1} = W^{n-1}$, so $P_a = WP_{a-1}W^{n-1} = W^{a-1}P_1W^{n-a+1}$.

5. There are 52 pieces of playing cards in a set of poker. The order of cards is changed in each shuffle, namely, the cards are made a permutation. If the shuffle is "strictly" done in the following rule: first separate the cards into two parts in equal number, then pick up one card from each part in order. After the strict shuffle the first and the last cards do not change their positions, while the remaining cards are rearranged. Try to find this permutation, to decompose it into a product of cycles without any common object, to write the cycle structure of the permutation, and to explain at least how many times of the strict shuffles will make the order of cards into its original one.

Solution. After the strict shuffle, the cards in the first part are arranged into the odd positions, while the cards in the second part are arranged into the even positions. In the mathematical language, after a shuffle, the nth card is changed to the $(2n - 1)$th position, while the $(26 + n)$th card is changed to the $(2n)$th position, where $1 \le n \le 26$. By this rule, we want to decompose the permutation corresponding to the strict shuffle into the product of cycles without any common object. First, this permutation contains two cycles (1) and (52) with length 1. Second, according to the rule, the shuffle transforms 2 into 3, 3 into 5, 5 into 9, 9 into 17, 17 into 33, 33 into 14, 14 into 27, and 27 into 2, so we obtain a cycle with length 8

$$(2\ 3\ 5\ 9\ 17\ 33\ 14\ 27).$$

We arbitrarily choose a card in the remaining cards, say the fourth card. The shuffle transforms 4 into 7, 7 into 13, 13 into 25, 25 into 49, 49 into 46, 46 into 40, 40 into 28, and 28 into 4, so we obtain another cycle with length 8

$$(4\ 7\ 13\ 25\ 49\ 46\ 40\ 28).$$

Similarly, in the remaining cards, we choose the sixth card. The shuffle transforms 6 into 11, 11 into 21, 21 into 41, 41 into 30, 30 into 8, 8 into 15,

15 into 29, and 29 into 6, so we obtain a cycle with length 8

$$(6\ 11\ 21\ 41\ 30\ 8\ 15\ 29).$$

In the remaining cards, we choose the tenth card. The shuffle transforms 10 into 19, 19 into 37, 37 into 22, 22 into 43, 43 into 34, 34 into 16, 16 into 31, and 31 into 10, so we obtain a cycle with length 8

$$(10\ 19\ 37\ 22\ 43\ 34\ 16\ 31).$$

In the remaining cards, we choose the twelfth card, the shuffle transforms 12 into 23, 23 into 45, 45 into 38, 38 into 24, 24 into 47, 47 into 42, 42 into 32, and 32 into 12, so we obtain a cycle with length 8

$$(12\ 23\ 45\ 38\ 24\ 47\ 42\ 32).$$

In the remaining cards, we choose the 18th card, the shuffle transforms 18 into 35, and 35 into 18. This time we obtain a cycle (18 35) with length 2. In the remaining cards, we choose the 20th card, the shuffle transforms 20 into 39, 39 into 26, 26 into 51, 51 into 50, 50 into 48, 48 into 44, 44 into 36, and 36 into 20, so we obtained a cycle with length 8

$$(20\ 39\ 26\ 51\ 50\ 48\ 44\ 36).$$

Now, 52 cards are exhausted, and the permutation corresponding to the shuffle is decomposed into the product of two cycles with length 1, one cycle with length 2 and six cycles with length 8. All those cycles do not contain any common digit. The cycle structure of this permutation is $(1^2, 2, 8^6)$. The least common multiple of the cycle lengths is 8, namely, after 8 times of the strict shuffles, the order of playing cards turns into its original one. In fact, the shuffle is not so strict. Since many accidental cases occur, there is no worry about.

6.2 Young Patterns, Young Tableaux and Young Operators

★ A class in the permutation group S_n is described by the cycle structure $(\ell) = (\ell_1, \ell_2, \ldots)$ of the permutations in the class. Since $\sum_j \ell_j = n$, (ℓ) is a partition number of n. For a finite group, the number of the inequivalent irreducible representations is equal to the number of the classes. Therefore, the irreducible representation of S_n can also be described by a partition number of n, denoted by $[\lambda] = [\lambda_1, \lambda_2, \cdots]$, $\lambda_1 \geq \lambda_2 \geq \cdots$, $\sum_j \lambda_j = n$. Usually, when some ℓ_j or λ_j are duplicated, one may denote them by an exponent, for example, $(2^3, 3^2) = (2, 2, 2, 3, 3)$.

★ From a partition number $[\lambda]$ of n, we define a Young pattern which consists of n boxes lined up on the top and on the left, where the jth row contains λ_j boxes. The Young pattern is also denoted by $[\lambda]$. Remind that in a Young pattern, the number of boxes in the upper row is not less than that in the lower row, and the number of boxes in the left column is not less than that in the right column.

A Young pattern $[\lambda]$ is said to be larger than a Young pattern $[\lambda']$ if $\lambda_i = \lambda'_i$, $1 \leq i < j$, and $\lambda_j > \lambda'_j$. There is no analytic formula for the number of different Young patterns with n boxes. However, one can list all different Young patterns with n boxes from the largest to the smallest.

★ Filling n digits 1, 2, ..., n arbitrarily into a given Young pattern $[\lambda]$ with n boxes, we obtain a Young tableau. A Young tableau is said to be standard if the digit in each column of the tableau increases downwards and the digit in each row increases from left to right.

For two standard Young tableaux with the same Young pattern, if the digits filled in the first $(i-1)$ rows and in the first $(j-1)$ columns of the ith row are the same as each other, respectively, but the digits filled in the box at the jth column of the ith row are different, the Young tableau with the larger digit in this box is said to be larger than the other. This increasing order for the standard Young tableaux is the so-called dictionary order. We can enumerate the standard Young tableaux from the smallest to the largest by an integer μ. The number of the standard Young tableaux with a given Young pattern $[\lambda]$ may be calculated by the hook rule,

$$d_{[\lambda]}(\mathbf{S}_n) = \frac{n!}{\prod h_{ij}} , \qquad (6.4)$$

where h_{ij} is the hook number of the box in the jth column of the ith row of the given Young pattern, which is equal to the number of boxes at its right in the ith row, plus the number of boxes below it in the jth column, and plus one. Two Young patterns related by a transpose are called the associated Young patterns. The corresponding standard Young tableaux for two associated Young patterns are related by a transpose such that the larger Young tableau for one Young pattern becomes the smaller for the associated Young pattern, but the numbers of the standard Young tableaux for two associated Young patterns are the same.

★ A permutation of the digits in the same row of a given Young tableau is called the horizontal permutation P of the Young tableau, and a permutation of the digits in the same column is called its vertical permutation Q. Generally, a horizontal permutation P may be a product of the horizontal permutations P_j, where P_j is a permutation of digits in the jth row. Similarly, Q may be a product of Q_k, where Q_k is a permutation of digits in the kth column. The sum of all horizontal permutations of a given Young tableau is called its horizontal operator $\mathcal{P} = \sum P = \prod_j \sum P_j$, and the sum of all vertical permutations multiplied by their permutation parities is called its vertical operator $\mathcal{Q} = \sum \delta(Q)Q = \prod_k \sum \delta(Q_k)Q_k$. The Young operator $\mathcal{Y}$ corresponding to the Young tableau is equal to the product of the horizontal operator and the vertical operator, $\mathcal{Y} = \mathcal{P}\mathcal{Q}$. A Young operator is said to be standard if its Young tableau is standard. Since for a given Young operator $\mathcal{Y}$, its Young tableau and its Young pattern have been determined, we often say the Young pattern $\mathcal{Y}$ and the Young tableau $\mathcal{Y}$ for convenience. P and Q are said to be the horizontal permutation and the vertical permutation of the Young operator $\mathcal{Y}$ (or the Young tableau $\mathcal{Y}$), respectively, and PQ is said to be a permutation belonging to the Young operator $\mathcal{Y}$ (or the Young tableau $\mathcal{Y}$). Except for the identity, no permutation can be both the horizontal permutation and the vertical permutation belonging to the same Young tableau.

★ The Young operator $\mathcal{Y}$ can be expressed as a combination of its permutations, $\mathcal{Y} = \sum \delta(Q)PQ$, and satisfies

$$P\mathcal{Y} = \mathcal{Y} = \delta(Q)\mathcal{Y}Q,$$

$$\left[E + \sum_{\mu=1}^{\lambda} (a_\mu \, b_\nu) \right] \mathcal{Y} = 0, \qquad \mathcal{Y} \left[E - \sum_{\mu=1}^{\tau} (c_\mu \, d_\nu) \right] = 0, \tag{6.5}$$

where the last two formulas are called the Fock conditions. The sum in the first Fock condition runs over all λ objects a_μ in a given row, while b_ν is any object in another row with less boxes, $\lambda' \le \lambda$. Similarly, the sum in the second Fock condition runs over all τ objects c_μ in a given column, while d_ν is any object in another column with less boxes, $\tau' \le \tau$.

If there exist two digits filled in the same row of a Young tableau $\mathcal{Y}$ and in the same column of another Young tableau $\mathcal{Y}'$, then $\mathcal{Y}'\mathcal{Y} = 0$. Two Young operators $\mathcal{Y}$ and $\mathcal{Y}'$ with different Young patterns are orthogonal to each other, $\mathcal{Y}\mathcal{Y}' = \mathcal{Y}'\mathcal{Y} = 0$. For a given Young pattern, if a standard Young tableau $\mathcal{Y}_\mu$ is larger than another standard Young tableau $\mathcal{Y}_\nu$, then

$\mathcal{Y}_\mu \mathcal{Y}_\nu = 0.$

★ If a permutation $R \in S_n$ transforms one Young tableau $\mathcal{Y}$ into another Young tableau $\mathcal{Y}'$ with the same Young pattern $[\lambda]$ for the permutation group S_n, the corresponding Young operators satisfy $\mathcal{Y}' = R\mathcal{Y}R^{-1}$. R can be calculated in the following way: Put the digits filled in the Young tableau $\mathcal{Y}$ to the first row of the two-row matrix for R, and the digits in the Young tableau $\mathcal{Y}'$ to the second row in the same order, namely, in each column of the two-row matrix for R, the upper digit and the lower digit are filled in the same box of two Young tableaux, respectively. The necessary and sufficient condition for $\mathcal{Y}'\mathcal{Y} \neq 0$ is that R belongs to the Young tableau $\mathcal{Y}$, namely, R is a product of the horizontal permutation P and the vertical permutation Q of the Young tableau $\mathcal{Y}$. Note that $P' = RPR^{-1}$ and $Q' = RQR^{-1}$ are respectively the horizontal one and the vertical one of the Young tableau $\mathcal{Y}'$, and $R = PQ = P'Q' = Q'P$. The square of a Young operator $\mathcal{Y}$ corresponding to the Young pattern $[\lambda]$ for S_n is

$$\mathcal{Y}^2 = \frac{n!}{d_{[\lambda]}(S_n)}\mathcal{Y}, \tag{6.6}$$

where $d_{[\lambda]}(S_n)$ is the number of the standard Young tableaux for the Young pattern, given in Eq. (6.4).

6. Write all Young patterns of the permutation groups S_5, S_6 and S_7 from the largest to the smallest, respectively.

Solution. For the permutation group S_5, its Young patterns are $[5]$, $[4,1]$, $[3,2]$, $[3,1^2]$, $[2^2,1]$, $[2,1^3]$, $[1^5]$.

For the permutation group S_6, its Young patterns are $[6]$, $[5,1]$, $[4,2]$, $[4,1^2]$, $[3^2]$, $[3,2,1]$, $[3,1^3]$, $[2^3]$, $[2^2,1^2]$, $[2,1^4]$, $[1^6]$.

For the permutation group S_7, its Young patterns are $[7]$, $[6,1]$, $[5,2]$, $[5,1^2]$, $[4,3]$, $[4,2,1]$, $[4,1^3]$, $[3^2,1]$, $[3,2^2]$, $[3,2,1^2]$, $[3,1^4]$, $[2^3,1]$, $[2^2,1^3]$, $[2,1^5]$, $[1^7]$.

7. Calculate the number $n(\ell)$ of elements in each class (ℓ) of the permutation groups S_4, S_5, S_6 and S_7, respectively.

Solution. The number $n(\ell)$ of elements in the class (ℓ) for S_4: $n(4) = 6$, $n(3,1) = 8$, $n(2^2) = 3$, $n(2,1^2) = 6$, $n(1^4) = 1$.

The number $n(\ell)$ of elements in the class (ℓ) for S_5: $n(5) = 24$, $n(4,1) = 30$, $n(3,2) = 20$, $n(3,1^2) = 20$, $n(2^2,1) = 15$, $n(2,1^3) = 10$, $n(1^5) = 1$.

The number $n(\ell)$ of elements in the class (ℓ) for S_6: $n(6) = 120$, $n(5,1) = 144$, $n(4,2) = 90$, $n(4,1^2) = 90$, $n(3^2) = 40$, $n(3,2,1) = 120$, $n(3,1^3) = 40$,

$n(2^3) = 15$, $n(2^2, 1^2) = 45$, $n(2, 1^4) = 15$, $n(1^6) = 1$.

The number $n(\ell)$ of elements in the class (ℓ) for S_7: $n(7) = 720$, $n(6, 1) = 840$, $n(5, 2) = 504$, $n(5, 1^2) = 504$, $n(4, 3) = 420$, $n(4, 2, 1) = 630$, $n(4, 1^3) = 210$, $n(3^2, 1) = 280$, $n(3, 2^2) = 210$, $n(3, 2, 1^2) = 420$, $n(3, 1^4) = 70$, $n(2^3, 1) = 105$, $n(2^2, 1^3) = 105$, $n(2, 1^5) = 21$, $n(1^7) = 1$.

8. Calculate the number $d_{[\lambda]}(S_n)$ of the standard Young tableaux for each Young pattern $[\lambda]$ of the permutation groups S_n, $5 \le n \le 9$.

Solution. We calculate the number $d_{[\lambda]}(S_n)$ of the standard Young tableaux for the Young pattern $[\lambda]$ by the hook rule (6.4). Remind that the associate Young patterns $[\lambda]$ and $[\widetilde{\lambda}]$ have the same number of the standard Young tableaux.

$[\lambda]$ of S_5	[5]	[4,1]	[3,2]	[3, 1^2]
$[\widetilde{\lambda}]$	$[1^5]$	$[2, 1^3]$	$[2^2, 1]$	
$d_{[\lambda]}(S_5)$	1	4	5	6

$[\lambda]$ of S_6	[6]	[5,1]	[4,2]	[4, 1^2]	[3^2]	[3,2,1]
$[\widetilde{\lambda}]$	$[1^6]$	$[2, 1^4]$	$[2^2, 1^2]$	$[3, 1^3]$	$[2^3]$	
$d_{[\lambda]}(S_6)$	1	5	9	10	5	16

$[\lambda]$ of S_7	[7]	[6,1]	[5,2]	[5, 1^2]	[4, 3]	[4,2,1]	[4, 1^3]	[3^2, 1]
$[\widetilde{\lambda}]$	$[1^7]$	$[2, 1^5]$	$[2^2, 1^3]$	$[3, 1^4]$	$[2^3, 1]$	$[3, 2, 1^2]$		$[3, 2^2]$
$d_{[\lambda]}(S_7)$	1	6	14	15	14	35	20	21

$[\lambda]$ of S_8	[8]	[7,1]	[6,2]	[6, 1^2]	[5, 3]	[5,2,1]
$[\widetilde{\lambda}]$	$[1^8]$	$[2, 1^6]$	$[2^2, 1^4]$	$[3, 1^5]$	$[2^3, 1^2]$	$[3, 2, 1^3]$
$d_{[\lambda]}(S_8)$	1	7	20	21	28	64
$[\lambda]$ of S_8	[5, 1^3]	[4^2]	[4, 3, 1]	[4, 2^2]	[4, 2, 1^2]	[3^2, 2]
$[\widetilde{\lambda}]$	$[4, 1^4]$	$[2^4]$	$[3, 2^2, 1]$	$[3^2, 1^1]$		
$d_{[\lambda]}(S_8)$	35	14	70	56	90	42

$[\lambda]$ of S_9	[9]	[8,1]	[7,2]	[7, 1^2]	[6, 3]	[6,2,1]	[6, 1^3]	[5, 4]
$[\widetilde{\lambda}]$	$[1^9]$	$[2, 1^7]$	$[2^2, 1^5]$	$[3, 1^6]$	$[2^3, 1^3]$	$[3, 2, 1^4]$	$[4, 1^5]$	$[2^4, 1]$
$d_{[\lambda]}(S_9)$	1	8	27	28	48	105	56	42
$[\lambda]$ of S_9	[5, 3, 1]	[5, 2^2]	[5, 2, 1^2]	[5, 1^4]	[4^2, 1]	[4, 3, 2]	[4, 3, 1^2]	[3^3]
$[\widetilde{\lambda}]$	$[3, 2^2, 1^2]$	$[3^2, 1^3]$	$[4, 2, 1^3]$		$[3, 2^3]$	$[3^2, 2, 1]$	$[4, 2^2, 1]$	
$d_{[\lambda]}(S_9)$	162	120	189	70	84	168	216	42

9. Write the Young operators corresponding to the following Young tableaux:

(1) :

1	2	3
4		

(2) :

1	2
3	4

(3) :

1	2	3	4
5			

Solution.

(1) : $[E + (1\ 2) + (1\ 3) + (2\ 3) + (1\ 2\ 3) + (3\ 2\ 1)]\,[E - (1\ 4)]$
$= E + (1\ 2) + (1\ 3) + (2\ 3) + (1\ 2\ 3) + (3\ 2\ 1)$
$-(1\ 4) - (2\ 1\ 4) - (3\ 1\ 4) - (2\ 3)(1\ 4) - (2\ 3\ 1\ 4) - (3\ 2\ 1\ 4).$

(2) : $[E + (1\ 2)]\,[E + (3\ 4)]\,[E - (1\ 3)]\,[E - (2\ 4)]$
$= E + (1\ 2) + (3\ 4) + (1\ 2)(3\ 4)$
$- (1\ 3) - (2\ 1\ 3) - (4\ 3\ 1) - (2\ 1\ 4\ 3)$
$- (2\ 4) - (1\ 2\ 4) - (3\ 4\ 2) - (1\ 2\ 3\ 4)$
$+ (1\ 3)(2\ 4) + (1\ 3\ 2\ 4) + (3\ 1\ 4\ 2) + (3\ 2)(1\ 4).$

(3) : $E + (1\ 2) + (1\ 3) + (1\ 4) + (2\ 3) + (2\ 4) + (3\ 4) + (1\ 2)(3\ 4)$
$+(1\ 3)(2\ 4) + (1\ 4)(2\ 3) + (1\ 2\ 3) + (1\ 3\ 2) + (1\ 2\ 4) + (1\ 4\ 2)$
$+(1\ 3\ 4) + (1\ 4\ 3) + (2\ 3\ 4) + (2\ 4\ 3) + (1\ 2\ 3\ 4) + (1\ 2\ 4\ 3)$
$+(1\ 3\ 4\ 2) + (1\ 3\ 2\ 4) + (1\ 4\ 2\ 3) + (1\ 4\ 3\ 2) - (1\ 5) - (2\ 1\ 5)$
$-(3\ 1\ 5) - (4\ 1\ 5) - (2\ 3)(1\ 5) - (2\ 4)(1\ 5) - (3\ 4)(1\ 5)$
$-(3\ 4)(2\ 1\ 5) - (2\ 4)(3\ 1\ 5) - (2\ 3)(4\ 1\ 5) - (2\ 3\ 1\ 5)$
$-(3\ 2\ 1\ 5) - (2\ 4\ 1\ 5) - (4\ 2\ 1\ 5) - (3\ 4\ 1\ 5) - (4\ 3\ 1\ 5)$
$-(2\ 3\ 4)(1\ 5) - (2\ 4\ 3)(1\ 5) - (2\ 3\ 4\ 1\ 5) - (2\ 4\ 3\ 1\ 5)$
$-(3\ 4\ 2\ 1\ 5) - (3\ 2\ 4\ 1\ 5) - (4\ 2\ 3\ 1\ 5) - (4\ 3\ 2\ 1\ 5).$

10. Write five standard Young tableaux $\mathcal{Y}_\mu$ corresponding to the Young pattern [3,2] of S_5 from the smallest to the largest, and calculate the permutations $R_{\mu\nu}$ transforming the Young tableau $\mathcal{Y}_\nu$ into Young tableau $\mathcal{Y}_\mu$.

Solution. The standard Young tableaux $\mathcal{Y}_\mu$ corresponding to the Young pattern [3,2] of S_5 are as follows:

$\mathcal{Y}_1$ $\mathcal{Y}_2$ $\mathcal{Y}_3$ $\mathcal{Y}_4$ $\mathcal{Y}_5$

1	2	3
4	5	

,

1	2	4
3	5	

,

1	2	5
3	4	

,

1	3	4
2	5	

,

1	3	5
2	4	

,

$$R_{12} = (3\ 4), \qquad R_{13} = (3\ 4\ 5), \qquad R_{14} = (2\ 4\ 3),$$

$$R_{15} = (2\ 4\ 5\ 3), \qquad R_{23} = (4\ 5), \qquad R_{24} = (2\ 3),$$

$$R_{25} = (2\ 3)(4\ 5), \qquad R_{34} = (2\ 3)(4\ 5), \qquad R_{35} = (2\ 3),$$

$$R_{45} = (4\ 5), \qquad R_{\mu\mu} = E, \qquad R_{\mu\nu} = R_{\nu\mu}^{-1}.$$

11. Calculate the permutation R_{12} transforming the following Young tableau $\mathcal{Y}_2$ to $\mathcal{Y}_1$, and check the formulas $\mathcal{P}_1 R_{12} = R_{12}\mathcal{P}_2$, $\mathcal{Q}_1 R_{12} = R_{12}\mathcal{Q}_2$ and $\mathcal{Y}_1 R_{12} = R_{12}\mathcal{Y}_2$, where

$$\mathcal{Y}_1: \quad \boxed{\begin{array}{ccc} 1 & 2 & 3 \end{array}} \atop \boxed{4} \quad, \qquad \mathcal{Y}_2: \quad \boxed{\begin{array}{ccc} 1 & 2 & 4 \end{array}} \atop \boxed{3} \quad,$$

Solution. $R_{12} = \begin{pmatrix} 1\ 2\ 4\ 3 \\ 1\ 2\ 3\ 4 \end{pmatrix} = (3\ 4).$

$$P_1 R_{12} = [E + (1\ 2) + (1\ 3) + (2\ 3) + (1\ 2\ 3) + (3\ 2\ 1)]\,(3\ 4)$$
$$= (3\ 4) + (1\ 2)(3\ 4) + (1\ 3\ 4) + (2\ 3\ 4) + (1\ 2\ 3\ 4) + (2\ 1\ 3\ 4),$$

$$R_{12} P_2 = (3\ 4)\,[E + (1\ 2) + (1\ 4) + (2\ 4) + (1\ 2\ 4) + (4\ 2\ 1)]$$
$$= (3\ 4) + (1\ 2)(3\ 4) + (3\ 4\ 1) + (3\ 4\ 2) + (3\ 4\ 1\ 2) + (3\ 4\ 2\ 1),$$

$$\mathcal{Q}_1 R_{12} = [E - (1\ 4)]\,(3\ 4) = (3\ 4) - (1\ 4\ 3),$$

$$R_{12} \mathcal{Q}_2 = (3\ 4)\,[E - (1\ 3)] = (3\ 4) - (4\ 3\ 1).$$

Thus,

$$\mathcal{P}_1 R_{12} = R_{12} \mathcal{P}_2, \qquad \mathcal{Q}_1 R_{12} = R_{12}\mathcal{Q}_2.$$

Since $\mathcal{Y}_1 = \mathcal{P}_1 \mathcal{Q}_1$ and $\mathcal{Y}_2 = \mathcal{P}_2 \mathcal{Q}_2$, we have $\mathcal{Y}_1 R_{12} = R_{12}\mathcal{Y}_2$.

12. It can be seen that there is no pair of digits filled in the same row of the Young tableau $\mathcal{Y}$ and in the same column of the Young tableau $\mathcal{Y}'$. Calculate the permutation R transforming the Young tableau $\mathcal{Y}$ into the Young tableau $\mathcal{Y}'$, and express R as PQ belonging to the Young tableau $\mathcal{Y}$, and as $P'Q'$ belonging to the Young tableau $\mathcal{Y}'$:

The Young tableau $\mathcal{Y}$ The Young tableau $\mathcal{Y}'$

1	2	4	7
3	5	9	
6	8		

1	2	3	4
5	6	7	
8	9		

Solution.

$$R = \begin{pmatrix} 1\ 2\ 4\ 7\ 3\ 5\ 9\ 6\ 8 \\ 1\ 2\ 3\ 4\ 5\ 6\ 7\ 8\ 9 \end{pmatrix} = (4\ 3\ 5\ 6\ 8\ 9\ 7)$$

$$= (7\ 4)(4\ 3\ 5\ 6\ 8\ 9) = (7\ 4)(3\ 5)(5\ 6\ 8\ 9\ 4)$$

$$= (7\ 4)(3\ 5)(5\ 6\ 8\ 9)(9\ 4) = (7\ 4)(3\ 5)(9\ 5)(5\ 6\ 8)(9\ 4)$$

$$= (7\ 4)(3\ 5\ 9)(6\ 8)(8\ 5)(9\ 4) = PQ,$$

$$P = (7\ 4)(3\ 5\ 9)(6\ 8), \qquad Q = (8\ 5)(9\ 4),$$

$$R = (4\ 3\ 5\ 6\ 8\ 9\ 7) = (4\ 3)(5\ 6\ 8\ 9\ 7\ 3) = (4\ 3)(5\ 6)(8\ 9\ 7\ 3\ 6)$$

$$= (4\ 3)(5\ 6)(8\ 9)(6\ 9\ 7\ 3) = (4\ 3)(8\ 9)(5\ 6)(7\ 6\ 9)(7\ 3)$$

$$= (4\ 3)(8\ 9)(5\ 6\ 7)(6\ 9)(7\ 3) = P'Q',$$

$$P' = (4\ 3)(5\ 6\ 7)(8\ 9) = RPR^{-1}, \qquad Q' = (9\ 6)(7\ 3) = RQR^{-1}.$$

13. Try to expand all non-standard Young tableaux $\mathcal{Y}$ of the Young pattern $[2,1]$ in terms of the standard Young tableaux $\mathcal{Y}_\mu$, $\mathcal{Y} = \sum_\mu t_\mu \mathcal{Y}_\mu$, where t_μ is the vector in the group space of S_3.

Solution. There are six Young tableaux corresponding to the Young pattern $[2,1]$, where two Young tableaux $\mathcal{Y}_1$ and $\mathcal{Y}_2$ are standard, and the remaining are non-standard.

$$\mathcal{Y}_1 = \begin{array}{|c|c|} \hline 1 & 2 \\ \hline 3 \\ \cline{1-1} \end{array}, \quad \mathcal{Y}_2 = \begin{array}{|c|c|} \hline 1 & 3 \\ \hline 2 \\ \cline{1-1} \end{array}, \quad \mathcal{Y}_a = \begin{array}{|c|c|} \hline 3 & 2 \\ \hline 1 \\ \cline{1-1} \end{array},$$

$$\mathcal{Y}_b = \begin{array}{|c|c|} \hline 2 & 3 \\ \hline 1 \\ \cline{1-1} \end{array}, \quad \mathcal{Y}_c = \begin{array}{|c|c|} \hline 2 & 1 \\ \hline 3 \\ \cline{1-1} \end{array}, \quad \mathcal{Y}_d = \begin{array}{|c|c|} \hline 3 & 1 \\ \hline 2 \\ \cline{1-1} \end{array}.$$

The symmetry and the Fock conditions of a Young operator lead to

$$\mathcal{Y}_a = (1\ 3)\mathcal{Y}_1(1\ 3) = -(1\ 3)\mathcal{Y}_1 = [E + (2\ 3)]\mathcal{Y}_1,$$

$$\mathcal{Y}_b = (1\ 2)\mathcal{Y}_2(1\ 2) = -(1\ 2)\mathcal{Y}_2 = [E + (2\ 3)]\mathcal{Y}_2,$$

$$\mathcal{Y}_c = \mathcal{Y}_c[(1\ 2) + (1\ 3)] = (1\ 2)\mathcal{Y}_1 + (1\ 3)\mathcal{Y}_b = \mathcal{Y}_1 - (2\ 3)\mathcal{Y}_2,$$

$$\mathcal{Y}_d = (2\ 3)\mathcal{Y}_c(2\ 3) = -(2\ 3)\mathcal{Y}_c = -(2\ 3)\mathcal{Y}_1 + \mathcal{Y}_2.$$

6.3 Primitive Idempotents in the Group Algebra

★ A vector e in the group algebra $\mathcal{L}$ of a finite group G is called an idempotent if $e^2 = e \in \mathcal{L}$. n vectors e_μ in $\mathcal{L}$ are called a set of the mutually orthogonal idempotents if $e_\mu e_\nu = \delta_{\mu\nu} e_\mu$. $\mathcal{L}e_\mu = \mathcal{L}_\mu$ and $e_\mu \mathcal{L} = \mathcal{R}_\mu$ are respectively called the left ideal and the right ideal of $\mathcal{L}$ generated by the idempotent e_μ. Two left (or right) ideals generated by two orthogonal idempotents contain no common vector.

For a complete set of the arbitrarily chosen bases x_ρ in a left ideal $\mathcal{L}_\mu = \mathcal{L}e_\mu$, the matrices $D(S)$ of the group elements S in the bases, $Sx_\rho = \sum_\tau x_\tau D_{\tau\rho}(S)$, form a representation of the group G, which is called the representation corresponding to the left ideal $\mathcal{L}_\mu$. Two left ideals, or two idempotents generating those two left ideals, are said to be equivalent if the representations corresponding to two left ideals are equivalent. A left ideal $\mathcal{L}_\mu$ is said to be minimum and its idempotent is said to be primitive if the representation corresponding to the left ideal is irreducible. The necessary and sufficient condition for an idempotent e_μ to be primitive is that $e_\mu t e_\mu = \lambda_t e_\mu$ holds for any vectors t in the group algebra $\mathcal{L}$, where λ_t is a constant depending on t, which is allowed to be zero. The necessary and sufficient condition for two primitive idempotents e_μ and e_ν to be equivalent is that there exists at least one group element R such that $e_\mu R e_\nu \neq 0$. These conclusions are also suitable for the right ideals, except for the calculation of the representation matrix of S in a right ideal $\mathcal{R}_\mu$ by $x_\rho S = \sum_\tau \overline{D}_{\rho\tau}(S) x_\tau$.

★ The Young operator is proportional to the primitive idempotent of the permutation group. Two Young operators corresponding to different Young patterns are orthogonal to each other. The representations generated by them are inequivalent and irreducible. Thus, an irreducible representation for a permutation group is described by a Young pattern.

When $n \leq 4$, those standard Young operators corresponding to the same Young pattern for the permutation group $\mathbf{S}_n$ are orthogonal to one another. This conclusion is not generally true. When $n \geq 5$, for a given Young pattern $[\lambda]$, we define $d_{[\lambda]}(\mathbf{S}_n)$ mutually orthogonal primitive idempotents in the following way. For simplicity, the index $[\lambda]$ and $\mathbf{S}_n$ are omitted. Denote by $R_{\mu\nu}$ the permutation transforming the standard Young tableau $\mathcal{Y}_\nu$ into the standard Young tableau $\mathcal{Y}_\mu$. If $\mathcal{Y}_\mu \mathcal{Y}_\nu \neq 0$, then $R_{\mu\nu} = P_\nu^{(\mu)} Q_\nu^{(\mu)} = P_\mu^{(\nu)} Q_\mu^{(\nu)}$, where $P_\nu^{(\mu)}$ and $Q_\nu^{(\mu)}$ are respectively the horizontal and the vertical permutations belonging to the Young tableau $\mathcal{Y}_\nu$, while $P_\mu^{(\nu)}$ and $Q_\mu^{(\nu)}$ are respectively the horizontal and the vertical permutations belonging to the Young tableau $\mathcal{Y}_\mu$. Let

$$P_{\mu\nu} = \begin{cases} P_\nu^{(\mu)} & \text{when } \mathcal{Y}_\mu \mathcal{Y}_\nu \neq 0 \\ 0 & \text{when } \mathcal{Y}_\mu \mathcal{Y}_\nu = 0, \end{cases}$$

$$y_\mu = E - \sum_{\rho=\mu+1}^{d} P_{\mu\rho} y_\rho, \qquad y_d = E, \qquad e_\mu = \frac{d}{n!} \mathcal{Y}_\mu y_\mu. \tag{6.7}$$

d primitive idempotents e_μ are mutually orthogonal. In another way, let

$$Q_{\mu\nu} = \begin{cases} \delta(Q_\mu^{(\nu)}) Q_\mu^{(\nu)} & \text{when } \mathcal{Y}_\mu \mathcal{Y}_\nu \neq 0 \\ 0 & \text{when } \mathcal{Y}_\mu \mathcal{Y}_\nu = 0, \end{cases}$$

$$y_\nu' = E - \sum_{\rho=1}^{\nu-1} y_\rho' Q_{\rho\nu}, \qquad y_1' = E, \qquad e_\nu' = \frac{d}{n!} y_\nu' \mathcal{Y}_\nu. \tag{6.8}$$

d primitive idempotents e_μ' are mutually orthogonal. The identity E of the permutation group may respectively expand in terms of two sets of the primitive idempotents

$$\begin{aligned} E &= \sum_{[\lambda],\mu} e_\mu^{[\lambda]} = \frac{1}{n!} \sum_{[\lambda]} d_{[\lambda]} \sum_\mu^{d_{[\lambda]}} \mathcal{Y}_\mu^{[\lambda]} y_\mu^{[\lambda]} \\ &= \sum_{[\lambda],\nu} e_\nu'^{[\lambda]} = \frac{1}{n!} \sum_{[\lambda]} d_{[\lambda]} \sum_\nu^{d_{[\lambda]}} y_\nu'^{[\lambda]} \mathcal{Y}_\nu^{[\lambda]}. \end{aligned} \tag{6.9}$$

Note that $n!/d_{[\lambda]}$ is nothing but the product of the hook numbers h_{ij} for all boxes in the Young pattern $[\lambda]$.

14. Expand explicitly the identity of the S_4 group in terms of the Young operators.

Solution. $E = \dfrac{A}{24} + \dfrac{B}{12} + \dfrac{C}{8}$, where

$$A = \mathcal{Y}_{\boxed{1\,2\,3\,4}} + \mathcal{Y}_{\boxed{\begin{matrix}1\\2\\3\\4\end{matrix}}}$$

$$= 2\{E + (1\ 2)(3\ 4) + (1\ 3)(2\ 4) + (1\ 4)(2\ 3) + (1\ 2\ 3) + (1\ 3\ 2)$$
$$+ (1\ 2\ 4) + (1\ 4\ 2) + (1\ 3\ 4) + (1\ 4\ 3) + (2\ 3\ 4) + (2\ 4\ 3)\},$$

$$B = \mathcal{Y}_{\boxed{\begin{matrix}1\,2\\3\,4\end{matrix}}} + \mathcal{Y}_{\boxed{\begin{matrix}1\,3\\2\,4\end{matrix}}}$$

$$= E + (1\ 2) + (3\ 4) + (1\ 2)(3\ 4) - (1\ 3) - (2\ 1\ 3)$$
$$- (4\ 3\ 1) - (2\ 1\ 4\ 3) - (2\ 4) - (1\ 2\ 4) - (3\ 4\ 2) - (1\ 2\ 3\ 4)$$
$$+ (1\ 3)(2\ 4) + (1\ 3\ 2\ 4) + (3\ 1\ 4\ 2) + (1\ 4)(3\ 2)$$
$$+ E + (1\ 3) + (2\ 4) + (1\ 3)(2\ 4) - (1\ 2) - (3\ 1\ 2)$$
$$- (4\ 2\ 1) - (3\ 1\ 4\ 2) - (3\ 4) - (1\ 3\ 4) - (2\ 4\ 3) - (1\ 3\ 2\ 4)$$
$$+ (1\ 2)(3\ 4) + (1\ 2\ 3\ 4) + (2\ 1\ 4\ 3) + (1\ 4)(2\ 3)$$
$$= 2\left\{E + (1\ 2)(3\ 4) + (1\ 3)(2\ 4) + (1\ 4)(3\ 2)\right\} - (1\ 2\ 3) - (1\ 3\ 2)$$
$$- (1\ 2\ 4) - (1\ 4\ 2) - (1\ 3\ 4) - (1\ 4\ 3) - (2\ 3\ 4) - (2\ 4\ 3),$$

$$= 2\left\{E + (1\ 2\ 3) + (1\ 3\ 2) - (2\ 3)(1\ 4)\right\} - (2\ 1\ 4) - (4\ 1\ 2)$$
$$- (3\ 1\ 4) - (4\ 1\ 3) - (2\ 3\ 1\ 4) + (4\ 1\ 2\ 3) - (3\ 2\ 1\ 4) + (4\ 1\ 3\ 2)$$
$$+ 2\left\{E + (1\ 2\ 4) + (1\ 4\ 2) - (2\ 4)(1\ 3)\right\} - (2\ 1\ 3) - (3\ 1\ 2)$$
$$- (4\ 1\ 3) - (3\ 1\ 4) - (2\ 4\ 1\ 3) + (3\ 1\ 2\ 4) - (4\ 2\ 1\ 3) + (3\ 1\ 4\ 2)$$
$$+ 2\left\{E + (1\ 4\ 3) + (1\ 3\ 4) - (4\ 3)(1\ 2)\right\} - (4\ 1\ 2) - (2\ 1\ 4)$$
$$- (3\ 1\ 2) - (2\ 1\ 3) - (4\ 3\ 1\ 2) + (2\ 1\ 4\ 3) - (3\ 4\ 1\ 2) + (2\ 1\ 3\ 4)$$
$$= 6E - 2(2\ 3)(1\ 4) - 2(2\ 4)(1\ 3) - 2(4\ 3)(1\ 2).$$

15. Calculate the orthogonal primitive idempotents for the Young pattern $[2, 2, 1]$ of the permutation group S_5.

Solution. There are five standard Young tableaux $\mathcal{Y}_\mu$ for the Young pattern $[2, 2, 1]$ of S_5:

After checking, the product of each pair of the standard Young operators is vanishing except for $\mathcal{Y}_1\mathcal{Y}_5 \neq 0$. The permutation transforming the Young tableau $\mathcal{Y}_5$ into the Young tableau $\mathcal{Y}_1$ is $R_{15} = (4\ 2\ 3\ 5) = (2\ 5)\ (5\ 4)(2\ 3)$. From Eq. (6.7) we obtain $y_1 = E - (2\ 5)$, and the remaining y_μ are E.

Thus, the orthogonal primitive idempotents are

$$e_1 = \frac{1}{24}\mathcal{Y}_1\left[E - (2\ 5)\right], \qquad e_\mu = \frac{1}{24}\mathcal{Y}_\mu, \qquad 2 \leq \mu \leq 5.$$

From Eq. (6.8) we may decompose R_{15} as $R_{15} = (3\ 4)\ (4\ 2)(3\ 5)$, and obtain another set of the orthogonal primitive idempotents

$$e_5' = \frac{1}{24}[E - (4\ 2)(3\ 5)]\mathcal{Y}_5, \qquad e_\nu' = \frac{1}{24}\mathcal{Y}_\nu, \qquad 1 \leq \nu \leq 4.$$

16. Calculate the orthogonal primitive idempotents for the Young pattern $[4, 2]$ of the permutation group S_6.

Solution. There are nine standard Young tableaux $\mathcal{Y}_\mu$ for the Young pattern $[4, 2]$ of S_6, listed from the smallest to the largest:

$$\begin{matrix} 1\ 2\ 3\ 4 & 1\ 2\ 3\ 5 & 1\ 2\ 3\ 6 & 1\ 2\ 4\ 5 & 1\ 2\ 4\ 6 \\ 5\ 6 & 4\ 6 & 4\ 5 & 3\ 6 & 3\ 5 \end{matrix}$$

$$\begin{matrix} 1\ 2\ 5\ 6 & 1\ 3\ 4\ 5 & 1\ 3\ 4\ 6 & 1\ 3\ 5\ 6 \\ 3\ 4 & 2\ 6 & 2\ 5 & 2\ 4 \end{matrix}$$

After checking, only the products of three pairs of the Young operators are non-vanishing:

$$\mathcal{Y}_1\mathcal{Y}_8 \neq 0, \qquad R_{18} = (2\ 5\ 6\ 4\ 3) = (6\ 4\ 3)(2\ 5)\ (5\ 3),$$
$$\mathcal{Y}_2\mathcal{Y}_9 \neq 0, \qquad R_{29} = (2\ 4\ 6\ 5\ 3) = (6\ 5\ 3)(2\ 4)\ (4\ 3),$$
$$\mathcal{Y}_3\mathcal{Y}_9 \neq 0, \qquad R_{39} = (2\ 4\ 5\ 3) = (5\ 3)(2\ 4)\ (4\ 3).$$

From Eq. (6.7) we obtain the orthogonal primitive idempotents as

$$e_1 = \frac{1}{80}\mathcal{Y}_1\left[E - (2\ 5)(3\ 6\ 4)\right], \qquad e_2 = \frac{1}{80}\mathcal{Y}_2\left[E - (2\ 4)(3\ 6\ 5)\right],$$
$$e_3 = \frac{1}{80}\mathcal{Y}_3\left[E - (2\ 4)(3\ 5)\right], \qquad e_\mu = \frac{1}{80}\mathcal{Y}_\mu, \qquad 4 \leq \mu \leq 9.$$

From Eq. (6.8) we may decompose the above $R_{\mu\nu}$ as $R_{18} = (4\ 3\ 2)(5\ 6)\ (6\ 2)$, $R_{29} = (5\ 3\ 2)(4\ 6)\ (6\ 2)$, $R_{39} = (3\ 2)(4\ 5)\ (5\ 2)$, and obtain another set of orthogonal primitive idempotents

$$e_8' = \frac{1}{80}[E + (2\ 6)]\,\mathcal{Y}_8, \qquad e_9' = \frac{1}{80}[E + (2\ 6) + (2\ 5)]\,\mathcal{Y}_9,$$
$$e_\nu' = \frac{1}{80}\mathcal{Y}_\nu, \qquad\qquad 1 \leq \nu \leq 7.$$

17. Calculate the orthogonal primitive idempotents for the Young pattern [3, 2, 1] of the permutation group S_6.

Solution. There are 16 standard Young tableaux $\mathcal{Y}_\mu$ for the Young pattern [3, 2, 1] of S_6, listed from the smallest to the largest:

```
1 2 3   1 2 3   1 2 4   1 2 4   1 2 5   1 2 5   1 2 6   1 2 6
4 5     4 6     3 5     3 6     3 4     3 6     3 4     3 5
6       5       6       5       6       4       5       4

1 3 4   1 3 4   1 3 5   1 3 5   1 3 6   1 3 6   1 4 5   1 4 6
2 5     2 6     2 4     2 6     2 4     2 5     2 6     2 5
6       5       6       4       5       4       3       3
```

After checking, there are 8 pairs of the Young operators whose products are non-vanishing:

$$\mathcal{Y}_1\mathcal{Y}_{11} \neq 0, \quad \mathcal{Y}_1\mathcal{Y}_{12} \neq 0, \quad \mathcal{Y}_2\mathcal{Y}_{13} \neq 0, \quad \mathcal{Y}_2\mathcal{Y}_{14} \neq 0,$$
$$\mathcal{Y}_3\mathcal{Y}_{15} \neq 0, \quad \mathcal{Y}_4\mathcal{Y}_{16} \neq 0, \quad \mathcal{Y}_5\mathcal{Y}_{15} \neq 0, \quad \mathcal{Y}_7\mathcal{Y}_{16} \neq 0.$$

$$R_{1,11} = (2\ 4\ 5\ 3) = (2\ 4)(5\ 3)\ (3\ 4) = (4\ 5)(3\ 2)\ (2\ 5),$$
$$R_{1,12} = (2\ 4\ 6\ 5\ 3) = (5\ 3)(6\ 2)\ (2\ 4)(6\ 3) = (3\ 2)(5\ 4)\ (4\ 6)(5\ 2),$$
$$R_{2,13} = (2\ 4\ 6\ 3) = (2\ 4)(6\ 3)\ (3\ 4) = (4\ 6)(3\ 2)\ (2\ 6),$$
$$R_{2,14} = (2\ 4\ 5\ 6\ 3) = (6\ 3)(5\ 2)\ (2\ 4)(5\ 3) = (3\ 2)(6\ 4)\ (4\ 5)(6\ 2),$$
$$R_{3,15} = (2\ 3\ 6\ 5\ 4) = (5\ 4)(6\ 2)\ (2\ 3)(6\ 4) = (4\ 2)(5\ 3)\ (3\ 6)(5\ 2),$$
$$R_{4,16} = (2\ 3\ 5\ 6\ 4) = (6\ 4)(5\ 2)\ (2\ 3)(5\ 4) = (4\ 2)(6\ 3)\ (3\ 5)(6\ 2),$$
$$R_{5,15} = (2\ 3\ 6\ 4) = (6\ 2)\ (2\ 3)(6\ 4) = (4\ 3)\ (3\ 6)(4\ 2),$$
$$R_{7,16} = (2\ 3\ 5\ 4) = (5\ 2)\ (2\ 3)(5\ 4) = (4\ 3)\ (3\ 5)(4\ 2).$$

Thus, we obtain the orthogonal primitive idempotents from Eq. (6.7):

$$e_1 = \frac{1}{45}\mathcal{Y}_1\left[E - (2\ 4)(3\ 5) - (2\ 6)(3\ 5)\right],$$

$$e_2 = \frac{1}{45}\mathcal{Y}_2\left[E - (2\ 4)(3\ 6) - (2\ 5)(3\ 6)\right],$$

$$e_3 = \frac{1}{45}\mathcal{Y}_3\left[E - (2\ 6)(4\ 5)\right], \quad e_4 = \frac{1}{45}\mathcal{Y}_4\left[E - (2\ 5)(4\ 6)\right],$$

$$e_5 = \frac{1}{45}\mathcal{Y}_5\left[E - (2\ 6)\right], \quad e_7 = \frac{1}{45}\mathcal{Y}_7\left[E - (2\ 5)\right],$$

$$e_\mu = \frac{1}{45}\mathcal{Y}_\mu, \quad \mu = 6 \ \text{ or } \ 8 \leq \mu \leq 16.$$

Another set of the orthogonal primitive idempotents is obtain from Eq. (6.8)

$$e'_{11} = \frac{1}{45}\,[E + (2\ 5)]\,\mathcal{Y}_{11}, \quad e'_{12} = \frac{1}{45}\,[E - (2\ 5)(4\ 6)]\,\mathcal{Y}_{12},$$

$$e'_{13} = \frac{1}{45}\,[E + (2\ 6)]\,\mathcal{Y}_{13}, \quad e'_{14} = \frac{1}{45}\,[E - (2\ 6)(4\ 5)]\,\mathcal{Y}_{14},$$

$$e'_{15} = \frac{1}{45}\,[E - (2\ 5)(3\ 6) - (2\ 4)(3\ 6)]\,\mathcal{Y}_{15},$$

$$e'_{16} = \frac{1}{45}\,[E - (2\ 6)(3\ 5) - (2\ 4)(3\ 5)]\,\mathcal{Y}_{16},$$

$$e'_{\nu} = \frac{1}{45}\mathcal{Y}_{\nu}, \qquad 1 \le \nu \le 10.$$

18. Calculate the orthogonal primitive idempotents for the Young patterns [3, 3] and [4,1,1] of the permutation group S_6, respectively.

Solution. There are five standard Young tableaux $\mathcal{Y}_\mu$ for the Young pattern [3, 3] of S_6:

$$\mathcal{Y}_1 \quad \mathcal{Y}_2 \quad \mathcal{Y}_3 \quad \mathcal{Y}_4 \quad \mathcal{Y}_5$$

$\mathcal{Y}_1$	$\mathcal{Y}_2$	$\mathcal{Y}_3$	$\mathcal{Y}_4$	$\mathcal{Y}_5$
1 2 3	1 2 4	1 2 5	1 3 4	1 3 5
4 5 6	3 5 6	3 4 6	2 5 6	2 4 6

The digit 6 can only be filled in the last box of the second row, so the standard Young tableaux for the Young pattern [3, 3] of S_6 are the same as those for the Young pattern [3, 2] of S_5, respectively, except for adding the last box filled with 6. The calculation for the orthogonal primitive idempotents is also the same.

After checking, the product of each pair of the standard Young operators is vanishing except for $\mathcal{Y}_1\mathcal{Y}_5 \ne 0$. The permutation transforming the Young tableau $\mathcal{Y}_5$ into the Young tableau $\mathcal{Y}_1$ is $R_{15} = (2\ 4\ 5\ 3) = (2\ 4)(5\ 3)\ (3\ 4) = (4\ 5)(3\ 2)\ (2\ 5)$. From Eqs. (6.7) and (6.8) we obtain two sets of the orthogonal primitive idempotents as

$$e_1 = \frac{1}{144}\mathcal{Y}_1\,[E - (2\ 4)(3\ 5)], \qquad e_\mu = \frac{1}{24}\mathcal{Y}_\mu, \qquad 2 \le \mu \le 5,$$

$$e'_5 = \frac{1}{144}\,[E + (2\ 5)]\mathcal{Y}_5, \qquad e'_\nu = \frac{1}{24}\mathcal{Y}_\nu, \qquad 1 \le \nu \le 4.$$

There are ten standard Young tableaux $\mathcal{Y}_\mu$ for the Young pattern [4, 1, 1] of S_6, listed from the smallest to the largest:

```
1 2 3 4   1 2 3 5   1 2 3 6   1 2 4 5   1 2 4 6
5         4         4         3         3
6         6         5         6         5


1 2 5 6   1 3 4 5   1 3 4 6   1 3 5 6   1 4 5 6
3         2         2         2         2
4         6         5         4         3
```

After checking, each pair of the Young operators are orthogonal. $\mathcal{Y}_\mu/72$ are ten orthogonal primitive idempotents for the Young pattern $[4,1,1]$ of S_6

19. Give an example to demonstrate that the primitive idempotent generating a minimum left ideal is not unique.

Solution. Take the S_3 group as an example. Let $\mathcal{Y}_1$ be the standard Young operator corresponding to the Young tableau $\begin{array}{|c|c|}\hline 1 & 2 \\\hline 3 \\\cline{1-1}\end{array}$. $\mathcal{Y}_1$ generates the minimal left ideal $\mathcal{L}_1 = \mathcal{L}\mathcal{Y}_1$. Let $\mathcal{Y}_2$ be the non-standard Young operator corresponding to Young tableau $\begin{array}{|c|c|}\hline 3 & 2 \\\hline 1 \\\cline{1-1}\end{array}$. It is easy to see that $\mathcal{Y}_2 = (1\ 3)\mathcal{Y}_1(1\ 3) = -(1\ 3)\mathcal{Y}_1$. Being a group algebra, $\mathcal{L} = -\mathcal{L}(1\ 3)$. Therefore, the left ideals generated by $\mathcal{Y}_1$ and $\mathcal{Y}_2$ are the same, $\mathcal{L}_1 = \mathcal{L}\mathcal{Y}_1 = \mathcal{L}\mathcal{Y}_2$. $\mathcal{Y}_1/3$ and $\mathcal{Y}_2/3$ are both the primitive idempotents generating the same minimal left ideal $\mathcal{L}_1$.

6.4 Irreducible Representations and Characters

In this section we will discuss the irreducible representation of S_n denoted by a given Young pattern $[\lambda]$. Sometimes, we neglect the index $[\lambda]$ for simplicity.

★ Let the permutation $R_{\mu\nu}$ transform the standard Young tableau $\mathcal{Y}_\nu$ into the standard Young tableau $\mathcal{Y}_\mu$, and d be the number of the standard Young tableaux corresponding to the Young pattern $[\lambda]$. Define the standard basis as

$$b_{\mu\nu} = \left(\frac{d}{n!}\right)^2 \mathcal{Y}_\mu y_\mu R_{\mu\nu} \mathcal{Y}_\nu y_\nu = \frac{d}{n!} R_{\mu\nu} \mathcal{Y}_\nu y_\nu, \qquad b_{\mu\rho} b_{\tau\nu} = \delta_{\rho\tau} b_{\mu\nu}. \quad (6.10)$$

For a given ν, the d bases $b_{\mu\nu}$ span the left ideal $\mathcal{L}_\nu$, while for a given μ, the d bases $b_{\mu\nu}$ span the right ideal $\mathcal{R}_\mu$. The representation matrix of the group element S in the standard bases take the same form both for the left ideal and for the right ideal:

$$Sb_{\mu\nu} = \sum_{\rho} b_{\rho\nu} D_{\rho\mu}(S), \qquad b_{\mu\nu}S = \sum_{\rho} D_{\nu\rho}(S)b_{\mu\rho}. \qquad (6.11)$$

Of course, another set of the standard bases can also be used:

$$b'_{\mu\nu} = (d/n!)^2 \, y'_{\mu} \mathcal{Y}_{\mu} R_{\mu\nu} y'_{\nu} \mathcal{Y}_{\nu} = \frac{d}{n!} y'_{\mu} \mathcal{Y}_{\mu} R_{\mu\nu}, \qquad b'_{\mu\rho} b'_{\tau\nu} = \delta_{\rho\tau} b'_{\mu\nu}.$$

For definiteness, we will only discuss the former set of bases.

★ The representation matrix of the group element S in the standard bases can be calculated by the tabular method. Denote by $\mathcal{Y}_{\nu}(S)$ the Young tableau transformed from the standard Young tableau $\mathcal{Y}_{\nu}$ by the permutation S. Let $y_{\mu} = \sum \delta_k T_k$, and denote by $\mathcal{Y}_{\mu k}$ the Young tableau transformed from the standard Young tableau $\mathcal{Y}_{\mu}$ by the permutation T_k^{-1}. Define the coefficient $A_{\mu k}^{\nu}(S)$ by comparing two Young tableaux $\mathcal{Y}_{\mu k}$ and $\mathcal{Y}_{\nu}(S)$. If there is a pair of digits filled in the same row of the Young tableau $\mathcal{Y}_{\nu}(S)$ and in the same column of the Young tableau $\mathcal{Y}_{\mu k}$, then $A_{\mu k}^{\nu}(S) = 0$. Otherwise, $A_{\mu k}^{\nu}(S)$ is defined to be the permutation parity of the vertical permutation R of the Young tableau $\mathcal{Y}_{\mu k}$ which transforms the Young tableau $\mathcal{Y}_{\mu k}$ into the Young tableau $\mathcal{Y}'$ such that the digits in each row of the Young tableau $\mathcal{Y}'$ are the same as those in the same row of the Young tableau $\mathcal{Y}_{\nu}(S)$. The representation matrix entry $D_{\mu\nu}(S)$ of the permutation S in the standard bases $b_{\mu\rho}$ is equal to $\sum_k \delta_k A_{\mu k}^{\nu}(S)$, which can be calculated by the tabular method. In fact, one may draw a table, where each row is designated by the algebraic sum of the Young tableaux, $\sum \delta_k \{$the Young tableau $\mathcal{Y}_{\mu k}\}$, and each column is designated by the Young tableau $\mathcal{Y}_{\nu}(S)$. $A_{\mu k}^{(\nu)}$ can be calculated by comparing the Young tableau $\mathcal{Y}_{\mu k}$ in the row and the Young tableau $\mathcal{Y}_{\nu}(S)$ in the column. The irreducible representation $D(S)$ in the standard bases is generally not unitary.

★ In the orthogonal basis Φ_{μ}, the representation matrix $\overline{D}(P_a)$ of a transposition P_a for two neighbored objects, $P_a = (a \; a + 1)$, is the direct sum of several one- or two-dimensional submatrices, which can be calculated as follows. If a and $a + 1$ are filled in the same row or the same column of the standard Young tableau $\mathcal{Y}_{\mu}$, we have a one-dimensional submatrix $\overline{D}_{\mu\mu}(P_a) = 1$ or -1, respectively. Otherwise, we have a two-dimensional submatrix. Without loss of generality, we assume that a is filled in the upper row of the standard Young tableau $\mathcal{Y}_{\mu}$ than $(a + 1)$, and the standard Young tableau $\mathcal{Y}_{\mu_a}$ is obtained from the standard Young tableau $\mathcal{Y}_{\mu}$

by interchanging a and $a + 1$. Since a and $a + 1$ are filled in the different rows and in the different columns of the standard Young tableau $\mathcal{Y}_\mu$, the Young tableau $\mathcal{Y}_{\mu_a}$ must be standard and larger than the standard Young tableau $\mathcal{Y}_\mu$. Now, the two-dimensional submatrix is

$$\begin{pmatrix} \overline{D}_{\mu\mu}(P_a) & \overline{D}_{\mu\mu_a}(P_a) \\ \overline{D}_{\mu_a\mu}(P_a) & \overline{D}_{\mu_a\mu_a}(P_a) \end{pmatrix} = \frac{1}{m} \begin{pmatrix} -1 & \sqrt{m^2 - 1} \\ \sqrt{m^2 - 1} & 1 \end{pmatrix},$$

$$m = m_\mu(a) - m_\mu(a + 1), \tag{6.12}$$

$$m_\mu(a) = c_\mu(a) - r_\mu(a),$$

$$m_\mu(a + 1) = c_\mu(a + 1) - r_\mu(a + 1),$$

where the digit a is filled in the $c_\mu(a)$th column of the $r_\mu(a)$th row of the standard Young tableau $\mathcal{Y}_\mu$, and similar for the digit $a + 1$. $m_\mu(a)$ is called the content of a in the Young tableau $\mathcal{Y}_\mu$. If one goes from the box filled with a in the Young tableau $\mathcal{Y}_\mu$ downwards or leftwards and at last reaches the box filled with $a + 1$, the number of steps is equal to m.

★ The merit of the representation $D(S)$ of S_n in the standard bases is that the bases are explicitly obtained, but the representation is not unitary. The merit of the representation $\overline{D}(P_a)$ of S_n in the orthogonal bases is that the representation is real orthogonal, but its bases have to be calculated from the standard bases. The similarity transformation matrix X from the standard bases to the orthogonal bases is an upper triangle matrix:

$$D(P_a)X = X\overline{D}(P_a), \qquad X_{\mu\nu} = 0, \qquad \text{when} \quad \mu > \nu. \tag{6.13}$$

If only one left ideal $\mathcal{L}_\nu$ is considered, the orthogonal basis $\phi_{\mu\nu}$ is

$$\phi_{\mu\nu} = \sum_\rho b_{\rho\nu} X_{\rho\mu}, \qquad S\phi_{\mu\nu} = \sum_\rho \phi_{\rho\nu}\overline{D}_{\rho\mu}(S). \tag{6.14}$$

If both left- and right-ideals are considered, we have to calculate the orthogonal bases $\Phi_{\mu\nu}$

$$\Phi_{\mu\nu} = \sum_\tau (X^{-1})_{\nu\tau}\phi_{\mu\tau} = \sum_{\rho\tau} (X^{-1})_{\nu\tau} b_{\rho\tau} X_{\rho\mu},$$

$$S\Phi_{\mu\nu} = \sum_\rho \Phi_{\rho\nu}\overline{D}_{\rho\mu}(S), \qquad \Phi_{\mu\nu}S = \sum_\rho \overline{D}_{\nu\rho}(S)\Phi_{\mu\rho}. \tag{6.15}$$

★ There is a graphic method for calculating the character of a class (ℓ) in the irreducible representation $[\lambda]$. Rearrange the partition number ℓ_j in an arbitrary order, but it is convenient to arrange ℓ_j from the smallest to the largest. According to the following rule, we fill n digits into the Young pattern $[\lambda]$ in order: first ℓ_1 digits 1, then ℓ_2 digits 2 and so on:

a) After each set of ℓ_j digits j is filled, the boxes which have been filled must constitute a sub-Young pattern, namely the boxes are lined up on the top and on the left, where the number of boxes on the upper row is not less than that on the lower row, and the number of boxes at the left column is not less than that at the right column. In each row, there is no unfilled box embedding between two filled boxes.

b) The boxes filled with the same digit are connected such that from the lowest and the leftist box one can go rightwards or upwards, without going leftwards or downwards, through all boxes filled with the same digit.

If all digits are filled into the Young pattern according to the rule, we say there is one regular application. The filling parity for the digit j is defined to be 1 if the number of rows of the boxes filled with j is odd, and to be -1 if that is even. The filling parity of a regular application is defined to be the product of the filling parities for all digits. The character $\chi^{[\lambda]}\{(\ell)\}$ of the class (ℓ) in the representation $[\lambda]$ is equal to the sum of the filling parities of all regular applications. If the digits cannot be all filled in the Young pattern satisfying the above rule, we say there is no regular application and obtain $\chi^{[\lambda]}\{(\ell)\} = 0$. For the class (1^n) composed of only the identity, each standard Young tableau corresponds to one regular application, so its character is just the dimension of the representation. Usually, the character of the identity is calculated by the hook rule instead of the graphic method.

★ Two Young patterns related by a transpose are called the associated Young patterns. Each regular application of the class (ℓ) in one Young pattern is a transpose of a regular application in its associated Young pattern. Since the sum of the number of rows and the number of columns of the boxes filled with j in a regular application is equal to $\ell_j + 1$. Note that $(-1)^{\ell_j+1}$ is just the permutation parity of the cycle with length ℓ_j. Therefore, the product of two filling parities of the digit j in the corresponding regular applications for two associated Young patterns is equal to the permutation parity of the cycle, and the characters of the class (ℓ) in two representations denoted by two associated Young patterns are the same if the permutation parity of the elements in the class is even, and different by a sign if that is odd. The Young pattern of the identical representation is $[n]$ and its associated Young pattern $[1^n]$ denotes the antisymmetric representation, where the representation matrix of S is equal to the permutation parity $\delta(S)$. A representation denoted by $[\lambda]$ is equivalent to the direct product of the antisymmetric representation $[1^n]$ and the representation denoted by the associated Young pattern $[\tilde{\lambda}]$.

20. Calculate the Clebsch-Gordan coefficients for the reduction of the self-product of the irreducible representation $[2, 1]$ of the S_3 group in the standard bases and in the orthogonal bases, respectively.

Solution. There are three inequivalent irreducible representations for S_3. $[3]$ is the symmetric (identical) representation. $[1^3]$ is the antisymmetric representation. Take the transpositions P_1 and P_2 of the neighboring objects as the generators of S_3. Their representation matrices are known:

$$D^{[3]}(P_1) = D^{[3]}(P_2) = 1, \qquad D^{[1^3]}(P_1) = D^{[1^3]}(P_2) = -1.$$

$[2,1]$ is the mixed representation in which the representation matrices of the generators can be calculated by the tabular method. There are two orthogonal standard Young operators for the Young tableau $[2,1]$. The rows in the table are designated by the standard Young tableaux $\mathcal{Y}_\mu$, listed in the first column of the table. Apply P_a, $a = 1$ or 2, to the standard Young tableau $\mathcal{Y}_\nu$, one obtains the Young tableau $\mathcal{Y}_\nu(P_a)$. The columns in the table are designated by the Young tableau $\mathcal{Y}_\nu(P_a)$, listed in the first row of the table. The coefficients $A^\nu_\mu(P_a)$ is determined by comparing the Young tableaux indicating the μth row and the νth column. $A^\nu_\mu(P_a)$ is nothing but the representation matrix entry of P_a, $a = 1$ or 2.

	P_1		P_2	
representation $[2,1]$	$\begin{matrix} 2\ 1 \\ 3 \end{matrix}$	$\begin{matrix} 2\ 3 \\ 1 \end{matrix}$	$\begin{matrix} 1\ 3 \\ 2 \end{matrix}$	$\begin{matrix} 1\ 2 \\ 3 \end{matrix}$
$\begin{matrix} 1\ 2 \\ 3 \end{matrix}$	1	-1	0	1
$\begin{matrix} 1\ 3 \\ 2 \end{matrix}$	0	-1	1	0

The representation matrices for the generators in the mixed representation $[2, 1]$ are

$$D^{[2,1]}(P_1) = \begin{pmatrix} 1 & -1 \\ 0 & -1 \end{pmatrix}, \quad D^{[2,1]}(P_2) = \begin{pmatrix} 0 & 1 \\ 1 & 0 \end{pmatrix}.$$

The reduction for the self-product of the representation [2,1] is

$$[2, 1] \times [2, 1] \simeq [3] \oplus [1^3] \oplus [2, 1].$$

Denote by X the similarity transformation matrix for reduction

$$\begin{pmatrix} 1 & -1 & -1 & 1 \\ 0 & -1 & 0 & 1 \\ 0 & 0 & -1 & 1 \\ 0 & 0 & 0 & 1 \end{pmatrix} X = X \begin{pmatrix} 1 & 0 & 0 & 0 \\ 0 & -1 & 0 & 0 \\ 0 & 0 & 1 & -1 \\ 0 & 0 & 0 & -1 \end{pmatrix},$$

$$\begin{pmatrix} 0 & 0 & 0 & 1 \\ 0 & 0 & 1 & 0 \\ 0 & 1 & 0 & 0 \\ 1 & 0 & 0 & 0 \end{pmatrix} X = X \begin{pmatrix} 1 & 0 & 0 & 0 \\ 0 & -1 & 0 & 0 \\ 0 & 0 & 0 & 1 \\ 0 & 0 & 1 & 0 \end{pmatrix}.$$

Due to $P_a^2 = E$, the eigenvalues for two matrices are both ± 1. The eigenvectors for the eigenvalue 1 of two representation matrices before transformation respectively are

$$P_1 : \begin{pmatrix} 1 \\ 0 \\ 0 \\ 0 \end{pmatrix}, \begin{pmatrix} 0 \\ 1 \\ 1 \\ 2 \end{pmatrix}; \qquad P_2 : \begin{pmatrix} 1 \\ 0 \\ 0 \\ 1 \end{pmatrix}, \begin{pmatrix} 0 \\ 1 \\ 1 \\ 0 \end{pmatrix}.$$

The transpose of the common eigenvector is $X_1^T = (2\ 1\ 1\ 2)$. The eigenvectors for the eigenvalue -1 respectively are

$$P_1 : \begin{pmatrix} 0 \\ 1 \\ -1 \\ 0 \end{pmatrix}, \begin{pmatrix} 1 \\ 1 \\ 1 \\ 0 \end{pmatrix}; P_2 : \begin{pmatrix} 1 \\ 0 \\ 0 \\ -1 \end{pmatrix}, \begin{pmatrix} 0 \\ 1 \\ -1 \\ 0 \end{pmatrix}.$$

The transpose of the common eigenvector is $X_2^T = (0\ 1\ -1\ 0)$. X_1 and X_2 are respectively the first and the second columns of X. Denote by X_3 and X_4 the third and the fourth columns of X. From the first formula of the similarity transformation, X_3 is an eigenvector with the eigenvalue 1 of the representation matrix before transformation, $X_3^T = (a\ b\ b\ 2b)$. Substituting X_3 into the second formula, we calculate the third column of the matrix, and obtain $X_4^T = (2b\ b\ b\ a)$. Substituting them into the first formula, we calculate the fourth column of matrix, and obtain

$$\begin{pmatrix} a \\ a-b \\ a-b \\ a \end{pmatrix} = \begin{pmatrix} -a-2b \\ -2b \\ -2b \\ -a-2b \end{pmatrix}.$$

Thus, $a = -b$. Taking $-a = b = 1$, we have

$$X = \begin{pmatrix} 2 & 0 & -1 & 2 \\ 1 & 1 & 1 & 1 \\ 1 & -1 & 1 & 1 \\ 2 & 0 & 2 & -1 \end{pmatrix}.$$

The reader is suggested to be familiar with this method. If not, he can calculate the matrix X by a direct way

$$X = \begin{pmatrix} 2 & 0 & a_1 & b_1 \\ 1 & 1 & a_2 & b_2 \\ 1 & -1 & a_3 & b_3 \\ 2 & 0 & a_4 & b_4 \end{pmatrix}.$$

Substituting it into the formulas of the similarity transformation, he will obtain the same result. It is worthy to mention that the column-matrices in X are not orthogonal to one another because the representation before transformation is not unitary.

Denoting by $|\mu\rangle|\nu\rangle$ the bases before transformation, and by $||[3]\rangle$, $||[1^3]\rangle$ and $||[2,1],\rho\rangle$ the bases after transformation, where μ, ν and ρ are respectively taken 1 and 2, we have

$$||[3]\rangle = 2|1\rangle|1\rangle + |1\rangle|2\rangle + |2\rangle|1\rangle + 2|2\rangle|2\rangle,$$

$$||[1^3]\rangle = |1\rangle|2\rangle - |2\rangle|1\rangle,$$

$$||[2,1],1\rangle = -|1\rangle|1\rangle + |1\rangle|2\rangle + |2\rangle|1\rangle + 2|2\rangle|2\rangle,$$

$$||[2,1],2\rangle = 2|1\rangle|1\rangle + |1\rangle|2\rangle + |2\rangle|1\rangle - |2\rangle|2\rangle.$$

For the representations [3] and $[1^3]$, the representation matrices in the orthogonal bases are the same as those in the standard bases. For the representation [2,1], the representation matrices of the generators in the orthogonal bases are

$$\overline{D}^{[2,1]}(P_1) = \begin{pmatrix} 1 & 0 \\ 0 & -1 \end{pmatrix}, \quad \overline{D}^{[2,1]}(P_2) = \frac{1}{2}\begin{pmatrix} -1 & \sqrt{3} \\ \sqrt{3} & 1 \end{pmatrix}.$$

Denote by Z the real orthogonal similarity transformation matrix in the reduction of the self-product of the representation [2,1]:

$$\begin{pmatrix} 1 & 0 & 0 & 0 \\ 0 & -1 & 0 & 0 \\ 0 & 0 & -1 & 0 \\ 0 & 0 & 0 & 1 \end{pmatrix} Z = Z \begin{pmatrix} 1 & 0 & 0 & 0 \\ 0 & -1 & 0 & 0 \\ 0 & 0 & 1 & 0 \\ 0 & 0 & 0 & -1 \end{pmatrix},$$

$$\frac{1}{4}\begin{pmatrix} 1 & -\sqrt{3} & -\sqrt{3} & 3 \\ -\sqrt{3} & -1 & 3 & \sqrt{3} \\ -\sqrt{3} & 3 & -1 & \sqrt{3} \\ 3 & \sqrt{3} & \sqrt{3} & 1 \end{pmatrix} Z = \frac{1}{2}Z \begin{pmatrix} 2 & 0 & 0 & 0 \\ 0 & -2 & 0 & 0 \\ 0 & 0 & -1 & \sqrt{3} \\ 0 & 0 & \sqrt{3} & 1 \end{pmatrix}.$$

Due to $P_a^2 = E$, the eigenvalues of two matrices are both ± 1. The eigenvectors for the eigenvalue 1 of two representation matrices before transformation respectively are

$$P_1 : \begin{pmatrix} 1 \\ 0 \\ 0 \\ 0 \end{pmatrix}, \begin{pmatrix} 0 \\ 0 \\ 0 \\ 1 \end{pmatrix}; \quad P_2 : \begin{pmatrix} 1 \\ 0 \\ 0 \\ 1 \end{pmatrix}, \begin{pmatrix} \sqrt{3} \\ -1 \\ -1 \\ \sqrt{1/3} \end{pmatrix}.$$

The normalized common eigenvector is $Z_1^T = (1\ 0\ 0\ 1)/\sqrt{2}$. The eigenvectors for the eigenvalue -1 respectively are

$$P_1 : \begin{pmatrix} 0 \\ 1 \\ 0 \\ 0 \end{pmatrix}, \begin{pmatrix} 0 \\ 0 \\ 1 \\ 0 \end{pmatrix}; P_2 : \begin{pmatrix} 0 \\ 1 \\ -1 \\ 0 \end{pmatrix}, \begin{pmatrix} \sqrt{3} \\ 1 \\ 1 \\ -\sqrt{3} \end{pmatrix}.$$

The normalized common eigenvector is $Z_2^T = (0\ 1\ -1\ 0)/\sqrt{2}$. Z_1 and Z_2 are the first and the second columns of Z, respectively. From the condition of real orthogonal matrix, one may write the matrix Z as

$$Z = \frac{1}{\sqrt{2}} \begin{pmatrix} 1 & 0 & a & 0 \\ 0 & 1 & 0 & 1 \\ 0 & -1 & 0 & 1 \\ 1 & 0 & -a & 0 \end{pmatrix}.$$

Substituting it into the formulas for the similarity transformation related to P_2, we obtain $a = -1$ from the first row of the fourth column. Denoting again by $|\mu\rangle|\nu\rangle$ the bases before transformation and by $||[3]\rangle$, $||[1^3]\rangle$ and

$||[2,1],\rho\rangle$ the bases after transformation, where μ, ν and ρ are respectively taken 1 and 2, we have

$$||[3]\rangle = \sqrt{1/2}\,\{|1\rangle|1\rangle + |2\rangle|2\rangle\}\,,$$
$$||[1^3]\rangle = \sqrt{1/2}\,\{|1\rangle|2\rangle - |2\rangle|1\rangle\}\,,$$
$$||[2,1],1\rangle = \sqrt{1/2}\,\{-|1\rangle|1\rangle + |2\rangle|2\rangle\}\,,$$
$$||[2,1],2\rangle = \sqrt{1/2}\,\{|1\rangle|2\rangle + |2\rangle|1\rangle\}\,.$$

21. Calculate the standard bases for the irreducible representation $[3,1]$ in the group space of S_4, and calculate the representation matrix of the transposition P_a for the neighbored objects in the standard bases by the tabular method.

Solution. There are three standard Young tableaux $\mathcal{Y}_1$, $\mathcal{Y}_2$, and $\mathcal{Y}_3$ for the Young pattern $[3,1]$ of the S_3 group

$$\mathcal{Y}_1: \begin{array}{|c|c|c|} \hline 1 & 2 & 3 \\ \hline 4 \\ \cline{1-1} \end{array}\,, \qquad \mathcal{Y}_2: \begin{array}{|c|c|c|} \hline 1 & 2 & 4 \\ \hline 3 \\ \cline{1-1} \end{array}\,, \qquad \mathcal{Y}_3: \begin{array}{|c|c|c|} \hline 1 & 3 & 4 \\ \hline 2 \\ \cline{1-1} \end{array}\,.$$

The corresponding Young operators $\mathcal{Y}_\mu$ are orthogonal to one another. The permutations between three standard Young tableaux respectively are

$$R_{12} = R_{21} = (3\ 4), \qquad R_{23} = R_{32} = (2\ 3), \qquad R_{13} = R_{31}^{-1} = (2\ 4\ 3).$$

Removing the factor $1/8$, we obtain the standard bases

$$\begin{array}{lll} b_{11} = \mathcal{Y}_1, & b_{12} = (3\ 4)\mathcal{Y}_2, & b_{13} = (2\ 4\ 3)\mathcal{Y}_3, \\ b_{21} = (3\ 4)\mathcal{Y}_1, & b_{22} = \mathcal{Y}_2, & b_{23} = (2\ 3)\mathcal{Y}_3, \\ b_{31} = (2\ 3\ 4)\mathcal{Y}_1, & b_{32} = (2\ 3)\mathcal{Y}_2, & b_{33} = \mathcal{Y}_3, \end{array}$$

where each column belongs to one left ideal, each row belongs to one right ideal, and

$$\mathcal{Y}_1 = E + (1\ 2) + (1\ 3) + (2\ 3) + (1\ 2\ 3) + (3\ 2\ 1) - (1\ 4)$$
$$- (2\ 1\ 4) - (3\ 1\ 4) - (2\ 3)(1\ 4) - (2\ 3\ 1\ 4) - (3\ 2\ 1\ 4),$$
$$\mathcal{Y}_2 = (3\ 4)\mathcal{Y}_1(3\ 4), \qquad \mathcal{Y}_3 = (2\ 4)\mathcal{Y}_1(2\ 4).$$

The representation matrices for P_a are calculated by the tabular method:

	P_1			P_2			P_3		
Rep. [3,1]	$\frac{2\,1\,3}{4}$	$\frac{2\,1\,4}{3}$	$\frac{2\,3\,4}{1}$	$\frac{1\,3\,2}{4}$	$\frac{1\,3\,4}{2}$	$\frac{1\,2\,4}{3}$	$\frac{1\,2\,4}{3}$	$\frac{1\,2\,3}{4}$	$\frac{1\,4\,3}{2}$
$\frac{1\,2\,3}{4}$	1	0	−1	1	0	0	0	1	0
$\frac{1\,2\,4}{3}$	0	1	−1	0	0	1	1	0	0
$\frac{1\,3\,4}{2}$	0	0	−1	0	1	0	0	0	1

Thus, the representation matrices of the transpositions P_a for the neighbored objects in the representation $[3,1]$ are:

$$D(P_1) = \begin{pmatrix} 1 & 0 & -1 \\ 0 & 1 & -1 \\ 0 & 0 & -1 \end{pmatrix}, \quad D(P_2) = \begin{pmatrix} 1 & 0 & 0 \\ 0 & 0 & 1 \\ 0 & 1 & 0 \end{pmatrix}, \quad D(P_3) = \begin{pmatrix} 0 & 1 & 0 \\ 1 & 0 & 0 \\ 0 & 0 & 1 \end{pmatrix}.$$

22. Calculate the real orthogonal representation matrix of the transpositions P_a for the neighbored objects in the irreducible representation $[3,1]$ of the S_4 group in the orthogonal bases by the formula (6.12). Calculate the similarity transformation matrix between two representations in the standard bases (Problem 21) and in the orthogonal bases and write explicitly the orthogonal bases $\Phi_{\mu\nu}$ in the group space of S_4.

Solution. The representation matrices of P_a in the orthogonal bases of the irreducible representation $[3,1]$ of S_4 can be calculated by Eq. (6.12),

$$\overline{D}(P_1) = \begin{pmatrix} 1 & 0 & 0 \\ 0 & 1 & 0 \\ 0 & 0 & -1 \end{pmatrix}, \quad \overline{D}(P_2) = \begin{pmatrix} 1 & 0 & 0 \\ 0 & -1/2 & \sqrt{3}/2 \\ 0 & \sqrt{3}/2 & 1/2 \end{pmatrix},$$

$$\overline{D}(P_3) = \begin{pmatrix} -1/3 & \sqrt{8}/3 & 0 \\ \sqrt{8}/3 & 1/3 & 0 \\ 0 & 0 & 1 \end{pmatrix}.$$

Denote by X the similarity transformation matrix transforming the representation $D(P_a)$ in the standard bases into the representation $\overline{D}(P_a)$ in the orthogonal bases. X is an upper triangular matrix with three column matrices X_1, X_2 and X_3:

$$D(P_a)X = X\overline{D}(P_a), \quad X = \begin{pmatrix} \sqrt{8} & a_1 & b_1 \\ 0 & a_2 & b_2 \\ 0 & 0 & b_3 \end{pmatrix},$$

where the coefficient $\sqrt{8}$ is added for simplicity. In calculation we neglect the rows where all entries are vanishing. From

$$D(P_3)X_1 = -\frac{1}{3}X_1 + \frac{\sqrt{8}}{3}X_2, \quad \begin{pmatrix} 0 \\ \sqrt{8} \end{pmatrix} = -\frac{1}{3}\begin{pmatrix} \sqrt{8} \\ 0 \end{pmatrix} + \frac{\sqrt{8}}{3}\begin{pmatrix} a_1 \\ a_2 \end{pmatrix},$$

we obtain $a_1 = 1$ and $a_2 = 3$. From

$$D(P_2)X_2 = -\frac{1}{2}X_2 + \frac{\sqrt{3}}{2}X_3, \quad \begin{pmatrix} 1 \\ 0 \\ 3 \end{pmatrix} = -\frac{1}{2}\begin{pmatrix} 1 \\ 3 \\ 0 \end{pmatrix} + \frac{\sqrt{3}}{2}\begin{pmatrix} b_1 \\ b_2 \\ b_3 \end{pmatrix},$$

we obtain $b_1 = b_2 = \sqrt{3}$ and $b_3 = 2\sqrt{3}$. Finally, the similarity transformation matrix X and its inverse matrix are

$$X = \begin{pmatrix} \sqrt{8} & 1 & \sqrt{3} \\ 0 & 3 & \sqrt{3} \\ 0 & 0 & 2\sqrt{3} \end{pmatrix}, \quad X^{-1} = \sqrt{1/8}\begin{pmatrix} 1 & -1/3 & -1/3 \\ 0 & \sqrt{8}/3 & -\sqrt{2}/3 \\ 0 & 0 & \sqrt{2/3} \end{pmatrix}.$$

According to Eqs. (6.14) and (6.15), we first calculate the bases $\phi_{\mu\nu}$ in a left ideal by X, then calculate the nine orthogonal bases $\Phi_{\mu\nu}([3,1])$ by X^{-1}. In the results the normalization factors $\sqrt{1/96}$ and $2\sqrt{3}$ are used.

$$
\begin{aligned}
\phi_{11} &= \sqrt{1/12}\,b_{11} \\
&= \sqrt{1/12}\,\{E + (1\ 2) + (1\ 3) + (2\ 3) + (1\ 2\ 3) + (3\ 2\ 1) - (1\ 4) \\
&\quad - (2\ 1\ 4) - (3\ 1\ 4) - (2\ 3)(1\ 4) - (2\ 3\ 1\ 4) - (3\ 2\ 1\ 4)\}, \\
\phi_{21} &= \sqrt{1/96}\,\{b_{11} + 3b_{21}\} = \sqrt{1/96}\,\{E + 3(3\ 4)\}\,b_{11} \\
&= \sqrt{1/96}\,\{E + (1\ 2) - 2(1\ 3) + (2\ 3) + (1\ 2\ 3) - 2(3\ 2\ 1) \\
&\quad - (1\ 4) - (2\ 1\ 4) + 2(3\ 1\ 4) - (2\ 3)(1\ 4) - (2\ 3\ 1\ 4) \\
&\quad + 2(3\ 2\ 1\ 4) + 3(3\ 4) + 3(3\ 4)(1\ 2) + 3(4\ 3\ 2) + 3(4\ 3\ 1\ 2) \\
&\quad - 3(3\ 4\ 1) - 3(3\ 4\ 2\ 1) - 3(3\ 2\ 4\ 1) - 3(3\ 1)(4\ 2)\}, \\
\phi_{31} &= \sqrt{1/32}\,\{b_{11} + b_{21} + 2b_{31}\} = \sqrt{1/32}\,\{E + (3\ 4) + 2(4\ 2)\}\,b_{11} \\
&= \sqrt{1/32}\,\{E - (1\ 2) + (2\ 3) - (1\ 2\ 3) - (1\ 4) + (2\ 1\ 4) \\
&\quad - (2\ 3)(1\ 4) + (2\ 3\ 1\ 4) + (3\ 4) - (3\ 4)(1\ 2) + (4\ 3\ 2) \\
&\quad - (4\ 3\ 1\ 2) - (3\ 4\ 1) + (3\ 4\ 2\ 1) - (3\ 2\ 4\ 1) + (3\ 1)(4\ 2) \\
&\quad + 2(4\ 2) + 2(4\ 2\ 3) - 2(2\ 4\ 1) - 2(2\ 3\ 4\ 1)\},
\end{aligned}
$$

$$\phi_{12} = \sqrt{1/12}\, b_{12} = \sqrt{1/12}(3\ 4)b_{22}$$
$$= \sqrt{1/12}\{(3\ 4) + (3\ 4)(1\ 2) + (3\ 4\ 1) + (3\ 4\ 2) + (3\ 4\ 1\ 2)$$
$$+ (3\ 4\ 2\ 1) - (4\ 3\ 1) - (4\ 3\ 2\ 1) - (4\ 1) - (4\ 2\ 3\ 1)$$
$$- (4\ 1)(3\ 2) - (4\ 2\ 1)\},$$

$$\phi_{22} = \sqrt{1/96}\{b_{12} + 3b_{22}\} = \sqrt{1/96}\{(3\ 4) + 3E\}b_{22}$$
$$= \sqrt{1/96}\{(3\ 4) + (3\ 4)(1\ 2) - 2(3\ 4\ 1) + (3\ 4\ 2) + (3\ 4\ 1\ 2)$$
$$- 2(3\ 4\ 2\ 1) - (4\ 3\ 1) - (4\ 3\ 2\ 1) + 2(4\ 1) - (4\ 2\ 3\ 1)$$
$$- (4\ 1)(3\ 2) + 2(4\ 2\ 1) + 3E + 3(1\ 2) + 3(2\ 4) + 3(1\ 2\ 4)$$
$$- 3(1\ 3) - 3(2\ 1\ 3) - 3(2\ 4)(1\ 3) - 3(2\ 4\ 1\ 3)\},$$

$$\phi_{32} = \sqrt{1/32}\{b_{12} + b_{22} + 2b_{32}\} = \sqrt{1/32}\{(3\ 4) + E + 2(2\ 3)\}b_{22}$$
$$= \sqrt{1/32}\{(3\ 4) - (3\ 4)(1\ 2) + (3\ 4\ 2) - (3\ 4\ 1\ 2) - (4\ 3\ 1)$$
$$+ (4\ 3\ 2\ 1) - (4\ 2\ 3\ 1) + (4\ 1)(3\ 2) + E - (1\ 2) + (2\ 4)$$
$$- (1\ 2\ 4) - (1\ 3) + (2\ 1\ 3) - (2\ 4)(1\ 3) + (2\ 4\ 1\ 3)$$
$$+ 2(2\ 3) + 2(3\ 2\ 4) - 2(2\ 3\ 1) - 2(2\ 4\ 3\ 1)\},$$

$$\phi_{13} = \sqrt{1/12}\, b_{13} = \sqrt{1/12}(2\ 4)b_{33}$$
$$= \sqrt{1/12}\{(2\ 4) + (2\ 4)(1\ 3) + (2\ 4\ 1) + (2\ 4\ 3) + (2\ 4\ 1\ 3)$$
$$+ (2\ 4\ 3\ 1) - (4\ 2\ 1) - (4\ 2\ 3\ 1) - (4\ 1) - (4\ 3\ 2\ 1)$$
$$- (4\ 1)(2\ 3) - (4\ 3\ 1)\},$$

$$\phi_{23} = \sqrt{1/96}\{b_{13} + 3b_{23}\} = \sqrt{1/96}\{(2\ 4) + 3(2\ 3)\}b_{33}$$
$$= \sqrt{1/96}\{(2\ 4) - 2(2\ 4)(1\ 3) + (2\ 4\ 1) + (2\ 4\ 3) - 2(2\ 4\ 1\ 3)$$
$$+ (2\ 4\ 3\ 1) - (4\ 2\ 1) + 2(4\ 2\ 3\ 1) - (4\ 1) - (4\ 3\ 2\ 1)$$
$$+ 2(4\ 1)(2\ 3) - (4\ 3\ 1) + 3(2\ 3) + 3(2\ 3\ 1) + 3(2\ 3\ 4)$$
$$+ 3(2\ 3\ 4\ 1) - 3(3\ 2\ 1) - 3(3\ 1) - 3(3\ 4\ 2\ 1) - 3(3\ 4\ 1)\},$$

$$\phi_{33} = \sqrt{1/32}\{b_{13} + b_{23} + 2b_{33}\} = \sqrt{1/32}\{(2\ 4) + (2\ 3) + 2E\}b_{33}$$
$$= \sqrt{1/32}\{(2\ 4) - (2\ 4\ 1) + (2\ 4\ 3) - (2\ 4\ 3\ 1) - (4\ 2\ 1)$$
$$+ (4\ 1) - (4\ 3\ 2\ 1) + (4\ 3\ 1) + (2\ 3) - (2\ 3\ 1) + (2\ 3\ 4)$$
$$- (2\ 3\ 4\ 1) - (3\ 2\ 1) + (3\ 1) - (3\ 4\ 2\ 1) + (3\ 4\ 1)$$
$$+ 2E + 2(3\ 4) - 2(1\ 2) - 2(3\ 4)(1\ 2)\},$$

$$\Phi_{11}([3,1]) = \sqrt{1/6}\,\{3\phi_{11} - \phi_{12} - \phi_{13}\}$$

$$= \sqrt{1/72}\,\{3b_{11} - (3\ 4)b_{22} - (2\ 4)b_{33}\}$$

$$= \sqrt{1/72}\,\{3E + 3(1\ 2) + 3(1\ 3) + 3(2\ 3) + 3(1\ 2\ 3) + 3(3\ 2\ 1)$$
$$- (1\ 4) - (2\ 1\ 4) - (3\ 1\ 4) - (2\ 3)(1\ 4) - (2\ 3\ 1\ 4) - (3\ 2\ 1\ 4)$$
$$- (3\ 4) - (3\ 4)(1\ 2) - (3\ 4\ 1) - (3\ 4\ 2) - (3\ 4\ 1\ 2) - (3\ 4\ 2\ 1)$$
$$- (2\ 4) - (2\ 4)(1\ 3) - (2\ 4\ 1) - (2\ 4\ 3) - (2\ 4\ 1\ 3) - (2\ 4\ 3\ 1)\},$$

$$\Phi_{21}([3,1]) = \sqrt{1/6}\,\{3\phi_{21} - \phi_{22} - \phi_{23}\}$$

$$= (1/24)\,\{3\,[E + 3(3\ 4)]\,b_{11} - [(3\ 4) + 3E]\,b_{22} - [(2\ 4) + 3(2\ 3)]\,b_{33}\}$$

$$= (1/6)\,\{-(1\ 4) - (2\ 1\ 4) - (2\ 3)(1\ 4) - (2\ 3\ 1\ 4) - (3\ 4\ 1)$$
$$- (3\ 4\ 2\ 1) - (3\ 2\ 4\ 1) - (3\ 1)(4\ 2) - (2\ 4) - (3\ 4\ 1\ 2)$$
$$- (1\ 2\ 4) - (2\ 3\ 4) + 2(3\ 1\ 4) + 2(3\ 2\ 1\ 4) + 2(3\ 4)$$
$$+ 2(3\ 4)(1\ 2) + 2(4\ 3\ 2) + 2(4\ 3\ 1\ 2)\},$$

$$\Phi_{31}([3,1]) = \sqrt{1/6}\,\{3\phi_{31} - \phi_{32} - \phi_{33}\}$$

$$= \sqrt{1/192}\,\{3\,[E + (3\ 4) + 2(4\ 2)]\,b_{11} - [(3\ 4) + E + 2(2\ 3)]\,b_{22}$$
$$- [(2\ 4) + (2\ 3) + 2E]\,b_{33}\}$$

$$= \sqrt{1/12}\,\{-(1\ 4) + (2\ 1\ 4) - (2\ 3)(1\ 4) + (2\ 3\ 1\ 4) - (3\ 4\ 1)$$
$$+ (3\ 4\ 2\ 1) - (3\ 2\ 4\ 1) + (3\ 1)(4\ 2) + (4\ 2) + (4\ 2\ 3)$$
$$- (2\ 4\ 1) - (2\ 3\ 4\ 1)\},$$

$$\Phi_{12}([3,1]) = \sqrt{1/3}\,\{2\phi_{12} - \phi_{13}\} = (1/6)\,\{2(3\ 4)b_{22} - (2\ 4)b_{33}\}$$

$$= (1/6)\,\{2(3\ 4) + 2(3\ 4)(1\ 2) + 2(3\ 4\ 1) + 2(3\ 4\ 2)$$
$$+ 2(3\ 4\ 1\ 2) + 2(3\ 4\ 2\ 1) - (4\ 3\ 1) - (4\ 3\ 2\ 1) - (4\ 1)$$
$$- (4\ 2\ 3\ 1) - (4\ 1)(3\ 2) - (4\ 2\ 1) - (2\ 4) - (2\ 4)(1\ 3)$$
$$- (2\ 4\ 1) - (2\ 4\ 3) - (2\ 4\ 1\ 3) - (2\ 4\ 3\ 1)\},$$

$$\Phi_{22}([3,1]) = \sqrt{1/3}\,\{2\phi_{22} - \phi_{23}\}$$

$$= \sqrt{1/288}\,\{2\,[(3\ 4) + 3E]\,b_{22} - [(2\ 4) + 3(2\ 3)]\,b_{33}\}$$

$$= \sqrt{1/288}\,\{2(3\ 4) + 2(3\ 4)(1\ 2) - (3\ 4\ 1) - (3\ 4\ 2)$$
$$- (3\ 4\ 1\ 2) - (3\ 4\ 2\ 1) - (4\ 3\ 1) - (4\ 3\ 2\ 1) + 5(4\ 1)$$
$$- 4(4\ 2\ 3\ 1) - 4(4\ 1)(3\ 2) + 5(4\ 2\ 1) + 6E + 6(1\ 2)$$
$$+ 5(2\ 4) + 5(1\ 2\ 4) - 3(1\ 3) - 3(2\ 1\ 3) - 4(2\ 4)(1\ 3)$$
$$- 4(2\ 4\ 1\ 3) - (2\ 4\ 3) - (2\ 4\ 3\ 1) - 3(2\ 3) - 3(2\ 3\ 1)\},$$

$$\Phi_{32}([3,1]) = \sqrt{1/3}\,\{2\phi_{32} - \phi_{33}\}$$
$$= \sqrt{1/96}\,\{2\,[(3\ 4) + E + 2(2\ 3)]\,b_{22}$$
$$- [(2\ 4) + (2\ 3) + 2E]\,b_{33}\}$$
$$= \sqrt{1/96}\,\{(3\ 4\ 2) - (3\ 4\ 1\ 2) + (2\ 4) - (1\ 2\ 4) + (4\ 2\ 1)$$
$$- (4\ 1) + (3\ 4\ 2\ 1) - (3\ 4\ 1) - 2(4\ 2\ 3\ 1) + 2(4\ 1)(3\ 2)$$
$$- 2(2\ 4)(1\ 3) + 2(2\ 4\ 1\ 3) - 3(4\ 3\ 1) - 3(1\ 3) + 3(2\ 1\ 3)$$
$$+ 3(4\ 3\ 2\ 1) - 3(2\ 3\ 1) - 3(2\ 4\ 3\ 1) + 3(2\ 3) + 3(3\ 2\ 4)\},$$

$$\Phi_{13}([3,1]) = \phi_{13} = \sqrt{1/12}(2\ 4)b_{33}$$
$$= \sqrt{1/12}\,\{(2\ 4) + (2\ 4)(1\ 3) + (2\ 4\ 1) + (2\ 4\ 3) + (2\ 4\ 1\ 3)$$
$$+ (2\ 4\ 3\ 1) - (4\ 2\ 1) - (4\ 2\ 3\ 1) - (4\ 1) - (4\ 3\ 2\ 1)$$
$$- (4\ 1)(2\ 3) - (4\ 3\ 1)\},$$

$$\Phi_{23}([3,1]) = \phi_{23} = \sqrt{1/96}\,\{(2\ 4) + 3(2\ 3)\}\,b_{33}$$
$$= \sqrt{1/96}\,\{(2\ 4) + (2\ 4\ 1) + (2\ 4\ 3) + (2\ 4\ 3\ 1) - (4\ 2\ 1)$$
$$- (4\ 1) - (4\ 3\ 2\ 1) - 2(2\ 4)(1\ 3) - 2(2\ 4\ 1\ 3) + 2(4\ 2\ 3\ 1)$$
$$+ 2(4\ 1)(2\ 3) - (4\ 3\ 1) + 3(2\ 3) + 3(2\ 3\ 1) + 3(2\ 3\ 4)$$
$$+ 3(2\ 3\ 4\ 1) - 3(3\ 2\ 1) - 3(3\ 1) - 3(3\ 4\ 2\ 1) - 3(3\ 4\ 1)\},$$

$$\Phi_{33}([3,1]) = \phi_{33} = \sqrt{1/32}\,\{(2\ 4) + (2\ 3) + 2E\}\,b_{33}$$
$$= \sqrt{1/32}\,\{(2\ 4) - (2\ 4\ 1) + (2\ 4\ 3) - (2\ 4\ 3\ 1) - (4\ 2\ 1)$$
$$+ (4\ 1) - (4\ 3\ 2\ 1) + (4\ 3\ 1) + (2\ 3) - (2\ 3\ 1) + (2\ 3\ 4)$$
$$- (2\ 3\ 4\ 1) - (3\ 2\ 1) + (3\ 1) - (3\ 4\ 2\ 1) + (3\ 4\ 1)$$
$$+ 2E + 2(3\ 4) - 2(1\ 2) - 2(3\ 4)(1\ 2)\}.$$

23. Calculate the representation matrices of the transpositions P_a for the neighbored objects both in the standard bases and in the orthogonal bases of the irreducible representation $[2, 1, 1]$ of the S_4 group. Calculate explicitly the orthogonal bases $\Phi_{\mu\nu}$ of the irreducible representation $[2, 1, 1]$ in the group space of S_4.

Solution. There are three standard Young tableaux $\mathcal{Y}_1$, $\mathcal{Y}_2$ and $\mathcal{Y}_3$ corresponding to the Young pattern $[2, 1, 1]$ in S_3:

$$\mathcal{Y}_1: \quad \begin{array}{|c|c|} \hline 1 & 2 \\ \hline 3 \\ \cline{1-1} 4 \\ \cline{1-1} \end{array} \quad , \quad \mathcal{Y}_2: \quad \begin{array}{|c|c|} \hline 1 & 3 \\ \hline 2 \\ \cline{1-1} 4 \\ \cline{1-1} \end{array} \quad , \quad \mathcal{Y}_3: \quad \begin{array}{|c|c|} \hline 1 & 4 \\ \hline 2 \\ \cline{1-1} 3 \\ \cline{1-1} \end{array} \quad .$$

The corresponding Young operators $\mathcal{Y}_\mu$ are orthogonal to one another. The permutations between three standard Young tableaux respectively are $R_{12} = (2\ 3)$, $R_{13} = (2\ 3\ 4)$ and $R_{23} = (3\ 4)$, so the standard bases are

$$
\begin{aligned}
b_{11} &= \mathcal{Y}_1, & b_{12} &= (2\ 3)\mathcal{Y}_2, & b_{13} &= (2\ 3\ 4)\mathcal{Y}_3 = -(4\ 2)\mathcal{Y}_3, \\
b_{21} &= (2\ 3)\mathcal{Y}_1, & b_{22} &= \mathcal{Y}_2, & b_{23} &= (3\ 4)\mathcal{Y}_3, \\
b_{31} &= (2\ 4\ 3)\mathcal{Y}_1 = -(2\ 4)\mathcal{Y}_1, & b_{32} &= (3\ 4)\mathcal{Y}_2, & b_{33} &= \mathcal{Y}_3.
\end{aligned}
$$

The representation matrices of the transpositions P_a of the neighbored objects in the representation $[2, 1, 1]$ can be calculated by the tabular method:

	P_1			P_2			P_3		
Rep. [2,1,1]	2 1 3 4	2 3 1 4	2 4 1 3	1 3 2 4	1 2 3 4	1 4 3 2	1 2 4 3	1 4 2 3	1 3 2 4
1 2 3 4	1	−1	1	0	1	0	−1	0	0
1 3 2 4	0	−1	0	1	0	0	0	0	1
1 4 2 3	0	0	−1	0	0	−1	0	1	0

Namely,

$$
D(P_1) = \begin{pmatrix} 1 & -1 & 1 \\ 0 & -1 & 0 \\ 0 & 0 & -1 \end{pmatrix}, \quad
D(P_2) = \begin{pmatrix} 0 & 1 & 0 \\ 1 & 0 & 0 \\ 0 & 0 & -1 \end{pmatrix}, \quad
D(P_3) = \begin{pmatrix} -1 & 0 & 0 \\ 0 & 0 & 1 \\ 0 & 1 & 0 \end{pmatrix}.
$$

The representation matrices of P_a in the orthogonal bases of the irreducible representation $[2, 1, 1]$ is calculated by the formula (6.12):

$$
\overline{D}(P_1) = \begin{pmatrix} 1 & 0 & 0 \\ 0 & -1 & 0 \\ 0 & 0 & -1 \end{pmatrix}, \quad
\overline{D}(P_2) = \begin{pmatrix} -1/2 & \sqrt{3}/2 & 0 \\ \sqrt{3}/2 & 1/2 & 0 \\ 0 & 0 & -1 \end{pmatrix},
$$

$$
\overline{D}(P_3) = \begin{pmatrix} -1 & 0 & 0 \\ 0 & -1/3 & \sqrt{8}/3 \\ 0 & \sqrt{8}/3 & 1/3 \end{pmatrix}.
$$

From $D(P_a)X = X\overline{D}(P_a)$, the similarity transformation matrix X and its inverse matrix can be calculated to be

$$X = \begin{pmatrix} 1 & \sqrt{1/3} & -\sqrt{1/6} \\ 0 & 2/\sqrt{3} & \sqrt{1/6} \\ 0 & 0 & \sqrt{3/2} \end{pmatrix}, \quad X^{-1} = \begin{pmatrix} 1 & -1/2 & 1/2 \\ 0 & \sqrt{3}/2 & -\sqrt{1/12} \\ 0 & 0 & \sqrt{2/3} \end{pmatrix}.$$

The orthogonal bases with the normalization factor $\sqrt{1/8}$ are calculated by Eqs. (6.13-14):

$$\begin{aligned}
\Phi_{11}([2,1,1]) &= \sqrt{1/32}\,\{2b_{11} - b_{12} + b_{13}\} \\
&= \sqrt{1/32}\,\{2b_{11} - (2\ 3)b_{22} - (4\ 2)b_{33}\} \\
&= \sqrt{1/32}\,\{-(1\ 3) - (1\ 4) + (1\ 3\ 4) + (4\ 3\ 1) - (2\ 1\ 3) - (2\ 1\ 4) \\
&\quad + (2\ 1\ 3\ 4) + (2\ 1\ 4\ 3) - (2\ 3) + (3\ 2\ 4) - (3\ 1) \\
&\quad + (3\ 1\ 2\ 4) - (4\ 2) + (4\ 2\ 3) - (2\ 4\ 1) + (4\ 1\ 2\ 3) \\
&\quad + 2E - 2(3\ 4) + 2(1\ 2) - 2(1\ 2)(3\ 4)\},
\end{aligned}$$

$$\begin{aligned}
\Phi_{21}([2,1,1]) &= \sqrt{1/96}\,\{2b_{11} + 4b_{21} - b_{12} - 2b_{22} + b_{13} + 2b_{23}\} \\
&= \sqrt{1/96}\,\{2\,[E + 2(2\ 3)]\,b_{11} - [(2\ 3) + 2E]\,b_{22} \\
&\quad + [-(4\ 2) + 2(3\ 4)]\,b_{33}\} \\
&= \sqrt{1/96}\,\{-(3\ 2\ 4) - (3\ 2\ 1\ 4) + (3\ 1\ 2\ 4) + (3\ 1\ 4) + (2\ 4) \\
&\quad - (1\ 2\ 4) - (1\ 4) + (4\ 2\ 1) - 2(2\ 3)(1\ 4) + 2(2\ 3\ 1\ 4) \\
&\quad - 2(3\ 2\ 4\ 1) + 2(3\ 1)(2\ 4) - 3(1\ 3) + 3(1\ 3\ 4) + 3(2\ 1\ 3) \\
&\quad - 3(2\ 1\ 3\ 4) + 3(2\ 3) - 3(2\ 3\ 1) - 3(2\ 3\ 4) + 3(2\ 3\ 4\ 1)\},
\end{aligned}$$

$$\begin{aligned}
\Phi_{31}([2,1,1]) &= \sqrt{1/192}\,\{-2b_{11} + 2b_{21} + 6b_{31} + b_{12} - b_{22} \\
&\quad - 3b_{32} - b_{13} + b_{23} + 3b_{33}\} \\
&= \sqrt{1/192}\,\{2\,[-E + (2\ 3) - 3(2\ 4)]\,b_{11} - [-(2\ 3) + E + 3(3\ 4)]\,b_{22} \\
&\quad + [(4\ 2) + (3\ 4) + 3E]\,b_{33}\} \\
&= \sqrt{1/12}\,\{(1\ 4) - (4\ 3\ 1) - (2\ 1\ 4) + (2\ 1\ 4\ 3) - (2\ 3)(1\ 4) \\
&\quad + (2\ 3\ 1\ 4) - (2\ 4) + (2\ 4)(1\ 3) + (2\ 4\ 1) \\
&\quad + (2\ 4\ 3) - (2\ 4\ 1\ 3) - (2\ 4\ 3\ 1)\},
\end{aligned}$$

$$\begin{aligned}
\Phi_{12}([2,1,1]) &= \sqrt{1/96}\,\{3b_{12} - b_{13}\} = \sqrt{1/96}\,\{3(2\ 3)b_{22} + (4\ 2)b_{33}\} \\
&= \sqrt{1/96}\,\{(4\ 2) - (4\ 2\ 1) - (4\ 2\ 3) + (4\ 2\ 1\ 3) + (2\ 4\ 1) \\
&\quad - (4\ 1) - (4\ 1\ 2\ 3) + (4\ 1\ 3) - 2(2\ 3)(1\ 4) + 2(3\ 2\ 4\ 1) \\
&\quad - 2(2\ 3\ 1\ 4) + 2(3\ 1)(2\ 4) + 3(2\ 3\ 1) - 3(3\ 1) - 3(3\ 1\ 2\ 4) \\
&\quad + 3(3\ 1\ 4) + 3(2\ 3) - 3(3\ 2\ 1) - 3(3\ 2\ 4) + 3(3\ 2\ 1\ 4)\},
\end{aligned}$$

$$\Phi_{22}([2,1,1]) = \sqrt{1/288}\left\{3b_{12} + 6b_{22} - b_{13} - 2b_{23}\right\}$$

$$= \sqrt{1/288}\left\{3\left[(2\ 3) + 2E\right]b_{22} - \left[-(4\ 2) + 2(3\ 4)\right]b_{33}\right\}$$

$$= \sqrt{1/288}\left\{-(3\ 2\ 4) + (3\ 2\ 1\ 4) - (4\ 2\ 3) + (4\ 2\ 1\ 3) + (4\ 1\ 2\ 3)\right.$$
$$- (4\ 1\ 3) + (3\ 1\ 2\ 4) - (3\ 1\ 4) - 2(3\ 4) + 2(3\ 4)(1\ 2)$$
$$+ 3(2\ 3) - 3(3\ 2\ 1) - 3(2\ 3\ 1) + 3(3\ 1) - 4(2\ 3)(1\ 4)$$
$$+ 4(3\ 2\ 4\ 1) + 4(2\ 3\ 1\ 4) - 4(3\ 1)(2\ 4) - 5(1\ 4) - 5(2\ 4)$$
$$\left. + 5(1\ 2\ 4) + 5(4\ 2\ 1) + 6E - 6(1\ 2)\right\},$$

$$\phi_{32}([2,1,1]) = (1/24)\left\{-3b_{12} + 3b_{22} + 9b_{32} + b_{13} - b_{23} - 3b_{33}\right\}$$

$$= (1/24)\left\{3\left[-(2\ 3) + E + 3(3\ 4)\right]b_{22} - \left[(4\ 2) + (3\ 4) + 3E\right]b_{33}\right\}$$

$$= (1/6)\left\{(2\ 3)(1\ 4) + (3\ 2\ 4) - (3\ 2\ 4\ 1) - (3\ 2\ 1\ 4) - (2\ 3\ 1\ 4)\right.$$
$$- (3\ 1\ 2\ 4) + (3\ 1)(2\ 4) + (3\ 1\ 4) - (1\ 4) - (2\ 4)$$
$$+ (1\ 2\ 4) + (4\ 2\ 1) + 2(3\ 4) - 2(3\ 4)(1\ 2) - 2(3\ 4\ 1)$$
$$\left. - 2(3\ 4\ 2) + 2(3\ 4\ 1\ 2) + 2(3\ 4\ 2\ 1)\right\},$$

$$\Phi_{13}([2,1,1]) = \sqrt{1/12}\,b_{13} = -\sqrt{1/12}(4\ 2)b_{33}$$

$$= \sqrt{1/12}\left\{-(4\ 2) + (4\ 2\ 1) + (4\ 2)(1\ 3) + (4\ 2\ 3) - (4\ 2\ 3\ 1)\right.$$
$$- (4\ 2\ 1\ 3) - (2\ 4\ 1) + (4\ 1) + (2\ 4\ 1\ 3) + (4\ 1\ 2\ 3)$$
$$\left. - (4\ 1)(2\ 3) - (4\ 1\ 3)\right\},$$

$$\Phi_{23}([2,1,1]) = (1/6)\left\{b_{13} + 2b_{23}\right\} = (1/6)\left\{-(4\ 2) + 2(3\ 4)\right\}b_{33}$$

$$= (1/6)\left\{-(4\ 2) + (4\ 2\ 1) + (4\ 2)(1\ 3) + (4\ 2\ 3) - (4\ 2\ 3\ 1)\right.$$
$$- (4\ 2\ 1\ 3) + (2\ 4\ 1) - (4\ 1) - (2\ 4\ 1\ 3) - (4\ 1\ 2\ 3)$$
$$+ (4\ 1)(2\ 3) + (4\ 1\ 3) + 2(3\ 4) - 2(3\ 4)(1\ 2) - 2(4\ 3\ 1)$$
$$\left. - 2(4\ 3\ 2) + 2(4\ 3\ 1\ 2) + 2(4\ 3\ 2\ 1)\right\},$$

$$\Phi_{33}([2,1,1]) = \sqrt{1/72}\left\{-b_{13} + b_{23} + 3b_{33}\right\}$$

$$= \sqrt{1/72}\left\{(4\ 2) + (3\ 4) + 3E\right\}b_{33}$$

$$= \sqrt{1/72}\left\{(4\ 2) - (4\ 2\ 1) - (4\ 2)(1\ 3) - (4\ 2\ 3) + (4\ 2\ 3\ 1)\right.$$
$$+ (4\ 2\ 1\ 3) - (2\ 4\ 1) + (4\ 1) + (2\ 4\ 1\ 3) + (4\ 1\ 2\ 3)$$
$$- (4\ 1)(2\ 3) - (4\ 1\ 3) + (3\ 4) - (3\ 4)(1\ 2) - (4\ 3\ 1)$$
$$- (4\ 3\ 2) + (4\ 3\ 1\ 2) + (4\ 3\ 2\ 1)$$
$$\left. + 3E - 3(1\ 2) - 3(1\ 3) - 3(2\ 3) + 3(1\ 2\ 3) + 3(3\ 2\ 1)\right\}.$$

24. Calculate the standard bases and the orthogonal bases in the group space of S_4 for the irreducible representation $[2, 2]$, and calculate the representation matrices of the transpositions P_a for the neighbored objects in these two sets of bases.

Solution. There are two standard Young tableaux $\mathcal{Y}_1$ and $\mathcal{Y}_2$ corresponding to the Young pattern $[2, 2]$ of S_4

$$\mathcal{Y}_1 : \quad \begin{array}{|c|c|} \hline 1 & 2 \\ \hline 3 & 4 \\ \hline \end{array}, \qquad \mathcal{Y}_2 : \quad \begin{array}{|c|c|} \hline 1 & 3 \\ \hline 2 & 4 \\ \hline \end{array}.$$

Two Young operators $\mathcal{Y}_1$ and $\mathcal{Y}_2$ are orthogonal to each other. The permutation relating two standard Young tableaux are $R_{12} = R_{21} = (2\ 3)$. Therefore, removing the normalization factor $1/12$, we obtain the standard bases

$$
\begin{aligned}
b_{11} = \mathcal{Y}_1 =\ & E + (1\ 2) + (3\ 4) + (1\ 2)(3\ 4) - (1\ 3) - (2\ 1\ 3) \\
& - (4\ 3\ 1) - (2\ 1\ 4\ 3) - (2\ 4) - (1\ 2\ 4) - (3\ 4\ 2) - (1\ 2\ 3\ 4) \\
& + (1\ 3)(2\ 4) + (1\ 3\ 2\ 4) + (3\ 1\ 4\ 2) + (3\ 2)(1\ 4),
\end{aligned}
$$

$$b_{21} = (2\ 3)\mathcal{Y}_1 \qquad b_{12} = (2\ 3)\mathcal{Y}_2, \qquad b_{22} = \mathcal{Y}_2 = (2\ 3)\mathcal{Y}_1(2\ 3).$$

The representation matrices of the transposition P_a of the neighbored objects in the standard bases for the representation $[2, 2]$ are calculated by the tabular method:

Rep. [2,2]	P_1		P_2		P_3	
	2 1 3 4	2 3 1 4	1 3 2 4	1 2 3 4	1 2 4 3	1 4 2 3
1 2 3 4	1	−1	0	1	1	−1
1 3 2 4	0	−1	1	0	0	−1

$$D(P_1) = D(P_3) = \begin{pmatrix} 1 & -1 \\ 0 & -1 \end{pmatrix}, \qquad D(P_2) = \begin{pmatrix} 0 & 1 \\ 1 & 0 \end{pmatrix}.$$

On the other hand, the representation matrices of P_a in the orthogonal bases of the irreducible representation $[2, 2]$ are calculated by Eq. (6.12)

$$\overline{D}(P_1) = \overline{D}(P_3) = \begin{pmatrix} 1 & 0 \\ 0 & -1 \end{pmatrix}, \qquad \overline{D}(P_2) = \frac{1}{2} \begin{pmatrix} -1 & \sqrt{3} \\ \sqrt{3} & 1 \end{pmatrix}.$$

We calculate the similarity transformation matrix X and its inverse matrix from $D(P_a)X = X\overline{D}(P_a)$:

$$X = \begin{pmatrix} 1 & \sqrt{1/3} \\ 0 & 2/\sqrt{3} \end{pmatrix}, \qquad X^{-1} = \begin{pmatrix} 1 & -1/2 \\ 0 & \sqrt{3}/2 \end{pmatrix}.$$

Thus, from Eqs. (6.13-14), we obtain the orthogonal bases with the normalization factor $\sqrt{1/12}$:

$$\Phi_{11}([2,2]) = \sqrt{1/48}\,\{2b_{11} - b_{12}\} = \sqrt{1/48}\,\{2b_{11} - (2\ 3)b_{22}\}$$
$$= \sqrt{1/48}\,\{2E + 2(1\ 2) + 2(3\ 4) + 2(1\ 2)(3\ 4) + 2(1\ 3)(2\ 4)$$
$$+ 2(1\ 3\ 2\ 4) + 2(3\ 1\ 4\ 2) + 2(3\ 2)(1\ 4) - (1\ 3) - (2\ 1\ 3)$$
$$- (4\ 3\ 1) - (2\ 1\ 4\ 3) - (2\ 4) - (1\ 2\ 4) - (3\ 4\ 2)$$
$$- (1\ 2\ 3\ 4) - (2\ 3) - (2\ 3\ 1) - (3\ 2\ 4) - (3\ 1\ 2\ 4)$$
$$- (2\ 1\ 3\ 4) - (3\ 4\ 1) - (2\ 1\ 4) - (1\ 4)\},$$

$$\Phi_{21}([2,2]) = \sqrt{1/144}\,\{2b_{11} + 4b_{21} - b_{12} - 2b_{22}\}$$
$$= \sqrt{1/144}\,\{2\,[E + 2(2\ 3)]\,b_{11} - [(2\ 3) + 2E]\,b_{22}\}$$
$$= (1/4)\,\{-(1\ 3) + (2\ 1\ 3) + (4\ 3\ 1) - (2\ 1\ 4\ 3) - (2\ 4)$$
$$+ (1\ 2\ 4) + (3\ 4\ 2) - (1\ 2\ 3\ 4) + (2\ 3) + (2\ 1\ 3\ 4) - (2\ 3\ 1)$$
$$- (2\ 1\ 4) - (3\ 2\ 4) - (3\ 4\ 1) + (3\ 1\ 2\ 4) + (1\ 4)\},$$

$$\Phi_{12}([2,2]) = (1/4)b_{12} = (1/4)(2\ 3)b_{22}$$
$$= (1/4)\,\{(2\ 3) + (2\ 3\ 1) + (3\ 2\ 4) + (3\ 1\ 2\ 4) - (3\ 2\ 1)$$
$$- (3\ 1) - (3\ 2\ 1\ 4) - (3\ 1\ 4) - (2\ 3\ 4) - (2\ 3\ 4\ 1) - (2\ 4)$$
$$- (2\ 4\ 1) + (2\ 1\ 3\ 4) + (3\ 4\ 1) + (2\ 1\ 4) + (1\ 4)\},$$

$$\Phi_{22}([2,2]) = \sqrt{1/48}\,\{b_{12} + 2b_{22}\} = \sqrt{1/48}\,\{(2\ 3) + 2E\}\,b_{22}$$
$$= \sqrt{1/48}\,\{(2\ 3) - (2\ 3\ 1) - (3\ 2\ 4) + (3\ 1\ 2\ 4) - (3\ 2\ 1) + (3\ 1)$$
$$+ (3\ 2\ 1\ 4) - (3\ 1\ 4) - (2\ 3\ 4) + (2\ 3\ 4\ 1) + (2\ 4) - (2\ 4\ 1)$$
$$+ (2\ 1\ 3\ 4) - (3\ 4\ 1) - (2\ 1\ 4) + (1\ 4) + 2E + 2(1\ 3)(2\ 4) - 2(1\ 2)$$
$$- 2(3\ 1\ 4\ 2) - 2(3\ 4) - 2(1\ 3\ 2\ 4) + 2(1\ 2)(3\ 4) + 2(2\ 3)(1\ 4)\}.$$

25. Calculate the symmetric bases of the oscillatory wave function of a molecule with the symmetry of a regular tetrahedron, for example, the methane CH_4.

Solution. The molecule CH_4 has the symmetry of a regular tetrahedron. The carbon atom is located at the center of the regular tetrahedron denoted by O, while four hydrogen atoms lie in four vertices of the tetrahedron denoted by A_j, $1 \le j \le 4$, respectively. The stretching oscillators of the C-H bonds OA_μ are described by the state $|a_1 a_2 a_3 a_4\rangle$ with four oscillatory quantum numbers. Due to the symmetry, the frequency ω of four oscillators are the same, and the energy for the state is $E = \sum_{j=1}^{4} (a_j + 2)\hbar\omega$.

The symmetry group for a regular tetrahedron is the $\mathbf{T}_d$ group, which is isomorphic onto the permutation group S_4. The correspondence between the elements of two groups can be established as follows. If an element R of $\mathbf{T}_d$ transform four vertices A_j of the tetrahedron respectively into the positions of A_{r_j}, R corresponds to the following permutation:

$$R \longleftrightarrow \begin{pmatrix} 1 & 2 & 3 & 4 \\ r_1 & r_2 & r_3 & r_4 \end{pmatrix}.$$

Especially, from Fig. 6.1 we see that three improper twofold rotations around the directions $(\mathbf{e}_1 + \mathbf{e}_2)/\sqrt{2}$, $(\mathbf{e}_1 - \mathbf{e}_3)/\sqrt{2}$, and $(\mathbf{e}_1 - \mathbf{e}_2)/\sqrt{2}$ correspond to the transpositions P_1, P_2, and P_3 of the neighbored objects, respectively. Since two groups are isomorphic, we will not distinguish the elements in the T_d group and in the permutation group S_4.

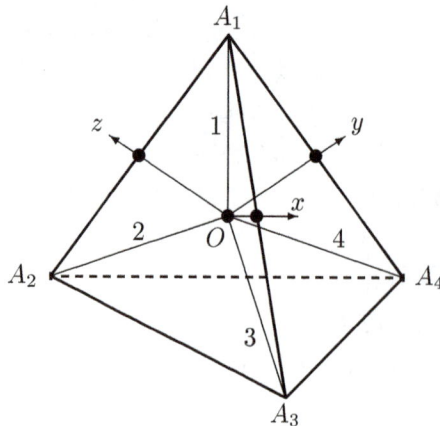

Fig. 6.1   Diagram of a molecule with the $\mathbf{T}_d$ symmetry.

We have obtained the orthogonal bases $\Phi_{\mu\nu}([\lambda])$ for the inequivalent irreducible representations of S_4 in Problems 21-24, in addition to the bases for two one-dimensional representations

$$\Phi([4]) = \sqrt{1/24}\,\{E + (1\ 2) + (1\ 3) + (1\ 4) + (2\ 3) + (2\ 4) + (3\ 4)$$
$$+ (1\ 2)(3\ 4) + (1\ 3)(2\ 4) + (1\ 4)(2\ 3) + (1\ 2\ 3) + (1\ 3\ 2) + (1\ 2\ 4)$$
$$+ (1\ 4\ 2) + (1\ 3\ 4) + (1\ 4\ 3) + (2\ 3\ 4) + (2\ 4\ 3) + (1\ 2\ 3\ 4)$$

$$+ (1\ 2\ 4\ 3) + (1\ 3\ 4\ 2) + (1\ 3\ 2\ 4) + (1\ 4\ 2\ 3) + (1\ 4\ 3\ 2)\},$$

$$\Phi([1^4]) = \sqrt{1/24}\,\{E - (1\ 2) - (1\ 3) - (1\ 4) - (2\ 3) - (2\ 4) - (3\ 4)$$
$$+ (1\ 2)(3\ 4) + (1\ 3)(2\ 4) + (1\ 4)(2\ 3) + (1\ 2\ 3) + (1\ 3\ 2) + (1\ 2\ 4)$$
$$+ (1\ 4\ 2) + (1\ 3\ 4) + (1\ 4\ 3) + (2\ 3\ 4) + (2\ 4\ 3) - (1\ 2\ 3\ 4)$$
$$- (1\ 2\ 4\ 3) - (1\ 3\ 4\ 2) - (1\ 3\ 2\ 4) - (1\ 4\ 2\ 3) - (1\ 4\ 3\ 2)\}.$$

Applying those operators to the oscillatory state $|a_1 a_2 a_3 a_4\rangle$, we obtain the symmetric bases of the oscillatory wave function of the molecule with T_d symmetry:

$$R\Phi_{\mu\nu}([\lambda])|a_1 a_2 a_3 a_4\rangle = \sum_{\rho} \overline{D}^{[\lambda]}_{\rho\mu}(R)\Phi_{\rho\nu}([\lambda])|a_1 a_2 a_3 a_4\rangle.$$

The application of an element R of S_4 to the oscillation state $|a_1 a_2 a_3 a_4\rangle$ is defined as follows. For the permutation R,

$$R = \begin{pmatrix} 1 & 2 & 3 & 4 \\ r_1 & r_2 & r_3 & r_4 \end{pmatrix} = \begin{pmatrix} s_1 & s_2 & s_3 & s_4 \\ 1 & 2 & 3 & 4 \end{pmatrix},$$

we have

$$R|a_1 a_2 a_3 a_4\rangle = |a_{s_1} a_{s_2} a_{s_3} a_{s_4}\rangle.$$

Recall that the application of a permutation R to the state only changes the order of four oscillatory quantum numbers a_j, but does not change their values. In fact, there exist 24 linearly independent symmetric bases belonging to the inequivalent irreducible representations if four oscillatory quantum numbers a_j all are different. The number of the linearly independent bases will decrease if some oscillatory quantum numbers a_j are equal to each other. For example, when two and only two quantum numbers coincide with each other, there exist 12 linearly independent symmetric bases belonging to five irreducible representations ($[4] \oplus [2,2] \oplus 2\,[3,1] \oplus [2,1,1]$).

26. Calculate the representation matrices of the generators (1 2) and (1 2 3 4 5) of the S_5 group in the irreducible representation $[2,2,1]$ by the tabular method.

Solution. From Problem 15, the orthogonal standard Young operators for the Young pattern $[2,2,1]$ of S_5 are $\mathcal{Y}_1\,[E - (2\ 5)]$ and $\mathcal{Y}_\mu$, $2 \leq \mu \leq 5$. Now we will calculate the representation matrices of the elements (1 2) and (1 2 3 4 5) in this representation $[2,2,1]$ by the tabular method.

Problems and Solutions in Group Theory

element (1 2)	2 1 3 4 5	2 1 3 5 4	2 3 1 4 5	2 3 1 5 4	2 4 1 5 3
1 2 1 5 3 4 − 3 4 5 2	1 − 0	0 − 0	−1 − 0	0 − 0	0 + 1
1 2 3 5 4	0	1	0	−1	1
1 3 2 4 5	0	0	−1	0	0
1 3 2 5 4	0	0	0	−1	0
1 4 2 5 3	0	0	0	0	−1

element (1 2 3 4 5)	2 3 4 5 1	2 3 4 1 5	2 4 3 5 1	2 4 3 1 5	2 5 3 1 4
1 2 1 5 3 4 − 3 4 5 2	1 − 0	−1 − 0	0 − 1	0 − 0	0 − 0
1 2 3 5 4	1	0	−1	0	0
1 3 2 4 5	1	−1	−1	1	0
1 3 2 5 4	1	0	0	0	1
1 4 2 5 3	0	0	1	0	0

From the tables, the representation matrices for the generators (1 2) and (1 2 3 4 5) in the representation $[2, 2, 1]$ respectively are

$$\begin{pmatrix} 1 & 0 & -1 & 0 & 1 \\ 0 & 1 & 0 & -1 & 1 \\ 0 & 0 & -1 & 0 & 0 \\ 0 & 0 & 0 & -1 & 0 \\ 0 & 0 & 0 & 0 & -1 \end{pmatrix}, \quad \begin{pmatrix} 1 & -1 & -1 & 0 & 0 \\ 1 & 0 & -1 & 0 & 0 \\ 1 & -1 & -1 & 1 & 0 \\ 1 & 0 & 0 & 0 & 1 \\ 0 & 0 & 1 & 0 & 0 \end{pmatrix}.$$

27. Respectively calculate the real orthogonal representation matrices of the transpositions P_a of the neighbored objects in two equivalent irreducible representations $[2^3] \simeq [1^6] \times [3, 3]$ of S_6, and find the similarity transformation matrix X between them.

Solution. The standard Young tableaux for the Young pattern $[2^3]$ and the Young pattern $[3,3]$ are listed from the smallest to the largest, respectively

$$
\begin{array}{llllll}
1\,2 & 1\,2 & 1\,3 & 1\,3 & 1\,4 \\
3\,4 & 3\,5 & 2\,4 & 2\,5 & 2\,5 \\
5\,6 & 4\,6 & 5\,6 & 4\,6 & 3\,6
\end{array}
\qquad
\begin{array}{lllll}
1\,2\,3 & 1\,2\,4 & 1\,2\,5 & 1\,3\,4 & 1\,3\,5 \\
4\,5\,6 & 3\,5\,6 & 3\,4\,6 & 2\,5\,6 & 2\,4\,6
\end{array}
$$

Obviously, each standard Young tableau in $[2^3]$ is the transpose of the corresponding one in $[3,3]$, but the larger in $[2^3]$ becomes the smaller in $[3,3]$. According to Eq. (6.12), the representation matrices of the transpositions P_a for the neighbored objects in the representations $[2^3]$ and $[1^6] \times [3,3]$ respectively are

representation $[2^3]$ representation $[1^6] \times [3,3]$

$P_1:$
$$
\begin{pmatrix}
1 & 0 & 0 & 0 & 0 \\
0 & 1 & 0 & 0 & 0 \\
0 & 0 & -1 & 0 & 0 \\
0 & 0 & 0 & -1 & 0 \\
0 & 0 & 0 & 0 & -1
\end{pmatrix},
\qquad
\begin{pmatrix}
-1 & 0 & 0 & 0 & 0 \\
0 & -1 & 0 & 0 & 0 \\
0 & 0 & -1 & 0 & 0 \\
0 & 0 & 0 & 1 & 0 \\
0 & 0 & 0 & 0 & 1
\end{pmatrix}
$$

$P_2: \frac{1}{2}$
$$
\begin{pmatrix}
-1 & 0 & \sqrt{3} & 0 & 0 \\
0 & -1 & 0 & \sqrt{3} & 0 \\
\sqrt{3} & 0 & 1 & 0 & 0 \\
0 & \sqrt{3} & 0 & 1 & 0 \\
0 & 0 & 0 & 0 & -2
\end{pmatrix},
\quad \frac{1}{2}
\begin{pmatrix}
-2 & 0 & 0 & 0 & 0 \\
0 & 1 & 0 & -\sqrt{3} & 0 \\
0 & 0 & 1 & 0 & -\sqrt{3} \\
0 & -\sqrt{3} & 0 & -1 & 0 \\
0 & 0 & -\sqrt{3} & 0 & -1
\end{pmatrix}
$$

$P_3: \frac{1}{3}$
$$
\begin{pmatrix}
3 & 0 & 0 & 0 & 0 \\
0 & -3 & 0 & 0 & 0 \\
0 & 0 & -3 & 0 & 0 \\
0 & 0 & 0 & -1 & \sqrt{8} \\
0 & 0 & 0 & \sqrt{8} & 1
\end{pmatrix},
\quad \frac{1}{3}
\begin{pmatrix}
1 & -\sqrt{8} & 0 & 0 & 0 \\
-\sqrt{8} & -1 & 0 & 0 & 0 \\
0 & 0 & -3 & 0 & 0 \\
0 & 0 & 0 & -3 & 0 \\
0 & 0 & 0 & 0 & 3
\end{pmatrix}
$$

$P_4: \frac{1}{2}$
$$
\begin{pmatrix}
-1 & \sqrt{3} & 0 & 0 & 0 \\
\sqrt{3} & 1 & 0 & 0 & 0 \\
0 & 0 & -1 & \sqrt{3} & 0 \\
0 & 0 & \sqrt{3} & 1 & 0 \\
0 & 0 & 0 & 0 & -2
\end{pmatrix},
\quad \frac{1}{2}
\begin{pmatrix}
-2 & 0 & 0 & 0 & 0 \\
0 & 1 & -\sqrt{3} & 0 & 0 \\
0 & -\sqrt{3} & -1 & 0 & 0 \\
0 & 0 & 0 & 1 & -\sqrt{3} \\
0 & 0 & 0 & -\sqrt{3} & -1
\end{pmatrix}
$$

$P_5:$
$$
\begin{pmatrix}
1 & 0 & 0 & 0 & 0 \\
0 & -1 & 0 & 0 & 0 \\
0 & 0 & 1 & 0 & 0 \\
0 & 0 & 0 & -1 & 0 \\
0 & 0 & 0 & 0 & -1
\end{pmatrix},
\qquad
\begin{pmatrix}
-1 & 0 & 0 & 0 & 0 \\
0 & -1 & 0 & 0 & 0 \\
0 & 0 & 1 & 0 & 0 \\
0 & 0 & 0 & -1 & 0 \\
0 & 0 & 0 & 0 & 1
\end{pmatrix}
$$

Each matrix in $[1^6] \times [3,3]$ can be obtained from the corresponding one in $[2^3]$ by a similarity transformation $X = YZ$. Y is a diagonal matrix diag$\{-1, 1, 1, -1, 1\}$. Since the non-diagonal matrix entries of P_a in the representation $[2^3]$ appear only in 12, 13, 24, 34 and 45 row-columns, Y changes the signs of all non-diagonal matrix entries of P_a in $[2^3]$. The matrix entries in Z are all vanishing except for those on the minor diagonal line which are 1. Z transpose the matrix in $[2^3]$ with respect to the minor diagonal line. Namely,

$$X^{-1}[2^3]X = [1^6] \times [3,3], \quad X = \begin{pmatrix} 0 & 0 & 0 & 0 & -1 \\ 0 & 0 & 0 & 1 & 0 \\ 0 & 0 & 1 & 0 & 0 \\ 0 & -1 & 0 & 0 & 0 \\ 1 & 0 & 0 & 0 & 0 \end{pmatrix}.$$

This method is suitable for other pair of the representations with the associated Young patterns.

28. Calculate the character of each class of the permutation group S_5 in the representation $[2,2,1]$ by the graphic method.

Solution. The characters of each class of S_5 in the representation $[2,2,1]$ are calculated in the following table.

class	(1^5)	$(1^3,2)$	$(1,2^2)$	$(1^2,3)$	$(2,3)$	$(1,4)$	(5)
Regular application	neglected	1 4 2 4 3	1 3 2 3 2	1 2 3 3 3	1 1 2 2 2	1 2 2 2 2	no
Application parity		-1	1	-1	-1	1	
$\chi^{[3,2]}[(\ell)]$	5	-1	1	-1	-1	1	0

29. Calculate the characters of each class of the permutation group S_6 in the representations $[3,2,1]$, $[3,3]$, and $[2^3]$ by the graphic method, respectively.

Solution. We calculate the characters in three representations of S_6 by the graphic method in the following table, where we also list the permitted regular application. The regular application of the identity (1^n) is neglected. The Young pattern $[2^3]$ is the one associated with the Young pattern $[3,3]$. The regular applications for $[2^3]$ are neglected because they can be obtained from those for $[3,3]$ by transpose. The Young pattern $[3,2,1]$ is a self-associated Young pattern, so the characters of the classes composed of the odd permutations are vanishing.

class	Characters in the irreducible representations of S_6			
	$[3,2,1]$	$[3,3]$	$[2^3]$	Regular application
(1^6)	16	5	5	neglected
$(2,1^4)$	0	1	-1	1 2 3 1 2 4 1 3 4 1 2 5 1 3 5 4 5 5 3 5 5 2 5 5 3 4 5 2 4 5
$(2^2,1^2)$	0	1	1	1 2 4 1 3 3 1 3 4 3 3 4 2 4 4 2 3 4
(2^3)	0	-3	3	1 1 3 1 2 2 1 2 3 2 2 3 1 3 3 1 2 3
$(3,1^3)$	-2	-1	-1	1 2 3 1 4 4 1 2 3 1 2 4 1 3 4 4 4 2 4 4 4 4 3 4 4 2 4 4 4 3
$(3,2,1)$	0	1	-1	1 2 2 1 3 3 1 2 2 3 3 2 3 3 3 3 3 2
(3^2)	-2	2	2	1 1 1 1 2 2 1 1 1 1 1 2 2 2 1 2 2 2 2 1 2 2 2 1
$(4,1^2)$	0	-1	1	1 2 3 3 3 3
$(4,2)$	0	-1	-1	1 1 2 2 2 2
$(5,1)$	1	0	0	1 2 2 2 2 2
(6)	0	0	0	

30. Calculate and fill in the character tables of S_3, S_4, S_5, S_6 and S_7 by the graphic method.

Solution. Here we will not give explicitly the regular applications, but only list the calculation results in the table. The character of each class in the antisymmetry representation $[1^n]$ shows whether the permutation in the class is even or odd. $n(\ell)$ denotes the number of elements in the class (ℓ). Instead of showing the pair of associated representations, only one of them is shown for simplicity.

Character Table of S_3			
class	$n(\ell)$	$[1^3]$	$[2,1]$
(1^3)	1	1	2
$(2,1)$	3	-1	0
(3)	2	1	-1

Character Table of S_4				
class	$n(\ell)$	$[1^4]$	$[3,1]$	$[2,2]$
(1^4)	1	1	3	2
$(2,1^2)$	6	-1	1	0
(2^2)	3	1	-1	2
$(3,1)$	8	1	0	-1
(4)	6	-1	-1	0

Character Table of S_5					
class	$n(\ell)$	$[1^5]$	$[4,1]$	$[3,2]$	$[3,1^2]$
(1^5)	1	1	4	5	6
$(2,1^3)$	10	-1	2	1	0
$(2^2,1)$	15	1	0	1	-2
$(3,1^2)$	20	1	1	-1	0
$(3,2)$	20	-1	-1	1	0
$(4,1)$	30	-1	0	-1	0
(5)	24	1	-1	0	1

Character Table of S_6							
class	$n(\ell)$	$[1^6]$	$[5,1]$	$[4,2]$	$[4,1^2]$	$[3,3]$	$[3,2,1]$
(1^6)	1	1	5	9	10	5	16
$(2,1^4)$	15	-1	3	3	2	1	0
$(2^2,1^2)$	45	1	1	1	-2	1	0
(2^3)	15	-1	-1	3	-2	-3	0
$(3,1^3)$	40	1	2	0	1	-1	-2
$(3,2,1)$	120	-1	0	0	-1	1	0
(3^2)	40	1	-1	0	1	2	-2
$(4,1^2)$	90	-1	1	-1	0	-1	0
$(4,2)$	90	1	-1	1	0	-1	0
$(5,1)$	144	1	0	-1	0	0	1
(6)	120	-1	-1	0	1	0	0

Character Table of S_7									
class	$n(\ell)$	$[1^7]$	$[6,1]$	$[5,2]$	$[5,1^2]$	$[4,3]$	$[4,2,1]$	$[4,1^3]$	$[3^2,1]$
(1^7)	1	1	6	14	15	14	35	20	21
$(2,1^5)$	21	-1	4	6	5	4	5	0	1
$(2^2,1^3)$	105	1	2	2	-1	2	-1	-4	1
$(2^3,1)$	105	-1	0	2	-3	0	1	0	-3
$(3,1^4)$	70	1	3	2	3	-1	-1	2	-3
$(3,2,1^2)$	420	-1	1	0	-1	1	-1	0	1
$(3,2^2)$	210	1	-1	2	-1	-1	-1	2	1
$(3^2,1)$	280	1	0	-1	0	2	-1	2	0
$(4,1^3)$	210	-1	2	0	1	-2	-1	0	-1
$(4,2,1)$	630	1	0	0	-1	0	1	0	-1
$(4,3)$	420	-1	-1	0	1	1	-1	0	-1
$(5,1^2)$	504	1	1	-1	0	-1	0	0	1
$(5,2)$	504	-1	-1	1	0	-1	0	0	1
$(6,1)$	840	-1	0	-1	0	0	1	0	0
(7)	720	1	-1	0	1	0	0	-1	0

6.5   The Inner and Outer Products of Representations

★ The direct product of two irreducible representations of a permutation group S_n is called the inner product of two representations for S_n, which is reducible in general. There is no special graphic rule for the reduction of the inner product of representations for the permutation group. The C-G series and the C-G coefficients for S_n can be calculated by the character method just like the usual finite group:

$$[\lambda] \times [\tau] \simeq \bigoplus_{[\omega]} a([\lambda], [\tau], [\omega])[\omega],$$

$$a([\lambda], [\tau], [\omega]) = \frac{1}{n!} \sum_{R \in S_n} \chi^{[\lambda]}(R)\chi^{[\tau]}(R)\chi^{[\omega]}(R). \tag{6.16}$$

Since the irreducible representations of S_n are real, the multiplicity $a([\lambda], [\tau], [\omega])$ is totally symmetric with respect to three representations. This symmetry leads to some common property for the C-G series of S_n. Letting $[\lambda]$ and $[\widetilde{\lambda}]$ be two associated Young patterns, we have

$$[n] \times [\lambda] = [\lambda], \qquad [1^n] \times [\lambda] \simeq [\widetilde{\lambda}], \qquad [\lambda] \times [\widetilde{\tau}] \simeq [\widetilde{\lambda}] \times [\tau],$$

$$[\lambda] \times [\lambda] \simeq [n] \oplus \ldots, \qquad [\lambda] \times [\widetilde{\lambda}] \simeq [1^n] \oplus \ldots, \tag{6.17}$$

$$a([\lambda], [\tau], [\omega]) = a([\widetilde{\lambda}], [\widetilde{\tau}], [\omega]) = a([\widetilde{\lambda}], [\tau], [\widetilde{\omega}]) = a([\lambda], [\widetilde{\tau}], [\widetilde{\omega}]).$$

In the reduction of the inner product of two irreducible representations of S_n, there is one identical representation $[n]$ if and only if two representations are equivalent, and there is one antisymmetric representation $[1^n]$ if and only if the Young patterns of two irreducible representations are associated.

★ We are going to apply the concepts of the subduced representation and the induced representation, discussed in §3.3, to the permutation group. The subgroup $H = S_n \times S_m$ of the permutation group $G = S_{n+m}$ is the direct product of two subgroups S_n and S_m, where S_n consists of the permutations among the first n objects in the $(n + m)$ objects and S_m consists of the permutations among the last m objects. The order of G is $g = (n+m)!$. The order of the subgroup H is $h = n!m!$ and its index is $n = g/h$. Denote by $R_j H$, $2 \leq j \leq n$, the left cosets of H in G where the element R_j is a permutation transforming the first n objects to n different positions in all $(n + m)$ objects. R_j is not unique, but we assume that R_j has been chosen. For convenience, we denote the subgroup H by $R_1 H$ with $R_1 = E$.

Let $\mathcal{L}$ be the group algebra of G. Denote by $e^{[\omega]}$ the primitive idempotent in $\mathcal{L}$, and by $\mathcal{L}e^{[\omega]}$ the minimal left ideal generated by $e^{[\omega]}$, where the Young pattern $[\omega]$ with $(n + m)$ boxes describes the irreducible representation of G. Let $\mathcal{L}^{nm} \subset \mathcal{L}$ be the group algebra of the subgroup H. Denote by $e^{[\lambda]}e^{[\mu]}$ the primitive idempotent in $\mathcal{L}^{nm}$, and by $\mathcal{L}^{nm}e^{[\lambda]}e^{[\mu]}$ the minimal left ideal generated by $e^{[\lambda]}e^{[\mu]}$, corresponding to the irreducible representation of H described by the direct product of two Young patterns $[\lambda] \times [\mu]$, where the numbers of the boxes in two Young patterns $[\lambda]$ and $[\mu]$ are n and m, respectively.

★ A subduced representation from an irreducible representation $[\omega]$ of G with respect to the subgroup H is generally reducible and can be reduced with respect to the irreducible representations $[\lambda] \times [\mu]$ of H,

$$
\begin{aligned}
X^{-1}[\omega]X &= \bigoplus a^{\omega}_{\lambda\mu}[\lambda] \times [\mu], \\
d_{[\omega]}(S_{n+m}) &= \sum a^{\omega}_{\lambda\mu}d_{[\lambda]}(S_n)d_{[\mu]}(S_m).
\end{aligned}
\tag{6.18}
$$

From the viewpoint of the group algebra, Eq. (6.18) shows that there are $a^{\omega}_{\lambda\mu}$ linear independent vectors $e^{[\lambda]}e^{[\mu]}t_\alpha e^{[\omega]}$, which map the left ideal $\mathcal{L}e^{[\omega]}$ onto the minimal left ideals for H, respectively,

$$
\mathcal{L}^{nm}e^{[\lambda]}e^{[\mu]}t_\alpha e^{[\omega]} \subset \mathcal{L}e^{[\omega]}.
\tag{6.19}
$$

Equation (6.19) gives the linear expansion of the new basis in the minimal left ideal in terms of the bases in $\mathcal{L}e^{[\omega]}$. The coefficients in the expansion constitute the column-matrices in the similarity transformation matrix X.

★ The primitive idempotent $e^{[\lambda]}e^{[\mu]}$ in $\mathcal{L}^{nm}$ is also an idempotent in $\mathcal{L}$, but generally not primitive. The left ideal $\mathcal{L}e^{[\lambda]}e^{[\mu]}$ generated by $e^{[\lambda]}e^{[\mu]}$ in $\mathcal{L}$ corresponds to a reducible representation of G, which is the induced representation from the representation $[\lambda] \times [\mu]$ of H with respect to G. For the permutation group, the induced representation is called the outer product of two representations, denoted by $[\lambda] \otimes [\mu]$. This representation can be reduced with respect to the irreducible representations $[\omega]$ of G,

$$
\begin{aligned}
Y^{-1}\left([\lambda] \otimes [\mu]\right)Y &= \bigoplus b^{\omega}_{\lambda\mu}[\omega], \\
\frac{(n+m)!}{n!m!}d_{[\lambda]}(S_n)d_{[\mu]}(S_m) &= \sum b^{\omega}_{\lambda\mu}d_{[\omega]}(S_{n+m}).
\end{aligned}
\tag{6.20}
$$

From the viewpoint of the group algebra, Eq. (6.20) shows that there are $b^{\omega}_{\lambda\mu}$ linear independent vectors $e^{[\omega]}t'_\alpha e^{[\lambda]}e^{[\mu]}$, which map the left ideal

$\mathcal{L}e^{[\lambda]}e^{[\mu]}$ onto the minimal left ideals for G, respectively,

$$\mathcal{L}e^{[\omega]}t'_\alpha e^{[\lambda]}e^{[\mu]} \subset \mathcal{L}e^{[\lambda]}e^{[\mu]}. \tag{6.21}$$

Equation (6.21) gives the linear expansion of the new basis in the minimal left ideal in terms of the bases in $\mathcal{L}e^{[\lambda]}e^{[\mu]}$. The coefficients in the expansion constitute the column-matrices in the similarity transformation matrix Y. Due to the symmetry between left- and right-ideals, the numbers of the linear independent vectors $e^{[\lambda]}e^{[\mu]}t_\alpha e^{[\omega]}$ and $e^{[\omega]}t'_\alpha e^{[\lambda]}e^{[\mu]}$ are equal, $a^\omega_{\lambda\mu} = b^\omega_{\lambda\mu}$. This is an example of the Frobenius theorem (see Chapter 3).

★ There is a graphic rule, called the Littlewood-Richardson rule, for calculating the multiplicity $b^\omega_{\lambda\mu}$ in the reduction of the outer product of two irreducible representations for the permutation group, although it can be calculated by the character method as usual. The calculation method is as follows.

Arbitrarily choose one of the Young patterns in an outer product $[\lambda] \otimes [\mu]$, say $[\mu]$. Usually, choose the Young pattern with the less boxes. Assign each box in the Young pattern $[\mu]$ with its row number, namely assign the box in the jth row with j. Now, attach the boxes assigned with 1 to another Young pattern $[\lambda]$ from its right or from its bottom, and then attach the boxes assigned with 2, and so on, in all possible way subject to the following rule. After attaching the boxes assigned with each digit, say j, the resultant diagram satisfies the following conditions:

> 1) The diagram constitutes a Young pattern, namely, the diagram is lined up on the top and on the left without unconnected boxes in each row such that the number of boxes in the upper row of the diagram is not less than that in the lower row, and the number of boxes in the left column is not less than that in the right column.
> 2) No two j appear in the same column of the diagram.
> 3) We read the attached boxes from right to left in the first row, then in the second row, etc., such that in each step of the reading process, the number of boxes assigned with the smaller digit must not be less than that assigned with the larger digit.

In each way subject to the above rule, the resultant diagram $[\omega]$ after attaching all boxes assigned with digits denotes an irreducible representation $[\omega]$ of S_{n+m} which appears in the reduction of the outer product $[\lambda] \otimes [\mu]$.

If the same Young pattern $[\omega]$ is obtained in different ways, the number of the ways is the multiplicity of the irreducible representation $[\omega]$ in the reduction.

★ Due to the Frobenius theorem, the Littlewood-Richardson rule can also be used for calculating the reduction of the subduced representation from the irreducible representation $[\omega]$ of S_{n+m} with respect to the subgroup $S_n \times S_m$. In the calculation, all the Young patterns $[\lambda]$ with n boxes and all the Young patterns $[\mu]$ with m boxes are taken to constitute the irreducible representation $[\omega]$ of S_{n+m} with the multiplicity $b^{\omega}_{\lambda\mu}$, if $[\omega]$ appears in the reduction of the outer product $[\lambda] \otimes [\mu]$ according to the Littlewood-Richardson rule with the same multiplicity.

31. Calculate the Clebsch-Gordan series of the inner product of each pair of the irreducible representations for the permutation groups S_3, S_4, S_5, S_6 and S_7 by the character method.

Solution. We can calculate the C-G series for the reduction of the inner product of two representations by Eq. (6.16). We only list some results for the C-G series, and neglect the remaining C-G series which can be obtained from them by Eq. (6.17).

For the S_3 group: $[2,1] \times [2,1] \simeq [3] \oplus [1^3] \oplus [2,1]$.

For the S_4 group:
$$[3,1] \times [3,1] \simeq [4] \oplus [3,1] \oplus [2^2] \oplus [2,1^2],$$
$$[3,1] \times [2^2] \simeq [3,1] \oplus [2,1^2],$$
$$[2^2] \times [2^2] \simeq [4] \oplus [2^2] \oplus [1^4].$$

For the S_5 group:

$[4,1] \times [4,1] \simeq [5] \oplus [4,1] \oplus [3,2] \oplus [3,1^2]$,

$[4,1] \times [3,2] \simeq [4,1] \oplus [3,2] \oplus [3,1^2] \oplus [2^2,1]$,

$[4,1] \times [3,1^2] \simeq [4,1] \oplus [3,2] \oplus [3,1^2] \oplus [2^2,1] \oplus [2,1^3]$,

$[3,2] \times [3,2] \simeq [5] \oplus [4,1] \oplus [3,2] \oplus [3,1^2] \oplus [2^2,1] \oplus [2,1^3]$,

$[3,2] \times [3,1^2] \simeq [4,1] \oplus [3,2] \oplus 2[3,1^2] \oplus [2^2,1] \oplus [2,1^3]$,

$[3,1^2] \times [3,1^2] \simeq [5] \oplus [4,1] \oplus 2[3,2] \oplus [3,1^2] \oplus 2[2^2,1] \oplus [2,1^3] \oplus [1^5]$.

For the S_6 group:

$[5,1] \times [5,1] \simeq [6] \oplus [5,1] \oplus [4,2] \oplus [4,1^2]$,

$[5,1] \times [4,2] \simeq [5,1] \oplus [4,2] \oplus [4,1^2] \oplus [3^2] \oplus [3,2,1]$,

$$[5,1] \times [4,1^2] \simeq [5,1] \oplus [4,2] \oplus [4,1^2] \oplus [3,2,1] \oplus [3,1^3],$$

$$[5,1] \times [3^2] \simeq [4,2] \oplus [3,2,1],$$

$$[5,1] \times [3,2,1] \simeq [4,2] \oplus [4,1^2] \oplus [3^2] \oplus 2[3,2,1] \oplus [2^3] \oplus [3,1^3]$$
$$\oplus [2^2,1^2],$$

$$[4,2] \times [4,2] \simeq [6] \oplus [5,1] \oplus 2[4,2] \oplus [4,1^2] \oplus 2[3,2,1]$$
$$\oplus [2^3] \oplus [3,1^3],$$

$$[4,2] \times [4,1^2] \simeq [5,1] \oplus [4,2] \oplus 2[4,1^2] \oplus [3^2] \oplus 2[3,2,1]$$
$$\oplus [3,1^3] \oplus [2^2,1^2],$$

$$[4,2] \times [3^2] \simeq [5,1] \oplus [4,1^2] \oplus [3^2] \oplus [3,2,1] \oplus [2^2,1^2],$$

$$[4,2] \times [3,2,1] \simeq [5,1] \oplus 2[4,2] \oplus 2[4,1^2] \oplus [3^2]$$
$$\oplus 3[3,2,1] \oplus [2^3] \oplus 2[3,1^3] \oplus 2[2^2,1^2] \oplus [2,1^4],$$

$$[4,1^2] \times [4,1^2] \simeq [6] \oplus [5,1] \oplus 2[4,2] \oplus [4,1^2] \oplus [3^2]$$
$$\oplus 2[3,2,1] \oplus [2^3] \oplus [3,1^3] \oplus [2^2,1^2] \oplus [2,1^4],$$

$$[4,1^2] \times [3^2] \simeq [4,2] \oplus [4,1^2] \oplus [3,2,1] \oplus [2^3] \oplus [3,1^3],$$

$$[4,1^2] \times [3,2,1] \simeq [5,1] \oplus 2[4,2] \oplus 2[4,1^2] \oplus [3^2] \oplus 4[3,2,1]$$
$$\oplus [2^3] \oplus 2[3,1^3] \oplus 2[2^2,1^2] \oplus [2,1^4],$$

$$[3^2] \times [3^2] \simeq [6] \oplus [4,2] \oplus [2^3] \oplus [3,1^3],$$

$$[3^2] \times [3,2,1] \simeq [5,1] \oplus [4,2] \oplus [4,1^2] \oplus 2[3,2,1] \oplus [3,1^3]$$
$$\oplus [2^2,1^2] \oplus [2,1^4],$$

$$[3,2,1] \times [3,2,1] \simeq [6] \oplus 2[5,1] \oplus 3[4,2] \oplus 4[4,1^2] \oplus 2[3^2]$$
$$\oplus 5[3,2,1] \oplus 2[2^3] \oplus 4[3,1^3] \oplus 3[2^2,1^2] \oplus 2[2,1^4] \oplus [1^6].$$

For the S_7 group:

$$[6,1] \times [6,1] = [7] \oplus [6,1] \oplus [5,2] \oplus [5,1^2],$$

$$[6,1] \times [5,2] = [6,1] \oplus [5,2] \oplus [5,1^2] \oplus [4,3] \oplus [4,2,1],$$

$$[6,1] \times [5,1^2] = [6,1] \oplus [5,2] \oplus [5,1^2] \oplus [4,2,1] \oplus [4,1^3],$$

$$[6,1] \times [4,3] = [5,2] \oplus [4,3] \oplus [4,2,1] \oplus [3^2,1]$$

$$[6,1] \times [4,2,1] = [5,2] \oplus [5,1^2] \oplus [4,3] \oplus 2[4,2,1] \oplus [4,1^3]$$
$$\oplus [3^2,1] \oplus [3,2,1^2] \oplus [3,2^2],$$

$$[6,1] \times [4,1^3] = [5,1^2] \oplus [4,2,1] \oplus [4,1^3] \oplus [3,2,1^2] \oplus [3,1^4],$$

$$[6,1] \times [3^2,1] = [4,3] \oplus [4,2,1] \oplus [3^2,1] \oplus [3,2,1^2] \oplus [3,2^2],$$

$[5,2] \times [5,2] = [7] \oplus [6,1] \oplus 2[5,2] \oplus [5,1^2] \oplus [4,3]$
$\qquad \oplus 2[4,2,1] \oplus [4,1^3] \oplus [3^2,1] \oplus [3,2^2],$

$[5,2] \times [5,1^2] = [6,1] \oplus [5,2] \oplus 2[5,1^2] \oplus [4,3] \oplus 2[4,2,1]$
$\qquad \oplus [4,1^3] \oplus [3^2,1] \oplus [3,2,1^2],$

$[5,2] \times [4,3] = [6,1] \oplus [5,2] \oplus [5,1^2] \oplus [4,3] \oplus 2[4,2,1]$
$\qquad \oplus [3^2,1] \oplus [3,2^2] \oplus [3,2,1^2],$

$[5,2] \times [4,2,1] = [6,1] \oplus 2[5,2] \oplus 2[5,1^2] \oplus 2[4,3] \oplus 4[4,2,1]$
$\qquad \oplus 2[4,1^3] \oplus 2[3^2,1] \oplus 2[3,2^2] \oplus 3[3,2,1^2] \oplus [2^3,1] \oplus [3,1^4],$

$[5,2] \times [4,1^3] = [5,2] \oplus [5,1^2] \oplus 2[4,2,1] \oplus 2[4,1^3] \oplus [3^2,1]$
$\qquad \oplus [3,2^2] \oplus 2[3,2,1^2] \oplus [3,1^4] \oplus [2^2,1^3],$

$[5,2] \times [3^2,1] = [5,2] \oplus [5,1^2] \oplus [4,3] \oplus 2[4,2,1] \oplus [4,1^3]$
$\qquad \oplus 2[3^2,1] \oplus [3,2^2] \oplus 2[3,2,1^2] \oplus [2^3,1] \oplus [2^2,1^3],$

$[5,1^2] \times [5,1^2] = [7] \oplus [6,1] \oplus 2[5,2] \oplus [5,1^2] \oplus [4,3]$
$\qquad \oplus 2[4,2,1] \oplus [4,1^3] \oplus [3,2^2] \oplus [3,2,1^2] \oplus [3,1^4],$

$[5,1^2] \times [4,3] = [5,2] \oplus [5,1^2] \oplus [4,3] \oplus 2[4,2,1] \oplus [4,1^3]$
$\qquad \oplus [3^2,1] \oplus [3,2^2] \oplus [3,2,1^2],$

$[5,1^2] \times [4,2,1] = [6,1] \oplus 2[5,2] \oplus 2[5,1^2] \oplus 2[4,3] \oplus 4[4,2,1]$
$\qquad \oplus 2[4,1^3] \oplus 3[3^2,1] \oplus 2[3,2^2] \oplus 3[3,2,1^2] \oplus [2^3,1] \oplus [3,1^4]$
$\qquad \oplus [2^2,1^3],$

$[5,1^2] \times [4,1^3] = [6,1] \oplus [5,2] \oplus [5,1^2] \oplus [4,3] \oplus 2[4,2,1]$
$\qquad \oplus [4,1^3] \oplus [3^2,1] \oplus [3,2^2] \oplus 2[3,2,1^2] \oplus [2^3,1] \oplus [3,1^4]$
$\qquad \oplus [2^2,1^3] \oplus [2,1^5],$

$[5,1^2] \times [3^2,1] = [5,2] \oplus [4,3] \oplus 3[4,2,1] \oplus [4,1^3] \oplus [3^2,1]$
$\qquad \oplus 2[3,2^2] \oplus 2[3,2,1^2] \oplus [2^3,1] \oplus [3,1^4],$

$[4,3] \times [4,3] = [7] \oplus [6,1] \oplus [5,2] \oplus [5,1^2] \oplus [4,3] \oplus [4,2,1]$
$\qquad \oplus [4,1^3] \oplus [3^2,1] \oplus [3,2^2] \oplus [3,2,1^2] \oplus [2^3,1],$

$[4,3] \times [4,2,1] = [6,1] \oplus 2[5,2] \oplus 2[5,1^2] \oplus [4,3] \oplus 4[4,2,1] \oplus 2[4,1^3]$
$\qquad \oplus 2[3^2,1] \oplus 2[3,2^2] \oplus 3[3,2,1^2] \oplus [2^3,1] \oplus [3,1^4] \oplus [2^2,1^3],$

$[4,3] \times [4,1^3] = [5,1^2] \oplus [4,3] \oplus 2[4,2,1] \oplus 2[4,1^3] \oplus [3^2,1]$
$\qquad \oplus [3,2^2] \oplus 2[3,2,1^2] \oplus [2^3,1] \oplus [3,1^4],$

$[4,3] \times [3^2,1] = [6,1] \oplus [5,2] \oplus [5,1^2] \oplus [4,3] \oplus 2[4,2,1] \oplus [4,1^3]$
$\qquad \oplus [3^2,1] \oplus [3,2^2] \oplus 2[3,2,1^2] \oplus [2^3,1] \oplus [3,1^4] \oplus [2^2,1^3],$

$$[4, 2, 1] \times [4, 2, 1] = [7] \oplus 2[6, 1] \oplus 4[5, 2] \oplus 4[5, 1^2] \oplus 4[4, 3]$$
$$\oplus 9[4, 2, 1] \oplus 5[4, 1^3] \oplus 5[3^2, 1] \oplus 5[3, 2^2] \oplus 8[3, 2, 1^2] \oplus 3[2^3, 1]$$
$$\oplus 3[3, 1^4] \oplus 3[2^2, 1^3] \oplus [2, 1^5],$$

$$[4, 2, 1] \times [4, 1^3] = [6, 1] \oplus 2[5, 2] \oplus 2[5, 1^2] \oplus 2[4, 3] \oplus 5[4, 2, 1]$$
$$\oplus 2[4, 1^3] \oplus 3[3^2, 1] \oplus 3[3, 2^2] \oplus 5[3, 2, 1^2] \oplus 2[2^3, 1] \oplus 2[3, 1^4]$$
$$\oplus 2[2^2, 1^3] \oplus [2, 1^5],$$

$$[4, 2, 1] \times [3^2, 1] = [6, 1] \oplus 2[5, 2] \oplus 3[5, 1^2] \oplus 2[4, 3] \oplus 5[4, 2, 1]$$
$$\oplus 3[4, 1^3] \oplus 3[3^2, 1] \oplus 3[3, 2^2] \oplus 5[3, 2, 1^2] \oplus 2[2^3, 1] \oplus 2[3, 1^4]$$
$$\oplus 2[2^2, 1^3] \oplus [2, 1^5],$$

$$[4, 1^3] \times [4, 1^3] = [7] \oplus [6, 1] \oplus 2[5, 2] \oplus [5, 1^2] \oplus 2[4, 3]$$
$$\oplus 2[4, 2, 1] \oplus [4, 1^3] \oplus 2[3^2, 1] \oplus 2[3, 2^2] \oplus 2[3, 2, 1^2] \oplus 2[2^3, 1]$$
$$\oplus [3, 1^4] \oplus 2[2^2, 1^3] \oplus [2, 1^5] \oplus [1^7],$$

$$[4, 1^3] \times [3^2, 1] = [5, 2] \oplus [5, 1^2] \oplus [4, 3] \oplus 3[4, 2, 1] \oplus 2[4, 1^3]$$
$$\oplus 2[3^2, 1] \oplus 2[3, 2^2] \oplus 3[3, 2, 1^2] \oplus [2^3, 1] \oplus [3, 1^4] \oplus [2^2, 1^3],$$

$$[3^2, 1] \times [3^2, 1] = [7] \oplus [6, 1] \oplus 2[5, 2] \oplus [5, 1^2] \oplus [4, 3] \oplus 3[4, 2, 1]$$
$$\oplus 2[4, 1^3] \oplus [3^2, 1] \oplus 2[3, 2^2] \oplus 3[3, 2, 1^2] \oplus [2^3, 1] \oplus 2[3, 1^4]$$
$$\oplus [2^2, 1^3] \oplus [2, 1^5].$$

32. Calculate the Clebsch-Gordan coefficients for the reduction of the self-product of the irreducible representation $[3, 2]$ of S_5.

Solution. This is a typical example for calculating the reduction of a reducible representation with a higher dimension for a finite group. Usually in the reduction of a reducible representation, one may try to find one or a few commutable elements in the group such that their eigenvalues can designate the bases for each irreducible representation without degeneration. This method has been used in the calculation for the Clebsch-Gordan coefficients of the groups **T**, **O** and **I** in Problems 25-27 of Chapter 3. For the permutation group S_n, when $n > 5$, the dimension of the irreducible representation is quite high. It is hard to find the eigenvalues of some commutable elements to designate the bases of each irreducible representation without degeneration. For example, one can designate the bases by the eigenvalues of two commutable elements (1 2) and (3 4 5), where the number of the different pairs of the eigenvalues is six. However, there is a 16-dimensional representation $[3, 2, 1]$ in S_6. In the present Problem we are going to introduce another method for calculating the C-G coefficients,

which will be suitable for the representations with higher dimension.
From Problem 31 we have

$$[3,2] \times [3,2] \simeq [5] \oplus [4,1] \oplus [3,2] \oplus [3,1^2] \oplus [2^2,1] \oplus [2,1^3].$$

We take the orthogonal bases where the representation matrix of each transposition P_a of two neighbored objects is known. The basis state before reduction is denoted by $|\mu\rangle|\nu\rangle$, where μ and ν both run from 1 to 5, respectively. Denote by $||[\lambda], \rho\rangle$ the basis state after reduction, where ρ runs from 1 to the dimension of $[\lambda]$. There is a characteristic in this example that two representations in the direct product are the same. The expansion of the new basis $||[\lambda], \rho\rangle$ can be chosen to be symmetric or antisymmetric in the permutation of two states in the product. For simplicity, we use the short notation for the basis state with the given symmetry in the permutation: $|\mu, \nu\rangle_S = |\mu\rangle|\nu\rangle + |\nu\rangle|\mu\rangle$ and $|\mu, \nu\rangle_A = |\mu\rangle|\nu\rangle - |\nu\rangle|\mu\rangle$. When $\mu = \nu$, only symmetric state exists, and we just use the basis state $|\mu\rangle|\mu\rangle$ without the subscript S. The method is outlined as follows. For a given representation, the representation matrix of P_1 is diagonal. The eigenvalue of P_1 is used to classify the basis states. The basis state with the eigenvalue 1 of P_1 is called the A-type basis, which has the following expansion

$$a_1|1\rangle|1\rangle + a_2|1\rangle|2\rangle + a_3|2\rangle|1\rangle + a_4|2\rangle|2\rangle + a_5|1\rangle|3\rangle + a_6|3\rangle|1\rangle + a_7|2\rangle|3\rangle$$
$$+ a_8|3\rangle|2\rangle + a_9|3\rangle|3\rangle + a_{10}|4\rangle|4\rangle + a_{11}|4\rangle|5\rangle + a_{12}|5\rangle|4\rangle + a_{13}|5\rangle|5\rangle, \tag{a}$$

The basis state with the eigenvalue -1 of P_1 is called the B-type basis, which has the following expansion

$$b_1|1\rangle|4\rangle + b_2|4\rangle|1\rangle + b_3|1\rangle|5\rangle + b_4|5\rangle|1\rangle + b_5|2\rangle|4\rangle + b_6|4\rangle|2\rangle + b_7|2\rangle|5\rangle \atop + b_8|5\rangle|2\rangle + b_9|3\rangle|4\rangle + b_{10}|4\rangle|3\rangle + b_{11}|3\rangle|5\rangle + b_{12}|5\rangle|3\rangle. \tag{b}$$

The action of the transposition P_a on the basis state is

$$P_a||[\lambda], \rho\rangle = \sum_\tau ||[\lambda], \tau\rangle D^{[\lambda]}_{\tau\rho}(P_a),$$
$$P_a|\mu\rangle|\nu\rangle = \sum_{\mu'\nu'} |\mu'\rangle|\nu'\rangle D^{[3,2]}_{\mu'\mu}(P_a) D^{[3,2]}_{\nu'\nu}(P_a), \tag{c}$$

where the representation matrix of P_a in $[\lambda]$ is known. The representation matrices of P_a in the representation $[3,2]$ are

$$P_1 : \begin{pmatrix} 1 & 0 & 0 & 0 & 0 \\ 0 & 1 & 0 & 0 & 0 \\ 0 & 0 & 1 & 0 & 0 \\ 0 & 0 & 0 & -1 & 0 \\ 0 & 0 & 0 & 0 & -1 \end{pmatrix}, \quad P_2 : \frac{1}{2} \begin{pmatrix} 2 & 0 & 0 & 0 & 0 \\ 0 & -1 & 0 & \sqrt{3} & 0 \\ 0 & 0 & -1 & 0 & \sqrt{3} \\ 0 & \sqrt{3} & 0 & 1 & 0 \\ 0 & 0 & \sqrt{3} & 0 & 1 \end{pmatrix},$$

$$P_3 : \frac{1}{3} \begin{pmatrix} -1 & \sqrt{8} & 0 & 0 & 0 \\ \sqrt{8} & 1 & 0 & 0 & 0 \\ 0 & 0 & 3 & 0 & 0 \\ 0 & 0 & 0 & 3 & 0 \\ 0 & 0 & 0 & 0 & -3 \end{pmatrix}, \quad P_4 : \frac{1}{2} \begin{pmatrix} 2 & 0 & 0 & 0 & 0 \\ 0 & -1 & \sqrt{3} & 0 & 0 \\ 0 & \sqrt{3} & 1 & 0 & 0 \\ 0 & 0 & 0 & -1 & \sqrt{3} \\ 0 & 0 & 0 & \sqrt{3} & 1 \end{pmatrix}.$$

(d)

For each irreducible representation $[\lambda]$, we first choose a basis state which is the common eigenstate of as many transpositions P_a as possible, and write its expansion according to its eigenvalue of P_1. Then, in terms of the representation matrices of P_a and the orthonormal condition we can determine the coefficients in the expansion. From the Schur theorem, two basis states belonging to two inequivalent irreducible representations or belonging to the different rows of a unitary irreducible representation are orthogonal. If the multiplicity of $[\lambda]$ in the reduction is one, the expansion must have a definite symmetry in the permutation. If it is larger than one, we can choose the expansions having definite symmetry. From the first basis state we are able to calculate the remaining basis states by Eq. (c), which have the same symmetry in the permutation as that the first basis state has. The orthonormal condition and the symmetry in permutation can be used for check.

First, we calculate the basis state in the representation [5], which is the identical representation with dimension one. The basis state $||[5]\rangle$ is given in Eq. (a). All representation matrices of P_a in [5] are 1. According to Eqs. (c) and (d), we have from the action of P_2

$$\begin{aligned} P_2||[5]\rangle &= ||[5]\rangle \\ &= a_1|1\rangle|1\rangle + a_2[(-1/2)|1\rangle|2\rangle + (\sqrt{3}/2)|1\rangle|4\rangle] \\ &\quad + a_3[(-1/2)|2\rangle|1\rangle + (\sqrt{3}/2)|4\rangle|1\rangle] + [a_4/4 + 3a_{10}/4]\,|2\rangle|2\rangle \\ &\quad + (\sqrt{3}/4)\,[-a_4 + a_{10}]\,(|2\rangle|4\rangle + |4\rangle|2\rangle) \\ &\quad + [3a_4/4 + a_{10}/4]\,|4\rangle|4\rangle + a_5[(-1/2)|1\rangle|3\rangle + (\sqrt{3}/2)|1\rangle|5\rangle] \\ &\quad + a_6[(-1/2)|3\rangle|1\rangle + (\sqrt{3}/2)|5\rangle|1\rangle] + [a_7/4 + 3a_{11}/4]\,|2\rangle|3\rangle \\ &\quad + (\sqrt{3}/4)\,[-a_7 + a_{11}]\,(|2\rangle|5\rangle + |4\rangle|3\rangle) \end{aligned}$$

$$+ [3a_7/4 + a_{11}/4] |4\rangle|5\rangle + [a_8/4 + 3a_{12}/4] |3\rangle|2\rangle$$
$$+ (\sqrt{3}/4) [-a_8 + a_{12}] (|5\rangle|2\rangle + |3\rangle|4\rangle)$$
$$+ [3a_8/4 + a_{12}/4] |5\rangle|4\rangle + [a_9/4 + 3a_{13}/4] |3\rangle|3\rangle$$
$$+ (\sqrt{3}/4) [-a_9 + a_{13}] (|3\rangle|5\rangle + |5\rangle|3\rangle)$$
$$+ [3a_9/4 + a_{13}/4] |5\rangle|5\rangle.$$

Comparing the coefficient of each term, especially the coefficients of the terms which do not appear in $||[5]\rangle$, we obtain $a_2 = a_3 = a_5 = a_6 = 0$, $a_4 = a_{10}$, $a_7 = a_{11}$, $a_8 = a_{12}$ and $a_9 = a_{13}$, namely

$$||[5]\rangle = a_1|1\rangle|1\rangle + a_4 (|2\rangle|2\rangle + |4\rangle|4\rangle) + a_7 (|2\rangle|3\rangle + |4\rangle|5\rangle)$$
$$+ a_8 (|3\rangle|2\rangle + |5\rangle|4\rangle) + a_9 (|3\rangle|3\rangle + |5\rangle|5\rangle). \qquad (e)$$

Applying P_3 to $||[5]\rangle$, we have

$$P_3||[5]\rangle = ||[5]\rangle$$
$$= [a_1/9 + 8a_4/9] |1\rangle|1\rangle + (\sqrt{8}/9) [-a_1 + a_4] (|1\rangle|2\rangle + |2\rangle|1\rangle)$$
$$+ [8a_1/9 + a_4/9] |2\rangle|2\rangle + a_4|4\rangle|4\rangle + a_9 (|3\rangle|3\rangle + |5\rangle|5\rangle)$$
$$+ a_7 [(1/3)|2\rangle|3\rangle + (\sqrt{8}/3)|1\rangle|3\rangle - |4\rangle|5\rangle]$$
$$+ a_8 [(1/3)|3\rangle|2\rangle + (\sqrt{8}/3)|3\rangle|1\rangle - |5\rangle|4\rangle].$$

Thus, $a_1 = a_4$, $a_7 = a_8 = 0$,

$$||[5]\rangle = a_1 (|1\rangle|1\rangle + |2\rangle|2\rangle + |4\rangle|4\rangle) + a_9 (|3\rangle|3\rangle + |5\rangle|5\rangle). \qquad (f)$$

Applying P_4 to $||[5]\rangle$, we have

$$P_4||[5]\rangle = ||[5]\rangle$$
$$= a_1|1\rangle|1\rangle + [a_1/4 + 3a_9/4] (|2\rangle|2\rangle + |4\rangle|4\rangle)$$
$$+ (\sqrt{3}/4) [-a_1 + a_9] (|2\rangle|3\rangle + |3\rangle|2\rangle + |4\rangle|5\rangle + |5\rangle|4\rangle)$$
$$+ [3a_1/4 + a_9/4] (|3\rangle|3\rangle + |5\rangle|5\rangle).$$

Thus, $a_1 = a_9$. After normalization we obtain

$$||[5]\rangle = \frac{1}{\sqrt{5}} \{|1\rangle|1\rangle + |2\rangle|2\rangle + |3\rangle|3\rangle + |4\rangle|4\rangle + |5\rangle|5\rangle\}.$$

Second, for the four-dimensional representation $[4, 1]$, we obtain the representation matrices of P_a from Eq. (12)

$$P_1 : \begin{pmatrix} 1 & 0 & 0 & 0 \\ 0 & 1 & 0 & 0 \\ 0 & 0 & 1 & 0 \\ 0 & 0 & 0 & -1 \end{pmatrix}, \qquad P_2 : \frac{1}{2}\begin{pmatrix} 2 & 0 & 0 & 0 \\ 0 & 2 & 0 & 0 \\ 0 & 0 & -1 & \sqrt{3} \\ 0 & 0 & \sqrt{3} & 1 \end{pmatrix},$$

$$P_3 : \frac{1}{3}\begin{pmatrix} 3 & 0 & 0 & 0 \\ 0 & -1 & \sqrt{8} & 0 \\ 0 & \sqrt{8} & 1 & 0 \\ 0 & 0 & 0 & 3 \end{pmatrix}, \qquad P_4 : \frac{1}{4}\begin{pmatrix} -1 & \sqrt{15} & 0 & 0 \\ \sqrt{15} & 1 & 0 & 0 \\ 0 & 0 & 1 & 0 \\ 0 & 0 & 0 & 1 \end{pmatrix}.$$

The basis state $||[4,1],1\rangle$ is an A-type basis. Applying P_1, P_2 and P_3 to $||[4,1],1\rangle$ one by one, we obtain a formula similar to Eq. (f):

$$||[4,1],1\rangle = a_1\left(|1\rangle|1\rangle + |2\rangle|2\rangle + |4\rangle|4\rangle\right) + a_9\left(|3\rangle|3\rangle + |5\rangle|5\rangle\right).$$

Since the basis state $||[4,1],1\rangle$ is orthogonal to $||[5]\rangle$, we have

$$||[4,1],1\rangle = (30)^{-1/2}\left\{2|1\rangle|1\rangle + 2|2\rangle|2\rangle - 3|3\rangle|3\rangle + 2|4\rangle|4\rangle - 3|5\rangle|5\rangle\right\}.$$

Applying P_4 to $||[4,1],1\rangle$, we have

$$\begin{aligned} P_4||[4,1],1\rangle &= (-1/4)||[4,1],1\rangle + (\sqrt{15}/4)||[4,1],2\rangle \\ &= (30)^{-1/2}\left\{2|1\rangle|1\rangle + [1/2 - 9/4]\left(|2\rangle|2\rangle + |4\rangle|4\rangle\right)\right. \\ &\quad + [3/2 - 3/4]\left(|3\rangle|3\rangle + |5\rangle|5\rangle\right) \\ &\quad \left. + \sqrt{3}\left[-1/2 - 3/4\right]\left(|2\rangle|3\rangle + |3\rangle|2\rangle + |4\rangle|5\rangle + |5\rangle|4\rangle\right)\right\}. \end{aligned}$$

The solution is

$$\begin{aligned} ||[4,1],2\rangle &= (4/\sqrt{15})\left\{P_4||[4,1],1\rangle + (1/4)||[4,1],1\rangle\right\} \\ &= (4/\sqrt{450})\left\{(2 + 1/2)|1\rangle|1\rangle + [-7/4 + 1/2]\left(|2\rangle|2\rangle + |4\rangle|4\rangle\right)\right. \\ &\quad + [3/4 - 3/4]\left(|3\rangle|3\rangle + |5\rangle|5\rangle\right) \\ &\quad \left. - (5\sqrt{3}/4)\left(|2\rangle|3\rangle + |3\rangle|2\rangle + |4\rangle|5\rangle + |5\rangle|4\rangle\right)\right\} \\ &= (3\sqrt{2})^{-1}\left\{2|1\rangle|1\rangle - |2\rangle|2\rangle - |4\rangle|4\rangle - \sqrt{3}|2,3\rangle_S - \sqrt{3}|4,5\rangle_S\right\}. \end{aligned}$$

The following calculation is similar. We only list the method and the results. Applying P_3 to $||[4,1],2\rangle$ gives

$$P_3||[4,1],2\rangle = (-1/3)||[4,1],2\rangle + (\sqrt{8}/3)||[4,1],3\rangle.$$

Thus,

$$\begin{aligned} ||[4,1],3\rangle &= (6)^{-1}\left\{2|2\rangle|2\rangle - 2|4\rangle|4\rangle - \sqrt{2}|1,2\rangle_S\right. \\ &\quad \left. - \sqrt{6}|1,3\rangle_S - \sqrt{3}|2,3\rangle_S + \sqrt{3}|4,5\rangle_S\right\}. \end{aligned}$$

Applying P_2 to $||[4, 1], 3\rangle$, we have

$$P_2||[4, 1], 3\rangle = (-1/2)||[4, 1], 3\rangle + (\sqrt{3}/2)||[4, 1], 4\rangle.$$

Thus,

$$
\begin{aligned}
||[4, 1], 4\rangle = (6)^{-1} \{ &-\sqrt{2}|1, 4\rangle_S - \sqrt{6}|1, 5\rangle_S - 2|2, 4\rangle_S \\
&+ \sqrt{3}|2, 5\rangle_S + \sqrt{3}|3, 4\rangle_S \}.
\end{aligned}
$$

Note that $||[4, 1], 4\rangle$ is a B-type basis state.

Third, we calculate the basis states in the representation $[3, 2]$ where the representation matrices of P_a were given in Eq. (d). $||[3, 2], 1\rangle$ is an A-type basis state. Applying P_1 and P_2 to $||[3, 2], 1\rangle$ one by one leads to a formula similar to Eq. (e):

$$
\begin{aligned}
||[3, 2], 1\rangle = \ &a_1|1\rangle|1\rangle + a_4 (|2\rangle|2\rangle + |4\rangle|4\rangle) + a_7 (|2\rangle|3\rangle + |4\rangle|5\rangle) \\
&+ a_8 (|3\rangle|2\rangle + |5\rangle|4\rangle) + a_9 (|3\rangle|3\rangle + |5\rangle|5\rangle).
\end{aligned}
$$

Applying P_4 to $||[3, 2], 1\rangle$, we have

$$
\begin{aligned}
P_4||[3, 2], 1\rangle &= ||[3, 2], 1\rangle \\
&= a_1|1\rangle|1\rangle + (1/4) \left[a_4 - \sqrt{3}a_7 - \sqrt{3}a_8 + 3a_9 \right] (|2\rangle|2\rangle + |4\rangle|4\rangle) \\
&+ (1/4) \left[3a_4 + \sqrt{3}a_7 + \sqrt{3}a_8 + a_9 \right] (|3\rangle|3\rangle + |5\rangle|5\rangle) \\
&+ (1/4) \left[-\sqrt{3}a_4 - a_7 + 3a_8 + \sqrt{3}a_9 \right] (|2\rangle|3\rangle + |4\rangle|5\rangle) \\
&+ (1/4) \left[-\sqrt{3}a_4 + 3a_7 - a_8 + \sqrt{3}a_9 \right] (|3\rangle|2\rangle + |5\rangle|4\rangle).
\end{aligned}
$$

Thus, $a_7 = a_8 = \sqrt{3}(a_9 - a_4)/2$. Since $||[3, 2], 1\rangle$ is orthogonal to $||[5]\rangle$ and $||[4, 1], 1\rangle$, we have

$$a_1 + 2a_4 + 2a_9 = 0, \qquad 2a_1 + 4a_4 - 6a_9 = 0.$$

Thus, $a_1 = -2a_4$ and $a_9 = 0$. The normalization condition leads to

$$||[3, 2], 1\rangle = (6)^{-1} \left\{ 4|1\rangle|1\rangle - 2|2\rangle|2\rangle - 2|4\rangle|4\rangle + \sqrt{3}|2, 3\rangle_S + \sqrt{3}|4, 5\rangle_S \right\}.$$

Applying P_3 to $||[3, 2], 1\rangle$, we have

$$P_3||[3, 2], 1\rangle = (-1/3)||[3, 2], 1\rangle + (\sqrt{8}/3)||[3, 2], 2\rangle.$$

Thus,

$$
\begin{aligned}
||[3, 2], 2\rangle = (6\sqrt{2})^{-1} \{ &4|2\rangle|2\rangle - 4|4\rangle|4\rangle - 2\sqrt{2}|1, 2\rangle_S + \sqrt{6}|1, 3\rangle_S \\
&+ \sqrt{3}|2, 3\rangle_S - \sqrt{3}|4, 5\rangle_S \}.
\end{aligned}
$$

Applying P_4 to $||[3,2],2\rangle$, we have

$$P_4||[3,2],2\rangle = (-1/2)||[3,2],2\rangle + (\sqrt{3}/2)||[3,2],3\rangle.$$

Thus,

$$||[3,2],3\rangle = (2\sqrt{6})^{-1}\left\{|2\rangle|2\rangle + 3|3\rangle|3\rangle - |4\rangle|4\rangle - 3|5\rangle|5\rangle + \sqrt{2}|1,2\rangle_S\right\}.$$

Applying P_2 to $||[3,2],2\rangle$, we have

$$P_2||[3,2],2\rangle = (-1/2)||[3,2],2\rangle + (\sqrt{3}/2)||[3,2],4\rangle.$$

Thus,

$$\begin{aligned}||[3,2],4\rangle = {} & (6\sqrt{2})^{-1}\left\{-2\sqrt{2}|1,4\rangle_S + \sqrt{6}|1,5\rangle_S - 4|2,4\rangle_S\right. \\ & \left. - \sqrt{3}|2,5\rangle_S - \sqrt{3}|3,4\rangle_S\right\}.\end{aligned}$$

$||[3,2],4\rangle$ is a B-type basis. Applying P_4 to $||[3,2],4\rangle$, we have

$$P_4||[3,2],4\rangle = (-1/2)||[3,2],4\rangle + (\sqrt{3}/2)||[3,2],5\rangle.$$

Thus,

$$||[3,2],5\rangle = (2\sqrt{6})^{-1}\left\{\sqrt{2}|1,4\rangle_S - |2,4\rangle_S - 3|3,5\rangle_S\right\}.$$

Fourth, we calculate the basis states in the six-dimensional representation $[3,1^2]$, where the representation matrices of P_a are

$$P_1: \begin{pmatrix} 1 & 0 & 0 & 0 & 0 & 0 \\ 0 & 1 & 0 & 0 & 0 & 0 \\ 0 & 0 & 1 & 0 & 0 & 0 \\ 0 & 0 & 0 & -1 & 0 & 0 \\ 0 & 0 & 0 & 0 & -1 & 0 \\ 0 & 0 & 0 & 0 & 0 & -1 \end{pmatrix}, \qquad P_2: \frac{1}{2}\begin{pmatrix} 2 & 0 & 0 & 0 & 0 & 0 \\ 0 & -1 & 0 & \sqrt{3} & 0 & 0 \\ 0 & 0 & -1 & 0 & \sqrt{3} & 0 \\ 0 & \sqrt{3} & 0 & 1 & 0 & 0 \\ 0 & 0 & \sqrt{3} & 0 & 1 & 0 \\ 0 & 0 & 0 & 0 & 0 & -1 \end{pmatrix},$$

$$P_3: \frac{1}{3}\begin{pmatrix} -1 & \sqrt{8} & 0 & 0 & 0 & 0 \\ \sqrt{8} & 1 & 0 & 0 & 0 & 0 \\ 0 & 0 & -3 & 0 & 0 & 0 \\ 0 & 0 & 0 & 3 & 0 & 0 \\ 0 & 0 & 0 & 0 & -1 & \sqrt{8} \\ 0 & 0 & 0 & 0 & \sqrt{8} & 1 \end{pmatrix}, \qquad P_4: \frac{1}{4}\begin{pmatrix} -4 & 0 & 0 & 0 & 0 & 0 \\ 0 & -1 & \sqrt{15} & 0 & 0 & 0 \\ 0 & \sqrt{15} & 1 & 0 & 0 & 0 \\ 0 & 0 & 0 & -1 & \sqrt{15} & 0 \\ 0 & 0 & 0 & \sqrt{15} & 1 & 0 \\ 0 & 0 & 0 & 0 & 0 & 4 \end{pmatrix}.$$

$||[3,1^2],1\rangle$ is an A-type basis. Applying P_1 and P_2 to $||[3,1^2],1\rangle$ one by one leads to a formula similar to Eq. (e):

$$\begin{aligned}||[3,1^2],1\rangle = {} & a_1|1\rangle|1\rangle + a_4\left(|2\rangle|2\rangle + |4\rangle|4\rangle\right) + a_7\left(|2\rangle|3\rangle + |4\rangle|5\rangle\right) \\ & + a_8\left(|3\rangle|2\rangle + |5\rangle|4\rangle\right) + a_9\left(|3\rangle|3\rangle + |5\rangle|5\rangle\right).\end{aligned}$$

Applying P_4 to $||[3, 1^2], 1\rangle$ changes its sign

$$
\begin{aligned}
P_4||[3, 1^2], 1\rangle &= -||[3, 1^2], 1\rangle \\
&= a_1|1\rangle|1\rangle + (1/4)\left[a_4 - \sqrt{3}a_7 - \sqrt{3}a_8 + 3a_9\right](|2\rangle|2\rangle + |4\rangle|4\rangle) \\
&\quad + (1/4)\left[3a_4 + \sqrt{3}a_7 + \sqrt{3}a_8 + a_9\right](|3\rangle|3\rangle + |5\rangle|5\rangle) \\
&\quad + (1/4)\left[-\sqrt{3}a_4 - a_7 + 3a_8 + \sqrt{3}a_9\right](|2\rangle|3\rangle + |4\rangle|5\rangle) \\
&\quad + (1/4)\left[-\sqrt{3}a_4 + 3a_7 - a_8 + \sqrt{3}a_9\right](|3\rangle|2\rangle + |5\rangle|4\rangle).
\end{aligned}
$$

Thus, $a_1 = 0$ and $a_4 = -a_9 = \sqrt{3}(a_7 + a_8)/2$. The orthogonal condition between $||[3, 1^2], 1\rangle$ and $||[2^2, 1], 1\rangle$ leads to $a_4 = a_9 = 0$. Then, $a_7 = -a_8$. The basis state $||[3, 1^2], 1\rangle$ is antisymmetric in the permutation. After normalization we have

$$
||[3, 1^2], 1\rangle = (2)^{-1}\left\{|2, 3\rangle_A + |4, 5\rangle_A\right\}.
$$

Applying P_3 to $||[3, 1^2], 1\rangle$, we have

$$
P_3||[3, 1^2], 1\rangle = (-1/3)||[3, 1^2], 1\rangle + (\sqrt{8}/3)||[3, 1^2], 2\rangle.
$$

Thus,

$$
||[3, 1^2], 2\rangle = (2\sqrt{2})^{-1}\left\{\sqrt{2}|1, 3\rangle_A + |2, 3\rangle_A - |4, 5\rangle_A\right\}.
$$

Applying P_4 to $||[3, 1^2], 2\rangle$, we have

$$
P_4||[3, 1^2], 2\rangle = (-1/4)||[3, 1^2], 2\rangle + (\sqrt{15}/4)||[3, 1^2], 3\rangle.
$$

Thus,

$$
\begin{aligned}
||[3, 1^2], 3\rangle = (2\sqrt{10})^{-1}\{&2\sqrt{2}|1, 2\rangle_A + \sqrt{6}|1, 3\rangle_A \\
&- \sqrt{3}|2, 3\rangle_A + \sqrt{3}|4, 5\rangle_A\}.
\end{aligned}
$$

Applying P_2 to $||[3, 1^2], 2\rangle$, we have

$$
P_2||[3, 1^2], 2\rangle = (-1/2)||[3, 1^2], 2\rangle + (\sqrt{3}/2)||[3, 1^2], 4\rangle.
$$

Thus,

$$
||[3, 1^2], 4\rangle = (2\sqrt{2})^{-1}\left\{\sqrt{2}|1, 5\rangle_A - |2, 5\rangle_A + |3, 4\rangle_A\right\}.
$$

$||[3, 1^2], 4\rangle$ is a B-type basis state. Applying P_4 to $||[3, 1^2], 4\rangle$, we have

$$
P_4||[3, 1^2], 4\rangle = (-1/4)||[3, 1^2], 4\rangle + (\sqrt{15}/4)||[3, 1^2], 5\rangle.
$$

Thus,

$$\begin{aligned}
\|[3,1^2],5\rangle = {}& (2\sqrt{10})^{-1}\,\{2\sqrt{2}|1,4\rangle_A + \sqrt{6}|1,5\rangle_A \\
& + \sqrt{3}|2,5\rangle_A - \sqrt{3}|3,4\rangle_A\}\,.
\end{aligned}$$

Applying P_3 to $\|[3,1^2],5\rangle$, we have

$$P_3\|[3,1^2],5\rangle = (-1/3)\|[3,1^2],5\rangle + (\sqrt{8}/3)\|[3,1^2],6\rangle.$$

Thus,

$$\|[3,1^2],6\rangle = (2\sqrt{5})^{-1}\left\{2|2,4\rangle_A - \sqrt{3}|2,5\rangle_A - \sqrt{3}|3,4\rangle_A\right\}.$$

Fifth, we calculate the basis states in the five-dimensional representation $[2^2,1]$, where the representation matrices of P_a are

$$P_1:\begin{pmatrix}1&0&0&0&0\\0&1&0&0&0\\0&0&-1&0&0\\0&0&0&-1&0\\0&0&0&0&-1\end{pmatrix},\qquad P_2:\frac{1}{2}\begin{pmatrix}-1&0&\sqrt{3}&0&0\\0&-1&0&\sqrt{3}&0\\\sqrt{3}&0&1&0&0\\0&\sqrt{3}&0&1&0\\0&0&0&0&-2\end{pmatrix},$$

$$P_3:\frac{1}{3}\begin{pmatrix}3&0&0&0&0\\0&-3&0&0&0\\0&0&-3&0&0\\0&0&0&-1&\sqrt{8}\\0&0&0&\sqrt{8}&1\end{pmatrix},\qquad P_4:\frac{1}{2}\begin{pmatrix}-1&\sqrt{3}&0&0&0\\\sqrt{3}&1&0&0&0\\0&0&-1&\sqrt{3}&0\\0&0&\sqrt{3}&1&0\\0&0&0&0&-2\end{pmatrix}.$$

The basis state $\|[2^2,1],5\rangle$ is the common eigenstate of P_1, P_2 and P_4. We begin the calculation with this state. $\|[2^2,1],5\rangle$ is a B-type basis with the form given in Eq. (b). Applying P_2 to $\|[2^2,1],5\rangle$ leads to

$$\begin{aligned}
P_2\|[2^2,1],5\rangle = {}& -\|[2^2,1],5\rangle \\
= {}& (b_1/2)\left[\sqrt{3}|1\rangle|2\rangle + |1\rangle|4\rangle\right] + (b_2/2)\left[\sqrt{3}|2\rangle|1\rangle + |4\rangle|1\rangle\right] \\
& + (b_3/2)\left[\sqrt{3}|1\rangle|3\rangle + |1\rangle|5\rangle\right] + (b_4/2)\left[\sqrt{3}|3\rangle|1\rangle + |5\rangle|1\rangle\right] \\
& + (\sqrt{3}/4)\,[b_5 + b_6]\,(-|2\rangle|2\rangle + |4\rangle|4\rangle) \\
& + (1/4)\,[-b_5 + 3b_6]\,|2\rangle|4\rangle + (1/4)\,[3b_5 - b_6]\,|4\rangle|2\rangle \\
& + (\sqrt{3}/4)\,[b_7 + b_{10}]\,(-|2\rangle|3\rangle + |4\rangle|5\rangle) \\
& + (1/4)\,[-b_7 + 3b_{10}]\,|2\rangle|5\rangle + (1/4)\,[3b_7 - b_{10}]\,|4\rangle|3\rangle \\
& + (\sqrt{3}/4)\,[b_8 + b_9]\,(-|3\rangle|2\rangle + |5\rangle|4\rangle) \\
& + (1/4)\,[-b_8 + 3b_9]\,|5\rangle|2\rangle + (1/4)\,[3b_8 - b_9]\,|3\rangle|4\rangle \\
& + (\sqrt{3}/4)\,[b_{11} + b_{12}]\,(-|3\rangle|3\rangle + |5\rangle|5\rangle)
\end{aligned}$$

$$+ (1/4) \left[-b_{11} + 3b_{12}\right] |3\rangle|5\rangle + (1/4) \left[3b_{11} - b_{12}\right] |5\rangle|3\rangle.$$

The solution is $b_1 = b_2 = b_3 = b_4 = 0$, $b_5 = -b_6$, $b_7 = -b_{10}$, $b_8 = -b_9$ and $b_{11} = -b_{12}$, namely

$$
\begin{aligned}
||[2^2, 1], 5\rangle =\ & b_5 |2, 4\rangle_A + b_7 \left(|2\rangle|5\rangle - |4\rangle|3\rangle\right) \\
& + b_8 \left(|5\rangle|2\rangle - |3\rangle|4\rangle\right) + b_{11} |3, 5\rangle_A.
\end{aligned}
\tag{g}
$$

Applying P_4 to $||[2^2, 1], 5\rangle$, we have

$$
\begin{aligned}
P_4 ||[2^2, 1], 5\rangle =\ & - ||[2^2, 1], 5\rangle \\
=\ & (1/4) \left[b_5 - \sqrt{3}b_7 + \sqrt{3}b_8 + 3b_{11}\right] |2, 4\rangle_A \\
& + (1/4) \left[-\sqrt{3}b_5 - b_7 - 3b_8 + \sqrt{3}b_{11}\right] |2\rangle|5\rangle \\
& + (1/4) \left[\sqrt{3}b_5 - 3b_7 - b_8 - \sqrt{3}b_{11}\right] |5\rangle|2\rangle \\
& + (1/4) \left[-\sqrt{3}b_5 + 3b_7 + b_8 + \sqrt{3}b_{11}\right] |3\rangle|4\rangle \\
& + (1/4) \left[\sqrt{3}b_5 + b_7 + 3b_8 - \sqrt{3}b_{11}\right] |4\rangle|3\rangle \\
& (1/4) \left[3b_5 + \sqrt{3}b_7 - \sqrt{3}b_8 + b_{11}\right] |3, 5\rangle_A.
\end{aligned}
$$

The solution is $b_5 = -b_{11} = \sqrt{3}(b_7 - b_8)/2$. The orthogonal condition between $||[2^2, 1], 5\rangle$ and $||[3, 1^2], 6\rangle$ leads to $b_5 = b_{11} = 0$ and $b_7 = b_8$. After normalization we obtain the symmetric expansion

$$||[2^2, 1], 5\rangle = (1/2) \left\{|3\,4\rangle_S - |2, 5\rangle_S\right\}.$$

Applying P_3 to $||[2^2, 1], 5\rangle$, we have

$$P_3 ||[2^2, 1], 5\rangle = (1/3) ||[2^2, 1], 5\rangle + (\sqrt{8}/3) ||[2^2, 1], 4\rangle.$$

The solution is

$$||[2^2, 1], 4\rangle = (2\sqrt{2})^{-1} \left\{\sqrt{2}|1, 5\rangle_S + |2, 5\rangle_S + |3, 4\rangle_S\right\}.$$

Applying P_4 to $||[2^2, 1], 4\rangle$, we have

$$P_4 ||[2^2, 1], 4\rangle = (1/2) ||[2^2, 1], 4\rangle + (\sqrt{3}/2) ||[2^2, 1], 3\rangle.$$

Thus,

$$||[2^2, 1], 3\rangle = (2\sqrt{2})^{-1} \left\{\sqrt{2}|1, 4\rangle_S - |2, 4\rangle_S + |3, 5\rangle_S\right\}.$$

Applying P_2 to $||[2^2, 1], 4\rangle$, we have

$$P_2 ||[2^2, 1], 4\rangle = (1/2) ||[2^2, 1], 4\rangle + (\sqrt{3}/2) ||[2^2, 1], 2\rangle.$$

Thus, $\quad ||[2^2, 1], 2\rangle = (2\sqrt{2})^{-1} \left\{ \sqrt{2}|1, 3\rangle_S - |2, 3\rangle_S + |4, 5\rangle_S \right\}.$

Applying P_4 to $||[2^2, 1], 2\rangle$, we have

$$P_4 ||[2^2, 1], 2\rangle = (1/2)||[2^2, 1], 2\rangle + (\sqrt{3}/2)||[2^2, 1], 1\rangle.$$

Thus,

$$||[2^2, 1], 1\rangle = (2\sqrt{2})^{-1} \left\{ |2\rangle|2\rangle - |3\rangle|3\rangle - |4\rangle|4\rangle + |5\rangle|5\rangle + \sqrt{2}|1, 2\rangle_S \right\}.$$

Finally, we calculate the basis states in the four-dimensional representation $[2, 1^3]$, where the representation matrices of P_a are

$$P_1 : \begin{pmatrix} 1 & 0 & 0 & 0 \\ 0 & -1 & 0 & 0 \\ 0 & 0 & -1 & 0 \\ 0 & 0 & 0 & -1 \end{pmatrix}, \quad P_2 : \frac{1}{2} \begin{pmatrix} -1 & \sqrt{3} & 0 & 0 \\ \sqrt{3} & 1 & 0 & 0 \\ 0 & 0 & -2 & 0 \\ 0 & 0 & 0 & -2 \end{pmatrix},$$

$$P_3 : \frac{1}{3} \begin{pmatrix} -3 & 0 & 0 & 0 \\ 0 & -1 & \sqrt{8} & 0 \\ 0 & \sqrt{8} & 1 & 0 \\ 0 & 0 & 0 & -3 \end{pmatrix}, \quad P_4 : \frac{1}{4} \begin{pmatrix} -4 & 0 & 0 & 0 \\ 0 & -4 & 0 & 0 \\ 0 & 0 & -1 & \sqrt{15} \\ 0 & 0 & \sqrt{15} & 1 \end{pmatrix}.$$

The basis state $||[2, 1^3], 4\rangle$ is the common eigenstate of P_1, P_2 and P_3. We start the calculation with this state. $||[2, 1^3], 4\rangle$ is a B-type basis. Applying P_2 to $||[2, 1^3], 4\rangle$, we obtain a similar formula to Eq. (g):

$$||[2, 1^3], 4\rangle = b_5|2, 4\rangle_A + b_7 (|2\rangle|5\rangle - |4\rangle|3\rangle)$$
$$+ b_8 (|5\rangle|2\rangle - |3\rangle|4\rangle) + b_{11}|3, 5\rangle_A.$$

Applying P_3 to $||[2, 1^3], 4\rangle$, we have

$$P_3 ||[2, 1^3], 4\rangle = -||[2, 1^3], 4\rangle$$
$$= (b_5/3) \left[\sqrt{8}|1, 4\rangle_A + |2, 4\rangle_A \right] - (b_7/3) \left[\sqrt{8}|1\rangle|5\rangle + |2\rangle|5\rangle \right]$$
$$- (b_8/3) \left[\sqrt{8}|5\rangle|1\rangle + |5\rangle|2\rangle \right] - b_7|4\rangle|3\rangle - b_8|3\rangle|4\rangle - b_{11}|3, 5\rangle_A.$$

Thus, $b_5 = b_7 = b_8 = 0$. After normalization we obtain the antisymmetric expansion

$$||[2, 1^3], 4\rangle = \sqrt{1/2}|3, 5\rangle_A.$$

Applying P_4 to $||[2, 1^3], 4\rangle$, we have

$$P_4 ||[2, 1^3], 4\rangle = (1/4)||[2, 1^3], 4\rangle + (\sqrt{15}/4)||[2, 1^3], 3\rangle.$$

The solution is

$$||[2,1^3],3\rangle = \sqrt{1/10}\left\{\sqrt{3}|2,4\rangle_A + |2,5\rangle_A + |3,4\rangle_A\right\}.$$

Applying P_3 to $||[2,1^3],3\rangle$, we have

$$P_3||[2,1^3],3\rangle = (1/3)||[2,1^3],3\rangle + (\sqrt{8}/3)||[2,1^3],2\rangle.$$

Thus,

$$||[2,1^3],2\rangle = (2\sqrt{5})^{-1}\left\{\sqrt{6}|1,4\rangle_A - \sqrt{2}|1,5\rangle_A - |2,5\rangle_A + |3,4\rangle_A\right\}.$$

Applying P_2 to $||[2,1^3],2\rangle$, we have

$$P_2||[2,1^3],2\rangle = (1/2)||[2,1^3],2\rangle + (\sqrt{3}/2)||[2,1^3],1\rangle.$$

Thus,

$$||[2,1^3],1\rangle = (2\sqrt{5})^{-1}\left\{\sqrt{6}|1,2\rangle_A - \sqrt{2}|1,3\rangle_A + |2,3\rangle_A - |4,5\rangle_A\right\}.$$

33. Calculate the reduction of the following outer products of the representations in the permutation groups by the Littlewood-Richardson rule:

(1) $[3,2,1] \otimes [3]$,   (2) $[3,2] \otimes [2,1]$,   (3) $[2,1] \otimes [4,2^3]$.

Solution. (1)

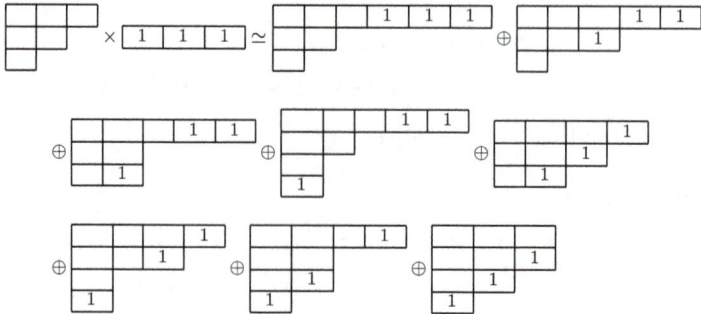

$$[3,2,1] \times [3] \simeq [6,2,1] \oplus [5,3,1] \oplus [5,2,2]$$
$$\oplus [5,2,1,1] \oplus [4,3,2] \oplus [4,3,1,1] \oplus [4,2,2,1] \oplus [3,3,2,1].$$

The dimensions of the representations on both sides are

$$\frac{9!}{6!\,3!} \times 16 \times 1 = 1344 = 105 + 162 + 120 + 189 + 168 + 216 + 216 + 168.$$

$$[3,2] \times [2,1] \simeq [5,3] \oplus [5,2,1] \oplus [4,4] \oplus 2[4,3,1]$$
$$\oplus [4,2,2] \oplus [4,2,1,1] \oplus [3,3,2] \oplus [3,3,1,1] \oplus [3,2,2,1].$$

The dimensions of the representations on both sides are

$$\frac{8!}{5!\,3!} \times 5 \times 2 = 560 = 28 + 64 + 14 + 2 \times 70 + 56 + 90 + 42 + 56 + 70.$$

$$[2,1] \times [4,2,2,2] \simeq [6,3,2,2] \oplus [6,2,2,2,1] \oplus [5,4,2,2]$$
$$\oplus [5,3,3,2] \oplus 2[5,3,2,2,1] \oplus [5,2,2,2,2] \oplus [5,2,2,2,1,1]$$
$$\oplus [4,4,3,2] \oplus [4,4,2,2,1] \oplus [4,3,3,2,1] \oplus [4,3,2,2,2]$$
$$\oplus [4,3,2,2,1,1] \oplus [4,2,2,2,2,1].$$

The dimensions of the representations on both sides are

$$\frac{13!}{3!\,10!} \times 2 \times 300 = 171600 = 12012 + 9009 + 12870 + 11583 + 2 \times 21450$$
$$+ 5005 + 10296 + 8580 + 12870 + 15015 + 8580 + 17160 + 5720.$$

34. Calculate the reduction of the subduced representations from the following irreducible representations of S_6 with respect to the subgroup $S_3 \otimes S_3$:

$$(1)\ [4,2]\,, \qquad (2)\ [2,2,1,1]\,, \qquad (3)\ [3,3]\,.$$

Solution. (1)

$$[4,2] \simeq [3] \times [3] \oplus [3] \times [2,1] \oplus [2,1] \times [3] \oplus [2,1] \times [2,1].$$

The dimensions of the representations on both sides are

$$9 = 1 \times 1 + 1 \times 2 + 2 \times 1 + 2 \times 2.$$

(2)

$$[2^2, 1^2] \simeq [2,1] \times [2,1] \oplus [2,1] \times [1^3] \oplus [1^3] \times [2,1] \oplus [1^3] \times [1^3].$$

The dimensions of the representations on both sides are

$$9 = 2 \times 2 + 2 \times 1 + 1 \times 2 + 1 \times 1.$$

(3)

$$\boxed{}\ \boxed{1}\ \boxed{1}\ \boxed{1} \oplus \begin{array}{cc} & \boxed{1} \\ \boxed{1} & \boxed{2} \end{array},$$

$$[3,3] \simeq [3] \times [3] \oplus [2,1] \times [2,1].$$

The dimensions of the representations on both sides are

$$5 = 1 \times 1 + 2 \times 2.$$

35. Calculate the similarity transformation matrix in the reduction of the subduced representation from the irreducible representation $[3,3]$ of S_6 with respect to the subgroup $S_3 \otimes S_3$.

Solution. The reduction of the subduced representation from $[3,3]$ of S_6 with respect to the subgroup $S_3 \otimes S_3$ was calculated in Problem 34:

$$[3,3] \simeq [3] \times [3] \oplus [2,1] \times [2,1].$$

The standard Young tableaux for the Young pattern $[3,3]$ were given in Problem 18. Denote by $\mathcal{Y}_1$ the Young operator for the standard Young tableau

1	2	3
4	5	6

The standard bases in the representation space $\mathcal{L}$ of $[3,3]$ are

$$b_1 = \mathcal{Y}_1, \qquad\qquad b_2 = (3\ 4)\mathcal{Y}_1,$$
$$b_3 = (3\ 5\ 4)\mathcal{Y}_1 = (3\ 5)\mathcal{Y}_1, \quad b_4 = (2\ 3\ 4)\mathcal{Y}_1 = (4\ 2)\mathcal{Y}_1,$$
$$b_5 = (2\ 3\ 5\ 4)\mathcal{Y}_1 = (5\ 2)\mathcal{Y}_1.$$

From the Fock condition (6.5) we have

$$\{E + (1\ 4) + (2\ 4) + (3\ 4)\}\mathcal{Y}_1 = 0,$$
$$\{E + (1\ 5) + (2\ 5) + (3\ 5)\}\mathcal{Y}_1 = 0,$$
$$\{E + (1\ 6) + (2\ 6) + (3\ 6)\}\mathcal{Y}_1 = 0,$$
$$\{E + (3\ 4) + (3\ 5) + (3\ 6)\}\mathcal{Y}_1 = 0.$$

The idempotent of the one-dimensional representation $[3] \times [3]$ of the subgroup $S_3 \otimes S_3$ is

$$\mathcal{Y}_2 = \mathcal{Y}\ \boxed{1\ \ 2\ \ 3}\ \mathcal{Y}\ \boxed{4\ \ 5\ \ 6}.$$

From Eq. (6.19), the basis state for the representation $[3] \times [3]$ is

$$\phi = \mathcal{Y}_2 \mathcal{Y}_1 = 36 b_1.$$

The idempotent of the four-dimensional representation $[2, 1] \times [2, 1]$ of the subgroup $S_3 \otimes S_3$ is

$$\mathcal{Y}_3 = \mathcal{Y}\begin{array}{|c|c|} \hline 1 & 2 \\ \hline 3 \\ \cline{1-1} \end{array} \mathcal{Y}\begin{array}{|c|c|} \hline 4 & 5 \\ \hline 6 \\ \cline{1-1} \end{array}.$$

Since $\mathcal{Y}_3 \mathcal{Y}_1 = 0$, from Eq. (6.19), we choose the basis state for the representation $[2, 1] \times [2, 1]$ as

$$
\begin{aligned}
\psi_1 &= \mathcal{Y}_3(3\ 4)\mathcal{Y}_1 \\
&= (3\ 4)\left\{E + (1\ 2) - (1\ 4) - (2\ 1\ 4)\right\}\left\{E + (3\ 5) - (3\ 6) - (5\ 3\ 6)\right\}\mathcal{Y}_1 \\
&= (3\ 4)\left\{2E - (1\ 4) - (4\ 2)\right\}\left\{E + (3\ 5) - 2(3\ 6)\right\}\mathcal{Y}_1 \\
&= \left\{2(3\ 4) + 2(4\ 3\ 5) - 4(4\ 3\ 6) - (3\ 4\ 1) - (4\ 1\ 3\ 5)\right. \\
&\quad \left. + 2(4\ 1\ 3\ 6) - (3\ 4\ 2) - (4\ 2\ 3\ 5) + 2(4\ 2\ 3\ 6)\right\}\mathcal{Y}_1 \\
&= 2b_2 + 2b_3 - 4(3\ 6)b_1 - (1\ 4)b_1 - (1\ 5)b_1 + 2(6\ 1)b_1 - b_4 - b_5 + 2(6\ 2)b_1 \\
&= 2b_2 + 2b_3 - b_4 - b_5 + [b_1 + b_2 + b_4] + [b_1 + b_3 + b_5] - 2b_1 \\
&\quad + 6[b_1 + b_2 + b_3] \\
&= 6b_1 + 9b_2 + 9b_3.
\end{aligned}
$$

Then, we have

$$
\begin{aligned}
\psi_2 &= (5\ 6)\psi_1 = 6b_1 + 9b_2 - 9[b_1 + b_2 + b_3] = -3b_1 - 9b_3, \\
\psi_3 &= (2\ 3)\psi_1 = 6b_1 + 9b_4 + 9b_5, \\
\psi_4 &= (2\ 3)(5\ 6)\psi_1 = (2\ 3)\psi_2 = -3b_1 - 9b_5.
\end{aligned}
$$

Thus, we obtain the similarity transformation matrix X:

$$
X = \begin{pmatrix}
36 & 6 & -3 & 6 & -3 \\
0 & 9 & 0 & 0 & 0 \\
0 & 9 & -9 & 0 & 0 \\
0 & 0 & 0 & 9 & 0 \\
0 & 0 & 0 & 9 & -9
\end{pmatrix}.
$$

The generators of $S_3 \otimes S_3$ are P_1, P_2, P_4, and P_5. Their representation matrices in $[3, 3]$ can be calculated by the tabular method:

$$D^{[3,3]}(P_1) = \begin{pmatrix} 1 & 0 & 0 & -1 & -1 \\ 0 & 1 & 0 & -1 & 0 \\ 0 & 0 & 1 & 0 & -1 \\ 0 & 0 & 0 & -1 & 0 \\ 0 & 0 & 0 & 0 & -1 \end{pmatrix}, \quad D^{[3,3]}(P_2) = \begin{pmatrix} 1 & 0 & 0 & 0 & 0 \\ 0 & 0 & 0 & 1 & 0 \\ 0 & 0 & 0 & 0 & 1 \\ 0 & 1 & 0 & 0 & 0 \\ 0 & 0 & 1 & 0 & 0 \end{pmatrix},$$

$$D^{[3,3]}(P_4) = \begin{pmatrix} 1 & 0 & 0 & 0 & 0 \\ 0 & 0 & 1 & 0 & 0 \\ 0 & 1 & 0 & 0 & 0 \\ 0 & 0 & 0 & 0 & 1 \\ 0 & 0 & 0 & 1 & 0 \end{pmatrix}, \quad D^{[3,3]}(P_5) = \begin{pmatrix} 1 & 0 & -1 & 0 & -1 \\ 0 & 1 & -1 & 0 & 0 \\ 0 & 0 & -1 & 0 & 0 \\ 0 & 0 & 0 & 1 & -1 \\ 0 & 0 & 0 & 0 & -1 \end{pmatrix}.$$

Their representation matrices in $[3] \times [3] \oplus [2,1] \times [2,1]$ are

$$\overline{D}(R) \equiv D^{[3] \times [3]}(R) \oplus D^{[2,1] \times [2,1]}(R),$$

$$\overline{D}(P_1) = 1 \oplus \begin{pmatrix} 1 & -1 \\ 0 & -1 \end{pmatrix} \times \begin{pmatrix} 1 & 0 \\ 0 & 1 \end{pmatrix} = \begin{pmatrix} 1 & 0 & 0 & 0 & 0 \\ 0 & 1 & 0 & -1 & 0 \\ 0 & 0 & 1 & 0 & -1 \\ 0 & 0 & 0 & -1 & 0 \\ 0 & 0 & 0 & 0 & -1 \end{pmatrix},$$

$$\overline{D}(P_2) = 1 \oplus \begin{pmatrix} 0 & 1 \\ 1 & 0 \end{pmatrix} \times \begin{pmatrix} 1 & 0 \\ 0 & 1 \end{pmatrix} = \begin{pmatrix} 1 & 0 & 0 & 0 & 0 \\ 0 & 0 & 0 & 1 & 0 \\ 0 & 0 & 0 & 0 & 1 \\ 0 & 1 & 0 & 0 & 0 \\ 0 & 0 & 1 & 0 & 0 \end{pmatrix},$$

$$\overline{D}(P_4) = 1 \oplus \begin{pmatrix} 1 & 0 \\ 0 & 1 \end{pmatrix} \times \begin{pmatrix} 1 & -1 \\ 0 & -1 \end{pmatrix} = \begin{pmatrix} 1 & 0 & 0 & 0 & 0 \\ 0 & 1 & -1 & 0 & 0 \\ 0 & 0 & -1 & 0 & 0 \\ 0 & 0 & 0 & 1 & -1 \\ 0 & 0 & 0 & 0 & -1 \end{pmatrix},$$

$$\overline{D}(P_5) = 1 \oplus \begin{pmatrix} 1 & 0 \\ 0 & 1 \end{pmatrix} \times \begin{pmatrix} 0 & 1 \\ 1 & 0 \end{pmatrix} = \begin{pmatrix} 1 & 0 & 0 & 0 & 0 \\ 0 & 0 & 1 & 0 & 0 \\ 0 & 1 & 0 & 0 & 0 \\ 0 & 0 & 0 & 0 & 1 \\ 0 & 0 & 0 & 1 & 0 \end{pmatrix}.$$

It is easy to check that X satisfies the similarity transformation

$$D^{[3,3]}(R)X = X\overline{D}(R) = X\left\{D^{[3]\times[3]}(R) \oplus D^{[2,1]\times[2,1]}(R)\right\},$$

$$R = P_a, \qquad a = 1, 2, 4, 5.$$

36. Calculate the representation matrices of the generators of S_4 in the induced representation from two-dimensional irreducible representation $[2, 1]$ of S_3 with respect to S_4.

Solution. For simplicity, let A and B respectively be the representation matrices of the generators of S_3 in the two-dimensional representation $[2, 1]$:

$$A = D^{[2,1]}(1\ 2) = \begin{pmatrix} 1 & 0 \\ 0 & -1 \end{pmatrix}, \qquad B = D^{[2,1]}(1\ 2\ 3) = \frac{1}{2}\begin{pmatrix} -1 & \sqrt{3} \\ -\sqrt{3} & -1 \end{pmatrix}.$$

Denoting by ϕ_μ, $\mu = 1$ and 2, the state bases in the representation space, we have

$$(1\ 2)\phi_\mu = \sum_{\nu=1}^{2} \phi_\nu A_{\nu\mu}, \qquad (1\ 2\ 3)\phi_\mu = \sum_{\nu=1}^{2} \phi_\nu B_{\nu\mu}.$$

This two-dimensional space is invariant for the S_3 group, but not for S_4. The S_3 group is a subgroup of S_4 with index 4. Extend the two-dimensional space:

$$\psi_{1\mu} = \phi_\mu, \qquad \psi_{2\mu} = (1\ 4)\phi_\mu, \qquad \psi_{3\mu} = (2\ 4)\phi_\mu, \qquad \psi_{4\mu} = (3\ 4)\phi_\mu.$$

The eight-dimensional space is invariant for S_4. In fact,

$$(1\ 2)\psi_{1\mu} = (1\ 2)\phi_\mu = \sum_{\nu=1}^{2} \psi_{1\nu} A_{\nu\mu},$$

$$(1\ 2)\psi_{2\mu} = (4\ 2)(2\ 1)\phi_\mu = \sum_{\nu=1}^{2} \psi_{3\nu} A_{\nu\mu},$$

$$(1\ 2)\psi_{3\mu} = (4\ 1)(1\ 2)\phi_\mu = \sum_{\nu=1}^{2} \psi_{2\nu} A_{\nu\mu},$$

$$(1\ 2)\psi_{4\mu} = (3\ 4)(1\ 2)\phi_\mu = \sum_{\nu=1}^{2} \psi_{4\nu} A_{\nu\mu},$$

$$(1\ 2\ 3\ 4)\psi_{1\mu} = (4\ 1)(1\ 2\ 3)\phi_\mu = \sum_{\nu=1}^{2} \psi_{2\nu} B_{\nu\mu},$$

$$(1\ 2\ 3\ 4)\psi_{2\mu} = (2\ 3\ 4)\phi_\mu = (4\ 2)(2\ 1)(1\ 2\ 3)\phi_\mu = \sum_{\nu=1}^{2} \psi_{3\nu} (AB)_{\nu\mu},$$

$$(1\ 2\ 3\ 4)\psi_{3\mu} = (3\ 4)(1\ 2)\phi_\mu = \sum_{\nu=1}^{2} \psi_{4\nu} A_{\nu\mu},$$

$$(1\ 2\ 3\ 4)\psi_{4\mu} = (1\ 2\ 3)\phi_\mu = \sum_{\nu=1}^{2} \psi_{1\nu} B_{\nu\mu}.$$

Therefore, the representation matrices of the generators $(1\ 2)$ and $(1\ 2\ 3\ 4)$ of S_4 in the induced representation are:

$$D(1\ 2) = \begin{pmatrix} A & 0 & 0 & 0 \\ 0 & 0 & A & 0 \\ 0 & A & 0 & 0 \\ 0 & 0 & 0 & A \end{pmatrix}, \qquad D(1\ 2\ 3\ 4) = \begin{pmatrix} 0 & 0 & 0 & B \\ B & 0 & 0 & 0 \\ 0 & AB & 0 & 0 \\ 0 & 0 & A & 0 \end{pmatrix}.$$

It is a reducible representation of S_4, which can be reduced into the direct sum of the representations $[3,1]$, $[2,2]$ and $[2,1^2]$.

37. Calculate the similarity transformation matrix in the reduction of the outer product representation $[2,1] \otimes [2]$ of the permutation group.

Solution. The outer product representation $[2,1] \otimes [2]$ is the induced representation from $[2,1] \times [2]$ of the subgroup $S_3 \times S_2$ with respect to the group S_5. Its reduction can be calculated by the Littlewood-Richardson rule:

$$[2,1] \otimes [2] \simeq [4,1] \oplus [3,2] \oplus [3,1^2] \oplus [2^2,1]$$

$$\frac{120}{6 \times 2} \times 2 \times 1 = 20 = 4 + 5 + 6 + 5.$$

The idempotent of the two-dimensional representation $[2,1] \times [2]$ of the subgroup $S_3 \otimes S_2$ is

$$\mathcal{Y}_1 = \mathcal{Y}\boxed{\begin{array}{|c|c|} \hline 1 & 2 \\ \hline 3 \\ \cline{1-1} \end{array}} \mathcal{Y}\boxed{\begin{array}{|c|c|} \hline 4 & 5 \\ \hline \end{array}}.$$

The standard bases in the representation space of $[2,1] \times [2]$ are

$$b_1 = \mathcal{Y}_1, \qquad b_2 = (2\ 3)\mathcal{Y}_1.$$

From the Fock condition (6.5) we have

$$(1\ 2)\mathcal{Y}_1 = (4\ 5)\mathcal{Y}_1 = \mathcal{Y}_1, \qquad (1\ 3)\mathcal{Y}_1 = -b_1 - b_2,$$

and the useful formula for the later calculation,

$$[E + (1\ 2)]\,[E - (1\ 3)]\,b_1 = [2E - (1\ 3) - (1\ 2)(1\ 3)]\,b_1 = 3b_1.$$

Extending the two-dimensional space, we obtain the bases

$$
\begin{aligned}
b_3 &= (1\ 4)b_1, & b_4 &= (1\ 4)b_2 = (1\ 4)(2\ 3)b_1, \\
b_5 &= (2\ 4)b_1, & b_6 &= (2\ 4)b_2 = (4\ 2\ 3)b_1, \\
b_7 &= (3\ 4)b_1, & b_8 &= (3\ 4)b_2 = (4\ 3\ 2)b_1, \\
b_9 &= (1\ 5)b_1, & b_{10} &= (1\ 5)b_2 = (1\ 5)(2\ 3)b_1, \\
b_{11} &= (2\ 5)b_1, & b_{12} &= (2\ 5)b_2 = (5\ 2\ 3)b_1, \\
b_{13} &= (3\ 5)b_1, & b_{14} &= (3\ 5)b_2 = (5\ 3\ 2)b_1, \\
b_{15} &= (1\ 4)(2\ 5)b_1, & b_{16} &= (1\ 4)(2\ 5)b_2 = (1\ 4)(5\ 2\ 3)b_1, \\
b_{17} &= (1\ 4)(3\ 5)b_1, & b_{18} &= (1\ 4)(3\ 5)b_2 = (1\ 4)(5\ 3\ 2)b_1, \\
b_{19} &= (2\ 4)(3\ 5)b_1, & b_{20} &= (2\ 4)(3\ 5)b_2 = (4\ 2\ 5\ 3)b_1.
\end{aligned}
$$

The representation matrices of the generators $P_1 = (1\ 2)$ and $W = (1\ 2\ 3\ 4\ 5)$ of S_5 in the out product representation $[2,1] \otimes [2]$ for the bases b_μ are

$$
P_1 b_\mu = \sum_{\nu=1}^{20} b_\nu D_{\nu\mu}^{[2,1]\otimes[2]}(P_1), \qquad
W b_\mu = \sum_{\nu=1}^{20} b_\nu D_{\nu\mu}^{[2,1]\otimes[2]}(W).
$$

Through a direct calculation we have

$$
\begin{aligned}
P_1 b_1 &= b_1, & P_1 b_2 &= -b_1 - b_2, & P_1 b_3 &= b_5, & P_1 b_4 &= -b_5 - b_6, \\
P_1 b_5 &= b_3, & P_1 b_6 &= -b_3 - b_4, & P_1 b_7 &= b_7, & P_1 b_8 &= -b_7 - b_8, \\
P_1 b_9 &= b_{11}, & P_1 b_{10} &= -b_{11} - b_{12}, & P_1 b_{11} &= b_9, & P_1 b_{12} &= -b_9 - b_{10}, \\
P_1 b_{13} &= b_{13}, & P_1 b_{14} &= -b_{13} - b_{14}, & P_1 b_{15} &= b_{15}, & P_1 b_{16} &= b_{16}, \\
P_1 b_{17} &= b_{19}, & P_1 b_{18} &= -b_{19} - b_{20}, & P_1 b_{19} &= b_{17}, & P_1 b_{20} &= -b_{17} - b_{18},
\end{aligned}
$$

$$
\begin{aligned}
W b_1 &= -b_3 - b_4, & W b_2 &= b_3, & W b_3 &= -b_{15} - b_{16}, & W b_4 &= b_{15}, \\
W b_5 &= -b_{17} - b_{18}, & W b_6 &= b_{17}, & W b_7 &= -b_9 - b_{10}, & W b_8 &= b_9, \\
W b_9 &= b_6, & W b_{10} &= b_5, & W b_{11} &= b_7, & W b_{12} &= -b_7 - b_8, \\
W b_{13} &= -b_1 - b_2, & W b_{14} &= b_1, & W b_{15} &= b_{20}, & W b_{16} &= -b_{19} - b_{20}, \\
W b_{17} &= -b_{11} - b_{12}, & W b_{18} &= b_{11}, & W b_{19} &= -b_{13} - b_{14}, & W b_{20} &= b_{13}.
\end{aligned}
$$

After the similarity transformation X, the out product representation $[2,1] \otimes [2]$ becomes a direct sum of four irreducible representations:

$$
\begin{aligned}
&\left\{ D^{[2,1]\otimes[2]}(R) \right\} X \\
&\quad = X \left\{ D^{[4,1]}(R) \oplus D^{[3,2]}(R) \oplus D^{[3,1^2]}(R) \oplus D^{[2^2,1]}(R) \right\},
\end{aligned}
$$

where $R = P_1$ and W. By the tabular method we obtain

$$D^{[4,1]}(P_1) = \begin{pmatrix} 1 & 0 & 0 & -1 \\ 0 & 1 & 0 & -1 \\ 0 & 0 & 1 & -1 \\ 0 & 0 & 0 & -1 \end{pmatrix}, \quad D^{[4,1]}(W) = \begin{pmatrix} -1 & 1 & 0 & 0 \\ -1 & 0 & 1 & 0 \\ -1 & 0 & 0 & 1 \\ -1 & 0 & 0 & 0 \end{pmatrix},$$

$$D^{[3,2]}(P_1) = \begin{pmatrix} 1 & 0 & 0 & -1 & -1 \\ 0 & 1 & 0 & -1 & 0 \\ 0 & 0 & 1 & 0 & -1 \\ 0 & 0 & 0 & -1 & 0 \\ 0 & 0 & 0 & 0 & -1 \end{pmatrix}, \quad D^{[3,2]}(W) = \begin{pmatrix} -1 & -1 & 1 & 1 & 0 \\ -1 & 0 & 0 & 0 & 1 \\ 0 & -1 & 0 & 0 & 0 \\ -1 & 0 & 0 & 1 & 0 \\ 0 & -1 & 0 & 1 & 0 \end{pmatrix},$$

$$D^{[3,1^2]}(P_1) = \begin{pmatrix} 1 & 0 & 0 & -1 & 1 & 0 \\ 0 & 1 & 0 & -1 & 0 & 1 \\ 0 & 0 & 1 & 0 & -1 & 1 \\ 0 & 0 & 0 & -1 & 0 & 0 \\ 0 & 0 & 0 & 0 & -1 & 0 \\ 0 & 0 & 0 & 0 & 0 & -1 \end{pmatrix}, \quad D^{[3,1^2]}(W) = \begin{pmatrix} 1 & -1 & 1 & 0 & 0 & 0 \\ 1 & 0 & 0 & -1 & 1 & 0 \\ 0 & 1 & 0 & -1 & 0 & 1 \\ 1 & 0 & 0 & 0 & 0 & 0 \\ 0 & 1 & 0 & 0 & 0 & 0 \\ 0 & 0 & 0 & 1 & 0 & 0 \end{pmatrix},$$

$$D^{[2^2,1]}(P_1) = \begin{pmatrix} 1 & 0 & -1 & 0 & 1 \\ 0 & 1 & 0 & -1 & 1 \\ 0 & 0 & -1 & 0 & 0 \\ 0 & 0 & 0 & -1 & 0 \\ 0 & 0 & 0 & 0 & -1 \end{pmatrix}, \quad D^{[2^2,1]}(W) = \begin{pmatrix} 1 & -1 & -1 & 0 & 0 \\ 1 & 0 & -1 & 0 & 0 \\ 1 & -1 & -1 & 1 & 0 \\ 1 & 0 & 0 & 0 & 1 \\ 0 & 0 & 1 & 0 & 0 \end{pmatrix}.$$

Now, we calculate the similarity transformation matrix X by Eq. (6.21). The idempotent of the four-dimensional representation $[4, 1]$ of S_5 is

$$\mathcal{Y}_2 = \mathcal{Y}_3^{[4,1]} = \mathcal{Y}\,\boxed{\begin{array}{|c|c|c|c|} \hline 1 & 2 & 4 & 5 \\ \hline 3 \\ \cline{1-1} \end{array}}.$$

From Eq. (6.21) we obtain the expansions of the basis states $\psi_\mu^{[4,1]}$ of $[4, 1]$ in the extending space:

$$\begin{aligned} \psi_3^{[4,1]} &= \sum_{\nu=1}^{20} b_\nu X_{\nu3} = \mathcal{Y}_2 \mathcal{Y}_1 \\ &= [E + (1\ 4) + (2\ 4) + (1\ 5) + (2\ 5) + (1\ 4)(2\ 5)] \\ &\quad \times [E + (1\ 2)]\,[E + (4\ 5)]\,[E - (1\ 3)]\,b_1 \\ &= 6\,[E + (1\ 4) + (2\ 4) + (1\ 5) + (2\ 5) + (1\ 4)(2\ 5)]\,b_1 \\ &= 6\,[b_1 + b_3 + b_5 + b_9 + b_{11} + b_{15}], \end{aligned}$$

$$\psi_1^{[4,1]} = \sum_{\nu=1}^{20} b_\nu X_{\nu 1} = (4\ 3\ 5)\psi_3^{[4,1]} = (3\ 5)\psi_3^{[4,1]}$$
$$= 6\left[b_{13} + b_{17} + b_{19} - b_9 - b_{10} + b_{12} + b_{16}\right],$$

$$\psi_2^{[4,1]} = \sum_{\nu=1}^{20} b_\nu X_{\nu 2} = (4\ 3)\psi_3^{[4,1]}$$
$$= 6\left[b_7 - b_3 - b_4 + b_6 - b_{17} - b_{18} + b_{20} - b_{15} - b_{16}\right],$$

$$\psi_4^{[4,1]} = \sum_{\nu=1}^{20} b_\nu X_{\nu 4} = (2\ 3)\psi_3^{[4,1]}$$
$$= 6\left[b_2 + b_4 + b_8 + b_{10} + b_{14} + b_{18}\right].$$

The idempotent of the five-dimensional representation $[3, 2]$ of S_5 is

$$\mathcal{Y}_3 = \mathcal{Y}_2^{[3,2]} = \mathcal{Y}\begin{array}{|c|c|c|}\hline 1 & 2 & 4 \\\hline 3 & 5 \\\cline{1-2}\end{array}.$$

From Eq. (6.21) we obtain the expansions of the basis states $\psi_\mu^{[3,2]}$ of $[3, 2]$ in the extending space:

$$\psi_2^{[3,2]} = \sum_{\nu=1}^{20} b_\nu X_{\nu 6} = \mathcal{Y}_3 \mathcal{Y}_1$$
$$= [E + (1\ 4) + (2\ 4)]\,[E + (1\ 2)]\,[E + (3\ 5)]$$
$$\times [E - (2\ 5)]\,[E - (1\ 3)]\,b_1$$
$$= [E + (1\ 4) + (2\ 4)]\,[E + (3\ 5)]\,[3b_1 - 2b_{11} - b_{12} - b_9 + b_{10}]$$
$$= [E + (1\ 4) + (2\ 4)]\,[3b_1 - 2b_{11} - b_{12} - b_9 + b_{10}$$
$$\qquad + 3b_{13} - b_{11} - 2b_{12} + b_9 + 2b_{10}]$$
$$= 3\,[b_1 - b_{11} - b_{12} + b_{10} + b_{13} + b_3 - b_{15} - b_{16} + b_{10} + b_{17}$$
$$\qquad + b_5 - b_{11} - b_{12} - b_{15} - b_{16} + b_{19}]$$
$$= 3\,[b_1 + b_3 + b_5 + 2b_{10} - 2b_{11} - 2b_{12} + b_{13} - 2b_{15}$$
$$\qquad - 2b_{16} + b_{17} + b_{19}]$$

$$\psi_1^{[3,2]} = \sum_{\nu=1}^{20} b_\nu X_{\nu 5} = (3\ 4)\psi_2^{[3,2]}$$
$$= 3\,[b_7 - b_3 - b_4 + b_6 + 2b_{18} - 2b_{19} - 2b_{20} + b_{13} + 2b_{15}$$
$$\qquad - b_9 - b_{10} + b_{12}],$$

$$\psi_3^{[3,2]} = \sum_{\nu=1}^{20} b_\nu X_{\nu 7} = (4\ 5)\psi_2^{[3,2]}$$
$$= 3\left[b_1 + b_9 + b_{11} + 2b_4 - 2b_5 - 2b_6 + b_7 + 2b_{16} - b_{17}\right.$$
$$\left. - b_{18} + b_{20}\right],$$

$$\psi_4^{[4,1]} = \sum_{\nu=1}^{20} b_\nu X_{\nu 8} = (2\ 3)\psi_2^{[3,2]}$$
$$= 3\left[b_2 + b_4 + b_8 + 2b_9 - 2b_{13} - 2b_{14} + b_{12}\right.$$
$$\left. - 2b_{18} - 2b_{17} + b_{16} + b_{19}\right],$$

$$\psi_5^{[4,1]} = \sum_{\nu=1}^{20} b_\nu X_{\nu 9} = (2\ 3)(4\ 5)\psi_2^{[3,2]}$$
$$= 3\left[b_2 + b_{10} + b_{14} + 2b_3 - 2b_7 - 2b_8 + b_6\right.$$
$$\left. + 2b_{17} - b_{15} - b_{16} + b_{20}\right].$$

The idempotent of the six-dimensional representation $[3, 1^2]$ of S_5 is

$$\mathcal{Y}_4 = \mathcal{Y}_2^{3,1^2]} = \mathcal{Y}\begin{array}{|c|c|c|}\hline 1 & 2 & 4 \\\hline 3 \\\cline{1-1} 5 \\\cline{1-1}\end{array}.$$

From Eq. (6.21) we obtain the expansions of the basis states $\psi_\mu^{[3,1^2]}$ of $[3, 1^2]$ in the extending space:

$$\psi_2^{[3,1^2]} = \sum_{\nu=1}^{20} b_\nu X_{\nu 11} = \mathcal{Y}_4\mathcal{Y}_1$$
$$= [E + (1\ 2) + (1\ 4) + (2\ 4) + (1\ 2\ 4) + (1\ 4\ 2)]$$
$$\times [E - (1\ 5) - (3\ 5)][E - (1\ 3)]b_1$$
$$= 2b_1 + b_2 - 2b_9 - b_{10} - 2b_{13} - b_{14} + b_1 - b_2 - b_{11} + b_{12}$$
$$- b_{13} + b_{14} + 2b_3 + b_4 - 2b_9 - b_{10} - 2b_{17} - b_{18} + 2b_5 + b_6$$
$$- b_{15} + b_{16} - 2b_{19} - b_{20} + b_3 - b_4 - b_{15} + b_{16} - b_{17} + b_{18}$$
$$+ b_5 - b_6 - b_{11} + b_{12} - b_{19} + b_{20}$$
$$= 3b_1 + 3b_3 + 3b_5 - 4b_9 - 2b_{10} - 2b_{11} + 2b_{12} - 3b_{13} - 2b_{15} + 2b_{16}$$
$$- 3b_{17} - 3b_{19},$$

$$\psi_1^{[3,1^2]} = \sum_{\nu=1}^{20} b_\nu X_{\nu 10} = (3\ 4)\psi_2^{[4,1]}$$
$$= 3b_7 - 3b_3 - 3b_4 + 3b_6 + 4b_{17} + 2b_{18} + 2b_{19} - 2b_{20} - 3b_{13} + 2b_{15}$$
$$+ 4b_{16} + 3b_9 + 3b_{10} - 3b_{12},$$

$$\psi_3^{[3,1^2]} = \sum_{\nu=1}^{20} b_\nu X_{\nu 12} = (4\ 5)\psi_1^{[3,1^2]}$$
$$= 3b_1 + 3b_9 + 3b_{11} - 4b_3 - 2b_4 - 2b_5 + 2b_6 - 3b_7 - 4b_{15} - 2b_{16}$$
$$+ 3b_{17} + 3b_{18} - 3b_{20},$$

$$\psi_4^{[3,1^2]} = \sum_{\nu=1}^{20} b_\nu X_{\nu 13} = (2\ 3)\psi_1^{[3,1^2]}$$
$$= 3b_2 + 3b_4 + 3b_8 - 2b_9 - 4b_{10} - 2b_{14} + 2b_{13} - 3b_{12} - 2b_{18} + 2b_{17}$$
$$- 3b_{16} - 3b_{19},$$

$$\psi_5^{[3,1^2]} = \sum_{\nu=1}^{20} b_\nu X_{\nu 14} = (2\ 3)(4\ 5)\psi_1^{[3,1^2]}$$
$$= 3b_2 + 3b_{10} + 3b_{14} - 2b_3 - 4b_4 - 2b_8 + 2b_7 - 3b_6 - 2b_{17} - 4b_{18}$$
$$+ 3b_{15} + 3b_{16} - 3b_{20},$$

$$\psi_6^{[3,1^2]} = \sum_{\nu=1}^{20} b_\nu X_{\nu 15} = (2\ 4\ 5\ 3)\psi_1^{[3,1^2]} = -(5\ 2)\psi_1^{[3,1^2]}$$
$$= -3b_{11} - 3b_{15} - 3b_5 + 2b_9 - 2b_{10} + 2b_1 - 2b_2 + 3b_{14} + 2b_3 - 2b_4$$
$$+ 3b_{18} + 3b_8.$$

The idempotent of the five-dimensional representation $[2^2, 1]$ of S_5 is

$$\mathcal{Y}_5 = \mathcal{Y}_1^{[2^2,1]} = \mathcal{Y}\begin{array}{|c|c|} \hline 1 & 2 \\ \hline 3 & 4 \\ \hline 5 \\ \cline{1-1} \end{array}.$$

From Eq. (6.21) we obtain the expansions of the basis states $\psi_\mu^{[2^2,1]}$ of $[2^2, 1]$ in the extending space:

$$\psi_1^{[2^2,1]} = \sum_{\nu=1}^{20} b_\nu X_{\nu 16} = \mathcal{Y}_5 \mathcal{Y}_1$$

$$= [E + (1\ 2)]\,[E + (3\ 4)]\,[E - (2\ 4)]$$
$$\times [E - (1\ 5) - (3\ 5)]\,[E - (1\ 3)]\,b_1$$
$$= [E + (1\ 2)]\,[E + (3\ 4)]\,[E - (2\ 4)]$$
$$\times [2b_1 + b_2 - 2b_9 - b_{10} - 2b_{13} - b_{14}]$$
$$= 2b_1 + b_2 - 2b_9 - b_{10} - 2b_{13} - b_{14} + b_1 - b_2 - b_{11} + b_{12}$$
$$\quad - b_{13} + b_{14} + 2b_7 + b_8 + 2b_{17} + b_{18} - 2b_{13} - b_{14} - 2b_5 - b_6$$
$$\quad + b_{15} - b_{16} + 2b_{19} + b_{20} + b_7 - b_8 + b_{19} - b_{20} - b_{13} + b_{14}$$
$$\quad - b_3 + b_4 + b_{15} - b_{16} + b_{17} - b_{18} - b_5 - 2b_6 - b_{15} - 2b_{16}$$
$$\quad + b_{11} + 2b_{12} + b_3 + 2b_4 - b_{15} - 2b_{16} - b_9 - 2b_{10}$$
$$= 3\,[b_1 + b_4 - b_5 - b_6 + b_7 - b_9 - b_{10} + b_{12}$$
$$\quad - 2b_{13} - 2b_{16} + b_{17} + b_{19}]\,,$$

$$\psi_3^{[2^2,1]} = \sum_{\nu=1}^{20} b_\nu X_{\nu 17} = (4\ 5)\psi_1^{[2^2,1]}$$
$$= 3\,[b_1 + b_{10} - b_{11} - b_{12} + b_{13} - b_3 - b_4 + b_6$$
$$\quad - 2b_7 + 2b_{15} + 2b_{16} - b_{17} - b_{18} + b_{20}]\,,$$

$$\psi_3^{[2^2,1]} = \sum_{\nu=1}^{20} b_\nu X_{\nu 18} = (2\ 3)\psi_1^{[2^2,1]}$$
$$= 3\,[b_2 + b_3 - b_7 - b_8 + b_6 - b_9 - b_{10} + b_{13}$$
$$\quad - 2b_{12} - 2b_{17} + b_{16} + b_{19}]\,,$$

$$\psi_4^{[2^2,1]} = \sum_{\nu=1}^{20} b_\nu X_{\nu 19} = (2\ 3)(4\ 5)\psi_1^{[2^2,1]}$$
$$= 3\,[b_2 + b_9 - b_{13} - b_{14} + b_{12} - b_3 - b_4 + b_7$$
$$\quad - 2b_6 + 2b_{17} + 2b_{18} - b_{15} - b_{16} + b_{20}]\,,$$

$$\psi_5^{[2^2,1]} = \sum_{\nu=1}^{20} b_\nu X_{\nu 20} = (2\ 4\ 5\ 3)\psi_1^{[2^2,1]} = (2\ 4)(5\ 3)\psi_1^{[2^2,1]}$$
$$= 3\,[b_{19} - b_{17} - b_{18} - b_{13} - b_{14} + b_8 + b_{15} + b_{11}$$
$$\quad - 2b_5 - 2b_9 + b_3 + b_1]\,.$$

Chapter 7

LIE GROUPS AND LIE ALGEBRAS

7.1 Classification of Semisimple Lie Algebras

★ An element R in a Lie group G can be described by a set of independent real parameters $(r_1, r_2, \ldots)$. Those parameters vary continuously in a given Euclidean region, called the group space. The number g of the parameters, which is equal to the dimension of the group space, is called the order of the group G. In this chapter we only discuss the Lie group whose group space is simply-connected.

A Lie group is said to be compact if its group space is a closed region in a Euclidean space. A Lie group is said to be simple if it does not contain any non-trivial invariant Lie subgroup. A Lie group is said to be semisimple if it does not contain any Abelian invariant Lie subgroup including the whole group. Any Lie group of order one is a simple Lie group, but not a semisimple Lie group. A simple Lie group whose order is more than one must be a semisimple Lie group.

★ The generators I_A in a representation $D(R)$ of a Lie group G are defined as

$$D(R) = \mathbf{1} - i \sum_A r_A I_A + \ldots, \qquad I_A = i \left. \frac{\partial D(R)}{\partial r_A} \right|_{R=E}. \qquad (7.1)$$

The generators I_A in any representation $D(R)$ of G satisfy the common commutation relation:

$$[I_A,\ I_B] = i \sum_D C_{AB}{}^D I_D, \qquad (7.2)$$

where the real parameters $C_{AB}{}^D$, independent of the representation, are called the structure constants of the Lie group.

269

★ The real linear space spanned by the bases $(-iI_A)$ is called the real Lie algebra if the product of two vectors X and Y in the space is defined by the Lie product:

$$X = \sum_A (-iI_A)\, x_A, \qquad Y = \sum_B (-iI_B)\, y_B,$$

$$[X,\, Y] = -\sum_{AB} x_A y_B\, [I_A,\, I_B] = \sum_D (-iI_D)\left(\sum_{AB} x_A y_B C_{AB}^{\ \ D}\right). \tag{7.3}$$

If the linear space is complex, the algebra is called the complex Lie algebra, or briefly, the Lie algebra. A real Lie algebra of a compact Lie group is compact. A Lie algebra of a simple or a semisimple Lie group is simple or semisimple, respectively. In fact, a simple Lie algebra does not contain any non-trivial ideal, and a semisimple Lie algebra does not contain any Abelian ideal, including the whole algebra. Therefore, a one-dimensional Lie algebra is simple, but not semisimple. In the rest of this chapter, if without special indication, "a simple Lie algebra with the dimension larger than one" will be simply called as "a simple Lie algebra" for simplicity. It is proved that a semisimple Lie algebra can be decomposed into the direct sum of some simple Lie algebras.

★ The Killing form g_{AB} of a Lie group or its Lie algebra is defined from its structure constants $C_{AB}^{\ \ D}$:

$$g_{AB} = \sum_{PQ} C_{AP}^{\ \ Q} C_{BQ}^{\ \ P} = -\operatorname{Tr}\left(I_A^{\mathrm{ad}} I_B^{\mathrm{ad}}\right). \tag{7.4}$$

g_{AB} is a real symmetric matrix of dimension g. When the parameters of a given Lie group are changed, the generators transform like a vector, and the structure constants as well as the Killing form transform like a tensor:

$$r_A' = \sum_B r_B \left(X^{-1}\right)_{BA}, \qquad \det X \neq 0,$$

$$I_A' = \sum_B X_{AB} I_B,$$

$$C_{AB}'^{\ \ D} = \sum_{A'B'D'} X_{AA'} X_{BB'} C_{A'B'}^{\ \ \ D'} \left(X^{-1}\right)_{D'D}, \tag{7.5}$$

$$g_{AB}' = \sum_{A'B'} X_{AA'} X_{BB'} g_{A'B'} = \sum_{A'B'} X_{AA'} g_{A'B'} \left(X^T\right)_{B'B}.$$

★ The Cartan Criteria says that a Lie algebra is semisimple if and only if its Killing form is non-singular, and a real semisimple Lie algebra is compact if and only if its Killing form is negative definite.

★ If $\{H_j,\ 1 \leq j \leq \ell\}$ is the largest set of the mutually commutable generators in a simple Lie algebra $\mathcal{L}$, the set constitutes the Cartan subalgebra and ℓ is called the rank of the Lie algebra $\mathcal{L}$. It is proved that the group parameters in a simple Lie algebra $\mathcal{L}$ can be chosen such that the generators H_j and the remaining generators E_α satisfy

$$[H_j,\ H_k] = 0, \qquad 1 \leq j \leq \ell,$$

$$[H_j,\ E_\alpha] = \alpha_j E_\alpha,$$

$$[E_\alpha,\ E_\beta] = \begin{cases} N_{\alpha,\beta} E_{\alpha+\beta} & \text{when } \vec{\alpha} + \vec{\beta} \text{ is a root,} \\ \sum_j \alpha_j H_j & \text{when } \vec{\alpha} + \vec{\beta} = 0, \\ 0 & \text{the remaining cases,} \end{cases} \tag{7.6}$$

and the Killing form becomes

$$g_{jk} = -\delta_{jk}, \qquad g_{\alpha\beta} = -\delta_{(-\alpha)\beta}, \qquad g_{j\alpha} = g_{\alpha j} = 0. \tag{7.7}$$

The real vector $\vec{\alpha}$ in the ℓ-dimensional space is called the root in $\mathcal{L}$, and the space is called the root space. The space spanned by the roots in a simple Lie algebra of rank ℓ is ℓ-dimensional. The roots $\pm\vec{\alpha}$ appear in pair. Except for the pair of roots, any two roots are not collinear. There is a one-to-one correspondence between the generator E_α and the root $\vec{\alpha}$. H_j and E_α are the Cartan-Weyl bases for the generators in the simple Lie algebra $\mathcal{L}$.

★ In a semisimple Lie algebra, we can constitute a root chain from a root $\vec{\alpha}$ by adding or subtracting another root $\vec{\beta}$ several times. Without loss of generality, if $\vec{\alpha} + n\vec{\beta}$, $-q_{\alpha\beta} \leq n \leq p_{\alpha\beta}$, are the roots, and $\vec{\alpha} + (p_{\alpha\beta} + 1)\vec{\beta}$ and $\vec{\alpha} - (q_{\alpha\beta} + 1)\vec{\beta}$ are not the roots, where $p_{\alpha\beta}$ and $q_{\alpha\beta}$ are both non-negative integer, it is proved that

$$\Gamma(\vec{\alpha}/\vec{\beta}) = \frac{\vec{\alpha} \cdot \vec{\beta}}{d_\beta} = q_{\alpha\beta} - p_{\alpha\beta}, \qquad d_\beta = \frac{1}{2}\vec{\beta} \cdot \vec{\beta},$$

$$N_{\alpha,\beta} = -N_{-\alpha,-\beta} = [p_{\alpha\beta}(q_{\alpha\beta} + 1)d_\beta]^{1/2}. \tag{7.8}$$

When $\vec{\alpha} = \vec{\beta}$, the root chain consists of $\vec{\alpha}$, 0 and $-\vec{\alpha}$.

★ A root is called positive if its first nonvanishing component is positive, otherwise it is negative. Sometimes, H_j is called the generator corresponding to the zero root, which is ℓ degeneracy. A positive root is called a simple root if it cannot be expressed as the non-negative integral combination of other positive roots. Therefore, any positive root is equal to a non-negative integral combination of the simple roots, and any negative root is equal to

a non-positive integral combination of the simple roots. The difference of
two simple roots is not a root. From Eq. (7.8), the simple roots are linearly
independent. Thus, the number of the simple roots in a ℓ-rank simple Lie
algebra is equal to ℓ. The angle between two simple roots has to be equal
to $5\pi/6$, $3\pi/4$, $2\pi/3$ or $\pi/2$, and correspondingly, the ratio of the length
squares of the two simple roots is 3, 2, 1, or no restriction, respectively.

★ There are four series of the classical simple Lie algebras (A_ℓ, B_ℓ, C_ℓ
and D_ℓ) and five exceptional simple Lie algebras (G_2, F_4, E_6, E_7 and E_8),
where the Lie group of A_ℓ is $SU(\ell+1)$, the Lie group of B_ℓ is $SO(2\ell+1)$,
the Lie group of C_ℓ is $Sp(2\ell)$, and the Lie group of D_ℓ is $SO(2\ell)$.

The Dynkin diagram for a simple Lie algebra is drawn by the following
rule. There are one or two different lengths among the simple roots in any
simple Lie algebra. Denote each longer simple root by a white circle, and
denote each shorter simple root, if it exists, by a black circle. Two circles
denoting two simple roots are connected by a single link, a double link, or
a triple link depending upon their angle to be $2\pi/3$, $3\pi/4$ or $5\pi/6$, respectively. Two circles are not connected if two simple roots are orthogonal and
the ratio of their lengths is not restricted. The Dynkin diagrams for the
simple Lie algebras are listed in Fig. 7.1.

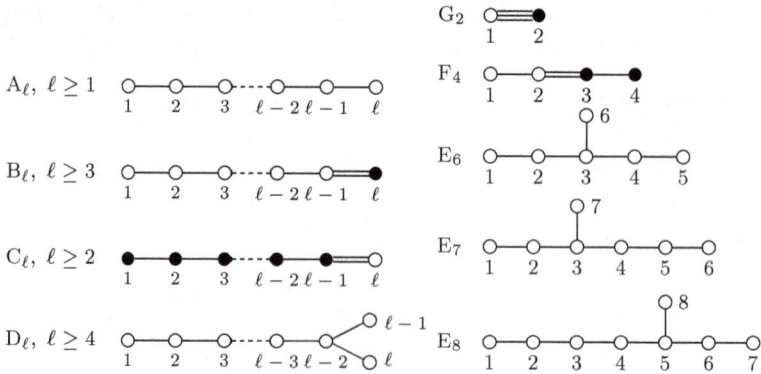

Fig. 7.1   The Dynkin diagrams of the simple Lie algebras.

★ The simple Lie algebra can also be described by the Cartan matrix.
The Cartan matrix for a simple Lie algebra with ℓ simple roots $\mathbf{r}_\mu$ is an
ℓ-dimensional matrix A:

$$A_{\mu\nu} = \Gamma\left(\mathbf{r}_\nu/\mathbf{r}_\mu\right) = \frac{2\mathbf{r}_\nu \cdot \mathbf{r}_\mu}{\mathbf{r}_\mu \cdot \mathbf{r}_\mu} = d_\mu^{-1}\left(\mathbf{r}_\nu \cdot \mathbf{r}_\mu\right). \tag{7.9}$$

The diagonal element of A is always equal to 2, and its non-diagonal element can be equal to 0, -1, -2 or -3. In comparison with the Dynkin diagram, if two simple roots $\mathbf{r}_\mu$ and $\mathbf{r}_\nu$ are disconnected, $A_{\mu\nu} = A_{\nu\mu} = 0$, and if they are connected by a single, double or triple link, and the length of $\mathbf{r}_\mu$ is not less than the length of $\mathbf{r}_\nu$, $A_{\mu\nu} = -1$ and $A_{\nu\mu} = -1$, -2 or -3, respectively.

The Dynkin diagram or the Cartan matrix gives all the information of the simple Lie algebra, such as the angles and the lengths of the simple roots as well as all roots. The roots can be calculated recursively as follows. Any positive root can be expressed as a sum of the simple roots:

$$\vec{\alpha} = \sum_{\nu=1}^{\ell} C_\nu \mathbf{r}_\nu, \qquad C_\nu \geq 0, \tag{7.10}$$

$\sum_\nu C_\nu$ is called the level of the root $\vec{\alpha}$. Any simple root is a root of level one. If all positive roots with the level less than n have been found, we want to judge whether $\vec{\alpha} + \mathbf{r}_\mu$ is a root where $\vec{\alpha}$ is a root of level $(n-1)$. Calculate

$$q - p = \Gamma\left(\vec{\alpha}/\mathbf{r}_\mu\right) = \sum_{\nu=1}^{\ell} A_{\mu\nu} C_\nu, \quad \text{if } \vec{\alpha} = \sum_{\nu=1}^{\ell} C_\nu \mathbf{r}_\nu. \tag{7.11}$$

Since q is known, one can calculate p from Eq. (7.11) to determine whether $\vec{\alpha} + \mathbf{r}_\mu$ is a root or not.

1. Prove the number $p_{\alpha\beta} + q_{\alpha\beta} + 1$ of roots in a root chain $\vec{\alpha} + n\vec{\beta}$, $-q_{\alpha\beta} \leq n \leq p_{\alpha\beta}$, in a simple Lie algebra is less than five.

Solution. Prove this problem by reduction to absurdity. If the length of the root chain is larger than four, one can redefine the root $\vec{\alpha}$ such that the following five vectors are all non-zero roots:

$$\vec{\alpha} - 2\vec{\beta}, \quad \vec{\alpha} - \vec{\beta}, \quad \vec{\alpha}, \quad \vec{\alpha} + \vec{\beta}, \quad \vec{\alpha} + 2\vec{\beta}.$$

However, the following four vectors are not roots:

$$(\vec{\alpha} \pm 2\vec{\beta}) + \vec{\alpha} = 2(\vec{\alpha} \pm \vec{\beta}), \qquad (\vec{\alpha} \pm 2\vec{\beta}) - \vec{\alpha} = \pm 2\vec{\beta}.$$

Thus,

$$0 = \Gamma[(\vec{\alpha} \pm 2\vec{\beta})/\vec{\alpha}] = d_\alpha^{-1}(\vec{\alpha} \pm 2\vec{\beta}) \cdot \vec{\alpha} = 2 \pm 2d_\alpha^{-1}\left(\vec{\beta} \cdot \vec{\alpha}\right).$$

Adding two equations leads to contradiction. This completes the proof.

Since $q_{\alpha\beta} + p_{\alpha\beta} + 1 \leq 4$, both $q_{\alpha\beta}$ and $p_{\alpha\beta}$ are less than four. Thus, $|\Gamma(\vec{\alpha}/\vec{\beta})| \leq 3$.

2. Prove that the difference of two simple roots is not a root, and there are ℓ simple roots in a simple Lie algebra of rank ℓ.

Solution. We first prove the difference of two simple roots $\mathbf{r}_\mu$ and $\mathbf{r}_\nu$ not to be a root by reduction to absurdity. Suppose $\mathbf{r}_\mu - \mathbf{r}_\nu = \vec{\alpha}$. If $\vec{\alpha}$ is a positive root, $\mathbf{r}_\mu = \mathbf{r}_\nu + \vec{\alpha}$ is the sum of two positive roos. If $\vec{\alpha}$ is a negative root, $\mathbf{r}_\nu = \mathbf{r}_\mu + (-\vec{\alpha})$ is the sum of two positive roos. It contradicts with the definition of a simple root. Thus,

$$\Gamma(\mathbf{r}_\mu/\mathbf{r}_\nu) = -p_{\mu\nu} \leq 0.$$

The angle of two simple roots is not an acute angle.

It is obvious that the number of the simple roots is not less than ℓ, because any root can be expressed as a linear combination of the simple roots. If the simple roots are linearly independent to each other, the number of the simple roots must be ℓ. Now, we prove it by reduction to absurdity. If there is a linear relation among simple roots, we separate the terms with the negative coefficients from those with the positive ones:

$$\mathbf{a} = \sum_\mu c_\mu \mathbf{r}_\mu = \sum_\nu d_\nu \mathbf{r}_\nu,$$

where c_μ and d_ν are all positive, and the values of μ and ν are different. Since $\mathbf{a} \neq 0$,

$$0 < \mathbf{a}^2 = \sum_{\mu\nu} c_\mu d_\nu \mathbf{r}_\mu \cdot \mathbf{r}_\nu \leq 0.$$

This completes the proof.

3. Prove that the angle between two simple roots has to be equal to $5\pi/6$, $3\pi/4$, $2\pi/3$ or $\pi/2$, and correspondingly, the ratio of the length squares of the two simple roots is 3, 2, 1, or no restriction, respectively.

Solution. Calculate the square of cosine of the angle between two simple roots:

$$4\cos^2\theta = \frac{4\left(\mathbf{r}_\mu \cdot \mathbf{r}_\nu\right)^2}{(\mathbf{r}_\mu \cdot \mathbf{r}_\mu)(\mathbf{r}_\nu \cdot \mathbf{r}_\nu)} = \Gamma(\mathbf{r}_\mu/\mathbf{r}_\nu)\Gamma(\mathbf{r}_\nu/\mathbf{r}_\mu) = \text{an integer}.$$

Since $\mathbf{r}_\mu$ and $\mathbf{r}_\nu$ is not collinear, $|\cos^2\theta| < 1$, and the integer can only be equal to 3, 2, 1 or 0. Namely, the angle θ can be $5\pi/6$, $3\pi/4$, $2\pi/3$ or $\pi/2$. Without loss of generality, let the length of $\mathbf{r}_\mu$ be not less than that of $\mathbf{r}_\nu$,

$$-\Gamma(\mathbf{r}_\mu/\mathbf{r}_\nu) \geq -\Gamma(\mathbf{r}_\nu/\mathbf{r}_\mu).$$

Due to the restriction for their product, $\Gamma(\mathbf{r}_\nu/\mathbf{r}_\mu)$ has to be equal to -1 or 0, and $\Gamma(\mathbf{r}_\mu/\mathbf{r}_\nu)$ is equal to -3, -2, -1 or 0. The ratio of those two number is

$$\frac{\mathbf{r}_\mu^2}{\mathbf{r}_\nu^2} = \frac{\Gamma(\mathbf{r}_\mu/\mathbf{r}_\nu)}{\Gamma(\mathbf{r}_\nu/\mathbf{r}_\mu)} = 3, \quad 2, \quad 1, \quad \text{or no restriction},$$

depending on the angle of two simple roots to be $5\pi/6$, $3\pi/4$, $2\pi/3$ or $\pi/2$.

4. Calculate the Cartan matrix of the Lie algebra E_6.

Solution. The Cartan matrix of E_6 is a six-dimensional matrix with the diagonal elements to be 2. According to the enumeration for the simple roots in its Dynkin diagram, the row-column numbers of the positions of the non-diagonal elements, which are -1, are 12, 23, 34, 36 and 45, or vice versa,

$$A = \begin{pmatrix} 2 & -1 & 0 & 0 & 0 & 0 \\ -1 & 2 & -1 & 0 & 0 & 0 \\ 0 & -1 & 2 & -1 & 0 & -1 \\ 0 & 0 & -1 & 2 & -1 & 0 \\ 0 & 0 & 0 & -1 & 2 & 0 \\ 0 & 0 & -1 & 0 & 0 & 2 \end{pmatrix}.$$

5. Draw the Dynkin diagram of a simple Lie algebra where its Cartan matrix is as follows, and indicate the enumeration for the simple roots.

$$A = \begin{pmatrix} 2 & -1 & 0 & 0 \\ -1 & 2 & -1 & 0 \\ 0 & -2 & 2 & -1 \\ 0 & 0 & -1 & 2 \end{pmatrix}, \qquad A_{\mu\nu} = \Gamma[\mathbf{r}_\nu/\mathbf{r}_\mu] = \frac{2\mathbf{r}_\nu \cdot \mathbf{r}_\mu}{\mathbf{r}_\mu \cdot \mathbf{r}_\mu}.$$

Solution. The Dynkin diagram consists of four circles. Since $A_{32} = -2$, the circle designated by 2 is white corresponding to a longer simple root, and the circle 3 is black corresponding to a shorter simple root. The circles 2 and 3 are connected by a double link. Since $A_{12} = A_{21} = A_{34} = A_{43} = -1$, the circles 1 and 2 are connected by a single link, so be the circles 3 and 4. Thus, the circle 1 is white and the circle 4 is black.

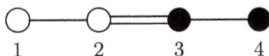

6. Calculate all positive roots in the C_2 Lie algebra.

Solution. There are two simple roots in C_2

$$\mathbf{r}_1 = \sqrt{1/2}\,(\mathbf{e}_1 - \mathbf{e}_2), \qquad \mathbf{r}_2 = \sqrt{2}\mathbf{e}_2,$$

where $\mathbf{r}_2$ is the longer simple root, and $\mathbf{r}_1$ is the shorter. Their angle is $3\pi/4$. Usually, the length square of the longer root is taken to be 2. The Cartan matrix of C_2 is

$$A = \begin{pmatrix} 2 & -2 \\ -1 & 2 \end{pmatrix}.$$

First, we calculate p_1 in the root chain $\mathbf{r}_1 + p_1\mathbf{r}_2$. From Eq. (7.11), we have

$$p_1 = -\Gamma(\mathbf{r}_1/\mathbf{r}_2) = -A_{21} = 1.$$

Thus, $\mathbf{r}_1 + \mathbf{r}_2 = \sqrt{1/2}\,(\mathbf{e}_1 + \mathbf{e}_2)$ is a root, but $\mathbf{r}_1 + 2\mathbf{r}_2$ is not a root. Then, we calculate p_2 in the root chain $\mathbf{r}_2 + p_2\mathbf{r}_1$. Since

$$p_2 = -A_{22} = 2,$$

in addition to $\mathbf{r}_1 + \mathbf{r}_2$, we know $2\mathbf{r}_1 + \mathbf{r}_2 = \sqrt{2}\mathbf{e}_1$ is a root. $3\mathbf{r}_1 + \mathbf{r}_2$ is not a root. $\mathbf{r}_1 + \mathbf{r}_2$ is the only root of level two, and $2\mathbf{r}_1 + \mathbf{r}_2$ is the only root of level three. Since $(2\mathbf{r}_1 + 2\mathbf{r}_2)$ is twice of $\mathbf{r}_1 + \mathbf{r}_2$, it is obviously not a root. Therefore, the C_2 Lie algebra contains four positive roots and four negative roots. The rank of C_2 is 2, and its order is ten. This result can be generalized to the C_ℓ Lie algebra. C_ℓ contains $2\ell^2$ roots: $\pm\sqrt{2}\mathbf{e}_j$, $\pm\sqrt{1/2}(\mathbf{e}_j + \mathbf{e}_k)$ and $\sqrt{1/2}(\mathbf{e}_j - \mathbf{e}_k)$. The rank of C_ℓ is ℓ and its order is $\ell(2\ell + 1)$.

7. Calculate all positive roots in the B_3 Lie algebra.

Solution. There are three simple roots in B_3:

$$\mathbf{r}_1 = \mathbf{e}_1 - \mathbf{e}_2, \qquad \mathbf{r}_2 = \mathbf{e}_2 - \mathbf{e}_3, \qquad \mathbf{r}_3 = \mathbf{e}_3,$$

where $\mathbf{r}_1$ and $\mathbf{r}_2$ are the longer roots with the length square 2, and $\mathbf{r}_3$ is the shorter root. The angle between $\mathbf{r}_1$ and $\mathbf{r}_2$ is $2\pi/3$. The angle between $\mathbf{r}_2$ and $\mathbf{r}_3$ is $3\pi/4$. $\mathbf{r}_1$ is orthogonal to $\mathbf{r}_3$. The Cartan matrix of B_3 is

$$A = \begin{pmatrix} 2 & -1 & 0 \\ -1 & 2 & -1 \\ 0 & -2 & 2 \end{pmatrix}.$$

We first calculate p_1, p_2 and p_3 in the root chains $\mathbf{r}_1 + p_1\mathbf{r}_2$, $\mathbf{r}_1 + p_2\mathbf{r}_3$, $\mathbf{r}_2 + p_3\mathbf{r}_3$. From Eq. (7.11), we have

$$p_1 = -A_{21} = 1, \qquad p_2 = -A_{31} = 0, \qquad p_3 = -A_{32} = 2.$$

Thus, there are two roots of level two: $\mathbf{r}_1 + \mathbf{r}_2$ and $\mathbf{r}_2 + \mathbf{r}_3$. We also know that $\mathbf{r}_2 + 2\mathbf{r}_3$ is a root of level three, and $\mathbf{r}_1 + 2\mathbf{r}_2$ and $\mathbf{r}_2 + 3\mathbf{r}_3$ are not the roots. Then, we calculate p_4, p_5, p_6 and p_7 in the root chains $(\mathbf{r}_1 + \mathbf{r}_2) + p_4\mathbf{r}_1$, $(\mathbf{r}_1 + \mathbf{r}_2) + p_5\mathbf{r}_3$, $(\mathbf{r}_2 + \mathbf{r}_3) + p_6\mathbf{r}_1$ and $(\mathbf{r}_2 + \mathbf{r}_3) + p_7\mathbf{r}_2$. Since

$$p_4 = 1 - A_{11} - A_{12} = 0, \qquad p_5 = -A_{31} - A_{32} = 2,$$
$$p_6 = -A_{12} - A_{13} = 1, \qquad p_7 = 1 - A_{22} - A_{23} = 0,$$

in addition to $\mathbf{r}_2 + 2\mathbf{r}_3$, there is another root of level three: $\mathbf{r}_1 + \mathbf{r}_2 + \mathbf{r}_3$. Besides, $\mathbf{r}_1 + \mathbf{r}_2 + 2\mathbf{r}_3$ is a root of level four. Adding $\mathbf{r}_2$ or $\mathbf{r}_3$ to the root $\mathbf{r}_2 + 2\mathbf{r}_3$ is not a root. We are going to calculate p_8 in the root chain $(\mathbf{r}_1 + \mathbf{r}_2 + \mathbf{r}_3) + p_8\mathbf{r}_2$. Since $\mathbf{r}_1 + \mathbf{r}_3$ is not a root, we have

$$p_8 = -A_{21} - A_{22} - A_{23} = 0.$$

Thus, $\mathbf{r}_1 + \mathbf{r}_2 + 2\mathbf{r}_3$ is the only root of level four. Adding $\mathbf{r}_3$ to it is not a root. We want to calculate p_9 and p_{10} in the root chains $(\mathbf{r}_1 + \mathbf{r}_2 + 2\mathbf{r}_3) + p_9\mathbf{r}_1$ and $(\mathbf{r}_1 + \mathbf{r}_2 + 2\mathbf{r}_3) + p_{10}\mathbf{r}_2$. Since

$$p_9 = 1 - A_{11} - A_{12} - 2A_{13} = 0,$$
$$p_{10} = -A_{21} - A_{22} - 2A_{23} = 1,$$

we obtain the only root of level five: $\mathbf{r}_1 + 2\mathbf{r}_2 + 2\mathbf{r}_3$. Besides, $\mathbf{r}_1 + 3\mathbf{r}_2 + 2\mathbf{r}_3$ is not a root. $2\mathbf{r}_1 + 2\mathbf{r}_2 + 2\mathbf{r}_3$ is twice of a known root, so it is not a root. Now, we are going to calculate p_{11} in the root chain $(\mathbf{r}_1 + 2\mathbf{r}_2 + 2\mathbf{r}_3) + p_{11}\mathbf{r}_3$. Since

$$p_{11} = -A_{31} - 2A_{32} - 2A_{33} = 0,$$

there is no root of level six.

In summary, the B_3 Lie algebra contains three roots of level one (three simple roots: $\mathbf{r}_1 = \mathbf{e}_1 - \mathbf{e}_2$, $\mathbf{r}_2 = \mathbf{e}_2 - \mathbf{e}_3$ and $\mathbf{r}_3 = \mathbf{e}_3$), two roots of level two ($\mathbf{r}_1 + \mathbf{r}_2 = \mathbf{e}_1 - \mathbf{e}_3$ and $\mathbf{r}_2 + \mathbf{r}_3 = \mathbf{e}_2$), two roots of level three ($\mathbf{r}_1 + \mathbf{r}_2 + \mathbf{r}_3 = \mathbf{e}_1$ and $\mathbf{r}_2 + 2\mathbf{r}_3 = \mathbf{e}_2 + \mathbf{e}_3$), one root of level four ($\mathbf{r}_1 + \mathbf{r}_2 + 2\mathbf{r}_3 = \mathbf{e}_1 + \mathbf{e}_3$), and one root of level five ($\mathbf{r}_1 + 2\mathbf{r}_2 + 2\mathbf{r}_3 = \mathbf{e}_1 + \mathbf{e}_2$). The rank of B_3 is 3, and its order is 21. This result can be generalized to the B_ℓ Lie algebra. B_ℓ contains $2\ell^2$ roots: $\pm\mathbf{e}_j$, $\pm(\mathbf{e}_j + \mathbf{e}_k)$ and $(\mathbf{e}_j - \mathbf{e}_k)$. The rank of B_ℓ is ℓ, and its order is $\ell(2\ell + 1)$.

8. Calculate all positive roots in the D_4 Lie algebra.

Solution. There are four simple roots in D_4:

$$\mathbf{r}_1 = \mathbf{e}_1 - \mathbf{e}_2, \qquad \mathbf{r}_2 = \mathbf{e}_2 - \mathbf{e}_3, \qquad \mathbf{r}_3 = \mathbf{e}_3 - \mathbf{e}_4, \qquad \mathbf{r}_4 = \mathbf{e}_3 + \mathbf{e}_4,$$

where the length squares of four simple roots are all 2. The angle between $\mathbf{r}_2$ and any other simple root $\mathbf{r}_a$, $a \neq 2$, is $2\pi/3$, and the other pair of simple roots is orthogonal to each other. The Cartan matrix of D_4 is

$$A = \begin{pmatrix} 2 & -1 & 0 & 0 \\ -1 & 2 & -1 & -1 \\ 0 & -1 & 2 & 0 \\ 0 & -1 & 0 & 2 \end{pmatrix}.$$

We first calculate p_1, p_2, p_3, p_4, p_5 and p_6 in the root chains $\mathbf{r}_1 + p_1\mathbf{r}_2$, $\mathbf{r}_1 + p_2\mathbf{r}_3$, $\mathbf{r}_1 + p_3\mathbf{r}_4$, $\mathbf{r}_2 + p_4\mathbf{r}_3$, $\mathbf{r}_2 + p_5\mathbf{r}_4$, and $\mathbf{r}_3 + p_6\mathbf{r}_4$. From Eq. (7.11) we have

$$p_1 = -A_{21} = 1, \qquad p_2 = -A_{31} = 0, \qquad p_3 = -A_{41} = 0,$$
$$p_4 = -A_{32} = 1, \qquad p_5 = -A_{42} = 1, \qquad p_6 = -A_{43} = 0.$$

There are three roots of level two: $\mathbf{r}_1 + \mathbf{r}_2 = \mathbf{e}_1 - \mathbf{e}_3$, $\mathbf{r}_2 + \mathbf{r}_3 = \mathbf{e}_2 - \mathbf{e}_4$, and $\mathbf{r}_2 + \mathbf{r}_4 = \mathbf{e}_2 + \mathbf{e}_4$. Besides, $\mathbf{r}_1 + 2\mathbf{r}_2$, $\mathbf{r}_2 + 2\mathbf{r}_3$ and $\mathbf{r}_2 + 2\mathbf{r}_4$ are not the roots. Then, we calculate p_7, p_8, p_9, p_{10}, p_{11}, p_{12}, p_{13}, p_{14} and p_{15} in the root chains $(\mathbf{r}_1 + \mathbf{r}_2) + p_7\mathbf{r}_1$, $(\mathbf{r}_1 + \mathbf{r}_2) + p_8\mathbf{r}_3$, $(\mathbf{r}_1 + \mathbf{r}_2) + p_9\mathbf{r}_4$, $(\mathbf{r}_2 + \mathbf{r}_3) + p_{10}\mathbf{r}_1$, $(\mathbf{r}_2 + \mathbf{r}_3) + p_{11}\mathbf{r}_2$, $(\mathbf{r}_2 + \mathbf{r}_3) + p_{12}\mathbf{r}_4$, $(\mathbf{r}_2 + \mathbf{r}_4) + p_{13}\mathbf{r}_1$, $(\mathbf{r}_2 + \mathbf{r}_4) + p_{14}\mathbf{r}_2$ and $(\mathbf{r}_2 + \mathbf{r}_4) + p_{15}\mathbf{r}_3$. Since

$$p_7 = 1 - A_{11} - A_{12} = 0, \qquad p_8 = -A_{31} - A_{32} = 1,$$
$$p_9 = -A_{41} - A_{42} = 1, \qquad p_{10} = -A_{12} - A_{13} = 1,$$
$$p_{11} = 1 - A_{22} - A_{23} = 0, \qquad p_{12} = -A_{42} - A_{43} = 1,$$
$$p_{13} = -A_{12} - A_{14} = 1, \qquad p_{14} = 1 - A_{22} - A_{24} = 0,$$
$$p_{15} = -A_{32} - A_{34} = 1,$$

we obtain three roots of level three: $\mathbf{r}_1 + \mathbf{r}_2 + \mathbf{r}_3 = \mathbf{e}_1 - \mathbf{e}_4$, $\mathbf{r}_1 + \mathbf{r}_2 + \mathbf{r}_4 = \mathbf{e}_1 + \mathbf{e}_4$, and $\mathbf{r}_2 + \mathbf{r}_3 + \mathbf{r}_4 = \mathbf{e}_2 + \mathbf{e}_3$. Based on them we want to calculate p_{16}, p_{17}, p_{18}, p_{19}, p_{20} and p_{21} in the root chains $(\mathbf{r}_1 + \mathbf{r}_2 + \mathbf{r}_3) + p_{16}\mathbf{r}_2$, $(\mathbf{r}_1 + \mathbf{r}_2 + \mathbf{r}_3) + p_{17}\mathbf{r}_4$, $(\mathbf{r}_1 + \mathbf{r}_2 + \mathbf{r}_4) + p_{18}\mathbf{r}_2$, $(\mathbf{r}_1 + \mathbf{r}_2 + \mathbf{r}_4) + p_{19}\mathbf{r}_3$, $(\mathbf{r}_2 +$

$\mathbf{r}_3 + \mathbf{r}_4) + p_{20}\mathbf{r}_1$ and $(\mathbf{r}_2 + \mathbf{r}_3 + \mathbf{r}_4) + p_{21}\mathbf{r}_2$. Since

$$p_{16} = -A_{21} - A_{22} - A_{23} = 0, \qquad p_{17} = A_{41} - A_{42} - A_{43} = 1,$$
$$p_{18} = -A_{21} - A_{22} - A_{24} = 0, \qquad p_{19} = A_{31} - A_{32} - A_{34} = 1,$$
$$p_{20} = -A_{12} - A_{13} - A_{14} = 1, \qquad p_{21} = -A_{22} - A_{23} - A_{24} = 0,$$

there is only one root of level four $\mathbf{r}_1 + \mathbf{r}_2 + \mathbf{r}_3 + \mathbf{r}_4 = \mathbf{e}_1 + \mathbf{e}_3$. Adding $\mathbf{r}_1$, $\mathbf{r}_3$ or $\mathbf{r}_4$ to the root of level four is not a root. We calculate p_{22} in the root chain $(\mathbf{r}_1 + \mathbf{r}_2 + \mathbf{r}_3 + \mathbf{r}_4) + p_{22}\mathbf{r}_2$. Since

$$p_{22} = -A_{21} - A_{22} - A_{23} - A_{24} = 1,$$

we obtain the only root of level five: $\mathbf{r}_1 + 2\mathbf{r}_2 + \mathbf{r}_3 + \mathbf{r}_4 = \mathbf{e}_1 + \mathbf{e}_2$. $\mathbf{r}_1 + 3\mathbf{r}_2 + \mathbf{r}_3 + \mathbf{r}_4$ is not a root. It is easy to know those vectors by adding $\mathbf{r}_1$, $\mathbf{r}_3$ or $\mathbf{r}_4$ to the root of level five is not a root, because

$$p_{23} = -A_{11} - 2A_{12} - A_{13} - A_{14} = 0,$$
$$p_{24} = -A_{31} - 2A_{32} - A_{33} - A_{34} = 0,$$
$$p_{25} = -A_{41} - 2A_{42} - A_{43} - A_{44} = 0.$$

Therefore, the D_4 Lie algebra contains four roots of level one, three roots of level two, three roots of level three, one root of level four, and one root of level five. The rank of D_4 is 4, and its order is 28. This result can be generalized to the D_ℓ Lie algebra. D_ℓ contains $2\ell(\ell - 1)$ roots: $\pm(\mathbf{e}_j + \mathbf{e}_k)$ and $(\mathbf{e}_j - \mathbf{e}_k)$. The rank of D_ℓ is ℓ, and its order is $\ell(2\ell - 1)$.

7.2 Irreducible Representations and the Chevalley Bases

★ Discuss an irreducible representation of a simple Lie algebra where the representation matrices of H_j are diagonal:

$$H_j|\mathbf{m}\rangle = m_j|\mathbf{m}\rangle, \tag{7.12}$$

where $|\mathbf{m}\rangle$ is the basis state in the representation space, and the vector $\mathbf{m}$ in an ℓ-dimensional space is called the weight of $|\mathbf{m}\rangle$. The ℓ-dimensional space is called the weight space. A weight $\mathbf{m}$ is said to be multiple if the number of the basis states $|\mathbf{m}\rangle$ satisfying Eq. (7.12) in the representation space is larger than one. Otherwise, it is a single weight. From Eq. (7.6) we have

$$H_j\left(E_\alpha|\mathbf{m}\rangle\right) = \left(m_j + \alpha_j\right)\left(E_\alpha|\mathbf{m}\rangle\right).$$

Therefore, E_α is called the raising operator and $E_{-\alpha}$ the lowering operator, if $\vec{\alpha}$ is a positive root.

★ In a given irreducible representation, one can construct a weight chain from a given weight **m** and a root $\vec{\alpha}$:

$$\mathbf{m} + n\vec{\alpha}, \quad -q \leq n \leq p, \quad \text{are weights}, \quad p \geq 0, \quad q \geq 0,$$
$$\mathbf{m} + (p+1)\vec{\alpha} \quad \text{and} \quad \mathbf{m} - (q+1)\vec{\alpha} \quad \text{are not the weights}. \tag{7.13}$$

It can be proved that

$$\Gamma\left(\mathbf{m}/\vec{\alpha}\right) = q - p = \text{integer}. \tag{7.14}$$

The weight $\mathbf{m}' = \mathbf{m} - \vec{\alpha}\Gamma\left(\mathbf{m}/\vec{\alpha}\right)$ is related to the weight **m** by a reflection with respect to the plane through the origin and orthogonal to the root $\vec{\alpha}$. **m** and $\mathbf{m}'$ have the same multiplicity and are called the equivalent weights. This reflection is called the Weyl reflection. All the mutually equivalent weights constitute the Weyl orbit, and their number is called the Weyl orbital size.

★ The Chevalley bases H_μ, E_μ and F_μ, $1 \leq \mu \leq \ell$, are another set of bases for the generators in a simple Lie algebra, which can be expressed as the linear combinations of the Cartan-Weyl bases H_j and E_α:

$$H_\mu = d_\mu^{-1} \sum_{j=1}^{\ell} (\mathbf{r}_\mu)_j H_j, \quad E_\mu = d_\mu^{-1/2} E_{\mathbf{r}_\mu}, \quad F_\mu = d_\mu^{-1/2} E_{-\mathbf{r}_\mu}. \tag{7.15}$$

They satisfy the following commutation relations:

$$[H_\mu, \ H_\nu] = 0, \qquad [H_\mu, \ E_\nu] = A_{\mu\nu} E_\nu,$$
$$[H_\mu, \ F_\nu] = -A_{\mu\nu} E_\nu, \qquad [E_\mu, \ F_\nu] = \delta_{\mu\nu} H_\mu, \tag{7.16}$$

and the Serre relations:

$$\underbrace{[E_\mu, \ [E_\mu, \ \cdots \ [E_\mu, \ E_\nu] \ \cdots \ ]]}_{1 - A_{\mu\nu}} = 0 \ ,$$
$$\underbrace{[F_\mu, \ [F_\mu, \ \cdots \ [F_\mu, \ F_\nu] \ \cdots \ ]]}_{1 - A_{\mu\nu}} = 0 \ . \tag{7.17}$$

The generator E_α corresponding to a non-simple root $\vec{\alpha}$ can be calculated by the commutator (7.6). The main merit of the Chevalley bases is that three Chevalley bases H_μ, E_μ and F_μ with the same subscript μ span a subalgebra, denoted by $\mathcal{A}_\mu$, which is isomorphic onto the Lie algebra A_1

(the SU(2) group). In fact, they satisfy the same commutation relations as those satisfied by the generators $2T_3$, T_+ and T_- in A_1:

$$[H_\mu, E_\mu] = 2E_\mu, \qquad [H_\mu, F_\mu] = -2F_\mu, \qquad [E_\mu, F_\mu] = H_\mu.$$

★ In the weight space, define a set of new bases $\mathbf{w}_\mu$, called the fundamental dominant weight, satisfying

$$\Gamma\left(\mathbf{w}_\mu/\mathbf{r}_\nu\right) = d_\nu^{-1}\left(\mathbf{w}_\mu \cdot \mathbf{r}_\nu\right) = \left(2\mathbf{w}_\mu \cdot \mathbf{r}_\nu\right)/\left(\mathbf{r}_\nu \cdot \mathbf{r}_\nu\right) = \delta_{\mu\nu}. \qquad (7.18)$$

In comparison with Eq. (7.9), we can relate the fundamental dominant weight $\mathbf{w}_\mu$ with the simple root $\mathbf{r}_\nu$ by the Cartan matrix

$$\mathbf{r}_\nu = \sum_{\mu=1}^{\ell} \mathbf{w}_\mu A_{\mu\nu}, \qquad \mathbf{w}_\mu = \sum_{\nu=1}^{\ell} \mathbf{r}_\nu (A^{-1})_{\nu\mu}. \qquad (7.19)$$

In the bases $\mathbf{w}_\mu$, all components of the roots and the weights are integers:

$$\mathbf{m} = \sum_{\mu=1}^{\ell} \mathbf{w}_\mu m_\mu, \qquad m_\mu = \Gamma(\mathbf{m}/\mathbf{r}_\mu),$$
$$H_\mu|\mathbf{m}\rangle = m_\mu|\mathbf{m}\rangle, \qquad H_\mu\left(E_\nu|\mathbf{m}\rangle\right) = \left(m_\mu + A_{\mu\nu}\right)\left(E_\nu|\mathbf{m}\rangle\right). \qquad (7.20)$$

The weight $\mathbf{m}'$ equivalent to $\mathbf{m}$ with respect to a simple root $\mathbf{r}_\mu$ can be expressed as

$$\mathbf{m}' = \mathbf{m} - \mathbf{r}_\mu\Gamma\left(\mathbf{m}/\mathbf{r}_\mu\right) = \mathbf{m} - m_\mu\mathbf{r}_\mu. \qquad (7.21)$$

★ Rewrite the weight chain in Eq. (7.13) with respect to a simple root $\mathbf{r}_\mu$ where $\mathbf{m} + p\mathbf{r}_\mu$ is redefined as $\mathbf{m}$:

$$\mathbf{m}, \quad \mathbf{m} - \mathbf{r}_\mu, \quad \dots, \mathbf{m} - q\mathbf{r}_\mu, \qquad q = \Gamma(\mathbf{m}/\mathbf{r}_\mu) = m_\mu. \qquad (7.22)$$

The states corresponding to the weight chain constitute a $(q+1)$-multiplet of the subalgebra A_μ. The basis states in the irreducible representation space constitute some multiplets of the different subalgebras A_μ. If all weights in the weight chain (7.22) are single, the representation matrix entry of F_μ for the multiplet is

$$F_\mu|\mathbf{m} - n\mathbf{r}_\mu\rangle = \sqrt{(q-n)(n+1)}\,|\mathbf{m} - (n+1)\mathbf{r}_\mu\rangle, \qquad 0 \leq n \leq q. \qquad (7.23)$$

The representation matrix of E_μ is the transpose of that of F_μ. If the multiple weight exists, the state in the multiplet of A_μ is a suitable combination of the basis states, calculated by the commutators (7.16) and the orthonormal condition.

★ For an irreducible representation with a finite dimension, there is a state $|\mathbf{M}\rangle$ with the highest weight $\mathbf{M}$:

$$H_\mu|\mathbf{M}\rangle = M_\mu|\mathbf{M}\rangle, \qquad E_\mu|\mathbf{M}\rangle = 0. \tag{7.24}$$

The highest weight $\mathbf{M}$ in an irreducible representation must be a single weight. Usually, an irreducible representation can be denoted by its highest weight $\mathbf{M}$. A representation is called fundamental if its highest weight is the fundamental dominant weight. Any basis state $|\mathbf{m}\rangle$ with the weight $\mathbf{m}$ in the irreducible representation space can be obtained from the highest weight state by the actions of the lowering operators. Thus,

$$\mathbf{m} = \mathbf{M} - \sum_\mu c_\mu \mathbf{r}_\mu, \qquad C_\mathbf{m} = \sum_\mu c_\mu. \tag{7.25}$$

$C_\mathbf{m}$ is called the height of the basis state $|\mathbf{m}\rangle$. The height of the highest weight state is zero.

★ Two irreducible representations are conjugate with each other if the lowest weight of one representation multiplied by -1 is equal to the highest weight of another representation. The irreducible representations is self-conjugate if its lowest weight multiplied by -1 is equal to its highest weight.

★ A weight is called a dominant weight if its components are non-negative integers. Due to Eq. (7.14), the highest weight must be a dominant weight, and any dominant weight can be the highest weight in one irreducible representation. However, for a given irreducible representation, there may exist a few dominant weights with multiplicities, among them only one dominant weight is the highest weight of this representation. The dimension of an irreducible representation is equal to the sum of the multiplicities of the dominant weights multiplied by their Weyl orbital sizes. Therefore, the dominant weights play a very important role in the calculation for the basis states and the representation matrices of the generators.

★ The method of the block weight diagram is effective for calculating the representation matrices of the generators in an irreducible representation. In a block weight diagram for an irreducible representation, each basis state corresponds to a block filled with its weight. For convenience, the negative component of the weight is denoted by a digit with a bar over it: $\overline{m} = -m$. The blocks corresponding to the multiple weight are enumerated. The block for the highest weight is located on the first row. The remaining blocks are arranged downwards as the height increases. The blocks for the weights with the same height are put in the same row.

Beginning with the highest weight, from each positive component $m_\nu > 0$, we draw m_μ blocks downwards. They constitute a weight chain of length $(m_\mu + 1)$ with respect to the simple root $\mathbf{r}_\mu$. The states corresponding to the weight chain constitute a multiplet of $\mathcal{A}_\mu$. Two neighbored blocks are related by a link denoting the application of the lowering operator F_μ. Usually, in the block weight diagram we connect two blocks by different links if their basis states are related by the different lowering operators. It is better to indicate the corresponding representation matrix entry of the lowering operator behind the link. Usually, the matrix entry is neglected when it is equal to one.

If all weights in a weight chain are single, the representation matrix entries of the lowering operator F_μ between two basis states corresponding to the neighbored weights can be calculated by Eq. (7.23). If a multiple weight appears in the weight chain, the state in the multiplet may be a combination of the basis states. The combination as well as the representation matrix entries of F_μ can be determined by Eq. (7.16) and the orthonormal condition.

When a dominant weight appears, we determine its multiplicity which is equal to the number of paths along which the dominant weight is obtained from the highest weight. If the representation matrix entry in a path is calculated to be zero, the multiplicity decreases. All weights equivalent to the dominant weight has the same multiplicity as the dominant weight has. Continue this method until all components of the weight are not positive. In the completed block weight diagram, the number of blocks with the same height first increases and then decreases as the height increases, symmetric up and down like a spindle.

★ For a two-rank Lie algebra the weights can be drawn in a planar diagram, called the planar weight diagram. The planar weight diagram of an irreducible representation is the inversion of that of its conjugate representation with respect to the origin. The planar weight diagram of a self-conjugate irreducible representations is symmetric with respect to the inversion. Note that the simple roots as well as the fundamental dominant weights are not required to be along the coordinate axes.

★ The representation in whose space the basis is the generator of the Lie algebra is the adjoint representation. Therefore, the root coincides with the weight in the adjoint representation, and the root space coincides with the weight space. To draw the block weight diagram or the planar weight diagram for the adjoint representation is another method for calculating all roots in a simple Lie algebra.

9. Draw the block weight diagrams and the planar weight diagrams of the representations $(1,0)$, $(0,1)$, $(1,1)$, $(3,0)$, and $(0,3)$ of the A_2 Lie algebra (the SU(3) group).

Solution. The Cartan matrix A and its inverse A^{-1} of A_2 are

$$A = \begin{pmatrix} 2 & -1 \\ -1 & 2 \end{pmatrix}, \qquad A^{-1} = \frac{1}{3}\begin{pmatrix} 2 & 1 \\ 1 & 2 \end{pmatrix}.$$

The relations between the simple roots and the fundamental dominant weights are

$$\mathbf{r}_1 = 2\mathbf{w}_1 - \mathbf{w}_2, \qquad\qquad \mathbf{r}_2 = -\mathbf{w}_1 + 2\mathbf{w}_2,$$

$$\mathbf{w}_1 = (1/3)\,(2\mathbf{r}_1 + \mathbf{r}_2), \qquad \mathbf{w}_2 = (1/3)\,(\mathbf{r}_1 + 2\mathbf{r}_2).$$

The Weyl orbits of the dominant weights $(1,0)$, $(0,1)$, $(1,1)$, $(3,0)$ and $(0,3)$ respectively are:

$$(1,0) \xrightarrow{\mathbf{r}_1} (\bar{1},1) \xrightarrow{\mathbf{r}_2} (0,\bar{1}),$$

$$(0,1) \xrightarrow{\mathbf{r}_2} (1,\bar{1}) \xrightarrow{\mathbf{r}_1} (\bar{1},0),$$

$$(1,1) \begin{array}{c} \xrightarrow{\mathbf{r}_1} (\bar{1},2) \xrightarrow{\mathbf{r}_2} (1,\bar{2}) \xrightarrow{\mathbf{r}_1} \\ \xrightarrow{\mathbf{r}_2} (2,\bar{1}) \xrightarrow{\mathbf{r}_1} (\bar{2},1) \xrightarrow{\mathbf{r}_2} \end{array} (\bar{1},\bar{1}),$$

$$(3,0) \xrightarrow{\mathbf{r}_1} (\bar{3},3) \xrightarrow{\mathbf{r}_2} (0,\bar{3}),$$

$$(0,3) \xrightarrow{\mathbf{r}_2} (3,\bar{3}) \xrightarrow{\mathbf{r}_1} (\bar{3},0).$$

The block weight diagrams of the representations $(1,0)$, $(0,1)$, $(1,1)$, $(3,0)$, and $(0,3)$ are given in Fig. 7.2. In the diagrams, the single link means the action of F_1 and the double link means that of F_2.

First, we discuss the fundamental representation $(1,0)$. From the highest weight $(1,0)$ we obtain a doublet of $\mathcal{A}_1$ with $(\bar{1},1)$. Then, from the weight $(\bar{1},1)$ we obtain a doublet of $\mathcal{A}_2$ with $(0,\bar{1})$. The weight $(0,\bar{1})$ contains no positive component. It is the lowest weight in the representation $(1,0)$. The dimension of the representation $(1,0)$ is three, where three weights are equivalent to each other. The representation matrix entries of the lowering operators are

$$F_1|(1,0),(1,0)\rangle = |(1,0),(\bar{1},1)\rangle, \qquad F_2|(1,0),(\bar{1},1)\rangle = |(1,0),(0,\bar{1})\rangle.$$

Second, we discuss the fundamental representation $(0,1)$. From the highest weight $(0,1)$ we obtain a doublet of $\mathcal{A}_2$ with $(1,\bar{1})$. Then, from the weight $(1,\bar{1})$ we obtain a doublet of $\mathcal{A}_1$ with $(\bar{1},0)$. The weight $(\bar{1},0)$ contains no positive component. It is the lowest weight in the representation $(0,1)$. The dimension of the representation $(0,1)$ is three, where three weights are equivalent to each other. The representation matrix entries of the lowering operators are

$$F_2|(0,1),(0,1)\rangle = |(0,1),(1,\bar{1})\rangle, \qquad F_1|(0,1),(1,\bar{1})\rangle = |(0,1),(\bar{1},0)\rangle.$$

It is worthy to notice that the block weight diagram of $(0,1)$ can be obtained from that of $(1,0)$ by turning upside down and changing the signs of all weights. In fact, the lowest weight $(0,\bar{1})$ in the representation $(1,0)$, times by -1, is the highest weight in the representation $(0,1)$. Two representations are conjugate to each other.

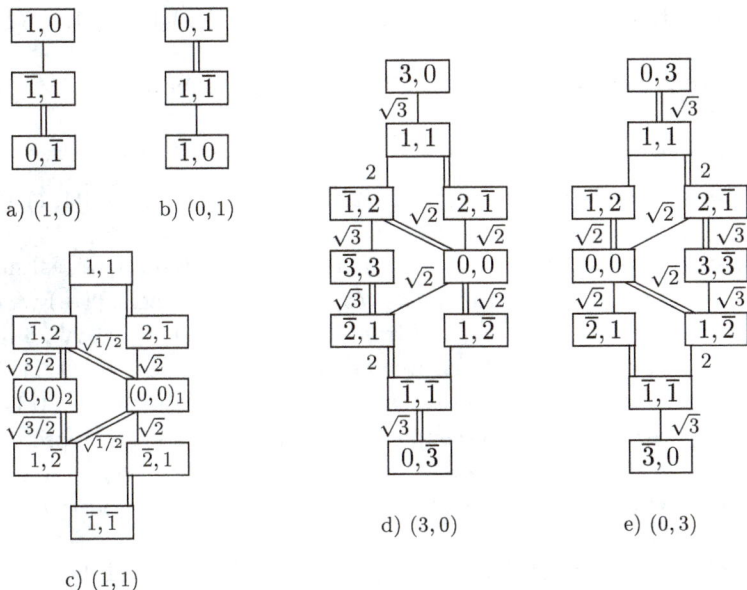

Fig. 7.2   The block weight diagrams of some representations of $\mathcal{A}_2$.

Third, we discuss the adjoint representation $(1,1)$ of A_2. From the highest weight $(1,1)$ we obtain a doublet of A_1 with $(\bar{1},2)$ and a doublet of A_2 with $(2,\bar{1})$. Then, from the weight $(2,\bar{1})$ we obtain a triplet of A_1 with $(0,0)$ and $(\bar{2},1)$, and from the weight $(\bar{1},2)$ we obtain a triplet of A_2 with $(0,0)$ and $(1,\bar{2})$. There are two paths to the dominant weight $(0,0)$, which may be a double weight. We define two orthonormal basis states $(0,0)_1$ and $(0,0)_2$, where $(0,0)_1$ belongs to the triplet of A_1 with $(2,\bar{1})$ and $(\bar{2},1)$, and $(0,0)_2$ is a singlet of A_1. The state $(0,0)$ in the triplet of A_2 with $(\bar{1},2)$ and $(1,\bar{2})$ has to be a combination of $(0,0)_1$ and $(0,0)_2$:

$$F_2|(1,1),(\bar{1},2)\rangle = a|(1,1),(0,0)_1\rangle + b|(1,1),(0,0)_2\rangle, \qquad a^2 + b^2 = 2.$$

Applying $E_1 F_2 = F_2 E_1$ to $|(1,1),(\bar{1},2)\rangle$, we have

$$E_1 F_2|(1,1),(\bar{1},2)\rangle = a\sqrt{2}|(1,1),(2,\bar{1})\rangle$$
$$= F_2 E_1|(1,1),(\bar{1},2)\rangle = F_2|(1,1),(1,1)\rangle = |(1,1),(2,\bar{1})\rangle.$$

Thus, $a = \sqrt{1/2}$. Choosing the phase of $|(1,1),(0,0)_2\rangle$ such that b is a positive real number, we have $b = \sqrt{2 - 1/2} = \sqrt{3/2}$. Let

$$F_2|(1,1),(0,0)_1\rangle = c|(1,1),(1,\bar{2})\rangle, \quad F_2|(1,1),(0,0)_2\rangle = d|(1,1),(1,\bar{2})\rangle.$$

Applying $E_2 F_2 = F_2 E_2 + H_2$ to $|(1,1),(0,0)_1\rangle$, we have

$$E_2 F_2|(1,1),(0,0)_1\rangle = c^2|(1,1),(0,0)_1\rangle + cd|(1,1),(0,0)_2\rangle$$
$$= (F_2 E_2 + H_2)|(1,1),(0,0)_1\rangle = (1/2)|(1,1),(0,0)_1\rangle + \sqrt{3/4}|(1,1),(0,0)_2\rangle.$$

Choosing the phase of $|(1,1),(1,\bar{2})\rangle$ such that c is a positive real number, we have $c = \sqrt{1/2}$, and then, $d = \sqrt{3/2}$. From the weight $(1,\bar{2})$ we obtain a doublet of A_1 with $(\bar{1},\bar{1})$, and from the weight $(\bar{2},1)$ we obtain a doublet of A_2 with $(\bar{1},\bar{1})$. Since the weight $(\bar{1},\bar{1})$ is equivalent to $(1,1)$, it is a single weight and is the lowest weight in the representation $(1,1)$. The representation $(1,1)$ is self-conjugate. The dimension of the representation $(1,1)$ is eight containing six equivalent single weights and one double weight $(0,0)$. The representation matrix entries of the lowering operators are

$$F_1|(1,1),(1,1)\rangle = |(1,1),(\bar{1},2)\rangle, \qquad F_2|(1,1),(1,1)\rangle = |(1,1),(2,\bar{1})\rangle,$$
$$F_1|(1,1),(2,\bar{1})\rangle = \sqrt{2}|(1,1),(0,0)_1\rangle, \quad F_1|(1,1),(0,0)_1\rangle = \sqrt{2}|(1,1),(\bar{2},1)\rangle,$$

$$F_2|(1,1),(\overline{1},2)\rangle = \sqrt{1/2}|(1,1),(0,0)_1\rangle + \sqrt{3/2}|(1,1),(0,0)_2\rangle,$$
$$F_2|(1,1),(0,0)_1\rangle = \sqrt{1/2}|(1,1),(1,\overline{2})\rangle,$$
$$F_2|(1,1),(0,0)_2\rangle = \sqrt{3/2}|(1,1),(1,\overline{2})\rangle,$$
$$F_1|(1,1),(1,\overline{2})\rangle = |(1,1),(\overline{1},\overline{1})\rangle, \qquad F_2|(1,1),(\overline{2},1)\rangle = |(1,1),(\overline{1},\overline{1})\rangle.$$

We see from the block weight diagram of the adjoint representation $(1,1)$ that the A_2 Lie algebra contains three positive roots:

$$2\mathbf{w}_1 - \mathbf{w}_2 = \mathbf{r}_1, \qquad -\mathbf{w}_1 + 2\mathbf{w}_2 = \mathbf{r}_2, \qquad \mathbf{w}_1 + \mathbf{w}_2 = \mathbf{r}_1 + \mathbf{r}_2.$$

Fourth, we discuss the representation $(3,0)$. From the highest weight $(3,0)$ we obtain a quartet of A_1 with $(1,1)$, $(\overline{1},2)$ and $(\overline{3},3)$, where the representation matrix entries of F_1 are $\sqrt{3}$, 2, and $\sqrt{3}$, respectively. The weight $(1,1)$ is a dominant weight, which is a single weight because there is only one path to $(1,1)$ lowered from the highest weight. All weights equivalent to $(3,0)$ or $(1,1)$ are single, too. Then, from the weight $(1,1)$ we obtain a doublet of A_2 with $(2,\overline{1})$. From the weight $(\overline{1},2)$ we obtain a triplet of A_2 with $(0,0)$ and $(1,\overline{2})$, and from the weight $(2,\overline{1})$ we obtain a triplet of A_1 with $(0,0)$ and $(\overline{2},1)$. There are two paths to the dominant weight $(0,0)$, which may be a double weight. We define two orthonormal basis states $(0,0)_1$ and $(0,0)_2$, where $(0,0)_1$ belongs to the triplet of A_1 with $(2,\overline{1})$ and $(\overline{2},1)$, and $(0,0)_2$ is a singlet of A_1. The state $(0,0)$ belonging the triplet of A_2 with $(\overline{1},2)$ and $(1,\overline{2})$ has to be a combination of $(0,0)_1$ and $(0,0)_2$:

$$F_2|(3,0),(\overline{1},2)\rangle = a|(3,0),(0,0)_1\rangle + b|(3,0),(0,0)_2\rangle, \qquad a^2 + b^2 = 2.$$

Applying $E_1 F_2 = F_2 E_1$ to $|(3,0),(\overline{1},2)\rangle$, we have

$$E_1 F_2|(3,0),(\overline{1},2)\rangle = a\sqrt{2}|(3,0),(2,\overline{1})\rangle$$
$$= F_2 E_1|(3,0),(\overline{1},2)\rangle = 2F_2|(3,0),(1,1)\rangle = 2|(3,0),(2,\overline{1})\rangle.$$

Thus, $a = \sqrt{2}$ and $b = 0$. It means that the weight $(0,0)$ is single. The basis state $|(3,0),(0,0)_2\rangle$ does not exist, and $|(3,0),(0,0)_1\rangle$ can be denoted by $|(3,0),(0,0)\rangle$.

From the weight $(1,\overline{2})$ we obtain a doublet of A_1 with $(\overline{1},\overline{1})$. From the weight $(\overline{3},3)$ we obtain a quartet of A_2 with $(\overline{2},1)$, $(\overline{1},\overline{1})$, and $(0,\overline{3})$, where the representation matrix entries of F_2 are $\sqrt{3}$, 2, and $\sqrt{3}$, respectively. The weights $(\overline{2},1)$ and $(\overline{1},\overline{1})$ are single because they are equivalent to $(1,1)$. Thus, the state $(\overline{2},1)$ coincides with the state in the triplet of A_1 with $(2,\overline{1})$ and $(0,0)$, and the state $(\overline{1},\overline{1})$ coincides with the state in the doublet of $\mathcal{A}_1$ with $(1,\overline{2})$. The block weight diagram of the representation $(3,0)$ is complete. The dimension of the representation $(3,0)$ is ten, containing three single dominant weights $(3,0)$, $(1,1)$ and $(0,0)$, with the Weyl orbital sizes 3, 6 and 0, respectively. The representation matrix entries of the lowering operators are

$$F_1|(3,0),(3,0)\rangle = \sqrt{3}|(3,0),(1,1)\rangle, \quad F_1|(3,0),(1,1)\rangle = 2|(3,0),(\overline{1},2)\rangle,$$
$$F_1|(3,0),(\overline{1},2)\rangle = \sqrt{3}|(3,0),(\overline{3},3)\rangle, \quad F_2|(3,0),(1,1)\rangle = |(3,0),(2,\overline{1})\rangle,$$
$$F_2|(3,0),(\overline{1},2)\rangle = \sqrt{2}|(3,0),(0,0)\rangle, \quad F_2|(3,0),(0,0)\rangle = \sqrt{2}|(3,0),(1,\overline{2})\rangle,$$
$$F_1|(3,0),(2,\overline{1})\rangle = \sqrt{2}|(3,0),(0,0)\rangle, \quad F_1|(3,0),(0,0)\rangle = \sqrt{2}|(3,0),(\overline{2},1)\rangle,$$
$$F_1|(3,0),(1,\overline{2})\rangle = |(3,0),(\overline{1},\overline{1})\rangle, \quad F_2|(3,0),(\overline{3},3)\rangle = \sqrt{3}|(3,0),(\overline{2},1)\rangle,$$
$$F_2|(3,0),(\overline{2},1)\rangle = 2|(3,0),(\overline{1},\overline{1})\rangle, \quad F_2|(3,0),(\overline{1},\overline{1})\rangle = \sqrt{3}|(3,0),(0,\overline{3})\rangle.$$

The lowest weight $(0,\overline{3})$ in the representation $(3,0)$, multiplied by -1, is the highest weight of the representation $(0,3)$. Therefore, the representation $(0,3)$ is conjugate with the representation $(3,0)$, and its block weight diagram is upside down of that of $(3,0)$, where all weights change their signs.

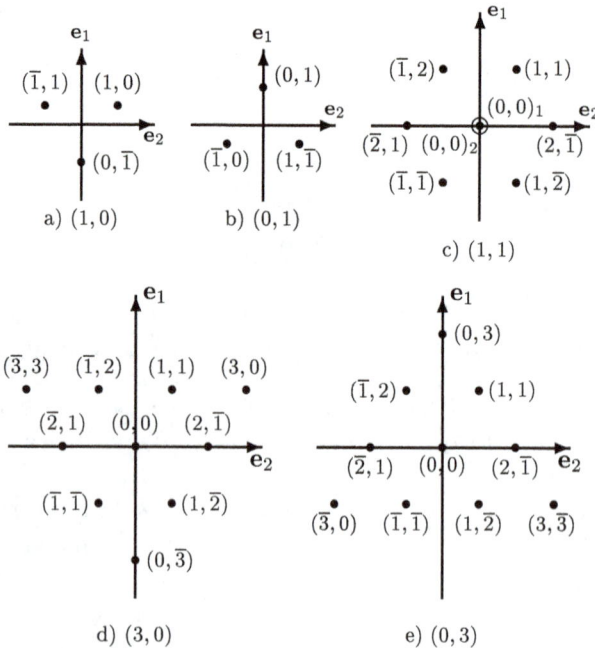

Fig. 7.3   The planar weight diagrams of some representations of A_2.

In the rectangular coordinate frame, the simple roots and the fundamental dominant weights of A_2 are

$$\mathbf{r}_1 = \sqrt{2}\mathbf{e}_2, \qquad\qquad \mathbf{r}_2 = \sqrt{1/2}\left(\sqrt{3}\mathbf{e}_1 - \mathbf{e}_2\right),$$

$$\mathbf{w}_1 = \sqrt{1/6}\left(\mathbf{e}_1 + \sqrt{3}\mathbf{e}_2\right), \qquad \mathbf{w}_2 = \sqrt{2/3}\mathbf{e}_1.$$

Physically, people prefer to put $\mathbf{e}_1$ along the ordinate axis, and $\mathbf{e}_2$ along the abscissa axis for A_2. The planar weight diagrams for the representations $(1,0)$, $(0,1)$, $(1,1)$, $(3,0)$, and $(0,3)$ are given in Fig. 7.3. Generally, the planar weight diagram for the representation $(0, M)$ of A_2 is a straight regular triangle where all weights are single, and that of $(M, 0)$ is a regular triangle upside down. The planar weight diagram of (M, M) of A_2 is a regular hexagon, where the weights on the boundary are single, and the multiplicity increases as the position of the weight goes to the origin.

10. Draw the block weight diagrams and the planar weight diagrams of two fundamental representations and the adjoint representation $(2, 0)$ of the C_2 Lie algebra.

Solution. The Cartan matrix A and its inverse A^{-1} in C_2 is

$$A = \begin{pmatrix} 2 & -2 \\ -1 & 2 \end{pmatrix}, \qquad A^{-1} = \frac{1}{2}\begin{pmatrix} 2 & 2 \\ 1 & 2 \end{pmatrix}.$$

The relations between the simple roots and the fundamental dominant weights are

$$\mathbf{r}_1 = 2\mathbf{w}_1 - \mathbf{w}_2, \qquad \mathbf{r}_2 = -2\mathbf{w}_1 + 2\mathbf{w}_2,$$

$$\mathbf{w}_1 = \mathbf{r}_1 + \mathbf{r}_2/2, \qquad \mathbf{w}_2 = \mathbf{r}_1 + \mathbf{r}_2.$$

The Weyl orbits of the dominant weights $(1, 0)$, $(0, 1)$, and $(2, 0)$ respectively are:

$$(1,0) \xrightarrow{\mathbf{r}_1} (\overline{1}, 1) \xrightarrow{\mathbf{r}_2} (1, \overline{1}) \xrightarrow{\mathbf{r}_1} (\overline{1}, 0),$$

$$(0,1) \xrightarrow{\mathbf{r}_2} (2, \overline{1}) \xrightarrow{\mathbf{r}_1} (\overline{2}, 1) \xrightarrow{\mathbf{r}_2} (0, \overline{1}),$$

$$(2,0) \xrightarrow{\mathbf{r}_1} (\overline{2}, 2) \xrightarrow{\mathbf{r}_2} (2, \overline{2}) \xrightarrow{\mathbf{r}_1} (\overline{2}, 0).$$

Their Weyl orbital sizes are all 4. The block weight diagrams of the representations $(1, 0)$, $(0, 1)$, and $(2, 0)$ are given in Fig. 7.4. In the diagrams, the single link means the action of F_1 and the double link means that of F_2.

$$
\begin{array}{ccc}
\boxed{1,0} & \boxed{0,1} & \boxed{2,0} \\
\quad 1\Big| & \quad 1\Big\| & \sqrt{2}\Big| \\
\boxed{\bar{1},1} & \boxed{2,\bar{1}} & \boxed{0,1} \\
\quad 1\Big\| & \sqrt{2}\Big\| & \sqrt{2}\Big| \qquad 1\Big\| \\
\boxed{1,\bar{1}} & \boxed{0,0} & \boxed{\bar{2},2}\ {}_1\ \boxed{2,\bar{1}} \\
\quad 1\Big| & \sqrt{2}\Big\| & 1\Big| \qquad \sqrt{2}\Big\| \\
\boxed{\bar{1},0} & \boxed{\bar{2},1} & \boxed{(0,0)_2}\ {}_1\ \boxed{(0,0)_1} \\
 & \quad 1\Big\| & 1\Big| \qquad \sqrt{2}\Big\| \\
 & \boxed{0,\bar{1}} & \boxed{2,\bar{2}} \quad \boxed{\bar{2},1} \\
 & & \sqrt{2}\Big| \qquad 1\Big\| \\
 & & \boxed{0,\bar{1}} \\
 & & \sqrt{2}\Big| \\
 & & \boxed{\bar{2},0}
\end{array}
$$

a) (1,0)   b) (0,1)   c) (2,0)

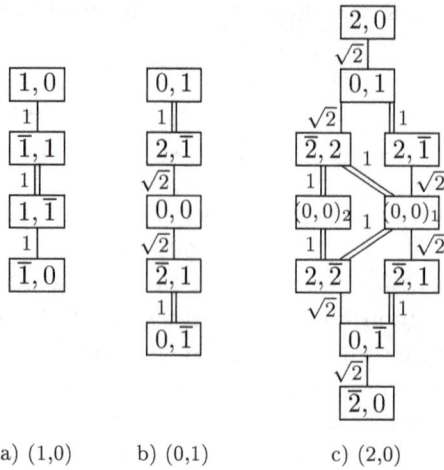

Fig. 7.4   The block weight diagrams of some representations of C_2.

First, we discuss the fundamental representation $(1,0)$. From the highest weight $(1,0)$ we obtain a doublet of $\mathcal{A}_1$ with $(\bar{1},1)$. Then, from the weight $(\bar{1},1)$ we obtain a doublet of $\mathcal{A}_2$ with $(1,\bar{1})$. From the weight $(1,\bar{1})$ we obtain a doublet of $\mathcal{A}_1$ with $(\bar{1},0)$, which contains no positive component. The dimension of the representation $(1,0)$ is four, where four weights are equivalent to each other. The representation matrix entries of the lowering operators are

$$F_1|(1,0),(1,0)\rangle = |(1,0),(\bar{1},1)\rangle, \quad F_2|(1,0),(\bar{1},1)\rangle = |(1,0),(1,\bar{1})\rangle,$$
$$F_1|(1,0),(1,\bar{1})\rangle = |(1,0),(\bar{1},0)\rangle.$$

Second, we discuss the fundamental representation $(0,1)$. From the highest weight $(0,1)$ we obtain a doublet of $\mathcal{A}_2$ with $(2,\bar{1})$. Then, from the weight $(2,\bar{1})$ we obtain a triplet of $\mathcal{A}_1$ with $(0,0)$ and $(\bar{2},1)$. The representation matrix entries of F_1 in the triplet are both $\sqrt{2}$. The weight $(0,0)$ is a dominant weight. There is only one path to $(0,0)$ lowered from the highest weight, so $(0,0)$ is single. From the weight $(\bar{2},1)$ we obtain a doublet of $\mathcal{A}_2$ with $(0,\bar{1})$, which contains no positive component. The five-dimensional representation $(0,1)$ contains two single dominant weights $(0,1)$ and $(0,0)$ with the Weyl orbital sizes four and one, respectively. The representation matrix entries of the lowering operators are

$$F_2|(0,1),(0,1)\rangle = |(0,1),(2,\bar{1})\rangle, \qquad F_1|(0,1),(2,\bar{1})\rangle = \sqrt{2}|(0,1),(0,0)\rangle,$$
$$F_1|(0,1),(0,0)\rangle = \sqrt{2}|(0,1),(\bar{2},1)\rangle, \quad F_2|(0,1),(\bar{2},1)\rangle = |(0,1),(0,\bar{1})\rangle.$$

At last, we discuss the adjoint representation $(2,0)$. From the highest weight $(2,0)$ we obtain a triplet of $\mathcal{A}_1$ with $(0,1)$ and $(\bar{2},2)$, where the representation matrix entries of F_1 are both $\sqrt{2}$. The dominant weight $(0,1)$ is single because there is only one path to $(0,1)$ lowered from the highest weight. From the weight $(0,1)$ we obtain a doublet of $\mathcal{A}_2$ with $(2,\bar{1})$. From the weight $(\bar{2},2)$ we obtain a triplet of $\mathcal{A}_2$ with $(0,0)$ and $(2,\bar{2})$, and from the weight $(2,\bar{1})$ we obtain a triplet of $\mathcal{A}_1$ with $(0,0)$ and $(\bar{2},1)$. There are two paths to the dominant weight $(0,0)$, which may be a double weight. We define two orthonormal basis states $(0,0)_1$ and $(0,0)_2$, where $(0,0)_1$ belongs to the triplet of $\mathcal{A}_1$ with $(2,\bar{1})$ and $(\bar{2},1)$, and $(0,0)_2$ is a singlet of $\mathcal{A}_1$. The state $(0,0)$ in the triplet of $\mathcal{A}_2$ with $(\bar{2},2)$ and $(2,\bar{2})$ has to be a combination of $(0,0)_1$ and $(0,0)_2$:

$$F_2|(2,0),(\bar{2},2)\rangle = a|(2,0),(0,0)_1\rangle + b|(2,0),(0,0)_2\rangle, \qquad a^2+b^2=2,$$

$$F_2|(2,0),(0,0)_1\rangle = c|(2,0),(2,\bar{2})\rangle, \quad F_2|(2,0),(0,0)_2\rangle = d|(2,0),(2,\bar{2})\rangle.$$

Applying $E_1 F_2 = F_2 E_1$ to $|(2,0),(\bar{2},2)\rangle$, we have

$$E_1 F_2|(2,0),(\bar{2},2)\rangle = a\sqrt{2}|(2,0),(2,\bar{1})\rangle$$

$$= F_2 E_1|(2,0),(\bar{2},2)\rangle = \sqrt{2}F_2|(2,0),(0,1)\rangle = \sqrt{2}|(2,0),(2,\bar{1})\rangle.$$

Thus, $a=1$. Choosing the phase of $|(2,0),(0,0)_2\rangle$ such that b is a positive real number, we have $b=\sqrt{2-1^2}=1$. Applying $E_2 F_2 = F_2 E_2 + H_2$ to $|(2,0),(0,0)_1\rangle$, we have

$$E_2 F_2|(2,0),(0,0)_1\rangle = c^2|(2,0),(0,0)_1\rangle + cd|(2,0),(0,0)_2\rangle$$

$$= (F_2 E_2 + H_2)|(2,0),(0,0)_1\rangle = |(2,0),(0,0)_1\rangle + |(2,0),(0,0)_2\rangle.$$

Choosing the phase of $|(2,0),(2,\bar{2})\rangle$ such that c is a positive real number, we have $c=1$, and then, $d=1$. All weights in the representation $(2,0)$ are single except for $(0,0)$. From the weight $(2,\bar{2})$ we obtain a triplet of $\mathcal{A}_1$ with $(0,\bar{1})$ and $(\bar{2},0)$, where the representation matrix entries of F_1 are both $\sqrt{2}$. From the weight $(\bar{2},1)$ we obtain a doublet of $\mathcal{A}_2$ with $(0,\bar{1})$. Since the weight $(0,\bar{1})$ is single, it also has to belong to the triplet of $\mathcal{A}_1$ with $(2,\bar{2})$ and $(\bar{2},0)$. Thus, the block weight diagram of the representation $(2,0)$ is complete. The representation $(2,0)$ contains two single dominant weights $(2,0)$ and $(0,1)$, and one double dominant weight $(0,0)$. The dimension of the representation $(2,0)$ is $1 \times 4 + 1 \times 4 + 2 \times 1 = 10$. The representation matrix entries of the lowering operators are

$$F_1|(2,0),(2,0)\rangle = \sqrt{2}|(2,0),(0,1)\rangle, \quad F_1|(2,0),(0,1)\rangle = \sqrt{2}|(2,0),(\bar{2},2)\rangle,$$

$$F_2|(2,0),(0,1)\rangle = |(2,0),(2,\bar{1})\rangle, \quad F_1|(2,0),(2,\bar{1})\rangle = \sqrt{2}|(2,0),(0,0)_1\rangle,$$

$$F_2|(2,0),(\bar{2},2)\rangle = |(2,0),(0,0)_1\rangle + |(2,0),(0,0)_2\rangle,$$

$$F_2|(2,0),(0,0)_1\rangle = |(2,0),(2,\bar{2})\rangle, \quad F_2|(2,0),(0,0)_2\rangle = |(2,0),(2,\bar{2})\rangle,$$

$$F_1|(2,0),(0,0)_1\rangle = \sqrt{2}|(2,0),(\bar{2},1)\rangle, \quad F_1|(2,0),(2,\bar{2})\rangle = \sqrt{2}|(2,0),(0,\bar{1})\rangle,$$

$$F_2|(2,0),(\bar{2},1)\rangle = |(2,0),(0,\bar{1})\rangle \quad F_1|(2,0),(0,\bar{1})\rangle = \sqrt{2}|(2,0),(\bar{2},0)\rangle.$$

We see from the block weight diagram of the adjoint representation $(2,0)$ that the C_2 Lie algebra contains four positive roots:

$$2\mathbf{w}_1 - \mathbf{w}_2 = \mathbf{r}_1, \qquad -2\mathbf{w}_1 + 2\mathbf{w}_2 = \mathbf{r}_2,$$
$$\mathbf{w}_2 = \mathbf{r}_1 + \mathbf{r}_2, \qquad 2\mathbf{w}_1 = 2\mathbf{r}_1 + \mathbf{r}_2.$$

In three representations $(1,0)$, $(0,1)$, and $(2,0)$ of C_2, the lowest weights are equal to the highest weight times by -1, respectively. It means that three representations are all self-conjugate.

In the rectangular coordinate frame, the simple roots (see Problem 6) and the fundamental dominant weights of C_2 are

$$\mathbf{r}_1 = \sqrt{1/2}\,(\mathbf{e}_1 - \mathbf{e}_2), \qquad \mathbf{r}_2 = \sqrt{2}\mathbf{e}_2,$$
$$\mathbf{w}_1 = \sqrt{1/2}\mathbf{e}_1, \qquad \mathbf{w}_2 = \sqrt{1/2}\,(\mathbf{e}_1 + \mathbf{e}_2).$$

Thus, we obtain the planar weight diagrams for the representations $(1,0)$, $(0,1)$ and $(2,0)$ as follows.

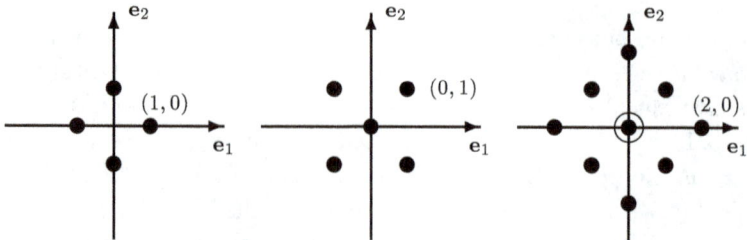

11. Draw the block weight diagrams and the planar weight diagrams of three representations $(0,1)$, $(1,0)$, and $(0,2)$ of the exceptional Lie algebra G_2.

Solution. The Cartan matrix A and its inverse A^{-1} in G_2 are

$$A = \begin{pmatrix} 2 & -1 \\ -3 & 2 \end{pmatrix}, \qquad A^{-1} = \begin{pmatrix} 2 & 1 \\ 3 & 2 \end{pmatrix}.$$

The relations between the simple roots and the fundamental dominant weights are

$$\mathbf{r}_1 = 2\mathbf{w}_1 - 3\mathbf{w}_2, \qquad \mathbf{r}_2 = -\mathbf{w}_1 + 2\mathbf{w}_2,$$

$$\mathbf{w}_1 = 2\mathbf{r}_1 + 3\mathbf{r}_2, \qquad \mathbf{w}_2 = \mathbf{r}_1 + 2\mathbf{r}_2.$$

The Weyl orbits of the dominant weights $(0, 1)$, $(1, 0)$, and $(0, 2)$ respectively are:

$$(0,1) \xrightarrow{\mathbf{r}_2} (1,\bar{1}) \xrightarrow{\mathbf{r}_1} (\bar{1},2) \xrightarrow{\mathbf{r}_2} (1,\bar{2}) \xrightarrow{\mathbf{r}_1} (\bar{1},1) \xrightarrow{\mathbf{r}_2} (0,\bar{1}),$$

$$(1,0) \xrightarrow{\mathbf{r}_1} (\bar{1},3) \xrightarrow{\mathbf{r}_2} (2,\bar{3}) \xrightarrow{\mathbf{r}_1} (\bar{2},3) \xrightarrow{\mathbf{r}_2} (1,\bar{3}) \xrightarrow{\mathbf{r}_1} (\bar{1},0),$$

$$(0,2) \xrightarrow{\mathbf{r}_2} (2,\bar{2}) \xrightarrow{\mathbf{r}_1} (\bar{2},4) \xrightarrow{\mathbf{r}_2} (2,\bar{4}) \xrightarrow{\mathbf{r}_1} (\bar{2},2) \xrightarrow{\mathbf{r}_2} (0,\bar{2}).$$

Their Weyl orbital sizes are all 6.

First, we discuss the fundamental representation $(0, 1)$. From the highest weight $(0, 1)$ we obtain a doublet of $\mathcal{A}_2$ with $(1, \bar{1})$. Then, from the weight $(1, \bar{1})$ we obtain a doublet of $\mathcal{A}_1$ with $(\bar{1}, 2)$. From the weight $(\bar{1}, 2)$ we obtain a triplet of $\mathcal{A}_2$ with $(0, 0)$ and $(1, \bar{2})$. The representation matrix entries of F_2 in the triplet are both $\sqrt{2}$. The weight $(0, 0)$ is a dominant weight. There is only one path to $(0, 0)$ lowered from the highest weight, so $(0, 0)$ is single. From the weight $(1, \bar{2})$ we obtain a doublet of $\mathcal{A}_1$ with $(\bar{1}, 1)$. From the weight $(\bar{1}, 1)$ we obtain a doublet of $\mathcal{A}_2$ with $(0, \bar{1})$, which contains no positive component. The dimension of the representation $(0, 1)$ is seven, where six weights are equivalent to each other and $(0, 0)$ is equivalent only to itself. The representation matrix entries of the lowering operators are

$$F_2|(0,1),(0,1)\rangle = |(0,1),(1,\bar{1})\rangle, \qquad F_1|(0,1),(1,\bar{1})\rangle = |(0,1),(\bar{1},2)\rangle,$$

$$F_2|(0,1),(\bar{1},2)\rangle = \sqrt{2}|(0,1),(0,0)\rangle, \quad F_2|(0,1),(0,0)\rangle = \sqrt{2}|(0,1),(1,\bar{2})\rangle,$$

$$F_1|(0,1),(1,\bar{2})\rangle = |(0,1),(\bar{1},1)\rangle, \qquad F_2|(0,1),(\bar{1},1)\rangle = |(0,1),(0,\bar{1})\rangle.$$

Second, we discuss the fundamental representation $(1, 0)$. From the highest weight $(1, 0)$ we obtain a doublet of $\mathcal{A}_1$ with $(\bar{1}, 3)$. Then, from the weight $(\bar{1}, 3)$ we obtain a quartet of $\mathcal{A}_2$ with $(0, 1)$, $(1, \bar{1})$ and $(2, \bar{3})$. The representation matrix entries of F_1 in the quartet are $\sqrt{3}$, 2, and $\sqrt{3}$, respectively. The weight $(0, 1)$ is a dominant weight. There is only one path to $(0, 1)$ lowered from the highest weight, so $(0, 1)$ is single. All weights equivalent to $(1,0)$ or $(0,1)$ are single, too. From the weight $(1,\bar{1})$ we obtain a doublet of $\mathcal{A}_1$ with $(\bar{1},2)$. From the weight $(2,\bar{3})$ we obtain a triplet of $\mathcal{A}_1$ with $(0,0)$ and $(\bar{2},3)$, and from the weight $(\bar{1},2)$ we obtain a triplet of $\mathcal{A}_2$ with $(0,0)$ and $(1,\bar{2})$. There are two paths to the dominant weight $(0,0)$, which may be a double weight. We define two orthonormal basis states $(0,0)_1$ and $(0,0)_2$, where $(0,0)_1$ belongs to the triplet of $\mathcal{A}_1$ with $(2,\bar{3})$ and $(\bar{2},3)$, and $(0,0)_2$ is a singlet of $\mathcal{A}_1$. The state $(0,0)$ in the triplet of $\mathcal{A}_2$ with $(\bar{1},2)$ and $(1,\bar{2})$ has to be a combination of $(0,0)_1$ and $(0,0)_2$:

$$F_2|(1,0),(\bar{1},2)\rangle = a|(1,0),(0,0)_1\rangle + b|(1,0),(0,0)_2\rangle, \qquad a^2 + b^2 = 2,$$

$$F_2|(1,0),(0,0)_1\rangle = c|(1,0),(1,\bar{2})\rangle, \quad F_2|(1,0),(0,0)_2\rangle = d|(1,0),(1,\bar{2})\rangle.$$

Applying $E_1 F_2 = F_2 E_1$ to $|(1,0),(\bar{1},2)\rangle$, we have

$$E_1 F_2|(1,0),(\bar{1},2)\rangle = a\sqrt{2}|(1,0),(2,\bar{3})\rangle$$

$$= F_2 E_1|(1,0),(\bar{1},2)\rangle = F_2|(1,0),(1,\bar{1})\rangle = \sqrt{3}|(1,0),(2,\bar{3})\rangle.$$

Thus, $a = \sqrt{3/2}$. Choosing the phase of $|(1,0),(0,0)_2\rangle$ such that b is a positive real number, we have $b = \sqrt{2-3/2} = \sqrt{1/2}$. Applying $E_2 F_2 = F_2 E_2 + H_2$ to $|(1,0),(0,0)_1\rangle$, we have

$$E_2 F_2|(1,0),(0,0)_1\rangle = c^2|(1,0),(0,0)_1\rangle + cd|(1,0),(0,0)_2\rangle$$

$$= (F_2 E_2 + H_2)|(1,0),(0,0)_1\rangle = (3/2)|(1,0),(0,0)_1\rangle + \sqrt{3/4}|(1,0),(0,0)_2\rangle.$$

Choosing the phase of $|(1,0),(1,\bar{2})\rangle$ such that c is a positive real number, we have $c = \sqrt{3/2}$, and then, $d = \sqrt{1/2}$. The remaining part of the block weight diagram is easy to calculate. We only list the results in the block diagram and the representation matrix entries of the lowering operators in the following. The representation $(1,0)$ contains two single dominant weights $(1,0)$ and $(0,1)$ with the Weyl orbital size 6 and one double dominant weight $(0,0)$. The dimension of $(1,0)$ is $1 \times 6 + 1 \times 6 + 2 \times 1 = 14$.

$$F_1|(1,0),(1,0)\rangle = |(1,0),(\bar{1},3)\rangle, \qquad F_2|(1,0),(\bar{1},3)\rangle = \sqrt{3}|(1,0),(0,1)\rangle,$$

$$F_2|(1,0),(0,1)\rangle = 2|(1,0),(1,\bar{1})\rangle, \qquad F_2|(1,0),(1,\bar{1})\rangle = \sqrt{3}|(1,0),(2,\bar{3})\rangle,$$

$$F_1|(1,0),(1,\bar{1})\rangle = |(1,0),(\bar{1},2)\rangle, \qquad F_1|(1,0),(2,\bar{3})\rangle = \sqrt{2}|(1,0),(0,0)_1\rangle,$$

$$F_2|(1,0),(\bar{1},2)\rangle = \sqrt{3/2}|(1,0),(0,0)_1\rangle + \sqrt{1/2}|(1,0),(0,0)_2\rangle,$$

$$F_2|(1,0),(0,0)_1\rangle = \sqrt{3/2}|(1,0),(1,\bar{2})\rangle,$$

$$F_2|(1,0),(0,0)_2\rangle = \sqrt{1/2}|(1,0),(1,\bar{2})\rangle,$$

$$F_1|(1,0),(0,0)_1\rangle = \sqrt{2}|(1,0),(\bar{2},3)\rangle, \qquad F_1|(1,0),(1,\bar{2})\rangle = |(1,0),(\bar{1},1)\rangle,$$

$$F_2|(1,0),(\overline{2},3)\rangle = \sqrt{3}|(1,0),(\overline{1},1)\rangle, \quad F_2|(1,0),(\overline{1},1)\rangle = 2|(1,0),(0,\overline{1})\rangle,$$

$$F_2|(1,0),(0,\overline{1})\rangle = \sqrt{3}|(1,0),(1,\overline{3})\rangle, \quad F_1|(1,0),(1,\overline{3})\rangle = |(1,0),(\overline{1},0)\rangle.$$

The representation $(1,0)$ is the adjoint representation of G_2. From the block weight diagram we obtain all positive roots of G_2 as follows:

$$2\mathbf{w}_1 - 3\mathbf{w}_2 = \mathbf{r}_1, \qquad\qquad -\mathbf{w}_1 + 2\mathbf{w}_2 = \mathbf{r}_2,$$

$$\mathbf{w}_1 - \mathbf{w}_2 = \mathbf{r}_1 + \mathbf{r}_2, \qquad\qquad \mathbf{w}_2 = \mathbf{r}_1 + 2\mathbf{r}_2,$$

$$-\mathbf{w}_1 + 3\mathbf{w}_2 = \mathbf{r}_1 + 3\mathbf{r}_2, \qquad \mathbf{w}_1 = 2\mathbf{r}_1 + 3\mathbf{r}_2.$$

At last, we discuss the representation $(0,2)$. From the highest weight $(0,2)$ we obtain a triplet of $\mathcal{A}_2$ with $(1,0)$ and $(2,\overline{2})$, where the representation matrix entries of F_2 are both $\sqrt{2}$. The weight $(1,0)$ is a dominant weight. There is only one path to $(1,0)$ lowered from the highest weight, so $(1,0)$ is single. From the weight $(1,0)$ we obtain a doublet of $\mathcal{A}_1$ with $(\overline{1},3)$. From the weight $(2,\overline{2})$ we obtain a triplet of $\mathcal{A}_1$ with $(0,1)$ and $(\overline{2},4)$, and from the weight $(\overline{1},3)$ we obtain a quartet of $\mathcal{A}_2$ with $(0,1)$, $(1,\overline{1})$, and $(2,\overline{3})$. There are two paths to the dominant weight $(0,1)$, which may be a double weight. We define two orthonormal basis states $(0,1)_1$ and $(0,1)_2$, where $(0,1)_1$ belongs to the triplet of $\mathcal{A}_1$ with $(2,\overline{2})$ and $(\overline{2},4)$, and $(0,1)_2$ is a singlet of $\mathcal{A}_1$. The state $(0,1)$ in the quartet of $\mathcal{A}_2$, composed of $(\overline{1},3)$, $(0,1)$, $(1,\overline{1})$, and $(2,\overline{3})$, has to be a combination of $(0,0)_1$ and $(0,0)_2$:

$$F_2|(0,2),(\overline{1},3)\rangle = a_1|(0,2),(0,1)_1\rangle + a_2|(0,2),(0,1)_2\rangle, \qquad a_1^2 + a_2^2 = 3.$$

Applying $E_1 F_2 = F_2 E_1$ to $|(0,2),(\overline{1},3)\rangle$, we have

$$E_1 F_2|(0,2),(\overline{1},3)\rangle = a_1\sqrt{2}|(0,2),(2,\overline{2})\rangle$$

$$= F_2 E_1|(0,2),(\overline{1},3)\rangle = F_2|(0,2),(1,0)\rangle = \sqrt{2}|(0,2),(2,\overline{2})\rangle.$$

Thus, $a_1 = 1$. Choosing the phase of $|(0,2),(0,1)_2\rangle$ such that a_2 is a positive real number, we have $a_2 = \sqrt{3-1^2} = \sqrt{2}$. Since the weight $(1,\overline{1})$ is equivalent to the dominant weight $(0,1)$, it is a double weight. Define two orthonormal basis states $|(0,2),(1,\overline{1})_1\rangle$ and $|(0,2),(1,\overline{1})_2\rangle$, where $|(0,2),(1,\overline{1})_1\rangle$ is proportional to $F_2|(0,2),(0,1)_1\rangle$ with a positive real coefficient b_1:

$$F_2|(0,2),(0,1)_1\rangle = b_1|(0,2),(1,\overline{1})_1\rangle,$$

$$F_2|(0,2),(0,1)_2\rangle = b_2|(0,2),(1,\overline{1})_1\rangle + b_3|(0,2),(1,\overline{1})_2\rangle.$$

Applying $E_2 F_2 = F_2 E_2 + H_2$ to $|(0,2),(0,1)_1\rangle$ and $|(0,2),(0,1)_2\rangle$, we have

$$E_2 F_2|(0,2),(0,1)_1\rangle = b_1^2|(0,2),(0,1)_1\rangle + b_1 b_2|(0,2),(0,1)_2\rangle$$
$$= (F_2 E_2 + H_2)|(0,2),(0,1)_1\rangle = (1+1)|(0,2),(0,1)_1\rangle + \sqrt{2}|(0,2),(0,1)_2\rangle$$
$$E_2 F_2|(0,2),(0,1)_2\rangle = b_1 b_2|(0,2),(0,1)_1\rangle + (b_2^2 + b_3^2)|(0,2),(0,1)_2\rangle$$
$$= (F_2 E_2 + H_2)|(0,2),(0,1)_2\rangle = \sqrt{2}|(0,2),(0,1)_1\rangle + (2+1)|(0,2),(0,1)_2\rangle.$$

Thus, $b_1 = \sqrt{2}$ and $b_2 = 1$. Choosing the phase of $|(0,2),(1,\bar{1})_2\rangle$ such that b_3 is a positive real number, we have $b_3 = \sqrt{2}$. Let

$$F_2|(0,2),(1,\bar{1})_1\rangle = b_4|(0,2),(2,\bar{3})\rangle, \quad F_2|(0,2),(1,\bar{1})_2\rangle = b_5|(0,2),(2,\bar{3})\rangle.$$

Applying $E_2 F_2 = F_2 E_2 + H_2$ to $|(0,2),(1,\bar{1})_1\rangle$, we have

$$E_2 F_2|(0,2),(1,\bar{1})_1\rangle = b_4^2|(0,2),(1,\bar{1})_1\rangle + b_4 b_5|(0,2),(1,\bar{1})_2\rangle$$
$$= (F_2 E_2 + H_2)|(0,2),(1,\bar{1})_1\rangle$$
$$= (2+1-1)|(0,2),(1,\bar{1})_1\rangle + \sqrt{2}|(0,2),(1,\bar{1})_2\rangle.$$

Choosing the phase of $|(0,2),(2,\bar{3})\rangle$ such that b_4 is a positive real number, we have $b_4 = \sqrt{2}$ and $b_5 = 1$.

From the weight $(\bar{2},4)$ we obtain a quintet of $\mathcal{A}_2$ with $(\bar{1},2)$, $(0,0)$, $(1,\bar{2})$, and $(2,\bar{4})$, where the representation matrix entries of F_2 are 2, $\sqrt{6}$, $\sqrt{6}$, and 2, respectively. Since the weights $(\bar{1},2)$ and $(1,\bar{2})$ are equivalent to the dominant weight $(0,1)$, they are the double weights. We define the orthonormal basis states $(\bar{1},2)_1$, $(\bar{1},2)_2$, $(1,\bar{2})_1$, and $(1,\bar{2})_2$, where $(\bar{1},2)_1$ and $(1,\bar{2})_1$ belong to the quintet of $\mathcal{A}_2$, and $(\bar{1},2)_2$ and $(1,\bar{2})_2$ belong to a triplet of $\mathcal{A}_2$ with $(0,0)$. Let

$$F_1|(0,2),(1,\bar{1})_1\rangle = c_1|(0,2),(\bar{1},2)_1\rangle + c_2|(0,2),(\bar{1},2)_2\rangle, \quad c_1^2 + c_2^2 = 1,$$
$$F_1|(0,2),(1,\bar{1})_2\rangle = c_3|(0,2),(\bar{1},2)_1\rangle + c_4|(0,2),(\bar{1},2)_2\rangle, \quad c_3^2 + c_4^2 = 1.$$

Applying $E_2 F_1 = F_1 E_2$ to $|(0,2),(1,\bar{1})_1\rangle$ and $|(0,2),(1,\bar{1})_2\rangle$, we have

$$E_2 F_1|(0,2),(1,\bar{1})_1\rangle = 2c_1|(0,2),(\bar{2},4)\rangle$$
$$= F_1 E_2|(0,2),(1,\bar{1})_1\rangle = \sqrt{2} F_1|(0,2),(0,1)_1\rangle + F_1|(0,2),(0,1)_2\rangle$$
$$= 2|(0,2),(\bar{2},4)\rangle,$$
$$E_2 F_1|(0,2),(1,\bar{1})_2\rangle = 2c_3|(0,2),(\bar{2},4)\rangle$$
$$= F_1 E_2|(0,2),(1,\bar{1})_2\rangle = \sqrt{2} F_1|(0,2),(0,1)_2\rangle = 0.$$

Thus, $c_1 = 1$ and $c_2 = c_3 = 0$. Choosing the phase of $|(0,2),(\bar{1},2)_2\rangle$ such that c_4 is a positive real number, we have $c_4 = 1$.

In addition to a quintet and a triplet of $\mathcal{A}_2$, which both contain the weight $(0,0)$, there is another triplet of $\mathcal{A}_1$ containing $(0,0)$. The triplet of $\mathcal{A}_1$ consists of $(2,\bar{3})$, $(0,0)$, $(\bar{2},3)$. There are three paths to the dominant weight $(0,0)$ lowered from the highest weight, so $(0,0)$ may be a triple weight. We define three orthonormal basis states $(0,0)_1$, $(0,0)_2$, and $(0,0)_3$, where $(0,0)_1$ belongs to the quintet of $\mathcal{A}_2$ composed of $(\bar{2},4)$, $(\bar{1},2)_1$, $(0,0)_1$, $(1,\bar{2})_1$ and $(2,\bar{4})$, $(0,0)_2$ belongs to the triplet of $\mathcal{A}_2$ composed of $(\bar{1},2)_2$, $(0,0)_2$ and $(1,\bar{2})_2$, and $(0,0)_3$ is a singlet of $\mathcal{A}_2$. Thus, we have

$$F_2|(0,2),(\bar{1},2)_1\rangle = \sqrt{6}|(0,2),(0,0)_1\rangle,$$
$$F_2|(0,2),(\bar{1},2)_2\rangle = \sqrt{2}|(0,2),(0,0)_2\rangle,$$
$$E_2|(0,2),(0,0)_3\rangle = 0, \quad F_2|(0,2),(0,0)_3\rangle = 0.$$

Let

$$F_1|(0,2),(2,\bar{3})\rangle = d_1|(0,2),(0,0)_1\rangle + d_2|(0,2),(0,0)_2\rangle + d_3|(0,2),(0,0)_3\rangle,$$

$$F_1|(0,2),(0,0)_1\rangle = d_4|(0,2),(\bar{2},3)\rangle, \quad F_1|(0,2),(0,0)_2\rangle = d_5|(0,2),(\bar{2},3)\rangle,$$

$$F_1|(0,2),(0,0)_3\rangle = d_6|(0,2),(\bar{2},3)\rangle, \quad d_1^2 + d_2^2 + d_3^2 = 2.$$

Applying $E_2 F_1 = F_1 E_2$ to $|(0,2),(2,\bar{3})\rangle$, we have

$$E_2 F_1|(0,2),(2,\bar{3})\rangle = \sqrt{6}d_1|(0,2),(\bar{1},2)_1\rangle + \sqrt{2}d_2|(0,2),(\bar{1},2)_2\rangle$$
$$= F_1 E_2|(0,2),(2,\bar{3})\rangle = \sqrt{2}F_1|(0,2),(1,\bar{1})_1\rangle + F_1|(0,2),(1,\bar{1})_2\rangle$$
$$= \sqrt{2}|(0,2),(\bar{1},2)_1\rangle + |(0,2),(\bar{1},2)_2\rangle.$$

Thus, $d_1 = \sqrt{1/3}$ and $d_2 = \sqrt{1/2}$. Choosing the phase of $|(0,2),(0,0)_3\rangle$ such that d_3 is a positive real number, we have $d_3 = \sqrt{7/6}$. Applying $E_1 F_1 = F_1 E_1 + H_1$ to $|(0,2),(0,0)_3\rangle$, we have

$$E_1 F_1|(0,2),(0,0)_3\rangle$$
$$= d_4 d_6|(0,2),(0,0)_1\rangle + d_5 d_6|(0,2),(0,0)_2\rangle + d_6^2|(0,2),(0,0)_3\rangle$$
$$= (F_1 E_1 + H_1)|(0,2),(0,0)_3\rangle$$
$$= \sqrt{7/18}|(0,2),(0,0)_1\rangle + \sqrt{7/12}|(0,2),(0,0)_2\rangle + (7/6)|(0,2),(0,0)_3\rangle.$$

Choosing the phase of $|(0,2),(\bar{2},3)\rangle$ such that d_6 is a positive real number, we have $d_6 = \sqrt{7/6}$, and then $d_4 = \sqrt{1/3}$ and $d_5 = \sqrt{1/2}$. It is easy to calculate the remaining part of the block weight diagram.

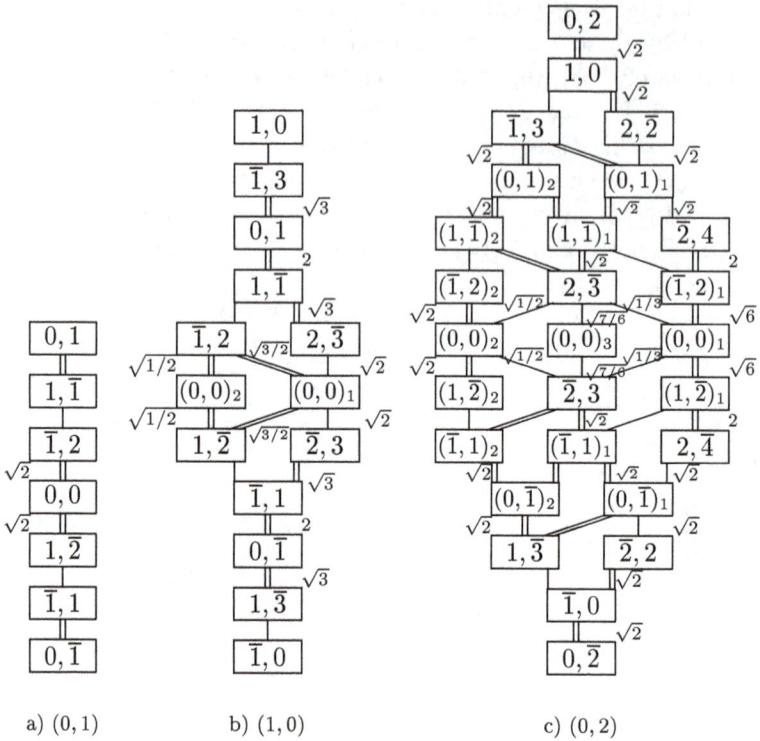

Fig. 7.5.　The block weight diagrams of some representations of G_2.

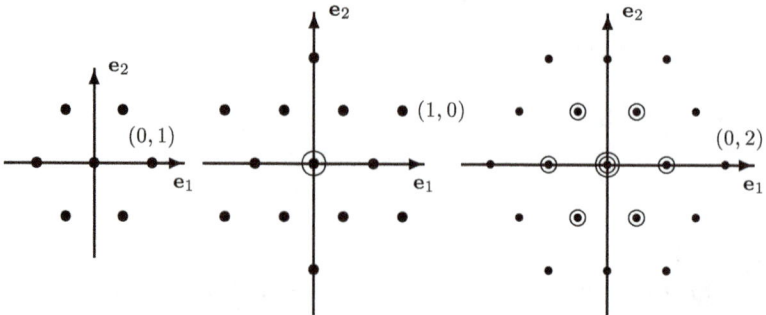

a) $(0,1)$　　　　b) $(1,0)$　　　　c) $(0,2)$

The results for the representation $(0,2)$ as well as $(1,0)$ and $(0,1)$ are given in the block weight diagrams (Fig. 7.5). In the diagrams, the single link means the action of F_1 and the double link means that of F_2. The planar weight diagrams are also given, where the simple roots and the fundamental dominant weights of G_2 in the rectangular coordinate frame are

$$\mathbf{r}_1 = \sqrt{2}\mathbf{e}_2, \qquad\qquad \mathbf{r}_2 = \sqrt{1/6}\mathbf{e}_1 - \sqrt{1/2}\mathbf{e}_2,$$

$$\mathbf{w}_1 = \sqrt{3/2}\mathbf{e}_1 + \sqrt{1/2}\mathbf{e}_2, \qquad \mathbf{w}_2 = \sqrt{2/3}\mathbf{e}_1.$$

7.3   Reduction of the Direct Product of Representations

★ The Clebsch-Gordan (CG) series for the reduction of the direct product $\mathbf{M}_1 \times \mathbf{M}_2$ of two irreducible representations of a simple Lie algebra can be calculated by the dominant weight diagrams. The main point is to list all dominant weights $\mathbf{N}$ contained in the product space, to count their multiplicities $n(\mathbf{M}_1 \times \mathbf{M}_2, \mathbf{N})$ in the space, to calculate their Weyl orbital sizes $OS(\mathbf{N})$, and to calculate the multiplicities $n(\mathbf{M}, \mathbf{N})$ of those dominant weights $\mathbf{N}$ in the representations $\mathbf{M}$ contained in the CG series, respectively. First, we list $OS(\mathbf{N})$ and $n(\mathbf{M}_1 \times \mathbf{M}_2, \mathbf{N})$ of all dominant weights $\mathbf{N}$ contained in the product space at the first column of the diagram. Let $\mathbf{M}_1 + \mathbf{M}_2 - \mathbf{N} = \sum_j c_j \mathbf{r}_j$, $\sum_j c_j$ is called the height of the dominant weight $\mathbf{N}$. Arrange the dominant weights $\mathbf{N}$ downwards as the height increases. The height of $\mathbf{M}_1 + \mathbf{M}_2$ is zero and is listed in the first row. The representation $\mathbf{M}_1 + \mathbf{M}_2$ is contained in the CG series with multiplicity one. Then, the second column of the diagram lists the dominant weights $\mathbf{N}$ in the representation $\mathbf{M}_1 + \mathbf{M}_2$ and their multiplicities $n(\mathbf{M}_1 + \mathbf{M}_2, \mathbf{N})$ (as the superscript). The difference $n(\mathbf{M}_1 \times \mathbf{M}_2, \mathbf{M}') - n(\mathbf{M}_1 + \mathbf{M}_2, \mathbf{M}')$ for the dominant weight $\mathbf{M}'$ with the next smallest height is equal to the multiplicity of the representation $\mathbf{M}'$ contained in the CG series. If it is not zero, the third column of the diagram lists the dominant weights $\mathbf{N}$ in the representation $\mathbf{M}'$ and their multiplicities $n(\mathbf{M}', \mathbf{N})$ (as the superscript). Continue the calculation to determine the multiplicities of the representations $\mathbf{N}$ in the CG series by comparing $n(\mathbf{M}_1 + \mathbf{M}_2, \mathbf{N})$ with the sum of the multiplicities of $\mathbf{N}$ in the representations, which we have known to be contained in the CG series, until all dominant weights $\mathbf{N}$ are calculated. At last, calculate the dimensions $d(\mathbf{M}')$ of the representations $\mathbf{M}'$ contained in the CG series by

$$d(\mathbf{M}') = \sum_{\mathbf{N}} OS(\mathbf{N}) n(\mathbf{M}', \mathbf{N}), \tag{7.26}$$

to see whether their sum is equal to the dimension of the direct product space:

$$d(\mathbf{M}_1 \times \mathbf{M}_2) = \sum_{\mathbf{N}} OS(\mathbf{N})n(\mathbf{M}_1 \times \mathbf{M}_2, \mathbf{N}) = \sum_{\mathbf{M}'} d(\mathbf{M}'). \qquad (7.27)$$

★ Denote by $|\mathbf{M}_1, \mathbf{m}_1\rangle|\mathbf{M}_2, \mathbf{m}_2\rangle$ the basis state before reduction, and by $\|\mathbf{M}, \mathbf{m}\rangle$ the basis state after reduction. Usually, the basis state $|\mathbf{M}_1, \mathbf{m}_1\rangle|\mathbf{M}_2, \mathbf{m}_2\rangle$ is briefly written as $|\mathbf{m}_1\rangle|\mathbf{m}_2\rangle$ for simplicity.

$$\|\mathbf{M}, \mathbf{m}\rangle = \sum_{\mathbf{m}_1', \mathbf{m}_2'} C^{\mathbf{M}_1 \mathbf{M}_2}_{\mathbf{m}_1 \mathbf{m}_2, \mathbf{M}\mathbf{m}} |\mathbf{m}_1\rangle|\mathbf{m}_2\rangle, \qquad (7.28)$$

where $C^{\mathbf{M}_1 \mathbf{M}_2}_{\mathbf{m}_1 \mathbf{m}_2, \mathbf{M}\mathbf{m}}$ are the Clebsch-Gordan (CG) coefficients. Applying the lowering operator F_μ to the basis state, we have:

$$F_\mu \|\mathbf{M}, \mathbf{m}\rangle = \sum_{\mathbf{m}'} D^{\mathbf{M}}_{\mathbf{m}'\mathbf{m}}(F_\mu)\|\mathbf{M}, \mathbf{m}'\rangle,$$
$$F_\mu |\mathbf{m}_1\rangle|\mathbf{m}_2\rangle = \sum_{\mathbf{m}_2'} D^{\mathbf{M}_2}_{\mathbf{m}_2'\mathbf{m}_2}(F_\mu)|\mathbf{m}_1\rangle|\mathbf{m}_2'\rangle + \sum_{\mathbf{m}_1'} D^{\mathbf{M}_1}_{\mathbf{m}_1'\mathbf{m}_1}(F_\mu)|\mathbf{m}_1'\rangle|\mathbf{m}_2\rangle.$$
$$(7.29)$$

The main steps of the calculation are as follows. First, the highest weight state in the representation $\mathbf{M}_1 + \mathbf{M}_2$ is

$$\|\mathbf{M}_1 + \mathbf{M}_2, \mathbf{M}_1 + \mathbf{M}_2\rangle = |\mathbf{M}_1\rangle|\mathbf{M}_2\rangle. \qquad (7.30)$$

In terms of Eq. (7.29) we are able to calculate the expansions of all basis states in the representation $\mathbf{M}_1 + \mathbf{M}_2$. Then, by the orthonormal condition or by Eq. (7.24) we calculate the highest weight state in the representation $\mathbf{M}'$ contained in the CG series. At last, we calculate the expansions of all basis states in the representation $\mathbf{M}'$ by Eq. (7.29).

When $\mathbf{M}_1 = \mathbf{M}_2$, the expansion of each basis state can be symmetric or antisymmetric with respect to the permutation between two basis states $|\mathbf{m}_1\rangle|\mathbf{m}_2\rangle$ and $|\mathbf{m}_2\rangle|\mathbf{m}_1\rangle$. For simplicity, we denote by (sym. terms) or (antisym. terms) the remaining terms obtained from the preceding terms by the permutation, multiplied by 1 or −1, respectively.

12. Calculate the Clebsch-Gordan series and the Clebsch-Gordan coefficients for the direct product representation $(1,0) \times (1,0)$ of the C_2 Lie algebra.

Solution. In Problem 10 we have calculated the block weight diagrams of the representations $(1,0)$, $(0,1)$ and $(2,0)$ of C_2. The representation $(1,0)$ contains only one single dominant weight $(1,0)$. The representation $(0,1)$

contains two single dominant weights $(0,1)$ and $(0,0)$. The representation $(2,0)$ contains two single dominant weights $(2,0)$ and $(0,1)$, and one double dominant weight $(0,0)$. The Weyl orbital sizes of the weights $(2,0)$, $(0,1)$, and $(0,0)$ are 4, 4, and 1, respectively. The linearly independent basis states for the dominant weights in the product space are

Dominant Weight	Basis State	No. of States
$(2,0)$:	$|(1,0)\rangle|(1,0)\rangle,$	1
$(0,1)$:	$|(1,0)\rangle|(\bar{1},1)\rangle,\ |(\bar{1},1)\rangle|(1,0)\rangle$	2
$(0,0)$:	$|(1,0)\rangle|(\bar{1},0)\rangle,\ |(\bar{1},1)\rangle|(1,\bar{1})\rangle,$	
	$|(1,\bar{1})\rangle|(\bar{1},1)\rangle,\ |(\bar{1},0)\rangle|(1,0)\rangle,$	4

Thus, we draw the diagram of the dominant weights as follows.

$OS \times n$

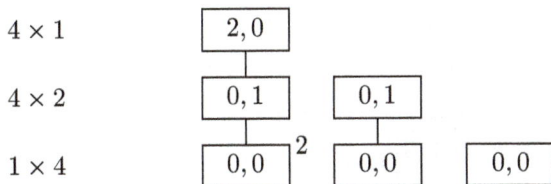

$$16 \quad = \quad 10 \quad + \quad 5 \quad + \quad 1$$

$$(1,0) \times (1,0) = (2,0) + (0,1) + (0,0)$$

Now, we calculate the expansions of the basis states. First, for the representation $(2,0)$, from Eq. (7.30) we have

$$\||(2,0),(2,0)\rangle = |(1,0)\rangle|(1,0)\rangle.$$

From the block weight diagram of $(2,0)$ of C_2 (see Fig. 7.4) and Eq. (7.29), we have

$$\begin{aligned}\||(2,0),(0,1)\rangle &= \sqrt{1/2}F_1\||(2,0),(2,0)\rangle \\ &= \sqrt{1/2}\left\{|(1,0)\rangle|(\bar{1},1)\rangle + |(\bar{1},1)\rangle|(1,0)\rangle\right\}.\end{aligned}$$

According to the orthonormal condition, we obtain the highest weight state in the representation $(0,1)$

$$\||(0,1),(0,1)\rangle = \sqrt{1/2}\left\{|(1,0)\rangle|(\bar{1},1)\rangle - |(\bar{1},1)\rangle|(1,0)\rangle\right\}.$$

Applying the lowering operators to the basis states continuously, we obtain

$$\|(2,0),(\overline{2},2)\rangle = \sqrt{1/2}F_1\|(2,0),(0,1)\rangle = |(\overline{1},1)\rangle|(\overline{1},1)\rangle,$$

$$\|(2,0),(2,\overline{1})\rangle = F_2\|(2,0),(0,1)\rangle$$
$$= \sqrt{1/2}\left\{|(1,0)\rangle|(1,\overline{1})\rangle + |(1,\overline{1})\rangle|(1,0)\rangle\right\},$$

$$\|(2,0),(0,0)_1\rangle = \sqrt{1/2}F_1\|(2,0),(2,\overline{1})\rangle$$
$$= (1/2)\left\{|(1,0)\rangle|(\overline{1},0)\rangle + |(\overline{1},1)\rangle|(1,\overline{1})\rangle + \text{(sym. terms)}\right\},$$

$$\|(2,0),(0,0)_2\rangle = F_2\|(2,0),(\overline{2},2)\rangle - \|(2,0)\rangle|(0,0)_1\rangle$$
$$= (1/2)\left\{-|(1,0)\rangle|(\overline{1},0)\rangle + |(\overline{1},1)\rangle|(1,\overline{1})\rangle + \text{(sym. terms)}\right\},$$

$$\|(2,0),(\overline{2},1)\rangle = \sqrt{1/2}F_1\|(2,0),(0,0)_1\rangle$$
$$= \sqrt{1/2}\left\{|(\overline{1},1)\rangle|(\overline{1},0)\rangle + |(\overline{1},0)\rangle|(\overline{1},1)\rangle\right\},$$

$$\|(2,0),(2,\overline{2})\rangle = F_2\|(2,0),(0,0)_1\rangle = |(1,\overline{1})\rangle|(1,\overline{1})\rangle,$$

$$\|(2,0),(0,\overline{1})\rangle = \sqrt{1/2}F_1\|(2,0),(2,\overline{2})\rangle$$
$$= \sqrt{1/2}\left\{|(1,\overline{1})\rangle|(\overline{1},0)\rangle + |(\overline{1},0)\rangle|(1,\overline{1})\rangle\right\},$$

$$\|(2,0),(\overline{2},0)\rangle = \sqrt{1/2}F_1\|(2,0),(0,\overline{1})\rangle = |(\overline{1},0)\rangle|(\overline{1},0)\rangle.$$

For the basis states in the representation $(0,1)$ we have

$$\|(0,1),(2,\overline{1})\rangle = F_2\|(0,1),(0,1)\rangle$$
$$= \sqrt{1/2}\left\{|(1,0)\rangle|(1,\overline{1})\rangle - |(1,\overline{1})\rangle|(1,0)\rangle\right\},$$

$$\|(0,1),(0,0)\rangle = \sqrt{1/2}F_1\|(0,1),(2,\overline{1})\rangle$$
$$= (1/2)\left\{|(1,0)\rangle|(\overline{1},0)\rangle + |(\overline{1},1)\rangle|(1,\overline{1})\rangle + \text{(antisym. terms)}\right\},$$

$$\|(0,1),(\overline{2},1)\rangle = \sqrt{1/2}F_1\|(0,1),(0,0)\rangle$$
$$= \sqrt{1/2}\left\{|(\overline{1},1)\rangle|(\overline{1},0)\rangle - |(\overline{1},0),(\overline{1},1)\rangle\right\},$$

$$\|(0,1),(0,\overline{1})\rangle = F_2\|(0,1)\rangle|(\overline{2},1)\rangle$$
$$= \sqrt{1/2}\left\{|(1,\overline{1})\rangle|(\overline{1},0)\rangle - |(\overline{1},0)\rangle|(1,\overline{1})\rangle\right\}.$$

The expansion of the basis state in the representation $(0,0)$ can be calculated by the orthonormal condition. But now, we prefer to calculate it by the condition (7.24) for the highest weight state. Let

$$\|(0,0),(0,0)\rangle = a|(1,0)\rangle|(\overline{1},0)\rangle + b|(\overline{1},1)\rangle|(1,\overline{1})\rangle$$
$$+ c|(1,\overline{1})\rangle|(\overline{1},1)\rangle + d|(\overline{1},0)\rangle|(1,0)\rangle.$$

Applying E_1 and E_2 to it, we have

$$\begin{aligned}
E_1||(0,0),(0,0)\rangle &= a|(1,0)\rangle|(1,\overline{1})\rangle + b|(1,0)\rangle|(1,\overline{1})\rangle \\
&\quad + c|(1,\overline{1})\rangle|(1,0)\rangle + d|(1,\overline{1})\rangle|(1,0)\rangle = 0, \\
E_2||(0,0),(0,0)\rangle &= b|(\overline{1},1)\rangle|(\overline{1},1)\rangle + c|(\overline{1},1)\rangle|(\overline{1},1)\rangle = 0.
\end{aligned}$$

Thus, $a = -b = c = -d$. After normalization we obtain

$$\begin{aligned}
||(0,0),(0,0)\rangle &= (1/2)\left\{|(1,0)\rangle|(\overline{1},0)\rangle - |(\overline{1},1)\rangle|(1,\overline{1})\rangle \right. \\
&\quad \left. +|(1,\overline{1})\rangle|(\overline{1},1)\rangle - |(\overline{1},0)\rangle|(1,0)\rangle\right\}.
\end{aligned}$$

It can be checked that four basis states with the weight $(0,0)$ are orthogonal to each other. In the permutation between $|\mathbf{m}_1\rangle$ and $|\mathbf{m}_2\rangle$, the representation $(2,0)$ is symmetric, but $(0,1)$ and $(0,0)$ are antisymmetric.

13. Calculate the Clebsch-Gordan series and the Clebsch-Gordan coefficients for the direct product representation $(0,1) \times (0,1)$ of the C_2 Lie algebra.

Solution. One may solve this problem following the way used in the preceding problem. But now, we prefer to calculate the block weight diagram and the CG coefficients with another method, where the CG coefficients for $(0,1) \times (0,1)$ and the block weight diagram for $(1,1)$ are calculated at the same time.

From the block weight diagram of $(0,1)$ given in Fig 7.4, we know that the linearly independent basis states for the dominant weights in the product space are as follows, where we neglect the basis state which can be obtained by the permutation of two states $|\mathbf{m}_1\rangle$ and $|\mathbf{m}_2\rangle$.

Dominant Weight	Basis State	No. of States
$(0,2)$:	$|(0,1)\rangle|(0,1)\rangle,$	1
$(2,0)$:	$|(0,1)\rangle|(2,\overline{1})\rangle,$	2
$(0,1)$:	$|(0,1)\rangle|(0,0)\rangle,$	2
$(0,0)$:	$|(0,1)\rangle|(0,\overline{1})\rangle, |(2,\overline{1})\rangle|(\overline{2},1)\rangle, |(0,0),(0,0)\rangle,$	5

We have calculated the block weight diagrams of the representations $(0,1)$ and $(2,0)$. However, we did not calculate the block weight diagram of the representation $(0,2)$. The representation $(0,1)$ contains two single dominant weights $(0,1)$ and $(0,0)$. The representation $(2,0)$ contains two single dominant weights $(2,0)$ and $(0,1)$, and one double dominant weight $(0,0)$. The Weyl orbital sizes of the weights $(2,0)$, $(0,1)$, and $(0,0)$ are 4,

4, and 1, respectively. The Weyl orbital of the weights $(0, 2)$ contains four weights: $(0, 2)$, $(4, \bar{2})$, $(\bar{4}, 2)$, and $(0, \bar{2})$.

From Eq. (7.30) we have the expansion of the highest weight state in the representation $(0, 2)$

$$\|(0, 2), (0, 2)\rangle = |(0, 1)\rangle|(0, 1)\rangle.$$

From the highest weight $(0, 2)$ we have a triplet of $\mathcal{A}_2$:

$$
\begin{aligned}
\|(0, 2), (2, 0)\rangle &= \sqrt{1/2} F_2 \|(0, 2), (0, 2)\rangle \\
&= \sqrt{1/2} \left\{ |(0, 1)\rangle|(2, \bar{1})\rangle + |(2, \bar{1})\rangle|(0, 1)\rangle \right\}, \\
\|(0, 2), (4, \bar{2})\rangle &= \sqrt{1/2} F_2 \|(0, 2), (2, 0)\rangle = |(2, \bar{1})\rangle|(2, \bar{1})\rangle.
\end{aligned}
$$

The dominant weight $(2, 0)$ is single in the representation $(0, 2)$ because there is only one path to it lowered from the highest weight. However, the multiplicity of the weight $(2, 0)$ in the product space is 2. It means that one representation $(2, 0)$ is contained in the CG series for $(0, 1) \times (0, 1)$.

From the weight $(2, 0)$ we have a triplet of $\mathcal{A}_1$,

$$
\begin{aligned}
\|(0, 2), (0, 1)\rangle &= \sqrt{1/2} F_1 \|(0, 2), (2, 0)\rangle \\
&= \sqrt{1/2} \left\{ |(0, 1)\rangle|(0, 0)\rangle + |(0, 0)\rangle|(0, 1)\rangle \right\} \\
\|(0, 2), (\bar{2}, 2)\rangle &= \sqrt{1/2} F_1 \|(0, 2), (0, 1)\rangle \\
&= \sqrt{1/2} \left\{ |(0, 1)\rangle|(\bar{2}, 1)\rangle + |(\bar{2}, 1)\rangle|(0, 1)\rangle \right\}.
\end{aligned}
$$

The dominant weight $(0, 1)$ is single in the representation $(0, 2)$ because there is only one path to it lowered from the highest weight. The representation $(2, 0)$ contains a single dominant weight $(0, 1)$, and the product space contains two basis states with $(0, 1)$. It means that the representation $(0, 1)$ is not contained in the CG series for $(0, 1) \times (0, 1)$.

From the weight $(4, \bar{2})$ we have a quintet of $\mathcal{A}_1$ with $(2, \bar{1})$, $(0, 0)$, $(\bar{2}, 1)$, and $(\bar{4}, 2)$, and from the weight $(\bar{2}, 2)$ we have a triplet of $\mathcal{A}_2$ with $(0, 0)$ and $(2, \bar{2})$. There are two paths to the dominant weight $(0, 0)$ lowered from the highest weight, so $(0, 0)$ may be a double dominant weight in the representation $(0, 2)$. We define two orthonormal basis states for the weight $(0, 0)$. $\|(0, 2), (0, 0)_1\rangle$ belongs to the quintet of $\mathcal{A}_1$, and $\|(0, 2), (0, 0)_2\rangle$ has to belong to a singlet of $\mathcal{A}_1$ because the weight $(2, \bar{1})$ is a single weight.

$$
\begin{aligned}
\|(0, 2), (2, \bar{1})\rangle &= (1/2) F_1 \|(0, 2), (4, \bar{2})\rangle \\
&= \sqrt{1/2} \left\{ |(2, \bar{1})\rangle|(0, 0)\rangle + |(0, 0)\rangle|(2, \bar{1})\rangle \right\},
\end{aligned}
$$

$$\|(0,2),(0,0)_1\rangle = \sqrt{1/6}F_1\|(0,2),(2,\overline{1})\rangle$$
$$= \sqrt{1/6}\left\{|(2,\overline{1})\rangle|(\overline{2},1)\rangle + 2|(0,0)\rangle|(0,0)\rangle\right.$$
$$\left. + |(\overline{2},1)\rangle|(2,\overline{1})\rangle\right\},$$

$$\|(0,2),(\overline{2},1)\rangle = \sqrt{1/6}F_1\|(0,2),(0,0)_1\rangle$$
$$= \sqrt{1/2}\left\{|(0,0)\rangle|(\overline{2},1)\rangle + |(\overline{2},1)\rangle|(0,0)\rangle\right\},$$

$$\|(0,2),(\overline{4},2)\rangle = (1/2)F_1\|(0,2),(\overline{2},1)\rangle = |(\overline{2},1)\rangle|(\overline{2},1)\rangle.$$

Applying F_2 to $\|(0,2),(\overline{2},2)\rangle$, we have

$$F_2\|(0,2),(\overline{2},2)\rangle = \sqrt{1/2}\left\{|(0,1)\rangle|(0,\overline{1})\rangle + |(2,\overline{1})\rangle|(\overline{2},1)\rangle\right.$$
$$\left. + |(\overline{2},1)\rangle|(2,\overline{1})\rangle + |(0,\overline{1})\rangle|(0,1)\rangle\right\}.$$

We obtain the inner product between this basis state and $\|(0,2),(0,0)_1\rangle$ to be $\sqrt{1/3}$. Thus,

$$a\|(0,2),(0,0)_2\rangle = F_2\|(0,2),(\overline{2},2)\rangle - \sqrt{1/3}\|(0,2),(0,0)_1\rangle$$
$$= \sqrt{1/18}\left\{3|(0,1)\rangle|(0,\overline{1})\rangle + 2|(2,\overline{1}),(\overline{2},1)\rangle\right.$$
$$\left. - 2|(0,0)\rangle|(0,0)\rangle + 2|(\overline{2},1)\rangle|(2,\overline{1})\rangle + 3|(0,\overline{1})\rangle|(0,1)\rangle\right\}.$$

After normalization we have $a = \sqrt{5/3}$. In fact, a can also be calculated by the property of a triplet: $a^2 + 1/3 = 2$. Hence,

$$F_2\|(0,2),(\overline{2},2)\rangle = \sqrt{1/3}\|(0,2),(0,0)_1\rangle + \sqrt{5/3}\|(0,2),(0,0)_2\rangle,$$
$$\|(0,2),(0,0)_2\rangle = \sqrt{1/30}\left\{3|(0,1)\rangle|(0,\overline{1})\rangle + 2|(2,\overline{1})\rangle|(\overline{2},1)\rangle\right.$$
$$\left. - 2|(0,0)\rangle|(0,0)\rangle + 2|(\overline{2},1)\rangle|(2,\overline{1})\rangle + 3|(0,\overline{1})\rangle|(0,1)\rangle\right\}.$$

It is easy to calculate the remaining expansions of the basis states in the representation $(0,2)$. We list the results as follows.

$$F_2\|(0,2),(0,0)_2\rangle = \sqrt{5/6}\left\{|(2,\overline{1})\rangle|(0,\overline{1})\rangle + |(0,\overline{1})\rangle|(2,\overline{1})\rangle\right\}$$
$$= \sqrt{5/3}\|(0,2),(2,\overline{2})\rangle,$$

$$\|(0,2),(2,\overline{2})\rangle = \sqrt{1/2}\left\{|(2,\overline{1})\rangle|(0,\overline{1})\rangle + |(0,\overline{1})\rangle|(2,\overline{1})\rangle\right\},$$

$$F_2\|(0,2),(0,0)_1\rangle = \sqrt{1/6}\left\{|(2,\overline{1})\rangle|(0,\overline{1})\rangle + |(0,\overline{1})\rangle|(2,\overline{1})\rangle\right\}$$
$$= \sqrt{1/3}\|(0,2),(2,\overline{2})\rangle,$$

$$\|(0,2),(0,\overline{1})\rangle = F_2\|(0,2),(\overline{2},1)\rangle$$
$$= \sqrt{1/2}\left\{|(0,0)\rangle|(0,\overline{1})\rangle + |(0,\overline{1})\rangle|(0,0)\rangle\right\},$$

$$||(0,2),(\bar{2},0)\rangle = \sqrt{1/2}F_2||(0,2),(\bar{4},2)\rangle$$
$$= \sqrt{1/2}\left\{|(\bar{2},1)\rangle|(0,\bar{1})\rangle + |(0,\bar{1})\rangle|(\bar{2},1)\rangle\right\},$$
$$||(0,2)\rangle|(0,\bar{2})\rangle = \sqrt{1/2}F_2||(0,2),(\bar{2},0)\rangle = |(0,\bar{1})\rangle|(0,\bar{1})\rangle.$$

In addition, we have

$$F_1||(0,2),(2,\bar{2})\rangle = \sqrt{2}||(0,2),(0,\bar{1})\rangle,$$
$$F_1||(0,2),(0,\bar{1})\rangle = \sqrt{2}||(0,2),(\bar{2},0)\rangle.$$

From the above calculation, we are able to draw the dominant weight diagram for $(0,1)\times(0,1)$ of C_2 and the block weight diagram and the planar weight diagram of the representation $(0,2)$ as follows.

$OS \times n$

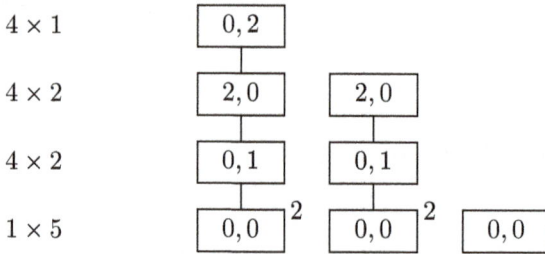

4×1	0, 2		
4×2	2, 0	2, 0	
4×2	0, 1	0, 1	
1×5	0, 0	0, 0	0, 0

$$25 \quad = \quad 14 \quad + \quad 10 \quad + \quad 1$$
$$(0,1)\times(0,1) \quad = \quad (0,2) \quad + \quad (2,0) \quad + \quad (0,0)$$

Now, we calculate the expansions of the basis states in the representation $(2,0)$. Since $||(2,0),(2,0)\rangle$ is orthogonal to $||(0,2),(2,0)\rangle$, we have

$$||(2,0),(2,0)\rangle = \sqrt{1/2}\left\{|(0,1)\rangle|(2,\bar{1})\rangle - |(2,\bar{1})\rangle|(0,1)\rangle\right\}.$$

The remaining basis states are

$$||(2,0),(0,1)\rangle = \sqrt{1/2}F_1||(2,0),(2,0)\rangle$$
$$= \sqrt{1/2}\left\{|(0,1)\rangle|(0,0)\rangle - |(0,0)\rangle|(0,1)\rangle\right\},$$
$$||(2,0),(\bar{2},2)\rangle = \sqrt{1/2}F_1||(2,0),(0,1)\rangle$$
$$= \sqrt{1/2}\left\{|(0,1)\rangle|(\bar{2},1)\rangle - |(\bar{2},1)\rangle|(0,1)\rangle\right\},$$
$$||(2,0),(2,\bar{1})\rangle = F_2||(2,0),(0,1)\rangle$$
$$= \sqrt{1/2}\left\{|(2,\bar{1})\rangle|(0,0)\rangle - |(0,0)\rangle|(2,\bar{1})\rangle\right\},$$

$$\|(2,0),(0,0)_1\rangle = \sqrt{1/2}F_1\|(2,0),(2,\overline{1})\rangle$$
$$= \sqrt{1/2}\left\{|(2,\overline{1})\rangle|(\overline{2},1)\rangle - |(\overline{2},1)\rangle|(2,\overline{1})\rangle\right\},$$

$$\|(2,0),(\overline{2},1)\rangle = \sqrt{1/2}F_1\|(2,0),(0,0)_1\rangle$$
$$= \sqrt{1/2}\left\{|(0,0)\rangle|(\overline{2},1)\rangle - |(\overline{2},1)\rangle|(0,0)\rangle\right\},$$

$$\|(2,0),(0,0)_2\rangle = F_2\|(2,0),(\overline{2},2)\rangle - \|(2,0)\rangle|(0,0)_1\rangle$$
$$= \sqrt{1/2}\left\{|(0,1)\rangle|(0,\overline{1})\rangle - |(0,\overline{1})\rangle|(0,1)\rangle\right\},$$

$$\|(2,0),(2,\overline{2})\rangle = F_2\|(2,0),(0,0)_2\rangle$$
$$= \sqrt{1/2}\left\{|(2,\overline{1})\rangle|(0,\overline{1})\rangle - |(0,\overline{1})\rangle|(2,\overline{1})\rangle\right\},$$

$$\|(2,0),(0,\overline{1})\rangle = \sqrt{1/2}F_1\|(2,0),(2,\overline{2})\rangle$$
$$= \sqrt{1/2}\left\{|(0,0)\rangle|(0,\overline{1})\rangle - |(0,\overline{1})\rangle|(0,0)\rangle\right\},$$

$$\|(2,0),(\overline{2},0)\rangle = \sqrt{1/2}F_1\|(2,0),(0,\overline{1})\rangle$$
$$= \sqrt{1/2}\left\{|(\overline{2},1)\rangle|(0,\overline{1})\rangle - |(0,\overline{1})\rangle|(\overline{2},1)\rangle\right\}.$$

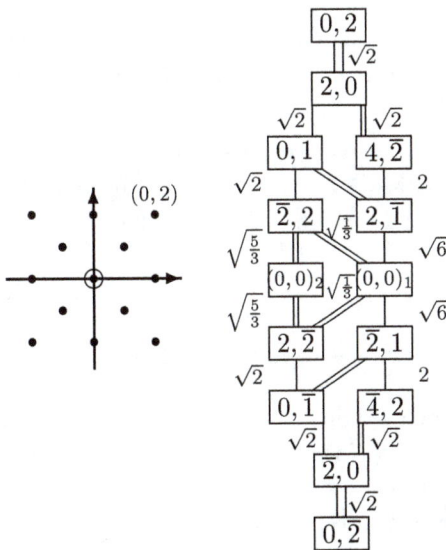

We may calculate the basis state $\|(0,0),(0,0)\rangle$ by the orthonormal condition to the preceding four basis states with the weight $(0,0)$. Here we calculate it by Eq. (7.24). Let

$$\|(0,0),(0,0)\rangle = a|(0,1)\rangle|(0,\overline{1})\rangle + b|(2,\overline{1})\rangle|(\overline{2},1)\rangle + c|(0,0),(0,0)\rangle$$
$$+ d|(\overline{2},1)\rangle|(2,\overline{1})\rangle + e|(0,\overline{1})\rangle|(0,1)\rangle.$$

From Eq. (7.29) we have

$$\begin{aligned}
E_1||(0,0),(0,0)\rangle &= \sqrt{2}b|(2,\bar{1})\rangle|(0,0)\rangle + \sqrt{2}c|(2,\bar{1})\rangle|(0,0)\rangle \\
&\quad + \sqrt{2}c|(0,0)\rangle|(2,\bar{1})\rangle + \sqrt{2}d|(0,0)\rangle|(2,\bar{1})\rangle = 0, \\
E_2||(0,0),(0,0)\rangle &= a|(0,1)\rangle|(\bar{2},1)\rangle + b|(0,1)\rangle|(\bar{2},1)\rangle \\
&\quad + d|(\bar{2},1)\rangle|(0,1)\rangle + e|(\bar{2},1)\rangle|(0,1)\rangle = 0.
\end{aligned}$$

Thus, $a = -b = c = -d = e$. After normalization, we have

$$\begin{aligned}
||(0,0),(0,0)\rangle = \sqrt{1/5}\,\big\{&|(0,1)\rangle|(0,\bar{1})\rangle - |(2,\bar{1})\rangle|(\bar{2},1)\rangle \\
+&|(0,0)\rangle|(0,0)\rangle - |(\bar{2},1)\rangle|(2,\bar{1})\rangle + |(0,\bar{1})\rangle|(0,1)\rangle\big\}.
\end{aligned}$$

It can be checked that five basis states with the weight $(0,0)$ are orthogonal to each other. In the reduction of the direct product representation $(0,1) \times (0,1)$ of C_2, the representations $(0,2)$ and $(0,0)$ are symmetric with respect to the permutation between $|\mathbf{m}_1\rangle$ and $|\mathbf{m}_2\rangle$, and the representation $(2,0)$ is antisymmetric.

14. Calculate the Clebsch-Gordan series for the direct product representation $(1,0) \times (0,1)$ of the C_2 Lie algebra and the expansion for the highest weight state of each irreducible representation in the CG series.

Solution. From the block weight diagrams for the representations $(1,0)$ and $(0,1)$ given in Fig 7.4, we know that the linearly independent basis states for the dominant weights in the product space are as follows.

Dominant Weight	Basis State	No. of States
$(1,1)$:	$\|(1,0)\rangle\|(0,1)\rangle,$	1
$(1,0)$:	$\|(1,0)\rangle\|(0,0)\rangle, \|(\bar{1},1)\rangle\|(2,\bar{1})\rangle, \|(1,\bar{1})\rangle\|(0,1)\rangle,$	3

There are a few methods to determine the dominant weights and their multiplicities contained in an irreducible representation. One method is to draw the upper part of the block weight diagram until all dominant weights in the representation appear. Another method is to look over the table book, for example, the table book [Bremner et al. (1985)] contains the information for the simple Lie algebras with rank less than 13. Anyway, we know that the representation $(1,1)$ contains one single dominant weight $(1,1)$ and one double dominant weight $(1,0)$. The representation $(1,0)$ contains only one single dominant weight $(1,0)$. The Weyl orbital size of the weight $(1,0)$ is 4. The Weyl orbital of the weights $(1,1)$ contains eight weights: $(1,1)$, $(\bar{1},2)$, $(3,\bar{1})$, $(3,\bar{2})$, $(\bar{3},2)$, $(\bar{3},1)$, $(1,\bar{2})$ and $(\bar{1},\bar{1})$. Now, we are able to draw the diagram of the dominant weights for the direct product representation $(1,0) \times (0,1)$ of C_2.

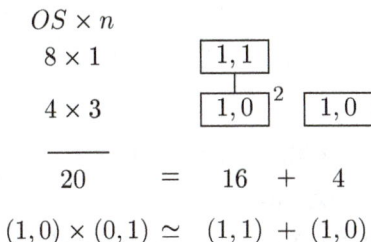

$$OS \times n$$

$$8 \times 1 \qquad \boxed{1,1}$$

$$4 \times 3 \qquad \boxed{1,0}\Big|^2 \quad \boxed{1,0}$$

$$\overline{20} \quad = \quad 16 \quad + \quad 4$$

$$(1,0) \times (0,1) \simeq \quad (1,1) + (1,0)$$

Fig. 7.6. The diagram of the dominant weights for $(1,0) \times (0,1)$ of C_2.

The highest weight states of the representations $(1,1)$ and $(1,0)$ can be calculated by Eqs. (7.30) and (7.24). The results are

$$||(1,1),(1,1)\rangle = |(1,0)\rangle|(0,1)\rangle,$$

$$||(1,0),(1,0)\rangle = \sqrt{1/5}\,\{|(1,0)\rangle|(0,0)\rangle - \sqrt{2}|(\bar{1},1)\rangle|(2,\bar{1})\rangle$$
$$+ \sqrt{2}|(1,\bar{1})\rangle|(0,1)\rangle\}.$$

15. Calculate the CG series for the following direct product representations of the G_2 Lie algebra and the expansion for the highest weight state of each irreducible representation in the CG series:

$$a): \quad (0,1) \times (0,1), \qquad b): \quad (0,1) \times (1,0), \qquad c): \quad (1,0) \times (1,0),$$

where the dimensions $d(\mathbf{M})$ of some representations $\mathbf{M}$, the Weyl orbital sizes $OS(\mathbf{M})$ of $\mathbf{M}$, and the multiplicities of the dominant weights in the representation $\mathbf{M}$ are listed in the following table.

$\mathbf{M}$	$d(\mathbf{M})$	$OS(\mathbf{M})$	The multiplicity of the dominant weight						
			$(0,0)$	$(0,1)$	$(1,0)$	$(0,2)$	$(1,1)$	$(0,3)$	$(2,0)$
$(0,0)$	1	1	1						
$(0,1)$	7	6	1	1					
$(1,0)$	14	6	2	1	1				
$(0,2)$	27	6	3	2	1	1			
$(1,1)$	64	12	4	4	2	2	1		
$(0,3)$	77	6	5	4	3	2	1	1	
$(2,0)$	77	6	5	3	3	2	1	1	1

Solution. In Problem 11 we have studied the block weight diagrams of the representations $(0,1)$ and $(1,0)$ of G_2. The representation $(0,1)$ contains seven single weights

$$(0,1), \quad (1,\bar{1}), \quad (\bar{1},2), \quad (0,0) \quad (1,\bar{2}), \quad (\bar{1},1), \quad (0,\bar{1}).$$

The representation $(1,0)$ contains 12 single weights and one double weight:

$$(1,0), \quad (\bar{1},3), \quad (0,1), \quad (1,\bar{1}), \quad (2,\bar{3}), \quad (\bar{1},2), \quad (0,0)_1,$$
$$(0,0)_2, \quad (1,\bar{2}), \quad (\bar{2},3), \quad (\bar{1},1), \quad (0,\bar{1}), \quad (1,\bar{3}), \quad (\bar{1},0).$$

In the following, we first list the linearly independent basis states $|\mathbf{m}_1\rangle|\mathbf{m}_2\rangle$ for each dominant weight in the product representation space, where we neglect the basis state which can be obtained by the permutation of two states $|\mathbf{m}_1\rangle$ and $|\mathbf{m}_2\rangle$. Then, comparing the number of the basis states of the dominant weight in the product representation space with the multiplicities of the dominant weight in the relevant representations, we are able to calculate the Clebsch-Gordan series. The expansion for the highest weight state of each irreducible representation in the CG series can be calculated by Eq. (7.24). The results are listed in the following.

a) The product representation $(0,1) \times (0,1)$.

Dominant Weight	Basis State	No. of States
$(0,2)$:	$|(0,1)\rangle|(0,1)\rangle$,	1
$(1,0)$:	$|(0,1)\rangle|(1,\bar{1})\rangle$,	2
$(0,1)$:	$|(0,1)\rangle|(0,0)\rangle, \ |(1,\bar{1})\rangle|(\bar{1},2)\rangle$,	4
$(0,0)$:	$|(0,1)\rangle|(0,\bar{1})\rangle, \ |(1,\bar{1})\rangle|(\bar{1},1)\rangle$,	
	$|(\bar{1},2)\rangle|(1,\bar{2})\rangle, \ |(0,0),(0,0)\rangle$,	7

The expansions for the highest weight states of the irreducible representations in the CG series are

$$\|(0,2),(0,2)\rangle = |(0,1)\rangle|(0,1)\rangle,$$
$$\|(1,0),(1,0)\rangle = \sqrt{1/2}\left\{|(1,\bar{1})|(0,1)\rangle\rangle - |(0,1)\rangle|(1,\bar{1})\rangle\right\},$$
$$\|(0,1)\rangle|(0,1)\rangle = \sqrt{1/6}\left\{|(0,1)\rangle|(0,0)\rangle - \sqrt{2}|(1,\bar{1})\rangle|(\bar{1},2)\rangle\right.$$
$$\left. + \sqrt{2}|(\bar{1},2)\rangle|(1,\bar{1})\rangle - |(0,0)\rangle|(0,1)\rangle\right\},$$
$$\|(0,0),(0,0)\rangle = \sqrt{1/7}\left\{|(0,1)\rangle|(0,\bar{1})\rangle - |(1,\bar{1})\rangle|(\bar{1},1)\rangle + |(\bar{1},2)\rangle|(1,\bar{2})\rangle\right.$$
$$\left. - |(0,0)\rangle|(0,0)\rangle + |(1,\bar{2})\rangle|(\bar{1},2)\rangle - |(\bar{1},1)\rangle|(1,\bar{1})\rangle + |(0,\bar{1})\rangle|(0,1)\rangle\right\}.$$

In the permutation between $|\mathbf{m}_1\rangle$ and $|\mathbf{m}_2\rangle$, the representation $(0,2)$ and $(0,0)$ are symmetric, but $(1,0)$ and $(0,1)$ are antisymmetric.

$OS \times n$

6×1	$\boxed{0,2}$			

| 6×2 | $\boxed{1,0}$ | $\boxed{1,0}$ | | |

| 6×4 | $\boxed{0,1}\ ^2$ | $\boxed{0,1}$ | $\boxed{0,1}$ | |

| 1×7 | $\boxed{0,0}\ ^3$ | $\boxed{0,0}\ ^2$ | $\boxed{0,0}$ | $\boxed{0,0}$ |

$$\frac{}{49} \quad = \quad 27 \quad + \quad 14 \quad + \quad 7 \quad + \quad 1$$

$$(0,1) \times (0,1) \simeq (0,2) \quad + \quad (1,0) \quad + \quad (0,1) \quad + \quad (0,0)$$

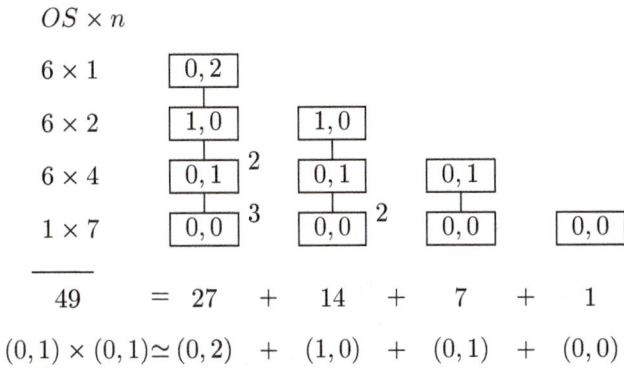

Fig. 7.7. The dominant weight diagram for $(0,1) \times (0,1)$ of G_2.

b) The product representation $(0,1) \times (1,0)$.

Dominant Weight	Basis State	No. of States
$(1,1):$	$\|(0,1)\rangle\|(1,0)\rangle,$	1
$(0,2):$	$\|(0,1)\rangle\|(0,1)\rangle,\ \|(1,\overline{1})\rangle\|(\overline{1},3)\rangle,\ \|(\overline{1},2)\rangle\|(1,0)\rangle,$	3
$(1,0):$	$\|(0,1)\rangle\|(1,\overline{1})\rangle,\ \|(1,\overline{1})\rangle\|(0,1)\rangle,\ \|(0,0)\rangle\|(1,0)\rangle,$	3
$(0,1):$	$\|(0,1)\rangle\|(0,0)_1\rangle,\ \|(0,1)\rangle\|(0,0)_2\rangle,\ \|(1,\overline{1})\rangle\|(\overline{1},2)\rangle,$	
	$\|(\overline{1},2)\rangle\|(1,\overline{1})\rangle,\ \|(0,0)\rangle\|(0,1)\rangle,$	
	$\|(1,\overline{2})\rangle\|(\overline{1},3)\rangle,\ \|(\overline{1},1)\rangle\|(1,0)\rangle,$	7
$(0,0):$	$\|(0,1)\rangle\|(0,\overline{1})\rangle,\ \|(1,\overline{1})\rangle\|(\overline{1},1)\rangle,\ \|(\overline{1},2)\rangle\|(1,\overline{2})\rangle,$	
	$\|(0,0)\rangle\|(0,0)_1\rangle,\ \|(0,0)\rangle\|(0,0)_2\rangle,\ \|(1,\overline{2})\rangle\|(\overline{1},2)\rangle,$	
	$\|(\overline{1},1)\rangle\|(1,\overline{1})\rangle,\ \|(0,\overline{1})\rangle\|(0,1)\rangle,$	8

The expansions for the highest weight states of the irreducible representations in the CG series are

$$\|(1,1),(1,1)\rangle = \|(0,1)\rangle\|(1,0)\rangle,$$

$$\|(0,2),(0,2)\rangle = \sqrt{1/7}\left\{\|(0,1)\rangle\|(0,1)\rangle - \sqrt{3}\|(1,\overline{1})\rangle\|(\overline{1},3)\rangle \right.$$
$$\left. + \sqrt{3}\|(\overline{1},2)\rangle\|(1,0)\rangle\right\},$$

$$\|(0,1),(0,1)\rangle = \sqrt{1/12}\left\{\sqrt{2}\|(0,1)\rangle\|(0,0)_2\rangle - \|(1,\overline{1})\rangle\|(\overline{1},2)\rangle \right.$$
$$+ \|(\overline{1},2)\rangle\|(1,\overline{1})\rangle - \sqrt{2}\|(0,0)\rangle\|(0,1)\rangle$$
$$\left. + \sqrt{3}\|(1,\overline{2})\rangle\|(\overline{1},3)\rangle - \sqrt{3}\|(\overline{1},1)\rangle\|(1,0)\rangle\right\}.$$

$$OS \times n$$

| 12×1 | $\boxed{1,1}$ | | | | |

6×3 $\boxed{0,2}$ ^{2} $\boxed{0,2}$

6×3 $\boxed{1,0}$ ^{2} $\boxed{1,0}$

6×7 $\boxed{0,1}$ ^{4} $\boxed{0,1}$ ^{2} $\boxed{0,1}$

1×8 $\boxed{0,0}$ ^{4} $\boxed{0,0}$ ^{3} $\boxed{0,0}$

$$98 \quad = 64 \; + \; 27 \; + \; 7$$

$$(0,1) \times (1,0) \simeq (1,1) \; + \; (0,2) \; + \; (0,1)$$

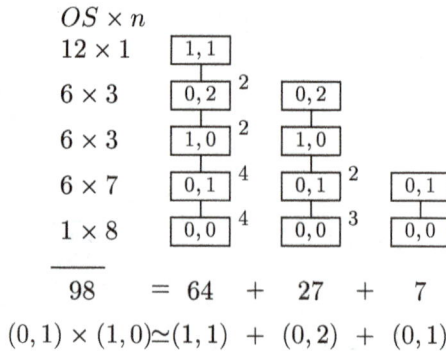

Fig. 7.8. The dominant weight diagram for $(0,1) \times (1,0)$ of G_2.

c) The product representation $(1,0) \times (1,0)$.

Dominant Weight	Basis State	No. of States
$(2,0)$:	$\lvert(1,0)\rangle\lvert(1,0)\rangle,$	1
$(0,3)$:	$\lvert(1,0)\rangle\lvert(\bar{1},3)\rangle,$	2
$(1,1)$:	$\lvert(1,0)\rangle\lvert(0,1)\rangle,$	2
$(0,2)$:	$\lvert(1,0)\rangle\lvert(\bar{1},2)\rangle,\ \lvert(\bar{1},3)\rangle\lvert(1,\bar{1})\rangle,\ \lvert(0,1)\rangle\lvert(0,1)\rangle,$	5
$(1,0)$:	$\lvert(1,0)\rangle\lvert(0,0)_1\rangle,\ \lvert(1,0)\rangle\lvert(0,0)_2\rangle,$ $\lvert(\bar{1},3)\rangle\lvert(2,\bar{3})\rangle,\ \lvert(0,1)\rangle\lvert(1,\bar{1})\rangle$	8
$(0,1)$:	$\lvert(1,0)\rangle\lvert(\bar{1},1)\rangle,\ \lvert(\bar{1},3)\rangle\lvert(1,\bar{2})\rangle,\ \lvert(0,1)\rangle\lvert(0,0)_1\rangle,$ $\lvert(0,1)\rangle\lvert(0,0)_1\rangle,\ \lvert(1,\bar{1})\rangle\lvert(\bar{1},2)\rangle,$	10
$(0,0)$:	12 pairs of the single weights, 4 pairs of the double weight $(0,0)$,	16

The expansions for the highest weight states of the irreducible representations in the CG series are

$$\lVert(2,0),(2,0)\rangle = \lvert(1,0)\rangle\lvert(1,0)\rangle,$$

$$\lVert(0,3),(0,3)\rangle = \sqrt{1/2}\left\{\lvert(1,0)\rangle\lvert(\bar{1},3)\rangle - \lvert(\bar{1},3)\rangle\lvert(1,0)\rangle\right\},$$

$$\lVert(0,2),(0,2)\rangle = (1/4)\left\{\sqrt{3}\lvert(1,0)\rangle\lvert(\bar{1},2)\rangle - \sqrt{3}\lvert(\bar{1},3)\rangle\lvert(1,\bar{1})\rangle\right.$$
$$\left. + (\text{sym. terms}) + 2\lvert(0,1)\rangle\lvert(0,1)\rangle\right\},$$

$$\lVert(1,0),(1,0)\rangle = (1/4)\left\{\lvert(1,0)\rangle\lvert(0,0)_1\rangle - \sqrt{3}\lvert(1,0)\rangle\lvert(0,0)_2\rangle\right.$$
$$\left. - \sqrt{2}\lvert(\bar{1},3)\rangle\lvert(2,\bar{3})\rangle + \sqrt{2}\lvert(0,1)\rangle\lvert(1,\bar{1})\rangle + (\text{antisym. terms})\right\},$$

$$\lVert(0,0),(0,0)\rangle = \sqrt{1/14}\left\{\lvert(1,0)\rangle\lvert(\bar{1},0)\rangle - \lvert(\bar{1},3)\rangle\lvert(1,\bar{3})\rangle + \lvert(0,1)\rangle\lvert(0,\bar{1})\rangle\right.$$
$$- \lvert(1,\bar{1})\rangle\lvert(\bar{1},1)\rangle + \lvert(\bar{1},2)\rangle\lvert(1,\bar{2})\rangle + \lvert(2,\bar{3})\rangle\lvert(\bar{2},3)\rangle + (\text{sym. terms})$$
$$\left. - \lvert(0,0)_1\rangle\lvert(0,0)_1\rangle - \lvert(0,0)_2\rangle\lvert(0,0)_2\rangle\right\}.$$

In the permutation between $|\mathbf{m}_1\rangle$ and $|\mathbf{m}_2\rangle$, the representation $(2,0)$, $(0,2)$ and $(0,0)$ are symmetric, but $(0,3)$ and $(1,0)$ are antisymmetric.

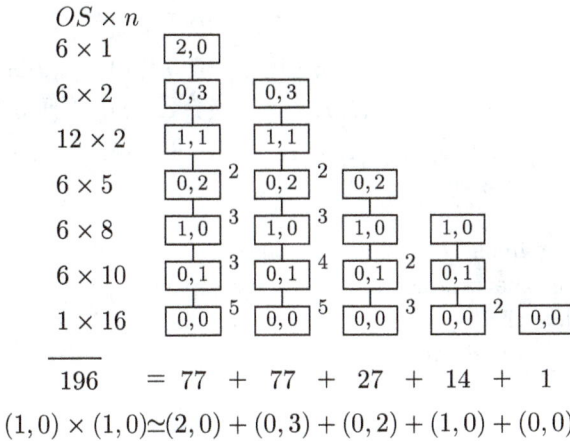

$$OS \times n$$

6×1 : $\boxed{2,0}$

6×2 : $\boxed{0,3}$ $\boxed{0,3}$

12×2 : $\boxed{1,1}$ $\boxed{1,1}$

6×5 : $\boxed{0,2}^{\,2}$ $\boxed{0,2}^{\,2}$ $\boxed{0,2}$

6×8 : $\boxed{1,0}^{\,3}$ $\boxed{1,0}^{\,3}$ $\boxed{1,0}$ $\boxed{1,0}$

6×10 : $\boxed{0,1}^{\,3}$ $\boxed{0,1}^{\,4}$ $\boxed{0,1}^{\,2}$ $\boxed{0,1}$

1×16 : $\boxed{0,0}^{\,5}$ $\boxed{0,0}^{\,5}$ $\boxed{0,0}^{\,3}$ $\boxed{0,0}^{\,2}$ $\boxed{0,0}$

$$196 = 77 + 77 + 27 + 14 + 1$$

$$(1,0) \times (1,0) \simeq (2,0) + (0,3) + (0,2) + (1,0) + (0,0)$$

Fig. 7.9. The dominant weight diagram for $(1,0) \times (1,0)$ of G_2.

16. Calculate the Clebsch-Gordan series for the direct product representation $(0,0,0,1) \times (0,0,0,1)$ of the F_4 Lie algebra and the expansion for the highest weight state of each irreducible representation in the Clebsch-Gordan series, where the dimensions $d(\mathbf{M})$ of some representations $\mathbf{M}$, the Weyl orbital sizes $OS(\mathbf{M})$ of $\mathbf{M}$, and the multiplicities of the dominant weights in the representation $\mathbf{M}$ are listed in the following table.

M	$d(\mathbf{M})$	OS	The multiplicity of the dominant weight				
			$(0,0,0,0)$	$(0,0,0,1)$	$(1,0,0,0)$	$(0,0,1,0)$	$(0,0,0,2)$
$(0,0,0,0)$	1	1	1				
$(0,0,0,1)$	26	24	2	1			
$(1,0,0,0)$	52	24	4	1	1		
$(0,0,1,0)$	273	96	9	5	2	1	
$(0,0,0,2)$	324	24	12	5	3	1	1

Solution. From the Cartan matrix of F_4 we obtain the relations between the simple roots $\mathbf{r}_j$ and the fundamental dominant weights $\mathbf{w}_j$

$$\mathbf{r}_1 = 2\mathbf{w}_1 - \mathbf{w}_2, \qquad \mathbf{r}_2 = -\mathbf{w}_1 + 2\mathbf{w}_2 - 2\mathbf{w}_3,$$
$$\mathbf{r}_3 = -\mathbf{w}_2 + 2\mathbf{w}_3 - \mathbf{w}_4, \qquad \mathbf{r}_4 = -\mathbf{w}_3 + 2\mathbf{w}_4.$$

The representation $(0,0,0,1)$ of F_4 contains 24 equivalent single weights and one doublet weight $(0,0,0,0)$:

$$(0,0,0,1), \quad (0,0,1,\bar{1}), \quad (0,1,\bar{1},0), \quad (1,\bar{1},1,0), \quad (\bar{1},0,1,0),$$
$$(1,0,\bar{1},1), \quad (\bar{1},1,\bar{1},1), \quad (1,0,0,\bar{1}), \quad (0,\bar{1},1,1), \quad (\bar{1},1,0,\bar{1}),$$
$$(0,0,\bar{1},2), \quad (0,\bar{1},2,\bar{1}), \quad (0,0,0,0)_1, \quad (0,0,0,0)_2, \quad (0,0,1,\bar{2}),$$
$$(0,1,\bar{2},1), \quad (0,1,\bar{1},\bar{1}), \quad (1,\bar{1},0,1), \quad (1,\bar{1},1,\bar{1}), \quad (\bar{1},0,0,1),$$
$$(\bar{1},0,1,\bar{1}), \quad (1,0,\bar{1},0), \quad (\bar{1},1,\bar{1},0), \quad (0,\bar{1},1,0), \quad (0,0,\bar{1},1),$$
$$(0,0,0,\bar{1}).$$

First, we list the linearly independent basis states $|\mathbf{m}_1\rangle|\mathbf{m}_2\rangle$ for each dominant weight in the product representation space $(0,0,0,1) \times (0,0,0,1)$, where we neglect the basis state which can be obtained by the permutation of two states $|\mathbf{m}_1\rangle$ and $|\mathbf{m}_2\rangle$.

Dominant Weight	Basis States	No. of States
$(0,0,0,2)$	$\|(0,0,0,1)\rangle\|(0,0,0,1)\rangle$	1
$(0,0,1,0)$	$\|(0,0,0,1)\rangle\|(0,0,1,\bar{1})\rangle$	2
$(1,0,0,0)$	$\|(0,0,0,1)\rangle\|(1,0,0,\bar{1})\rangle, \ \|(0,0,1,\bar{1})\rangle\|(1,0,\bar{1},1)\rangle$	
	$\|(0,1,\bar{1},0)\rangle\|(1,\bar{1},1,0)\rangle,$	6
$(0,0,0,1)$	$\|(0,0,0,1)\rangle\|(0,0,0,0)_1\rangle, \ \|(0,0,0,1)\rangle\|(0,0,0,0)_2\rangle$	
	$\|(0,0,1,\bar{1})\rangle\|(0,0,\bar{1},2)\rangle, \ \|(0,1,\bar{1},0)\rangle\|(0,\bar{1},1,1)\rangle,$	
	$\|(1,\bar{1},1,0)\rangle\|(\bar{1},1,\bar{1},1)\rangle, \ \|(\bar{1},0,1,0)\rangle\|(1,0,\bar{1},1)\rangle,$	12
$(0,0,0,0)$	24 pairs of the single weights,	
	4 pairs of the double weight $(0,0,0,0)$,	28

Then, comparing the number of the basis states of the dominant weight in the product representation space with the multiplicities of the dominant weight in the relevant representations, we are able to calculate the Clebsch-Gordan series, as given in the dominant weight diagram.

The expansion for the highest weight state of each irreducible representation in the CG series can be calculated by Eq. (7.24). The results are listed in the following.

$$\|(0,0,0,2),(0,0,0,2)\rangle = |(0,0,0,1)\rangle|(0,0,0,1)\rangle,$$
$$\|(0,0,1,0),(0,0,1,0)\rangle = \sqrt{1/2}\left\{|(0,0,0,1)\rangle|(0,0,1,\bar{1})\rangle \right.$$
$$\left. - |(0,0,1,\bar{1})\rangle|(0,0,0,1)\rangle\right\},$$

$$\||(1,0,0,0),(1,0,0,0)\rangle = \sqrt{1/6}\,\{|(0,0,0,1)\rangle|(1,0,0,\overline{1})\rangle$$
$$-\,|(0,0,1,\overline{1})\rangle|(1,0,\overline{1},1)\rangle + |(0,1,\overline{1},0)\rangle|(1,\overline{1},1,0)\rangle + (\text{antisym. terms})\},$$
$$\||(0,0,0,1),(0,0,0,1)\rangle = \sqrt{3/28}\,\{\sqrt{2/3}|(0,0,0,1)\rangle|(0,0,0,0)_2\rangle$$
$$-\,|(0,0,1,\overline{1})\rangle|(0,0,\overline{1},2)\rangle + |(0,1,\overline{1},0)\rangle|(0,\overline{1},1,1)\rangle$$
$$-\,|(1,\overline{1},1,0)\rangle|(\overline{1},1,\overline{1},1)\rangle + |(\overline{1},0,1,0)\rangle|(1,0,\overline{1},1)\rangle + (\text{sym. terms})\},$$
$$\||(0,0,0,0),(0,0,0,0)\rangle = \sqrt{1/26}\,\{|(0,0,0,1)\rangle|(0,0,0,\overline{1})\rangle$$
$$-\,|(0,0,1,\overline{1})\rangle|(0,0,\overline{1},1)\rangle + |(0,1,\overline{1},0)\rangle|(0,\overline{1},1,0)\rangle$$
$$-\,|(1,\overline{1},1,0)\rangle|(\overline{1},1,\overline{1},0)\rangle + |(\overline{1},0,1,0)\rangle|(1,0,\overline{1},0)\rangle$$
$$+\,|(1,0,\overline{1},1)\rangle|(\overline{1},0,1,\overline{1})\rangle - |(\overline{1},1,\overline{1},1)\rangle|(1,\overline{1},1,\overline{1})\rangle$$
$$-\,|(1,0,0,\overline{1})\rangle|(\overline{1},0,0,1)\rangle + |(0,\overline{1},1,1)\rangle|(0,1,\overline{1},\overline{1})\rangle$$
$$+\,|(\overline{1},1,0,\overline{1})\rangle|(1,\overline{1},0,1)\rangle - |(0,0,\overline{1},2)\rangle|(0,0,1,\overline{2})\rangle$$
$$-\,|(0,\overline{1},2,\overline{1})\rangle|(0,1,\overline{2},1)\rangle + (\text{sym. terms})$$
$$+\,|(0,0,0,0)_1\rangle|(0,0,0,0)_1\rangle + |(0,0,0,0)_2\rangle|(0,0,0,0)_2\rangle\},$$
$$F_3\,|(0,\overline{1},2,\overline{1})\rangle = \sqrt{2}\,|(0,0,0,0)_1\rangle,$$
$$F_4\,|(0,0,\overline{1},2)\rangle = \sqrt{1/2}\,|(0,0,0,0)_1\rangle + \sqrt{3/2}\,|(0,0,0,0)_2\rangle.$$

In the permutation between $|\mathbf{m}_1\rangle$ and $|\mathbf{m}_2\rangle$, the representation $(0,0,0,2)$, $(0,0,0,1)$ and $(0,0,0,0)$ are symmetric, but $(0,0,1,0)$ and $(1,0,0,0)$ are antisymmetric.

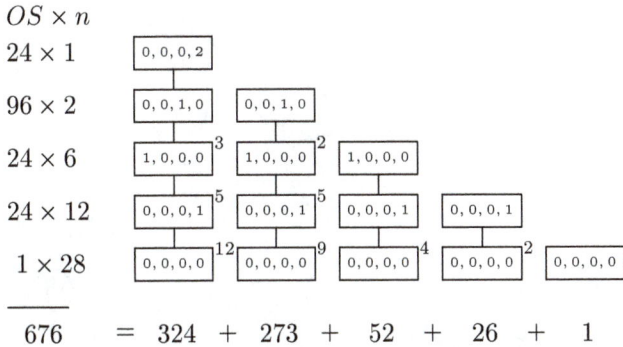

$$676 \quad = \quad 324 \;+\; 273 \;+\; 52 \;+\; 26 \;+\; 1$$

$$(0,0,0,1)\times(0,0,0,1)\simeq(0,0,0,2)+(0,0,1,0)+(1,0,0,0)+(0,0,0,1)+(0,0,0,0)$$

Fig. 7.10   The dominant weight diagram for $(0,0,0,1)\times(0,0,0,1)$ of F_4.

Chapter 8

UNITARY GROUPS

8.1 The SU(N) Group and Its Lie Algebra

★ The set of all N-dimensional unimodular unitary matrices, in the multiplication rule of matrices, constitutes the SU(N) group. It is a simply-connected compact Lie group with order $(N^2 - 1)$ and rank $(N - 1)$. The generators in its self-representation are traceless hermitian matrices:

$$
\begin{aligned}
\left(T_{ab}^{(1)}\right)_{cd} &= \frac{1}{2}\left(\delta_{ac}\delta_{bd} + \delta_{ad}\delta_{bc}\right), \\
\left(T_{ab}^{(2)}\right)_{cd} &= -\frac{i}{2}\left(\delta_{ac}\delta_{bd} - \delta_{ad}\delta_{bc}\right),
\end{aligned}
\qquad 1 \le a < b \le N,
$$

$$
\left(T_a^{(3)}\right)_{cd} = \begin{cases} \delta_{cd}\left[2a(a-1)\right]^{-1/2} & \text{when } c < a, \\ -\delta_{cd}\left[(a-1)/(2a)\right]^{1/2} & \text{when } c = a, \quad 2 \le a \le N, \\ 0 & \text{when } c > a, \end{cases}
$$

$$(8.1)$$

where the subscripts a and b are the ordinal indices of the generator, while c and d are the row and column indices of the matrix. The matrix $T_{ab}^{(1)}$ is symmetric with respect to both ab and cd, but $T_{ab}^{(2)}$ is antisymmetric. $T_a^{(3)}$ is diagonal. Usually, three types of the generators can be enumerated uniformly in the following order $T_1 = T_{12}^{(1)}$, $T_2 = T_{12}^{(2)}$, $T_3 = T_2^{(3)}$, $T_4 = T_{13}^{(1)}$, $T_5 = T_{13}^{(2)}$, $T_6 = T_{23}^{(1)}$, $T_7 = T_{23}^{(2)}$, $T_8 = T_3^{(3)}$, and so on. The generators T_A satisfy the normalization condition: $\mathrm{Tr}(T_A T_B) = \delta_{AB}/2$. Any element u in SU(N) can be expressed in the exponential matrix form:

$$
u = e^{-iH}, \quad H = \sum_{A=1}^{N^2-1} \omega_A T_A = \sum_{a<b}\left\{\omega_{ab}^{(1)}T_{ab}^{(1)} + \omega_{ab}^{(2)}T_{ab}^{(2)}\right\} + \sum_{a=2}^{N}\omega_a^{(3)}T_a^{(3)}.
$$

$$(8.2)$$

★ Let $N \equiv \ell + 1$, and define $(\ell + 1)$ vectors $\mathbf{V}_a$ in the ℓ-dimensional space:

$$(\mathbf{V}_a)_j = \left(T^{(3)}_{\ell-j+2} \right)_{aa}, \qquad \begin{array}{l} j = \ell,\ (\ell - 1),\ \ldots,\ 1, \\ \ell - j + 2 = 2,\quad 3,\qquad \ldots,\ N, \end{array}$$

$$\mathbf{V}_a \cdot \mathbf{V}_b = -\frac{1}{2(\ell+1)} + \frac{1}{2}\delta_{ab}, \qquad \mathbf{V}_N = -\sum_{a=1}^{\ell} \mathbf{V}_a. \tag{8.3}$$

$$\mathbf{V}_1 = \left\{ \frac{1}{\sqrt{2\ell(\ell+1)}},\ \frac{1}{\sqrt{2\ell(\ell-1)}},\ \ldots,\ \frac{1}{\sqrt{2a(a-1)}},\ \ldots,\ \frac{1}{\sqrt{12}},\ \frac{1}{2} \right\},$$

$$\mathbf{V}_2 = \left\{ \frac{1}{\sqrt{2\ell(\ell+1)}},\ \frac{1}{\sqrt{2\ell(\ell-1)}},\ \ldots,\ \frac{1}{\sqrt{2a(a-1)}},\ \ldots,\ \frac{1}{\sqrt{12}},\ -\frac{1}{2} \right\},$$

$$\mathbf{V}_3 = \left\{ \frac{1}{\sqrt{2\ell(\ell+1)}},\ \frac{1}{\sqrt{2\ell(\ell-1)}},\ \ldots,\ \frac{1}{\sqrt{2a(a-1)}},\ \ldots,\ -\frac{1}{\sqrt{3}},\ 0 \right\},$$

$$\cdots$$

$$\mathbf{V}_a = \left\{ \frac{1}{\sqrt{2\ell(\ell+1)}},\ \frac{1}{\sqrt{2\ell(\ell-1)}},\ \ldots,\ -\sqrt{\frac{a-1}{2a}},\ 0,\ \ldots,\ \ldots,\ 0 \right\},$$

$$\cdots$$

$$\mathbf{V}_\ell = \left\{ \frac{1}{\sqrt{2\ell(\ell+1)}},\ -\sqrt{\frac{\ell-1}{2\ell}},\ 0,\ \ldots\ \ldots\ \ldots\ \ldots\ \ldots,\ 0 \right\},$$

$$\mathbf{V}_N = \left\{ -\sqrt{\frac{\ell}{2(\ell+1)}},\ 0,\ 0,\ \ldots\ \ldots\ \ldots\ \ldots\ \ldots,\ 0 \right\}.$$

The Cartan-Weyl bases for the generators in the self-representation of $SU(N)$ can be expressed as

$$H_j = \sqrt{2}\, T^{(3)}_{\ell-j+2}, \qquad E_{\vec{\alpha}_{ab}} = T^{(1)}_{ab} + i T^{(2)}_{ab}, \tag{8.4}$$

$$[H_j,\ E_{\vec{\alpha}_{ab}}] = \begin{cases} -\left[\dfrac{\ell-j+1}{\ell-j+2} \right]^{1/2} E_{\vec{\alpha}_{ab}} & \text{when } \ell - j + 2 = a, \\[2ex] \left[\dfrac{1}{(\ell-j+2)(\ell-j+1)} \right]^{1/2} E_{\vec{\alpha}_{ab}} & \text{when } a < \ell - j + 2 < b, \\[2ex] \left\{ \left[\dfrac{1}{(\ell-j+2)(\ell-j+1)} \right]^{1/2} + \left[\dfrac{\ell-j+1}{\ell-j+2} \right]^{1/2} \right\} E_{\vec{\alpha}_{ab}} \\ \hspace{4cm} \text{when } \ell - j + 2 = b, \\[2ex] 0 \hspace{1.5cm} \text{when } \ell - j + 2 < a,\ \text{or } \ell - j + 2 > b, \end{cases}$$

where $a < b$. Namely,

$$[H_j,\ E_{\vec{\alpha}_{ab}}] = \sqrt{2}\left\{(V_a)_{jj} - (V_b)_{jj}\right\} E_{\vec{\alpha}_{ab}}. \qquad (8.5)$$

When $a = b - 1 = \mu$, $E_{\mathbf{r}_\mu} = E_{\vec{\alpha}_{\mu(\mu+1)}}$ corresponds to the simple root $\mathbf{r}_\mu$,

$$\mathbf{r}_\mu = \sqrt{2}\,(\mathbf{V}_\mu - \mathbf{V}_{\mu+1}),\quad \mathbf{r}_\mu \cdot \mathbf{r}_\nu = 2\delta_{\mu\nu} - \delta_{\mu(\nu-1)},\quad 1 \le \mu \le \nu \le \ell,$$

$$(\mathbf{V}_\mu - \mathbf{V}_{\mu+1})_\tau = \delta_{\mu(\ell-\tau+1)}\sqrt{\frac{\mu+1}{2\mu}} - \delta_{\mu(\ell-\tau+2)}\sqrt{\frac{\mu-1}{2\mu}}.$$
$$(8.6)$$

When $a < b$, $E_{\vec{\alpha}_{ab}}$ corresponds to the positive root $\vec{\alpha}_{ab} = \sqrt{2}\left\{\mathbf{V}_a - \mathbf{V}_b\right\} = \sum_{\mu=a}^{b-1} \mathbf{r}_\mu$. Therefore, the Lie algebra of the SU$(\ell+1)$ group is A_ℓ. The largest root in A_ℓ is $\vec{\omega} = \sum_{\mu=1}^{\ell} \mathbf{r}_\mu = \mathbf{w}_1 + \mathbf{w}_\ell$, which is the highest weight of the adjoint representation of A_ℓ.

1. Calculate the anti-commutation relations for the generators T_A in the self-representation of SU(N).

Solution. The anti-commutator for two generators T_A and T_B is hermitian, but not traceless. Let

$$\{T_A,\ T_B\} = T_A T_B + T_B T_A = \sum_{P=1}^{N^2-1} d_{ABP} T_P + c\mathbf{1}.$$

Taking the trace, we have $cN = 2\mathrm{Tr}\,(T_A T_B) = \delta_{AB}$. Thus,

$$\{T_A,\ T_B\} = T_A T_B + T_B T_A = \sum_{P=1}^{N^2-1} d_{ABP} T_P + \frac{1}{N}\delta_{AB}\mathbf{1}. \qquad (8.7)$$

2. Calculate the structure constants and d_{ABD} in Eq. (8.7) for the SU(3) group.

Solution. Write the commutation relations for the generators T_A in the self-representation of SU(N):

$$[T_A,\ T_B] = T_A T_B - T_B T_A = i\sum_{P=1}^{N^2-1} C_{AB}^{\ P} T_P, \qquad (8.8)$$

where $C_{AB}^{\ P}$ are the structure constants. Multiplying Eqs. (8.7) and (8.8) by T_D and taking the trace, we have

$$C_{AB}^{\ D} = -2i\mathrm{Tr}\,(T_A T_B T_D - T_B T_A T_D),$$
$$d_{ABD} = 2\mathrm{Tr}\,(T_A T_B T_D + T_B T_A T_D). \qquad (8.9)$$

C_{AB}^{D} are totally antisymmetric with respect to three indices, but d_{ABD} are totally symmetric. Substituting the generators T_A into Eq. (8.9), we obtain the nonvanishing structure constants C_{AB}^{D} and the coefficients d_{ABD} of SU(3) as listed in Table 8.1.

Table 8.1 The nonvanishing structure constants C_{AB}^{D} and the coefficients d_{ABD} of SU(3).

ABD	123	147	156	246	257	345	367	458	678
C_{AB}^{D}	1	$\dfrac{1}{2}$	$\dfrac{-1}{2}$	$\dfrac{1}{2}$	$\dfrac{1}{2}$	$\dfrac{1}{2}$	$\dfrac{-1}{2}$	$\dfrac{\sqrt{3}}{2}$	$\dfrac{\sqrt{3}}{2}$

ABD	118	146	157	228	247	256	338	344
d_{ABD}	$\dfrac{1}{\sqrt{3}}$	$\dfrac{1}{2}$	$\dfrac{1}{2}$	$\dfrac{1}{\sqrt{3}}$	$\dfrac{-1}{2}$	$\dfrac{1}{2}$	$\dfrac{1}{\sqrt{3}}$	$\dfrac{1}{2}$

ABD	355	366	377	448	558	668	778	888
d_{ABD}	$\dfrac{1}{2}$	$\dfrac{-1}{2}$	$\dfrac{-1}{2}$	$\dfrac{-1}{2\sqrt{3}}$	$\dfrac{-1}{2\sqrt{3}}$	$\dfrac{-1}{2\sqrt{3}}$	$\dfrac{-1}{2\sqrt{3}}$	$\dfrac{-1}{\sqrt{3}}$

3. Calculate the Killing form for SU(N) group.

Solution. The generator I_A^{adj} in the adjoint representation is related to the structure constant C_{AB}^{D} by

$$\left(I_A^{\text{adj}}\right)_{DB} = iC_{AB}^{D}. \tag{8.10}$$

Thus, the Killing form g_{AB} is

$$g_{AB} = \sum_{PQ} C_{AP}^{Q} C_{BQ}^{P} = -\text{Tr}\left(I_A^{\text{adj}} I_B^{\text{adj}}\right). \tag{8.11}$$

Being an $(N^2 - 1)$-dimensional matrix, g is commutable with any generator I_A^{adj} in the adjoint representation of SU(N):

$$
\begin{aligned}
\left[g, I_A^{\text{adj}}\right]_{BD} &= \sum_P \left\{-\text{Tr}\left(I_B^{\text{adj}} I_P^{\text{adj}}\right)\left(I_A^{\text{adj}}\right)_{PD} + \left(I_A^{\text{adj}}\right)_{BP}\text{Tr}\left(I_P^{\text{adj}} I_D^{\text{adj}}\right)\right\} \\
&= -i\sum_P \left\{C_{AD}^{P}\text{Tr}\left(I_B^{\text{adj}} I_P^{\text{adj}}\right) + C_{AB}^{P}\text{Tr}\left(I_P^{\text{adj}} I_D^{\text{adj}}\right)\right\} \\
&= -\text{Tr}\left\{I_B^{\text{adj}}\left(I_A^{\text{adj}} I_D^{\text{adj}} - I_D^{\text{adj}} I_A^{\text{adj}}\right) + \left(I_A^{\text{adj}} I_B^{\text{adj}} - I_B^{\text{adj}} I_A^{\text{adj}}\right) I_D^{\text{adj}}\right\} = 0.
\end{aligned}
$$

Due to the Schur theorem, g_{AB} is a constant matrix, $g_{AB} = c\delta_{AB}$. Calculate the constant c for the SU(3) group by taking $A = 3$:

$$c = g_{33} = 2\left\{ C_{31}{}^{2}C_{32}{}^{1} + C_{34}{}^{5}C_{35}{}^{4} + C_{36}{}^{7}C_{37}{}^{6} \right\} = -2\left\{ 1 + 2 \times (1/4) \right\} = 3.$$

Note that $T_4 = T_{13}^{(1)}$ and $T_6 = T_{23}^{(1)}$. When generalizing the result to the SU(N) group, the subscript 3 in $T_{13}^{(1)}$ and $T_{23}^{(1)}$ becomes a, $3 \leq a \leq N$, so that the number in the curve brackets becomes $1 + 2(N-2)/4$. Therefore,

$$g_{AB} = -N\delta_{AB}, \qquad \text{for SU}(N). \qquad (8.12)$$

8.2 Irreducible Tensor Representations of SU(N)

★ The Chevalley bases for the generators in the self-representation of SU(N) are

$$
\begin{aligned}
(H_\mu)_{cd} &= \sum_{j=1}^{\ell} (\mathbf{r}_\mu)_j \, (H_j)_{cd} = -\sqrt{\frac{2(\mu-1)}{\mu}} \left(T_\mu^{(3)} \right)_{cd} \\
&\quad + \sqrt{\frac{2(\mu+1)}{\mu}} \left(T_{\mu+1}^{(3)} \right)_{cd} = \delta_{cd} \left(\delta_{\mu c} - \delta_{(\mu+1)c} \right), \\
(E_\mu)_{cd} &= \left(E_{\mathbf{r}_\mu} \right)_{cd} = \delta_{\mu c}\delta_{(\mu+1)d}, \qquad (F_\mu)_{cd} = \left(E_{-\mathbf{r}_\mu} \right)_{cd} = \delta_{(\mu+1)c}\delta_{\mu d}, \\
& 1 \leq \mu \leq \ell = N - 1.
\end{aligned}
$$
$$(8.13)$$

Note that for a given μ, three generators H_μ, E_μ and F_μ constitute the subalgebra $\mathcal{A}_\mu$, which is isomorphic onto the A_1 algebra.

★ A covariant tensor $\mathbf{T}_{a_1 \ldots a_n}$ of rank n with respect to SU(N) transforms as

$$(O_u\mathbf{T})_{a_1 \ldots a_n} = \sum_{d_1 \ldots d_n} u_{a_1 d_1} \cdots u_{a_n d_n} \mathbf{T}_{d_1 \ldots d_n}, \qquad u \in \text{SU}(N). \qquad (8.14)$$

The tensor basis $(\Theta_{b_1 \ldots b_n})_{a_1 \ldots a_n} = (\Theta_{b_1})_{a_1} \cdots (\Theta_{b_n})_{a_n}$ transforms as

$$
\begin{aligned}
(\Theta_{b_1 \ldots b_n})_{a_1 \ldots a_n} &= \delta_{a_1 b_1} \cdots \delta_{a_n b_n}, \qquad (\Theta_{b_1})_{a_1} = \delta_{a_1 b_1}, \\
(O_u\Theta_{b_1 \ldots b_n})_{a_1 \ldots a_n} &= \sum_{d_1 \ldots d_n} u_{a_1 d_1} \cdots u_{a_n d_n} (\Theta_{b_1 \ldots b_n})_{d_1 \ldots d_n} \\
&= u_{a_1 b_1} \cdots u_{a_n b_n} = \sum_{d_1 \ldots d_n} (\Theta_{d_1 \ldots d_n})_{a_1 \ldots a_n} u_{d_1 b_1} \cdots u_{d_n b_n}.
\end{aligned}
$$
$$(8.15)$$

Since the matrix entries $u_{a_i d_i}$ in Eqs. (8.14) and (8.15) are commutable, the symmetry among the indices of a tensor keeps invariant in the SU(N) transformation. Therefore, the tensor space is reducible, and can be reduced into the direct sum of some irreducible tensor subspaces with the definite permutation symmetry among the indices.

★ In order to describe the symmetry of the indices of a tensor, we define the permutation operators R for the tensor. R is a linear operator by which a tensor $\mathbf{T}$ is transformed into a new tensor $\mathbf{T}'$. The product of two permutation operators should satisfy the multiplication rule of elements in the permutation group. Let us begin with a simple example. If

$$R = \begin{pmatrix} 1\,2\,3 \\ 2\,3\,1 \end{pmatrix} = \begin{pmatrix} 3\,1\,2 \\ 1\,2\,3 \end{pmatrix} = (1\,2\,3) = (1\,2)(2\,3),$$

then a tensor transforms as

$$\mathbf{T}'_{a_1 a_2 a_3} = (2\,3)\mathbf{T}_{a_1 a_2 a_3} = \mathbf{T}_{a_1 a_3 a_2},$$
$$R\mathbf{T}_{a_1 a_2 a_3} = (1\,2)\mathbf{T}'_{a_1 a_2 a_3} = \mathbf{T}'_{a_2 a_1 a_3} = \mathbf{T}_{a_2 a_3 a_1} \neq \mathbf{T}_{a_3 a_1 a_2}.$$

On the other hand, a tensor basis transforms as

$$(2\,3)\Theta_{b_1 b_2 b_3} = \Theta_{b_1 b_3 b_2},$$
$$R\Theta_{b_1 b_2 b_3} = (1\,2)\Theta_{b_1 b_3 b_2} = \Theta_{b_3 b_1 b_2} \neq \Theta_{b_2 b_3 b_1}.$$

They transform in the different but compatible ways. In fact, expanding a tensor $\mathbf{T}$ in the tensor basis Θ_{abc}, we have

$$
\begin{aligned}
(R\mathbf{T})_{a_1 a_2 a_3} &= R \sum_{b_1, b_2, b_3} T_{b_1 b_2 b_3} \left(\Theta_{b_1 b_2 b_3}\right)_{a_1 a_2 a_3} \\
&= \sum_{b_1, b_2, b_3} T_{b_1 b_2 b_3} \left(R\Theta_{b_1 b_2 b_3}\right)_{a_1 a_2 a_3} \\
&= \sum_{b_1, b_2, b_3} T_{b_1 b_2 b_3} \left(\Theta_{b_3 b_1 b_2}\right)_{a_1 a_2 a_3} \\
&= T_{a_2 a_3 a_1} = (\mathbf{T})_{a_2 a_3 a_1}.
\end{aligned}
$$

In general, if

$$R = \begin{pmatrix} 1 & 2 & \cdots & n \\ r_1 & r_2 & \cdots & r_n \end{pmatrix} = \begin{pmatrix} \bar{r}_1 & \bar{r}_2 & \cdots & \bar{r}_n \\ 1 & 2 & \cdots & n \end{pmatrix},$$

we have

$$(R\mathbf{T})_{a_1 \cdots a_n} = \mathbf{T}_{a_{r_1} \cdots a_{r_n}}, \qquad R\Theta_{b_1 \cdots b_n} = \Theta_{b_{\bar{r}_1} \cdots b_{\bar{r}_n}}. \tag{8.16}$$

It is easy to show that the product order of an $SU(N)$ transformation O_u and a permutation R among indices of a tensor are commutable

$$RO_u\mathbf{T} = O_uR\mathbf{T}. \qquad (8.17)$$

This property is called the Weyl reciprocity. It is the theoretical foundation for decomposing a tensor space by the Young operators.

★ In general, the tensor space $\mathcal{T}$ is reducible. The minimal tensor subspace $\mathcal{T}_\mu^{[\lambda]}$ corresponding to an irreducible representation $[\lambda]$ of $SU(N)$ can be obtained by the projection of a Young operator, $\mathcal{T}_\mu^{[\lambda]} = \mathcal{Y}_\mu^{[\lambda]}\mathcal{T}$. Two tensor subspaces denoted by different Young patterns are inequivalent. Two tensor subspaces denoted by different standard Young tableaux of the same Young pattern are equivalent, but are linearly independent.

★ An arbitrary tensor basis $\mathcal{Y}_\mu^{[\lambda]}\Theta_{a_1\ldots a_n}$ in a tensor subspace $\mathcal{T}_\mu^{[\lambda]}$ can be denoted by a tensor Young tableau with the Young pattern $[\lambda]$, where the box filled with j in the standard Young tableau $\mathcal{Y}_\mu^{[\lambda]}$ is now filled with the subscript a_j. For example, for the Young operator $\mathcal{Y}_1^{[2,1]}$ corresponding to the standard Young tableau $\boxed{\begin{array}{cc}1&2\\3\end{array}}$, the tensor Young tableau of the tensor basis $\mathcal{Y}_1^{[2,1]}\Theta_{a_1a_2a_3}$ is $\boxed{\begin{array}{cc}a_1&a_2\\a_3\end{array}}$. For convenience we write

$$\mathcal{Y}_1^{[2,1]}\Theta_{a_1a_2a_3} = \boxed{\begin{array}{cc}a_1&a_2\\a_3\end{array}} \quad \text{in } \mathcal{Y}_1^{[2,1]}\mathcal{T}. \qquad (8.18)$$

Note that the same tensor Young tableaux in different tensor subspaces denote the different tensor bases. For example, if the Young operator $\mathcal{Y}_2^{[2,1]}$ corresponds to the standard Young tableau $\boxed{\begin{array}{cc}1&3\\2\end{array}}$, the tensor Young tableau $\boxed{\begin{array}{cc}a&b\\c\end{array}}$ denotes the different tensors in two tensor subspaces:

$$\mathcal{Y}_1^{[2,1]}\Theta_{abc} = \Theta_{abc} + \Theta_{bac} - \Theta_{cba} - \Theta_{bca} \quad \text{in } \mathcal{Y}_1^{[2,1]}\mathcal{T},$$
$$\mathcal{Y}_2^{[2,1]}\Theta_{acb} = \Theta_{acb} + \Theta_{bca} - \Theta_{cab} - \Theta_{bac} \quad \text{in } \mathcal{Y}_2^{[2,1]}\mathcal{T}.$$

Due to the symmetry and the Fock condition (6.5) satisfied by the Young operator, the tensor Young tableaux in a given tensor subspace satisfy some linear relations. For example, for the Young operator $\mathcal{Y}_1^{[2,1]}$ we have

$$\mathcal{Y}_1^{[2,1]}\Theta_{abc} = -\mathcal{Y}_1^{[2,1]}(1\ 3)\Theta_{abc} = -\mathcal{Y}_1^{[2,1]}\Theta_{cba},$$
$$\mathcal{Y}_1^{[2,1]}\Theta_{abc} = \mathcal{Y}_1^{[2,1]}\left\{(2\ 1) + (2\ 3)\right\}\Theta_{abc} = \mathcal{Y}_1^{[2,1]}\Theta_{bac} + \mathcal{Y}_1^{[2,1]}\Theta_{acb},$$

$$\begin{array}{|c|c|}\hline a & b \\\hline c \\\cline{1-1}\end{array} = -\begin{array}{|c|c|}\hline c & b \\\hline a \\\cline{1-1}\end{array}, \qquad \begin{array}{|c|c|}\hline a & b \\\hline c \\\cline{1-1}\end{array} = \begin{array}{|c|c|}\hline b & a \\\hline c \\\cline{1-1}\end{array} + \begin{array}{|c|c|}\hline a & c \\\hline b \\\cline{1-1}\end{array}. \qquad (8.19)$$

The relations similar to Eq. (8.19) hold for the general cases. Those relations obviously depend upon the Young pattern, but are independent of the Young operators with the same Young pattern as well as the tensor subspaces. The first equality says that the tensor Young tableau is antisymmetric with respect to any two digits filled in the same column. The tensor Young tableau is equal to zero if there are two equal digits filled in its same column. We may rearrange the digits in each column of the tensor Young tableau by the antisymmetric relation such that the digit increases downwards. The second equality is also called the Fock condition.

★ Any tensor in a tensor subspace $T_\mu^{[\lambda]}$ is a combination of the tensor bases $\mathcal{Y}_\mu^{[\lambda]}\Theta_{a_1\ldots a_n}$. $\mathcal{Y}_\mu^{[\lambda]}\Theta_{a_1\ldots a_n}$ can be expressed as

$$\mathcal{Y}_\mu^{[\lambda]}\Theta_{a_1\ldots a_n} = \mathcal{Y}_\mu^{[\lambda]}S\Theta_{b_1\ldots b_n}, \qquad b_1 \leq b_2 \leq \ldots \leq b_n.$$

Since $\mathcal{Y}_\mu^{[\lambda]}R_{\mu\nu}$ are the standard bases in the minimal right ideal of the permutation group produced by the Young operator $\mathcal{Y}_\mu^{[\lambda]}$, the following tensor bases

$$\mathcal{Y}_\mu^{[\lambda]}R_{\mu\nu}\Theta_{b_1\ldots b_n}, \qquad b_1 \leq b_2 \leq \ldots \leq b_n, \qquad (8.20)$$

constitute a complete set of the tensor bases in the tensor subspace $T_\mu^{[\lambda]}$, where $R_{\mu\nu}$ is the permutation transforming the standard Young tableau $\mathcal{Y}_\nu^{[\lambda]}$ to the standard Young tableau $\mathcal{Y}_\mu^{[\lambda]}$. We are going to discuss the characteristic of the basis (8.20). Since

$$\mathcal{Y}_\mu^{[\lambda]}R_{\mu\nu}\Theta_{b_1\ldots b_n} = R_{\mu\nu}\left\{\mathcal{Y}_\nu^{[\lambda]}\Theta_{b_1\ldots b_n}\right\}, \qquad b_1 \leq b_2 \leq \ldots \leq b_n, \qquad (8.21)$$

the tensor Young tableau for the tensor basis $\mathcal{Y}_\mu^{[\lambda]}R_{\mu\nu}\Theta_{b_1\ldots b_n}$ in the tensor subspace $T_\mu^{[\lambda]}$ is the same as that for the tensor basis $\mathcal{Y}_\nu^{[\lambda]}\Theta_{b_1\ldots b_n}$ in the tensor subspace $T_\nu^{[\lambda]}$. In fact, if R permutates $\bar{r}_j$ to j, $R_{\mu\nu}\Theta_{b_1\ldots b_n} = \Theta_{b_{\bar{r}_1}\ldots b_{\bar{r}_n}}$. Therefore, the box filled with j in the standard Young tableau $\mathcal{Y}_\mu^{[\lambda]}$ is filled with $\bar{r}_j$ in the standard Young tableau $\mathcal{Y}_\nu^{[\lambda]}$, and is filled with $b_{\bar{r}_j}$ both in the tensor Young tableau $\mathcal{Y}_\mu^{[\lambda]}R_{\mu\nu}\Theta_{b_1\ldots b_n}$ and in the tensor Young tableau $\mathcal{Y}_\nu^{[\lambda]}\Theta_{b_1\ldots b_n}$.

★ A tensor Young tableau is called standard if the digit in each column of the tableau increases downwards and in each row does not decrease rightwards. Since $\mathcal{Y}_\nu^{[\lambda]}$ is a standard Young tableau and $b_1 \le b_2 \le \dots \le b_n$, the tensor Young tableau $\mathcal{Y}_\nu^{[\lambda]}\Theta_{b_1\dots b_n}$ as well as the tensor Young tableau $\mathcal{Y}_\mu^{[\lambda]}R_{\mu\nu}\Theta_{b_1\dots b_n}$ is standard. The standard tensor Young tableaux constitute a complete set of the tensor bases in the tensor subspace $\mathcal{T}_\mu^{[\lambda]}$. The standard tensor Young tableau is the eigenstate of H_μ where, due to Eq. (8.13), the eigenvalue is equal to the number of the digit μ filled in the tableau, subtracting the number of the digit $(\mu+1)$. Therefore, the representation matrix of the Chevalley basis H_μ is diagonal in the tensor bases. The action of F_μ on the standard tensor Young tableau is equal to the sum of all possible tensor Young tableaux, each of which is obtained from the original one by replacing one filled digit μ with the digit $(\mu+1)$.

If without special indication, we use the notation for the inner product of tensors that the bases $\Theta_{a_1\dots a_n}$ are orthonormal. Two standard tensor Young tableaux with the different sets of the filled digits are orthogonal, but may not be normalized to the same number. Two standard tensor Young tableaux with the equal set of the filled digits but different filling order may not be orthogonal. To find a set of the orthonormal bases, one need to make suitable combination and normalization from the standard tensor Young tableaux.

★ Since $R_{\mu\nu}\mathcal{Y}_\nu^{[\lambda]}$ are the standard bases in the minimal left ideal of the permutation group corresponding to the irreducible representation $[\lambda]$, Eq. (8.21) also shows that the same tensor Young tableaux in all tensor subspace $\mathcal{T}_\nu^{[\lambda]}$ with the same Young pattern $[\lambda]$ constitutes a complete set of the standard bases in the irreducible representation space of the permutation group, denoted by $[\lambda]$. Therefore, in the decomposition of the tensor space

$$\mathcal{T} = \frac{1}{n!}\sum_{[\lambda]} d_{[\lambda]}(S_n) \sum_\mu \mathcal{Y}_\mu^{[\lambda]}y_\mu^{[\lambda]}\mathcal{T} = \bigoplus_{[\lambda],\mu} \mathcal{T}_\mu^{[\lambda]},$$
$$\frac{d_{[\lambda]}(S_n)}{n!}\mathcal{Y}_\mu^{[\lambda]}y_\mu^{[\lambda]}\mathcal{T} = \mathcal{Y}_\mu^{[\lambda]}\mathcal{T} = \mathcal{T}_\mu^{[\lambda]},$$

(8.22)

the tensor subspace $\mathcal{T}_\mu^{[\lambda]}$ corresponds to the irreducible representation $[\lambda]$ of SU(N), and for a given Young pattern $[\lambda]$, the tensor bases with the same tensor Young tableau in the $d_{[\lambda]}(S_n)$ subspaces $\mathcal{T}_\mu^{[\lambda]}$ span the irreducible representation space of the permutation group S_n denoted by $[\lambda]$.

★ The dimension $d_{[\lambda]}(\text{SU}(N))$ of the representation $[\lambda]$ of SU(N) can be calculated by the hook rule. In this rule the dimension is expressed by a quotient, where the numerator is the product of $(N + m_{ij})$, and the denominator is the product of h_{ij}. $m_{ij} = j - i$ is called the content of the box located in the jth column of the ith row of the Young pattern $[\lambda]$. h_{ij} is called the hook number of that box in the Young pattern, which is equal to the number of the boxes on its right in the ith row of the Young pattern, plus the number of the boxes below it in the jth column, and plus one.

$$d_{[\lambda]}(\mathrm{SU}(N)) = \prod_{ij} \frac{N + j - i}{h_{ij}} = \frac{Y_A^{[\lambda]}}{Y_h^{[\lambda]}}, \qquad (8.23)$$

where the symbol $Y_A^{[\lambda]}$ is a tableau obtained from the Young pattern $[\lambda]$ by filling $(N + m_{ij})$ into the box located in its ith row and jth column, and the symbol $Y_h^{[\lambda]}$ is a tableau obtained from the Young pattern $[\lambda]$ by filling h_{ij} into that box. The symbol $Y_A^{[\lambda]}$ means the product of the filled digits in it, so does $Y_h^{[\lambda]}$.

★ The one-row Young pattern $[n]$ denotes the totally symmetric tensor representation. The one-column Young pattern $[1^n]$ denotes the totally antisymmetric tensor representation. The totally antisymmetric tensor space of rank N, denoted by $[1^N]$, contains only one standard tensor Young tableau, corresponds to the tensor basis

$$(\mathbf{E})_{a_1 \dots a_N} \equiv \left(\mathcal{Y}^{[1^N]} \Theta_{12 \dots N} \right)_{a_1 \dots a_N} = \epsilon_{a_1 \dots a_N}. \qquad (8.24)$$

The tensor $\mathbf{E}$ is an invariant tensor in the $\mathrm{SU}(N)$ transformation, so $[1^N]$ denotes the identical representation of $\mathrm{SU}(N)$. The tensor subspace described by a Young pattern with more than N rows is a null space. The reduction of the direct product of two irreducible representations described by two Young patterns $[\lambda]$ and $[\mu]$ can be calculated by the Littlewood-Richardson rule. However, the Young patterns with more than N rows contained in the reduction should be removed. The representation denoted by the Young pattern $[\lambda]$ with N rows is equivalent to the representation denoted by the Young pattern $[\lambda']$ obtained from $[\lambda]$ by removing its first column.

★ A contravariant tensor $\mathbf{T}^{a_1 \dots a_n}$ of rank n with respect to $\mathrm{SU}(N)$ transforms as

$$\begin{aligned}
(O_u \mathbf{T})^{a_1 \dots a_n} &= \sum_{d_1 \dots d_n} u_{a_1 d_1}^* \dots u_{a_n d_n}^* \mathbf{T}^{d_1 \dots d_n} \\
&= \sum_{d_1 \dots d_n} \mathbf{T}^{d_1 \dots d_n} \left(u^{-1} \right)_{d_1 a_1} \dots \left(u^{-1} \right)_{d_n a_n}, \qquad u \in \mathrm{SU}(N).
\end{aligned} \qquad (8.25)$$

The tensor basis $\Theta^{b_1 \ldots b_n} = \left(\Theta^{b_1} \right)^{a_1} \ldots \left(\Theta^{b_n} \right)^{a_n}$ transforms as

$$\left(\Theta^{b_1 \ldots b_n} \right)^{a_1 \ldots a_n} = \delta_{a_1 b_1} \ldots \delta_{a_n b_n}, \qquad \left(\Theta^{b_1} \right)^{a_1} = \delta_{a_1 b_1},$$

$$O_u \Theta^{b_1 \ldots b_n} = \sum_{d_1 \ldots d_n} \Theta^{d_1 \ldots d_n} u^*_{d_1 b_1} \ldots u^*_{d_n b_n}$$

$$= \sum_{d_1 \ldots d_n} \left(u^{-1} \right)_{b_1 d_1} \ldots \left(u^{-1} \right)_{b_n d_n} \Theta^{d_1 \ldots d_n}.$$

$$(8.26)$$

The contravariant tensor space $\mathcal{T}^*$ can also be decomposed by the projection of the Young operator, $\mathcal{T}^* = \bigoplus \mathcal{Y}_\mu^{[\lambda]} \mathcal{T}^* = \bigoplus \mathcal{T}_\mu^{[\lambda]*}$. The tensor subspace $\mathcal{T}_\mu^{[\lambda]*}$ corresponds to the representation $[\lambda]^*$ of SU(N), which is the conjugate one of the representation $[\lambda]$. The basis in the tensor subspace $\mathcal{T}_\mu^{[\lambda]*}$ is denoted by the standard tensor Young tableau with star, for example

$$\mathcal{Y}_1^{[2,1]} \Theta^{a_1 a_2 a_3} = \boxed{\begin{array}{cc} a_1 & a_2 \\ \hline a_3 \end{array}}^{\,*} \qquad \text{in } \mathcal{Y}_1^{[2,1]} \mathcal{T}^*. \qquad (8.27)$$

★ A mixed tensor $\mathbf{T}^{b_1 \ldots b_m}_{a_1 \ldots a_n}$ of rank (n, m) with respect to SU(N) transforms as

$$\left(O_u \mathbf{T} \right)^{b_1 \ldots b_m}_{a_1 \ldots a_n} = \sum_{c_1 \ldots c_n} \sum_{d_1 \ldots d_n} u_{a_1 c_1} \ldots u_{a_n d_n} \mathbf{T}^{d_1 \ldots d_n}_{c_1 \ldots c_n} \left(u^{-1} \right)_{d_1 b_1} \ldots \left(u^{-1} \right)_{d_n b_n}.$$

$$(8.28)$$

Obviously, the mixed tensor $(\mathbf{D})^b_a = \delta^b_a$ is an invariant tensor of rank $(1, 1)$ in the SU(N) transformation.

A mixed tensor is said to be made a contraction (or the trace) of its one pair of indices, one covariant and one contravariant, if to take the pair of indices to be equal and then to sum over the indices. The trace tensor of a mixed tensor of rank (n, m) is a tensor of rank $(n - 1, m - 1)$. A mixed tensor can be decomposed into the sum of a series of traceless tensors with different ranks. For example,

$$T^b_a = \left\{ T^b_a - \frac{1}{N} \delta^b_a \sum_d T^d_d \right\} + \delta^b_a \left[\frac{1}{N} \sum_d T^d_d \right],$$

$$\mathbf{T} = \left\{ \mathbf{T} - \mathbf{D} \frac{1}{N} \sum_d T^d_d \right\} + \mathbf{D} \left[\frac{1}{N} \delta^b_a \sum_d T^d_d \right],$$

$$(8.29)$$

where the tensor in the curve brackets is a traceless mixed tensor of rank $(1, 1)$, and the tensor in the square brackets is a scalar [the tensor of rank $(0, 0)$]. Therefore, a mixed tensor space can be decomposed into the direct sum of a series of traceless tensor subspaces with different ranks, each of which is invariant in the $SU(N)$ transformation.

The traceless tensor space can be decomposed into the direct sum of some minimal tensor subspace by the projection of the Young operator $\mathcal{Y}_\mu^{[\lambda]}$ on the covariant indices and the projection of the Young operator $\mathcal{Y}_\nu^{[\tau]}$ on the contravariant indices, where the Young patterns $[\lambda]$ and $[\tau]$ contain n and m boxes, respectively. The minimal tensor subspace corresponds to an irreducible representation of $SU(N)$, denoted by $[\lambda]\backslash[\tau]^*$. It can be shown that the tensor subspace corresponding to $[\lambda]\backslash[\tau]^*$ is null space if the sum of the numbers of rows of $[\lambda]$ and $[\tau]$ is larger than N, and the representation $[\lambda]\backslash[\tau]^*$ of $SU(N)$ is equivalent to the representation $[\lambda']\backslash[\tau']^*$ if $[\tau]$ contains t rows, $[\tau']$ is obtained from $[\tau]$ by removing its first column, and $[\lambda']$ is obtained from $[\lambda]$ by attaching a new column with $(N-t)$ boxes from its left. Thus, the representation $[\tau]^*$ of $SU(N)$ is equivalent to the representation $[\lambda]$ if by rotating the Young pattern $[\tau]$ upside down and then by attaching it to the Young pattern $[\lambda]$ from below, it just forms a rectangle with N row.

4. Calculate the dimensions of the irreducible representations denoted by the following Young patterns for the $SU(3)$ group and for the $SU(6)$ group, respectively:

$$[3], \quad [2,1], \quad [3,3], \quad [4,2], \quad [5,1].$$

Solution.

$$d_{[3]}(SU(3)) = \frac{3\ 4\ 5}{3\ 2\ 1} = 10, \quad d_{[3]}(SU(6)) = \frac{6\ 7\ 8}{3\ 2\ 1} = 56,$$

$$d_{[2,1]}(SU(3)) = \frac{\begin{array}{cc}3\ 4\\2\end{array}}{\begin{array}{cc}3\ 1\\1\end{array}} = 8, \quad d_{[2,1]}(SU(6)) = \frac{\begin{array}{cc}6\ 7\\5\end{array}}{\begin{array}{cc}3\ 1\\1\end{array}} = 70,$$

$$d_{[3,3]}(SU(3)) = \frac{\begin{array}{ccc}3\ 4\ 5\\2\ 3\ 4\end{array}}{\begin{array}{ccc}4\ 3\ 2\\3\ 2\ 1\end{array}} = 10, \quad d_{[3,3]}(SU(6)) = \frac{\begin{array}{ccc}6\ 7\ 8\\5\ 6\ 7\end{array}}{\begin{array}{ccc}4\ 3\ 2\\3\ 2\ 1\end{array}} = 490,$$

$$d_{[4,2]}(SU(3)) = \frac{\begin{array}{cccc}3\ 4\ 5\ 6\\2\ 3\end{array}}{\begin{array}{cccc}5\ 4\ 2\ 1\\2\ 1\end{array}} = 27, \quad d_{[4,2]}(SU(6)) = \frac{\begin{array}{cccc}6\ 7\ 8\ 9\\5\ 6\end{array}}{\begin{array}{cccc}5\ 4\ 2\ 1\\2\ 1\end{array}} = 1134,$$

$$d_{[5,1]}(SU(3)) = \frac{\begin{array}{c}3\,4\,5\,6\,7\\2\end{array}}{\begin{array}{c}6\,4\,3\,2\,1\\1\end{array}} = 35, \quad d_{[5,1]}(SU(6)) = \frac{\begin{array}{c}6\,7\,8\,9\,10\\5\end{array}}{\begin{array}{c}6\,4\,3\,2\,1\\1\end{array}} = 1050.$$

5. Calculate the Clebsch-Gordan series for the following direct product representations, and compare their dimensions by Eq. (8.23) for the SU(3) group and for the SU(6) group, respectively:

(a) $[2,1] \otimes [3,0]$, b) $[3,0] \otimes [3,0]$, c) $[3,0] \otimes [3,3]$, (d) $[4,2] \otimes [2,1]$.

Solution. a): $\begin{array}{c}0\,0\\0\end{array} \otimes 1\,1\,1 \simeq \begin{array}{c}0\,0\,1\,1\,1\\0\end{array} \oplus \begin{array}{c}0\,0\,1\,1\\0\,1\end{array} \oplus \begin{array}{c}0\,0\,1\,1\\0\\1\end{array} \oplus \begin{array}{c}0\,0\,1\\0\,1\\1\end{array}$,

$SU(3) : 8 \times 10 = 80 = 35 + 27 + 10 + 8,$

$SU(6) : 70 \times 56 = 3920 = 1050 + 1134 + 840 + 896.$

b): $0\,0\,0 \otimes 1\,1\,1 \simeq 0\,0\,0\,1\,1\,1 \oplus \begin{array}{c}0\,0\,0\,1\,1\\1\end{array} \oplus \begin{array}{c}0\,0\,0\,1\\1\,1\end{array} \oplus \begin{array}{c}0\,0\,0\\1\,1\,1\end{array}$,

$SU(3) : 10 \times 10 = 100 = 28 + 35 + 27 + 10,$

$SU(6) : 56 \times 56 = 3136 = 462 + 1050 + 1134 + 490.$

c): $1\,1\,1 \otimes \begin{array}{c}0\,0\,0\\0\,0\,0\end{array} \simeq \begin{array}{c}0\,0\,0\,1\,1\,1\\0\,0\,0\end{array} \oplus \begin{array}{c}0\,0\,0\,1\,1\\0\,0\,0\\1\end{array} \oplus \begin{array}{c}0\,0\,0\,1\\0\,0\,0\\1\,1\end{array} \oplus \begin{array}{c}0\,0\,0\\0\,0\,0\\1\,1\,1\end{array}$,

$SU(3) : 10 \times 10 \;=\; 100 = 64 + 27 + 8 + 1,$

$SU(6) : 56 \times 490 = 27440 = 9240 + 11340 + 5880 + 980.$

d): $\begin{array}{c}0\,0\,0\,0\\0\,0\end{array} \otimes \begin{array}{c}1\,1\\2\end{array} \simeq \begin{array}{c}0\,0\,0\,0\,1\,1\\0\,0\,2\end{array} \oplus \begin{array}{c}0\,0\,0\,0\,1\,1\\0\,0\\2\end{array} \oplus \begin{array}{c}0\,0\,0\,0\,1\\0\,0\,1\,2\end{array}$

$\oplus \begin{array}{c}0\,0\,0\,0\,1\\0\,0\,1\\2\end{array} \oplus \begin{array}{c}0\,0\,0\,0\,1\\0\,0\,2\\1\end{array} \oplus \begin{array}{c}0\,0\,0\,0\,1\\0\,0\\1\,2\end{array}$

$$
\begin{array}{c}
0\,0\,0\,0\,1 \\
0\,0 \\
\oplus\quad 1 \\
2
\end{array}
\quad
\begin{array}{c}
0\,0\,0\,0 \\
\oplus\,0\,0\,1\,1 \\
2
\end{array}
\quad
\begin{array}{c}
0\,0\,0\,0 \\
\oplus\,0\,0\,1 \\
1\,2
\end{array}
\quad
\begin{array}{c}
0\,0\,0\,0 \\
0\,0\,1 \\
\oplus\quad 1 \\
2
\end{array}
\quad
\begin{array}{c}
0\,0\,0\,0 \\
0\,0 \\
\oplus\quad 1\,1 \\
2
\end{array}\quad ,
$$

$SU(3):\ 27 \times 8 = 216$
$$= 64 + 35 + 35 + 2 \times 27 + 10 + 0 + 10 + 8 + 0 + 0,$$
$SU(6):\ 1134 \times 70 = 79380$
$$= 9240 + 11550 + 5880 + 2 \times 11340 + 6000 + 5670$$
$$+ 4704 + 5880 + 4536 + 3240.$$

6. Write all tensor Young tableaux in the tensor subspace $\mathcal{T}_1^{[2,1]} = \mathcal{Y}_1^{[2,1]}\mathcal{T}$ of SU(3), and prove that the standard tensor Young tableaux constitute the complete bases in this subspace.

Solution. The standard Young tableau $\mathcal{Y}_1^{[2,1]}$ is $\begin{array}{|c|c|}\hline 1 & 2 \\\hline 3 \\\cline{1-1}\end{array}$. Thus,

$$
\mathcal{Y}_1^{[2,1]}\Theta_{abc} = \{E + (1\,2) - (1\,3) - (1\,2)(1\,3)\}\,\Theta_{abc}
$$
$$
= \Theta_{abc} + \Theta_{bac} - \Theta_{cba} - \Theta_{bca}.
$$

If there is a pair of the filled digits to be equal in the tensor Young tableau, from Eq. (8.19) we have

$$
\begin{array}{|c|c|}\hline 1 & 1 \\\hline 2 \\\cline{1-1}\end{array} = -\,\begin{array}{|c|c|}\hline 2 & 1 \\\hline 1 \\\cline{1-1}\end{array} = \mathcal{Y}_1^{[2,1]}\Theta_{112} = 2\Theta_{112} - \Theta_{211} - \Theta_{121},
$$

$$
\begin{array}{|c|c|}\hline 1 & 1 \\\hline 3 \\\cline{1-1}\end{array} = -\,\begin{array}{|c|c|}\hline 3 & 1 \\\hline 1 \\\cline{1-1}\end{array} = \mathcal{Y}_1^{[2,1]}\Theta_{113} = 2\Theta_{113} - \Theta_{311} - \Theta_{131},
$$

$$
\begin{array}{|c|c|}\hline 1 & 2 \\\hline 2 \\\cline{1-1}\end{array} = -\,\begin{array}{|c|c|}\hline 2 & 2 \\\hline 1 \\\cline{1-1}\end{array} = \mathcal{Y}_1^{[2,1]}\Theta_{122} = \Theta_{122} + \Theta_{212} - 2\Theta_{221},
$$

$$
\begin{array}{|c|c|}\hline 2 & 2 \\\hline 3 \\\cline{1-1}\end{array} = -\,\begin{array}{|c|c|}\hline 3 & 2 \\\hline 2 \\\cline{1-1}\end{array} = \mathcal{Y}_1^{[2,1]}\Theta_{223} = 2\Theta_{223} - \Theta_{322} - \Theta_{232},
$$

$$
\begin{array}{|c|c|}\hline 1 & 3 \\\hline 3 \\\cline{1-1}\end{array} = -\,\begin{array}{|c|c|}\hline 3 & 3 \\\hline 1 \\\cline{1-1}\end{array} = \mathcal{Y}_1^{[2,1]}\Theta_{133} = \Theta_{133} + \Theta_{313} - 2\Theta_{331},
$$

$$
\begin{array}{|c|c|}\hline 2 & 3 \\\hline 3 \\\cline{1-1}\end{array} = -\,\begin{array}{|c|c|}\hline 3 & 3 \\\hline 2 \\\cline{1-1}\end{array} = \mathcal{Y}_1^{[2,1]}\Theta_{233} = \Theta_{233} + \Theta_{323} - 2\Theta_{332},
$$

where the left tensor Young tableau in each equality is standard. If all three filled digits in the tensor Young tableau are different, from Eq. (8.19) we have

$$\young(12,3) = -\;\young(32,1) = \mathcal{Y}_1^{[2,1]}\Theta_{123} = \Theta_{123} + \Theta_{213} - \Theta_{321} - \Theta_{231},$$

$$\young(13,2) = -\;\young(23,1) = \mathcal{Y}_1^{[2,1]}\Theta_{132} = \Theta_{132} + \Theta_{312} - \Theta_{231} - \Theta_{321},$$

$$\young(21,3) = -\;\young(31,2) = \young(12,3) - \young(13,2)$$

$$= \mathcal{Y}_1^{[2,1]}\Theta_{213} = \Theta_{213} + \Theta_{123} - \Theta_{312} - \Theta_{132},$$

where the left tensor Young tableaux in the first two equalities are standard. The dimension of the subspace $\mathcal{T}_1^{[2,1]}$ is eight. There are eight standard tensor Young tableaux. The first six standard tensor Young tableaux are normalized to 6, but the last two to 4. In addition, the last two standard tensor Young tableaux are not orthogonal to each other.

7. Try to express each nonzero tensor Young tableau for the irreducible representation $[3,1]$ of SU(3) as the linear combination of the standard tensor Young tableaux.

Solution. For the SU(3) group, the digits filled in the tensor Young tableaux are 1, 2 and 3. The equal digits cannot be filled in the same column. There are three types of nonzero tensor Young tableaux. A tensor Young tableau in the first type contains three equal digits. From Eq. (8.19) we have

$$\young(aaa,b) = -\;\young(baa,a)\,.$$

When $a < b$ the left tensor Young tableaux is standard, while when $a > b$ the right one is standard. There are $3 \times 2 = 6$ standard tensor Young tableaux in the first type. A tensor Young tableau in the second type contains two pairs of equal digits. From Eq. (8.19) we have

$$\young(aab,b) = \young(aba,b) = -\;\young(bab,a) = -\;\young(bba,a)\,.$$

When $a < b$, the first tensor Young tableau is standard, while when $a > b$ the fourth one is standard. There are 3 standard tensor Young tableaux in the second type. A tensor Young tableau in the third type contains one pair of equal digits. There are six standard tensor Young tableaux in the third type. The other tensor Young tableaux can be expanded in the standard tensor Young tableaux by Eq. (8.19),

$$\young(112,3) = \young(121,3) = -\young(312,1) = -\young(321,1),$$

$$\young(113,2) = \young(131,2) = -\young(213,1) = -\young(231,1),$$

$$\young(211,3) = -\young(311,2) = \young(112,3) - \young(113,2),$$

$$\young(123,2) = \young(132,2) = -\young(223,1) = -\young(232,1),$$

$$\young(122,3) = -\young(322,1),$$

$$\young(221,3) = \young(212,3) = -\young(321,2) = -\young(312,2)$$
$$= \young(122,3) - \young(123,2),$$

$$\young(133,2) = -\young(233,1),$$

$$\young(123,3) = \young(132,3) = -\young(323,1) = -\young(332,1),$$

$$\young(313,2) = \young(331,2) = -\young(213,3) = -\young(231,3)$$
$$= \young(133,2) - \young(123,3).$$

The dimension of the irreducible representation [3, 1] of SU(3) is 15. There are 15 standard tensor Young tableaux.

8. Write the explicit expansion of each standard tensor Young tableau in the tensor subspace $\mathcal{Y}_2^{[3,1]}\mathcal{T}$, where $\mathcal{T}$ is the tensor space of rank four for the SU(3) group and the standard Young tableau of the Young operator $\mathcal{Y}_2^{[3,1]}$ is $\young(124,3)$.

Solution. First, we write the explicit formula for the application of the

Young operator $\mathcal{Y}_2^{[3,1]}$ to the tensor basis Θ_{abcd}:

$$\mathcal{Y}_2^{[3,1]}\Theta_{abcd} = \{E + (1\ 2) + (1\ 4) + (2\ 4) + (1\ 2\ 4) + (1\ 4\ 2) - (1\ 3)$$
$$- (2\ 1\ 3) - (4\ 1\ 3) - (2\ 4)(1\ 3) - (2\ 4\ 1\ 3) - (4\ 2\ 1\ 3)\}\,\Theta_{abcd}$$
$$= \Theta_{abcd} + \Theta_{bacd} + \Theta_{dbca} + \Theta_{adcb} + \Theta_{dacb} + \Theta_{bdca}$$
$$- \Theta_{cbad} - \Theta_{bcad} - \Theta_{dbac} - \Theta_{cdab} - \Theta_{dcab} - \Theta_{bdac}.$$

Similar to Problem 7, there are three types of standard tensor Young tableaux. In the first type there are six standard tensor Young tableaux, each of which contains three equal digits.

$$\begin{array}{|c|c|c|}\hline a & a & a \\\hline b \\\cline{1-1}\end{array} = - \begin{array}{|c|c|c|}\hline b & a & a \\\hline a \\\cline{1-1}\end{array}$$
$$= \mathcal{Y}_2^{[3,1]}\Theta_{aaba} = 6\Theta_{aaba} - 2\Theta_{baaa} - 2\Theta_{abaa} - 2\Theta_{aaab}.$$

When $a < b$ the left tensor Young tableau is standard, and when $a > b$ the right one is standard. The module square of the standard tensor Young tableaux in the first type is 48. In the second type there are three standard tensor Young tableaux, each of which contains two pairs of equal digits:

$$\begin{array}{|c|c|c|}\hline a & a & b \\\hline b \\\cline{1-1}\end{array} = \mathcal{Y}_2^{[3,1]}\Theta_{aabb} = 2\Theta_{aabb} + 2\Theta_{baba} + 2\Theta_{abba}$$
$$- 2\Theta_{baab} - 2\Theta_{abab} - 2\Theta_{bbaa}.$$

The module square of the standard tensor Young tableau in the second type is 24. In the third type there are six standard tensor Young tableaux, each of which contains one pair of equal digits:

$$\begin{array}{|c|c|c|}\hline 1 & a & a \\\hline b \\\cline{1-1}\end{array} = \mathcal{Y}_2^{[3,1]}\Theta_{1aba} = 2\Theta_{1aba} + 2\Theta_{a1ba} + 2\Theta_{aab1}$$
$$- 2\Theta_{ba1a} - 2\Theta_{ab1a} - 2\Theta_{aa1b},$$

$$\begin{array}{|c|c|c|}\hline 1 & 1 & a \\\hline b \\\cline{1-1}\end{array} = \mathcal{Y}_2^{[3,1]}\Theta_{11ba} = 2\Theta_{11ba} + 2\Theta_{a1b1} + 2\Theta_{1ab1} - \Theta_{b11a}$$
$$- \Theta_{1b1a} - \Theta_{a11b} - \Theta_{ba11} - \Theta_{ab11} - \Theta_{1a1b},$$

$$\begin{array}{|c|c|c|}\hline 1 & 2 & 3 \\\hline 2 \\\cline{1-1}\end{array} = \mathcal{Y}_2^{[3,1]}\Theta_{1223} = \Theta_{1223} + \Theta_{2123} + \Theta_{3221} + \Theta_{1322}$$
$$+ \Theta_{3122} + \Theta_{2321} - 2\Theta_{2213} - 2\Theta_{3212} - 2\Theta_{2312},$$

$$\boxed{\begin{array}{|c|c|c|}\hline 1 & 2 & 3 \\\hline 3 \\\cline{1-1}\end{array}} = \mathcal{Y}_2^{[3,1]}\Theta_{1233} = \Theta_{1233} + \Theta_{2133} + \Theta_{3231} + \Theta_{1332}$$

$$+ \Theta_{3132} + \Theta_{2331} - 2\Theta_{3213} - 2\Theta_{2313} - 2\Theta_{3312},$$

where a and b are respectively taken 2 and 3. The module square of the first form, which includes two standard tensor Young tableaux, is 24, and that of the remaining standard tensor Young tableaux is 18. Two standard tensor Young tableaux with the equal set of the filled digits are not orthogonal.

9. Transform the following traceless mixed tensor representations of the SU(6) group into the covariant tensor representations, respectively, and calculate their dimensions:

      (1) $[3, 2, 1]^*$,   (2) $[3, 2, 1]\backslash[3, 3]^*$,   (3) $[4, 3, 1]\backslash[3, 2]^*$.

Solution. (1) $[3, 2, 1]^* \simeq [1^3]\backslash[2, 1]^* \simeq [2^3, 1]\backslash[1]^* \simeq [3^3, 2, 1]$,

$$d_{[3,2,1]}(SU(6)) = \frac{\begin{array}{l} 6\ 7\ 8 \\ 5\ 6 \\ 4 \end{array}}{\begin{array}{l} 5\ 3\ 1 \\ 3\ 1 \\ 1 \end{array}} = 896, \qquad d_{[3^3,2,1]}(SU(6)) = \frac{\begin{array}{l} 6\ 7\ 8 \\ 5\ 6\ 7 \\ 4\ 5\ 6 \\ 3\ 4 \\ 2 \end{array}}{\begin{array}{l} 7\ 5\ 3 \\ 6\ 4\ 2 \\ 5\ 3\ 1 \\ 3\ 1 \\ 1 \end{array}} = 896.$$

(2) $[3, 2, 1]\backslash[3, 3]^* \simeq [4, 3, 2, 1]\backslash[2, 2]^* \simeq [5, 4, 3, 2]\backslash[1, 1]^* \simeq [6, 5, 4, 3]$,

$$d_{[6,5,4,3]}(SU(6)) = \frac{\begin{array}{l} 6\ 7\ 8\ 9\ 10\ 11 \\ \cdot\ 5\ 6\ 7\ 8\ \ 9 \\ 4\ 5\ 6\ 7 \\ 3\ 4\ 5 \end{array}}{\begin{array}{l} 9\ 8\ 7\ 5\ 3\ 1 \\ 7\ 6\ 5\ 3\ 1 \\ 5\ 4\ 3\ 1 \\ 3\ 2\ 1 \end{array}} = 147840.$$

(3) $[4, 3, 1]\backslash[3, 2]^* \simeq [5, 4, 2, 1]\backslash[2, 1]^* \simeq [6, 5, 3, 2]\backslash[1]^* \simeq [7, 6, 4, 3, 1],$

$$d_{[7,6,4,3,1]}(SU(6)) = \frac{\begin{matrix} 6\ 7\ 8\ 9\ 10\ 11\ 12 \\ 5\ 6\ 7\ 8\ 9\ \ 10 \\ 4\ 5\ 6\ 7 \\ 3\ 4\ 5 \\ 2 \end{matrix}}{\begin{matrix} 11\ 9\ 8\ 6\ 4\ 3\ 1 \\ 9\ 7\ 6\ 4\ 2\ 1 \\ 6\ 4\ 3\ 1 \\ 4\ 2\ 1 \\ 1 \end{matrix}} = 612500.$$

10. Prove the identity:

$$\sum_{A=1}^{N^2-1} (T_A)_{ac}(T_A)_{bd} = \frac{1}{2}\delta_a^d\delta_b^c - \frac{1}{2N}\delta_a^c\delta_b^d,$$

where $(T_A)_{ac}$ are the generators in the self-representation of the $SU(N)$ group.

Solution. Define a mixed tensor **T** of rank $(2, 2)$ with respect to $SU(N)$

$$\mathbf{T}_{ab}^{cd} = \sum_{A=1}^{N^2-1} (T_A)_{ac}(T_A)_{bd}.$$

It is easy to show that **T** is invariant in the $SU(N)$ transformation,

$$
\begin{aligned}
(O_u\mathbf{T})_{ab}^{cd} &= \sum_{A=1}^{N^2-1} \left(uT_A u^{-1}\right)_{ac}\left(uT_A u^{-1}\right)_{bd} \\
&= \sum_{A,B,C=1}^{N^2-1} (T_B)_{ac}(T_C)_{bd}D_{BA}^{\mathrm{adj}}(u)D_{CA}^{\mathrm{adj}}(u) \\
&= \sum_{B,C=1}^{N^2-1} (T_B)_{ac}(T_C)_{bd}\sum_{A=1}^{N^2-1} D_{BA}^{\mathrm{adj}}(u)D_{AC}^{\mathrm{adj}}(u^{-1}) = \mathbf{T}_{ab}^{cd}.
\end{aligned}
$$

Therefore, **T** can be expressed as the product of the invariant mixed tensors δ_a^c of $SU(N)$,

$$\mathbf{T}_{ab}^{cd} = \sum_{A=1}^{N^2-1} (T_A)_{ac}(T_A)_{bd} = p\delta_a^c\delta_b^d + q\delta_a^d\delta_b^c.$$

Since T_A is a traceless matrix, $0 = \sum_a \mathbf{T}_{ab}^{ad} = (Np + q)\,\delta_b^d$, we have $q = -Np$. As

$$(T_A)_{NN} = \begin{cases} 0 & \text{when } A \neq N^2 - 1 \\ -\left(\dfrac{N-1}{2N}\right)^{1/2} & \text{when } A = N^2 - 1, \end{cases}$$

we have

$$\frac{N-1}{2N} = \mathbf{T}_{NN}^{NN} = p + q = -(N-1)p.$$

The solution is $p = -(2N)^{-1}$ and $q = 1/2$. This completes the proof.

8.3   Orthonormal Bases for Irreducible Representations

★ Recall that the standard tensor Young tableau is the eigenstate of the Chevalley basis H_μ where the eigenvalue is equal to the number of the digit μ filled in the tableau, subtracting the number of the digit $(\mu + 1)$. The action of the lowering operator F_μ on the standard tensor Young tableau is equal to the sum of all possible tensor Young tableaux, each of which is obtained from the original one by replacing one filled digit μ with the digit $(\mu + 1)$. Two standard tensor Young tableaux with different sets of the filled digits are orthogonal, but may not be normalized to the same number. Two standard tensor Young tableaux with the equal set of the filled digits but different filling order may not be orthogonal.

The merit of the method of Young operators is that the complete tensor bases in an irreducible representation space of SU(N) have been explicitly known. The shortcoming is that the tensor bases are not orthonormal, and the obtained representation is not unitary. On the other hand, the method of the block weight diagram gives the representaion matrices of the Chevalley bases of generators in the unitary irreducible representation, but the orthonormal bases are given only symbolically. To combine two methods will offset one's shortcoming by the other's strong point.

★ In the standard tensor Young tableau corresponding to the dominant weight, the number of each digit μ filled in the tableau must not be less than the number of the digit $(\mu+1)$ filled there. It is the criterion for determining which standard tensor Young tableau corresponds to the dominant weight in the representation. The states for the multiple weight correspond to the standard tensor Young tableaux with the equal set of the filled digits but in different filling order. In the standard tensor Young tableau corresponding to the highest weight, each box located in the jth row of the tableau is filled with the digit j, because each raising operator E_μ eliminates it. Therefore, the component M_μ of the highest weight $\mathbf{M}$ in the representation $[\lambda]$ is

$$M_\mu = \lambda_\mu - \lambda_{\mu+1}. \tag{8.30}$$

★ Gelfand presented the symbols for the orthonormal basis states of an irreducible representation $[\lambda]$ of $SU(N)$, usually called the Gelfand bases, and calculated the analytic formulas for the representation matrix entries of the generators (the Chevalley bases) of $SU(N)$ in the Gelfand bases. For a multiple weight, he chose the orthonormal basis states first according to the multiplets of the subalgebra $\mathcal{A}_1$, then, according to the multiplets of the subalgebra $\mathcal{A}_2$ if there still exist the multiple states, and so on. Gelfand described the basis state with the weight $\mathbf{m}$ in the representation $[\lambda]$ of $SU(N)$ by $N(N+1)/2$ parameters ω_{ab}, $1 \le a \le b \le N$, arranged as a regular triangle upside down:

$$|\omega_{ab}\rangle = \begin{vmatrix} \omega_{1N} & & \omega_{2N} & \cdots & \omega_{(N-1)N} & & \omega_{NN} \\ & \omega_{1(N-1)} & & \cdots & \cdots & \omega_{(N-1)(N-1)} & \\ & & \cdots & \cdots & \cdots & & \\ & & \omega_{12} & & \omega_{22} & & \\ & & & \omega_{11} & & & \end{vmatrix} \Bigg\rangle,$$

where $\omega_{aN} = \lambda_a$, $\omega_{NN} = \lambda_N = 0$, and

$$\omega_{ab} \ge \omega_{a(b-1)} \ge \omega_{(a+1)b}, \qquad 1 \le a \le b \le N. \tag{8.31}$$

The representation entries of the generators are

$$H_\mu |\omega_{ab}\rangle = m_\mu |\omega_{ab}\rangle, \qquad m_\mu = -\sum_{d=1}^{\mu+1} \omega_{d(\mu+1)} + 2\sum_{d=1}^{\mu} \omega_{d\mu} - \sum_{d=1}^{\mu-1} \omega_{d(\mu-1)},$$

$$E_\mu |\omega_{ab}\rangle = \sum_{\nu=1}^{\mu} A_{\nu\mu}(\omega_{ab}) |\omega_{ab} + \delta_{a\nu}\delta_{b\mu}\rangle,$$

$$F_\mu |\omega_{ab}\rangle = \sum_{\nu=1}^{\mu} A_{\nu\mu}(\omega_{ab} - \delta_{a\nu}\delta_{b\mu}) |\omega_{ab} - \delta_{a\nu}\delta_{b\mu}\rangle,$$

$A_{\nu\mu}(\omega_{ab})$

$$= \left\{ -\prod_{t=1}^{\mu-1} \left(\omega_{t(\mu-1)} - \omega_{\nu\mu} - t + \nu - 1\right) \prod_{p=1}^{\mu+1} \left(\omega_{p(\mu+1)} - \omega_{\nu\mu} - p + \nu\right) \right\}^{1/2}$$

$$\times \left\{ \prod_{d\neq\nu,d=1}^{\mu} \left(\omega_{d\mu} - \omega_{\nu\mu} - d + \nu\right)\left(\omega_{d\mu} - \omega_{\nu\mu} - d + \nu - 1\right) \right\}^{-1/2} .$$

$$(8.32)$$

Namely, the eigenvalue of H_μ is equal to twice the sum of the parameters in the μth row from the bottom, minus the sum of the parameters in the $(\mu - 1)$th row and the $(\mu+1)$th row from the bottom. The application of E_μ (F_μ) to a Gelfand basis leads to a linear combination of the possible basis states, each of which is obtained by increasing (decreasing) one parameter $\omega_{a\mu}$ in the μth row from the bottom under the condition (8.31). The parameters of the highest weight state in the Gelfand bases satisfy

$$\omega_{ab} = \lambda_a, \qquad a \le b \le N, \qquad \lambda_N = 0. \qquad (8.33)$$

11. Expand the Gelfand bases in the irreducible representation [3, 0] of the SU(3) group with respect to the standard tensor Young tableaux by making use of its block weight diagram given in Fig. 7.2.

Solution. The tensor subspace described by the representation $[\lambda] = [3, 0]$ consists of the completely symmetric tensors of rank three. There is no multiple weight in the representation. Its highest weight is $\mathbf{M} = (3, 0)$, which corresponds to the standard tensor Young tableau and the Gelfand basis are

$$|(3,0), (3,0)\rangle = \boxed{1 \mid 1 \mid 1} = \left| \begin{matrix} 3 & \;0\; & \;0\; \\ & 3 & \;0\; \\ & & 3 \end{matrix} \right) .$$

There are three types of the standard tensor Young tableaux which are normalized to 6, 12 and 36, respectively:

$$\boxed{a \mid b \mid c} = \sqrt{6}\left\{6^{-1/2}\left(\Theta_{abc} + \Theta_{acb} + \Theta_{bac} + \Theta_{bca} + \Theta_{cab} + \Theta_{cba}\right)\right\},$$

$$\boxed{a \mid b \mid b} = 2\sqrt{3}\left\{3^{-1/2}\left(\Theta_{abb} + \Theta_{bab} + \Theta_{bba}\right)\right\},$$

$$\boxed{a \mid a \mid a} = 6\left\{\Theta_{aaa}\right\},$$

where a, b and c are all different. In the application of the lowering operator F_μ, all possible terms, which may be non-standard tensor Young tableaux, need to be considered. For example

$$F_1 \,\boxed{1\,|\,1\,|\,1}\, = 3\,\boxed{1\,|\,1\,|\,2}\,, \quad F_1\,\boxed{1\,|\,1\,|\,2}\, = 2\,\boxed{1\,|\,2\,|\,2}\,.$$

Now, the orthogonal bases can be calculated as follows.

$$|(3,0),(1,1)\rangle = \sqrt{1/3}F_1|(3,0),(3,0)\rangle = \sqrt{1/3}F_1\,\boxed{1\,|\,1\,|\,1}$$
$$= \sqrt{3}\,\boxed{1\,|\,1\,|\,2}\, = \left|\begin{smallmatrix}3 & & 0 & & 0\\ & 3 & & 0 & \\ & & 2 & & \end{smallmatrix}\right\rangle,$$

$$|(3,0),(\bar{1},2)\rangle = (1/2)F_1|(3,0),(1,1)\rangle = (1/2)F_1\left\{\sqrt{3}\,\boxed{1\,|\,1\,|\,2}\right\}$$
$$= \sqrt{3}\,\boxed{1\,|\,2\,|\,2}\, = \left|\begin{smallmatrix}3 & & 0 & & 0\\ & 3 & & 0 & \\ & & 1 & & \end{smallmatrix}\right\rangle,$$

$$|(3,0),(\bar{3},3)\rangle = \sqrt{1/3}F_1|(3,0),(\bar{1},2)\rangle = \sqrt{1/3}F_1\left\{\sqrt{3}\,\boxed{1\,|\,2\,|\,2}\right\}$$
$$= \boxed{2\,|\,2\,|\,2}\, = \left|\begin{smallmatrix}3 & & 0 & & 0\\ & 3 & & 0 & \\ & & 0 & & \end{smallmatrix}\right\rangle,$$

$$|(3,0),(2,\bar{1})\rangle = F_2|(3,0),(1,1)\rangle = F_2\left\{\sqrt{3}\,\boxed{1\,|\,1\,|\,2}\right\}$$
$$= \sqrt{3}\,\boxed{1\,|\,1\,|\,3}\, = \left|\begin{smallmatrix}3 & & 0 & & 0\\ & 2 & & 0 & \\ & & 2 & & \end{smallmatrix}\right\rangle,$$

$$|(3,0),(0,0)\rangle = \sqrt{1/2}F_2|(3,0),(\bar{1},2)\rangle = \sqrt{1/2}F_2\left\{\sqrt{3}\,\boxed{1\,|\,2\,|\,2}\right\}$$
$$= \sqrt{6}\,\boxed{1\,|\,2\,|\,3}\, = \left|\begin{smallmatrix}3 & & 0 & & 0\\ & 2 & & 0 & \\ & & 1 & & \end{smallmatrix}\right\rangle,$$

$$|(3,0),(1,\bar{2})\rangle = \sqrt{1/2}F_2|(3,0),(0,0)\rangle = \sqrt{1/2}F_2\left\{\sqrt{6}\,\boxed{1\,|\,2\,|\,3}\right\}$$
$$= \sqrt{3}\,\boxed{1\,|\,3\,|\,3}\, = \left|\begin{smallmatrix}3 & & 0 & & 0\\ & 1 & & 0 & \\ & & 1 & & \end{smallmatrix}\right\rangle,$$

$$|(3,0),(\bar{2},1)\rangle = \sqrt{1/3}F_2|(3,0),(\bar{3},3)\rangle = \sqrt{1/3}F_2\,\boxed{2\,|\,2\,|\,2}$$
$$= \sqrt{3}\,\boxed{2\,|\,2\,|\,3}\, = \left|\begin{smallmatrix}3 & & 0 & & 0\\ & 2 & & 0 & \\ & & 0 & & \end{smallmatrix}\right\rangle,$$

$$|(3,0),(\bar{1},\bar{1})\rangle = (1/2)F_2|(3,0),(\bar{2},1)\rangle = (1/2)F_2\left\{\sqrt{3}\,\boxed{2\,|\,2\,|\,3}\right\}$$
$$= \sqrt{3}\,\boxed{2\,|\,3\,|\,3}\, = \left|\begin{smallmatrix}3 & & 0 & & 0\\ & 1 & & 0 & \\ & & 0 & & \end{smallmatrix}\right\rangle,$$

$$|(3,0),(0,\bar{3})\rangle = \sqrt{1/3}F_2|(3,0),(\bar{1},\bar{1})\rangle = \sqrt{1/3}F_2\left\{\sqrt{3}\,\boxed{2\,|\,3\,|\,3}\right\}$$

$$= \boxed{3\,|\,3\,|\,3} = \left|3 \begin{array}{ccc} & 0 & 0 & 0 \\ & 0 & 0 \\ & 0 & \end{array}\right\rangle.$$

The expansions of the orthonormal basis states with respect to the standard tensor Young tableaux give the wave functions of the hadrons. Substituting the expansions into the planar weight diagram of $(3,0)$ given in Fig. 7.3, we obtained the wave functions for the baryon decuplet. In the planar weight diagram, the ordinate axis and the abscissa axis are the eigenvalues of T_3 and T_8, respectively (see the end of Problem 14).

12. Expand the Gelfand bases in the irreducible representation $[3,3]$ of the SU(3) group with respect to the standard tensor Young tableaux by making use of its block weight diagram given in Fig. 7.2.

Solution. The representation $[\lambda] = [3,3]$ of SU(3) is the conjugate representation of $[3,0]$, $[3,3] \simeq [3,0]^*$. This problem can be solved in the same way as that used in Problem 11. However, in the present problem we prefer to calculate the standard tensor Young tableaux of the contravariant tensor instead of those of the covariant tensor. Both results for two kinds of the standard tensor Young tableaux are given here. Note that the action of the generator on the basis state of the conjugate representation is different from that on the basis state of the original representation. Denote by $D(I_A)$ and $D'(I_A)$ the representation matrices of the generator I_A in the representations $D(R)$ and those in its conjugate one $D(R)^*$, respectively:

$$D(R) = 1 - i\sum_A D(I_A)\omega_A,$$

$$D(R)^* = 1 + i\sum_A D(I_A)^*\omega_A = 1 - i\sum_A D'(I_A)\omega_A.$$

Thus, $D'(I_A) = -D(I_A)^*$. For the Chevalley bases we have

$$D'(H_\mu) = -D(H_\mu)^*, \quad D'(E_\mu) = -D(F_\mu)^*, \quad D'(F_\mu) = -D(E_\mu)^*. \quad (8.34)$$

Namely, the standard tensor Young tableau with star is the eigenstate of the Chevalley basis H_μ where the eigenvalue is equal to the number of the digit $(\mu+1)$ filled in the tableau, subtracting the number of the digit μ. The action of the lowering operator F_μ on the standard tensor Young tableau with star is equal to the sum multiplied by -1, of all possible tensor Young tableaux with star, each of which is obtained from the original one by replacing one filled digit $(\mu+1)$ with digit μ. The order of the filled digits in the standard tensor Young tableau changed to 3, 2, 1 instead of 1, 2, 3.

The highest weight of the representation [3,3] is $\mathbf{M} = (0,3)$, which corresponds to the standard tensor Young tableau and the Gelfand basis are

$$|(0,3),(0,3)\rangle = \boxed{3\;3\;3}^{\,*} = \boxed{\begin{array}{ccc}1&1&1\\2&2&2\end{array}} = \left|\begin{array}{ccc}3 & & \\ & 3 & \\ 3 & & 3 \\ & 3 & \\ & & 0\end{array}\right\rangle .$$

The normalization factor for the standard tensor Young tableau with star is similar to that for the standard tensor Young tableau. Now, the orthogonal bases can be calculated as follows.

$$|(0,3),(1,1)\rangle = \sqrt{1/3}\,F_2|(0,3),(0,3)\rangle = \sqrt{1/3}\,F_2\boxed{3\;3\;3}^{\,*}$$
$$= -\sqrt{3}\,\boxed{3\;3\;2}^{\,*} = \sqrt{3}\,\boxed{\begin{array}{ccc}1&1&1\\2&2&3\end{array}} = \left|\begin{array}{ccc}3 & & \\ & 3 & \\ 3 & & 2 \\ & 3 & \\ & & 0\end{array}\right\rangle ,$$

$$|(0,3),(2,\bar{1})\rangle = (1/2)F_2|(0,3),(1,1)\rangle = (1/2)F_2\left\{-\sqrt{3}\,\boxed{3\;3\;2}^{\,*}\right\}$$
$$= \sqrt{3}\,\boxed{3\;2\;2}^{\,*} = \sqrt{3}\,\boxed{\begin{array}{ccc}1&1&1\\2&3&3\end{array}} = \left|\begin{array}{ccc}3 & & \\ & 3 & \\ 3 & & 1 \\ & 3 & \\ & & 0\end{array}\right\rangle ,$$

$$|(0,3),(3,\bar{3})\rangle = \sqrt{1/3}\,F_2|(0,3),(2,\bar{1})\rangle = \sqrt{1/3}\,F_2\left\{\sqrt{3}\,\boxed{3\;2\;2}^{\,*}\right\}$$
$$= -\boxed{2\;2\;2}^{\,*} = \boxed{\begin{array}{ccc}1&1&1\\3&3&3\end{array}} = \left|\begin{array}{ccc}3 & & \\ & 3 & \\ 3 & & 0 \\ & 3 & \\ & & 0\end{array}\right\rangle ,$$

$$|(0,3),(\bar{1},2)\rangle = F_1|(0,3),(1,1)\rangle = F_1\left\{-\sqrt{3}\,\boxed{3\;3\;2}^{\,*}\right\}$$
$$= \sqrt{3}\,\boxed{3\;3\;1}^{\,*} = \sqrt{3}\,\boxed{\begin{array}{ccc}1&1&2\\2&2&3\end{array}} = \left|\begin{array}{ccc}3 & & \\ & 3 & \\ 3 & & 2 \\ & 2 & \\ & & 0\end{array}\right\rangle ,$$

$$|(0,3),(0,0)\rangle = \sqrt{1/2}\,F_1|(0,3),(2,\bar{1})\rangle = \sqrt{1/2}\,F_1\left\{\sqrt{3}\,\boxed{3\;2\;2}^{\,*}\right\}$$
$$= -\sqrt{6}\,\boxed{3\;2\;1}^{\,*} = \sqrt{6}\,\boxed{\begin{array}{ccc}1&1&2\\2&3&3\end{array}} = \left|\begin{array}{ccc}3 & & \\ & 3 & \\ 3 & & 1 \\ & 2 & \\ & & 0\end{array}\right\rangle ,$$

$$|(0,3),(\bar{2},1)\rangle = \sqrt{1/2}\,F_1|(0,3),(0,0)\rangle = \sqrt{1/2}\,F_1\left\{-\sqrt{6}\,\boxed{3\;2\;1}^{\,*}\right\}$$
$$= \sqrt{3}\,\boxed{3\;1\;1}^{\,*} = \sqrt{3}\,\boxed{\begin{array}{ccc}1&2&2\\2&3&3\end{array}} = \left|\begin{array}{ccc}3 & & \\ & 3 & \\ 3 & & 1 \\ & 1 & \\ & & 0\end{array}\right\rangle ,$$

$$|(0,3),(1,\overline{2})\rangle = \sqrt{1/3}F_1|(0,3),(3,\overline{3})\rangle = \sqrt{1/3}F_1\left\{-\boxed{\begin{array}{ccc} 2 & 2 & 2 \end{array}}^*\right\}$$

$$= \sqrt{3}\,\boxed{\begin{array}{ccc} 2 & 2 & 1 \end{array}}^* = \sqrt{3}\,\boxed{\begin{array}{ccc} 1 & 1 & 2 \\ 3 & 3 & 3 \end{array}} = \left|\begin{array}{cccc} 3 & & 3 & 0 \\ & 3 & & 0 \\ & & 2 & \end{array}\right\rangle,$$

$$|(0,3),(\overline{1},\overline{1})\rangle = (1/2)F_1|(0,3),(1,\overline{2})\rangle = (1/2)F_1\left\{\sqrt{3}\,\boxed{\begin{array}{ccc} 2 & 2 & 1 \end{array}}^*\right\}$$

$$= -\sqrt{3}\,\boxed{\begin{array}{ccc} 2 & 1 & 1 \end{array}}^* = \sqrt{3}\,\boxed{\begin{array}{ccc} 1 & 2 & 2 \\ 3 & 3 & 3 \end{array}} = \left|\begin{array}{cccc} 3 & & 3 & 0 \\ & 3 & & 0 \\ & & 1 & \end{array}\right\rangle,$$

$$|(0,3),(\overline{3},0)\rangle = \sqrt{1/3}F_1|(0,3),(\overline{1},\overline{1})\rangle = \sqrt{1/3}F_1\left\{-\sqrt{3}\,\boxed{\begin{array}{ccc} 2 & 1 & 1 \end{array}}^*\right\}$$

$$= \boxed{\begin{array}{ccc} 1 & 1 & 1 \end{array}}^* = \boxed{\begin{array}{ccc} 2 & 2 & 2 \\ 3 & 3 & 3 \end{array}} = \left|\begin{array}{cccc} 3 & & 3 & 0 \\ & 3 & & 0 \\ & & 0 & \end{array}\right\rangle.$$

13. Expand the Gelfand bases in the irreducible representation $[2,1]$ of the SU(3) group with respect to the standard tensor Young tableaux by making use of its block weight diagram given in Fig. 7.2.

Solution. The representation $[\lambda] = [2,1]$ is the adjoint representation of SU(3). For the tensor subspace $\mathcal{T}_1^{[2,1]} = \mathcal{Y}_1^{[2,1]}\mathcal{T}$, the standard Young operator is $\mathcal{Y}_1^{[2,1]} = E + (1\,2) - (1\,3) - (2\,1\,3)$, corresponding to the Young tableau $\boxed{\begin{array}{cc} 1 & 2 \\ 3 & \end{array}}$. There are six single weight and one double weight $(0,0)$ in the representation $[2,1]$. The typical standard tensor Young tableaux and their normalization factors are

$$\boxed{\begin{array}{cc} a & b \\ b & \end{array}} = \mathcal{Y}_1^{[2,1]}\Theta_{abb} = 6^{1/2}\left\{6^{-1/2}\left(\Theta_{abb} + \Theta_{bab} - 2\Theta_{bba}\right)\right\},$$

$$\boxed{\begin{array}{cc} a & b \\ c & \end{array}} = \mathcal{Y}_1^{[2,1]}\Theta_{abc} = 2\left\{2^{-1}\left(\Theta_{abc} + \Theta_{bac} - \Theta_{cba} - \Theta_{bca}\right)\right\}.$$

Due to the Fock condition (6.5), we have

$$\boxed{\begin{array}{cc} 2 & 1 \\ 3 & \end{array}} = \boxed{\begin{array}{cc} 1 & 2 \\ 3 & \end{array}} + \boxed{\begin{array}{cc} 2 & 3 \\ 1 & \end{array}} = \boxed{\begin{array}{cc} 1 & 2 \\ 3 & \end{array}} - \boxed{\begin{array}{cc} 1 & 3 \\ 2 & \end{array}},$$

   In particle physics, a hadron is composed of three quarks (baryon) or a pair of one quark and one antiquark (meson). The expansions of the orthonormal basis states in $[2,1]$ with respect to the standard tensor Young tableaux give the flavor wave functions for the baryon octet or for the meson octet. For the meson octet, the standard tensor Young tableau should be transformed into the traceless mixed tensor by the totally antisymmetric tensor ϵ_{abc}:

$$\begin{array}{|c|c|}\hline 1 & a \\\hline 2 \\\cline{1-1}\end{array} \longrightarrow \boxed{a}\!\!\diagdown\!\!\boxed{3}^{\,*}, \qquad a = 1 \text{ or } 2,$$

$$\begin{array}{|c|c|}\hline 1 & a \\\hline 3 \\\cline{1-1}\end{array} \longrightarrow -\boxed{a}\!\!\diagdown\!\!\boxed{2}^{\,*}, \qquad a = 1 \text{ or } 3,$$

$$\begin{array}{|c|c|}\hline 2 & a \\\hline 3 \\\cline{1-1}\end{array} \longrightarrow \boxed{a}\!\!\diagdown\!\!\boxed{1}^{\,*}, \qquad a = 2 \text{ or } 3,$$

$$\begin{aligned}
\begin{array}{|c|c|}\hline 1 & 3 \\\hline 2 \\\cline{1-1}\end{array} &= \frac{1}{3}\left\{ 2\,\begin{array}{|c|c|}\hline 1 & 3 \\\hline 2 \\\cline{1-1}\end{array} - \begin{array}{|c|c|}\hline 2 & 1 \\\hline 3 \\\cline{1-1}\end{array} + \begin{array}{|c|c|}\hline 1 & 2 \\\hline 3 \\\cline{1-1}\end{array} \right\} \\
&\longrightarrow \frac{1}{3}\left\{ 2\,\boxed{3}\!\!\diagdown\!\!\boxed{3}^{\,*} - \boxed{1}\!\!\diagdown\!\!\boxed{1}^{\,*} - \boxed{2}\!\!\diagdown\!\!\boxed{2}^{\,*} \right\} \\
\begin{array}{|c|c|}\hline 1 & 2 \\\hline 3 \\\cline{1-1}\end{array} &= \frac{1}{3}\left\{ \begin{array}{|c|c|}\hline 2 & 1 \\\hline 3 \\\cline{1-1}\end{array} + 2\,\begin{array}{|c|c|}\hline 1 & 2 \\\hline 3 \\\cline{1-1}\end{array} + \begin{array}{|c|c|}\hline 1 & 3 \\\hline 2 \\\cline{1-1}\end{array} \right\} \\
&\longrightarrow \frac{1}{3}\left\{ \boxed{1}\!\!\diagdown\!\!\boxed{1}^{\,*} - 2\,\boxed{2}\!\!\diagdown\!\!\boxed{2}^{\,*} + \boxed{3}\!\!\diagdown\!\!\boxed{3}^{\,*} \right\}.
\end{aligned} \tag{8.35}$$

The highest weight in $[2,1]$ is $\mathbf{M} = (1,1)$, which corresponds to the standard tensor Young tableau and the Gelfand basis are

$$|(1,1),(1,1)\rangle = \begin{array}{|c|c|}\hline 1 & 1 \\\hline 2 \\\cline{1-1}\end{array} = \boxed{3}\!\!\diagdown\!\!\boxed{1}^{\,*} = \left|\begin{array}{ccc} 2 & & 1 \quad 0 \\ & 2 & \\ & 2 & \quad 1 \end{array}\right\rangle.$$

The remaining orthonormal bases can be calculated from the block weight diagram given in Fig. 7.2:

$$|(1,1),(\bar{1},2)\rangle = F_1|(1,1),(1,1)\rangle = F_1\,\begin{array}{|c|c|}\hline 1 & 1 \\\hline 2 \\\cline{1-1}\end{array}$$

$$= \begin{array}{|c|c|}\hline 1 & 2 \\\hline 2 \\\cline{1-1}\end{array} = \boxed{2}\!\!\diagdown\!\!\boxed{3}^{\,*} = \left|\begin{array}{ccc} 2 & & 1 \quad 0 \\ & 2 & \\ & 1 & \quad 1 \end{array}\right\rangle,$$

$$|(1,1),(2,\bar{1})\rangle = F_2|(1,1),(1,1)\rangle = F_2\,\begin{array}{|c|c|}\hline 1 & 1 \\\hline 2 \\\cline{1-1}\end{array}$$

$$= \begin{array}{|c|c|}\hline 1 & 1 \\\hline 3 \\\cline{1-1}\end{array} = -\boxed{1}\!\!\diagdown\!\!\boxed{2}^{\,*} = \left|\begin{array}{ccc} 2 & & 1 \quad 0 \\ & 2 & \\ & 2 & \quad 0 \end{array}\right\rangle,$$

$$|(1,1),(0,0)_1\rangle = \sqrt{1/2}F_1|(1,1),(2,\bar{1})\rangle$$

$$= \sqrt{1/2}F_1\;\boxed{\begin{smallmatrix}1&1\\3\end{smallmatrix}} \;=\; \sqrt{1/2}\left\{\boxed{\begin{smallmatrix}1&2\\3\end{smallmatrix}} + \boxed{\begin{smallmatrix}2&1\\3\end{smallmatrix}}\right\}$$

$$= \sqrt{2}\;\boxed{\begin{smallmatrix}1&2\\3\end{smallmatrix}} - \sqrt{1/2}\;\boxed{\begin{smallmatrix}1&3\\2\end{smallmatrix}}$$

$$= \sqrt{1/2}\left\{\boxed{1}\!\!\diagdown\!\!\boxed{1}^{\,*} - \boxed{2}\!\!\diagdown\!\!\boxed{2}^{\,*}\right\} = \left|\begin{smallmatrix}2&&1&&0\\&2&&0&\\&&1&&\end{smallmatrix}\right\rangle ,$$

$$|(1,1),(\bar{2},1)\rangle = \sqrt{1/2}F_1|(1,1),(0,0)_1\rangle$$

$$= \sqrt{1/2}F_1\left\{\sqrt{2}\;\boxed{\begin{smallmatrix}1&2\\3\end{smallmatrix}} - \sqrt{1/2}\;\boxed{\begin{smallmatrix}1&3\\2\end{smallmatrix}}\right\}$$

$$= \boxed{\begin{smallmatrix}2&2\\3\end{smallmatrix}} = \boxed{2}\!\!\diagdown\!\!\boxed{1}^{\,*} = \left|\begin{smallmatrix}2&&1&&0\\&2&&0&\\&&0&&\end{smallmatrix}\right\rangle ,$$

$$|(1,1),(0,0)_2\rangle = \sqrt{2/3}\left\{F_2|(1,1),(\bar{1},2)\rangle - \sqrt{1/2}|(1,1),(0,0)_1\rangle\right\}$$

$$= \sqrt{2/3}\left\{F_2\;\boxed{\begin{smallmatrix}1&2\\2\end{smallmatrix}} - \boxed{\begin{smallmatrix}1&2\\3\end{smallmatrix}} + (1/2)\boxed{\begin{smallmatrix}1&3\\2\end{smallmatrix}}\right\} = \sqrt{3/2}\;\boxed{\begin{smallmatrix}1&3\\2\end{smallmatrix}}$$

$$= \sqrt{1/6}\left\{2\,\boxed{3}\!\!\diagdown\!\!\boxed{3}^{\,*} - \boxed{1}\!\!\diagdown\!\!\boxed{1}^{\,*} - \boxed{2}\!\!\diagdown\!\!\boxed{2}^{\,*}\right\}$$

$$= \left|\begin{smallmatrix}2&&1&&0\\&1&&1&\\&&1&&\end{smallmatrix}\right\rangle ,$$

$$|(1,1),(1,\bar{2})\rangle = \sqrt{2/3}F_2|(1,1),(0,0)_2\rangle = \sqrt{2/3}F_2\left\{\sqrt{3/2}\;\boxed{\begin{smallmatrix}1&3\\2\end{smallmatrix}}\right\}$$

$$= \boxed{\begin{smallmatrix}1&3\\3\end{smallmatrix}} = -\boxed{3}\!\!\diagdown\!\!\boxed{2}^{\,*} = \left|\begin{smallmatrix}2&&1&&0\\&1&&0&\\&&1&&\end{smallmatrix}\right\rangle ,$$

$$|(1,1),(\bar{1},\bar{1})\rangle = F_1|(1,1),(1,\bar{2})\rangle = F_1\;\boxed{\begin{smallmatrix}1&3\\3\end{smallmatrix}}$$

$$= \boxed{\begin{smallmatrix}2&3\\3\end{smallmatrix}} = \boxed{3}\!\!\diagdown\!\!\boxed{1}^{\,*} = \left|\begin{smallmatrix}2&&1&&0\\&1&&0&\\&&0&&\end{smallmatrix}\right\rangle ,$$

where we use the formula calculated from Eq. (8.32):

$$F_2\left|\begin{smallmatrix}2&&1&&0\\&2&&1&\\&&1&&\end{smallmatrix}\right\rangle = \sqrt{\frac{3}{2}}\left|\begin{smallmatrix}2&&1&&0\\&1&&1&\\&&1&&\end{smallmatrix}\right\rangle + \sqrt{\frac{1}{2}}\left|\begin{smallmatrix}2&&1&&0\\&2&&0&\\&&1&&\end{smallmatrix}\right\rangle .$$

Each basis state in the representation $[2,1]$ is normalized to 6. Especially, the basis state

$$|(1,1),(0,0)_1\rangle = \sqrt{2}\,\begin{array}{|c|c|}\hline 1 & 2 \\\hline 3 \\\cline{1-1}\end{array} - \sqrt{1/2}\,\begin{array}{|c|c|}\hline 1 & 3 \\\hline 2 \\\cline{1-1}\end{array}$$

$$= \sqrt{2}\,\{\Theta_{123} + \Theta_{213} - \Theta_{321} - \Theta_{231}\}$$

$$\quad - \sqrt{1/2}\,\{\Theta_{132} + \Theta_{312} - \Theta_{231} - \Theta_{321}\}$$

$$= \sqrt{1/2}\,\{2\Theta_{123} + 2\Theta_{213} - \Theta_{321} - \Theta_{231} - \Theta_{132} - \Theta_{312}\}.$$

14. Express each Gelfand basis in the irreducible representation $[4,0]$ of the SU(3) group by the standard tensor Young tableau and calculate the nonvanishing matrix entries for the lowering operators F_μ. Draw the block weight diagram and the planar weight diagram for the representation $[4,0]$ of SU(3).

Solution. The tensor subspace described by the representation $[\lambda] = [4,0]$ consists of the completely symmetric tensors of rank four. There are four single dominant weights in the representation:

$$(4,0) \qquad\qquad (2,1) \qquad\qquad (0,2) \qquad\qquad (1,0)$$

$$\begin{array}{|c|c|c|c|}\hline 1 & 1 & 1 & 1 \\\hline\end{array}, \quad \begin{array}{|c|c|c|c|}\hline 1 & 1 & 1 & 2 \\\hline\end{array}, \quad \begin{array}{|c|c|c|c|}\hline 1 & 1 & 2 & 2 \\\hline\end{array}, \quad \begin{array}{|c|c|c|c|}\hline 1 & 1 & 2 & 3 \\\hline\end{array}.$$

The highest weight of the representation $[4,0]$ is $\mathbf{M} = (4,0)$, which corresponds to the standard tensor Young tableau and the Gelfand basis:

$$|(4,0),(4,0)\rangle = \begin{array}{|c|c|c|c|}\hline 1 & 1 & 1 & 1 \\\hline\end{array} = \left|\begin{matrix} 4 & & 0 & & 0 \\ & 4 & & 0 & \\ & & 4 & & \end{matrix}\right\rangle.$$

There are four types of the standard tensor Young tableaux which normalize to 48, 96, 144, and 576, respectively. The normalization factors for them are $\sqrt{12}$, $\sqrt{6}$, 2, and 1, respectively.

$$\sqrt{12}\,\begin{array}{|c|c|c|c|}\hline a & b & c & c \\\hline\end{array} = 24\,\{12^{-1/2}\,(\Theta_{abcc} + \Theta_{bacc} + \Theta_{acbc} + \Theta_{bcac}$$

$$+ \Theta_{cabc} + \Theta_{cbac} + \Theta_{cacb} + \Theta_{cbca}$$

$$+ \Theta_{accb} + \Theta_{bcca} + \Theta_{ccab} + \Theta_{ccba})\},$$

$$\sqrt{6}\,\begin{array}{|c|c|c|c|}\hline a & a & c & c \\\hline\end{array} = 24\,\{6^{-1/2}\,(\Theta_{aacc} + \Theta_{acac} + \Theta_{caac} + \Theta_{caca}$$

$$+ \Theta_{acca} + \Theta_{ccaa})\},$$

$$2\,\begin{array}{|c|c|c|c|}\hline a & b & b & b \\\hline\end{array} = 24\,\{2^{-1}\,(\Theta_{abbb} + \Theta_{babb} + \Theta_{bbab} + \Theta_{bbba})\},$$

$$\begin{array}{|c|c|c|c|}\hline a & a & a & a \\\hline\end{array} = 24\,\{\Theta_{aaaa}\},$$

where a, b, and c all are different.

Now, we calculate the remaining basis states. From the highest weight $[\lambda] = [4, 0]$ we have a quintet of $\mathcal{A}_1$:

$$|(4,0),(2,1)\rangle = (1/2)F_1|(4,0),(4,0)\rangle = (1/2)F_1 \boxed{1\,|\,1\,|\,1\,|\,1}$$

$$= 2\,\boxed{1\,|\,1\,|\,1\,|\,2} = \left|\begin{matrix} 4 & & 0 & & 0 \\ & 4 & & 0 & \\ & & 3 & & \end{matrix}\right\rangle ,$$

$$|(4,0),(0,2)\rangle = \sqrt{1/6}\,F_1|(4,0),(2,1)\rangle = \sqrt{1/6}\,F_1\left\{2\,\boxed{1\,|\,1\,|\,1\,|\,2}\right\}$$

$$= \sqrt{6}\,\boxed{1\,|\,1\,|\,2\,|\,2} = \left|\begin{matrix} 4 & & 0 & & 0 \\ & 4 & & 0 & \\ & & 2 & & \end{matrix}\right\rangle ,$$

$$|(4,0),(\bar{2},3)\rangle = \sqrt{1/6}\,F_1|(4,0),(0,2)\rangle = \sqrt{1/6}\,F_1\left\{\sqrt{6}\,\boxed{1\,|\,1\,|\,2\,|\,2}\right\}$$

$$= 2\,\boxed{1\,|\,2\,|\,2\,|\,2} = \left|\begin{matrix} 4 & & 0 & & 0 \\ & 4 & & 0 & \\ & & 1 & & \end{matrix}\right\rangle ,$$

$$|(4,0),(\bar{4},4)\rangle = (1/2)F_1|(4,0),(\bar{2},3)\rangle = (1/2)F_1\left\{2\,\boxed{1\,|\,2\,|\,2\,|\,2}\right\}$$

$$= \boxed{2\,|\,2\,|\,2\,|\,2} = \left|\begin{matrix} 4 & & 0 & & 0 \\ & 4 & & 0 & \\ & & 0 & & \end{matrix}\right\rangle .$$

Then, from $|(4,0),(2,1)\rangle$, $|(4,0),(0,2)\rangle$, $|(4,0),(\bar{2},3)\rangle$, and $|(4,0),(\bar{4},4)\rangle$ we have a doublet, triplet, quartet and quintet of $\mathcal{A}_2$, respectively.

$$|(4,0),(3,\bar{1})\rangle = F_2|(4,0),(2,1)\rangle = F_2\left\{2\,\boxed{1\,|\,1\,|\,1\,|\,2}\right\}$$

$$= 2\,\boxed{1\,|\,1\,|\,1\,|\,3} = \left|\begin{matrix} 4 & & 0 & & 0 \\ & 3 & & 0 & \\ & & 3 & & \end{matrix}\right\rangle ,$$

$$|(4,0),(1,0)\rangle = \sqrt{1/2}\,F_2|(4,0),(0,2)\rangle = \sqrt{1/2}\,F_2\left\{\sqrt{6}\,\boxed{1\,|\,1\,|\,2\,|\,2}\right\}$$

$$= \sqrt{12}\,\boxed{1\,|\,1\,|\,2\,|\,3} = \left|\begin{matrix} 4 & & 0 & & 0 \\ & 3 & & 0 & \\ & & 2 & & \end{matrix}\right\rangle ,$$

$$|(4,0),(2,\bar{2})\rangle = \sqrt{1/2}\,F_2|(4,0),(1,0)\rangle = \sqrt{1/2}\,F_2\left\{\sqrt{12}\,\boxed{1\,|\,1\,|\,2\,|\,3}\right\}$$

$$= \sqrt{6}\,\boxed{1\,|\,1\,|\,3\,|\,3} = \left|\begin{matrix} 4 & & 0 & & 0 \\ & 2 & & 0 & \\ & & 2 & & \end{matrix}\right\rangle ,$$

$$|(4,0),(\bar 1,1)\rangle = \sqrt{1/3}\,F_2|(4,0),(\bar 2,3)\rangle = \sqrt{1/3}\,F_2\left\{2\;\boxed{1\;|\;2\;|\;2\;|\;2}\right\}$$

$$= \sqrt{12}\;\boxed{1\;|\;2\;|\;2\;|\;3} = \left|\begin{smallmatrix}4 & 0 & 0 \\ 3 & & 0 & 0 \\ & 1 & \end{smallmatrix}\right\rangle,$$

$$|(4,0),(0,\bar 1)\rangle = (1/2)F_2|(4,0),(\bar 1,1)\rangle = (1/2)F_2\left\{\sqrt{12}\;\boxed{1\;|\;2\;|\;2\;|\;3}\right\}$$

$$= \sqrt{12}\;\boxed{1\;|\;2\;|\;3\;|\;3} = \left|\begin{smallmatrix}4 & 0 & 0 \\ 2 & & 0 & 0 \\ & 1 & \end{smallmatrix}\right\rangle,$$

$$|(4,0),(1,\bar 3)\rangle = \sqrt{1/3}\,F_2|(4,0),(0,\bar 1)\rangle = \sqrt{1/3}\,F_2\left\{\sqrt{12}\;\boxed{1\;|\;2\;|\;3\;|\;3}\right\}$$

$$= 2\;\boxed{1\;|\;3\;|\;3\;|\;3} = \left|\begin{smallmatrix}4 & 0 & 0 \\ 1 & & 0 & 0 \\ & 1 & \end{smallmatrix}\right\rangle,$$

$$|(4,0),(\bar 3,2)\rangle = (1/2)F_2|(4,0),(\bar 4,4)\rangle = (1/2)F_2\;\boxed{2\;|\;2\;|\;2\;|\;2}$$

$$= 2\;\boxed{2\;|\;2\;|\;2\;|\;3} = \left|\begin{smallmatrix}4 & 0 & 0 \\ 3 & & 0 & 0 \\ & 0 & \end{smallmatrix}\right\rangle,$$

$$|(4,0),(\bar 2,0)\rangle = \sqrt{1/6}\,F_2|(4,0),(\bar 3,2)\rangle = \sqrt{1/6}\,F_2\left\{2\;\boxed{2\;|\;2\;|\;2\;|\;3}\right\}$$

$$= \sqrt{6}\;\boxed{2\;|\;2\;|\;3\;|\;3} = \left|\begin{smallmatrix}4 & 0 & 0 \\ 2 & & 0 & 0 \\ & 0 & \end{smallmatrix}\right\rangle,$$

$$|(4,0),(\bar 1,\bar 2)\rangle = \sqrt{1/6}\,F_2|(4,0),(\bar 2,0)\rangle = \sqrt{1/6}\,F_2\left\{\sqrt{6}\;\boxed{2\;|\;2\;|\;3\;|\;3}\right\}$$

$$= 2\;\boxed{2\;|\;3\;|\;3\;|\;3} = \left|\begin{smallmatrix}4 & 0 & 0 \\ 1 & & 0 & 0 \\ & 0 & \end{smallmatrix}\right\rangle,$$

$$|(4,0),(0,4)\rangle = (1/2)F_2|(4,0),(\bar 1,\bar 2)\rangle = (1/2)F_2\left\{2\;\boxed{2\;|\;3\;|\;3\;|\;3}\right\}$$

$$= \boxed{3\;|\;3\;|\;3\;|\;3} = \left|\begin{smallmatrix}4 & 0 & 0 \\ 0 & & 0 & 0 \\ & 0 & \end{smallmatrix}\right\rangle.$$

It is easy to check that

$$F_1|3,\bar 1\rangle = \sqrt{3}|1,0\rangle, \qquad F_1|1,0\rangle = 2|\bar 1,1\rangle, \qquad F_1|\bar 1,1\rangle = \sqrt{3}|\bar 3,2\rangle,$$
$$F_1|2,\bar 2\rangle = \sqrt{2}|0,\bar 1\rangle, \qquad F_1|0,\bar 1\rangle = \sqrt{2}|\bar 2,0\rangle, \qquad F_1|1,\bar 3\rangle = |\bar 1,\bar 2\rangle.$$

The block weight diagram and the planar weight diagram are now drawn as follows. From Eqs. (8.3) and (8.6), the simple roots and the fundamental dominant weights of the SU(3) group in the rectangular coordinate frame are

$$\mathbf{r}_1 = \sqrt{2}\,(\mathbf{V}_1 - \mathbf{V}_2) = \sqrt{2}\mathbf{e}_2,$$
$$\mathbf{r}_2 = \sqrt{2}\,(\mathbf{V}_2 - \mathbf{V}_3) = \sqrt{1/2}\left(\sqrt{3}\mathbf{e}_1 - \mathbf{e}_2\right),$$
$$\mathbf{w}_1 = (1/3)\,(2\mathbf{r}_1 + \mathbf{r}_2) = \sqrt{1/6}\left(\mathbf{e}_1 + \sqrt{3}\mathbf{e}_2\right),$$
$$\mathbf{w}_2 = (1/3)\,(\mathbf{r}_1 + 2\mathbf{r}_2) = \sqrt{2/3}\mathbf{e}_1.$$

From Eq. (8.4) we know that the first and the second components of the weight $\mathbf{m}$ in the rectangular coordinate frame are related to the eigenvalues of $T_8 = T_3^{(3)}$ and $T_3 = T_2^{(3)}$ in SU(3), respectively. Namely, the ordinate axis and the abscissa axis are the eigenvalues of T_3 and T_8, respectively.

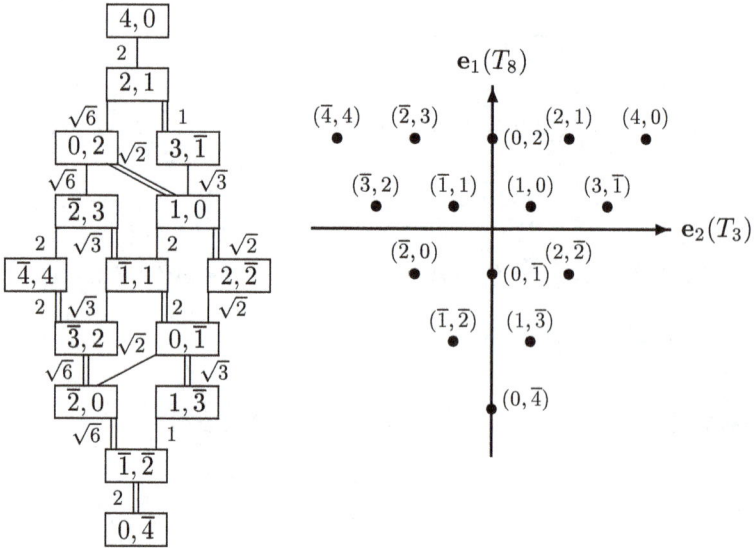

15. Express each Gelfand basis in the irreducible representation [3, 1] of the SU(3) group by the standard tensor Young tableau and calculate the nonvanishing matrix entries for the lowering operators F_μ. Draw the block weight diagram and the planar weight diagram for the representation [3, 1] of SU(3).

Solution. There are two single dominant weights $(2, 1)$ and $(0, 2)$, and one double dominant weight $(1, 0)$ in the representation $[\lambda] = [3, 1]$ of SU(3):

$$
\overset{(2,1)}{\boxed{\begin{array}{|c|c|c|}\hline 1 & 1 & 1 \\\hline 2 \\\cline{1-1}\end{array}}}, \quad
\overset{(0,2)}{\boxed{\begin{array}{|c|c|c|}\hline 1 & 1 & 2 \\\hline 2 \\\cline{1-1}\end{array}}}, \quad
\overset{(1,0)}{\boxed{\begin{array}{|c|c|c|}\hline 1 & 1 & 2 \\\hline 3 \\\cline{1-1}\end{array}}}, \quad
\overset{(1,0)}{\boxed{\begin{array}{|c|c|c|}\hline 1 & 1 & 3 \\\hline 2 \\\cline{1-1}\end{array}}}.
$$

The highest weight of the representation $[3,1]$ is $\mathbf{M} = (2,1)$, which corresponds to the standard tensor Young tableau and the Gelfand basis:

$$
|(2,1),(2,1)\rangle = \boxed{\begin{array}{|c|c|c|}\hline 1 & 1 & 1 \\\hline 2 \\\cline{1-1}\end{array}} = \left| \begin{array}{cccc} 3 & & & \\ & 3 & 1 & 0 \\ & & 3 & \end{array} \right\rangle .
$$

In the tensor subspace $\mathcal{T}_1^{[3,1]} = \mathcal{Y}_1^{[3,1]} \mathcal{T}$, the standard Young tableau is $\boxed{\begin{array}{|c|c|c|}\hline 1 & 2 & 3 \\\hline 4 \\\cline{1-1}\end{array}}$, and the action of the Young operator $\mathcal{Y}_1^{[3,1]}$ on the tensor basis Θ_{abcd} is:

$$
\begin{aligned}
\mathcal{Y}_1^{[3,1]}\Theta_{abcd} = {}& \Theta_{abcd} + \Theta_{acbd} + \Theta_{bacd} + \Theta_{bcad} + \Theta_{cabd} + \Theta_{cbad} \\
& - \Theta_{dbca} - \Theta_{dcba} - \Theta_{bdca} - \Theta_{bcda} - \Theta_{cdba} - \Theta_{cbda} .
\end{aligned}
$$

Note that

$$
\boxed{\begin{array}{|c|c|c|}\hline 2 & 1 & a \\\hline 3 \\\cline{1-1}\end{array}} = \boxed{\begin{array}{|c|c|c|}\hline 1 & 2 & a \\\hline 3 \\\cline{1-1}\end{array}} - \boxed{\begin{array}{|c|c|c|}\hline 1 & 3 & a \\\hline 2 \\\cline{1-1}\end{array}},
$$

where a may be filled in the second column or the third column.

There are four types of standard tensor Young tableaux which normalize to 18, 24, 24, and 48, respectively. The normalization factors for them are $\sqrt{8/3}$, $\sqrt{2}$, $\sqrt{2}$, and 1, respectively.

$$
\begin{aligned}
\sqrt{8/3}\,\boxed{\begin{array}{|c|c|c|}\hline a & a & c \\\hline d \\\cline{1-1}\end{array}} = {}& \sqrt{48}\Big\{ \sqrt{1/18}\,(2\Theta_{aacd} + 2\Theta_{acad} + 2\Theta_{caad} - \Theta_{daca} \\
& - \Theta_{dcaa} - \Theta_{adca} - \Theta_{acda} - \Theta_{cdaa} - \Theta_{cada})\Big\},
\end{aligned}
$$

$$
\begin{aligned}
\sqrt{2}\,\boxed{\begin{array}{|c|c|c|}\hline a & c & c \\\hline d \\\cline{1-1}\end{array}} = {}& \sqrt{48}\Big\{ \sqrt{1/6}\,(\Theta_{accd} + \Theta_{cacd} + \Theta_{ccad} \\
& - \Theta_{dcca} - \Theta_{cdca} - \Theta_{ccda})\Big\},
\end{aligned}
$$

$$
\begin{aligned}
\sqrt{2}\,\boxed{\begin{array}{|c|c|c|}\hline a & a & d \\\hline d \\\cline{1-1}\end{array}} = {}& \sqrt{48}\Big\{ \sqrt{1/6}\,(\Theta_{aadd} + \Theta_{adad} + \Theta_{daad} \\
& - \Theta_{dada} - \Theta_{ddaa} - \Theta_{adda})\Big\},
\end{aligned}
$$

$$
\boxed{\begin{array}{|c|c|c|}\hline a & a & a \\\hline d \\\cline{1-1}\end{array}} = \sqrt{48}\Big\{ \sqrt{1/12}\,(3\Theta_{aaad} - \Theta_{daaa} - \Theta_{adaa} - \Theta_{aada})\Big\}.
$$

From the highest weight state $(2,1)$ we have a triplet of $\mathcal{A}_1$ with $(0,2)$ and $(\bar{2},3)$, and a doublet of $\mathcal{A}_2$ with $(3,\bar{1})$.

$$|(2,1),(0,2)\rangle = \sqrt{1/2}\,F_1|(2,1),(2,1)\rangle = \sqrt{1/2}\,F_1\;\boxed{\begin{array}{ccc}1&1&1\\2&&\end{array}}$$

$$= \sqrt{2}\;\boxed{\begin{array}{ccc}1&1&2\\2&&\end{array}} = \left|\begin{array}{cccc}3&&1&0\\&3&1&\\&&2&\end{array}\right\rangle,$$

$$|(2,1),(\bar{2},3)\rangle = \sqrt{1/2}\,F_1|(2,1),(0,2)\rangle = \sqrt{1/2}\,F_1\left\{\sqrt{2}\;\boxed{\begin{array}{ccc}1&1&2\\2&&\end{array}}\right\}$$

$$= \boxed{\begin{array}{ccc}1&2&2\\2&&\end{array}} = \left|\begin{array}{cccc}3&&1&0\\&3&1&\\&&1&\end{array}\right\rangle,$$

$$|(2,1),(3,\bar{1})\rangle = F_2|(2,1),(2,1)\rangle = F_2\;\boxed{\begin{array}{ccc}1&1&1\\2&&\end{array}}$$

$$= \boxed{\begin{array}{ccc}1&1&1\\3&&\end{array}} = \left|\begin{array}{cccc}3&&1&0\\&3&0&\\&&3&\end{array}\right\rangle.$$

From the weight $(3,\bar{1})$ we have a quartet of $\mathcal{A}_1$ with $(1,0)$, $(\bar{1},1)$, and $(\bar{3},2)$, where the dominant weight $(1,0)$ and its equivalent weight $(\bar{1},1)$ are the double weights. We define one basis state for $(1,0)$ or for $(\bar{1},1)$ belongs to the quartet of $\mathcal{A}_1$, and the other belongs to the doublet.

$$|(2,1),(1,0)_1\rangle = \sqrt{1/3}\,F_1|(2,1),(3,\bar{1})\rangle = \sqrt{1/3}\,F_1\;\boxed{\begin{array}{ccc}1&1&1\\3&&\end{array}}$$

$$= \sqrt{1/3}\left\{2\;\boxed{\begin{array}{ccc}1&1&2\\3&&\end{array}} + \boxed{\begin{array}{ccc}2&1&1\\3&&\end{array}}\right\}$$

$$= \sqrt{3}\;\boxed{\begin{array}{ccc}1&1&2\\3&&\end{array}} - \sqrt{1/3}\;\boxed{\begin{array}{ccc}1&1&3\\2&&\end{array}} = \left|\begin{array}{cccc}3&&1&0\\&3&0&\\&&2&\end{array}\right\rangle,$$

$$|(2,1),(\bar{1},1)_1\rangle = (1/2)F_1|(2,1),(1,0)_1\rangle$$

$$= (1/2)F_1\left\{\sqrt{3}\;\boxed{\begin{array}{ccc}1&1&2\\3&&\end{array}} - \sqrt{1/3}\;\boxed{\begin{array}{ccc}1&1&3\\2&&\end{array}}\right\}$$

$$= \sqrt{3/4}\;\boxed{\begin{array}{ccc}1&2&2\\3&&\end{array}} + \sqrt{3/4}\;\boxed{\begin{array}{ccc}2&1&2\\3&&\end{array}} - \sqrt{1/12}\;\boxed{\begin{array}{ccc}1&2&3\\2&&\end{array}}$$

$$= \sqrt{3}\;\boxed{\begin{array}{ccc}1&2&2\\3&&\end{array}} - \sqrt{4/3}\;\boxed{\begin{array}{ccc}1&2&3\\2&&\end{array}} = \left|\begin{array}{cccc}3&&1&0\\&3&0&\\&&1&\end{array}\right\rangle,$$

$$|(2,1),(\bar{3},2)\rangle = \sqrt{1/3}F_1|(2,1),(\bar{1},1)_1\rangle$$

$$= \sqrt{1/3}F_1\left\{\sqrt{3}\;\begin{array}{|c|c|c|}\hline 1 & 2 & 2 \\\hline 3 \\\cline{1-1}\end{array} - \sqrt{4/3}\;\begin{array}{|c|c|c|}\hline 1 & 2 & 3 \\\hline 2 \\\cline{1-1}\end{array}\right\}$$

$$= \begin{array}{|c|c|c|}\hline 2 & 2 & 2 \\\hline 3 \\\cline{1-1}\end{array} = \left|\begin{array}{ccc} 3 & & \\ 3 & 1 & 0 \\ & 0 & \end{array}\right\rangle,$$

$$F_1\,|(2,1),(1,0)_2\rangle = |(2,1),(\bar{1},1)_2\rangle, \qquad E_1\,|(2,1),(1,0)_2\rangle = 0.$$

From the weight $(0,2)$ we have a triplet of $\mathcal{A}_2$ with $(1,0)$ and $(2,\bar{2})$, where the basis state with the weight $(1,0)$ is a combination of $|(1,0)_1\rangle$ and $|(1,0)_2\rangle$. Letting

$$F_2\,|(2,1),(0,2)\rangle = a_1\,|(2,1),(1,0)_1\rangle + a_2\,|(2,1),(1,0)_2\rangle,$$
$$F_2\,|(2,1),(1,0)_1\rangle = a_3\,|(2,1),(2,\bar{2})\rangle,$$
$$F_2\,|(2,1),(1,0)_2\rangle = a_4\,|(2,1),(2,\bar{2})\rangle,$$
$$a_1^2 + a_2^2 = a_3^2 + a_4^2 = 2,$$

we have

$$E_1F_2\,|(2,1),(0,2)\rangle = a_1\sqrt{3}\,|(2,1),(3,\bar{1})\rangle$$
$$= F_2E_1\,|(2,1),(0,2)\rangle = \sqrt{2}F_2\,|(2,1),(2,1)\rangle = \sqrt{2}\,|(2,1),(3,\bar{1})\rangle.$$

Thus, $a_1 = \sqrt{2/3}$, and $a_2 = \sqrt{2 - 2/3} = \sqrt{4/3}$. Applying $E_2F_2 = F_2E_2 + H_2$ to the basis state $|(2,1),(1,0)_1\rangle$, we have

$$E_2F_2\,|(2,1),(1,0)_1\rangle = a_3^2\,|(2,1),(1,0)_1\rangle + a_3a_4\,|(2,1),(1,0)_2\rangle$$
$$= (F_2E_2 + H_2)\,|(2,1),(1,0)_1\rangle$$
$$= (2/3)\,|(2,1),(1,0)_1\rangle + \sqrt{8/9}\,|(2,1),(1,0)_2\rangle.$$

Choosing the phase of the basis state $|(2,1),(2,\bar{2})\rangle$ such that a_3 is real positive, we obtain $a_3 = \sqrt{2/3}$, and then $a_4 = \sqrt{4/3}$.

$$|(2,1),(1,0)_2\rangle = \sqrt{3/4}\left\{F_2|(2,1),(0,2)\rangle - \sqrt{2/3}|(2,1),(1,0)_1\rangle\right\}$$

$$= \sqrt{3/4}F_2\left\{\sqrt{2}\;\begin{array}{|c|c|c|}\hline 1 & 1 & 2 \\\hline 2 \\\cline{1-1}\end{array}\right\}$$

$$- \sqrt{1/2}\left\{\sqrt{3}\;\begin{array}{|c|c|c|}\hline 1 & 1 & 2 \\\hline 3 \\\cline{1-1}\end{array} - \sqrt{1/3}\;\begin{array}{|c|c|c|}\hline 1 & 1 & 3 \\\hline 2 \\\cline{1-1}\end{array}\right\}$$

$$= \sqrt{8/3}\;\begin{array}{|c|c|c|}\hline 1 & 1 & 3 \\\hline 2 \\\cline{1-1}\end{array} = \left|\begin{array}{ccc} 3 & & \\ 2 & 1 & 0 \\ & 2 & \end{array}\right\rangle,$$

$$|(2,1),(\bar{1},1)_2\rangle = F_1|(2,1),(1,0)_2\rangle$$

$$= F_1\left\{\sqrt{8/3}\;\boxed{\begin{smallmatrix}1&1&3\\2&\end{smallmatrix}}\right\} = \sqrt{8/3}\;\boxed{\begin{smallmatrix}1&2&3\\2&\end{smallmatrix}} = \left|\begin{smallmatrix}3&&1&&0\\&2&&1\\&&1&\end{smallmatrix}\right\rangle,$$

$$|(2,1),(2,\bar{2})\rangle = \sqrt{3/4}F_2|(2,1),(1,0)_2\rangle = \sqrt{3/2}F_2|(2,1),(1,0)_1\rangle$$

$$= \sqrt{3/4}F_2\left\{\sqrt{8/3}\;\boxed{\begin{smallmatrix}1&1&3\\2&\end{smallmatrix}}\right\} = \sqrt{2}\;\boxed{\begin{smallmatrix}1&1&3\\3&\end{smallmatrix}} = \left|\begin{smallmatrix}3&&1&&0\\&2&&0\\&&2&\end{smallmatrix}\right\rangle.$$

From the weight $(2,\bar{2})$ we have a triplet of $\mathcal{A}_1$ with $(0,\bar{1})$ and $(\bar{2},0)$, where the weight $(0,\bar{1})$ is a double weight. We define one state basis $|(0,\bar{1})_1\rangle$ belongs to the triplet of $\mathcal{A}_1$, and the other $|(0,\bar{1})_2\rangle$ belongs to the singlet.

$$|(2,1),(0,\bar{1})_1\rangle = \sqrt{1/2}F_1|(2,1),(2,\bar{2})\rangle = \sqrt{1/2}F_1\left\{\sqrt{2}\;\boxed{\begin{smallmatrix}1&1&3\\3&\end{smallmatrix}}\right\}$$

$$= \boxed{\begin{smallmatrix}1&2&3\\3&\end{smallmatrix}} + \boxed{\begin{smallmatrix}2&1&3\\3&\end{smallmatrix}}$$

$$= 2\;\boxed{\begin{smallmatrix}1&2&3\\3&\end{smallmatrix}} - \boxed{\begin{smallmatrix}1&3&3\\2&\end{smallmatrix}} = \left|\begin{smallmatrix}3&&1&&0\\&2&&0\\&&1&\end{smallmatrix}\right\rangle,$$

$$|(2,1),(\bar{2},0)\rangle = \sqrt{1/2}F_1|(2,1),(0,\bar{1})_1\rangle$$

$$= \sqrt{1/2}F_1\left\{2\;\boxed{\begin{smallmatrix}1&2&3\\3&\end{smallmatrix}} - \boxed{\begin{smallmatrix}1&3&3\\2&\end{smallmatrix}}\right\}$$

$$= \sqrt{2}\;\boxed{\begin{smallmatrix}2&2&3\\3&\end{smallmatrix}} = \left|\begin{smallmatrix}3&&1&&0\\&2&&0\\&&0&\end{smallmatrix}\right\rangle.$$

From the weight $(\bar{2},3)$ we have a quartet of $\mathcal{A}_2$ with $(\bar{1},1)$, $(0,\bar{1})$ and $(1,\bar{3})$, where the states with the weights $(\bar{1},1)$ and $(0,\bar{1})$ are the combinations of the basis states. Letting

$$F_2\,|(2,1),(\bar{2},3)\rangle = b_1\,|(2,1),(\bar{1},1)_1\rangle + b_2\,|(2,1),(\bar{1},1)_2\rangle,$$

$$F_2\,|(2,1),(\bar{1},1)_1\rangle = c_1\,|(2,1),(0,\bar{1})_1\rangle + c_2\,|(2,1),(0,\bar{1})_2\rangle,$$

$$F_2\,|(2,1),(\bar{1},1)_2\rangle = c_3\,|(2,1),(0,\bar{1})_1\rangle + c_4\,|(2,1),(0,\bar{1})_2\rangle,$$

$$F_2\,|(2,1),(0,\bar{1})_1\rangle = d_1\,|(2,1),(1,\bar{3})\rangle,$$

$$F_2\,|(2,1),(0,\bar{1})_2\rangle = d_2\,|(2,1),(1,\bar{3})\rangle,$$

we have

$$E_1 F_2\,|(2,1),(\bar{2},3)\rangle = 2b_1\,|(2,1),(1,0)_1\rangle + b_2\,|(2,1),(1,0)_2\rangle$$

$$= F_2 E_1 \,|(2,1),(\overline{2},3)\rangle = \sqrt{4/3}\,|(2,1),(1,0)_1\rangle + \sqrt{8/3}\,|(2,1),(1,0)_2\rangle,$$

$$E_1 F_2 \,|(2,1),(\overline{1},1)_1\rangle = c_1\sqrt{2}\,|(2,1),(2,\overline{2})\rangle$$

$$= F_2 E_1 \,|(2,1),(\overline{1},1)_1\rangle = 2F_2\,|(2,1),(1,0)_1\rangle = \sqrt{8/3}\,|(2,1),(2,\overline{2})\rangle,$$

$$E_1 F_2 \,|(2,1),(\overline{1},1)_2\rangle = c_3\sqrt{2}\,|(2,1),(2,\overline{2})\rangle$$

$$= F_2 E_1 \,|(2,1),(\overline{1},1)_2\rangle = F_2\,|(2,1),(1,0)_2\rangle = \sqrt{4/3}\,|(2,1),(2,\overline{2})\rangle.$$

Thus, $b_1 = \sqrt{1/3}$, $b_2 = \sqrt{8/3}$, $c_1 = \sqrt{4/3}$, and $c_3 = \sqrt{2/3}$. Applying $E_2 F_2 = F_2 E_2 + H_2$ to the basis state $|(2,1),(\overline{1},1)_2\rangle$, we have

$$E_2 F_2 \,|(2,1),(\overline{1},1)_2\rangle$$
$$= \left(\sqrt{8/9} + c_2 c_4\right)|(2,1),(\overline{1},1)_1\rangle + \left(2/3 + c_4^2\right)|(2,1),(\overline{1},1)_2\rangle$$
$$= (F_2 E_2 + H_2)\,|(2,1),(\overline{1},1)_2\rangle$$
$$= \sqrt{8/9}\,|(2,1),(\overline{1},1)_1\rangle + (8/3 + 1)\,|(2,1),(\overline{1},1)_2\rangle.$$

Choosing the phase of the basis state $|(2,1),(0,\overline{1})_2\rangle$ such that c_4 is real positive, we obtain $c_4 = \sqrt{3}$, and then $c_2 = 0$. Applying $E_2 F_2 = F_2 E_2 + H_2$ to the basis state $|(2,1),(0,\overline{1})_2\rangle$, we have

$$E_2 F_2 \,|(2,1),(0,\overline{1})_2\rangle = d_1 d_2\,|(2,1),(0,\overline{1})_1\rangle + d_2^2\,|(2,1),(0,\overline{1})_2\rangle$$
$$= (F_2 E_2 + H_2)\,|(2,1),(0,\overline{1})_2\rangle$$
$$= \sqrt{2}\,|(2,1),(0,\overline{1})_1\rangle + (3-1)\,|(2,1),(0,\overline{1})_2\rangle.$$

Choosing the phase of the basis state $|(2,1),(1,\overline{3})\rangle$ such that d_2 is real positive, we obtain $d_2 = \sqrt{2}$, and then $d_1 = 1$. From the above calculation, we have

$$|(2,1),(0,\overline{1})_2\rangle = \sqrt{1/3}\left\{ F_2|(2,1),(\overline{1},1)_2\rangle - \sqrt{2/3}|(2,1),(0,\overline{1})_1\rangle \right\}$$

$$= \sqrt{1/3}F_2\left\{ \sqrt{8/3}\;\begin{array}{|c|c|c|}\hline 1 & 2 & 3 \\\hline 2 \\\cline{1-1}\end{array} \right\}$$

$$- \sqrt{2/9}\left\{ 2\;\begin{array}{|c|c|c|}\hline 1 & 2 & 3 \\\hline 3 \\\cline{1-1}\end{array} - \begin{array}{|c|c|c|}\hline 1 & 3 & 3 \\\hline 2 \\\cline{1-1}\end{array} \right\}$$

$$= \sqrt{2}\;\begin{array}{|c|c|c|}\hline 1 & 3 & 3 \\\hline 2 \\\cline{1-1}\end{array} = \left| \begin{matrix} 3 & & \\ & 1 & \\ & 1 & \end{matrix}\;\begin{matrix} 1 & 0 \\ 1 & \end{matrix} \right\rangle,$$

$$|(2,1),(1,\overline{3})\rangle = \sqrt{1/2}F_2|(2,1),(0,\overline{1})_2\rangle = F_2|(2,1),(0,\overline{1})_1\rangle$$

$$= \sqrt{1/2}F_2\left\{ \sqrt{2}\;\begin{array}{|c|c|c|}\hline 1 & 3 & 3 \\\hline 2 \\\cline{1-1}\end{array} \right\} = \begin{array}{|c|c|c|}\hline 1 & 3 & 3 \\\hline 3 \\\cline{1-1}\end{array} = \left| \begin{matrix} 3 & & \\ & 1 & \\ & 1 & \end{matrix}\;\begin{matrix} 1 & 0 & 0 \\ & & \end{matrix} \right\rangle,$$

$$|(2,1),(\bar{1},\bar{2})\rangle = F_1|(2,1),(1,\bar{3})\rangle = F_1 \;\begin{array}{|c|c|c|}\hline 1 & 3 & 3\\\hline 3\\\cline{1-1}\end{array}$$

$$= \;\begin{array}{|c|c|c|}\hline 2 & 3 & 3\\\hline 3\\\cline{1-1}\end{array}\; = \left|\begin{matrix}3 & 1 & 0\\ & 1 & 0\\ & & 0\end{matrix}\right\rangle .$$

Note that those representation matrix entries of F_2 related to the basis states with the multiple weights can also be calculated with the formula of Gelfand.

$$F_2\left|\begin{matrix}3 & 1 & 0\\ 3 & 1\\ 2\end{matrix}\right\rangle = A_{12}\left|\begin{matrix}3 & 1 & 0\\ 2 & 1\\ 2\end{matrix}\right\rangle + A_{22}\left|\begin{matrix}3 & 1 & 0\\ 3 & 0\\ 2\end{matrix}\right\rangle ,$$

$$F_2\left|\begin{matrix}3 & 1 & 0\\ 3 & 0\\ 2\end{matrix}\right\rangle = B_{12}\left|\begin{matrix}3 & 1 & 0\\ 2 & 0\\ 2\end{matrix}\right\rangle ,$$

$$F_2\left|\begin{matrix}3 & 1 & 0\\ 2 & 1\\ 2\end{matrix}\right\rangle = B_{22}\left|\begin{matrix}3 & 1 & 0\\ 2 & 0\\ 2\end{matrix}\right\rangle ,$$

$$F_2\left|\begin{matrix}3 & 1 & 0\\ 3 & 1\\ 1\end{matrix}\right\rangle = C_{12}\left|\begin{matrix}3 & 1 & 0\\ 2 & 1\\ 1\end{matrix}\right\rangle + C_{22}\left|\begin{matrix}3 & 1 & 0\\ 3 & 0\\ 1\end{matrix}\right\rangle ,$$

$$F_2\left|\begin{matrix}3 & 1 & 0\\ 2 & 1\\ 1\end{matrix}\right\rangle = D_{12}\left|\begin{matrix}3 & 1 & 0\\ 1 & 1\\ 1\end{matrix}\right\rangle + D_{22}\left|\begin{matrix}3 & 1 & 0\\ 2 & 0\\ 1\end{matrix}\right\rangle ,$$

$$F_2\left|\begin{matrix}3 & 1 & 0\\ 3 & 0\\ 1\end{matrix}\right\rangle = A'_{12}\left|\begin{matrix}3 & 1 & 0\\ 2 & 0\\ 1\end{matrix}\right\rangle ,$$

$$F_2\left|\begin{matrix}3 & 1 & 0\\ 2 & 0\\ 1\end{matrix}\right\rangle = B'_{12}\left|\begin{matrix}3 & 1 & 0\\ 1 & 0\\ 1\end{matrix}\right\rangle ,$$

$$F_2\left|\begin{matrix}3 & 1 & 0\\ 1 & 1\\ 1\end{matrix}\right\rangle = B'_{22}\left|\begin{matrix}3 & 1 & 0\\ 1 & 0\\ 1\end{matrix}\right\rangle .$$

From Eq. (8.32) we have

$$A_{12} = \{-(-1)(1)(-2)(-4)/[(-2)(-3)]\}^{1/2} = \sqrt{4/3},$$
$$A_{22} = \{-(2)(4)(1)(-1)/[(4)(3)]\}^{1/2} = \sqrt{2/3},$$
$$B_{12} = \{-(-1)(1)(-2)(-4)/[(-3)(-4)]\}^{1/2} = \sqrt{2/3},$$
$$B_{22} = \{-(2)(4)(1)(-1)/[(3)(2)]\}^{1/2} = \sqrt{4/3},$$
$$C_{12} = \{-(-2)(1)(-2)(-4)/[(-2)(-3)]\}^{1/2} = \sqrt{8/3},$$

$$C_{22} = \{-(1)(4)(1)(-1)/[(4)(3)]\}^{1/2} = \sqrt{1/3},$$
$$D_{12} = \{-(-1)(2)(-1)(-3)/[(-1)(-2)]\}^{1/2} = \sqrt{3},$$
$$D_{22} = \{-(1)(4)(1)(-1)/[(3)(2)]\}^{1/2} = \sqrt{2/3},$$
$$A'_{12} = \{-(-2)(1)(-2)(-4)/[(-3)(-4)]\}^{1/2} = \sqrt{4/3},$$
$$B'_{12} = \{-(-1)(2)(-1)(-3)/[(-2)(-3)]\}^{1/2} = 1,$$
$$B'_{22} = \{-(1)(4)(1)(-1)/[(2)(1)]\}^{1/2} = \sqrt{2}.$$

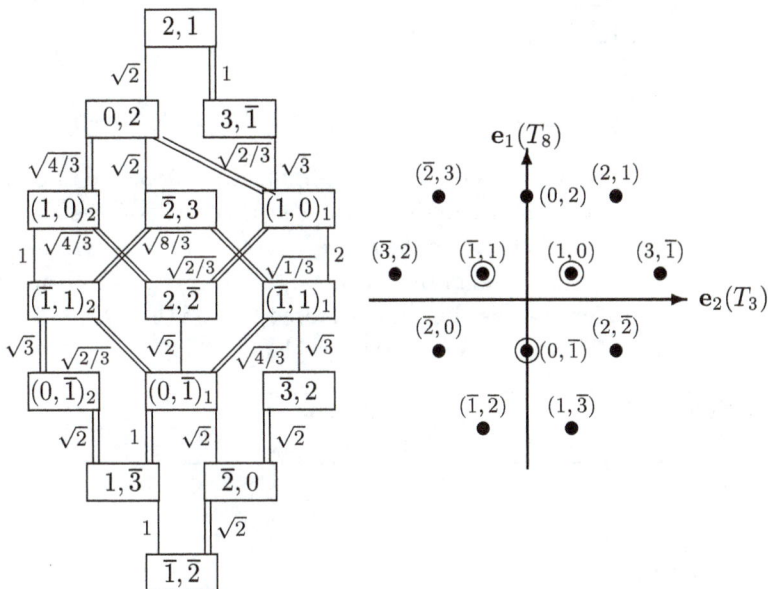

The block weight diagram and the planar weight diagram are given in the figure. The planar weight diagram of the representation $[\lambda_1, \lambda_2]$ of SU(3) is a hexagon with the lengths of two neighbored sides to be $(\lambda_1 - \lambda_2)$ (i.e. M_1) and λ_2 (i.e. M_2). The weights on the boundary is single, and the multiplicity increases as the position of the weight in the diagram goes inside until the hexagon reduces to a triangle.

16. Calculate the Clebsch-Gordan series for the direct product representation $[2, 1] \times [2, 1]$ of the SU(3) group, and expand the highest weight state of each irreducible representation in the CG series with respect to the standard tensor Young tableaux.

Solution. The representation $[2, 1]$ is the adjoint representation of SU(3) and was studied in Problem 13. The highest weight of $[2, 1]$ of SU(3) is $(1, 1)$.

Usually, the Clebsch-Gordan series for the direct product representation of the SU(N) group is best calculated with the Littlewood-Richardson rule, as used in Problem 5. For $[2,1] \times [2,1]$ of SU(3) we have

$$[2,1] \times [2,1] \simeq [4,2] \oplus [3,0] \oplus [3,3] \oplus [2,1] \oplus [2,1] \oplus [0,0] .$$

However, the method of the dominant weight diagram discussed in Chapter 7 is also a good method for it. In this method we need to know the dominant weights and their multiplicities in the relevant representations, which can be obtained, say, by the method of the block weight diagram, or by consulting the table book. They can also be obtained directly from the standard tensor Young tableaux. This is the aim of this problem.

Dominant weight	Standard tensor Young tableaux	No. of states
(2,2)		1
(3,0)		2
(0,3)		2
(1,1)		6
(0,0)		10

We have known that the weight for a standard tensor Young tableau is dominant if and only if the number of each digit μ filled in the tableau is not less than the number of the digit $(\mu + 1)$ filled there. We can easily count by this criterion which and how many standard tensor Young tableaux correspond to the dominant weights in the direct product representation space and in the relevant representation spaces. First, we list the standard tensor Young tableaux with the dominant weights in the direct product representation space $[2,1] \times [2,1]$, where only one standard tensor Young tableau in the pair related by a permutation of two product factors is listed for simplicity. Then, we calculate the standard tensor Young tableaux with the dominant weights in the representation $(2,2)$ which is contained once in the CG series because the weight $(2,2)$ is the highest weight in the product space.

Dominant weight	Standard tensor Young tableaux	No. of states
(2,2)	$\begin{array}{\|c\|c\|c\|c\|}\hline 1&1&1&1\\\hline 2&2\\\cline{1-2}\end{array}$	1
(3,0)	$\begin{array}{\|c\|c\|c\|c\|}\hline 1&1&1&1\\\hline 2&3\\\cline{1-2}\end{array}$	1
(0,3)	$\begin{array}{\|c\|c\|c\|c\|}\hline 1&1&1&2\\\hline 2&2\\\cline{1-2}\end{array}$	1
(1,1)	$\begin{array}{\|c\|c\|c\|c\|}\hline 1&1&1&2\\\hline 2&3\\\cline{1-2}\end{array}$ $\begin{array}{\|c\|c\|c\|c\|}\hline 1&1&1&3\\\hline 2&2\\\cline{1-2}\end{array}$	2
(0,0)	$\begin{array}{\|c\|c\|c\|c\|}\hline 1&1&2&2\\\hline 3&3\\\cline{1-2}\end{array}$ $\begin{array}{\|c\|c\|c\|c\|}\hline 1&1&2&3\\\hline 2&3\\\cline{1-2}\end{array}$	3
	$\begin{array}{\|c\|c\|c\|c\|}\hline 1&1&3&3\\\hline 2&2\\\cline{1-2}\end{array}$	

By comparing respectively the multiplicities of the dominant weights $(3,0)$ and $(0,3)$ in the product space with those in the representation $(2,2)$, $2-1 = 1$, we conclude that there is one representation $(3,0)$ and one representation $(0,3)$ contained in the CG series. Third, we calculate the standard tensor Young tableaux with the dominant weights in the representations $(3,0)$ and $(0,3)$. Comparing the numbers of the standard tensor Young tableaux with the dominant weight $(1,1)$ in the product space and in the three representations $(2,2)$, $(3,0)$ and $(0,3)$, $6 - 2 - 1 - 1 = 2$, we conclude that the representation $(1,1)$ is contained in the CG series twice.

Dominant weight	Standard tensor Young tableaux		No. of states	
	in (3,0)	in (0,3)	in (3,0)	in (0,3)
(3,0)	1 1 1		1	0
(0,3)		1 1 1 / 2 2 2	0	1
(1,1)	1 1 2	1 1 1 / 2 2 3	1	1
(0,0)	1 2 3	1 1 2 / 2 3 3	1	1

Fourth, we calculate the tensor Young tableaux with the dominant weights in the representation $(1,1)$. Comparing the numbers of the standard tensor Young tableaux with the dominant weight $(0,0)$ in the product space and in the representations $(2,2)$, $(3,0)$, $(0,3)$, and two $(1,1)$, $10 - 3 - 1 - 1 - 2 \times 2 = 1$, we conclude that the representation $(0,0)$ is contained once in the CG series. Therefore, we complete the diagram of the dominant weights for the direct product representation $[2,1] \times [2,1]$, where the subscript S or A denotes the symmetric or antisymmetric combination in the permutation of two factors in the product, respectively.

Dominant weight	Standard tensor Young tableaux	No. of states
(1,1)	1 1 / 2	1
(0,0)	1 2 / 3 , 1 3 / 2	2

$OS \times n$

6×1 $\boxed{2,2}$

3×2 $\boxed{3,0}$ $\boxed{3,0}$

3×2 $\boxed{0,3}$ $\boxed{0,3}$

6×6 $\boxed{1,1}^2$ $\boxed{1,1}$ $\boxed{1,1}$ $\boxed{1,1}^2$

1×10 $\boxed{0,0}^3$ $\boxed{0,0}$ $\boxed{0,0}$ $\boxed{0,0}^2$ $\boxed{0,0}$

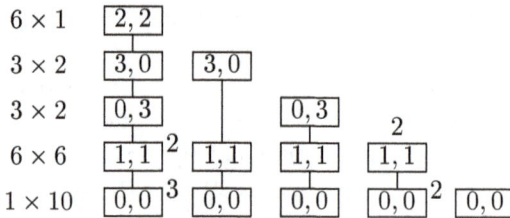

$\overline{64} = 27 + 10 + 10^* + 2 \times 8 + 1$

$[2,1] \times [2,1] \simeq [4,2] \oplus [3,0] \oplus [0,3] \oplus [2,1]_S \oplus [2,1]_A \oplus [0,0]$

At last, we calculate the expansion for the highest weight state of each representation contained in the CG series by Eq. (7.24), namely, each raising operator E_μ annihilates the highest weight state. The results are

$$\begin{array}{|c|c|c|c|}\hline 1&1&1&1\\\hline 2&2\\\cline{1-2}\end{array} \sim \begin{array}{|c|c|}\hline 1&1\\\hline 2\\\cline{1-1}\end{array} \times \begin{array}{|c|c|}\hline 1&1\\\hline 2\\\cline{1-1}\end{array},$$

$$\begin{array}{|c|c|c|c|}\hline 1&1&1&1\\\hline 2\\\cline{1-1}3\\\cline{1-1}\end{array} \sim \begin{array}{|c|c|}\hline 1&1\\\hline 2\\\cline{1-1}\end{array} \times \begin{array}{|c|c|}\hline 1&1\\\hline 3\\\cline{1-1}\end{array} - \begin{array}{|c|c|}\hline 1&1\\\hline 3\\\cline{1-1}\end{array} \times \begin{array}{|c|c|}\hline 1&1\\\hline 2\\\cline{1-1}\end{array},$$

$$\begin{array}{|c|c|c|}\hline 1&1&1\\\hline 2&2&2\\\hline\end{array} \sim \begin{array}{|c|c|}\hline 1&1\\\hline 2\\\cline{1-1}\end{array} \times \begin{array}{|c|c|}\hline 1&2\\\hline 2\\\cline{1-1}\end{array} - \begin{array}{|c|c|}\hline 1&2\\\hline 2\\\cline{1-1}\end{array} \times \begin{array}{|c|c|}\hline 1&1\\\hline 2\\\cline{1-1}\end{array},$$

$$\begin{array}{|c|c|c|}\hline 1&1&1\\\hline 2&2\\\cline{1-2}3\\\cline{1-1}\end{array}_{S} \sim \begin{array}{|c|c|}\hline 1&1\\\hline 2\\\cline{1-1}\end{array} \times \begin{array}{|c|c|}\hline 1&2\\\hline 3\\\cline{1-1}\end{array} - \begin{array}{|c|c|}\hline 1&2\\\hline 2\\\cline{1-1}\end{array} \times \begin{array}{|c|c|}\hline 1&1\\\hline 3\\\cline{1-1}\end{array} + \text{(sym. terms)},$$

$$\begin{array}{|c|c|c|}\hline 1&1&1\\\hline 2&2\\\cline{1-2}3\\\cline{1-1}\end{array}_{A} \sim \begin{array}{|c|c|}\hline 1&1\\\hline 2\\\cline{1-1}\end{array} \times \begin{array}{|c|c|}\hline 1&2\\\hline 3\\\cline{1-1}\end{array} - 2\begin{array}{|c|c|}\hline 1&1\\\hline 2\\\cline{1-1}\end{array} \times \begin{array}{|c|c|}\hline 1&3\\\hline 2\\\cline{1-1}\end{array}$$

$$- \begin{array}{|c|c|}\hline 1&2\\\hline 2\\\cline{1-1}\end{array} \times \begin{array}{|c|c|}\hline 1&1\\\hline 3\\\cline{1-1}\end{array} + \text{(antisym. terms)},$$

$$\begin{array}{|c|c|}\hline 1&1\\\hline 2&2\\\hline 3&3\\\hline\end{array} \sim \begin{array}{|c|c|}\hline 1&1\\\hline 2\\\cline{1-1}\end{array} \times \begin{array}{|c|c|}\hline 2&3\\\hline 3\\\cline{1-1}\end{array} - \begin{array}{|c|c|}\hline 1&1\\\hline 3\\\cline{1-1}\end{array} \times \begin{array}{|c|c|}\hline 2&2\\\hline 3\\\cline{1-1}\end{array}$$

$$- \begin{array}{|c|c|}\hline 1&2\\\hline 2\\\cline{1-1}\end{array} \times \begin{array}{|c|c|}\hline 1&3\\\hline 3\\\cline{1-1}\end{array} - \begin{array}{|c|c|}\hline 1&2\\\hline 3\\\cline{1-1}\end{array} \times \begin{array}{|c|c|}\hline 1&3\\\hline 2\\\cline{1-1}\end{array} + \text{(sym. terms)}$$

$$+ 2\begin{array}{|c|c|}\hline 1&2\\\hline 3\\\cline{1-1}\end{array} \times \begin{array}{|c|c|}\hline 1&2\\\hline 3\\\cline{1-1}\end{array} + 2\begin{array}{|c|c|}\hline 1&3\\\hline 2\\\cline{1-1}\end{array} \times \begin{array}{|c|c|}\hline 1&3\\\hline 2\\\cline{1-1}\end{array}.$$

In terms of the orthonormal basis states we have

$$\|(2,2),(2,2)\rangle = |1,1\rangle\,|1,1\rangle\,,$$

$$\|(3,0),(3,0)\rangle = \sqrt{1/2}\,\{\,|1,1\rangle\,|2,\bar{1}\rangle - |2,\bar{1}\rangle\,|1,1\rangle\,\}\,,$$

$$\|(0,3),(0,3)\rangle = \sqrt{1/2}\,\{\,|1,1\rangle\,|\bar{1},2\rangle - |\bar{1},2\rangle\,|1,1\rangle\,\}\,,$$

$$\|(1,1),(1,1)\rangle_S = \sqrt{1/20}\,\{\,\sqrt{3}\,|1,1\rangle\,|(0,0)_1\rangle + |1,1\rangle\,|(0,0)_2\rangle$$
$$- \sqrt{6}\,|\bar{1},2\rangle\,|2,\bar{1}\rangle + \text{(sym. terms)}\,\}\,,$$

$$\|(1,1),(1,1)\rangle_A = 6^{-1}\,\{\,\sqrt{3}\,|1,1\rangle\,|(0,0)_1\rangle - 3\,|1,1\rangle\,|(0,0)_2\rangle$$
$$- \sqrt{6}\,|\bar{1},2\rangle\,|2,\bar{1}\rangle + \text{(antisym. terms)}\,\}\,,$$

$$||(0,0),(0,0)\rangle = \sqrt{1/8} \left\{ |1,1\rangle \, |\bar{1},\bar{1}\rangle - |2,\bar{1}\rangle \, |\bar{2},1\rangle - |\bar{1},2\rangle \, |1,\bar{2}\rangle \right.$$

$$\left. +(\text{sym. terms}) + |(0,0)_1\rangle \, |(0,0)_1\rangle + |(0,0)_2\rangle \, |(0,0)_2\rangle \right\}.$$

17. A neutron is composed of one u quark and two d quarks. Construct the wave function of a neutron with spin $S_3 = -1/2$, satisfying the correct permutation symmetry among the identical particles.

Solution. In particle physics, a hadron is composed of three quarks (baryon) or a pair of one quark and one antiquark (meson). The strong interaction among quarks is the gauge interaction of the colored SU(3) group. In addition to the color, a quark brings the flavor quantum number. The u and d quarks span the self-representation space of the flavor SU(2) group, the isospin symmetry group. The isospin is conserved in the strong interaction. The next quark is s quark, which is heavier than the u and d quarks. u, d and s quarks span the self-representation space of the flavor SU(3) group, which describes an approximate symmetry, but played an important role in the history of particle physics. Even now, it is still useful in the rough estimation. The expansions of the orthonormal basis states with respect to the standard tensor Young tableaux give the flavor wave functions of the hadrons.

The flavor SU(3) group is a compact simple Lie group with order 8 and rank 2. The two commutable generators have important physical meanings. In the self-representation, T_3 describes the third component of the isospin of a quark, and T_8 is proportional to the supercharge Y of a quark:

$$T_3 = \frac{1}{2} \begin{pmatrix} 1 & 0 & 0 \\ 0 & -1 & 0 \\ 0 & 0 & 0 \end{pmatrix}, \qquad Y = \frac{2}{\sqrt{3}} T_8 = \frac{1}{3} \begin{pmatrix} 1 & 0 & 0 \\ 0 & 1 & 0 \\ 0 & 0 & -2 \end{pmatrix}. \tag{8.36}$$

The basis states in an irreducible representation space of the flavor SU(3) group can be described in the planar weight diagram with the coordinate axes T_3 and Y, as shown in Fig. 7.3 and Problems 14 and 15 in this Chapter. Since the planar weight diagram is intuitive, it has widely applied in particle physics.

A neutron consists of three quarks. It is a system of three identical fermions. According to the Fermi statistics, the wave function has to be totally antisymmetric in the permutation of quarks. The internal orbital angular momentum of quarks is assumed to be zero. The total wave function is composed of three parts: the color part, the flavor part, and the spinor part. Since the neutron has to be the color singlet, its wave function of the color part is totally antisymmetric in the permutation. Thus, the product of the wave functions of the flavor part and the spinor part should be combined into the totally symmetric state in the permutation. The neutron belongs to a flavor octet with $T = -T_3 = 1/2$ and $Y = 1$, and to a spinor doublet. In this problem we discuss the state with $S_3 = -1/2$. The Young patterns corresponding to two parts both are $[2, 1]$. Choosing the standard Young tableau $\boxed{\begin{array}{cc} 1 & 2 \\ 3 \end{array}}$, we obtain two basis states of the flavor wave function by the standard tensor Young tableaux:

$$\boxed{\begin{array}{cc} u & d \\ d \end{array}} = udd + dud - 2ddu, \qquad (23)\ \boxed{\begin{array}{cc} u & d \\ d \end{array}} = udd + ddu - 2dud .$$

There are also two basis states of the spinor wave function:

$$\boxed{\begin{array}{cc} + & - \\ - \end{array}} = (+--) + (-+-) - 2(--+) ,$$

$$(23)\ \boxed{\begin{array}{cc} + & - \\ - \end{array}} = (+--) + (--+) - 2(-+-) .$$

In this set of the standard bases, the representation matrices of the generators of the permutation group S_3 can be calculated by the tabular method (see Problem 20 in Chap. 6),

$$D^{[2,1]}(1\ 2) = \begin{pmatrix} 1 & -1 \\ 0 & -1 \end{pmatrix}, \qquad D^{[2,1]}(2\ 3) = \begin{pmatrix} 0 & 1 \\ 1 & 0 \end{pmatrix}.$$

In the direct product representation,

$$D^{[2,1]}(1\ 2) \times D^{[2,1]}(1\ 2) = \begin{pmatrix} 1 & -1 & -1 & 1 \\ 0 & -1 & 0 & 1 \\ 0 & 0 & -1 & 1 \\ 0 & 0 & 0 & 1 \end{pmatrix},$$

$$D^{[2,1]}(2\ 3) \times D^{[2,1]}(2\ 3) = \begin{pmatrix} 0 & 0 & 0 & 1 \\ 0 & 0 & 1 & 0 \\ 0 & 1 & 0 & 0 \\ 1 & 0 & 0 & 0 \end{pmatrix}.$$

Their eigenvectors for the eigenvalue 1 respectively

$$(1\ 2): \begin{pmatrix} a \\ b \\ b \\ 2b \end{pmatrix}, \qquad (2\ 3): \begin{pmatrix} c \\ d \\ d \\ c \end{pmatrix}.$$

The totally symmetric state corresponds to the common eigenvector with eigenvalue 1, which is $(2\ 1\ 1\ 2)^T$. Thus, the total wave function of the flavor part and the spinor part for the neutron with $S_3 = -1/2$ is

$$N_{-1/2} = \boxed{\begin{smallmatrix} u & d \\ d & \end{smallmatrix}} \left\{ 2\,\boxed{\begin{smallmatrix} + & - \\ - & \end{smallmatrix}} + \left[(2\ 3)\,\boxed{\begin{smallmatrix} + & - \\ - & \end{smallmatrix}} \right] \right\}$$

$$+ \left[(2\ 3)\,\boxed{\begin{smallmatrix} u & d \\ d & \end{smallmatrix}} \right] \left\{ \boxed{\begin{smallmatrix} + & - \\ - & \end{smallmatrix}} + 2\left[(2\ 3)\,\boxed{\begin{smallmatrix} + & - \\ - & \end{smallmatrix}} \right] \right\}$$

$$= \{udd + dud - 2ddu\} \cdot 3\{(+--) - (--+)\}$$
$$+ \{udd + ddu - 2dud\} \cdot 3\{(+--) - (-+-)\}$$
$$= 3\{2u_+d_-d_- - d_+d_-u_- - d_+u_-d_- - u_-d_-d_+ + 2d_-d_-u_+$$
$$- d_-u_-d_+ - u_-d_+d_- - d_-d_+u_- + 2d_-u_+d_-\} .$$

For normalization, one may replace the factor 3 with $\sqrt{1/18}$.

8.4   Subduced Representations

★ Now, we discuss the reduction problem of the subduced representation of an irreducible representation of the SU(N) group with respect to its subgroup. There are three kinds of important subgroups for the SU(N) group. One is SU($N-1$) ⊂SU(N), the second is SU(M)×SU(N/M) ⊂SU(N), and the third is SU(M)×SU($N-M$) ⊂SU(N).

★ The reduction of the subduced representation from an irreducible representation $[\omega]$ of SU(N) with respect to the subgroup SU($N-1$) can be solved directly by the standard tensor Young tableau. The tensor indices taken the value N keep invariant in the subgroup SU($N-1$). In a standard tensor Young tableau, the box filled with N can only be located in the lowest position of each column, where the neighbored right position has to be empty or a box filled with N. Removing the boxes filled with N in a standard tensor Young tableau of SU(N), we obtain a basis state of an irreducible representation of the subgroup $SU(N-1)$. Thus, by removing several (including zero) boxes located in the lowest positions in some columns from the Young pattern $[\lambda]$, we obtain the Young pattern $[\mu]$, denoting an irreducible representation of $SU(N-1)$ in the Clebsch-Gordan series for the reduction of the subduced representation:

$$[\lambda] \simeq \bigoplus [\mu], \qquad d_{[\lambda]}(SU(N)) = \sum_{[\mu]} d_{[\mu]}(SU(N-1)),$$

$$\lambda_N \leq \mu_{N-1} \leq \lambda_{N-1} \leq \mu_{N-2} \leq \cdots \leq \mu_2 \leq \lambda_2 \leq \mu_1 \leq \lambda_1. \tag{8.37}$$

This method may be applied continuously to the reduction for the subgroups in the group chain

$$SU(N) \supset SU(N-1) \supset \cdots \supset SU(3) \supset SU(2) \ .$$

Thus, the subduced representation of an irreducible representation $[\lambda]$ of $SU(N)$ with respect to its subgroup $SU(N')$ with $N' < N$ can be reduced.

★ For the $SU(NM)$ group, containing the subgroup $SU(N) \times SU(M)$, each tensor index α consists of two indices (ai), where a is the tensor index of the subgroup $SU(N)$, and i is the tensor index of the subgroup $SU(M)$. In the reduction of a subduced representation from an irreducible representation $[\omega]$ of $SU(NM)$ with respect to the irreducible representations $[\lambda] \otimes [\mu]$ of the subgroup $SU(N) \times SU(M)$, the numbers of the boxes in three Young patterns are the same n. The three Young patterns respectively describe the permutation symmetry among n indices α, n indices a, and n indices i. When n indices i are fixed, the n indices a have the permutation symmetry denoted by $[\lambda]$. When n indices a are fixed, the n indices i have the permutation symmetry denoted by $[\mu]$. When both sets of indices are transformed, n indices α have the permutation symmetry denoted by $[\omega]$. Therefore, the representation $[\omega]$ of the permutation group S_n should be contained in the reduction of the inner product of two representations $[\lambda]$ and $[\mu]$:

$$[\lambda] \times [\mu] \simeq [\omega] \oplus \cdots \ .$$

Now, being an irreducible representation of $SU(NM)$, the number of rows of the Young pattern $[\omega]$ cannot be larger than NM. Correspondingly, the number of rows of the Young pattern $[\lambda]$ for $SU(N)$ cannot be larger than N, and that of the Young pattern $[\mu]$ for $SU(M)$ cannot be larger than M.

★ For the $SU(N+M)$ group, containing the subgroup $SU(N) \times SU(M)$, each tensor index α takes two possible kinds of values, either the tensor index a of the subgroup $SU(N)$, or the tensor index i of the subgroup $SU(M)$. In the reduction of the subduced representation from the irreducible representation $[\omega]$ of the $SU(N + M)$ group with respect to the irreducible representation $[\lambda] \otimes [\mu]$ of the subgroup $SU(N) \times SU(M)$, the number n_1 of boxes in the Young pattern $[\lambda]$ and the number n_2 of boxes in the Young pattern $[\mu]$ are not fixed, but the sum $n_1 + n_2$ of their numbers of boxes is fixed and equal to the number n of boxes in the Young pattern $[\omega]$. In the permutation among the tensor indices, all indices are divided into two sets. The n_1 indices a in the first set have the permutation symmetry denoted by $[\lambda]$, which contains n_1 boxes, and the n_2 indices i in the second set have the permutation symmetry denoted by $[\mu]$, which contains n_2 boxes. Altogether, they have the permutation symmetry denoted by $[\omega]$, which contains $n = n_1 + n_2$ boxes. Therefore, the representation $[\omega]$ of the permutation group S_n should be contained in the reduction of the outer product of two representations $[\lambda]$ and $[\mu]$ of the permutation group:

$$[\lambda] \otimes [\mu] \simeq a^\omega_{\lambda\mu} [\omega] \oplus \cdots .$$

The reduction of the subduced representation can be calculated by the Littlewood-Richardson rule.

$$[\omega] \simeq \bigoplus a^\omega_{\lambda\mu} \{[\lambda] \otimes [\mu]\} ,$$

$$d_{[\omega]}(SU(N + M)) = \sum a^\omega_{\lambda\mu} d_{[\lambda]}(SU(N))d_{[\mu]}(SU(M)) .$$

Being an irreducible representation of $SU(N + M)$, the number of rows of the Young pattern $[\omega]$ cannot be larger than $N + M$. Correspondingly, the number of rows of the Young pattern $[\lambda]$ for $SU(N)$ cannot be larger than N, and that of the Young pattern $[\mu]$ for $SU(M)$ cannot be larger than M.

18. Reduce the subduced representation from $[3, 1]$ of the $SU(4)$ group with respect to the subgroup $SU(3)$ and list the standard tensor Young tableaux as the basis states in each irreducible representation space of $SU(3)$.

Solution. The dimension of the representation $[3, 1]$ of $SU(4)$ is

$$d_{[3,1]}(SU(4)) = \frac{\begin{array}{l} 4\,5\,6 \\ 3 \end{array}}{\begin{array}{l} 4\,2\,1 \\ 1 \end{array}} = 45.$$

Its subduced representation with respect to SU(3) is reduced into the direct sum of six irreducible representations of SU(3). They and their basis states are listed as follows.

1). The representation [3, 1] of SU(3) is 15-dimensional with the following standard tensor Young tableaux as the basis states

$$
\begin{array}{lllll}
\dfrac{1\,1\,1}{2} & \dfrac{1\,1\,1}{3} & \dfrac{1\,1\,2}{2} & \dfrac{1\,1\,2}{3} & \dfrac{1\,1\,3}{2} \\[2mm]
\dfrac{1\,1\,3}{3} & \dfrac{1\,2\,2}{2} & \dfrac{1\,2\,2}{3} & \dfrac{1\,2\,3}{2} & \dfrac{1\,2\,3}{3} \\[2mm]
\dfrac{1\,3\,3}{2} & \dfrac{1\,3\,3}{3} & \dfrac{2\,2\,2}{3} & \dfrac{2\,2\,3}{3} & \dfrac{2\,3\,3}{3}
\end{array}
$$

2). The representation [3, 0] of SU(3) is 10-dimensional with the following standard tensor Young tableaux as the basis states

$$
\begin{array}{lllll}
\dfrac{1\,1\,1}{4} & \dfrac{1\,1\,2}{4} & \dfrac{1\,1\,3}{4} & \dfrac{1\,2\,2}{4} & \dfrac{1\,2\,3}{4} \\[2mm]
\dfrac{1\,3\,3}{4} & \dfrac{2\,2\,2}{4} & \dfrac{2\,2\,3}{4} & \dfrac{2\,3\,3}{4} & \dfrac{3\,3\,3}{4}
\end{array}
$$

3). The representation [2, 1] of SU(3) is 8-dimensional with the following standard tensor Young tableaux as the basis states

$$
\begin{array}{llll}
\dfrac{1\,1\,4}{2} & \dfrac{1\,1\,4}{3} & \dfrac{1\,2\,4}{2} & \dfrac{1\,2\,4}{3} \\[2mm]
\dfrac{1\,3\,4}{2} & \dfrac{1\,3\,4}{3} & \dfrac{2\,2\,4}{3} & \dfrac{2\,3\,4}{3}
\end{array}
$$

4). The representation [2, 0] of SU(3) is 6-dimensional with the following standard tensor Young tableaux as the basis states

$$
\begin{array}{llllll}
\dfrac{1\,1\,4}{4} & \dfrac{1\,2\,4}{4} & \dfrac{1\,3\,4}{4} & \dfrac{2\,2\,4}{4} & \dfrac{2\,3\,4}{4} & \dfrac{3\,3\,4}{4}
\end{array}
$$

5). The representation [1, 1] of SU(3) is 3-dimensional with the following standard tensor Young tableaux as the basis states

$$
\begin{array}{lll}
\dfrac{1\,4\,4}{2} & \dfrac{1\,4\,4}{3} & \dfrac{2\,4\,4}{3}
\end{array}
$$

6). The representation [1, 0] of SU(3) is 3-dimensional with the following standard tensor Young tableaux as the basis states

$$\begin{array}{ccc} \begin{array}{ccc}1&4&4\end{array} & \begin{array}{ccc}2&4&4\end{array} & \begin{array}{ccc}3&4&4\end{array} \\ 4 & 4 & 4 \end{array}$$
$$,\qquad,\qquad.$$

19. Reduce the subduced representations from the following irreducible representations of the SU(6) group with respect to the subgroup SU(3)⊗SU(2), respectively:

$$[1],\quad [1,1],\quad [2],\quad [3],\quad [2,1],\quad [1,1,1],\quad [2,1^4].$$

Solution. In particle physics, a representation of the flavor SU(3) group describes a flavor multiplet, and a representation of the spinor SU(2) group describes a spinor multiplet. All those representations are denoted by the state numbers contained in the multiplets.

$$6 \simeq (3,\ 2)$$

$$15 \simeq (3^*,\ 3)\ \oplus\ (6,\ 1)$$

$$21 \simeq (3^*,\ 1)\ \oplus\ (6,\ 3)$$

$$56 \simeq (10,\ 4)\ \oplus\ (8,\ 2)$$

$$70 \simeq (10,\,2) \oplus (8,\,4) \oplus (8,\,2) \oplus (1,\,2)$$

$$20 \simeq (1,\,4) \oplus (8,\,2)$$

$$35 \simeq (1,\,3) \oplus (8,\,1) \oplus (8,\,3)$$

Notice the reductions of the subduced representations from [3] and $[2, 1^4] \simeq [1] \backslash [1]^*$. The former gives the observed multiplets of the low-energy baryons in the sixties of the 20th century, namely, a decuplet with spin 3/2 and an octet with spin 1/2. Since they belong to the same representation of SU(6), their parities have to be the same (to be positive relative to the parity of a proton). The latter gives the observed multiplets of the low-energy mesons in that years, namely, a singlet with spin 1, an octet with spin 0, and an octet with spin 1, whose parities are the same (to be negative from experiments). It is the success of the SU(6) symmetry theory for hadrons. In the SU(6) symmetry theory, the spin is described by the SU(2) group, so that this theory is only a nonrelativitic theory. This theory cannot explain many phenomena in high energy particle physics.

20. Reduce the subduced representations from the following irreducible representations of the SU(5) group with respect to the subgroup SU(3)⊗SU(2), respectively:

$$[1], \quad [1, 1], \quad [2], \quad [3], \quad [2, 1], \quad [1, 1, 1], \quad [2, 1^4].$$

Solution. In the grand unified model proposed in 1974, one needs to study the reduction of the subduced representations from some irreducible representations of the SU(5) group with respect to the subgroup SU(3)⊗SU(2). There is a U(1) subgroup in the SU(5) group with the generator $Y = \mathrm{diag}\{-1/3,\ -1/3,\ -1/3,\ 1/2,\ 1/2\}$, which is commutable with all elements in the subgroup SU(3)⊗SU(2). However, since there are five common elements in two subgroups in addition to the identity, the SU(5) group is not the direct product of two subgroups SU(3)⊗SU(2) and U(1). The color subgroup SU(3) describes the strong interaction of quarks, and the subgroup SU(2)⊗U(1) describes the weak-electric interaction. Each representation in the reduction is denoted by two multiplets of SU(3) and SU(2) and by the eigenvalue of Y as the subscript. In the SU(5) grand unified model, the electric charge Q is equal to $T_3 + Y$, where T_3 is the third generator of the subgroup SU(2):

$$Q = T_3 + Y = \mathrm{diag}\{0,\ 0,\ 0,\ 1/2,\ -1/2\}$$
$$+\ \mathrm{diag}\{-1/3,\ -1/3,\ -1/3,\ 1/2,\ 1/2\}$$
$$=\ \mathrm{diag}\{-1/3,\ -1/3,\ -1/3,\ 1,\ 0\}\ .$$

$\square \simeq \square \times [0] \oplus [0] \otimes \square\ ,$

$5 \simeq (3,\ 1)_{-1/3} \oplus (1,\ 2)_{1/2}\ ,$

$\square\!\square \simeq \square\!\square \times [0] \oplus \square \otimes \square \oplus [0] \otimes \square\!\square\ ,$

$10 \simeq (3^*,\ 1)_{-2/3} \oplus (3,\ 2)_{1/6} \oplus (1,\ 1)_1\ ,$

$\square\square \simeq \square\square \times [0] \oplus \square \otimes \square \oplus [0] \otimes \square\square\ ,$

$15 \simeq (6,\ 1)_{-2/3} \oplus (3,\ 2)_{1/6} \oplus (1,\ 3)_1\ ,$

$\square\square\square \simeq \square\square\square \times [0] \oplus \square\square \otimes \square$
$\oplus\ \square \otimes \square\square \oplus [0] \otimes \square\square\square\ ,$

$35 \simeq (10,\ 1)_{-1} \oplus (6,\ 2)_{-1/6} \oplus (3,\ 3)_{2/3} \oplus (1,\ 4)_{3/2}\ ,$

$\square\square \simeq \square\square \times [0] \oplus \square\square \otimes \square \oplus \square \otimes \square$
$\oplus\ \square \otimes \square\square \oplus \square \otimes \square\square \oplus [0] \otimes \square\square\ ,$

$$40 \simeq (8,\ 1)_{-1} \oplus (6,\ 2)_{-1/6} \oplus (3^*,\ 2)_{-1/6}$$
$$\oplus\ (3,\ 3)_{2/3} \oplus (3,\ 1)_{2/3} \oplus (1,\ 2)_{3/2}\ ,$$

$$10^* \simeq (1,\ 1)_{-1} \oplus (3^*,\ 2)_{-1/6} \oplus (3,\ 1)_{2/3}\ ,$$

$$24 \simeq (3,\ 2)_{-5/6} \oplus (1,\ 3)_0 \oplus (1,\ 1)_0 \oplus (8,\ 1)_0 \oplus (3^*,\ 2)_{5/6}\ .$$

In the SU(5) grand unified model, the state of a particle is divided into the left-hand state and the right-hand state. They are filled into the 5-dimensional and 10-dimensional representations, respectively. The first generation of particles are filled as follows:

$$\begin{pmatrix} d_1 \\ d_2 \\ d_3 \\ e^c \\ \nu^c \end{pmatrix}_R, \qquad \begin{pmatrix} 0 & u_3^c & -u_2^c & u_1 & d_1 \\ & 0 & u_1^c & u_2 & d_2 \\ & & 0 & u_3 & d_3 \\ & & & 0 & e^c \\ & & & & 0 \end{pmatrix}_L,$$

where the superscript c denotes the charge conjugate state, and the subscript R and L denote the right-hand state and left-hand state, respectively. The 10-dimensional representation is the antisymmetric tensor representation, so only the upper half matrix is needed to be filled with the particle states. The 24-dimensional representation is the adjoint representation, describing the gauge particles, which is not listed here.

8.5 Casimir Invariants of SU(N)

★ The generators T_A in the self-representation of SU(N) satisfy the commutation relations and the anticommutation relationship

$$[T_A, \ T_B] = i \sum_{D=1}^{N^2-1} C_{AB}^{D} T_D,$$

$$\{T_A, \ T_B\} = \sum_{D=1}^{N^2-1} d_{ABD} T_D + \frac{1}{N}\delta_{AB}\mathbf{1}.$$

Due to the normalization condition, $\mathrm{Tr}(T_A T_B) = \delta_{AB}/2$, the coefficients C_{AB}^{D} and d_{ABD} can be expressed by T_A as given in Eq. (8.9)

$$C_{AB}^{D} = -2i\mathrm{Tr}\left(T_A T_B T_D - T_B T_A T_D\right),$$
$$d_{ABD} = 2\mathrm{Tr}\left(T_A T_B T_D + T_B T_A T_D\right). \tag{8.9}$$

Recall that the adjoint representation $D^{\mathrm{adj}}(u)$ of SU(N) satisfy

$$D^{[\lambda]}(u)^{-1} I_A^{[\lambda]} D^{[\lambda]}(u) = \sum_{A'=1}^{N^2-1} D_{AA'}^{\mathrm{adj}}(u) I_{A'}^{[\lambda]}, \quad u \in \mathrm{SU}(N), \tag{8.38}$$

where $D^{[\lambda]}(u)$ and $I_A^{[\lambda]}$ are the representation matrices of the element u and the generator I_A in the representation $[\lambda]$ of SU(N), respectively. When $[\lambda]$ is the self-representation $[1]$ of SU(N), $D^{[\lambda]}(u) = u$ and $I_A^{[\lambda]} = T_A$. Let C_{AB}^{D} and d_{ABD} be the totally symmetric and antisymmetric tensors of rank three with respect to the adjoint representation. We are going to show that they are the invariant tensors. Prove it with C_{AB}^{D} as example.

$$C_{AB}^{D} \longrightarrow \sum_{A'B'D'} D_{AA'}^{\mathrm{ad}} D_{BB'}^{\mathrm{ad}} D_{DD'}^{\mathrm{ad}} C_{A'B'}^{D'}$$
$$= -2i\mathrm{Tr}\left\{u^{-1}T_A u u^{-1}T_B u u^{-1}T_D u - u^{-1}T_B u u^{-1}T_A u u^{-1}T_D u\right\}$$
$$= C_{AB}^{D}. \tag{8.39}$$

According to the Littlewood-Richardson rule, except for $N = 2$, the C-G series for the direct product of two adjoint representations of SU(N) contains, in addition to the other representations, only one identical representation and two adjoint representations. Therefore, the identical representation only appears twice in the reduction of the direct product of three adjoint representations of SU(N). Equation (8.39) says that C_{AB}^{D} and d_{ABD} are the only invariant tensors of rank three, up to a constant factor, with respect to the adjoint representation of SU(N). For the SU(2) group, since there is only one adjoint representation contained in the CG series for the direct product of two adjoint representations, and then, there is only one identical representation contained in the CG series for the direct product of three adjoint representations. On the other hand, $d_{ABD} = 0$ for SU(2), and the invariant tensor of rank three with respect to the adjoint representation of SU(2) has to be proportional to C_{AB}^{D}.

Now, we replace T_A on the right hand side of Eq. (8.9) with $I_A^{[\lambda]}$. They are still the symmetric or antisymmetric invariant tensors of rank three with respect to the adjoint representation. Thus, they are proportional to C_{AB}^{D} or d_{ABD}, respectively

$$-2i\text{Tr}\left(I_A^{[\lambda]}I_B^{[\lambda]}I_D^{[\lambda]} - I_B^{[\lambda]}I_A^{[\lambda]}I_D^{[\lambda]}\right) = 2T_2([\lambda])C_{AB}^{D},$$

$$2\text{Tr}\left(I_A^{[\lambda]}I_B^{[\lambda]}I_D^{[\lambda]} + I_B^{[\lambda]}I_A^{[\lambda]}I_D^{[\lambda]}\right) = A([\lambda])d_{ABD}.$$

(8.40)

From the first equality we obtain

$$\text{Tr}\left(I_A^{[\lambda]}I_D^{[\lambda]}\right) = \delta_{AD}T_2([\lambda]).$$

(8.41)

$T_2([\lambda])$ and $A([\lambda])$ are respectively related to the Casimir operators of rank two and rank three of SU(N), and sometimes are simply called the Casimir invariants of rank two and rank three in the literature. Both $T_2([\lambda])$ and $A([\lambda])$ are independent of the similarity transformation of the representation. It is easy to show from Eq. (8.40) that

$$T_2([\lambda]^*) = T_2([\lambda]), \qquad A([\lambda]^*) = -A([\lambda]),$$
$$T_2([\lambda] \oplus [\mu]) = T_2([\lambda]) + T_2([\mu]),$$
$$A([\lambda] \oplus [\mu]) = A([\lambda]) + A([\mu]),$$
$$T_2([\lambda] \otimes [\mu]) = d_{[\mu]}T_2([\lambda]) + d_{[\lambda]}T_2([\mu]),$$
$$A([\lambda] \otimes [\mu]) = d_{[\mu]}A([\lambda]) + d_{[\lambda]}A([\mu]).$$

(8.42)

In the calculation of $T_2([\lambda])$ with Eq. (8.41), we take $A = D = 3$ such that both generators contained in Eq. (8.41) belong to the subgroup SU(2). Denote $T_2([\lambda])$ in SU(2) by $T_2^{(0)}([\lambda])$, which is calculated to be

$$T_2^{(0)}([\lambda]) = \text{Tr}\left\{\left(I_3^{[\lambda]}\right)^2\right\} = \frac{1}{12}\,\lambda(\lambda+1)(\lambda+2) .$$

(8.43)

Calculate $T_2([\lambda])$ in the subduced representation from the irreducible representation $[\lambda]$ of SU(N) with respect to the subgroup SU($N-2$)⊗SU(2), which is equivalent to a direct sum of some irreducible representation $[\nu] \otimes [\mu]$ of the subgroup,

$$T_2([\lambda]) = \sum d_{[\nu]}(\text{SU}(N-2))T_2^{(0)}([\mu]) .$$

(8.44)

In the calculation of $A([\lambda])$ with Eq. (8.40), we take $A = B = 3$ and $D = 8$ such that three generators are contained in Eq. (8.40) all belong to the subgroup SU(3), where $d_{338} = \sqrt{1/3}$. Denote $A([\lambda])$ in SU(3) by $A^{(0)}([\lambda])$,

$$A^{(0)}([\lambda]) = 4\sqrt{3}\mathrm{Tr}\left\{T_8^{[\lambda]}(T_3^{[\lambda]})^2\right\} .$$

In the planar weight diagram, the states located in the same horizontal level have the same eigenvalues of $T_8^{[\lambda]}$, and the sum of $\left(T_3^{[\lambda]}\right)^2$ can be calculated by Eq. (8.43). Then, we sum up the contribution for the states with different $T_8^{[\lambda]}$. For a one-row Young pattern, we have

$$
\begin{aligned}
A^{(0)}([\lambda, 0]) &= 2\sum_{m=0}^{\lambda} (\lambda - 3m)T_2^{(0)}([\lambda - m]) \\
&= (1/6)\sum_{m=0}^{\lambda} (\lambda - 3m)(\lambda - m)(\lambda - m + 1)(\lambda - m + 2) \\
&= (1/120)\lambda(\lambda + 1)(\lambda + 2)(\lambda + 3)(2\lambda + 3).
\end{aligned}
\tag{8.45}
$$

Generally,

$$
\begin{aligned}
A^{(0)}([\lambda + \lambda', \lambda']) &= A^{(0)}([\lambda] \otimes [\lambda']^*) - A^{(0)}([\lambda - 1] \otimes [\lambda' - 1]^*) \\
&= d_{[\lambda']}(SU(3))A^{(0)}([\lambda]) - d_{[\lambda]}(SU(3))A^{(0)}([\lambda']) \\
&\quad - d_{[\lambda'-1]}(SU(3))A^{(0)}([\lambda - 1]) + d_{[\lambda-1]}(SU(3))A^{(0)}([\lambda' - 1]) \\
&= \frac{1}{120}(\lambda + 1)(\lambda' + 1)(\lambda - \lambda')(\lambda + \lambda' + 2)(\lambda + 2\lambda' + 3)(2\lambda + \lambda' + 3).
\end{aligned}
\tag{8.46}
$$

Calculate $A([\lambda])$ in the subduced representation from the irreducible representation $[\lambda]$ of SU(N) with respect to the subgroup SU($N - 3$)$\otimes$SU(3), which is equivalent to a direct sum of some irreducible representation $[\nu] \otimes [\mu]$ of the subgroup,

$$A([\lambda]) = \sum d_{[\nu]}(SU(N - 3))A^{(0)}([\mu]). \tag{8.47}$$

21. Calculate the Casimir invariants $T_2([1^r])$ and $A([1^r])$ in the representation denoted by a one-column Young pattern ($[1^r]$) of the SU(N) group.

Solution. From Eqs. (8.44) and (8.47), we obtain

$$T_2([1^r]) = d_{[1^{r-1}]}(SU(N-2))T_2^{(0)}([1]) = \frac{1}{2}\binom{N-2}{r-1},$$

$$
\begin{aligned}
A([1^r]) &= \{d_{[1^{r-1}]}(SU(N-3)) - d_{[1^{r-2}]}(SU(N-3))\} A^{(0)}([1]) \\
&= \frac{(N-3)!}{(r-1)!(N-r-2)!} - \frac{(N-3)!}{(r-2)!(N-r-1)!} \\
&= \frac{(N-3)!\{(N-r-1)-(r-1)\}}{(r-1)!(N-r-1)!} \\
&= \frac{(N-2r)}{(r-1)}\binom{N-3}{r-2}.
\end{aligned}
$$

22. Calculate the Casimir invariants $T_2([3])$ and $A([3])$ in the irreducible representation [3] of the SU(6) group.

Solution. From Eq. (8.43) we have

$$T_2^{(0)}([3]) = 5, \quad T_2^{(0)}([2]) = 2, \quad T_2^{(0)}([1]) = 1/2 \quad T_2^{(0)}([0]) = 0.$$

From the reduction of the subduced representation from [3] of SU(6) with respect to the subgroup SU(4)⊗SU(2)

$$[3] \longrightarrow [3] \otimes [0] \oplus [2] \otimes [1] \oplus [1] \otimes [2] \oplus [0] \otimes [3],$$

we obtain

$$
\begin{aligned}
T_2([3]) &= d_{[3]}(SU(4))T_2^{(0)}([0]) + d_{[2]}(SU(4))T_2^{(0)}([1]) \\
&\quad + d_{[1]}(SU(4))T_2^{(0)}([2]) + d_{[0]}(SU(4))T_2^{(0)}([3]) \\
&= 0 + 10 \times (1/2) + 4 \times 2 + 1 \times 5 = 18.
\end{aligned}
$$

From Eq. (8.45) we have

$$A^{(0)}([3]) = 27, \quad A^{(0)}([2]) = 7, \quad A^{(0)}([1]) = 1 \quad A^{(0)}([0]) = 0.$$

From the reduction of the subduced representation from [3] of SU(6) with respect to the subgroup SU(3)⊗SU(3)

$$[3] \longrightarrow [3] \otimes [0] \oplus [2] \otimes [1] \oplus [1] \otimes [2] \oplus [0] \otimes [3].$$

we obtain

$$A([\lambda]) = d_{[3]}(SU(3))A^{(0)}([0]) + d_{[2]}(SU(3))A^{(0)}([1])$$
$$+ d_{[1]}(SU(3))A^{(0)}([2]) + d_{[0]}(SU(3))A^{(0)}([3])$$
$$= 0 + 6 \times 1 + 3 \times 7 + 1 \times 27 = 54.$$

23. Calculate the Casimir invariants $T_2([\lambda])$ and $A([\lambda])$ in the representation $([\lambda] = [\lambda, 0, 0, 0])$ of the SU(5) group.

Solution. The calculation method is the same as Problem 22. Here we only list the result.

$$T_2^{(0)}([\lambda]) = \lambda(\lambda + 1)(\lambda + 2)/12,$$
$$A^{(0)}([\lambda]) = \lambda(\lambda + 1)(\lambda + 2)(\lambda + 3)(2\lambda + 3)/120,$$
$$d_{[\lambda]}(SU(3)) = (\lambda + 2)(\lambda + 1)/2, \quad d_{[\lambda]}(SU(2)) = \lambda + 1.$$

Hence

$$T_2([\lambda]) = \sum_{n=0}^{\lambda} d_{[\lambda-n]}(SU(3))T_2^{(0)}([n])$$
$$= \sum_{n=0}^{\lambda} (\lambda - n + 2)(\lambda - n + 1)n(n + 1)(n + 2)/24$$
$$= \lambda(\lambda + 1)(\lambda + 2)(\lambda + 3)(\lambda + 4)(\lambda + 5)/1440,$$

$$A([\lambda]) = \sum_{n=0}^{\lambda} d_{[\lambda-n]}(SU(2))A^{(0)}([n])$$
$$= \sum_{n=0}^{\lambda} (\lambda - n + 1)n(n + 1)(n + 2)(n + 3)(2n + 3)/120$$
$$= \lambda(\lambda + 1)(\lambda + 2)(\lambda + 3)(\lambda + 4)(\lambda + 5)(2\lambda + 5)/5040.$$

Generalizing this result to the SU(N) group, we obtain

$$T_2([\lambda]) = \frac{1}{2} \prod_{n=0}^{N} \frac{\lambda + n}{n + 1} = \frac{\lambda(\lambda + 1)\cdots(\lambda + N)}{2(N + 1)!},$$

$$A([\lambda]) = \frac{2\lambda + N}{N + 2} \prod_{n=0}^{N} \frac{\lambda + n}{n + 1} = \frac{\lambda(\lambda + 1)\cdots(\lambda + N)(2\lambda + N)}{(N + 2)!}.$$

Chapter 9

REAL ORTHOGONAL GROUPS

9.1 Tensor Representations of SO(N)

★ The set of all N-dimensional unimodular real orthogonal matrices, in the multiplication rule of matrices, constitutes SO(N) group. In this chapter we discuss only the case of $N \geq 3$, because SO(2) group is an Abelian group which is isomorphic onto the U(1) group. The SO(N) group is a doubly-connected compact Lie group with order $N(N-1)/2$. SO(N) is a subgroup of SU(N). The generators in the self-representation of SO(N) is usually taken as the second type of generators in the self-representation of SU(N) multiplied by two

$$
\begin{aligned}
(T_{ab})_{rs} = 2 \left(T_{ab}^{(2)} \right)_{rs} &= -i\delta_{ar}\delta_{bs} + i\delta_{as}\delta_{br}, \qquad a < b, \\
\mathrm{Tr}\,(T_{ab}T_{cd}) &= 2\delta_{ac}\delta_{bd} - 2\delta_{ad}\delta_{bc}, \\
[T_{ab},\ T_{cd}] &= -i\left\{ \delta_{bc}T_{ad} + \delta_{ad}T_{bc} - \delta_{ac}T_{bd} - \delta_{bd}T_{ac} \right\},
\end{aligned}
\tag{9.1}
$$

where the indices a and b are the ordinal indices of the generator, while r and s are the row and column indices of the matrix. $N(N-1)/2$ matrices T_{ab} all are antisymmetric, and they are also antisymmetric with respect to the ordinal indices. Any matrix R in SO(N) can be expressed in the form of exponential matrix function:

$$
R = \exp\left\{ -i \sum_{a<b} \omega_{ab} T_{ab} \right\}.
\tag{9.2}
$$

★ The SO(N) group with $N = 2\ell$ or $N = 2\ell + 1$ contains ℓ mutually commutable generators,

$$
H_1 = T_{12}, \qquad H_2 = T_{34}, \qquad \ldots, \qquad H_\ell = T_{(2\ell-1)(2\ell)},
\tag{9.3}
$$

which span the Cartan subalgebra of the Lie algebra of $SO(N)$. Thus, the ranks of both $SO(2\ell)$ and $SO(2\ell + 1)$ are ℓ. We want to find the common eigenvectors of H_j, $[H_j, E_\alpha] = \alpha_j E_\alpha$. When $N = 2\ell$, the eigenvectors are

$$
\begin{aligned}
E_{ab}^{(1)} &= \frac{1}{2} \left[T_{(2a)(2b-1)} - iT_{(2a-1)(2b-1)} - iT_{(2a)(2b)} - T_{(2a-1)(2b)} \right], \\
E_{ab}^{(2)} &= \frac{1}{2} \left[T_{(2a)(2b-1)} + iT_{(2a-1)(2b-1)} + iT_{(2a)(2b)} - T_{(2a-1)(2b)} \right], \\
E_{ab}^{(3)} &= \frac{1}{2} \left[T_{(2a)(2b-1)} - iT_{(2a-1)(2b-1)} + iT_{(2a)(2b)} + T_{(2a-1)(2b)} \right], \\
E_{ab}^{(4)} &= \frac{1}{2} \left[T_{(2a)(2b-1)} + iT_{(2a-1)(2b-1)} - iT_{(2a)(2b)} + T_{(2a-1)(2b)} \right],
\end{aligned}
\tag{9.4}
$$

with the eigenvalues $\{\mathbf{e}_a - \mathbf{e}_b\}_j$, $\{-\mathbf{e}_a + \mathbf{e}_b\}_j$, $\{\mathbf{e}_a + \mathbf{e}_b\}_j$, and $\{-\mathbf{e}_a - \mathbf{e}_b\}_j$, respectively, where $a < b$, and $\{\mathbf{e}_a\}_j = \delta_{aj}$. When $N = 2\ell + 1$, in addition to Eq. (9.4), there are another two types of eigenvectors:

$$
\begin{aligned}
E_{ab}^{(5)} &= \sqrt{\frac{1}{2}} \left[T_{(2a)(2\ell+1)} - iT_{(2a-1)(2\ell+1)} \right], \\
E_{ab}^{(6)} &= \sqrt{\frac{1}{2}} \left[T_{(2a)(2\ell+1)} + iT_{(2a-1)(2\ell+1)} \right],
\end{aligned}
\tag{9.5}
$$

with the eigenvalues $\{\mathbf{e}_a\}_j$ and $-\{\mathbf{e}_a\}_j$, respectively.

The simple roots for the $SO(2\ell + 1)$ group are

$$
\mathbf{r}_\mu = \mathbf{e}_\mu - \mathbf{e}_{\mu+1}, \qquad \mathbf{r}_\ell = \mathbf{e}_\ell, \qquad 1 \le \mu \le \ell - 1.
\tag{9.6}
$$

The remaining positive roots are

$$
\mathbf{e}_a - \mathbf{e}_b = \sum_{\mu=a}^{b-1} \mathbf{r}_\mu, \quad \mathbf{e}_a + \mathbf{e}_b = \sum_{\mu=a}^{b-1} \mathbf{r}_\mu + 2\sum_{\nu=b}^{\ell} \mathbf{r}_\nu, \quad \mathbf{e}_a = \sum_{\mu=a}^{\ell} \mathbf{r}_\mu,
$$

where $a < b$. The largest root, which is the highest weight of the adjoint representation of $SO(2\ell + 1)$, is

$$
\vec{\omega} = \mathbf{e}_1 + \mathbf{e}_2 = \mathbf{r}_1 + 2\sum_{\nu=2}^{\ell} \mathbf{r}_\nu.
\tag{9.7}
$$

Therefore, the Lie algebra of $SO(2\ell + 1)$ is B_ℓ, and $\vec{\omega} = \mathbf{w}_2$. The simple roots for the $SO(2\ell)$ group are

$$
\mathbf{r}_\mu = \mathbf{e}_\mu - \mathbf{e}_{\mu+1}, \qquad \mathbf{r}_\ell = \mathbf{e}_{\ell-1} + \mathbf{e}_\ell, \qquad 1 \le \mu \le \ell - 1.
\tag{9.8}
$$

The remaining positive root are

$$\mathbf{e}_a - \mathbf{e}_b = \sum_{\mu=a}^{b-1} \mathbf{r}_\mu, \qquad \mathbf{e}_a + \mathbf{e}_b = \sum_{\mu=a}^{b-1} \mathbf{r}_\mu + 2\sum_{\nu=b}^{\ell-2} \mathbf{r}_\nu + \mathbf{r}_{\ell-1} + \mathbf{r}_\ell,$$

where $a < b$. The largest root, which is the highest weight of the adjoint representation of $SO(2\ell)$, is

$$\vec{\omega} = \mathbf{e}_1 + \mathbf{e}_2 = \mathbf{r}_1 + 2\sum_{\nu=2}^{\ell-2} \mathbf{r}_\nu + \mathbf{r}_{\ell-1} + \mathbf{r}_\ell. \tag{9.9}$$

Therefore, the Lie algebra of $SO(2\ell)$ is D_ℓ, and $\vec{\omega} = \mathbf{w}_2$.

★ The Chevalley bases in the self-representation of the $SO(2\ell+1)$ group (the B_ℓ Lie algebra) are

$$
\begin{aligned}
H_\mu &= T_{(2\mu-1)(2\mu)} - T_{(2\mu+1)(2\mu+2)}, \\
E_\mu &= \tfrac{1}{2}\left\{ T_{(2\mu)(2\mu+1)} - iT_{(2\mu-1)(2\mu+1)} - iT_{(2\mu)(2\mu+2)} - T_{(2\mu-1)(2\mu+2)} \right\}, \\
F_\mu &= \tfrac{1}{2}\left\{ T_{(2\mu)(2\mu+1)} + iT_{(2\mu-1)(2\mu+1)} + iT_{(2\mu)(2\mu+2)} - T_{(2\mu-1)(2\mu+2)} \right\}, \\
H_\ell &= 2T_{(2\ell-1)(2\ell)}, \\
E_\ell &= T_{(2\ell)(2\ell+1)} - iT_{(2\ell-1)(2\ell+1)}, \\
F_\ell &= T_{(2\ell)(2\ell+1)} + iT_{(2\ell-1)(2\ell+1)},
\end{aligned}
$$
$$\tag{9.10}$$

where $1 \le \mu < \ell$. The Chevalley bases in the self-representation of the $SO(2\ell)$ group (the D_ℓ Lie algebra) are the same as those of the $SO(2\ell+1)$ group except for those when $\mu = \ell$, which are

$$
\begin{aligned}
H_\ell &= T_{(2\ell-3)(2\ell-2)} + T_{(2\ell-1)(2\ell)}, \\
E_\ell &= \tfrac{1}{2}\left\{ T_{(2\ell-2)(2\ell-1)} - iT_{(2\ell-3)(2\ell-1)} + iT_{(2\ell-2)(2\ell)} + T_{(2\ell-3)(2\ell)} \right\}, \\
F_\ell &= \tfrac{1}{2}\left\{ T_{(2\ell-2)(2\ell-1)} + iT_{(2\ell-3)(2\ell-1)} - iT_{(2\ell-2)(2\ell)} + T_{(2\ell-3)(2\ell)} \right\}.
\end{aligned}
$$
$$\tag{9.11}$$

★ The definition for a tensor with respect to the $SO(N)$ group is the same as that to the $SU(N)$ group, except that the transformation matrix $u \in SU(N)$ is replaced with $R \in SO(N)$. The Weyl reciprocity still holds for the tensors of $SO(N)$. However, this replacement makes some new characteristics for the tensors of $SO(N)$. Since the transformation matrix $R \in SO(N)$ is real, there is no difference between the covariant tensor and the contravariant tensor for the $SO(N)$ group. As a result, first, the tensor space of $SO(N)$ group can be decomposed into a direct sum of a series of the traceless tensor subspaces $\mathcal{T}$, which are invariant in $SO(N)$. Each $\mathcal{T}$ can be projected by the Young operators into the subspace $\mathcal{T}_\mu^{[\lambda]} = \mathcal{Y}_\mu^{[\lambda]}\mathcal{T}$.

Second, since the tensor is traceless, the tensor subspace $\mathcal{T}_\mu^{[\lambda]}$ is a null space if the sum of the numbers of boxes in the first and the second columns of the Young pattern $[\lambda]$ is larger than N. Third, since there is no difference between the covariant tensor and the contravariant tensor, the representation described by a Young pattern $[\lambda]$ with the row number m larger than $N/2$ is equivalent to that by $[\lambda']$, where $[\lambda']$ is obtained from $[\lambda]$ by replacing its first column with a column containing $(N - m)$ boxes. Fourth, when the row number of $[\lambda]$ is equal to $N/2$, the representation space $\mathcal{T}_\mu^{[\lambda]}$ is reduced into two minimal invariant subspaces with the same dimension. They are the self-dual and the anti-self-dual tensor subspaces, whose representations are denoted by $[+\lambda]$ and $[-\lambda]$, respectively.

★ A tensor of rank two for $\mathrm{SO}(N)$ is easy to be decomposed into the sum of the traceless tensors:

$$
\mathbf{T}_{ab} = \left\{ \mathbf{T}_{ab} - \delta_{ab} N^{-1} \sum_d \mathbf{T}_{dd} \right\} + \delta_{ab} \left[N^{-1} \sum_d \mathbf{T}_{dd} \right],
$$

where the tensor in the curve bracket is a traceless tensor of rank two, and the tensor in the square bracket is a tensor of rank zero. However, generally speaking, the decomposition of a tensor of rank n into a sum of some traceless tensors of different ranks is straightforward, but tedious. As discussed in Chap. 8, the irreducible tensor bases for $\mathrm{SU}(N)$ are obtained explicitly. In fact, they are the standard tensor Young tableaux. The bases can be orthonormalized by making use of the block weight diagram. However, due to the traceless condition, some standard tensor Young tableaux for $\mathrm{SO}(N)$ become linearly dependent. It becomes a new problem how to find the complete set of the independent irreducible tensor bases for $\mathrm{SO}(N)$ explicitly. Another problem is that the tensor basis $\Theta_{a_1 \ldots a_n}$ is not the common eigenstate of the Chevalley bases H_μ. Since a tensor basis $\Theta_{a_1 \ldots a_n}$ is a direct product of n vector bases $\Theta_{a_1} \cdots \Theta_{a_n}$, we only need to find the new vector bases such that H_μ are diagonal. Namely, we want to diagonalize H_μ in the self-representation of $\mathrm{SO}(N)$ by a similarity transformation. Both problems are well solved for the $\mathrm{SO}(3)$ group. Now, we generalize the solution to the $\mathrm{SO}(N)$ group.

★ In the self-representation space of $\mathrm{SO}(2\ell + 1)$ we defined a set of new orthonormal bases, called the spherical harmonic bases,

$$\Phi_\alpha = \begin{cases} (-1)^{\ell-\alpha+1}\sqrt{1/2}\,(\Theta_{2\alpha-1} + i\Theta_{2\alpha}) & \text{when } 1 \le \alpha \le \ell, \\ \Theta_{2\ell+1} & \text{when } \alpha = \ell+1, \\ \sqrt{1/2}\,(\Theta_{4\ell-2\alpha+3} - i\Theta_{4\ell-2\alpha+4}) & \text{when } \ell+2 \le \alpha \le 2\ell+1. \end{cases}$$
(9.12)

The tensor basis $\Phi_{\alpha_1 \dots \alpha_n}$ of rank n for $SO(2\ell+1)$ is the direct product of n vector bases $\Phi_{\alpha_1} \cdots \Phi_{\alpha_n}$. In the spherical harmonic bases Φ_α, the nonvanishing matrix entries of the Chevalley bases are

$$\begin{aligned} H_\mu \Phi_\mu &= \Phi_\mu, & H_\mu \Phi_{\mu+1} &= -\Phi_{\mu+1}, \\ H_\mu \Phi_{2\ell-\mu+1} &= \Phi_{2\ell-\mu+1}, & H_\mu \Phi_{2\ell-\mu+2} &= -\Phi_{2\ell-\mu+2}, \\ H_\ell \Phi_\ell &= 2\Phi_\ell, & H_\ell \Phi_{\ell+2} &= -2\Phi_{\ell+2}, \\[4pt] E_\mu \Phi_{\mu+1} &= \Phi_\mu, & E_\mu \Phi_{2\ell-\mu+2} &= \Phi_{2\ell-\mu+1}, \\ E_\ell \Phi_{\ell+1} &= \sqrt{2}\Phi_\ell, & E_\ell \Phi_{\ell+2} &= \sqrt{2}\Phi_{\ell+1}, \\[4pt] F_\mu \Phi_\mu &= \Phi_{\mu+1}, & F_\mu \Phi_{2\ell-\mu+1} &= \Phi_{2\ell-\mu+2}, \\ F_\ell \Phi_\ell &= \sqrt{2}\Phi_{\ell+1}, & F_\ell \Phi_{\ell+1} &= \sqrt{2}\Phi_{\ell+2}, \end{aligned}$$
(9.13)

where $1 \le \mu \le \ell-1$. In the spherical harmonic bases, the standard tensor Young tableau $\mathcal{Y}_\mu^{[\lambda]} \Phi_{\alpha_1 \dots \alpha_n}$ in the representation $[\lambda]$ of $SO(2\ell+1)$ corresponds to the highest weight if each box in its jth row is filled with the digit j, because each raising operator E_μ annihilates it. Therefore, the relation between the highest weight $\mathbf{M}$ and the Young pattern $[\lambda]$ for $SO(2\ell+1)$ is

$$M_\mu = \lambda_\mu - \lambda_{\mu+1}, \qquad 1 \le \mu < \ell, \qquad M_\ell = 2\lambda_\ell. \tag{9.14}$$

Although the standard tensor Young tableaux is generally not traceless, the standard tensor Young tableau with the highest weight in $[\lambda]$ of $SO(2\ell+1)$ must be traceless, because it only contains Φ_α with $\alpha < \ell+1$. For example, the tensor basis $\Theta_1 \Theta_1$ is not traceless, but $\Phi_1 \Phi_1$ is traceless. The remaining traceless tensor bases in an irreducible representation space $[\lambda]$ of $SO(2\ell+1)$ can be calculated from the basis with the highest weight by the lowering operators F_μ. In the calculation the block weight diagram is helpful.

★ In the self-representation space of $SO(2\ell)$ we defined a set of new orthonormal bases, called the spherical harmonic bases,

$$\Phi_\alpha = \begin{cases} (-1)^{\ell-\alpha}\sqrt{1/2}\,(\Theta_{2\alpha-1} + i\Theta_{2\alpha}) & \text{when } 1 \le \alpha \le \ell, \\ \sqrt{1/2}\,(\Theta_{4\ell-2\alpha+1} - i\Theta_{4\ell-2\alpha+2}) & \text{when } \ell+1 \le \alpha \le 2\ell. \end{cases}$$
(9.15)

The tensor basis $\Phi_{\alpha_1 \ldots \alpha_n}$ of rank n for $SO(2\ell)$ is the direct product of n vector bases $\Phi_{\alpha_1} \cdots \Phi_{\alpha_n}$. In the spherical harmonic bases Φ_α, the nonvanishing matrix entries of the Chevalley bases are

$$
\begin{aligned}
H_\mu \Phi_\mu &= \Phi_\mu, & H_\mu \Phi_{\mu+1} &= -\Phi_{\mu+1}, \\
H_\mu \Phi_{2\ell-\mu} &= \Phi_{2\ell-\mu}, & H_\mu \Phi_{2\ell-\mu+1} &= -\Phi_{2\ell-\mu+1}, \\
H_\ell \Phi_{\ell-1} &= \Phi_{\ell-1}, & H_\ell \Phi_\ell &= \Phi_\ell, \\
H_\ell \Phi_{\ell+1} &= -\Phi_{\ell+1}, & H_\ell \Phi_{\ell+2} &= -\Phi_{\ell+2}, \\[4pt]
E_\mu \Phi_{\mu+1} &= \Phi_\mu, & E_\mu \Phi_{2\ell-\mu+1} &= \Phi_{2\ell-\mu}, \\
E_\ell \Phi_{\ell+1} &= \Phi_{\ell-1}, & E_\ell \Phi_{\ell+2} &= \Phi_\ell, \\[4pt]
F_\mu \Phi_\mu &= \Phi_{\mu+1}, & F_\mu \Phi_{2\ell-\mu} &= \Phi_{2\ell-\mu+1}, \\
F_\ell \Phi_{\ell-1} &= \Phi_{\ell+1}, & F_\ell \Phi_\ell &= \Phi_{\ell+2},
\end{aligned}
\tag{9.16}
$$

where $1 \leq \mu \leq \ell - 1$. In the spherical harmonic bases, the standard tensor Young tableau $\mathcal{Y}_\mu^{[\lambda]} \Phi_{\alpha_1 \ldots \alpha_n}$ in the representation $[\lambda]$ of $SO(2\ell)$ corresponds to the highest weight if each box in its jth row, $j < \ell$, is filled with the digit j and if all boxes located in its ℓth row, if $[\lambda]$ contains ℓ row, are filled with the same digit ℓ or $\ell + 1$, depending on whether the representation is self-dual $[+\lambda]$ or anti-self-dual $[-\lambda]$. Therefore, the relation between the highest weight $\mathbf{M}$ and the Young pattern $[\lambda]$ for $SO(2\ell)$ is

$$
\begin{aligned}
M_\mu &= \lambda_\mu - \lambda_{\mu+1}, & &\text{when } 1 \leq \mu < \ell - 1, \\
M_{\ell-1} &= M_\ell = \lambda_{\ell-1}, & &\text{for } [\lambda] \text{ with } \lambda_\ell = 0, \\
M_{\ell-1} &= \lambda_{\ell-1} - \lambda_\ell, \quad M_\ell = \lambda_{\ell-1} + \lambda_\ell, & &\text{for } [+\lambda], \\
M_{\ell-1} &= \lambda_{\ell-1} + \lambda_\ell, \quad M_\ell = \lambda_{\ell-1} - \lambda_\ell, & &\text{for } [-\lambda].
\end{aligned}
\tag{9.17}
$$

Since the standard tensor Young tableau with the highest weight contains either Φ_ℓ or $\Phi_{\ell+1}$, in addition to Φ_α with $\alpha < \ell$, it is a traceless tensor. The remaining traceless tensor bases in an irreducible representation space $[\lambda]$ of $SO(2\ell)$ can be calculated from the basis with the highest weight by the lowering operators F_μ.

★ The dimension $d_{[\lambda]}(SO(N))$ of the representation $[\lambda]$ of $SO(N)$ can be calculated by the hook rule. In this rule the dimension is expressed as a quotient, where the numerator and the denominator are denoted by the symbols $Y_T^{[\lambda]}$ and $Y_h^{[\lambda]}$, respectively:

$$d_{[\lambda]}(\mathrm{SO}(N)) = \frac{Y_T^{[\lambda]}}{Y_h^{[\lambda]}} \qquad \text{when} \quad \lambda_\ell = 0,$$

$$d_{[\pm\lambda]}(\mathrm{SO}(N)) = \frac{Y_T^{[\lambda]}}{2Y_h^{[\lambda]}} \qquad \text{when} \quad \lambda_\ell \neq 0. \tag{9.18}$$

Let us explain the meaning of two symbols $Y_h^{[\lambda]}$ and $Y_T^{[\lambda]}$. The hook path $(i,\,j)$ in the Young pattern $[\lambda]$ is defined to be a path which enters the Young pattern at the rightist of the ith row, goes leftwards in the i row, turns downward at the j column, goes downwards in the j column, and leaves from the Young pattern at the bottom of the j column. The inverse path $\overline{(i,\,j)}$ is the same path as the hook path $(i,\,j)$ but with the opposite direction. The number of boxes contained in the path $(i,\,j)$, as well as in its inverse, is the hook number h_{ij}. $Y_h^{[\lambda]}$ is a tableau of the Young pattern $[\lambda]$ where the box in the jth column of the ith row is filled with the hook number h_{ij}. Define a series of the tableaux $Y_{T_a}^{[\lambda]}$ recursively by the rule given below. $Y_T^{[\lambda]}$ is a tableau of the Young pattern $[\lambda]$ where each box is filled with the sum of the digits which are respectively filled in the same box of each tableau $Y_{T_a}^{[\lambda]}$ in the series. The symbol $Y_T^{[\lambda]}$ means the product of the filled digits in it, so does the symbol $Y_h^{[\lambda]}$.

The tableaux $Y_{T_a}^{[\lambda]}$ are defined by the following rule:

(a) $Y_{T_0}^{[\lambda]}$ is a tableau of the Young pattern $[\lambda]$ where the box in the jth column of the ith row is filled with the digit $(N + j - i)$.

(b) Let $[\lambda^{(1)}] = [\lambda]$. Beginning with $[\lambda^{(1)}]$, we define recursively the Young pattern $[\lambda^{(a)}]$ by removing the first row and the first column of the Young pattern $[\lambda^{(a-1)}]$ until $[\lambda^{(a)}]$ contains less than two columns.

(c) If $[\lambda^{(a)}]$ contains more than one column, define $Y_{T_a}^{[\lambda]}$ to be a tableau of the Young pattern $[\lambda]$ where the boxes in the first $(a - 1)$ row and in the first $(a - 1)$ column are filled with 0, and the remaining part of the Young pattern is nothing but $[\lambda^{(a)}]$. Let $[\lambda^{(a)}]$ have r rows. Fill the first r boxes along the hook path $(1,\,1)$ of the Young pattern $[\lambda^{(a)}]$, beginning with the box on the rightmost, with the digits $(\lambda_1^{(a)} - 1)$, $(\lambda_2^{(a)} - 1)$, $\cdots$, $(\lambda_r^{(a)} - 1)$, box by box, and fill the first $(\lambda_i^{(a)} - 1)$ boxes in each inverse path $\overline{(i,\ 1)}$ of the Young pattern $[\lambda^{(a)}]$, $1 \le i \le r$, with -1. The remaining boxes are filled with 0. If a few -1 are filled in the same box, the digits are summed. The sum of all filled digits in the pattern $Y_{T_a}^{[\lambda]}$ is zero.

1. Calculate the dimensions of the irreducible representations of the SO(8) group denoted by the following Young patterns: (a) [4, 2], (b) [3, 2], (c) [4, 4], (d) [+3, 2, 1, 1], (e) [+3, 3, 1, 1].

Solution. (a). $d_{[4,2]}(SO(8)) = \dfrac{\begin{array}{l}8\ 9\ 10\ 11 \\ 7\ 8\end{array} + \begin{array}{l}-1\ -1\ 1\ 3 \\ -2\ \ \ 0\end{array}}{\begin{array}{l}5\ 4\ 2\ 1 \\ 2\ 1\end{array}} = \dfrac{\begin{array}{l}7\ 8\ 11\ 14 \\ 5\ 8\end{array}}{\begin{array}{l}5\ 4\ 2\ 1 \\ 2\ 1\end{array}} = 4312.$

(b). $d_{[3,2]}(SO(8)) = \dfrac{\begin{array}{l}8\ 9\ 10 \\ 7\ 8\end{array} + \begin{array}{l}-1\ 1\ 2 \\ -2\ 0\end{array}}{\begin{array}{l}4\ 3\ 1 \\ 2\ 1\end{array}} = \dfrac{\begin{array}{l}7\ 10\ 12 \\ 5\ \ 8\end{array}}{\begin{array}{l}4\ 3\ 1 \\ 2\ 1\end{array}} = 1400.$

(c). $d_{[4,4]}(SO(8)) = \dfrac{\begin{array}{l}8\ 9\ 10\ 11 \\ 7\ 8\ 9\ 10\end{array} + \begin{array}{l}-1\ -1\ \ \ 3\ \ \ 3 \\ -2\ -1\ -1\ \ \ 0\end{array} + \begin{array}{l}0\ \ \ 0\ \ \ \ 0\ \ \ \ 0 \\ 0\ -1\ -1\ \ \ 2\end{array}}{\begin{array}{l}5\ 4\ 3\ 2 \\ 4\ 3\ 2\ 1\end{array}} = \dfrac{\begin{array}{l}7\ 8\ 13\ 14 \\ 5\ 6\ \ 7\ 12\end{array}}{\begin{array}{l}5\ 4\ 3\ 2 \\ 4\ 3\ 2\ 1\end{array}}$

$\qquad\qquad\qquad = 8918.$

(d). $d_{[+3,2,1,1]}(SO(8)) = \dfrac{\begin{array}{l}8\ 9\ 10 \\ 7\ 8 \\ 6 \\ 5\end{array} + \begin{array}{l}0\ \ \ 1\ 2 \\ 0\ \ \ 0 \\ -1 \\ -2\end{array}}{\begin{array}{l}6\ 3\ 1 \\ 2\times\begin{array}{l}4\ 1 \\ 2\end{array} \\ 1\end{array}} = \dfrac{\begin{array}{l}8\ 10\ 12 \\ 7\ 8 \\ 5 \\ 3\end{array}}{\begin{array}{l}6\ 3\ 1 \\ 2\times\begin{array}{l}4\ 1 \\ 2\end{array} \\ 1\end{array}} = 2800.$

(e). $d_{[+3,3,1,1]}(SO(6)) = \dfrac{\begin{array}{l}8\ 9\ 10 \\ 7\ 8\ 9 \\ 6 \\ 5\end{array} + \begin{array}{l}0\ \ 2\ 2 \\ 0\ \ 0\ 0 \\ -2 \\ -2\end{array} + \begin{array}{l}0\ \ \ 0\ \ \ 0 \\ 0\ -1\ 1 \\ 0 \\ 0\end{array}}{\begin{array}{l}6\ 3\ 2 \\ 2\times\begin{array}{l}5\ 2\ 1 \\ 2\end{array} \\ 1\end{array}} = \dfrac{\begin{array}{l}8\ 11\ 12 \\ 7\ 7\ 10 \\ 4 \\ 3\end{array}}{\begin{array}{l}6\ 3\ 2 \\ 2\times\begin{array}{l}5\ 2\ 1 \\ 2\end{array} \\ 1\end{array}} = 4312.$

2. Calculate the Clebsch-Gordan series for the subduced representations of the following irreducible representations of the SU(N) group with respect to the subgroup SO(N), and then check the results by their dimensions for $N = 7$:

$[2]$, $[3]$, $[2,1]$, $[4]$, $[3,1]$, $[2,2]$

$[2,1^2]$, $[5]$, $[4,1]$, $[3,2]$, $[3,1,1]$, $[2,2,1]$,

$[2,1^3]$, $[6]$, $[5,1]$, $[4,2]$, $[4,1,1]$, $[3,3]$,

$[3,2,1]$, $[3,1^3]$, $[2^3]$, $[2^2,1^2]$, $[2,1^4]$.

Solution. The following results for the reductions of the subduced representations of $\mathrm{SU}(N)$ with respect to the subgroup $\mathrm{SO}(N)$ are independent of N, except for the Young patterns in the CG series with the row number not less than $N/2$, or with the digit number in the first two column larger than N.

$\square\square \longrightarrow \square\square \oplus 1$, $28 = 27 + 1$.

$\square\square\square \longrightarrow \square\square\square \oplus \square$, $84 = 77 + 7$.

(diagram) $\longrightarrow$ (diagram) $\oplus \square$, $112 = 105 + 7$.

$\square\square\square\square \longrightarrow \square\square\square\square \oplus \square\square \oplus 1$, $210 = 182 + 27 + 1$.

(diagram) $\longrightarrow$ (diagram) $\oplus \square\square \oplus$ (diagram), $378 = 330 + 27 + 21$.

(diagram) $\longrightarrow$ (diagram) $\oplus \square\square \oplus 1$, $196 = 168 + 27 + 1$.

(diagram) $\longrightarrow$ (diagram) $\oplus$ (diagram), $210 = 189 + 21$.

$\square\square\square\square\square \longrightarrow \square\square\square\square\square \oplus \square\square\square \oplus \square$, $462 = 378 + 77 + 7$.

(diagram) $\longrightarrow$ (diagram) $\oplus \square\square\square \oplus$ (diagram) $\oplus \square$,

$1008 = 819 + 77 + 105 + 7$.

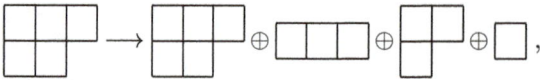

$$882 = 693 + 77 + 105 + 7.$$

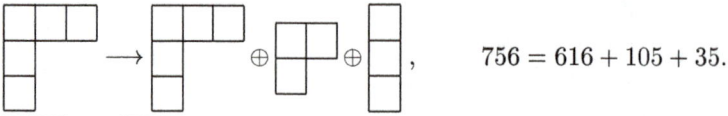

$$756 = 616 + 105 + 35.$$

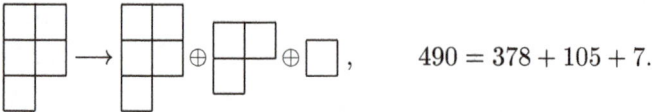

$$490 = 378 + 105 + 7.$$

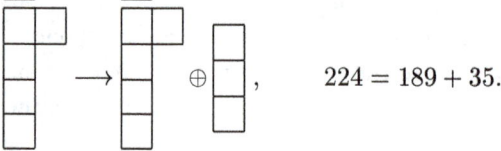

$$224 = 189 + 35.$$

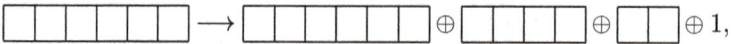

$$924 = 714 + 182 + 27 + 1.$$

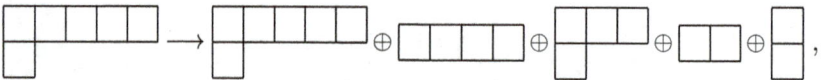

$$2310 = 1750 + 182 + 330 + 27 + 21.$$

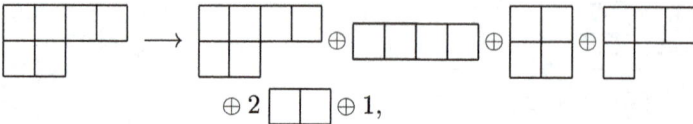

$$2646 \ = \ 1911 + 182 + 168 + 330 + 2 \times 27 + 1.$$

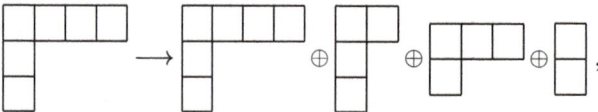

$$2100 = 1560 + 189 + 330 + 21.$$

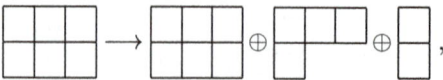

$$1176 = 825 + 330 + 21.$$

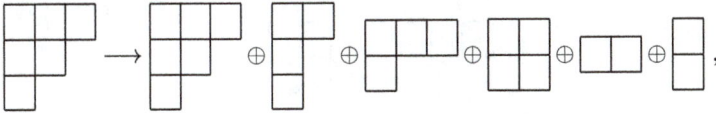

$$2352 = 1617 + 189 + 330 + 168 + 27 + 21.$$

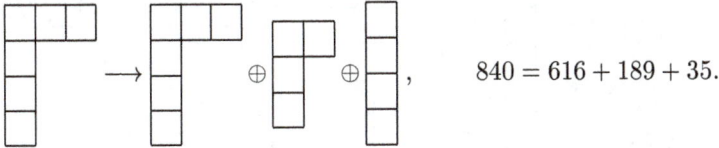

$$840 = 616 + 189 + 35.$$

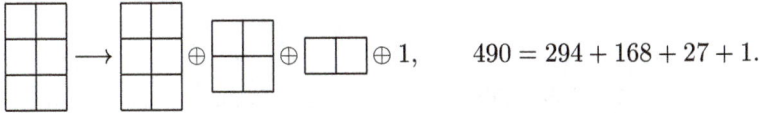

$$490 = 294 + 168 + 27 + 1.$$

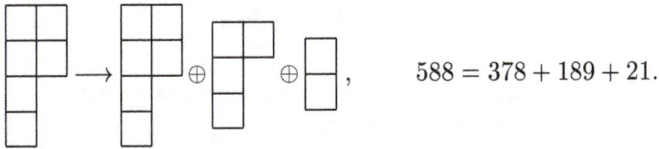

$$588 = 378 + 189 + 21.$$

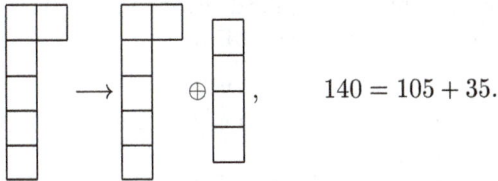

$$140 = 105 + 35.$$

3. Calculate the Clebsch-Gordan series in the reductions of the following direct products of the irreducible tensor representations, and check the results by their dimensions for the SO(7) group:

$$(1)\ [2] \otimes [2], \quad (2)\ [2] \otimes [1,1], \quad (3)\ [3] \otimes [2,1].$$

Solution. (1):

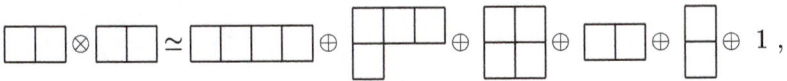

$$27 \times 27 = 729 = 182 + 330 + 168 + 27 + 21 + 1.$$

(2):

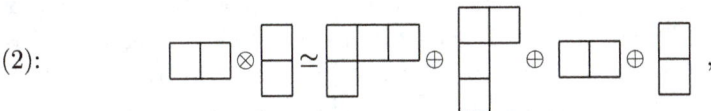

$$27 \times 21 = 567 = 330 + 189 + 27 + 21.$$

(3):

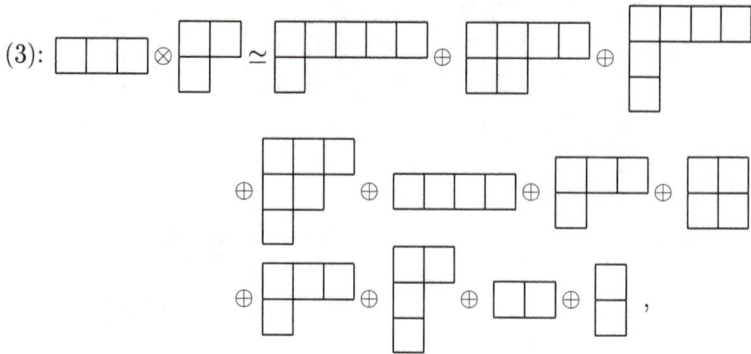

$$77 \times 105 = 8085 = 1750 + 1911 + 1560 + 1617$$
$$+ 182 + 330 + 168 + 330 + 189 + 27 + 21.$$

4. Calculate the orthonormal bases in the irreducible representation space $[2,2]$ of SO(5) by the method of the block weight diagram, and then express the orthonormal bases by the standard tensor Young tableaux in the traceless tensor space of rank four for SO(5).

Solution. The Lie algebra of SO(5) is B_2. The relations between the simple roots $\mathbf{r}_\mu$ and the fundamental dominant weights $\mathbf{w}_\mu$ are

$$\mathbf{r}_1 = 2\mathbf{w}_1 - 2\mathbf{w}_2, \qquad \mathbf{r}_2 = -\mathbf{w}_1 + 2\mathbf{w}_2,$$

$$\mathbf{w}_1 = \mathbf{r}_1 + \mathbf{r}_2, \qquad \mathbf{w}_2 = \frac{1}{2}\mathbf{r}_1 + \mathbf{r}_2.$$

The highest weight of the representation $[2,2]$ of SO(5) is $(0,4)$, and its dimension is

$$d_{[2,2]}(SO(5)) = \frac{\begin{array}{cc} 5\,6 \\ 4\,5 \end{array} + \begin{array}{cc} 1\,1 \\ -2\,0 \end{array} \quad \begin{array}{cc} 6\,7 \\ 2\,5 \end{array}}{\begin{array}{cc} 3\,2 \\ 2\,1 \end{array}} = \frac{\begin{array}{cc} 6\,7 \\ 2\,5 \end{array}}{\begin{array}{cc} 3\,2 \\ 2\,1 \end{array}} = 35.$$

From the highest weight state $(0,4)$ we have a quartet of $\mathcal{A}_2$ with $(1,2)$, $(2,0)$, $(3,\bar{2})$ and $(4,\bar{4})$, where the matrix entries of F_2 are 2, $\sqrt{6}$, $\sqrt{6}$, and 2, respectively. The weights $(1,2)$ and $(2,0)$ are the single dominant weights. From $(1,2)$ we have a doublet of $\mathcal{A}_1$ with $(\bar{1},4)$. From the weight $(2,0)$ we have a triplet of $\mathcal{A}_1$ with $(0,2)$ and $(\bar{2},4)$, and from the weight $(\bar{1},4)$ we have a quintet of $\mathcal{A}_2$ with $(0,2)$, $(1,0)$, $(2,\bar{2})$ and $(3,\bar{4})$. The weight $(0,2)$ may be a double dominant weight. Define the first basis state $|(0,2)_1\rangle$ belonging to the triplet of $\mathcal{A}_1$ and the other $|(0,2)_2\rangle$ belonging to a singlet of $\mathcal{A}_1$. Let

$$F_1 \, |(2,0)\rangle = \sqrt{2} \, |(0,2)_1\rangle, \quad F_1 \, |(0,2)_1\rangle = \sqrt{2} \, |(\bar{2},4)\rangle,$$
$$F_2 \, |(\bar{1},4)\rangle = a_1 \, |(0,2)_1\rangle + a_2 \, |(0,2)_2\rangle, \quad a_1^2 + a_2^2 = 4.$$

Applying $E_1 F_2 = F_2 E_1$ to the basis state $|(\bar{1},4)\rangle$, we have

$$E_1 F_2 \, |(\bar{1},4)\rangle = \sqrt{2} \, a_1 \, |(2,0)\rangle$$
$$= F_2 E_1 \, |(\bar{1},4)\rangle = \ F_2 \, |(1,2)\rangle = \sqrt{6} \, |(2,0)\rangle.$$

Thus, $a_1 = \sqrt{3}$, and then, $a_2 = \sqrt{4-3} = 1$. From the weight $(3,\bar{2})$ we have a quartet of $\mathcal{A}_1$ with $(1,0)$, $(\bar{1},2)$, and $(\bar{3},4)$. On the other hand, applying F_2 to two states $|(0,2)_1\rangle$ and $|(0,2)_2\rangle$, we obtain another two states with the weight $(1,0)$. Thus, the dominant weight $(1,0)$ may be a triple weight. We define the first basis state $(1,0)_1$ belonging to the quartet of $\mathcal{A}_1$ and the other two belonging to two doublets of $\mathcal{A}_1$. The action of F_2 on $|(0,2)_1\rangle$ is the combination of $|(1,0)_1\rangle$ and $|(1,0)_2\rangle$. The weight $(2,\bar{2})$ is the double weight because it is equivalent to $(0,2)$. The weight $(\bar{1},2)$ is equivalent to $(1,0)$. We have

$$F_1 \, |(3,\bar{2})\rangle = \sqrt{3} \, |(1,0)_1\rangle, \quad F_1 \, |(1,0)_1\rangle = 2 \, |(\bar{1},2)_1\rangle,$$
$$F_1 \, |(\bar{1},2)_1\rangle = \sqrt{3} \, |(\bar{3},4)\rangle, \quad F_1 \, |(1,0)_2\rangle = \ |(\bar{1},2)_2\rangle,$$
$$F_1 \, |(1,0)_3\rangle = \ |(\bar{1},2)_3\rangle.$$

Let

$$F_2 \, |(0,2)_1\rangle = a_3 \, |(1,0)_1\rangle + a_4 \, |(1,0)_2\rangle,$$
$$F_2 \, |(0,2)_2\rangle = a_5 \, |(1,0)_1\rangle + a_6 \, |(1,0)_2\rangle + a_7 \, |(1,0)_3\rangle,$$

Applying $E_1 F_2 = F_2 E_1$ to the basis state $|(0,2)_1\rangle$ and $|(0,2)_2\rangle$, we have

$$E_1 F_2 \, |(0,2)_1\rangle = \sqrt{3} \, a_3 \, |(3,\bar{2})\rangle$$
$$= F_2 E_1 \, |(0,2)_1\rangle = \ \sqrt{2} F_2 \, |(2,0)\rangle = \sqrt{12} \, |(3,\bar{2})\rangle$$
$$E_1 F_2 \, |(0,2)_2\rangle = \sqrt{3} \, a_5 \, |(3,\bar{2})\rangle$$
$$= F_2 E_1 \, |(0,2)_1\rangle = 0.$$

Thus, $a_3 = 2$ and $a_5 = 0$. Applying $E_2 F_2 = F_2 E_2 + H_2$ to the basis state $|(0,2)_1\rangle$ and $|(0,2)_2\rangle$, we have

$$E_2 F_2 \, |(0,2)_1\rangle = \left(4 + a_4^2\right) \, |(0,2)_1\rangle + a_4 a_6 \, |(0,2)_2\rangle$$
$$= (F_2 E_2 + H_2) \, |(0,2)_1\rangle = (3+2) \, |(0,2)_1\rangle + \sqrt{3} \, |(0,2)_2\rangle$$
$$E_2 F_2 \, |(0,2)_2\rangle = a_4 a_6 \, |(0,2)_1\rangle + \left(a_6^2 + a_7^2\right) \, |(0,2)_2\rangle$$
$$= (F_2 E_2 + H_2) \, |(0,2)_2\rangle = \sqrt{3} \, |(0,2)_1\rangle + (1+2) \, |(0,2)_2\rangle .$$

Choosing the phases of the basis states $|(1,0)_2\rangle$ and $|(1,0)_3\rangle$ such that a_4 and a_7 are real positive, we have $a_4 = 1$, $a_6 = \sqrt{3}$, and $a_7 = 0$. It means that the dominant weight $(1,0)$ as well as its equivalent weight $(\bar{1},2)$ in $[2,2]$ are the double weights, not triple. The states $|(1,0)_3\rangle$ and $|(\bar{1},2)_3\rangle$ do not exist in $[2,2]$. From the weight $(4,\bar{4})$ we have a quintet of $\mathcal{A}_1$ with the weights $(2,\bar{2})_1$, $(0,0)_1$, $(\bar{2},2)_1$ and $(\bar{4},4)$, where the matrix entries of F_1 are 2, $\sqrt{6}$, $\sqrt{6}$ and 2, respectively. Another basis state $|(2,\bar{2})_2\rangle$ belongs to a triplet of $\mathcal{A}_1$ with $(0,0)_2$ and $(\bar{2},2)_2$, where both matrix entries of F_1 are $\sqrt{2}$. Let

$$F_2 \, |(1,0)_1\rangle = a_8 \, |(2,\bar{2})_1\rangle + a_9 \, |(2,\bar{2})_2\rangle ,$$
$$F_2 \, |(1,0)_2\rangle = a_{10} \, |(2,\bar{2})_1\rangle + a_{11} \, |(2,\bar{2})_2\rangle ,$$
$$F_2 \, |(2,\bar{2})_1\rangle = a_{12} \, |(3,\bar{4})\rangle , \qquad F_2 \, |(2,\bar{2})_2\rangle = a_{13} \, |(3,\bar{4})\rangle .$$

Applying $E_1 F_2 = F_2 E_1$ to the basis state $|(1,0)_1\rangle$ and $|(1,0)_2\rangle$, we have

$$E_1 F_2 \, |(1,0)_1\rangle = 2 a_8 |(4,\bar{4})\rangle$$
$$= F_2 E_1 \, |(1,0)_1\rangle = \sqrt{3} F_2 \, |(3,\bar{2})\rangle = 2\sqrt{3} \, |(4,\bar{4})\rangle ,$$
$$E_1 F_2 \, |(1,0)_2\rangle = 2 a_{10} |(4,\bar{4})\rangle$$
$$= F_2 E_1 \, |(1,0)_2\rangle = 0 .$$

Thus, $a_8 = \sqrt{3}$ and $a_{10} = 0$. Applying $E_2 F_2 = F_2 E_2 + H_2$ to the basis state $|(1,0)_1\rangle$, we have

$$E_2 F_2 \, |(1,0)_1\rangle = \left(3 + a_9^2\right) \, |(1,0)_1\rangle + a_9 a_{11} \, |(1,0)_2\rangle$$
$$= (F_2 E_2 + H_2) \, |(1,0)_1\rangle = 4 \, |(1,0)_1\rangle + 2 \, |(1,0)_2\rangle .$$

Choosing the phase of the basis state $|(2,\bar{2})_2\rangle$ such that a_9 is real positive, we have $a_9 = 1$ and $a_{11} = 2$. Applying $E_2 F_2 = F_2 E_2 + H_2$ to the basis state $|(2,\bar{2})_1\rangle$, we have

$$E_2 F_2 \, |(2,\bar{2})_1\rangle = a_{12}^2 \, |(2,\bar{2})_1\rangle + a_{12} a_{13} \, |(2,\bar{2})_2\rangle$$
$$= (F_2 E_2 + H_2) \, |(2,\bar{2})_1\rangle = (3-2) \, |(2,\bar{2})_1\rangle + \sqrt{3} \, |(2,\bar{2})_2\rangle .$$

Choosing the phase of the basis state $|(3, \overline{4})\rangle$ such that a_{12} is real positive, we have $a_{12} = 1$ and $a_{13} = \sqrt{3}$.

Applying F_2 to the states with the double weight $(\overline{1}, 2)$, we have two basis states with the weight $(0, 0)$, in addition to the state bases $|(0, 0)_1\rangle$ in the quintet and $|(0, 0)_2\rangle$ in the triplet of $\mathcal{A}_1$. Thus, the dominant weight $(0, 0)$ may be a quadruple weight. Both $|(0, 0)_3\rangle$ and $|(0, 0)_4\rangle$ are singlet of $\mathcal{A}_1$. Let

$$F_2 |(\overline{2}, 4)\rangle = b_1 |(\overline{1}, 2)_1\rangle + b_2 |(\overline{1}, 2)_2\rangle,$$
$$F_2 |(\overline{1}, 2)_1\rangle = b_3 |(0, 0)_1\rangle + b_4 |(0, 0)_2\rangle + b_5 |(0, 0)_4\rangle$$
$$F_2 |(\overline{1}, 2)_2\rangle = b_6 |(0, 0)_1\rangle + b_7 |(0, 0)_2\rangle + b_8 |(0, 0)_3\rangle + b_9 |(0, 0)_4\rangle.$$

Applying $E_1 F_2 = F_2 E_1$ to the basis states $|(\overline{2}, 4)\rangle$, $|(\overline{1}, 2)_1\rangle$ and $|(\overline{1}, 2)_2\rangle$, we have

$$E_1 F_2 |(\overline{2}, 4)\rangle = 2b_1 |(1, 0)_1\rangle + b_2 |(1, 0)_2\rangle$$
$$= F_2 E_1 |(\overline{2}, 4)\rangle = \sqrt{2} F_2 |(0, 2)_1\rangle = \sqrt{8} |(1, 0)_1\rangle + \sqrt{2} |(1, 0)_2\rangle,$$
$$E_1 F_2 |(\overline{1}, 2)_1\rangle = \sqrt{6} b_3 |(2, \overline{2})_1\rangle + \sqrt{2} b_4 |(2, \overline{2})_2\rangle$$
$$= F_2 E_1 |(\overline{1}, 2)_1\rangle = 2 F_2 |(1, 0)_1\rangle = \sqrt{12} |(2, \overline{2})_1\rangle + 2 |(2, \overline{2})_2\rangle,$$
$$E_1 F_2 |(\overline{1}, 2)_2\rangle = \sqrt{6} b_6 |(2, \overline{2})_1\rangle + \sqrt{2} b_7 |(2, \overline{2})_2\rangle$$
$$= F_2 E_1 |(\overline{1}, 2)_2\rangle = F_2 |(1, 0)_2\rangle = 2 |(2, \overline{2})_2\rangle.$$

Thus, we have $b_1 = b_2 = b_3 = b_4 = b_7 = \sqrt{2}$ and $b_6 = 0$. Applying $E_2 F_2 = F_2 E_2 + H_2$ to the state basis $|(\overline{1}, 2)_1\rangle$, we have

$$E_2 F_2 |(\overline{1}, 2)_1\rangle = (2 + 2 + b_5^2) |(\overline{1}, 2)_1\rangle + (2 + b_5 b_9) |(\overline{1}, 2)_2\rangle$$
$$= (F_2 E_2 + H_2) |(\overline{1}, 2)_1\rangle = (2 + 2) |(\overline{1}, 2)_1\rangle + 2 |(\overline{1}, 2)_2\rangle.$$

Thus, $b_5 = 0$, and the basis state $|(0, 0)_4\rangle$ does not exist. So, $b_9 = 0$. Applying $E_2 F_2 = F_2 E_2 + H_2$ to the basis state $|(\overline{1}, 2)_2\rangle$, we have

$$E_2 F_2 |(\overline{1}, 2)_2\rangle = 2 |(\overline{1}, 2)_1\rangle + (2 + b_8^2) |(\overline{1}, 2)_2\rangle$$
$$= (F_2 E_2 + H_2) |(\overline{1}, 2)_2\rangle = 2 |(\overline{1}, 2)_1\rangle + (2 + 2) |(\overline{1}, 2)_2\rangle.$$

Choosing the phase of the basis state $|(0, 0)_3\rangle$ such that b_8 is real positive, we have $b_8 = \sqrt{2}$.

The remaining basis states and the matrix entries of the lowering operators F_μ can be calculated similarly. The results are given in Fig. 9.1.

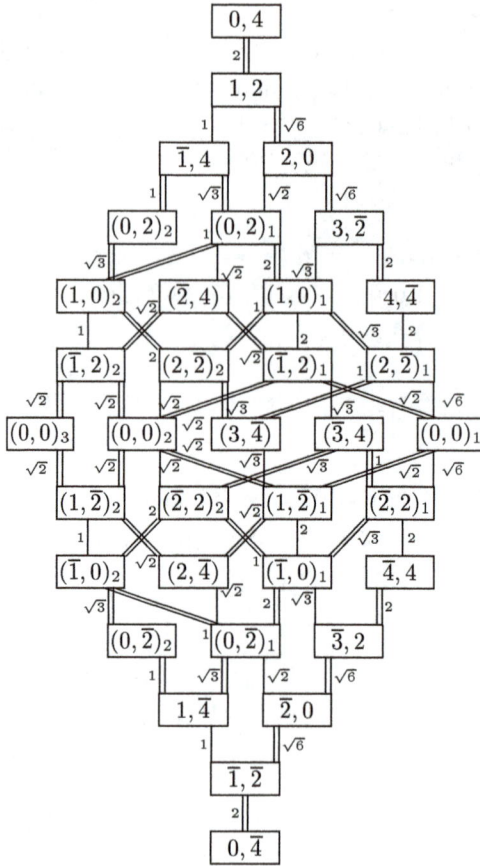

Fig. 9.1   The block weight diagram for $[2,2]$ of $SO(5)$.

Now, we expand the orthonormal basis states in the standard tensor Young tableaux. The Young tableau $\mathcal{Y}^{[2,2]}_1$ is $\begin{array}{|c|c|}\hline 1 & 2 \\\hline 3 & 4 \\\hline\end{array}$. From the Fock condition (8.19) we have

$$\begin{array}{|c|c|}\hline a & b \\\hline c & d \\\hline\end{array} = -\begin{array}{|c|c|}\hline a & d \\\hline c & b \\\hline\end{array} = -\begin{array}{|c|c|}\hline c & b \\\hline a & d \\\hline\end{array} = \begin{array}{|c|c|}\hline b & a \\\hline d & c \\\hline\end{array} = \begin{array}{|c|c|}\hline a & c \\\hline b & d \\\hline\end{array} + \begin{array}{|c|c|}\hline a & b \\\hline d & c \\\hline\end{array}.$$

The normalization factors for the typical tensor Young tableaux are

$$\young(aa,bb) = y_1^{[2,2]}\Phi_{aabb} = \sqrt{48}\left\{\sqrt{1/12}\left[2\Phi_{aabb} + 2\Phi_{bbaa} - \Phi_{baab}\right.\right.$$
$$\left.\left. - \Phi_{baba} - \Phi_{abab} - \Phi_{abba}\right]\right\},$$

$$\young(aa,bc) = y_1^{[2,2]}\Phi_{aabc} = \sqrt{24}\left\{\sqrt{1/24}\left[2\Phi_{aabc} + 2\Phi_{aacb} + 2\Phi_{bcaa}\right.\right.$$
$$+ 2\Phi_{cbaa} - \Phi_{baac} - \Phi_{baca} - \Phi_{abac} - \Phi_{abca}$$
$$\left.\left. - \Phi_{acba} - \Phi_{acab} - \Phi_{caba} - \Phi_{caab}\right]\right\},$$

$$\young(ab,cd) = y_1^{[2,2]}\Phi_{abcd} = A - C, \qquad \young(ac,bd) = y_1^{[2,2]}\Phi_{acbd} = B - C,$$

$$A = \Phi_{abcd} + \Phi_{abdc} + \Phi_{bacd} + \Phi_{badc} + \Phi_{cdab} + \Phi_{cdba} + \Phi_{dcab} + \Phi_{dcba},$$
$$B = \Phi_{acbd} + \Phi_{acdb} + \Phi_{cabd} + \Phi_{cadb} + \Phi_{bdac} + \Phi_{bdca} + \Phi_{dbac} + \Phi_{dbca},$$
$$C = \Phi_{cbad} + \Phi_{cbda} + \Phi_{bcad} + \Phi_{bcda} + \Phi_{adcb} + \Phi_{adbc} + \Phi_{dacb} + \Phi_{dabc}.$$

From Problem 2, we know that the number of linearly independent trace tensors in the tensor space of rank four projected by $y_1^{[2,2]}$ is $50 - 35 = 15$. In fact, they belong to the representation $[2,0]$ and $[0,0]$ of SO(5). We can calculate the trace tensors from the highest weight condition (7.24) and in terms of the lowering operators. In the following we only list the trace tensors with the dominant weights.

$$|(2,0),(2,0)\rangle = 2\,\young(11,24) - \young(11,33),$$

$$|(2,0),(0,2)\rangle = \sqrt{2}\left\{\young(12,24) + \young(11,25) - \young(12,33)\right\},$$

$$|(2,0),(1,0)\rangle = \sqrt{2}\left\{2\,\young(13,24) - \young(12,34) + \young(11,35)\right\},$$

$$|(2,0),(0,0)_1\rangle = \young(13,35) + \young(23,34) - \young(22,44) + \young(11,55),$$

$$|(2,0),(0,0)_2\rangle = \sqrt{1/5}\left\{2\,\young(12,45) + 2\,\young(14,25) + 3\,\young(13,35)\right.$$
$$\left. - 3\,\young(23,34) - \young(11,55) - \young(22,44)\right\},$$

$$|(0,0),(0,0)\rangle = 2\begin{array}{|c|c|}\hline 1 & 2\\\hline 4 & 5\\\hline\end{array} + 2\begin{array}{|c|c|}\hline 1 & 4\\\hline 2 & 5\\\hline\end{array} - 2\begin{array}{|c|c|}\hline 1 & 3\\\hline 3 & 5\\\hline\end{array} + 2\begin{array}{|c|c|}\hline 2 & 3\\\hline 3 & 4\\\hline\end{array}$$

$$- \begin{array}{|c|c|}\hline 2 & 2\\\hline 4 & 4\\\hline\end{array} - \begin{array}{|c|c|}\hline 1 & 1\\\hline 5 & 5\\\hline\end{array}.$$

Beginning with the highest weight state, we calculate the expansions of the basis states in terms of Eq. (9.13) and the block weight diagram. Due to the Wigner-Eckart theorem, the state basis belonging to the representation $[2,2]$ of $SO(5)$ is orthogonal to the basis states belonging to $[2,0]$ and $[0,0]$, so it is a traceless tensor. For simplicity we neglect the index $(0,4)$ for the highest weight in the basis state $|(0,4),(m_1,m_2)\rangle$.

$$|(0,4)\rangle = \begin{array}{|c|c|}\hline 1 & 1\\\hline 2 & 2\\\hline\end{array},$$

$$|(1,2)\rangle = (1/2)F_2|(0,4)\rangle = \sqrt{2}\begin{array}{|c|c|}\hline 1 & 1\\\hline 2 & 3\\\hline\end{array},$$

$$|(2,0)\rangle = \sqrt{1/6}F_2|(1,2)\rangle = \sqrt{2/3}\left\{\begin{array}{|c|c|}\hline 1 & 1\\\hline 2 & 4\\\hline\end{array} + \begin{array}{|c|c|}\hline 1 & 1\\\hline 3 & 3\\\hline\end{array}\right\},$$

$$|(3,\overline{2})\rangle = \sqrt{1/6}F_2|(2,0)\rangle = \sqrt{2}\begin{array}{|c|c|}\hline 1 & 1\\\hline 3 & 4\\\hline\end{array},$$

$$|(4,\overline{4})\rangle = (1/2)F_2|(3,\overline{2})\rangle = \begin{array}{|c|c|}\hline 1 & 1\\\hline 4 & 4\\\hline\end{array},$$

$$|(\overline{1},4)\rangle = F_1|(1,2)\rangle = \sqrt{2}\begin{array}{|c|c|}\hline 1 & 2\\\hline 2 & 3\\\hline\end{array},$$

$$|(0,2)_1\rangle = \sqrt{1/2}F_1|(2,0)\rangle = \sqrt{1/3}\left\{\begin{array}{|c|c|}\hline 1 & 1\\\hline 2 & 5\\\hline\end{array} + \begin{array}{|c|c|}\hline 1 & 2\\\hline 2 & 4\\\hline\end{array} + 2\begin{array}{|c|c|}\hline 1 & 2\\\hline 3 & 3\\\hline\end{array}\right\},$$

$$|(0,2)_2\rangle = F_2|(\overline{1},4)\rangle - \sqrt{3}|(0,2)_1\rangle = \begin{array}{|c|c|}\hline 1 & 2\\\hline 2 & 4\\\hline\end{array} - \begin{array}{|c|c|}\hline 1 & 1\\\hline 2 & 5\\\hline\end{array},$$

$$|(1,0)_1\rangle = \sqrt{1/3}F_1|(3,\overline{2})\rangle = \sqrt{2/3}\left\{\begin{array}{|c|c|}\hline 1 & 1\\\hline 3 & 5\\\hline\end{array} + 2\begin{array}{|c|c|}\hline 1 & 2\\\hline 3 & 4\\\hline\end{array} - \begin{array}{|c|c|}\hline 1 & 3\\\hline 2 & 4\\\hline\end{array}\right\},$$

$$|(\overline{1},2)_1\rangle = (1/2)F_1|(1,0)_1\rangle = \sqrt{2/3}\left\{2\begin{array}{|c|c|}\hline 1 & 2\\\hline 3 & 5\\\hline\end{array} - \begin{array}{|c|c|}\hline 1 & 3\\\hline 2 & 5\\\hline\end{array} + \begin{array}{|c|c|}\hline 2 & 2\\\hline 3 & 4\\\hline\end{array}\right\},$$

$$|(\overline{3},4)\rangle = \sqrt{1/3}F_1|(\overline{1},2)_1\rangle = \sqrt{2}\begin{array}{|c|c|}\hline 2 & 2\\\hline 3 & 5\\\hline\end{array},$$

$$|(1,0)_2\rangle = \sqrt{1/3}F_2|(0,2)_2\rangle = \sqrt{2/3}\left\{\begin{array}{|c|c|}\hline 1 & 3\\\hline 2 & 4\\\hline\end{array} + \begin{array}{|c|c|}\hline 1 & 2\\\hline 3 & 4\\\hline\end{array} - \begin{array}{|c|c|}\hline 1 & 1\\\hline 3 & 5\\\hline\end{array}\right\},$$

$$|(\bar{2},4)\rangle = \sqrt{1/2}F_1|(0,2)_1\rangle = \sqrt{2/3}\left\{\begin{array}{|c|c|}\hline 1 & 2 \\\hline 2 & 5 \\\hline\end{array} + \begin{array}{|c|c|}\hline 2 & 2 \\\hline 3 & 3 \\\hline\end{array}\right\},$$

$$|(2,\bar{2})_1\rangle = (1/2)F_1|(4,\bar{4})\rangle = \begin{array}{|c|c|}\hline 1 & 1 \\\hline 4 & 5 \\\hline\end{array} + \begin{array}{|c|c|}\hline 1 & 2 \\\hline 4 & 4 \\\hline\end{array},$$

$$|(0,0)_1\rangle = \sqrt{1/6}F_1|(2,\bar{2})_1\rangle$$
$$= \sqrt{1/6}\left\{4\begin{array}{|c|c|}\hline 1 & 2 \\\hline 4 & 5 \\\hline\end{array} - 2\begin{array}{|c|c|}\hline 1 & 4 \\\hline 2 & 5 \\\hline\end{array} + \begin{array}{|c|c|}\hline 1 & 1 \\\hline 5 & 5 \\\hline\end{array} + \begin{array}{|c|c|}\hline 2 & 2 \\\hline 4 & 4 \\\hline\end{array}\right\},$$

$$|(\bar{2},2)_1\rangle = \sqrt{1/6}F_1|(0,0)_1\rangle = \begin{array}{|c|c|}\hline 1 & 2 \\\hline 5 & 5 \\\hline\end{array} + \begin{array}{|c|c|}\hline 2 & 2 \\\hline 4 & 5 \\\hline\end{array},$$

$$|(\bar{4},4)\rangle = (1/2)F_1|(\bar{2},2)_1\rangle = \begin{array}{|c|c|}\hline 2 & 2 \\\hline 5 & 5 \\\hline\end{array},$$

$$|(2,\bar{2})_2\rangle = (1/2)F_2|(1,0)_2\rangle = \sqrt{1/3}\left\{2\begin{array}{|c|c|}\hline 1 & 3 \\\hline 3 & 4 \\\hline\end{array} + \begin{array}{|c|c|}\hline 1 & 2 \\\hline 4 & 4 \\\hline\end{array} - \begin{array}{|c|c|}\hline 1 & 1 \\\hline 4 & 5 \\\hline\end{array}\right\},$$

$$|(\bar{1},2)_2\rangle = F_1|(1,0)_2\rangle = \sqrt{2/3}\left\{2\begin{array}{|c|c|}\hline 1 & 3 \\\hline 2 & 5 \\\hline\end{array} - \begin{array}{|c|c|}\hline 1 & 2 \\\hline 3 & 5 \\\hline\end{array} + \begin{array}{|c|c|}\hline 2 & 2 \\\hline 3 & 4 \\\hline\end{array}\right\},$$

$$|(3,\bar{4})\rangle = F_2|(2,\bar{2})_1\rangle = \sqrt{2}\begin{array}{|c|c|}\hline 1 & 3 \\\hline 4 & 4 \\\hline\end{array},$$

$$|(0,0)_2\rangle = \sqrt{1/2}F_1|(2,\bar{2})_2\rangle$$
$$= \sqrt{1/6}\left\{2\begin{array}{|c|c|}\hline 1 & 3 \\\hline 3 & 5 \\\hline\end{array} + 2\begin{array}{|c|c|}\hline 2 & 3 \\\hline 3 & 4 \\\hline\end{array} + \begin{array}{|c|c|}\hline 2 & 2 \\\hline 4 & 4 \\\hline\end{array} - \begin{array}{|c|c|}\hline 1 & 1 \\\hline 5 & 5 \\\hline\end{array}\right\},$$

$$|(0,0)_3\rangle = \sqrt{1/2}F_2|(\bar{1},2)_2\rangle - |(0,0)_2\rangle$$
$$= \sqrt{1/6}\left\{4\begin{array}{|c|c|}\hline 1 & 4 \\\hline 2 & 5 \\\hline\end{array} - 2\begin{array}{|c|c|}\hline 1 & 2 \\\hline 4 & 5 \\\hline\end{array} + \begin{array}{|c|c|}\hline 1 & 1 \\\hline 5 & 5 \\\hline\end{array} + \begin{array}{|c|c|}\hline 2 & 2 \\\hline 4 & 4 \\\hline\end{array}\right\},$$

$$|(1,\bar{2})_1\rangle = \sqrt{1/3}F_1|(3,\bar{4})\rangle = \sqrt{2/3}\left\{2\begin{array}{|c|c|}\hline 1 & 3 \\\hline 4 & 5 \\\hline\end{array} - \begin{array}{|c|c|}\hline 1 & 4 \\\hline 3 & 5 \\\hline\end{array} + \begin{array}{|c|c|}\hline 2 & 3 \\\hline 4 & 4 \\\hline\end{array}\right\},$$

$$|(\bar{1},0)_1\rangle = (1/2)F_1|(1,\bar{2})_1\rangle = \sqrt{2/3}\left\{\begin{array}{|c|c|}\hline 1 & 3 \\\hline 5 & 5 \\\hline\end{array} + 2\begin{array}{|c|c|}\hline 2 & 3 \\\hline 4 & 5 \\\hline\end{array} - \begin{array}{|c|c|}\hline 2 & 4 \\\hline 3 & 5 \\\hline\end{array}\right\},$$

$$|(\bar{3},2)\rangle = \sqrt{1/3}F_1|(\bar{1},0)_1\rangle = \sqrt{2}\begin{array}{|c|c|}\hline 2 & 3 \\\hline 5 & 5 \\\hline\end{array},$$

$$|(\bar{2},2)_2\rangle = \sqrt{1/2}F_1|(0,0)_2\rangle = \sqrt{1/3}\left\{2\begin{array}{|c|c|}\hline 2 & 3 \\\hline 3 & 5 \\\hline\end{array} + \begin{array}{|c|c|}\hline 2 & 2 \\\hline 4 & 5 \\\hline\end{array} - \begin{array}{|c|c|}\hline 1 & 2 \\\hline 5 & 5 \\\hline\end{array}\right\},$$

$$|(1,\bar{2})_2\rangle = \sqrt{1/2}F_2|(0,0)_3\rangle = \sqrt{2/3}\left\{2\begin{array}{|c|c|}\hline 1 & 4 \\\hline 3 & 5 \\\hline\end{array} - \begin{array}{|c|c|}\hline 1 & 3 \\\hline 4 & 5 \\\hline\end{array} + \begin{array}{|c|c|}\hline 2 & 3 \\\hline 4 & 4 \\\hline\end{array}\right\},$$

$$|(2,\overline{4})\rangle = \sqrt{1/2}\,F_2|(1,\overline{2})_2\rangle = \sqrt{2/3}\left\{\begin{array}{|c|c|}\hline 1 & 4 \\\hline 4 & 5 \\\hline\end{array} + \begin{array}{|c|c|}\hline 3 & 3 \\\hline 4 & 4 \\\hline\end{array}\right\},$$

$$|(\overline{1},0)_2\rangle = F_1|(1,\overline{2})_2\rangle = \sqrt{2/3}\left\{\begin{array}{|c|c|}\hline 2 & 4 \\\hline 3 & 5 \\\hline\end{array} + \begin{array}{|c|c|}\hline 2 & 3 \\\hline 4 & 5 \\\hline\end{array} - \begin{array}{|c|c|}\hline 1 & 3 \\\hline 5 & 5 \\\hline\end{array}\right\},$$

$$|(0,\overline{2})_1\rangle = \sqrt{1/2}\,F_1|(2,\overline{4})\rangle = \sqrt{1/3}\left\{\begin{array}{|c|c|}\hline 1 & 4 \\\hline 5 & 5 \\\hline\end{array} + \begin{array}{|c|c|}\hline 2 & 4 \\\hline 4 & 5 \\\hline\end{array} + 2\begin{array}{|c|c|}\hline 3 & 3 \\\hline 4 & 5 \\\hline\end{array}\right\},$$

$$|(\overline{2},0)\rangle = \sqrt{1/2}\,F_1|(0,\overline{2})_1\rangle = \sqrt{2/3}\left\{\begin{array}{|c|c|}\hline 2 & 4 \\\hline 5 & 5 \\\hline\end{array} + \begin{array}{|c|c|}\hline 3 & 3 \\\hline 5 & 5 \\\hline\end{array}\right\},$$

$$|(0,\overline{2})_2\rangle = \sqrt{1/3}\left\{F_2|(\overline{1},0)_2\rangle - |(0,\overline{2})_1\rangle\right\} = \begin{array}{|c|c|}\hline 2 & 4 \\\hline 4 & 5 \\\hline\end{array} - \begin{array}{|c|c|}\hline 1 & 4 \\\hline 5 & 5 \\\hline\end{array},$$

$$|(1,\overline{4})\rangle = F_2|(0,\overline{2})_2\rangle = \sqrt{2}\,\begin{array}{|c|c|}\hline 3 & 4 \\\hline 4 & 5 \\\hline\end{array},$$

$$|(\overline{1},\overline{2})\rangle = \sqrt{1/6}\,F_2|(\overline{2},0)\rangle = \sqrt{2}\,\begin{array}{|c|c|}\hline 3 & 4 \\\hline 5 & 5 \\\hline\end{array},$$

$$|(0,\overline{4})\rangle = (1/2)F_2|(\overline{1},\overline{2})\rangle = \begin{array}{|c|c|}\hline 4 & 4 \\\hline 5 & 5 \\\hline\end{array}.$$

5. Calculate the orthonormal bases in the irreducible representation space $[2,0,0,0]$ of $SO(8)$ by the method of the block weight diagram, and then, express the orthonormal bases by the standard tensor Young tableaux in the traceless symmetric tensor space of rank two for $SO(8)$.

Solution. The Lie algebra of $SO(8)$ is D_4. The relations between the simple roots $\mathbf{r}_\mu$ and the fundamental dominant weights $\mathbf{w}_\mu$ are

$$\mathbf{r}_1 = 2\mathbf{w}_1 - \mathbf{w}_2, \qquad\qquad \mathbf{r}_2 = -\mathbf{w}_1 + 2\mathbf{w}_2 - \mathbf{w}_3 - \mathbf{w}_4,$$

$$\mathbf{r}_3 = -\mathbf{w}_2 + 2\mathbf{w}_3, \qquad\qquad \mathbf{r}_4 = -\mathbf{w}_2 + 2\mathbf{w}_4,$$

$$\mathbf{w}_1 = \mathbf{r}_1 + \mathbf{r}_2 + \frac{1}{2}\mathbf{r}_3 + \frac{1}{2}\mathbf{r}_4, \quad \mathbf{w}_2 = \mathbf{r}_1 + 2\mathbf{r}_2 + \mathbf{r}_3 + \mathbf{r}_4,$$

$$\mathbf{w}_3 = \frac{1}{2}\mathbf{r}_1 + \mathbf{r}_2 + \mathbf{r}_3 + \frac{1}{2}\mathbf{r}_4, \quad \mathbf{w}_4 = \frac{1}{2}\mathbf{r}_1 + \mathbf{r}_2 + \frac{1}{2}\mathbf{r}_3 + \mathbf{r}_4.$$

The highest weight of the representation $[2,0,0,0]$ of $SO(8)$ is $(2,0,0,0)$, and its dimension is

$$d_{[2,0,0,0]}(SO(8)) = \frac{\begin{array}{|c|c|}\hline 8 & 9 \\\hline\end{array} + \begin{array}{|c|c|}\hline -1 & 1 \\\hline\end{array}}{\begin{array}{|c|c|}\hline 2 & 1 \\\hline\end{array}} = \frac{7 \times 10}{2 \times 1} = 35.$$

From the highest weight state $(2,0,0,0)$ we have a triplet of $\mathcal{A}_1$ with

$(0, 1, 0, 0)$, and $(\bar{2}, 2, 0, 0)$, where both matrix entries of F_1 are $\sqrt{2}$. The weight $(0, 1, 0, 0)$ is a single dominant weight. The weights equivalent to two dominant weights are

$$(2, 0, 0, 0), \quad (\bar{2}, 2, 0, 0), \quad (0, \bar{2}, 2, 2), \quad (0, 0, \bar{2}, 2),$$
$$(0, 0, 2, \bar{2}), \quad (0, 2, \bar{2}, \bar{2}), \quad (2, \bar{2}, 0, 0), \quad (\bar{2}, 0, 0, 0).$$

$$(0, 1, 0, 0), \quad (1, \bar{1}, 1, 1), \quad (\bar{1}, 0, 1, 1), \quad (1, 0, \bar{1}, 1), \quad (1, 0, 1, \bar{1}),$$
$$(\bar{1}, 1, \bar{1}, 1), \quad (\bar{1}, 1, 1, \bar{1}), \quad (1, 1, \bar{1}, \bar{1}), \quad (\bar{1}, 2, \bar{1}, \bar{1}), \quad (2, \bar{1}, 0, 0),$$
$$(0, \bar{1}, 2, 0), \quad (0, \bar{1}, 0, 2), \quad (1, \bar{2}, 1, 1), \quad (\bar{2}, 1, 0, 0), \quad (0, 1, \bar{2}, 0),$$
$$(0, 1, 0, \bar{2}), \quad (\bar{1}, \bar{1}, 1, 1), \quad (1, \bar{1}, \bar{1}, 1), \quad (1, \bar{1}, 1, \bar{1}), \quad (\bar{1}, 0, \bar{1}, 1),$$
$$(\bar{1}, 0, 1, \bar{1}), \quad (1, 0, \bar{1}, \bar{1}), \quad (\bar{1}, 1, \bar{1}, \bar{1}), \quad (0, \bar{1}, 0, 0).$$

Therefore, the representation $[2, 0, 0, 0]$ contains two simple dominant weights $(2, 0, 0, 0)$ and $(0, 1, 0, 0)$ and one triple dominant weight $(0, 0, 0, 0)$. The block weight diagram for the representation $[2, 0, 0, 0]$ is very easy except for the part related to the triple weight $(0, 0, 0, 0)$, which will be explained below. In Fig. 9.2 we give the block weight diagram for the representation $[2, 0, 0, 0]$ where only those matrix entries which are not equal to 1 are indicated.

Let $|(2, \bar{1}, 0, 0)\rangle$, $|(0, 0, 0, 0)_1\rangle$ and $|(\bar{2}, 1, 0, 0)\rangle$ constitute a triplet of $\mathcal{A}_1$, and the other two basis states $|(0, 0, 0, 0)_2\rangle$ and $|(0, 0, 0, 0)_3\rangle$ are the singlets of $\mathcal{A}_1$. We assume

$$F_1 |(2, \bar{1}, 0, 0)\rangle = \sqrt{2} |(0, 0, 0, 0)_1\rangle,$$
$$F_3 |(0, \bar{1}, 2, 0)\rangle = a_1 |(0, 0, 0, 0)_1\rangle + a_2 |(0, 0, 0, 0)_2\rangle,$$
$$F_4 |(0, \bar{1}, 0, 2)\rangle = b_1 |(0, 0, 0, 0)_1\rangle + b_2 |(0, 0, 0, 0)_2\rangle + b_3 |(0, 0, 0, 0)_3\rangle,$$
$$F_2 |(\bar{1}, 2, \bar{1}, \bar{1})\rangle = c_1 |(0, 0, 0, 0)_1\rangle + c_2 |(0, 0, 0, 0)_2\rangle + c_3 |(0, 0, 0, 0)_3\rangle,$$
$$a_1^2 + a_2^2 = b_1^2 + b_2^2 + b_3^2 = c_1^2 + c_2^2 + c_3^2 = 2,$$

where we neglect the index $(2, 0, 0, 0)$ for the highest weight in the basis state $|(2, 0, 0, 0), (m_1, m_2, m_3, m_4)\rangle$ for simplicity. Applying $E_1 F_2 = F_2 E_1$ to the basis state $|(\bar{1}, 2, \bar{1}, \bar{1})\rangle$, $E_1 F_3 = F_3 E_1$ to the basis state $|(0, \bar{1}, 2, 0)\rangle$, and $E_1 F_4 = F_4 E_1$ to the basis state $|(0, \bar{1}, 0, 2)\rangle$, we have

$$E_1 F_2 |(\bar{1}, 2, \bar{1}, \bar{1})\rangle = \sqrt{2} c_1 |(2, \bar{1}, 0, 0)\rangle$$
$$= F_2 E_1 |(\bar{1}, 2, \bar{1}, \bar{1})\rangle = F_2 |(1, 1, \bar{1}, \bar{1})\rangle = |(2, \bar{1}, 0, 0)\rangle,$$

$$E_1 F_3 \; |(0,\bar{1},2,0)\rangle = \sqrt{2}a_1 \; |(2,\bar{1},0,0)\rangle$$
$$= F_3 E_1 \; |(0,\bar{1},2,0)\rangle = 0,$$
$$E_1 F_4 \; |(0,\bar{1},0,2)\rangle = \sqrt{2}b_1 \; |(2,\bar{1},0,0)\rangle$$
$$= F_4 E_1 \; |(0,\bar{1},0,2)\rangle = 0.$$

Thus, $c_1 = \sqrt{1/2}$ and $a_1 = b_1 = 0$. Choosing the phase of the basis state $|(0,0,0,0)_2\rangle$ such that a_2 is real positive, we have $a_2 = \sqrt{2}$. Applying $E_3 F_2 = F_2 E_3$ to the basis state $|(\bar{1},2,\bar{1},\bar{1})\rangle$, and $E_3 F_4 = F_4 E_3$ to the basis state $|(0,\bar{1},0,2)\rangle$, we have

$$E_3 F_2 \; |(\bar{1},2,\bar{1},\bar{1})\rangle = \sqrt{2}c_2 \; |(0,\bar{1},2,0)\rangle$$
$$= F_2 E_3 \; |(\bar{1},2,\bar{1},\bar{1})\rangle = F_2 \; |(\bar{1},1,1,\bar{1})\rangle = \; |(0,\bar{1},2,0)\rangle,$$
$$E_3 F_4 \; |(0,\bar{1},0,2)\rangle = \sqrt{2}b_2 \; |(0,\bar{1},2,0)\rangle$$
$$= F_4 E_3 \; |(0,\bar{1},0,2)\rangle = \sqrt{2}F_4 \; |(0,\bar{2},2,2)\rangle = 2 \; |(0,\bar{1},2,0)\rangle.$$

Thus, $c_2 = \sqrt{1/2}$ and $b_2 = \sqrt{2}$. Then, $b_3 = 0$. Choosing the phase of the basis state $|(0,0,0,0)_3\rangle$ such that c_3 is real positive, we have $c_3 = 1$.

Now, we expand the orthonormal basis states in the standard tensor Young tableaux. The Young tableau $\mathcal{Y}_1^{[2,0,0,0]}$ is $\boxed{1\;2}$. The normalization factors for two typical tensor Young tableaux are

$$\boxed{a\;a} = \mathcal{Y}_1^{[2,0,0,0]}\Phi_{aa} = 2\,\Phi_{aa},$$
$$\boxed{a\;b} = \mathcal{Y}_1^{[2,0,0,0]}\Phi_{ab} = \sqrt{2}\left\{\sqrt{1/2}\,[\Phi_{ab} + \Phi_{ba}]\right\}.$$

There is only one trace tensor in the tensor space of rank two projected by $\mathcal{Y}_1^{[2,0,0,0]}$. It belongs to the representation $[0,0,0,0]$ of SO(8):

$$|(0,0,0,0),(0,0,0,0)\rangle = -\,\boxed{1\;8} + \boxed{2\;7} - \boxed{3\;6} + \boxed{4\;5}.$$

Beginning with the highest weight state, we calculate the expansions of the basis states in terms of Eq. (9.16) and the block weight diagram. Due to the Wigner-Eckart theorem, the basis state belonging to the representation $[2,0,0,0]$ of SO(8) is orthogonal to the basis state belonging to $[0,0,0,0]$, so it is a traceless tensor.

$$|(2,0,0,0)\rangle = \boxed{1\;1},$$
$$|(0,1,0,0)\rangle = \sqrt{1/2}F_1|(2,0,0,0)\rangle = \sqrt{2}\,\boxed{1\;2},$$

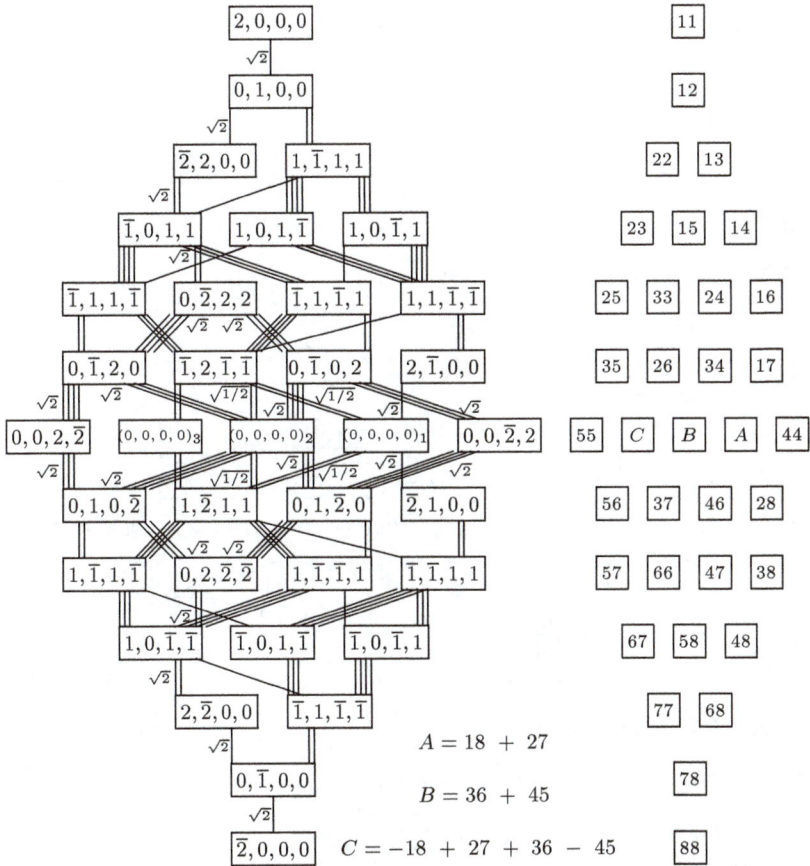

Fig. 9.2 The block weight diagram for $[2,0,0,0]$ of SO(8).

$$|(\overline{2},2,0,0)\rangle = \sqrt{1/2}F_1|(0,1,0,0)\rangle = \boxed{\begin{array}{|c|c|} 2 & 2 \end{array}} \,,$$

$$|(1,\overline{1},1,1)\rangle = F_2|(0,1,0,0)\rangle = \sqrt{2}\,\boxed{\begin{array}{|c|c|} 1 & 3 \end{array}} \,,$$

$$|(\overline{1},0,1,1)\rangle = F_1|(1,\overline{1},1,1)\rangle = \sqrt{2}\,\boxed{\begin{array}{|c|c|} 2 & 3 \end{array}} \,,$$

$$|(1,0,\overline{1},1)\rangle = F_3|(1,\overline{1},1,1)\rangle = \sqrt{2}\,\boxed{\begin{array}{|c|c|} 1 & 4 \end{array}} \,,$$

$$|(1,0,1,\overline{1})\rangle = F_4|(1,\overline{1},1,1)\rangle = \sqrt{2}\,\boxed{\begin{array}{|c|c|} 1 & 5 \end{array}} \,,$$

$$|(0,\overline{2},2,2)\rangle = \sqrt{1/2}F_2|(\overline{1},0,1,1)\rangle = \boxed{\begin{array}{|c|c|} 3 & 3 \end{array}} \,,$$

$$|(\overline{1},1,\overline{1},1)\rangle = F_3|(\overline{1},0,1,1)\rangle = \sqrt{2}\,\boxed{\begin{array}{|c|c|} 2 & 4 \end{array}} \,,$$

$$|(\bar{1},1,1,\bar{1})\rangle = F_4|(\bar{1},0,1,1)\rangle = \sqrt{2}\,\boxed{2\ 5}\ ,$$

$$|(1,1,\bar{1},\bar{1})\rangle = F_3|(1,0,1,\bar{1})\rangle = \sqrt{2}\,\boxed{1\ 6}\ ,$$

$$|(0,\bar{1},2,0)\rangle = F_2|(\bar{1},1,1,\bar{1})\rangle = \sqrt{2}\,\boxed{3\ 5}\ ,$$

$$|(\bar{1},2,\bar{1},\bar{1})\rangle = F_3|(\bar{1},1,1,\bar{1})\rangle = \sqrt{2}\,\boxed{2\ 6}\ ,$$

$$|(0,\bar{1},0,2)\rangle = F_2|(\bar{1},1,\bar{1},1)\rangle = \sqrt{2}\,\boxed{3\ 4}\ ,$$

$$|(2,\bar{1},0,0)\rangle = F_2|(1,1,\bar{1},\bar{1})\rangle = \sqrt{2}\,\boxed{1\ 7}\ ,$$

$$|(0,0,\bar{2},2)\rangle = \sqrt{1/2}F_3|(0,\bar{1},0,2)\rangle = \boxed{4\ 4}\ ,$$

$$|(0,0,2,\bar{2})\rangle = \sqrt{1/2}F_4|(0,\bar{1},2,0)\rangle = \boxed{5\ 5}\ ,$$

$$|(0,0,0,0)_1\rangle = \sqrt{1/2}F_1|(2,\bar{1},0,0)\rangle = \boxed{1\ 8} + \boxed{2\ 7}\ ,$$

$$|(0,0,0,0)_2\rangle = \sqrt{1/2}F_3|(0,\bar{1},2,0)\rangle = \boxed{3\ 6} + \boxed{4\ 5}\ ,$$

$$|(0,0,0,0)_3\rangle = F_2|(\bar{1},2,\bar{1},\bar{1})\rangle - \sqrt{1/2}|(0,0,0,0)_1\rangle - \sqrt{1/2}|(0,0,0,0)_2\rangle$$

$$= \sqrt{1/2}\left\{-\boxed{1\ 8} + \boxed{2\ 7} + \boxed{3\ 6} - \boxed{4\ 5}\right\}\ ,$$

$$|(\bar{2},1,0,0)\rangle = \sqrt{1/2}F_1|(0,0,0,0)_1\rangle = \sqrt{2}\,\boxed{2\ 8}\ ,$$

$$|(0,1,0,\bar{2})\rangle = \sqrt{1/2}F_4|(0,0,0,0)_2\rangle = \sqrt{2}\,\boxed{5\ 6}\ ,$$

$$|(1,\bar{2},1,1)\rangle = F_2|(0,0,0,0)_3\rangle = \sqrt{2}\,\boxed{3\ 7}\ ,$$

$$|(0,1,\bar{2},0)\rangle = \sqrt{1/2}F_4|(0,0,\bar{2},2)\rangle = \sqrt{2}\,\boxed{4\ 6}\ ,$$

$$|(1,\bar{1},1,\bar{1})\rangle = F_2|(0,1,0,\bar{2})\rangle = \sqrt{2}\,\boxed{5\ 7}\ ,$$

$$|(0,2,\bar{2},\bar{2})\rangle = \sqrt{1/2}F_3|(0,1,0,\bar{2})\rangle = \boxed{6\ 6}\ ,$$

$$|(1,\bar{1},\bar{1},1)\rangle = F_3|(1,\bar{2},1,1)\rangle = \sqrt{2}\,\boxed{4\ 7}\ ,$$

$$|(\bar{1},\bar{1},1,1)\rangle = F_2|(\bar{2},1,0,0)\rangle = \sqrt{2}\,\boxed{3\ 8}\ ,$$

$$|(\bar{1},0,1,\bar{1})\rangle = F_4|(\bar{1},\bar{1},1,1)\rangle = \sqrt{2}\,\boxed{5\ 8}\ ,$$

$$|(\bar{1},0,\bar{1},1)\rangle = F_3|(\bar{1},\bar{1},1,1)\rangle = \sqrt{2}\,\boxed{4\ 8}\ ,$$

$$|(1,0,\bar{1},\bar{1})\rangle = F_3|(1,\bar{1},1,1)\rangle = \sqrt{2}\,\boxed{6\ 7}\ ,$$

$$|(2,\bar{2},0,0)\rangle = \sqrt{1/2}F_2|(1,0,\bar{1},\bar{1})\rangle = \boxed{7\ 7}\ ,$$

$$|(\bar{1},1,\bar{1},\bar{1})\rangle = F_1|(1,0,\bar{1},\bar{1})\rangle = \sqrt{2}\,\boxed{6\ 8}\ ,$$

$$|(0,\bar{1},0,0)\rangle = F_2|(\bar{1},1,\bar{1},\bar{1})\rangle = \sqrt{2}\,\boxed{7\ 8}\ ,$$

$$|(\bar{2},0,0,0)\rangle = \sqrt{1/2}F_1|(0,\bar{1},0,0)\rangle = \boxed{8\ 8}\ .$$

6. Calculate the spherical harmonic functions $Y_{\mathbf{m}}^{[\lambda]}$ in an N-dimensional space.

Solution. The relations between the rectangular coordinates x_a and the spherical coordinates r and θ_b in an N-dimensional space are

$$
\begin{aligned}
x_1 &= r\cos\theta_1\sin\theta_2\ldots\sin\theta_{N-1}, \\
x_2 &= r\sin\theta_1\sin\theta_2\ldots\sin\theta_{N-1}, \\
x_b &= r\cos\theta_{b-1}\sin\theta_b\ldots\sin\theta_{N-1}, \quad 3\le b\le N-1, \\
x_N &= r\cos\theta_{N-1}, \\
\sum_{a=1}^{N} x_a^2 &= r^2.
\end{aligned}
\tag{9.19}
$$

The unit vector along $\mathbf{x}$ is usually denoted by $\hat{\mathbf{x}} = \mathbf{x}/r$. The volume element of the configuration space is

$$
\prod_{a=1}^{N} dx_a = r^{N-1}dr\,d\Omega, \qquad d\Omega = \prod_{a=1}^{N-1}(\sin\theta_a)^{a-1}\,d\theta_a,
\tag{9.20}
$$

$$
0\le r<\infty, \quad -\pi\le\theta_1\le\pi, \quad 0\le\theta_b\le\pi, \quad 2\le b\le N-1.
$$

The orbital angular momentum operators L_{ab} are the generators of the transformation operators P_R for the scalar function, $R\in\mathrm{SO}(N)$,

$$
L_{ab} = -L_{ba} = -ix_a\frac{\partial}{\partial x_b} + ix_b\frac{\partial}{\partial x_a}, \qquad L^2 = \sum_{a<b=2}^{N} L_{ab}^2.
\tag{9.21}
$$

From the second Lie theorem, L_{ab} satisfy the same commutation relations as those of the generators T_{ab} in the self representation of $\mathrm{SO}(N)$. The Chevalley bases $H_\mu(L)$, $E_\mu(L)$ and $F_\mu(L)$ can also be expressed by Eqs. (9.10) and (9.11), where T_{ab} are replaced with L_{ab}. Because

$$
L_{ab}x_d = \sum_{c=1}^{N} x_c\,(T_{ab})_{cd}, \qquad J_{ab}\Theta_d = \sum_{c=1}^{N} \Theta_c\,(T_{ab})_{cd},
$$

where J_{ab} are the generators of O_R. The common eigenfunctions X_α of $H_\mu(L)$ can be obtained from Φ_α given in Eq. (9.12) for $\mathrm{SO}(2\ell+1)$ and in Eq. (9.15) for $\mathrm{SO}(2\ell)$ by replacing the vector basis Θ_a with the rectangular coordinate x_a/r, where the factor r^{-1} stands for removing the dimension of length.

The spherical harmonic function $Y_{\mathbf{m}}^{[\lambda]}(\hat{\mathbf{x}})$ is the eigenfunction of the orbital angular momentum for a single particle. Since there is only one coordinate vector $\mathbf{x}$, the representation $[\lambda]$ for the spherical harmonic function $Y_{\mathbf{m}}^{[\lambda]}(\hat{\mathbf{x}})$ has to be the totally symmetric representation denoted by the one-row Young pattern $[\lambda] = [\lambda, 0, \ldots, 0]$. X_α, multiplied by a normalization factor, is nothing but the spherical harmonic function $Y_{\mathbf{m}}^{[1]}(\hat{\mathbf{x}})$ in the self-representation $[\lambda] = [1]$. The component m_μ of the weight $\mathbf{m}$ is the eigenvalue of $H_\mu(L)$ in the eigenfunction X_α. Generally, for the highest weight state $\mathbf{M} = (\lambda, 0, \ldots, 0)$ we have

$$Y_{\mathbf{M}}^{[\lambda]}(\hat{\mathbf{x}}) = C_{N,\lambda} X_1^\lambda = C_{N,\lambda} \left\{ \frac{(-1)^t (x_1 + ix_2)}{r\sqrt{2}} \right\}^\lambda, \tag{9.22}$$

where t is equal to ℓ when $N = 2\ell + 1$ and to $(\ell - 1)$ when $N = 2\ell$. The remaining spherical harmonic function $Y_{\mathbf{m}}^{[\lambda]}(\hat{\mathbf{x}})$ with the weight $\mathbf{m}$ can be calculated by the lowering operators $F_\mu(L)$, just like we have done in Problems 4 and 5. Here we have to calculate the normalization factor $C_{N,\lambda}$ and the eigenvalue of the angular momentum square L^2. Both problems can be solved from the highest weight state $Y_{\mathbf{M}}^{[\lambda]}(\hat{\mathbf{x}})$.

Since $x_1 + ix_2 = re^{i\theta_1} \sin\theta_2 \ldots \sin\theta_{D-1}$ and

$$A_n \equiv \int_0^\pi d\theta \, (\sin\theta)^n = \begin{cases} \pi \cdot \dfrac{1 \cdot 3 \cdot 5 \cdots (n-1)}{2 \cdot 4 \cdot 6 \cdots (n)} & \text{when } n \text{ is even,} \\[2mm] 2 \cdot \dfrac{2 \cdot 4 \cdot 6 \cdots (n-1)}{1 \cdot 3 \cdot 5 \cdots (n)} & \text{when } n \text{ is odd,} \end{cases}$$

we have

$$1 = \int d\Omega \, |Y_{\mathbf{M}}^{[\lambda]}(\hat{\mathbf{x}})|^2 = C_{N,\lambda}^2 2^{1-\lambda} \pi \prod_{a=2}^{N-1} A_{2\lambda+a-1}.$$

For $SO(2\ell + 1)$ group, we have

$$C_{(2\ell+1),\lambda}^{-2} = 2^{1-\lambda}\pi \, (A_{2\lambda+1} A_{2\lambda+2})(A_{2\lambda+3} A_{2\lambda+4}) \ldots (A_{2\lambda+2\ell-3} A_{2\lambda+2\ell-2})$$

$$\times A_{2\lambda+2\ell-1}$$

$$= 2^{1-\lambda}\pi \left(\frac{\pi}{\lambda+1} \right) \left(\frac{\pi}{\lambda+2} \right) \cdots \left(\frac{\pi}{\lambda+\ell-1} \right) 2 \cdot \frac{2 \cdot 4 \cdot 6 \cdots (2\lambda+2\ell-2)}{1 \cdot 3 \cdot 5 \cdots (2\lambda+2\ell-1)}$$

$$= \frac{2^{\lambda+2\ell}\pi^\ell \lambda!(\lambda+\ell-1)!}{(2\lambda+2\ell-1)!}. \tag{9.23}$$

For SO(2ℓ) group, we have

$$C^{-2}_{(2\ell),\lambda} = 2^{1-\lambda}\pi \left(A_{2\lambda+1}A_{2\lambda+2}\right)\left(A_{2\lambda+3}A_{2\lambda+4}\right)\dots\left(A_{2\lambda+2\ell-3}A_{2\lambda+2\ell-2}\right)$$

$$= 2^{1-\lambda}\pi \left(\frac{\pi}{\lambda+1}\right)\left(\frac{\pi}{\lambda+2}\right)\cdots\left(\frac{\pi}{\lambda+\ell-1}\right) = \frac{\pi^\ell\lambda!}{2^{\lambda-1}(\lambda+\ell-1)!}.$$

$$(9.24)$$

Assuming both b and d are not equal to 1 and 2, we have

$$L^2_{12}X^\ell_1 = \ell^2 X^\ell_1, \qquad \left(L^2_{1b} + L^2_{2b}\right)X^\ell_1 = \ell X^\ell_1, \qquad L^2_{bd}X^\ell_1 = 0.$$

Thus, the eigenvalue of L^2 in the spherical harmonic function $Y^{[\lambda]}_{\mathbf{m}}(\hat{\mathbf{x}})$ is $\ell(\ell+N-2)$.

7. Calculate the complete set of the independent bases for the eigenfunctions of angular momentum in a three-body system which is spherically symmetric in an N-dimensional space.

Solution. After separating the motion of the center of mass, there are two Jacobi coordinate vectors in a three-body system, denoted by $\mathbf{R}_1 = \mathbf{x}$ and $\mathbf{R}_2 = \mathbf{y}$. Thus, the eigenfunction of angular momentum in the system belongs to an irreducible representation, denoted by a Young pattern with only one or two rows, $[\lambda_1, \lambda_2, 0, \dots, 0] = [\lambda_1, \lambda_2]$. In a three-dimensional space, the eigenfunction of angular momentum is described by the representation D^ℓ and the magnetic quantum number m. But, in the N-dimensional space, $N > 3$, the eigenfunction of angular momentum is described by the representation $[\lambda_1, \lambda_2]$ and the weight $\mathbf{m}$. Due to the spherical symmetry, we only need to calculate the eigenfunction of angular momentum with the highest weight, namely, $m = \ell$ in the three-dimensional space and $\mathbf{m} = \mathbf{M} = (\lambda_1 - \lambda_2, \lambda_2, 0, \dots, 0)$ in the N-dimensional space. The remaining eigenfunctions can be calculated by the lowering operators $F_\mu(L)$, just like we have done in Problems 4 and 5. Therefore, the eigenfunction of angular momentum with the highest weight in the N-dimensional space is described by two parameters λ_1 and λ_2.

Denote by $Q^{\lambda_1\lambda_2}_q(\mathbf{x}, \mathbf{y})$ the independent bases of the eigenfunctions of angular momentum with the highest weight, where q is understood as an ordinal parameter. What the complete set means is that any eigenfunction $\psi^{[\lambda_1,\lambda_2]}_{\mathbf{M}}(\mathbf{x}, \mathbf{y})$ with the angular momentum $[\lambda_1, \lambda_2]$ and the highest weight $\mathbf{M}$ can be expanded with respect to $Q^{\lambda_1\lambda_2}_q(\mathbf{x}, \mathbf{y})$,

$$\psi^{[\lambda_1,\lambda_2]}_{\mathbf{M}}(\mathbf{x}, \mathbf{y}) = \sum_q \phi_q(\xi)Q^{\lambda_1\lambda_2}_q(\mathbf{x}, \mathbf{y}), \qquad (9.25)$$

where the coefficients $\phi_q(\xi)$ only depend upon the internal invariants ξ. In the three-body system there are three internal variables: $\xi_1 = \mathbf{x} \cdot \mathbf{x}$, $\xi_2 = \mathbf{y} \cdot \mathbf{y}$ and $\xi_3 = \mathbf{x} \cdot \mathbf{y}$. What the independent bases means is that any basis cannot be expanded with respect to the other bases like Eq. (9.25). Obviously, a basis is not a new independent basis if it is a product of another basis and a function of the internal variables.

Consider the product of two spherical harmonic functions $Y_{\mathbf{m}}^{[q]}(\hat{\mathbf{x}})$ and $Y_{\mathbf{m}'}^{[p]}(\hat{\mathbf{y}})$. It belongs to the direct product $[q] \times [p]$ of two irreducible representations, which is usually a reducible one. Its reduction can be calculated by the Littlewood-Richardson rule and the contraction of the components of $\mathbf{x}$ and $\mathbf{y}$. In the contraction the factor of the internal variables appears. Therefore, the independent bases of angular momentum come from the calculation result by the Littlewood-Richardson rule.

$$[q] \times [p] \simeq \bigoplus_{t=0}^{\min\{q,p\}} \bigoplus_{\mu=0}^{\min\{q,p\}-t} [q + p - \mu - 2t, \mu].$$

The representations with $t > 0$ correspond to the bases which are not independent. Conversely, any independent basis of angular momentum $[\lambda_1, \lambda_2]$ can be calculated by the product $Y_{\mathbf{m}}^{[q]}(\hat{\mathbf{x}})Y_{\mathbf{m}'}^{[p]}(\hat{\mathbf{y}})$ where $\lambda_2 \leq q \leq \lambda_1$ and $p = \lambda_1 + \lambda_2 - q$. Note that the highest weight state corresponds to the standard tensor Young tableau where each box in the first row is filled with 1 and each box in the second row is filled with 2, and the tensor indices in the same column of the tensor Young tableau are antisymmetric. Now, the digit 1 means $X_1 = c(x_1 + ix_2)$ or $Y_1 = c(y_1 + iy_2)$ and the digit 2 means $X_2 = -c(x_3 + ix_4)$ or $Y_2 = -c(y_3 + iy_4)$, where c is a constant factor. Up to the normalization factor, the independent basis state $Q_q^{\lambda_1\lambda_2}(\mathbf{x}, \mathbf{y})$ is expressed as a product of three parts:

$$Q_q^{\lambda_1\lambda_2}(\mathbf{x}, \mathbf{y}) = (X_1Y_2 - X_2Y_1)^{\lambda_2} \cdot X_1^{q-\lambda_2} \cdot Y_1^{\lambda_1-q}, \qquad \lambda_2 \leq q \leq \lambda_1. \quad (9.26)$$

For SO(4) group the irreducible representation denoted by a two-row Young pattern can be self-dual or anti-self-dual. The standard tensor Young tableau for the highest weight state of the self-dual representation is still in the form (9.26), but that of the anti-self-dual representation is different. In fact, each box in the second row of that standard tensor Young tableau for the highest weight state of the anti-self-dual representation is filled with 3, not 2. In this case, the independent basis state $Q_q^{\lambda_1 \lambda_2}(\mathbf{x}, \mathbf{y})$ is expressed as

$$Q_q^{\lambda_1 \lambda_2}(\mathbf{x}, \mathbf{y}) = (X_1 Y_3 - X_3 Y_1)^{\lambda_2} \cdot X_1^{q - \lambda_2} \cdot Y_1^{\lambda_1 - q}, \qquad \lambda_2 \leq q \leq \lambda_1$$

$$X_1 = c(x_1 + ix_2) \quad Y_1 = c(y_1 + iy_2), \tag{9.27}$$

$$X_3 = -c(x_3 - ix_4), \qquad Y_3 = -c(y_3 - iy_4).$$

For SO(3) group, due to the traceless condition, the sum of the numbers of boxes in the first two columns of the Young pattern describing the irreducible representation is not larger than 3, namely, $\lambda_2 = 0$ or 1. On the other hand, the representation $[\lambda, 1]$ for SO(3) is equivalent to the representation $[\lambda, 0]$. But, the bases of angular momentum for $[\lambda, 0]$ and $[\lambda, 1]$ have different parities.

9.2 Spinor Representations of SO(N)

★ Let N matrices γ_a satisfy the anti-commutation relations,

$$\{\gamma_a, \ \gamma_b\} = \gamma_a \gamma_b + \gamma_b \gamma_a = 2\delta_{ab}\mathbf{1}, \qquad a, \ b \leq N. \tag{9.28}$$

The set of all their products constitutes a finite group, denoted by Γ_N. Γ_N is a matrix group. Usually, the matrix γ_a is defined by its representation matrix in a faithful unitary irreducible representation of Γ_N. The matrices γ_a, called the irreducible γ_a matrices, are unitary and hermitian. When $N = 2\ell$ is even, define

$$\gamma_f = (-i)^{N/2} \gamma_1 \gamma_2 \ldots \gamma_N. \tag{9.29}$$

γ_f is also unitary and hermitian, and γ_f is anticommutable with any matrix γ_a. Thus, the matrix γ_f and N matrices γ_a satisfy the anti-commutation relations (9.28). They are regarded as the definition of γ_a in Γ_{N+1}. Note that

$$\gamma_1 \gamma_2 \ldots \gamma_N \gamma_f = (i)^{N/2} \mathbf{1}. \tag{9.30}$$

The left-hand side of Eq. (9.30) changes its sign if one transposes arbitrary two matrices γ_a and γ_b, or to change the sign of arbitrary one matrix γ_a. When $N = 4m + 1$, the right-hand side of Eq. (9.30) is not a new element, the group Γ_{4m+1} is isomorphic onto the group Γ_{4m}. When $N = 4m - 1$, the right-hand side of Eq. (9.30) is a new element, the set Γ_{4m-2} and its product with i is isomorphic onto the group Γ_{4m-1}.

$$\Gamma_{4m+1} \approx \Gamma_{4m}, \qquad \Gamma_{4m-1} \approx \{\Gamma_{4m-2},\, i\Gamma_{4m-2}\}. \tag{9.31}$$

When $N = 2\ell$, there are $2^{2\ell+1}$ elements in Γ_N. For any element of Γ_N, except for the constant matrix, there exists at least one element which is anti-commutable with the given element. Thus, the trace of the element is zero. The dimension $d^{(2\ell)}$ of the irreducible γ_a matrices which satisfy Eq. (9.28) can be calculated to be 2^ℓ by the character formula. The matrices in $\Gamma_{2\ell}$, neglecting their signs, are linear independent, and form a complete set of bases for the 2^ℓ-dimensional matrices.

The equivalent theorem says that any two sets of irreducible γ_a matrices which respectively satisfy Eq. (9.28) are equivalent to each other. Namely, they can be related one-to-one by a similarity transformation X:

$$\overline{\gamma}_a = X^{-1}\gamma_a X, \qquad 1 \le a \le 2\ell. \tag{9.32}$$

When $N = 2\ell + 1$, the dimension of matrix γ_a is still 2^ℓ. The equivalent condition for two sets of the irreducible γ_a matrices should include, in addition to Eq. (9.28), that two products $\gamma_1 \ldots \gamma_N$ are equal to each other.

★ The equivalent theorem of γ_a plays an important role in the spinor theory of $SO(N)$. The charge conjugate matrix C and the space-time inversion matrix B in particle physics can be defined based on it. When $N = 2\ell$,

$$
\begin{aligned}
C^{-1}\gamma_a C &= -(\gamma_a)^T, & C^\dagger C &= \mathbf{1}, \\
\det C &= 1, & C^T &= (-1)^{\ell(\ell+1)/2} C, \\
B^{-1}\gamma_a B &= (\gamma_a)^T, & B^\dagger B &= \mathbf{1}, \\
\det B &= 1, & B^T &= (-1)^{\ell(\ell-1)/2} B.
\end{aligned}
$$

Due to the condition (9.30), only matrix C can be defined when $N = 4m-1$, and only matrix B can be defined when $N = 4m + 1$.

The fundamental spinor representation $D^{[s]}(R)$ of $SO(N)$, which is simply called the spinor representation in literature, is also defined based on the equivalent theorem:

$$
\begin{aligned}
D^{[s]}(R^{-1})\gamma_a D^{[s]}(R) &= \sum_{b=1}^{N} R_{ab}\gamma_b, & R &\in SO(N), \\
C^{-1}D^{[s]}(R)C = \left\{D^{[s]}(R^{-1})\right\}^T = D^{[s]}(R)^*, & & \text{when } N \ne 4m+1, \\
B^{-1}D^{[s]}(R)B = \left\{D^{[s]}(R^{-1})\right\}^T = D^{[s]}(R)^*, & & \text{when } N = 4m+1.
\end{aligned}
\tag{9.33}
$$

Its generators $S_{ab} = -i\gamma_a\gamma_b/2$, $a < b$, are called the spinor operators. The fundamental spinor representation, being a group, is homomorphic onto $\mathrm{SO}(N)$ by 2:1 correspondence. When N is odd, the fundamental spinor representation $D^{[s]}(R)$, or denoted by $[s]$, is irreducible. $[s]$ is real when $N = 8k \pm 1$, but self-conjugate when $N = 8k \pm 3$. Its dimension is $d_{[s]} = 2^{(N-1)/2}$. The fundamental spinor representation $[s]$ is reducible when N is even. It can be reduced into two inequivalent representations, denoted by $D^{[\pm s]}(R)$ or $[\pm s]$, $[s] \simeq [+s] \oplus [-s]$, with the same dimension $d_{[\pm s]} = 2^{(N/2)-1}$. $[\pm s]$ are conjugate to each other when $N = 4k + 2$. $[\pm s]$ are real when $N = 8k$ and self-conjugate when $N = 8k + 4$.

★ In $\mathrm{SO}(N)$ transformation, Ψ is called the fundamental spinor of $\mathrm{SO}(N)$ if it transforms by the fundamental spinor representation $D^{[s]}(R)$:

$$O_R\Psi = D^{[s]}(R)\Psi, \qquad R \in \mathrm{SO}(N),$$

where Ψ is a column matrix with $d_{[s]}$ components. When N is even, Ψ can be decomposed into two parts by the project operators $P_\pm = (1/2)(1 \pm \gamma_f)$, $\Psi_\pm = P_\pm\Psi$.

★ The spin-tensor is a spinor with the tensor subscripts,

$$O_R\Psi_{a_1\ldots a_n} = \sum_{b_a\ldots b_n} R_{a_1 b_1} \ldots R_{a_n b_n} D(R)\Psi_{b_1\ldots b_n}.$$

The spin-tensor space is reducible. In its minimal invariant subspace, the spin-tensor has to satisfy the usual traceless condition and the second traceless condition:

$$\sum_b \psi_{a\cdots b\cdots b\cdots c} = 0, \qquad \sum_b \gamma_b\psi_{a\cdots b\cdots c} = 0, \tag{9.34}$$

and is projected by the Young operator. The Young pattern for the Young operator has the row number not larger than $N/2$, otherwise the subspace corresponds to the null space. The representation corresponding to the minimal invariant subspace is called the irreducible spinor-tensor representation or simply the spinor representation, which are described by a Young diagram with a symbol s: $[s, \lambda] = [s, \lambda_1, \lambda_2, \cdots]$ for an odd N or $[\pm s, \lambda] = [\pm s, \lambda_1, \lambda_2, \cdots]$ for an even N. The spinor representation is double-valued one of $\mathrm{SO}(N)$.

★ The dimension of a spinor representation $[s, \lambda]$ of $\mathrm{SO}(2\ell + 1)$ or $[\pm s, \lambda]$ of $\mathrm{SO}(2\ell)$ can be calculated by the hook rule. In this rule the dimension is expressed as a quotient, where the numerator and the denominator are denoted by the symbols $Y_S^{[\lambda]}$ and $Y_h^{[\lambda]}$, respectively:

$$d_{[s,\lambda]}(\mathrm{SO}(2\ell+1)) = d_{[s]}(\mathrm{SO}(2\ell+1)) \, \frac{Y_S^{[\lambda]}}{Y_h^{[\lambda]}},$$

$$d_{[\pm s,\lambda]}(\mathrm{SO}(2\ell)) = d_{[\pm s]}(\mathrm{SO}(2\ell)) \, \frac{Y_S^{[\lambda]}}{Y_h^{[\lambda]}}. \qquad (9.35)$$

We still use the concept of the hook path (i, j) in the Young pattern $[\lambda]$, which enters the Young pattern at the rightmost of the ith row, goes leftwards in the i row, turns downwards at the j column, goes downwards in the j column, and leaves from the Young pattern at the bottom of the j column. The inverse path $\overline{(i, j)}$ is the same as the hook path (i, j) except for the opposite direction. The number of boxes contained in the hook path (i, j) is the hook number h_{ij} of the box in the jth column of the ith row. $Y_h^{[\lambda]}$ is a tableau of the Young pattern $[\lambda]$ where the box in the jth column of the ith row is filled with the hook number h_{ij}. Define a series of the tableaux $Y_{S_a}^{[\lambda]}$ recursively by the rule given below. $Y_S^{[\lambda]}$ is a tableau of the Young pattern $[\lambda]$ where each box is filled with the sum of the digits which are respectively filled in the same box of each tableau $Y_{S_a}^{[\lambda]}$ in the series. The symbol $Y_S^{[\lambda]}$ means the product of the filled digits in it, so does the symbol $Y_h^{[\lambda]}$.

The tableaux $Y_{S_a}^{[\lambda]}$ are defined by the following rule:

(a) $Y_{S_0}^{[\lambda]}$ is a tableau of the Young pattern $[\lambda]$ where the box in the jth column of the ith row is filled with the digit $(N - 1 + j - i)$.

(b) Let $[\lambda^{(1)}] = [\lambda]$. Beginning with $[\lambda^{(1)}]$, we define recursively the Young pattern $[\lambda^{(a)}]$ by removing the first row and the first column of the Young pattern $[\lambda^{(a-1)}]$ until $[\lambda^{(a)}]$ contains less than two rows.

(c) If $[\lambda^{(a)}]$ contains more than one row, define $Y_{S_a}^{[\lambda]}$ to be a tableau of the Young pattern $[\lambda]$ where the boxes in the first $(a - 1)$ row and in the first $(a - 1)$ column are filled with 0, and the remaining part of the Young pattern is nothing but $[\lambda^{(a)}]$. Let $[\lambda^{(a)}]$ be r row. Fill the first $(r - 1)$ boxes along the hook path $(1, 1)$ of the Young pattern $[\lambda^{(a)}]$, beginning with the box on the rightmost, with the digits $\lambda_2^{(a)}$, $\lambda_3^{(a)}$,

$\cdots$, $\lambda_r^{(a)}$, box by box, and fill the first $\lambda_i^{(a)}$ boxes in each inverse path $\overline{(i,\ 1)}$ of the Young pattern $[\lambda^{(a)}]$, $2 \le i \le r$, with -1. The remaining boxes are filled with 0. If a few -1 are filled in the same box, the digits are summed. The sum of all filled digits in the pattern $Y_{S_a}^{[\lambda]}$ is zero.

8. Calculate the dimension of the irreducible spinor representation of the SO(7) group denoted by the following Young patterns:

$$(1)\ [s,4,2], \qquad (2)\ [s,3,2], \qquad (3)\ [s,4,4],$$
$$(4)\ [s,3,1,1], \qquad (5)\ [s,3,2,2].$$

Solution.

(1): $d_{[s,4,2]}(SO(7)) = 8 \times \left\{ \dfrac{\begin{smallmatrix} 6\ 7\ 8\ 9 \\ 5\ 6 \end{smallmatrix} + \begin{smallmatrix} 0 \quad 0 \quad 0\ 2 \\ -1\ -1 \end{smallmatrix}}{\begin{smallmatrix} 5\ 4\ 2\ 1 \\ 2\ 1 \end{smallmatrix}} \right\} = 8 \times \left\{ \dfrac{\begin{smallmatrix} 6\ 7\ 8\ 11 \\ 4\ 5 \end{smallmatrix}}{\begin{smallmatrix} 5\ 4\ 2\ 1 \\ 2\ 1 \end{smallmatrix}} \right\} = 7392.$

(2): $d_{[s,3,2]}(SO(7)) = 8 \times \left\{ \dfrac{\begin{smallmatrix} 6\ 7\ 8 \\ 5\ 6 \end{smallmatrix} + \begin{smallmatrix} 0 \quad 0 \quad 2 \\ -1\ -1 \end{smallmatrix}}{\begin{smallmatrix} 4\ 3\ 1 \\ 2\ 1 \end{smallmatrix}} \right\} = 8 \times \left\{ \dfrac{\begin{smallmatrix} 6\ 7\ 10 \\ 4\ 5 \end{smallmatrix}}{\begin{smallmatrix} 4\ 3\ 1 \\ 2\ 1 \end{smallmatrix}} \right\} = 2800.$

(3): $d_{[s,4,3]}(SO(7)) = 8 \times \left\{ \dfrac{\begin{smallmatrix} 6\ 7\ 8\ 9 \\ 5\ 6\ 7 \end{smallmatrix} + \begin{smallmatrix} 0 \quad 0 \quad 0\ 3 \\ -1\ -1\ -1 \end{smallmatrix}}{\begin{smallmatrix} 5\ 4\ 3\ 1 \\ 3\ 2\ 1 \end{smallmatrix}} \right\} = 8 \times \left\{ \dfrac{\begin{smallmatrix} 6\ 7\ 8\ 12 \\ 4\ 5\ 6 \end{smallmatrix}}{\begin{smallmatrix} 5\ 4\ 3\ 1 \\ 3\ 2\ 1 \end{smallmatrix}} \right\} = 10752.$

(4): $d_{[s,3,1,1]}(SO(7)) = 8 \times \left\{ \dfrac{\begin{smallmatrix} 6\ 7\ 8 \\ 5 \\ 4 \end{smallmatrix} + \begin{smallmatrix} 0\ 1\ 1 \\ 0 \\ -2 \end{smallmatrix}}{\begin{smallmatrix} 5\ 2\ 1 \\ 2 \\ 1 \end{smallmatrix}} \right\} = 8 \times \left\{ \dfrac{\begin{smallmatrix} 6\ 8\ 9 \\ 5 \\ 2 \end{smallmatrix}}{\begin{smallmatrix} 5\ 2\ 1 \\ 2 \\ 1 \end{smallmatrix}} \right\} = 1728.$

(5): $d_{[s,3,2,2]}(SO(7)) = 8 \times \left\{ \dfrac{\begin{smallmatrix} 6\ 7\ 8 \\ 5\ 6 \\ 4\ 5 \end{smallmatrix} + \begin{smallmatrix} 0\ 2\ 2 \\ -1\ 0 \\ -2\ -1 \end{smallmatrix} + \begin{smallmatrix} 0\ 0\ 0 \\ +0\ 1 \\ 0\ -1 \end{smallmatrix}}{\begin{smallmatrix} 5\ 4\ 1 \\ 3\ 2 \\ 2\ 1 \end{smallmatrix}} \right\} = 8 \times \left\{ \dfrac{\begin{smallmatrix} 6\ 9\ 10 \\ 4\ 7 \\ 2\ 3 \end{smallmatrix}}{\begin{smallmatrix} 5\ 4\ 1 \\ 3\ 2 \\ 2\ 1 \end{smallmatrix}} \right\} = 3024.$

9. Calculate the highest weight of the fundamental spinor representations of the SO(N)

Solution. From the second Lie theorem, the spinor operators S_{ab} satisfy the same commutation relations as those of the generators T_{ab} in the self representation of SO(N). The Chevalley bases $H_\mu(S)$, $E_\mu(S)$ and $F_\mu(S)$ in the fundamental spinor representation can also be expressed by Eqs. (9.10) and (9.11), where T_{ab} are replaced with S_{ab}. It is convenient to choose an explicit forms of the γ_a matrices for calculating the highest weight of the fundamental spinor representations, although the calculated result is independent of the chosen forms. Since the dimension of γ_a for the SO(2ℓ) group is 2^ℓ, we may define the γ_a matrices of $\Gamma_{2\ell}$ to be the direct product of ℓ Pauli matrices such that Eq. (9.28) is satisfied:

$$\gamma_{2\mu-1} = \underbrace{\mathbf{1} \times \ldots \times \mathbf{1}}_{\mu-1} \times \sigma_1 \times \underbrace{\sigma_3 \times \ldots \times \sigma_3}_{\ell-\mu} \,,$$

$$\gamma_{2\mu} = \underbrace{\mathbf{1} \times \ldots \times \mathbf{1}}_{\mu-1} \times \sigma_2 \times \underbrace{\sigma_3 \times \ldots \times \sigma_3}_{\ell-\mu} \,, \qquad (9.36)$$

$$\gamma_f = \underbrace{\sigma_3 \times \cdots \times \sigma_3}_{\ell} \,.$$

In fact, they are also γ_a matrices of $\Gamma_{2\ell+1}$ with $\gamma_{2\ell+1} = \gamma_f$. Thus, the Chevalley bases in the fundamental spinor representation of SO($2\ell + 1$), whose Lie algebra is B_ℓ, are

$$H_\mu = \underbrace{\mathbf{1} \times \ldots \times \mathbf{1}}_{\mu-1} \times \mathrm{diag}\,\{0,\ 1,\ -1,\ 0\} \times \underbrace{\mathbf{1} \times \ldots \times \mathbf{1}}_{\ell-\mu-1} \,,$$

$$E_\mu = \underbrace{\mathbf{1} \times \ldots \times \mathbf{1}}_{\mu-1} \times \{\sigma_+ \times \sigma_-\} \times \underbrace{\mathbf{1} \times \ldots \times \mathbf{1}}_{\ell-\mu-1} \,,$$

$$F_\mu = \underbrace{\mathbf{1} \times \ldots \times \mathbf{1}}_{\mu-1} \times \{\sigma_- \times \sigma_+\} \times \underbrace{\mathbf{1} \times \ldots \times \mathbf{1}}_{\ell-\mu-1} \,, \qquad (9.37)$$

$$H_\ell = \underbrace{\mathbf{1} \times \ldots \times \mathbf{1}}_{\ell-1} \times \sigma_3 \,,$$

$$E_\ell = \underbrace{\sigma_3 \times \cdots \times \sigma_3}_{\ell-1} \times \sigma_+ \,, \qquad F_\ell = \underbrace{\sigma_3 \times \cdots \times \sigma_3}_{\ell-1} \times \sigma_- \,,$$

where $1 \leq \mu < \ell$. The Chevalley bases in the fundamental spinor representation of the SO(2ℓ) group, whose Lie algebra is D_ℓ, are the same as those of the SO($2\ell + 1$) group except for those when $\mu = \ell$, which are

$$H_\ell = \underbrace{\mathbf{1} \times \ldots \times \mathbf{1}}_{\ell-2} \times \mathrm{diag}\,\{1,\ 0,\ 0,\ -1\} \,,$$

$$E_\ell = -\underbrace{\mathbf{1} \times \ldots \times \mathbf{1}}_{\ell-2} \times \{\sigma_+ \times \sigma_+\} \,, \qquad (9.38)$$

$$F_\ell = -\underbrace{\mathbf{1} \times \ldots \times \mathbf{1}}_{\ell-2} \times \{\sigma_- \times \sigma_-\} \,.$$

Let α and β be two eigenfunctions of σ_3, respectively,

$$\alpha = \begin{pmatrix} 1 \\ 0 \end{pmatrix}, \qquad \beta = \begin{pmatrix} 0 \\ 1 \end{pmatrix},$$

$$\sigma_3\alpha = \alpha, \qquad \sigma_+\alpha = 0, \qquad \sigma_-\alpha = \beta,$$

$$\sigma_3\beta = -\beta, \qquad \sigma_+\beta = \alpha, \qquad \sigma_-\beta = 0.$$

Denote by $\chi(\mathbf{m})$ the basis states with the weight $\mathbf{m}$ in the fundamental spinor representation $[s]$ of $SO(2\ell+1)$ and by $\chi_\pm(\mathbf{m})$ the basis states with the weight $\mathbf{m}$ in the fundamental spinor representation $[\pm s]$ of $SO(2\ell)$. From the condition that each raising operator annihilates the highest weight state, we obtain the basis state $\chi(\mathbf{M})$ for the highest weight for $[s]$ of $SO(2\ell+1)$ and the basis state $\chi_\pm(\mathbf{M})$ for the highest weight for $[\pm s]$ of $SO(2\ell)$ as follows:

$$\chi(\mathbf{M}) = \underbrace{\alpha \times \cdots \times \alpha}_{\ell},$$

$$\mathbf{M} = (0,\, 0,\, \ldots,\, 0, 1), \qquad \text{for } [s] \text{ of } SO(2\ell+1),$$

$$\chi_+(\mathbf{M}) = \underbrace{\alpha \times \cdots \times \alpha}_{\ell}, \tag{9.39}$$

$$\mathbf{M} = (0,\, 0,\, \ldots,\, 0, 1), \qquad \text{for } [+s] \text{ of } SO(2\ell),$$

$$\chi_-(\mathbf{M}) = \underbrace{\alpha \times \cdots \times \alpha}_{\ell-1} \times \beta,$$

$$\mathbf{M} = (0,\, 0,\, \ldots,\, 1, 0), \qquad \text{for } [-s] \text{ of } SO(2\ell).$$

10. Calculate the highest weight of the spinor representation $[s, \lambda]$ of $SO(2\ell+1)$, and the highest weights of the spinor representations $[\pm s, \lambda]$ of $SO(2\ell)$

Solution. The spin-tensor of $SO(N)$ corresponds to the irreducible representation $[s, \lambda]$ when $N = 2\ell+1$ or to $[\pm s, \lambda]$ when $N = 2\ell$ if the spin-tensor satisfy the usual traceless condition, and the second traceless condition, given in Eq. (9.34). The generators in the irreducible spinor representation are the total angular momentum operators $J_{ab} = L_{ab} + S_{ab}$, $a < b$. The orbital angular momentum operators L_{ab} and the spinor angular momentum operators S_{ab} are respectively applied to the tensor part and the spinor part of the spin-tensor. The total angular momentum operators J_{ab} satisfy the same commutation relations as those of L_{ab} and S_{ab}. The Chevalley bases $H_\mu(J)$, $E_\mu(J)$ and $F_\mu(J)$ in the irreducible representation $[s, \lambda]$ can also be expressed by Eqs. (9.10) and (9.11), where T_{ab} are replaced with

J_{ab}. Therefore, the weight of a spin-tensor basis is the sum of the weights of the tensor part and the spinor part. Let $[\lambda] = [\lambda_1, \lambda_2, \ldots, \lambda_\ell]$. The highest weight $\mathbf{M}$ of the spinor representation $[s, \lambda]$ of $SO(2\ell + 1)$ is

$$M_\mu = \lambda_\mu - \lambda_{\mu+1}, \qquad 1 \leq \mu < \ell, \qquad M_\ell = 2\lambda_\ell + 1.$$

The highest weight $\mathbf{M}$ of the spinor representation $[+s, \lambda]$ of $SO(2\ell)$ is

$$M_\mu = \lambda_\mu - \lambda_{\mu+1}, \qquad 1 \leq \mu \leq \ell - 2,$$
$$M_{\ell-1} = \lambda_{\ell-1} - \lambda_\ell, \qquad M_\ell = \lambda_{\ell-1} + \lambda_\ell + 1.$$

The highest weight $\mathbf{M}$ of the spinor representation $[-s, \lambda]$ of $SO(2\ell)$ is

$$M_\mu = \lambda_\mu - \lambda_{\mu+1}, \qquad 1 \leq \mu \leq \ell - 2,$$
$$M_{\ell-1} = \lambda_{\ell-1} + \lambda_\ell + 1, \qquad M_\ell = \lambda_{\ell-1} - \lambda_\ell.$$

11. Calculate the Clebsch-Gordan series for the direct product of the tensor representation $[\lambda]$ and the fundamental spinor representation $[s]$ of $SO(2\ell + 1)$ or $[\pm s]$ of $SO(2\ell)$, where $[\lambda]$ is a single row Young pattern or a single column Young pattern.

Solution. If the tensor part of a spin-tensor $\Psi_{a_1 \ldots a_n}$ belongs to the tensor representation $[\lambda]$ but it does not satisfy the second traceless condition, namely, its tensor part satisfies the traceless condition for each pair of the subscripts a_i and a_j and is projected by a Young operator, then, the spin-tensor space corresponds to the direct product of the tensor representation $[\lambda]$ and the fundamental spinor representation. The method for reducing the direct product representation is to decompose the spin-tensor space into a series of subspaces satisfying the second traceless condition (9.34). In fact, $\sum_b \gamma_b \Psi_{a_1 \ldots b \ldots a_n}$ is a new spin-tensor where the tensor rank decreases by one. If $[\lambda]$ is a single row Young pattern, the tensor part of the spin-tensor is a totally symmetric traceless tensor such that twice application of the second traceless condition is trivial, $\sum_{bc} \gamma_b \gamma_c \Psi_{a_1 \ldots b \ldots c \ldots a_n} = 0$ due to Eq. (9.28). If $[\lambda]$ is a single column Young pattern, the tensor part of the spin-tensor is a totally antisymmetric tensor. Note that the antisymmetric tensor space of rank n of $SO(N)$ is equivalent to that of rank $N - n$. Thus, the reductions of the direct product representations for $SO(2\ell + 1)$ are

$$[\lambda, 0, \ldots, 0] \times [s] \simeq [s, \lambda, 0, \ldots, 0] \oplus [s, (\lambda - 1), 0, \ldots, 0],$$
$$[1^n] \times [s] \simeq [s, 1^n] \oplus [s, 1^{n-1}] \oplus \ldots \oplus [s], \qquad n \leq \ell \,, \tag{9.40}$$

and those for $SO(2\ell)$ are

$$[\lambda, 0, \ldots, 0] \times [\pm s] \simeq [\pm s, \lambda, 0, \ldots, 0] \oplus [\mp s, (\lambda - 1), 0, \ldots, 0],$$

$$[1^n] \times [\pm s] \simeq [\pm s, 1^n] \oplus [\mp s, 1^{n-1}] \oplus [\pm s, 1^{n-2}] \oplus \ldots, \quad n < \ell,$$

$$[\pm 1^\ell] \times [\pm s] \simeq [\pm s, 1^\ell] \oplus [\pm s, 1^{\ell-2}] \oplus [\pm s, 1^{\ell-4}] \oplus \ldots,$$

$$[\pm 1^\ell] \times [\mp s] \simeq [\pm s, 1^{\ell-1}] \oplus [\pm s, 1^{\ell-3}] \oplus [\pm s, 1^{\ell-5}] \oplus \ldots.$$

(9.41)

The sign in $\pm s$ changes because $P_\pm \gamma_\mu = \gamma_\mu P_\mp$ where $P_\pm = (\mathbf{1} \pm \gamma_f)/2$. Those formulas can be shown with the method of the dominant weight diagram. The reader is encouraged to check them for the groups SO(7) and SO(8). Equations (9.40) and (9.41) can be partly checked by their dimensions. From Eqs. (9.18) and (9.35), the dimensions of $[\lambda]$ and $[s, \lambda]$ of SO($2\ell + 1$), for example, are

$$d_{[\lambda,0,\ldots,0]} = \frac{(2\ell)(2\ell + 1) \ldots (2\ell + \lambda - 2)(2\ell + 2\lambda - 1)}{\lambda!}$$

$$= \frac{(2\ell + \lambda - 2)!(2\ell + 2\lambda - 1)}{(2\ell - 1)!\lambda!},$$

$$d_{[1^n]} = \frac{(2\ell + 1)!}{(2\ell - n + 1)!n!}, \quad \text{when } n \leq \ell,$$

$$d_{[s,\lambda,0,\ldots,0]} = 2^\ell \frac{(2\ell)(2\ell + 1) \ldots (2\ell + \lambda - 1)}{\lambda!} = 2^\ell \frac{(2\ell + \lambda - 1)!}{(2\ell - 1)!\lambda!},$$

$$d_{[s,1^n]}(\text{SO}(2\ell + 1)) = 2^\ell \frac{(2\ell + 1)(2\ell) \ldots (2\ell - n + 3)(2\ell - 2n + 2)}{n!}$$

$$= 2^\ell \frac{(2\ell + 1)!(2\ell - 2n + 2)}{(2\ell - n + 2)!n!}, \quad \text{when } n \leq \ell.$$

Then, for Eq. (9.40) we have

$$2^\ell \frac{(2\ell + \lambda - 2)!(2\ell + 2\lambda - 1)}{(2\ell - 1)!\lambda!} = 2^\ell \frac{(2\ell + \lambda - 1)!}{(2\ell - 1)!\lambda!} + 2^\ell \frac{(2\ell + \lambda - 2)!}{(2\ell - 1)!(\lambda - 1)!},$$

$$2^\ell \frac{(2\ell + 1)!}{(2\ell + 1 - n)!n!} = 2^\ell \sum_{a=0}^{n} \frac{(2\ell + 1)!(2\ell - 2a + 2)}{(2\ell - a + 2)!a!}, \quad \text{when } n \leq \ell.$$

12. Calculate the basis states in the fundamental spinor representation $[s]$ of SO(7).

Solution. The Lie algebra of the SO(7) group is B_3. The highest weight of the fundamental spinor representation $[s]$ of SO(7) is $(0, 0, 1)$. Its dimension is 8. Its block weight diagram is as follows. All nonvanishing matrix entries of the lowering operators F_μ are 1.

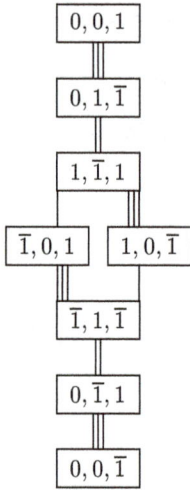

Fig. 9.3 The block weight
diagram of $[s]$ of B_3.

By the notation used in Problem 9,
we have

$$\chi(0,0,1) = \alpha \times \alpha \times \alpha,$$
$$\chi(0,1,\bar{1}) = \alpha \times \alpha \times \beta,$$
$$\chi(1,\bar{1},1) = \alpha \times \beta \times \alpha,$$
$$\chi(\bar{1},0,1) = \beta \times \alpha \times \alpha,$$
$$\chi(1,0,\bar{1}) = -\alpha \times \beta \times \beta,$$
$$\chi(\bar{1},1,\bar{1}) = -\beta \times \alpha \times \beta,$$
$$\chi(0,\bar{1},1) = -\beta \times \beta \times \alpha,$$
$$\chi(0,0,\bar{1}) = -\beta \times \beta \times \beta.$$

13. Calculate the basis states in the fundamental spinor representations $[\pm s]$ of SO(8).

Solution. The Lie algebra of SO(8) group is D_4. The highest weights of the fundamental spinor representations $[+s]$ and $[-s]$ of SO(8) are $(0,0,0,1)$ and $(0,0,1,0)$, respectively. The dimensions of both representations are 8. Their block weight diagrams are as follows. All nonvanishing matrix entries of the lowering operators F_μ are 1.

By the notation used in Problem 9, we have

$$\chi_+(0,0,0,1) = \alpha \times \alpha \times \alpha \times \alpha, \qquad \chi_-(0,0,1,0) = \alpha \times \alpha \times \alpha \times \beta,$$
$$\chi_+(0,1,0,\bar{1}) = -\alpha \times \alpha \times \beta \times \beta, \qquad \chi_-(0,1,\bar{1},0) = \alpha \times \alpha \times \beta \times \alpha,$$
$$\chi_+(1,\bar{1},1,0) = -\alpha \times \beta \times \alpha \times \beta, \qquad \chi_-(1,\bar{1},0,1) = \alpha \times \beta \times \alpha \times \alpha,$$
$$\chi_+(\bar{1},0,1,0) = -\beta \times \alpha \times \alpha \times \beta, \qquad \chi_-(\bar{1},0,0,1) = \beta \times \alpha \times \alpha \times \alpha,$$
$$\chi_+(1,0,\bar{1},0) = -\alpha \times \beta \times \beta \times \alpha, \qquad \chi_-(1,0,0,\bar{1}) = -\alpha \times \beta \times \beta \times \beta,$$
$$\chi_+(\bar{1},1,\bar{1},0) = -\beta \times \alpha \times \beta \times \alpha, \qquad \chi_-(\bar{1},1,0,\bar{1}) = -\beta \times \alpha \times \beta \times \beta,$$
$$\chi_+(0,\bar{1},0,1) = -\beta \times \beta \times \alpha \times \alpha, \qquad \chi_-(0,\bar{1},1,0) = -\beta \times \beta \times \alpha \times \beta,$$
$$\chi_+(0,0,0,\bar{1}) = \beta \times \beta \times \beta \times \beta, \qquad \chi_-(0,0,\bar{1},0) = -\beta \times \beta \times \beta \times \alpha.$$

$$\boxed{0,0,0,1}$$

$$\boxed{0,1,0,\bar{1}}$$

$$\boxed{1,\bar{1},1,0}$$

$$\boxed{\bar{1},0,1,0} \quad \boxed{1,0,\bar{1},0}$$

$$\boxed{\bar{1},1,\bar{1},0}$$

$$\boxed{0,\bar{1},0,1}$$

$$\boxed{0,0,0,\bar{1}}$$

a) $[+s]$

$$\boxed{0,0,1,0}$$

$$\boxed{0,1,\bar{1},0}$$

$$\boxed{1,\bar{1},0,1}$$

$$\boxed{\bar{1},0,0,1} \quad \boxed{1,0,0,\bar{1}}$$

$$\boxed{\bar{1},1,0,\bar{1}}$$

$$\boxed{0,\bar{1},1,0}$$

$$\boxed{0,0,\bar{1},0}$$

b) $[-s]$

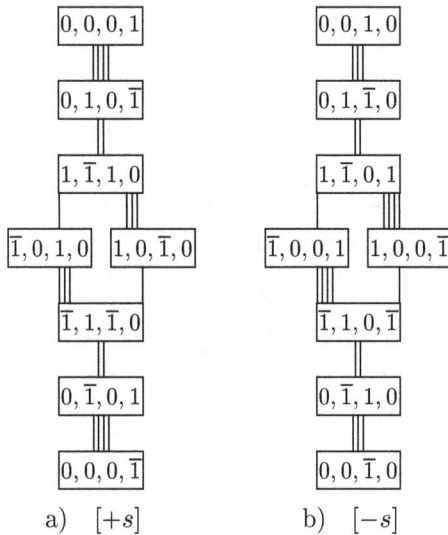

Fig. 9.4 The block weight diagrams of $[\pm s]$ of D_4.

14. Expand the eigenfunction of the total angular momentum with the highest weight in terms of the product of the spherical harmonic function $Y_{\mathbf{m}}^{[\lambda]}(\hat{\mathbf{x}})$ and the spinor basis $\chi(\mathbf{m})$.

Solution. In Problem 11 the Clebsch-Gordan series for the direct product representation $[\lambda, 0, \ldots, 0] \times [s]$ is given. For the $SO(2\ell + 1)$ group, we express the representations in Eq. (9.40) by their highest weights,

$$(\lambda, 0 \ldots, 0) \times (0 \ldots, 0, 1) \simeq (\lambda, 0 \ldots, 0, 1) \oplus (\lambda - 1, 0 \ldots, 0, 1).$$

Let $(J) = (\lambda, 0 \ldots, 0, 1)$, $(L) = (\lambda, 0 \ldots, 0)$, $(L + 1) = (\lambda + 1, 0 \ldots, 0)$, and $(S) = (0 \ldots, 0, 1)$. There are two ways to obtain the total angular momentum state (J): $(L) \times (S)$ and $(L + 1) \times (S)$. The highest weight state can be calculated by the condition that it should be annihilated by each raising operator. The highest weight state for the first case is easy to calculate:

$$\Psi_{(J)}^{(J)(L)(S)}(\hat{\mathbf{x}}) = Y_{(L)}^{(L)}(\hat{\mathbf{x}})\chi[(S)] = C_{(2\ell+1),\lambda} \left\{ \frac{(-1)^\ell (x_1 + ix_2)}{r\sqrt{2}} \right\}^\lambda \chi[(S)],$$

where $C_{(2\ell+1),\lambda}$ is given in Eq. (9.23). The highest weight state for the second case is

$$\Psi_{(J)}^{(J)(L+1)(S)}(\hat{\mathbf{x}}) = \sqrt{\frac{1}{2\ell + 2\lambda + 1}} \left\{ Y_{(\lambda,0,\ldots,0)}^{(L+1)}(\hat{\mathbf{x}})\chi[(S)] \right.$$
$$- \sqrt{2}Y_{(\lambda,0,\ldots,\bar{1},2)}^{(L+1)}(\hat{\mathbf{x}})\chi[(0,\ldots,0,1,\bar{1})]$$
$$+ \sqrt{2}Y_{(\lambda,0,\ldots,0,\bar{1},1,0)}^{(L+1)}(\hat{\mathbf{x}})\chi[(0,\ldots,0,1,\bar{1},1)] - + \ldots$$
$$- (-1)^{\ell}\sqrt{2}Y_{(\lambda-1,1,0,\ldots,0)}^{(L+1)}(\hat{\mathbf{x}})\chi[(1,\bar{1},0,\ldots,0,1)]$$
$$\left. + (-1)^{\ell}\sqrt{2(\lambda+1)}Y_{(\lambda+1,0,\ldots,0)}^{(L+1)}(\hat{\mathbf{x}}) \right\} \chi[(\bar{1},0,\ldots,0,1)]$$
$$= \frac{C_{(2\ell+1),(\lambda+1)}(-1)^{\ell\lambda}\sqrt{\lambda+1}}{(r)^{\lambda+1}2^{\lambda/2}\sqrt{2\ell+2\lambda+1}}(x_1+ix_2)^{\lambda}$$
$$\times \left\{ x_{2\ell+1}\chi[(S)] + (x_{2\ell-1}+ix_{2\ell})\chi[(0,\ldots,0,1,\bar{1})] \right.$$
$$+ (x_{2\ell-3}+ix_{2\ell-2})\chi[(0,\ldots,0,1,\bar{1},1)] + \ldots$$
$$+ (x_3+ix_4)\chi[(1,\bar{1},0,\ldots,0,1)]$$
$$\left. + (x_1+ix_2)\chi[(\bar{1},0,\ldots,0,1)] \right\}.$$

For the $SO(2\ell)$ group, we express the representations in Eq. (9.41) by their highest weights,

$$(\lambda,0\ldots,0) \times (0\ldots,0,1) \simeq (\lambda,0\ldots,0,1) \oplus (\lambda-1,0\ldots,0,1,0),$$
$$(\lambda,0\ldots,0) \times (0\ldots,0,1,0) \simeq (\lambda,0\ldots,0,1,0) \oplus (\lambda-1,0\ldots,0,1).$$

Let $(J_+) = (\lambda,0\ldots,0,1)$, $(J_-) = (\lambda,0\ldots,0,1,0)$, $(L) = (\lambda,0\ldots,0)$, $(L+1) = (\lambda+1,0\ldots,0)$, $(+S) = (0\ldots,0,1)$, and $(-S) = (0\ldots,0,1,0)$. There are two ways to obtain the total angular momentum state $(J_{\pm})$: $(L) \times (\pm S)$ and $(L+1) \times (\mp S)$. It is easy to calculate the highest weight state for the first case:

$$\Psi_{(J_{\pm})}^{(J_{\pm})(L)(\pm S)}(\hat{\mathbf{x}}) = Y_{(L)}^{(L)}(\hat{\mathbf{x}})\chi_{\pm}[(\pm S)]$$
$$= C_{(2\ell),\lambda} \left\{ \frac{(-1)^{\ell-1}(x_1+ix_2)}{r\sqrt{2}} \right\}^{\lambda} \chi_{\pm}[(\pm S)],$$

where $C_{(2\ell),\lambda}$ is given in Eq. (9.24). The highest weight state for the second case is

$$\Psi_{(J_+)}^{(J_+)(L+1)(-S)}(\hat{\mathbf{x}}) = \sqrt{\frac{1}{\ell+\lambda}} \left\{ Y_{(\lambda,0,\ldots,0,\bar{1},1)}^{(L+1)}(\hat{\mathbf{x}})\chi_{-}[(-S)] \right.$$
$$\left. - Y_{(\lambda,0,\ldots,0,\bar{1},1,1)}^{(L+1)}(\hat{\mathbf{x}})\chi_{-}[(0,\ldots,0,1,\bar{1},0)] \right.$$

$$+ Y^{(L+1)}_{(\lambda,0,\ldots,0,\bar{1},1,0,0)}(\hat{\mathbf{x}})\chi_-[(0,\ldots,0,1,\bar{1},0,1)] - + \ldots$$

$$+ (-1)^\ell Y^{(L+1)}_{(\lambda-1,1,0,\ldots,0)}(\hat{\mathbf{x}})\chi_-[(1,\bar{1},0,\ldots,0,1)]$$

$$- (-1)^\ell \sqrt{\lambda+1}\, Y^{(L+1)}_{(\lambda+1,0,\ldots,0)}(\hat{\mathbf{x}})\chi_-[(\bar{1},0,\ldots,0,1)]\Big\}$$

$$= \frac{C_{(2\ell),(\lambda+1)}(-1)^{(\ell-1)\lambda}\sqrt{\lambda+1}}{(r\sqrt{2})^{\lambda+1}\sqrt{\ell+\lambda}}(x_1+ix_2)^\lambda$$

$$\times \Big\{ (x_{2\ell-1}+ix_{2\ell})\chi_-[(-S)] + (x_{2\ell-3}+ix_{2\ell-2})\chi_-[(0,\ldots,0,1,\bar{1},0)]$$

$$+ (x_{2\ell-5}+ix_{2\ell-4})\chi_-[(0,\ldots,0,1,\bar{1},0,1)] + \ldots$$

$$+ (x_3+ix_4)\chi_-[(1,\bar{1},0,\ldots,0,1)]$$

$$+ (x_1+ix_2)\chi_-[(\bar{1},0,\ldots,0,1)]\Big\}.$$

$$\Psi^{(J_-)(L+1)(+S)}_{(J_-)}(\hat{\mathbf{x}}) = \sqrt{\frac{1}{\ell+\lambda}}\Big\{ Y^{(L+1)}_{(\lambda,0,\ldots,0,1,\bar{1})}(\hat{\mathbf{x}})\chi_+[(+S)]$$

$$- Y^{(L+1)}_{(\lambda,0,\ldots,0,\bar{1},1,1)}(\hat{\mathbf{x}})\chi_+[(0,\ldots,0,1,0,\bar{1})]$$

$$+ Y^{(L+1)}_{(\lambda,0,\ldots,0,\bar{1},1,0,0)}(\hat{\mathbf{x}})\chi_+[(0,\ldots,0,1,\bar{1},1,0)] - + \ldots$$

$$+ (-1)^\ell Y^{(L+1)}_{(\lambda-1,1,0,\ldots,0)}(\hat{\mathbf{x}})\chi_+[(1,\bar{1},0,\ldots,0,1,0)]$$

$$- (-1)^\ell \sqrt{\lambda+1}\, Y^{(L+1)}_{(\lambda+1,0,\ldots,0)}(\hat{\mathbf{x}})\chi_+[(\bar{1},0,\ldots,0,1,0)]\Big\}$$

$$= \frac{C_{(2\ell),(\lambda+1)}(-1)^{(\ell-1)\lambda}\sqrt{\lambda+1}}{(r\sqrt{2})^{\lambda+1}\sqrt{\ell+\lambda}}(x_1+ix_2)^\lambda$$

$$\times \Big\{ (x_{2\ell-1}-ix_{2\ell})\chi_+[(+S)] + (x_{2\ell-3}+ix_{2\ell-2})\chi_+[(0,\ldots,0,1,0,\bar{1})]$$

$$+ (x_{2\ell-5}+ix_{2\ell-4})\chi_+[(0,\ldots,0,1,\bar{1},1,0)] + \ldots$$

$$+ (x_3+ix_4)\chi_+[(1,\bar{1},0,\ldots,0,1,0)]$$

$$+ (x_1+ix_2)\chi_+[(\bar{1},0,\ldots,0,1,0)]\Big\}.$$

9.3 SO(4) Group and the Lorentz Group

★ The inequivalent irreducible representations of the Lorentz group can be obtained from those of SO(4) group. This is the general method for studying the inequivalent irreducible representations of a non-compact group.

★ Find new bases of generators of SO(4) from six generators T_{ab},

$$T_a^{(\pm)} = \frac{1}{2}\left(\sum_{b<c=2}^{3}\epsilon_{abc}T_{bc} \pm T_{a4}\right),$$

$$\left[T_a^{(\pm)}, T_b^{(\pm)}\right] = i\sum_{c=1}^{3}\epsilon_{abc}T_c^{(\pm)}, \qquad \left[T_a^{(+)}, T_b^{(-)}\right] = 0. \tag{9.42}$$

The new bases are separated into two sets. Two generators in different sets are commutable. Express them in the form of the direct product of the Pauli matrices:

$$T_1^{(+)} = \frac{1}{2}\sigma_2 \times \sigma_1, \quad T_2^{(+)} = \frac{-1}{2}\sigma_2 \times \sigma_3, \quad T_3^{(+)} = \frac{1}{2}\mathbf{1}_2 \times \sigma_2,$$

$$T_1^{(-)} = \frac{-1}{2}\sigma_1 \times \sigma_2, \quad T_2^{(-)} = \frac{-1}{2}\sigma_2 \times \mathbf{1}_2, \quad T_3^{(-)} = \frac{1}{2}\sigma_3 \times \sigma_2.$$

By the similarity transformation N, we have

$$N^{-1}T_a^{(+)}N = (\sigma_a/2) \times \mathbf{1}_2, \qquad N^{-1}T_a^{(-)}N = \mathbf{1}_2 \times (\sigma_a/2).$$

$$N = \frac{1}{\sqrt{2}}\begin{pmatrix} -1 & 0 & 0 & 1 \\ -i & 0 & 0 & -i \\ 0 & 1 & 1 & 0 \\ 0 & i & -i & 0 \end{pmatrix}.$$

Therefore, any element R of SO(4) can be expressed in the direct product of two matrices

$$\begin{aligned} R &= \exp\left(-i\sum_{a<b=2}^{4}\omega_{ab}T_{ab}\right) \\ &= \exp\left\{-i\sum_{a=1}^{3}\left(\omega_a^{(+)}T_a^{(+)} + \omega_a^{(-)}T_a^{(-)}\right)\right\} \\ &= \exp\left\{-i\omega^{(+)}\hat{\mathbf{n}}^{(+)} \cdot \mathbf{T}^{(+)}\right\}\exp\left\{-i\omega^{(-)}\hat{\mathbf{n}}^{(-)} \cdot \mathbf{T}^{(-)}\right\} \\ &= N\left\{u(\hat{\mathbf{n}}^{(+)}, \omega^{(+)}) \times u(\hat{\mathbf{n}}^{(-)}, \omega^{(-)})\right\}N^{-1}, \end{aligned} \tag{9.43}$$

where

$$\omega_a^{(\pm)} = \sum_{b<c=2}^{3}\epsilon_{abc}\omega_{bc} \pm \omega_{a4} = \omega^{(\pm)}n_a^{(\pm)}, \qquad \omega^{(\pm)} = \left\{\sum_{a=1}^{3}\left(\omega_a^{(\pm)}\right)^2\right\}^{1/2}.$$

Thus, matrix R is explicitly expressed in the product $u_1 \times u_2$ of two-dimensional unimodular unitary matrices. If two matrices u_1 and u_2 change their signs at the same time, R keeps invariant. Therefore, Eq. (9.43) gives a one-to-two correspondence between an element R in SO(4) and two elements u_1 and u_2 in SU(2). Furthermore, this correspondence keeps invariant in the multiplication of the group elements. Thus,

$$SO(4) \sim SU(2) \times SU(2)'. \tag{9.44}$$

We restrict the varied area of the parameters of the SO(4) group such that there is a one-to-one correspondence between a set of parameters and a group element, at least in the region with nonzero measure:

$$0 \le \omega^{(+)} \le 2\pi, \qquad 0 \le \omega^{(-)} \le \pi,$$
$$0 \le \theta^{(\pm)} \le \pi, \qquad -\pi \le \varphi^{(\pm)} \le \pi , \tag{9.45}$$

where $\theta^{(\pm)}$ and $\varphi^{(\pm)}$ are the polar angle and the azimuthal angle of the direction $\hat{\mathbf{n}}^{(\pm)}$. Since the group space of the second SU(2)' group is reduced to that similar to the group space of the SO(3) group, namely, the two ends of a diameter in the group space correspond to the same element, the group space of SO(4) is doubly-connected.

Any irreducible representation of SO(4) can be expressed as the direct product of two irreducible representations of two SU(2) groups, denoted by D^{jk}:

$$D^{jk}\left(\hat{\mathbf{n}}^{(+)}, \omega^{(+)}; \hat{\mathbf{n}}^{(-)}, \omega^{(-)}\right) = D^j\left(\hat{\mathbf{n}}^{(+)}, \omega^{(+)}\right) \times D^k\left(\hat{\mathbf{n}}^{(-)}, \omega^{(-)}\right). \tag{9.46}$$

The row (column) index of D^{jk} is denoted by two indices $(\mu\nu)$. The dimension of D^{jk} is $(2j+1)(2k+1)$. Its generator $I_a^{jk(\pm)}$ is expressed by the generator I_a^j of SU(2):

$$I_a^{jk(+)} = I_a^j \times \mathbf{1}_{2k+1}, \qquad I_a^{jk(-)} = \mathbf{1}_{2j+1} \times I_a^k,$$
$$I_{ab}^{jk} = \sum_{c=1}^3 \epsilon_{abc}\left(I_c^{jk(+)} + I_c^{jk(-)}\right)$$
$$= \sum_{c=1}^3 \epsilon_{abc}\left(I_c^j \times \mathbf{1}_{2k+1} + \mathbf{1}_{2j+1} \times I_c^k\right), \tag{9.47}$$
$$I_{a4}^{jk} = I_a^{jk(+)} - I_a^{jk(-)} = I_a^j \times \mathbf{1}_{2k+1} - \mathbf{1}_{2j+1} \times I_a^k.$$

D^{jk} with different superscripts jk are inequivalent to each other. When $(j+k)$ is an integer, D^{jk} is a single-valued representation of SO(4), i.e., the irreducible tensor representation. When $(j+k-1/2)$ is an integer, D^{jk} is a double-valued representation of SO(4), i.e., the spinor representation. The correspondence between D^{jk} and the irreducible representation denoted by a Young pattern $[\lambda_1, \lambda_2]$ is as follows. When $j + k$ is an integer, we have

$$D^{jj} \simeq [2j, 0], \qquad\qquad \text{when } j = k,$$
$$D^{jk} \simeq [+(j+k), (j-k)] \qquad \text{when } j > k,$$
$$D^{jk} \simeq [-(j+k), (k-j)] \qquad \text{when } j < k,$$

and when $j + k - 1/2$ is an integer, we have

$$D^{jk} \simeq [+s, (j+k-1/2), (j-k-1/2)], \qquad \text{when } j > k,$$
$$D^{jk} \simeq [-s, (j+k-1/2), (k-j-1/2)], \qquad \text{when } j < k.$$

Note that $[+\lambda_1, \lambda_2]$ is the self-dual representation, and $[-\lambda_1, \lambda_2]$ is the anti-self-dual one. D^{00} is the identical representation, $D^{\frac{1}{2}0}$ and $D^{0\frac{1}{2}}$ respectively are two fundamental spinor representations $[+s]$ and $[-s]$, and $D^{\frac{1}{2}\frac{1}{2}}$ is equivalent to the self-representation.

★ The Lorentz transformation matrix A between two inertial frames is a 4×4 orthogonal matrix:

$$A^T A = A A^T = \mathbf{1}. \qquad (9.48)$$

The matrix entries of A satisfy the following condition:

$$\left.\begin{array}{l} A_{ab} \;\; \text{and} \;\; A_{44} \;\; \text{are real} \\[2mm] A_{a4} \;\; \text{and} \;\; A_{4a} \;\; \text{are imaginary} \end{array}\right\} \quad a \text{ and } b = 1,\, 2,\, 3.$$

This condition keeps invariant in the product of two matrices A. The set of all such orthogonal matrices A, in the multiplication rule of matrices, constitutes the homogeneous Lorentz group, denoted by O(3,1) or L_h. The orthogonal condition (9.48) gives

$$\det A = \pm 1, \qquad A_{44}^2 = 1 + \sum_{a=1}^{3} |A_{a4}|^2 \geq 1.$$

These two discontinuous constraints separate the group space of O(3,1) into four unconnected sections. The element in the section where the identity belongs to satisfies

$$\det A = 1, \qquad A_{44} \geq 1. \qquad (9.49)$$

Those elements satisfying Eq. (9.49) constitute a simple Lie subgroup, called the proper Lorentz group, denoted by L_p. Since there is no upper limit of A_{44}, the group space of L_p is an open area in the Euclidean space. Thus, L_p is a non-compact Lie group. The representative elements in three cosets of L_p are usually chosen to be the space inversion σ, the time inversion τ and the whole inversion ρ:

$$\sigma = \operatorname{diag}(-1, \ -1, \ -1, \ 1),$$
$$\tau = \operatorname{diag}(1, \ 1, \ 1, \ -1),$$
$$\rho = \operatorname{diag}(-1, \ -1, \ -1, \ -1).$$

★ Let A be an infinitesemal element of L_p:

$$A = 1 - i\alpha X, \qquad\qquad A^T = 1 - i\alpha X^T,$$
$$1 = A^T A = 1 - i\alpha\left(X + X^T\right), \quad X^T = -X,$$
$$1 = \det A = 1 - i\alpha \operatorname{Tr} X, \qquad \operatorname{Tr} X = 0.$$

Thus, X is a traceless antisymmetric matrix. Expand X with respect to the generators in the self-representation of SO(4):

$$
\begin{aligned}
A &= 1 - i\sum_{a<b=2}^{3} \omega_{ab} T_{ab} - i\sum_{a=1}^{3} \omega_{a4} T_{a4}\\
&= 1 - i\sum_{a=1}^{3}\left(\Omega_a T_a^{(+)} + \Omega_a^* T_a^{(-)}\right),
\end{aligned}
\tag{9.50}
$$

where ω_{ab} is real, ω_{a4} is imaginary, and

$$\Omega_a = \sum_{b<c=2}^{3} \epsilon_{abc}\omega_{bc} + \omega_{a4}. \tag{9.51}$$

Except for the imaginary parameters, the generators in the self-representations of SO(4) and L_p are completely the same, so are the generators in the corresponding irreducible representations of two groups. The finite-dimensional inequivalent irreducible representations of L_p are also denoted by $D^{jk}(L_p)$, whose generators are given in Eq. (9.47). Since the parameters for SO(4) and L_p are different, the global properties of the two groups are very different. Any finite-dimensional irreducible representation D^{jk} of L_p, except for the identical representation, is not unitary. There exist infinite-dimensional unitary representations for L_p.

★ The transformation generated by T_{ab} is obviously a pure rotation, belonging to the subgroup SO(3). Let us study the transformation generated by T_{34} with the parameter $\omega_{34} = i\omega$:

$$A(\vec{e}_3, i\omega) \equiv \exp\left\{-i(i\omega)T_{34}\right\} = \begin{pmatrix} 1 & 0 & 0 & 0 \\ 0 & 1 & 0 & 0 \\ 0 & 0 & \cosh\omega & -i\sinh\omega \\ 0 & 0 & i\sinh\omega & \cosh\omega \end{pmatrix},$$

$$v = c\tanh\omega, \qquad \cosh\omega = \left(1 - v^2/c^2\right)^{-1/2},$$

$$\sinh\omega = (v/c)\left(1 - v^2/c^2\right)^{-1/2}.$$

(9.52)

It describes the Lorentz boost along the direction of z axis with the relative velocity $v = c\tanh\omega$.

Any Lorentz transformation A satisfying Eqs. (9.48) and (9.49) can be decomposed into the product of some rotational transformation and the Lorentz boost along the z direction. Let $A_{44} = \cosh\omega$, from which ω is determined. Extracting a factor $-i\sinh\omega$ from A_{a4}, we obtain a unit vector $\hat{n}(\theta, \varphi)$ in the three-dimensional space, whose rectangular coordinates are $(i/\sinh\omega)(A_{14}, A_{24}, A_{34})$:

$$A_{44} = \cosh\omega, \qquad\qquad iA_{14}/\sinh\omega = \sin\theta\cos\varphi,$$
$$iA_{24}/\sinh\omega = \sin\theta\sin\varphi, \qquad iA_{34}/\sinh\omega = \cos\theta.$$

(9.53)

Extending the rotational matrix into a 4×4 matrix, we have

$$R(\vec{e}_3, \varphi)R(\vec{e}_2, \theta) = \begin{pmatrix} \cos\varphi\cos\theta & -\sin\varphi & \cos\varphi\sin\theta & 0 \\ \sin\varphi\cos\theta & \cos\varphi & \sin\varphi\sin\theta & 0 \\ -\sin\theta & 0 & \cos\theta & 0 \\ 0 & 0 & 0 & 1 \end{pmatrix}.$$

Left-multiplying to A the inverse matrix of $R(\vec{e}_3, \varphi)R(\vec{e}_2, \theta)A(\vec{e}_3, i\omega)$, we obtain a rotational matrix $R(\alpha, \beta, \gamma)$,

$$\left\{R(\vec{e}_3, \varphi)R(\vec{e}_2, \theta)A(\vec{e}_3, i\omega)\right\}^{-1} A = R(\alpha, \beta, \gamma),$$

from which the Euler angles α, β and γ can be calculated. Thus,

$$A = R(\varphi, \theta, 0)A(\vec{e}_3, i\omega)R(\alpha, \beta, \gamma).$$

(9.54)

The geometrical meaning of the decomposition is evident. Two rotations in two-sides of $A(\vec{e}_3, i\omega)$ rotate two z axes in two frames before and after the Lorentz transformation A to the direction of the relative motion, and the remaining axes to be parallel to each other, respectively. After two rotations, the Lorentz transformation A is simplified into $A(\vec{e}_3, i\omega)$.

★ Since all inequivalent irreducible representations of L_p have been known, the inequivalent irreducible representations of the homogeneous Lorentz group O(3,1) can be obtained by determining the representation matrices of τ and ρ. The representation matrix of σ can be calculated by the formula $\sigma = \tau\rho$. Since ρ is commutable with any element in O(3,1) and the ρ^2 is equal to the identity, the representation matrix of ρ in an irreducible representation of O(3,1) is a constant matrix, where the constant is ± 1 in a single-valued representation. The property of the representation matrix of τ can be determined by the relations in the self-representation

$$\tau T_{ab}\tau^{-1} = T_{ab}, \quad \tau T_{a4}\tau^{-1} = -T_{a4}, \quad \tau T_a^{(\pm)}\tau^{-1} = T_a^{(\mp)}.$$

Let $V^{(\lambda)}$, $\lambda = 1, 2, 3$ and 4, be four inequivalent irreducible representations of the four-order inversion group V_4. When $j = k$, The irreducible representation D^{jj} of L_p induces four inequivalent irreducible representation $\Delta^{jj\lambda}$ of O(3,1):

$$\begin{aligned}
\Delta^{jj\lambda}(A) &= D^{jj}(A), \qquad A \in L_p, \\
\Delta^{jj\lambda}_{\mu\nu,\mu'\nu'}(\tau) &= V^{(\lambda)}(\tau)\delta_{\mu\nu'}\delta_{\nu\mu'}, \\
\Delta^{jj\lambda}(\rho) &= V^{(\lambda)}(\rho)\,\mathbf{1}, \qquad 1 \le \lambda \le 4.
\end{aligned} \tag{9.55}$$

When $j \ne k$, while $j + k$ is an integer, the representation $D^{jk} \oplus D^{kj}$ of L_p induces two inequivalent irreducible representation $\Delta^{jk\pm}$ of O(3,1):

$$\begin{aligned}
\Delta^{jk\pm}(\rho) &= \pm\mathbf{1}, \quad \Delta^{jk\pm}(A) = D^{jk}(A) \oplus D^{kj}(A), \qquad A \in L_p, \\
\Delta^{jk\pm}_{\mu\nu\alpha,\mu'\nu'\beta}(\tau) &= \delta_{(-\alpha)\beta}\delta_{\mu\nu'}\delta_{\nu\mu'}, \qquad \alpha, \beta = \pm 1,
\end{aligned} \tag{9.56}$$

where α and β are used to identify two representation spaces of D^{jk} and D^{kj}. Since $\Delta^{jk\pm}(\tau)$ has no diagonal entries, changing its sign leads to an equivalent representation.

When $j + k$ is a half-odd number, we only discuss the Dirac spinor representation $D(O(3,1))$ here. For definiteness, the index μ runs from 1 to 4, while the index a runs from 1 to 3. Let γ_μ be four anti-commutable matrices and C be the charge-conjugate transformation matrix:

$$C^{-1}\gamma_\mu C = -\gamma_\mu^T, \quad C^\dagger C = \mathbf{1}, \quad C^T = -C, \quad \det C = 1.$$

Thus, $D(A)$ satisfies

$$D(A)^{-1}\gamma_\mu D(A) = \sum_{\nu=1}^{4} A_{\mu\nu}\gamma_\nu, \qquad \det D(A) = 1,$$

$$C^{-1}D(A)C = \frac{A_{44}}{|A_{44}|}\left\{D(A^{-1})\right\}^T. \tag{9.57}$$

The additional factor $A_{44}/|A_{44}|$ comes from the physical reason. Its generators are

$$I_{\mu\nu} = \frac{-i}{4}\left(\gamma_\mu\gamma_\nu - \gamma_\nu\gamma_\mu\right). \tag{9.58}$$

The representation matrices of the representative elements in the cosets of L_p are

$$D(\sigma) = \pm i\gamma_4, \qquad D(\tau) = \pm\gamma_4\gamma_5, \qquad D(\rho) = \pm i\gamma_5. \tag{9.59}$$

15. Discuss the classes in the SO(4) group and calculate their characters in the irreducible representation D^{jk}.

Solution. Any element R of SO(4) can be transformed into the standard form by a real orthogonal similarity transformation X:

$$X^{-1}RX = \begin{pmatrix} \cos\varphi_1 & -\sin\varphi_1 & 0 & 0 \\ \sin\varphi_1 & \cos\varphi_1 & 0 & 0 \\ 0 & 0 & \cos\varphi_2 & -\sin\varphi_2 \\ 0 & 0 & \sin\varphi_2 & \cos\varphi_2 \end{pmatrix}$$

$$= N\left\{u(\vec{e}_3, \varphi_1 + \varphi_2) \times u(\vec{e}_3, \varphi_1 - \varphi_2)\right\}N^{-1},$$

$$u(\vec{e}_3, \alpha) = \begin{pmatrix} \exp\{-i\alpha/2\} & 0 \\ 0 & \exp\{i\alpha/2\} \end{pmatrix},$$

where Eq. (9.43) is used. Therefore, the class of SO(4) is described by two parameters φ_1 and φ_2. Its character in the representation $D^{jk}(R)$ is

$$\chi^{jk}(\varphi_1, \varphi_2) = \frac{\sin\left[(j+1/2)(\varphi_1 + \varphi_2)\right]\ \sin\left[(k+1/2)(\varphi_1 - \varphi_2)\right]}{\sin\left[(\varphi_1 + \varphi_2)/2\right]\ \sin\left[(\varphi_1 - \varphi_2)/2\right]}.$$

16. Calculate six parameters of the following proper Lorentz transformation A, and write its representation matrix in the irreducible representation $D^{jk}(A)$ of the proper Lorentz group L_p:

$$A(\varphi,\theta,\omega,\alpha,\beta,\gamma) = \begin{pmatrix} 1 & 0 & 0 & 0 \\ 0 & \sqrt{3}/2 & (\cosh\omega)/2 & -i(\sinh\omega)/2 \\ 0 & -1/2 & \sqrt{3}(\cosh\omega)/2 & -i\sqrt{3}(\sinh\omega)/2 \\ 0 & 0 & i\sinh\omega & \cosh\omega \end{pmatrix}.$$

Solution. From the fourth column of $A(\varphi,\theta,\omega,\alpha,\beta,\gamma)$, the boost parameter is ω, and $\theta = \pi/6$ and $\varphi = \pi/2$ because $\sin\theta\cos\varphi = 0$, $\sin\theta\sin\varphi = 1/2$ and $\cos\theta = \sqrt{3}/2$. The matrix form of the rotation $R(\pi/2,\pi/6,0)$ is

$$R(\pi/2,\pi/6,0) = \begin{pmatrix} 0 & -1 & 0 & 0 \\ 1 & 0 & 0 & 0 \\ 0 & 0 & 1 & 0 \\ 0 & 0 & 0 & 1 \end{pmatrix} \begin{pmatrix} \sqrt{3}/2 & 0 & 1/2 & 0 \\ 0 & 1 & 0 & 0 \\ -1/2 & 0 & \sqrt{3}/2 & 0 \\ 0 & 0 & 0 & 1 \end{pmatrix}$$

$$R(\pi/2,\pi/6,0)^{-1} = \begin{pmatrix} 0 & \sqrt{3}/2 & -1/2 & 0 \\ -1 & 0 & 0 & 0 \\ 0 & 1/2 & \sqrt{3}/2 & 0 \\ 0 & 0 & 0 & 1 \end{pmatrix}.$$

Thus,

$$A(\vec{e}_3,i\omega)^{-1}R(\pi/2,\pi/6,0)^{-1}A(\varphi,\theta,\omega,\alpha,\beta,\gamma) = \begin{pmatrix} 0 & 1 & 0 & 0 \\ -1 & 0 & 0 & 0 \\ 0 & 0 & 1 & 0 \\ 0 & 0 & 0 & 1 \end{pmatrix}.$$

From this we obtain $\alpha = -\pi/2$ and $\beta = \gamma = 0$. Thus, we obtain the representation matrix of A in $D^{jk}(L_p)$

$$\begin{aligned}
D^{jk}_{\mu\nu,\rho\lambda}(A) &= \left\{\left[D^j(\varphi,\theta,0) \times D^k(\varphi,\theta,0)\right] \cdot \left[e^{\omega I_3^j} \times e^{-\omega I_3^k}\right]\right. \\
&\quad \left. \cdot \left[D^j(\alpha,\beta,\gamma) \times D^k(\alpha,\beta,\gamma)\right]\right\}_{\mu\nu,\rho\lambda} \\
&= \sum_{\tau,\sigma} e^{-i(\mu+\nu)\varphi} d^j_{\mu\tau}(\theta) d^k_{\nu\sigma}(\theta) e^{\omega(\tau-\sigma)} e^{-i(\tau+\sigma)\alpha} \\
&\quad \cdot d^j_{\tau\rho}(\beta) d^k_{\sigma\lambda}(\beta) e^{-i(\rho+\lambda)\gamma} \\
&= \sum_{\tau,\sigma} e^{-i(\mu+\nu)\pi/2} d^j_{\mu\tau}(\pi/6) d^k_{\nu\sigma}(\pi/6) e^{\omega(\tau-\sigma)} e^{i(\tau+\sigma)\pi/2}.
\end{aligned}$$

17. Prove that the Dirac spinor representation satisfy

$$\gamma_4 D(A)^\dagger \gamma_4 = \frac{A_{44}}{|A_{44}|} D(A)^{-1}.$$

Solution. The Dirac spinor representation is not unitary. The aim of this problem is to study the property of its conjugate representation. From the definition (9.57) of the Dirac spinor representation we have:

$$D(A)^\dagger \gamma_a D(A^{-1})^\dagger = \sum_{b=1}^{3} A_{ab}\gamma_b - A_{a4}\gamma_4,$$
$$D(A)^\dagger \gamma_4 D(A^{-1})^\dagger = -\sum_{b=1}^{3} A_{4b}\gamma_b + A_{44}\gamma_4.$$

To make the formulas uniformly, we make a similarity transformation γ_4:

$$\left\{\gamma_4 D(A)^\dagger \gamma_4\right\} \gamma_\mu \left\{\gamma_4 D(A)^\dagger \gamma_4\right\}^{-1} = \sum_{\nu=1}^{4} A_{\mu\nu}\gamma_\nu.$$

From the equivalent theorem of γ_a, we obtain

$$\gamma_4 D(A)^\dagger \gamma_4 = cD(A)^{-1}.$$

Substituting Eq. (9.59) into it, we determine the constant c :

$$\gamma_4 D(A)^\dagger \gamma_4 = \frac{A_{44}}{|A_{44}|} D(A)^{-1}.$$

This is the reason why the matrix γ_4 should be inserted between two spinors $\psi^\dagger$ and ψ to construct an invariant quantity of the Lorentz transformation in the quantum field theory:

$$\overline{\psi}\psi = \psi^\dagger \gamma_4 \psi.$$

18. Discuss the classes of the proper Lorentz group L_p.

Solution. In the self-representation of L_p we write the pure rotation $R(\alpha, \beta, \gamma)$ and the Lorentz boost $A(\vec{e}_3, i\omega)$ with the relative velocity along the z axis in terms of Eq. (9.43)

$$R(\alpha, \beta, \gamma) = N \left\{u(\alpha, \beta, \gamma) \times u(\alpha, \beta, \gamma)\right\} N^{-1},$$
$$A(\vec{e}_3, i\omega) = N \left\{\exp(\omega\sigma_3/2) \times \exp(-\omega\sigma_3/2)\right\} N^{-1}.$$

Thus, any element $A(\varphi, \theta, \omega, \alpha, \beta, \gamma)$ in L_p can be expressed as

$$A(\varphi, \theta, \omega, \alpha, \beta, \gamma) = N\{M \times (\sigma_2 M^* \sigma_2)\} N^{-1}, \qquad (9.60)$$

where

$$
\begin{aligned}
M &= u(\varphi, \theta, 0) \exp(\omega \sigma_3 / 2) u(\alpha, \beta, \gamma), \\
\sigma_2 M^* \sigma_2 &= u(\varphi, \theta, 0) \exp(-\omega \sigma_3 / 2) u(\alpha, \beta, \gamma).
\end{aligned} \qquad (9.61)
$$

The matrix M is a two-dimensional matrix with determinant $+1$, which belongs to the two-dimensional unimodular complex matrix group SL(2,C). We are going to show the homomorphism between SL(2,C) and L_p with a 2:1 correspondence. First, we have known from Eq. (9.60) that an arbitrary Lorentz transformation $A = A(\varphi, \theta, \omega, \alpha, \beta, \gamma)$ corresponds to a matrix $M \in$ SL(2,C). Conversely, any M given in Eq. (9.61) determines a Lorentz transformation A. We will show that any element P in SL(2,C) can be expressed in the form (9.61) later. If two matrices M and M' correspond to the same Lorentz transformation A, then

$$N\{M \times (\sigma_2 M^* \sigma_2)\} N^{-1} = A = N\{M' \times [\sigma_2 (M')^* \sigma_2]\} N^{-1}.$$

Thus, $M^{-1} M' = c\mathbf{1}$. Since $\det M = 1$ and $\det M' = 1$, $c^2 = 1$ and $c = \pm 1$. There is a 2:1 correspondence between $\pm M$ given in Eq. (9.61) and the Lorentz transformation A. This correspondence keeps invariant in the multiplication of elements. In fact, if

$$A = N\{M \times (\sigma_2 M^* \sigma_2)\} N^{-1}, \qquad A' = N\{M' \times [\sigma_2 (M')^* \sigma_2]\} N^{-1},$$

we have

$$AA' = N\{MM' \times [\sigma_2 (MM')^* \sigma_2]\} N^{-1}.$$

Now, we have proved that the two-dimensional unimodular complex matrix group SL(2,C) is homomorphic onto the proper Lorentz group L_p with a 2:1 correspondence. Thus, we can determine the class of L_p by SL(2,C). Recall that two matrices $\pm M \in$ SL(2,C) correspond to the same Lorentz transformation $A \in L_p$. Two classes in SL(2,C) different by a sign correspond to the same class in L_p. If two eigenvalues of $M \in$ SL(2,C) are different, M can be diagonalized through a similarity transformation $X \in$ SL(2,C)

$$X^{-1} M X = \pm \begin{pmatrix} e^{(\omega - i\varphi)/2} & 0 \\ 0 & e^{-(\omega - i\varphi)/2} \end{pmatrix} = \pm e^{-i(\varphi + i\omega)\sigma_3 / 2},$$

$$-\pi \le \varphi \le \pi, \qquad 0 \le \omega < \infty.$$

Thus, we have

$$A = \begin{pmatrix} \cos\varphi & -\sin\varphi & 0 & 0 \\ \sin\varphi & \cos\varphi & 0 & 0 \\ 0 & 0 & \cosh\omega & -i\sinh\omega \\ 0 & 0 & i\sinh\omega & \cosh\omega \end{pmatrix} \quad \begin{array}{l} -\pi \le \varphi \le \pi, \\[2mm] 0 \le \omega < \infty. \end{array}$$

The class is described by two parameters ω and φ.

If two eigenvalues of M are equal, the eigenvalues have to be ± 1. $\pm\mathbf{1}$ correspond to the identity element of L_p, which constitutes one class in L_p. If M is not equal to $\pm\mathbf{1}$, M can be transformed into the Jordan form by a similarity transformation $Y \in SL(2,C)$ (see Problem 16 in Chap. 1):

$$Y^{-1}MY = \pm \begin{pmatrix} 1 & -2 \\ 0 & 1 \end{pmatrix}.$$

Since

$$\begin{pmatrix} 1 & -2 \\ 0 & 1 \end{pmatrix} = u(-\pi, \pi/4, 0) \begin{pmatrix} \sqrt{2}+1 & 0 \\ 0 & \sqrt{2}-1 \end{pmatrix} u(\pi, 3\pi/4, 0)$$

$$= \exp\{-\sigma_1 - i\sigma_2\} = \exp\left\{-i\sum_{a=1}^{3}\Omega_a\sigma_a/2\right\},$$

we obtain $\Omega_1 = -2i$, $\Omega_2 = 2$, $\Omega_3 = 0$. In order to calculate the six parameters of A, we make a transformation

$$\begin{pmatrix} 1 & -2 \\ 0 & 1 \end{pmatrix} = u(-\pi, \pi/4, 0) \begin{pmatrix} \sqrt{2}+1 & 0 \\ 0 & \sqrt{2}-1 \end{pmatrix} u(\pi, 3\pi/4, 0).$$

The solutions are $\varphi = -\pi$, $\theta = \pi/4$, $\cosh\omega = 3$, $\alpha = \pi$, $\beta = 3\pi/4$, $\gamma = 0$, and

$$A = \begin{pmatrix} 1 & 0 & 2 & 2i \\ 0 & 1 & 0 & 0 \\ -2 & 0 & -1 & -2i \\ -2i & 0 & -2i & 3 \end{pmatrix} = e^{-i2T_{31}-2T_{14}}.$$

It is easy to check that A is a proper Lorentz transformation with all four eigenvalues to be 1. A has only two linearly independent eigenvectors for the eigenvalue 1. A is transformed into the Jordan form by the similarity transformation Z:

$$Z = \begin{pmatrix} 0 & 0 & 2 & -1 \\ 1 & 0 & 0 & 0 \\ 0 & -4 & 0 & 1 \\ 0 & -4i & 0 & 0 \end{pmatrix}, \qquad Z^{-1} = \begin{pmatrix} 0 & 1 & 0 & 0 \\ 0 & 0 & 0 & i/4 \\ 1/2 & 0 & 1/2 & i/2 \\ 0 & 0 & 1 & i \end{pmatrix},$$

$$Z^{-1}AZ = \begin{pmatrix} 1 & 0 & 0 & 0 \\ 0 & 1 & 1 & 0 \\ 0 & 0 & 1 & 1 \\ 0 & 0 & 0 & 1 \end{pmatrix}.$$

Finally, we are going to show that, any two-dimensional unimodual complex matrix $P \in SL(2,C)$ can be expressed into the form of Eq. (9.61), namely

$$P = \begin{pmatrix} A & B \\ C & D \end{pmatrix} = u(\varphi,\theta,0) \begin{pmatrix} e^{\omega/2} & 0 \\ 0 & e^{-\omega/2} \end{pmatrix} u(\alpha,\beta,\gamma).$$

$$u(\alpha,\beta,\gamma) = \begin{pmatrix} \cos(\beta/2)e^{-i(\alpha+\gamma)/2} & -\sin(\beta/2)e^{-i(\alpha-\gamma)/2} \\ \sin(\beta/2)e^{i(\alpha-\gamma)/2} & \cos(\beta/2)e^{i(\alpha+\gamma)/2} \end{pmatrix}.$$

Solving the equation, we have

$$A = c_\theta c_\beta e^{\omega/2}e^{-i(\varphi+\alpha+\gamma)/2} - s_\theta s_\beta e^{-\omega/2}e^{-i(\varphi-\alpha+\gamma)/2},$$
$$B = -c_\theta s_\beta e^{\omega/2}e^{-i(\varphi+\alpha-\gamma)/2} - s_\theta c_\beta e^{-\omega/2}e^{-i(\varphi-\alpha-\gamma)/2},$$
$$C = s_\theta c_\beta e^{\omega/2}e^{i(\varphi-\alpha-\gamma)/2} + c_\theta s_\beta e^{-\omega/2}e^{i(\varphi+\alpha-\gamma)/2},$$
$$D = -s_\theta s_\beta e^{\omega/2}e^{i(\varphi-\alpha+\gamma)/2} + c_\theta c_\beta e^{-\omega/2}e^{i(\varphi+\alpha+\gamma)/2},$$

where $c_\theta = \cos(\theta/2)$, $s_\theta = \sin(\theta/2)$ and $c_\beta = \cos(\beta/2)$ and so on. The direct calculation leads to

$$|A|^2 = -(\sin\theta \ \sin\beta \ \cos\alpha)/2 + c_\theta^2 c_\beta^2 e^\omega + s_\theta^2 s_\beta^2 e^{-\omega},$$
$$|B|^2 = (\sin\theta \ \sin\beta \ \cos\alpha)/2 + c_\theta^2 s_\beta^2 e^\omega + s_\theta^2 c_\beta^2 e^{-\omega},$$
$$|C|^2 = (\sin\theta \ \sin\beta \ \cos\alpha)/2 + s_\theta^2 c_\beta^2 e^\omega + c_\theta^2 s_\beta^2 e^{-\omega},$$
$$|D|^2 = -(\sin\theta \ \sin\beta \ \cos\alpha)/2 + s_\theta^2 s_\beta^2 e^\omega + c_\theta^2 c_\beta^2 e^{-\omega},$$
$$AB = e^{-i\varphi}\left\{-\cos\beta\sin\theta + \sin\beta\left(-c_\theta^2 e^\omega e^{-i\alpha} + s_\theta^2 e^{-\omega}e^{i\alpha}\right)\right\}/2,$$
$$AC = e^{-i\gamma}\left\{\cos\theta \ \sin\beta + \sin\theta\left(c_\beta^2 e^\omega e^{-i\alpha} - s_\beta^2 e^{-\omega}e^{i\alpha}\right)\right\}/2,$$
$$AD = s_\theta^2 s_\beta^2 + c_\theta^2 c_\beta^2 - s_\theta c_\theta s_\beta c_\beta \left(e^\omega e^{-i\alpha} - e^{-\omega}e^{i\alpha}\right),$$
$$|A|^2 + |B|^2 + |C|^2 + |D|^2 = e^\omega + e^{-\omega},$$

$$|A|^2 + |B|^2 - |C|^2 - |D|^2 = \cos\theta \left(e^\omega - e^{-\omega}\right),$$
$$|A|^2 - |B|^2 + |C|^2 - |D|^2 = \cos\beta \left(e^\omega - e^{-\omega}\right).$$

ω, θ and β can be respectively calculated from the last three formulas. Then, α is determined by the formula for AD. At last, γ and φ can be calculated from the formulas for AC and AB. Here, the calculated method is not unique due to $\det P = 1$. This completes the proof.

For example, we calculate the parameters of a matrix P:

$$P = \frac{1}{4}\begin{pmatrix} \sqrt{3}\left[\sqrt{2} - 1 - i\left(\sqrt{2}+1\right)\right] & -\sqrt{2}+1-i3\left(\sqrt{2}+1\right) \\ \sqrt{2}+1-i3\left(\sqrt{2}-1\right) & \sqrt{3}\left[\sqrt{2}+1+i\left(\sqrt{2}-1\right)\right] \end{pmatrix}.$$

From $|A|^2 = |D|^2 = 9/8$, $|B|^2 = (15/8) + \sqrt{2}$ and $|C|^2 = (15/8) - \sqrt{2}$, we obtain

$$\cosh\omega = 3, \quad \sinh\omega = \sqrt{8}, \quad e^{\omega/2} = \sqrt{2}+1, \quad e^{-\omega/2} = \sqrt{2}-1$$
$$\cos\theta = 1/2, \quad \theta = \pi/3, \quad \cos\beta = -1/2, \quad \beta = 2\pi/3.$$

Then, from $AD = 3(1 - i\sqrt{8})/8$, we obtain

$$e^\omega e^{-i\alpha} + e^{-\omega}e^{i\alpha} = 6\cos\alpha - i4\sqrt{2}\sin\alpha = \frac{16}{3}\left\{\frac{3}{8} - AD\right\} = 4\sqrt{2}i.$$

The solution is $\alpha = -\pi/2$. Finally, from

$$AB = \frac{\sqrt{3}[-2(3+\sqrt{2}) - i]}{8} = e^{-i\varphi}\left\{\frac{\sqrt{3}}{8} - i\frac{\sqrt{3}}{4}\left(3+\sqrt{2}\right)\right\},$$
$$AC = \frac{\sqrt{3}[-1 - 2i(3-\sqrt{2})]}{8} = e^{-i\gamma}\left\{\frac{\sqrt{3}}{8} + i\frac{\sqrt{3}}{4}\left(3-\sqrt{2}\right)\right\},$$

we have $e^{-i\varphi} = -i$ and $e^{-i\gamma} = -1$, and determine $\varphi = \pi/2$ and $\gamma = \pi$. At last, we have

$$P = \frac{e^{i\pi/4}}{2}\begin{pmatrix} -i\sqrt{3} & i \\ 1 & \sqrt{3} \end{pmatrix}\begin{pmatrix} \sqrt{2}+1 & 0 \\ 0 & \sqrt{2}-1 \end{pmatrix}\frac{e^{-i\pi/4}}{2}\begin{pmatrix} 1 & \sqrt{3} \\ -i\sqrt{3} & i \end{pmatrix}.$$

19. Express an arbitrary element in the proper Lorentz group L_p in the form of exponential matrix function.

Solution. In Problem 18, we have shown that $L_p \sim \mathrm{SL}(2,C)$ with the correspondence (9.60):

$$\pm M = \pm u(\varphi, \theta, 0)u(\vec{e}_3, i\omega)u(\alpha, \beta, \gamma) \longrightarrow A(\varphi, \theta, \omega, \alpha, \beta, \gamma),$$

$$\pm u(\vec{e}_3, \alpha) = \pm e^{-i\alpha\sigma_3/2} = \pm \begin{pmatrix} e^{-i\alpha/2} & 0 \\ 0 & e^{i\alpha/2} \end{pmatrix} \longrightarrow R(\vec{e}_3, \alpha),$$

$$\pm u(\vec{e}_2, \beta) = \pm e^{-i\beta\sigma_2/2} = \pm \begin{pmatrix} \cos(\beta/2) & -\sin(\beta/2) \\ \sin(\beta/2) & \cos(\beta/2) \end{pmatrix} \longrightarrow R(\vec{e}_2, \beta),$$

$$\pm u(\vec{e}_3, i\omega) = \pm e^{\omega\sigma_3/2} = \pm \begin{pmatrix} e^{\omega/2} & 0 \\ 0 & e^{-\omega/2} \end{pmatrix} \longrightarrow A(\vec{e}_3, i\omega).$$

An arbitrary element A in L_p can be written in the form of exponential matrix function if its corresponding elements $\pm M$ in $\mathrm{SL}(2, C)$ can. Since $\det M = 1$, its two eigenvalues have the same sign. If its eigenvalues are positive, M can be transformed into the form of exponential matrix function by a unimodular similarity transformation. If its eigenvalues are negative, $-M$ can be transformed into the form of exponential matrix function by a unimodular similarity transformation. Ignoring the irrelevant sign, we can express an arbitrary element M in $\mathrm{SL}(2, C)$ as

$$M = \pm \exp\left(-i\vec{\Omega} \cdot \vec{\sigma}/2\right), \qquad \sigma_2 M^* \sigma_2 = \pm \exp\left(-i\vec{\Omega}^* \cdot \vec{\sigma}/2\right).$$

By Eq. (9.60), the proper Lorentz transformation A corresponding to M can be expressed as

$$
\begin{aligned}
A &= N\left\{\exp\left(-i\vec{\Omega} \cdot \vec{\sigma}/2\right) \times \exp\left(-i\vec{\Omega}^* \cdot \vec{\sigma}/2\right)\right\} N^{-1} \\
&= \exp\left(-i\vec{\Omega} \cdot \vec{T}^{(+)}\right) \exp\left(-i\vec{\Omega}^* \cdot \vec{T}^{(-)}\right) \\
&= \exp\left\{-i \sum_{a=1}^{3} \left(\Omega_a T_a^{(+)} + \Omega_a^* T_a^{(-)}\right)\right\} \\
&= \exp\left\{-i \sum_{a<b} \omega_{ab} T_{ab} - i \sum_{a=1}^{3} \omega_{a4} T_{a4}\right\},
\end{aligned}
$$

where $T_a^{(\pm)}$, T_{ab} and T_{a4} are the generators in the self-representation of $\mathrm{SO}(4)$, and the parameters ω_{ab} and ω_{c4} are the real part and the imaginary part of Ω_c, respectively:

$$\omega_{ab} = \frac{1}{2} \sum_{c=1}^{3} \epsilon_{abc} \left(\Omega_c + \Omega_c^*\right), \qquad \omega_{c4} = \frac{1}{2i} \left(\Omega_c - \Omega_c^*\right).$$

Thus, an arbitrary Lorentz transformation A can be expressed in the form of exponential matrix function, where the parameters ω_{ab} are real, and ω_{a4} are imaginary. The set of parameters $\omega_{\mu\nu}$ is mainly for the theoretical study, and the set of parameters $(\varphi, \theta, \omega, \alpha, \beta, \gamma)$ is convenient for calculation. The relation between two sets of parameters are given by Eq. (9.60), namely

$$M = u(\varphi, \theta, 0)u(\vec{e}_3, i\omega)u(\alpha, \beta, \gamma) = \exp\left(-i\vec{n} \cdot \vec{\sigma}/2\right).$$

20. Let

$$X = -ix_4 \mathbf{1} + \sum_{a=1}^{3} x_a \sigma_a = \begin{pmatrix} x_3 - ix_4 & x_1 - ix_2 \\ x_1 + ix_2 & -x_3 - ix_4 \end{pmatrix},$$

where x_a is real, $x_4 = ict$ is imaginary, and X is a hermitian matrix. Let $M \in SL(2,C)$,

$$MXM^\dagger = X' = -ix_4' \mathbf{1} + \sum_{a=1}^{3} x_a' \sigma_a.$$

(a) Calculate Tr X and det X.   (b) Prove $x_\mu' = \sum_\nu A_{\mu\nu} x_\nu$, where A is a proper Lorentz transformation.   (c) Prove that $L_p \sim SL(2,C)$. (d) Calculate the corresponding matrix M, when A is equal to $R(\vec{e}_3, \varphi)$, $R(\vec{e}_2, \theta)$, and $A(\vec{e}_3, i\omega)$, respectively.   (e) Calculate the matrix M corresponding to an arbitrary proper Lorentz transformation $A(\varphi, \theta, \omega, \alpha, \beta, \gamma)$.

Solution. (a) Tr$X = -i2x_4$,   det $X = -\sum_{\mu=1}^{4} x_\mu^2$.

(b) Since $X' = MXM^\dagger$ is hermitian, it can be expanded with respect to the Pauli matrices and $\mathbf{1}$, where the coefficients x_a' are real, and x_4' is imaginary. Further, x_μ' can be expressed as the linear combinations of x_μ,

$$x_\mu' = \sum_{\nu=1}^{4} A_{\mu\nu} x_\nu, \qquad 1 \le \mu \le 4,$$

where A_{ab} and A_{44} are real, and A_{4a} and A_{a4} are imaginary. Due to det $X' = $ det X,

$$\sum_{\mu=1}^{4} x_\mu'^{\,2} = \sum_{\mu=1}^{4} x_\mu^2.$$

Therefore, A is a Lorentz transformation matrix, which is an orthogonal matrix with the matrix entries depending upon M. When $M = 1$, A is the identity. Since the group space of SL(2,C) is connected, the set of A can be continuously extended from the identity, namely, A is a proper Lorentz transformation matrix.

(c) From a given matrix M, we can calculate $X' = MXM^\dagger$, and then, determine a proper Lorentz transformation matrix A uniquely. If the matrices X' calculated from M and M' are the same, so are the matrices A, then $M^{-1}M'$ is commutable with three Pauli matrices σ_a. Thus, it is a constant matrix. Due to det$M$=det$M' = 1$, $M' = \pm M$. Namely, there is a 2:1 correspondence between $\pm M$ and A. This mapping is obviously invariant in the multiplication of elements. Therefore, SL(2,C) is homomorphic onto L_p, $L_p \sim SL(2,C)$.

(d) Ignoring the sign in front of the matrix M, we can express the matrix M and A by the same parameters. If $A \in$SO(3) is a pure rotation, M obviously belongs to SU(2),

$$M(\vec{e}_3, \varphi) = u(\vec{e}_3, \varphi) = 1 \cos(\varphi/2) - i\sigma_3 \sin(\varphi/2),$$
$$M(\vec{e}_2, \theta) = u(\vec{e}_2, \theta) = 1 \cos(\theta/2) - i\sigma_2 \sin(\theta/2),$$

In general, $M(\alpha, \beta, \gamma) = u(\alpha, \beta, \gamma)$,

$$u(\alpha, \beta, \gamma) = \begin{pmatrix} \cos(\beta/2)e^{-i(\alpha+\gamma)/2} & -\sin(\beta/2)e^{-i(\alpha-\gamma)/2} \\ \sin(\beta/2)e^{i(\alpha-\gamma)/2} & \cos(\beta/2)e^{i(\alpha+\gamma)/2} \end{pmatrix}.$$

For a boost $A = A(\vec{e}_3, i\omega)$, we have

$$M(\vec{e}_3, i\omega)XM(\vec{e}_3, i\omega)^\dagger = -i1(ix_3 \sinh\omega + x_4 \cosh\omega) + \sigma_3(x_3 \cosh\omega - ix_4 \sinh\omega) + \sigma_1 x_1 + \sigma_2 x_2.$$

If only one component x_μ is nonvanishing, we have

$$M(\vec{e}_3, i\omega)\sigma_1 M(\vec{e}_3, i\omega)^\dagger = \sigma_1, \qquad\qquad x_1 = 1,$$
$$M(\vec{e}_3, i\omega)\sigma_2 M(\vec{e}_3, i\omega)^\dagger = \sigma_2, \qquad\qquad x_2 = 1,$$
$$M(\vec{e}_3, i\omega)\sigma_3 M(\vec{e}_3, i\omega)^\dagger = 1 \sinh\omega + \sigma_3 \cosh\omega, \quad x_3 = 1,$$
$$M(\vec{e}_3, i\omega)M(\vec{e}_3, i\omega)^\dagger = 1 \cosh\omega + \sigma_3 \sinh\omega, \quad x_4 = i.$$

Let $M(\vec{e}_3, i\omega) = 1M_0 + \sigma_1 M_1 + \sigma_2 M_2 + \sigma_3 M_3$. Extracting the Pauli matrix from the left-hand side of the first two formulas, and then, moving it to their right-hand side, we have

$$\{\mathbf{1}M_0 + \sigma_1 M_1 - \sigma_2 M_2 - \sigma_3 M_3\} M^\dagger = \mathbf{1},$$
$$\{\mathbf{1}M_0 - \sigma_1 M_1 + \sigma_2 M_2 - \sigma_3 M_3\} M^\dagger = \mathbf{1}.$$

Adding and subtracting two equations, we obtain $M_1 = M_2 = 0$, $|M_0|^2 - |M_3|^2 = 1$, and $M_0 M_3^*$ are real. Thus, the last two formulas become the same, and turn to $|M_0|^2 + |M_3|^2 = \cosh\omega$ and $2M_0 M_3^* = \sinh\omega$. Recall that $\det M = 1$. The solution is $M_0 = \cosh\omega/2$ and $M_3 = \sinh\omega/2$, namely

$$M(\vec{\mathbf{e}}_3, i\omega) = \begin{pmatrix} e^{\omega/2} & 0 \\ 0 & e^{-\omega/2} \end{pmatrix} = u(\vec{\mathbf{e}}_3, i\omega).$$

(e) For an arbitrary proper Lorentz transformation $A(\varphi, \theta, \omega, \alpha, \beta, \gamma)$, we have

$$\pm M(\varphi, \theta, \omega, \alpha, \beta, \gamma) = \pm u(\varphi, \theta, 0)u(\vec{\mathbf{e}}_3, i\omega)u(\alpha, \beta, \gamma).$$

Chapter 10

THE SYMPLECTIC GROUPS

10.1 The Groups Sp(2ℓ, R) and USp(2ℓ)

★ Take the subscript a of a vector in a (2ℓ)-dimensional space to be μ or $\bar{\mu}$, $1 \le \mu \le \ell$, in the following order:

$$a = 1,\ \bar{1},\ 2,\ \bar{2},\ \cdots,\ \ell,\ \bar{\ell}. \tag{10.1}$$

Define a (2ℓ)-dimensional antisymmetric matrix J:

$$J_{ab} = \begin{cases} 1 & \text{when } a = \mu, \quad b = \bar{\mu} \\ -1 & \text{when } a = \bar{\mu}, \quad b = \mu \\ 0 & \text{otherwise,} \end{cases} \tag{10.2}$$
$$J = 1_\ell \times (i\sigma_2) = -J^{-1} = -J^T, \qquad \det J = 1\ .$$

The set of all $(2\ell) \times (2\ell)$ real matrices R satisfying

$$R^T J R = J, \qquad R^* = R, \tag{10.3}$$

in the multiplication rule of matrices, constitutes the real symplectic group $\mathrm{Sp}(2\ell, R)$. Both the inverse R^{-1} and the transpose R^T of any element R of $\mathrm{Sp}(2\ell, R)$ satisfy Eq. (10.3):

$$R^{-1} = -J R^T J, \qquad \left(R^{-1}\right)^T J R^{-1} = J, \tag{10.4}$$
$$-J \left(-J R^T J\right)^T J \left(-J R^T J\right) J = R J R^T = J.$$

$\mathrm{Sp}(2\ell, R)$ is not a compact Lie group, because the general form for a diagonal matrix $R_0 \in \mathrm{Sp}(2\ell, R)$ contains the parameters ω_μ without an upper limit:

$$R_0 = \mathrm{diag}\left\{e^{\omega_1},\ e^{-\omega_1},\ e^{\omega_2},\ e^{-\omega_2},\ \ldots, e^{\omega_\ell},\ e^{-\omega_\ell}\right\}. \tag{10.5}$$

★ The set of all $(2\ell) \times (2\ell)$ unitary matrices u satisfying

$$u^T J u = J, \qquad u^\dagger = u^{-1}, \qquad (10.6)$$

in the multiplication rule of matrices, constitutes the unitary symplectic group $\mathrm{USp}(2\ell)$. Similarly, both the inverse u^{-1} and the transpose u^T of any element u of $\mathrm{USp}(2\ell)$ satisfy Eq. (10.6). Since any matrix entry of a unitary matrix is finite, $\mathrm{USp}(2\ell)$ is a compact Lie group. As an example, the general form of a diagonal matrix $u_0 \in \mathrm{USp}(2\ell)$ is:

$$u_0 = \mathrm{diag}\left\{ e^{-i\varphi_1},\ e^{i\varphi_1},\ e^{-i\varphi_2},\ e^{i\varphi_2},\ \ldots, e^{-i\varphi_\ell},\ e^{i\varphi_\ell} \right\}. \qquad (10.7)$$

It can be shown that the determinant of $R \in \mathrm{Sp}(2\ell, R)$ or $u \in \mathrm{USp}(2\ell)$ is $+1$. The group spaces of both $\mathrm{Sp}(2\ell, R)$ and $\mathrm{USp}(2\ell)$ are connected.

★ For the infinitesimal elements $R \in \mathrm{Sp}(2\ell, R)$ and $u \in \mathrm{USP}(2\ell)$,

$$\begin{aligned}
R = \mathbf{1} - i\alpha X, \qquad X^T = JXJ, \qquad X^* = -X, \\
u = \mathbf{1} - i\beta Y, \qquad Y^T = JYJ, \qquad Y^\dagger = Y.
\end{aligned} \qquad (10.8)$$

An imaginary matrix X of 2ℓ dimensions contains $4\ell^2$ real parameters, so does a hermitian matrix Y. The component form of the condition $X^T = JXJ$ is

$$X_{\nu\mu} = -X_{\overline{\mu}\overline{\nu}}, \qquad X_{\overline{\nu}\mu} = X_{\overline{\mu}\nu}, \qquad X_{\nu\overline{\mu}} = X_{\mu\overline{\nu}},$$

which gives $\ell^2 + \ell(\ell - 1)$ real constraints. The first equation gives $\mathrm{Tr}X = 0$, which means $\det R = 1$. The component form of $Y^T = JYJ$ is

$$Y_{\nu\mu} = -Y_{\overline{\mu}\overline{\nu}}, \qquad Y_{\overline{\nu}\mu} = Y_{\overline{\mu}\nu}.$$

Due to $Y^\dagger = Y$, the first equation contains ℓ^2 real constraints, including $\mathrm{Tr}Y = 0$, and the second equation contains $\ell(\ell - 1)$ complex constraints. Therefore, there are $\ell(2\ell + 1)$ real parameters both for R and for u. The order of $\mathrm{Sp}(2\ell, R)$ and the order of $\mathrm{USp}(2\ell)$ are both $\ell(2\ell + 1)$.

★ We choose the generators T_A in the self-representation of $\mathrm{USp}(2\ell)$

$$\begin{aligned}
T_{ab}^{(2)} \times \mathbf{1}_2, \qquad T_{ab}^{(1)} \times \sigma_p, \qquad T_{aa}^{(1)} \times \sigma_p / \sqrt{2}, \\
1 \le p \le 3, \qquad a < b, \qquad 1 \le a \le \ell,
\end{aligned} \qquad (10.9)$$

where $T_{ab}^{(1)}$ and $T_{ab}^{(2)}$ are the generators in the self-representation of the $\mathrm{SU}(\ell)$ group, and $\left(T_{aa}^{(1)} \right)_{bd} = \delta_{bd}\delta_{ab}$. The generators T_A are hermitian and normalized:

$$\mathrm{Tr}\,(T_A T_B) = \delta_{AB}. \tag{10.10}$$

Thus, the structure constants of $\mathrm{USp}(2\ell)$ are totally antisymmetric with respect to all three indices.

The diagonal generators span the Cartan subalgebra of $\mathrm{USp}(2\ell)$:

$$H_\mu = T_{\mu\mu}^{(1)} \times \sigma_3/\sqrt{2}, \qquad 1 \le \mu \le \ell. \tag{10.11}$$

Their simultaneous eigenvectors in the Lie algebra, $[H_\mu,\ E_\alpha] = \alpha_\mu E_\alpha$, are

$$
\begin{aligned}
E_{ab}^{(1)} &= \left\{ T_{ab}^{(1)} \times \sigma_3 + i T_{ab}^{(2)} \times \mathbf{1}_2 \right\}/\sqrt{2}, \\
E_{ab}^{(2)} &= \left\{ T_{ab}^{(1)} \times \sigma_3 - i T_{ab}^{(2)} \times \mathbf{1}_2 \right\}/\sqrt{2}, \\
E_{ab}^{(3)} &= T_{ab}^{(1)} \times (\sigma_1 + i\sigma_2)/\sqrt{2}, \\
E_{ab}^{(4)} &= T_{ab}^{(1)} \times (\sigma_1 - i\sigma_2)/\sqrt{2}, \\
E_{a}^{(5)} &= T_{aa}^{(1)} \times (\sigma_1 + i\sigma_2)/2, \\
E_{a}^{(6)} &= T_{aa}^{(1)} \times (\sigma_1 - i\sigma_2)/2,
\end{aligned}
$$

where $a < b$ and the eigenvalues (roots) are $\sqrt{1/2}\,(\mathbf{e}_a - \mathbf{e}_b)$, $-\sqrt{1/2}\,(\mathbf{e}_a - \mathbf{e}_b)$, $\sqrt{1/2}\,(\mathbf{e}_a + \mathbf{e}_b)$, $-\sqrt{1/2}\,(\mathbf{e}_a + \mathbf{e}_b)$, $\sqrt{2}\mathbf{e}_a$, and $-\sqrt{2}\mathbf{e}_a$, $a < b$, respectively. The simple roots are

$$\mathbf{r}_\mu = \sqrt{1/2}\,(\mathbf{e}_\mu - \mathbf{e}_{\mu+1}), \qquad 1 \le \mu \le \ell - 1, \qquad \mathbf{r}_\ell = \sqrt{2}\mathbf{e}_\ell. \tag{10.12}$$

The first $(\ell - 1)$ simple roots are shorter, $d_\mu = (\mathbf{r}_\mu \cdot \mathbf{r}_\mu)/2 = 1/2$, and the last one is longer, $d_\ell = (\mathbf{r}_\ell \cdot \mathbf{r}_\ell)/2 = 1$. The remaining positive roots are

$$
\begin{aligned}
\sqrt{1/2}\,(\mathbf{e}_a - \mathbf{e}_b) &= \sum_{\mu=a}^{b-1} \mathbf{r}_\mu, \\
\sqrt{1/2}\,(\mathbf{e}_a + \mathbf{e}_b) &= \sum_{\mu=a}^{b-1} \mathbf{r}_\mu + 2\sum_{\mu=b}^{\ell-1} \mathbf{r}_\mu + \mathbf{r}_\ell, \\
\sqrt{2}\mathbf{e}_a &= 2\sum_{\mu=a}^{\ell-1} \mathbf{r}_\mu + \mathbf{r}_\ell.
\end{aligned}
$$

Thus, the Lie algebra of $\mathrm{USp}(2\ell)$ is C_ℓ. The largest root, which is the highest weight of the adjoint representation of C_ℓ, is

$$\vec{\omega} = \sqrt{2}\mathbf{e}_1 = 2\sum_{\mu=1}^{\ell-1} \mathbf{r}_\mu + \mathbf{r}_\ell = 2\mathbf{w}_1. \tag{10.13}$$

★ The Chevalley bases of USp(2ℓ) are

$$H_\mu = \left\{ T^{(1)}_{\mu\mu} - T^{(1)}_{(\mu+1)(\mu+1)} \right\} \times \sigma_3, \qquad H_\ell = T^{(1)}_{\ell\ell} \times \sigma_3,$$

$$E_\mu = \left\{ T^{(1)}_{\mu(\mu+1)} \times \sigma_3 + i T^{(2)}_{\mu(\mu+1)} \times 1_2 \right\},$$

$$F_\mu = \left\{ T^{(1)}_{\mu(\mu+1)} \times \sigma_3 - i T^{(2)}_{\mu(\mu+1)} \times 1_2 \right\}, \tag{10.14}$$

$$E_\ell = T^{(1)}_{\ell\ell} \times (\sigma_1 + i\sigma_2)/2, \qquad F_\ell = T^{(1)}_{\ell\ell} \times (\sigma_1 - i\sigma_2)/2,$$

$$1 \le \mu \le \ell - 1.$$

★ The generators in the self-representation of the Sp($2\ell, R$) group are pure imaginary, $X^* = -X$. The explicit forms of the generators can be obtained from Eq. (10.9) by replacing σ_a with τ_a:

$$\tau_1 = i\sigma_1, \qquad \tau_2 = \sigma_2, \qquad \tau_3 = i\sigma_3. \tag{10.15}$$

Namely, some generators change by a factor i so that the Sp($2\ell, R$) group is a non-compact Lie group. USp(2ℓ) and Sp($2\ell, R$) have the same complex Lie algebra, but different real Lie algebras. In another viewpoint, one may change the relevant parameters of Sp($2\ell, R$) to be imaginary such that the generators in the corresponding irreducible representations of both USp(2ℓ) and Sp($2\ell, R$) are the same.

1. Prove that the determinant of R in Sp($2\ell, R$) and the determinant of u in USp(2ℓ) are both $+1$.

Solution. In the following proof for the determinant of R in Sp($2\ell, R$), we did not use the conjugate operation so that the proof is also effective for u in USp(2ℓ).

Let $R \in$ Sp($2\ell, R$) be the transformation matrix for a vector x_a in the 2ℓ-dimensional real space:

$$x_a \xrightarrow{R} x'_a = \sum_b R_{ab} x_b . \tag{10.16}$$

Define a pseudo-product for two real vectors x and y:

$$\{x, \, y\}_J \equiv \sum_{ab} x_a J_{ab} y_b = \sum_{\mu=1}^{\ell} \left(x_\mu y_{\bar\mu} - x_{\bar\mu} y_\mu \right) . \tag{10.17}$$

Obviously, the pseudo-product is invariant in the transformation R:

$$\{x, \, y\}_J = \{Rx, \, Ry\}_J . \tag{10.18}$$

The self-pseudo-product of one vector is vanishing, $\{x,\ x\}_J = 0$.

For the totally antisymmetric tensor of rank 2ℓ, $\epsilon_{a_1 \ldots a_{2\ell}}$, we can prove the following identity by Eq. (10.2)

$$\sum_{a_1 \ldots a_{2\ell}} \epsilon_{a_1 \ldots a_{2\ell}} J_{a_1 a_2} \ldots J_{a_{2\ell-1} a_{2\ell}} = 2^\ell \ell! \ .$$

In fact, the nonvanishing terms in the sum come from the permutations among J's ($\ell!$ terms) and the transpositions between two subscripts of any J (2^ℓ terms), and all are equal to one. Moving the column index of a 2ℓ-dimensional matrix X to be a superscript, we rewrite the determinant of X as

$$\det X = \sum_{b_1 \ldots b_{2\ell}} \epsilon_{b_1 \ldots b_{2\ell}} X_1^{b_1} \ldots X_{2\ell}^{b_{2\ell}} .$$

Permutating the order of X_a^b, and changing the summing indices to restore the order of the superscripts, we obtain

$$\epsilon_{a_1 \ldots a_{2\ell}} \det X = \sum_{b_1 \ldots b_{2\ell}} \epsilon_{b_1 \ldots b_{2\ell}} X_{a_1}^{b_1} \ldots X_{a_{2\ell}}^{b_{2\ell}} .$$

Thus, we have

$$\det X = \left(2^\ell \ell! \right)^{-1} \sum_{a_1 \ldots a_{2\ell}} (\det X) \epsilon_{a_1 \ldots a_{2\ell}} J_{a_1 a_2} \ldots J_{a_{2\ell-1} a_{2\ell}}$$

$$= \left(2^\ell \ell! \right)^{-1} \sum_{a_1 \ldots a_{2\ell}} J_{a_1 a_2} \ldots J_{a_{2\ell-1} a_{2\ell}} \sum_{b_1 \ldots b_{2\ell}} \epsilon_{b_1 \ldots b_{2\ell}} X_{a_1}^{b_1} \ldots X_{a_{2\ell}}^{b_{2\ell}}$$

$$= \left(2^\ell \ell! \right)^{-1} \sum_{b_1 \ldots b_{2\ell}} \epsilon_{b_1 \ldots b_{2\ell}} \left\{ X^{b_1},\ X^{b_2} \right\}_J \ldots \left\{ X^{b_{2\ell-1}},\ X^{b_{2\ell}} \right\}_J \ .$$

Since the right-hand side is invariant in the transformation $R \in \mathrm{Sp}(2\ell, R)$, we have

$$\det (RX) = \det X, \qquad \det R = 1 \ .$$

In the same reason, $\det u = 1$. Both $\mathrm{Sp}(2\ell, R)$ and $\mathrm{USp}(2\ell)$ are the simply-connected Lie groups.

2. Count the number of the independent real parameters of R in $\mathrm{Sp}(2\ell, R)$ and u in $\mathrm{USp}(2\ell)$ directly from their definitions (10.3) and (10.6).

Solution. A (2ℓ)-dimensional real matrix contains $(2\ell)^2$ real parameters. Equations (10.3) and (10.17) show that the pseudo-product of two column matrices of $R \in \mathrm{Sp}(2\ell, R)$ is equal to J_{ab}:

$$\{R_{\cdot a}, \ R_{\cdot b}\}_J = \sum_{\mu=1}^{\ell} (R_{\mu a} R_{\overline{\mu} b} - R_{\overline{\mu} a} R_{\mu b}) = J_{ab}.$$

It is an identity when $a = b$. It gives $\ell(2\ell - 1)$ independent real constraints when $a \neq b$, so that the order of $\mathrm{Sp}(2\ell, R)$ is $(2\ell)^2 - \ell(2\ell - 1) = \ell(2\ell + 1)$.

A (2ℓ)-dimensional complex matrix contains $2(2\ell)^2$ real parameters. From $u^{-1} = -J u^T J$, we have $u^* = -JuJ$,

$$u^*_{\mu\nu} = u_{\overline{\mu}\overline{\nu}}, \qquad u^*_{\mu\overline{\nu}} = -u_{\overline{\mu}\nu}.$$

Each gives $2\ell^2$ independent real constraints. From the unitary condition $\sum_d u^*_{da} u_{db} = \delta_{ab}$, we have the following constraints. The constraints for $a = b = \mu$ are the same as those for $a = b = \overline{\mu}$. There are ℓ real independent constraints:

$$a = b = \mu, \quad \sum_{\nu=1}^{\ell} (u_{\overline{\nu}\overline{\mu}} u_{\nu\mu} - u_{\nu\overline{\mu}} u_{\overline{\nu}\mu}) = \sum_{\nu=1}^{\ell} \left(|u_{\nu\mu}|^2 + |u_{\overline{\nu}\mu}|^2\right) = 1,$$

$$a = b = \overline{\mu}, \quad \sum_{\nu=1}^{\ell} (-u_{\overline{\nu}\mu} u_{\nu\overline{\mu}} + u_{\nu\mu} u_{\overline{\nu}\overline{\mu}}) = 1.$$

When $\mu \neq \nu$, the constraints for $a = \mu$ and $b = \nu$ are the same as those for $a = \overline{\nu}$, $b = \overline{\mu}$ as well as those for exchanging a and b. There are $\ell(\ell - 1)/2$ complex independent constraints:

$$a = \mu, \ b = \nu \neq \mu, \quad \sum_{\tau=1}^{\ell} (u_{\overline{\tau}\overline{\mu}} u_{\tau\nu} - u_{\tau\overline{\mu}} u_{\overline{\tau}\nu}) = 0,$$

$$a = \overline{\nu}, \ b = \overline{\mu} \neq \overline{\nu}, \quad \sum_{\tau=1}^{\ell} (-u_{\overline{\tau}\nu} u_{\tau\overline{\mu}} + u_{\tau\nu} u_{\overline{\tau}\overline{\mu}}) = 0.$$

When $\mu \neq \nu$, the constraints for $a = \mu$, $b = \overline{\nu}$ are the same as those for $a = \nu$, $b = \overline{\mu}$ as well as those for exchanging a and b. There are $\ell(\ell - 1)/2$ complex independent constraints:

$$a = \mu, \ b = \overline{\nu} \neq \overline{\mu}, \quad \sum_{\tau=1}^{\ell} (u_{\overline{\tau}\overline{\mu}} u_{\tau\overline{\nu}} - u_{\tau\overline{\mu}} u_{\overline{\tau}\overline{\nu}}) = 0,$$

$$a = \nu, \ b = \overline{\mu} \neq \overline{\nu}, \quad \sum_{\tau=1}^{\ell} (u_{\overline{\tau}\nu} u_{\tau\overline{\mu}} - u_{\tau\nu} u_{\overline{\tau}\overline{\mu}}) = 0.$$

One complex constraint is equivalent to two real constraints. Altogether there are $(2\ell)^2 + \ell + 2\ell(\ell-1) = 6\ell^2 - \ell$ real constraints. The number of the independent real parameters of u in $\mathrm{USp}(2\ell)$ is $8\ell^2 - (6\ell^2 - \ell) = \ell(2\ell+1)$, which is the order of the $\mathrm{USp}(\ell)$ group.

3. Express the simple roots of $\mathrm{USp}(2\ell)$ by the vectors $\mathbf{V}_a$ given in Eq. (8.3) for the $\mathrm{SU}(\ell+1)$ group, and then, write their Cartan-Weyl bases of generators in the self-representation of $\mathrm{USp}(2\ell)$

Solution. From the Cartan-Weyl bases (10.11), we have

$$
\begin{aligned}
&\left[H_\nu,\, E_{\mathbf{r}_\mu}\right] = (\mathbf{r}_\mu)_\nu\, E_{\mathbf{r}_\mu}, \\
&\mathbf{r}_\rho = \sqrt{1/2}\,(\mathbf{e}_\rho - \mathbf{e}_{\rho+1}), \qquad \mathbf{r}_\ell = \sqrt{2}\,\mathbf{e}_\ell \\
&H_\nu = T^{(1)}_{\nu\nu} \times \sigma_3/\sqrt{2}, \\
&E_{\mathbf{r}_\rho} = \left\{ T^{(1)}_{\rho(\rho+1)} \times \sigma_3 + iT^{(2)}_{\rho(\rho+1)} \times \mathbf{1}_2 \right\}/\sqrt{2}, \\
&E_{\mathbf{r}_\ell} = T^{(1)}_{\ell\ell} \times (\sigma_1 + i\sigma_2)/2, \\
&1 \le \rho \le \ell-1, \quad 1 \le \mu \le \ell, \quad 1 \le \nu \le \ell.
\end{aligned}
\tag{10.19}
$$

If we choose another set of the normalized bases H'_ν in the Cartan subalgebra:

$$
\begin{aligned}
H'_1 &= \mathbf{1}_\ell \times \sigma_3/\sqrt{2\ell} = \sqrt{1/\ell} \sum_{\rho=1}^{\ell} H_\rho, \\
H'_\tau &= T^{(3)}_{\ell-\tau+2} \times \sigma_3 \\
&= \sqrt{\frac{1}{(\ell-\tau+2)(\ell-\tau+1)}} \sum_{\rho=1}^{\ell-\tau+1} H_\rho - \sqrt{\frac{\ell-\tau+1}{\ell-\tau+2}} H_{\ell-\tau+2}, \\
&2 \le \tau \le \ell,
\end{aligned}
\tag{10.20}
$$

the component $(\mathbf{r}'_\mu)_\nu$ of the simple root can be calculated from the formula H'_ν by replacing H_ρ with $\sqrt{1/2}\,(\delta_{\rho\mu} - \delta_{\rho(\mu+1)})$ when $\mu < \ell$ and $\sqrt{2}\delta_{\rho\ell}$ when $\mu = \ell$.

$$
\begin{aligned}
(\mathbf{r}_\mu)_1 &= \delta_{\mu\ell}\sqrt{2/\ell}, \\
(\mathbf{r}_\mu)_\tau &= \delta_{\mu(\ell-\tau+1)}\left\{ \sqrt{\frac{1}{2(\ell-\tau+2)(\ell-\tau+1)}} + \sqrt{\frac{\ell-\tau+1}{2(\ell-\tau+2)}} \right\} \\
&\quad - \delta_{\mu(\ell-\tau+2)}\sqrt{\frac{\ell-\tau+1}{2(\ell-\tau+2)}} - \delta_{\mu\ell}\delta_{\tau 2}\sqrt{\frac{\ell-1}{2\ell}} \\
&= (\mathbf{V}_\mu - \mathbf{V}_{\mu+1})_\tau + \delta_{\mu\ell}\delta_{\tau 2}\,(\mathbf{V}_\ell)_\tau.
\end{aligned}
$$

Namely,

$$\mathbf{r}_\mu = \mathbf{V}_\mu - \mathbf{V}_{\mu+1}, \quad 1 \le \mu < \ell, \quad \mathbf{r}_\ell = 2\mathbf{V}_\ell - \frac{2}{\ell}\left(\sqrt{\ell+1}-1\right)\mathbf{V}_{\ell+1}.$$

(10.21)

The remaining positive roots are

$$\sum_{\mu=a}^{b-1} \mathbf{r}_\mu = \mathbf{V}_a - \mathbf{V}_b,$$

$$\sum_{\mu=a}^{b-1} \mathbf{r}_\mu + 2\sum_{\mu=b}^{\ell-1} \mathbf{r}_\mu + \mathbf{r}_\ell = \mathbf{V}_a + \mathbf{V}_b - \frac{2}{\ell}\left(\sqrt{\ell+1}-1\right)\mathbf{V}_{\ell+1},$$

$$2\sum_{\mu=a}^{\ell-1} \mathbf{r}_\mu + \mathbf{r}_\ell = 2\mathbf{V}_a - \frac{2}{\ell}\left(\sqrt{\ell+1}-1\right)\mathbf{V}_{\ell+1}.$$

10.2   Irreducible Representations of Sp(2ℓ)

★ Since the generators in the corresponding irreducible representations of USp(2ℓ) and Sp(2ℓ, R), as well as the orthonormal bases in the representation space are the same, we will only discuss the irreducible representations of Sp(2ℓ, R). In the following, Sp(2ℓ) stands for both USp(2ℓ) and Sp(2ℓ, R) for simplicity.

★ Let $R \in \mathrm{Sp}(2\ell, R)$ be the coordinate transformation in a real (2ℓ)-dimensional space:

$$x_a \xrightarrow{R} x_a' = \sum_b R_{ab}x_b .$$

(10.22)

From Eq. (10.22) we can define the tensors for Sp(2ℓ, R) transformations:

$$\mathbf{T}_{a_1\dots a_n} \xrightarrow{R} (O_R\mathbf{T})_{a_1\dots a_n} = \sum_{b_1\dots b_n} R_{a_1b_1}\dots R_{a_nb_n}\mathbf{T}_{b_1\dots b_n}.$$

(10.23)

Similar to the tensors of SU(N) and SO(N), the Weyl reciprocity holds for the tensors of Sp(2ℓ, R). The tensor space can be reduced by the Young operators. In addition, there are two invariant tensors for Sp(2ℓ, R). One is the antisymmetric tensor J_{ab} of rank two

$$(O_RJ)_{ab} = \sum_{cd} R_{ac}R_{bd}J_{cd} = \left(RJR^T\right)_{ab} = J_{ab}.$$

(10.24)

The contraction $\sum_{ab} J_{ab} T_{abc...}$ decreases the rank of the tensor $T_{abc...}$ by two:

$$
\begin{aligned}
\sum_{ab} O_R J_{ab} T_{abc...} &= \sum_{ab} \sum_{rsa'b'c'...} R_{ar} R_{bs} R_{aa'} R_{bb'} R_{cc'} \ldots J_{rs} T_{a'b'c'...} \\
&= \sum_{ab} \sum_{a'b'c'...} R_{aa'} R_{bb'} R_{cc'} \ldots J_{ab} T_{a'b'c'...} \\
&= \sum_{c'...} R_{cc'} \ldots \left(\sum_{a'b'} J_{a'b'} T_{a'b'c'...} \right).
\end{aligned}
$$

(10.25)

Sometimes, this operation is also called the trace. The trace subspace is invariant in $\mathrm{Sp}(2\ell, R)$. In order to reduce the tensor space, we have to decompose the tensor space into the direct sum of a series of traceless tensor subspaces with decreasing rank two by two. Applying the Young operator $\mathcal{Y}_\mu^{[\lambda]}$ to a traceless tensor space $\mathcal{T}$, we obtain the minimal tensor subspace $\mathcal{Y}_\mu^{[\lambda]} \mathcal{T}$, corresponding to an irreducible representation $[\lambda]$ of $\mathrm{Sp}(2\ell, R)$.

The other invariant tensor of $\mathrm{Sp}(2\ell, R)$ is the totally antisymmetric tensor $\epsilon_{a_1 \cdots a_{2\ell}}$ of rank 2ℓ. However, this tensor violates the new traceless condition. It is irrelevant to the reduction of the tensor space for $\mathrm{Sp}(2\ell, R)$.

★ The traceless tensor subspace denoted by a Young pattern $[\lambda]$ with the row number larger than ℓ is empty. The irreducible representation of the $\mathrm{Sp}(2\ell, R)$ group is described by a Young pattern with at most ℓ rows. The standard tensor Young tableau is still the common eigenstate of the Chevalley bases H_μ. The problem is that it is not necessary traceless. Fortunately, the contraction occurs only when a pair subscripts μ and $\overline{\mu}$ appears. It is convenient to rearrange the bases Θ_a in the vector space:

$$
\Phi_\mu = \Theta_\mu, \quad \Phi_{\ell+\mu} = \Theta_{\overline{\ell-\mu+1}}, \quad 1 \le \mu \le \ell. \tag{10.26}
$$

From Eq. (10.14) the nonvanishing matrix entries of the Chevalley bases in the bases Φ_a are

$$
\begin{aligned}
&H_\mu \Phi_\mu = \Phi_\mu, &&H_\mu \Phi_{\mu+1} = -\Phi_{\mu+1}, \\
&H_\mu \Phi_{2\ell-\mu} = \Phi_{2\ell-\mu}, &&H_\mu \Phi_{2\ell-\mu+1} = -\Phi_{2\ell-\mu+1}, \\
&H_\ell \Phi_\ell = \Phi_\ell, &&H_\ell \Phi_{\ell+1} = -\Phi_{\ell+1}, \\
&E_\mu \Phi_{\mu+1} = \Phi_\mu, &&E_\mu \Phi_{2\ell-\mu+1} = -\Phi_{2\ell-\mu}, \\
&F_\mu \Phi_\mu = \Phi_{\mu+1}, &&F_\mu \Phi_{2\ell-\mu} = -\Phi_{2\ell-\mu+1}, \\
&E_\ell \Phi_{\ell+1} = \Phi_\ell, &&F_\ell \Phi_\ell = \Phi_{\ell+1}.
\end{aligned}
$$

(10.27)

In the bases Φ_α, $1 \le \alpha \le 2\ell$, the standard tensor Young tableau

$\mathcal{Y}_\mu^{[\lambda]} \Phi_{\alpha_1 \dots \alpha_n}$ in the representation $[\lambda]$ of $\mathrm{Sp}(2\ell, R)$ corresponds to the highest weight if each box in its jth row, $j \leq \ell$, is filled with the digit j. The standard tensor Young tableau with the highest weight is a traceless tensor because all subscripts take the values less than $\ell + 1$, but the trace occurs only when a pair of subscripts μ and $2\ell - \mu + 1$ appears. The relation between the highest weight $\mathbf{M}$ and the Young pattern $[\lambda]$ for $\mathrm{Sp}(2\ell, R)$ is

$$M_\mu = \lambda_\mu - \lambda_{\mu+1}, \quad \text{when } 1 \leq \mu < \ell, \quad M_\ell = \lambda_\ell. \quad (10.28)$$

The remaining traceless tensor bases in an irreducible representation space $[\lambda]$ of $\mathrm{Sp}(2\ell, R)$ can be calculated from the basis with the highest weight by the lowering operators F_μ.

★ The dimension of an irreducible representation $[\lambda]$ of $\mathrm{Sp}(2\ell)$ can be calculated by the hook rule. In this rule, the dimension is expressed as a quotient, where the numerator and the denominator are denoted by the symbols $Y_P^{[\lambda]}$ and $Y_h^{[\lambda]}$, respectively:

$$d_{[\lambda]}(\mathrm{Sp}(2\ell)) = \frac{Y_P^{[\lambda]}}{Y_h^{[\lambda]}}. \quad (10.29)$$

We still use the concept of the hook path (i, j) in the Young pattern $[\lambda]$, which enters the Young pattern at the rightmost of the ith row, goes leftwards in the i row, turns downwards at the j column, goes downwards in the j column, and leaves from the Young pattern at the bottom of the j column. The inverse path $\overline{(i, j)}$ is the same path as the hook path (i, j) except for the opposite direction. The number of boxes contained in the hook path (i, j) is the hook number h_{ij} of the box in the jth column of the ith row. $Y_h^{[\lambda]}$ is a tableau of the Young pattern $[\lambda]$ where the box in the jth column of the ith row is filled with the hook number h_{ij}. Define a series of the tableaux $Y_{P_a}^{[\lambda]}$ recursively by the rule given below. $Y_P^{[\lambda]}$ is a tableau of the Young pattern $[\lambda]$ where each box is filled with the sum of the digits which are respectively filled in the same box of each tableau $Y_{P_a}^{[\lambda]}$ in the series. The symbol $Y_P^{[\lambda]}$ means the product of the filled digits in it, so does the symbol $Y_h^{[\lambda]}$.

The tableaux $Y_{P_a}^{[\lambda]}$ are defined by the following rule:

(a) $Y_{P_0}^{[\lambda]}$ is a tableau of the Young pattern $[\lambda]$ where the box in the jth column of the ith row is filled with the digit $(2\ell + j - i)$.

(b) Let $[\lambda^{(1)}] = [\lambda]$. Beginning with $[\lambda^{(1)}]$, we define recursively the Young pattern $[\lambda^{(a)}]$ by removing the first row and the first column of the Young pattern $[\lambda^{(a-1)}]$ until $[\lambda^{(a)}]$ contains less than two rows.

(c) If $[\lambda^{(a)}]$ contains more than one row, define $Y_{P_a}^{[\lambda]}$ to be a tableau of the Young pattern $[\lambda]$ where the boxes in the first $(a-1)$ rows and in the first $(a-1)$ columns are filled with 0, and the remaining part of the Young pattern is nothing but $[\lambda^{(a)}]$. Let $[\lambda^{(a)}]$ have r rows. Fill the first $(r-1)$ boxes along the hook path $(1, 1)$ of the Young pattern $[\lambda^{(a)}]$, beginning with the box on the rightmost, with the digits $\lambda_2^{(a)}$, $\lambda_3^{(a)}$, $\cdots$, $\lambda_r^{(a)}$, box by box, and fill the first $\lambda_i^{(a)}$ boxes in each inverse path $\overline{(i, 1)}$ of the Young pattern $[\lambda^{(a)}]$, $2 \le i \le r$, with -1. The remaining boxes are filled with 0. If a few -1 are filled in the same box, the digits are summed. The sum of all filled digits in the pattern $Y_{S_a}^{[\lambda]}$ is zero.

4. Calculate the dimensions of the irreducible representations of the Sp(6) group denoted by the following Young patterns:

(1) $[4, 2]$, (2) $[3, 2]$, (3) $[4, 4]$, (4) $[3, 3, 2]$, (5) $[4, 4, 3]$.

Solution.

(1): $d_{[4,2]}(\mathrm{Sp}(6)) = \left\{ \dfrac{\begin{matrix} 6\ 7\ 8\ 9 \\ 5\ 6 \end{matrix} + \begin{matrix} 0\ \ 0\ \ \ 0\ 2 \\ -1\ -1 \end{matrix}}{\begin{matrix} 5\ 4\ 2\ 1 \\ 2\ 1 \end{matrix}} \right\} = \left\{ \dfrac{\begin{matrix} 6\ 7\ 8\ 11 \\ 4\ 5 \end{matrix}}{\begin{matrix} 5\ 4\ 2\ 1 \\ 2\ 1 \end{matrix}} \right\} = 924.$

(2): $d_{[3,2]}(\mathrm{Sp}(6)) = \left\{ \dfrac{\begin{matrix} 6\ 7\ 8 \\ 5\ 6 \end{matrix} + \begin{matrix} 0\ \ 0\ \ 2 \\ -1\ -1 \end{matrix}}{\begin{matrix} 4\ 3\ 1 \\ 2\ 1 \end{matrix}} \right\} = \left\{ \dfrac{\begin{matrix} 6\ 7\ 10 \\ 4\ 5 \end{matrix}}{\begin{matrix} 4\ 3\ 1 \\ 2\ 1 \end{matrix}} \right\} = 350.$

(3): $d_{[4,4]}(\mathrm{Sp}(6)) = \left\{ \dfrac{\begin{matrix} 6\ 7\ 8\ 9 \\ 5\ 6\ 7\ 8 \end{matrix} + \begin{matrix} 0\ \ \ 0\ \ \ 0\ \ \ 4 \\ -1\ -1\ -1\ -1 \end{matrix}}{\begin{matrix} 5\ 4\ 3\ 2 \\ 4\ 3\ 2\ 1 \end{matrix}} \right\} = \left\{ \dfrac{\begin{matrix} 6\ 7\ 8\ 13 \\ 4\ 5\ 6\ 7 \end{matrix}}{\begin{matrix} 5\ 4\ 3\ 2 \\ 4\ 3\ 2\ 1 \end{matrix}} \right\} = 1274.$

(4): $d_{[3,3,2]}(\mathrm{Sp}(6)) = \left\{ \dfrac{\begin{matrix} 6\ 7\ 8 \\ 5\ 6\ 7 \\ 4\ 5 \end{matrix} + \begin{matrix} 0\ \ \ 2\ \ \ 3 \\ -1\ -1\ \ 0 \\ -2\ -1 \end{matrix} + \begin{matrix} 0\ \ 0\ \ 0 \\ 0\ \ 0\ \ 1 \\ 0\ -1 \end{matrix}}{\begin{matrix} 5\ 4\ 2 \\ 4\ 3\ 1 \\ 2\ 1 \end{matrix}} \right\} = \left\{ \dfrac{\begin{matrix} 6\ 9\ 11 \\ 4\ 5\ 8 \\ 2\ 3 \end{matrix}}{\begin{matrix} 5\ 4\ 2 \\ 4\ 3\ 1 \\ 2\ 1 \end{matrix}} \right\} = 594.$

$$(5): \ d_{[4,4,3]}(\mathrm{Sp}(6)) = \left\{ \begin{array}{c} \left. \begin{array}{ccccccc} 6\ 7\ 8\ 9 & 0 & 0\ 3\ 4 & 0\ 0\ 0\ 0 \\ 5\ 6\ 7\ 8 + & -1 - 1 - 1 & 0 + 0 & 0\ 0\ 2 \\ 4\ 5\ 6 & -2 - 1 - 1 & & 0 - 1 - 1 \\ \hline 6\ 5\ 4\ 2 \\ 5\ 4\ 3\ 1 \\ 3\ 2\ 1 \end{array} \right. \end{array} \right\} = \left\{ \begin{array}{c} 6\ 7\ 11\ 13 \\ 4\ 5\ 6\ 10 \\ 2\ 3\ 4 \\ 6\ 5\ 4\ 2 \\ 5\ 4\ 3\ 1 \\ 3\ 2\ 1 \end{array} \right\}$$

$$= 2002.$$

5. Calculate the orthonormal bases in the irreducible representation $[1, 1, 0]$ of the $\mathrm{Sp}(6)$ group by the method of the block weight diagram, and then, express the orthonormal bases by the standard tensor Young tableaux in the traceless tensor space of rank two for $\mathrm{Sp}(6)$.

Solution. The Lie algebra of $\mathrm{Sp}(6)$ is C_3. The relations between the simple roots $\mathbf{r}_\mu$ and the fundamental dominant weights $\mathbf{w}_\mu$ are

$$\mathbf{r}_1 = 2\mathbf{w}_1 - \mathbf{w}_2, \qquad \mathbf{r}_2 = -\mathbf{w}_1 + 2\mathbf{w}_2 - \mathbf{w}_3, \qquad \mathbf{r}_3 = -2\mathbf{w}_2 + 2\mathbf{w}_3,$$

$$\mathbf{w}_1 = \mathbf{r}_1 + \mathbf{r}_2 + \frac{1}{2}\mathbf{r}_3, \qquad \mathbf{w}_2 = \mathbf{r}_1 + 2\mathbf{r}_2 + \mathbf{r}_3, \qquad \mathbf{w}_3 = \mathbf{r}_1 + 2\mathbf{r}_2 + \frac{3}{2}\mathbf{r}_3.$$

The highest weight of the representation $[1, 1, 0]$ of $\mathrm{Sp}(6)$ is $\mathbf{M} = (0, 1, 0)$. The dimension of $[1, 1, 0]$ is $d_{[1,1,0]} = (7 \cdot 4)/(2 \cdot 1) = 14$.

The weights equivalent to the highest weight are

$$(0, 1, 0), \quad (1, \overline{1}, 1), \quad (\overline{1}, 0, 1), \quad (1, 1, \overline{1}), \quad (\overline{1}, 2, \overline{1}), \quad (2, \overline{1}, 0),$$
$$(1, \overline{2}, 1), \quad (\overline{2}, 1, 0), \quad (1, 0, \overline{1}), \quad (\overline{1}, \overline{1}, 1), \quad (\overline{1}, 1, \overline{1}), \quad (0, \overline{1}, 0).$$

Therefore, the representation $[1, 1, 0]$ contains one simple dominant weight $(0, 1, 0)$ and one double dominant weight $(0, 0, 0)$.

From the highest weight $(0, 1, 0)$ we have a doublet of $\mathcal{A}_2$ with $(1, \overline{1}, 1)$. From $(1, \overline{1}, 1)$ we have a doublet of $\mathcal{A}_1$ with $(\overline{1}, 0, 1)$ and a doublet of $\mathcal{A}_3$ with $(1, 1, \overline{1})$. From $(1, 1, \overline{1})$ we have a doublet of $\mathcal{A}_1$ with $(\overline{1}, 2, \overline{1})$ and a doublet of $\mathcal{A}_2$ with $(2, \overline{1}, 0)$. From $(\overline{1}, 0, 1)$ we have a doublet of $\mathcal{A}_3$ with $(\overline{1}, 2, \overline{1})$. Although $(\overline{1}, 2, \overline{1})$ comes from two paths, it is still a single weight because it is equivalent to $(0, 1, 0)$. Now, we meet the double weight $(0, 0, 0)$ by applying F_1 to $|(2, \overline{1}, 0)\rangle$ and applying F_2 to $|(\overline{1}, 2, \overline{1})\rangle$. Letting $|(2, \overline{1}, 0)\rangle$, $|(0, 0, 0)_1\rangle$ and $|(\overline{2}, 1, 0)\rangle$ constitute a triplet of $\mathcal{A}_1$, and the other basis state $|(0, 0, 0)_2\rangle$ is a singlet of $\mathcal{A}_1$, we have

$$F_1 |(2, \overline{1}, 0)\rangle = \sqrt{2} |(0, 0, 0)_1\rangle, \qquad F_1 |(0, 0, 0)_1\rangle = \sqrt{2} |(\overline{2}, 1, 0)\rangle,$$
$$F_2 |(\overline{1}, 2, \overline{1})\rangle = a |(0, 0, 0)_1\rangle + b |(0, 0, 0)_2\rangle, \qquad a^2 + b^2 = 2,$$
$$F_2 |(0, 0, 0)_1\rangle = c |(1, \overline{2}, 1)\rangle, \qquad F_2 |(0, 0, 0)_2\rangle = d |(1, \overline{2}, 1)\rangle,$$

where we neglect the first index $(0,1,0)$, which denotes the representation, in the basis state $|(0,1,0),(m_1,m_2,m_3)\rangle$ for simplicity. Applying $E_1F_2 = F_2E_1$ to the basis state $|(\bar{1},2,\bar{1})\rangle$, we have

$$E_1F_2\,|(\bar{1},2,\bar{1})\rangle = \sqrt{2}a\,|(2,\bar{1},0)\rangle$$
$$= F_2E_1\,|(\bar{1},2,\bar{1})\rangle = F_2\,|(1,1,\bar{1})\rangle = |(2,\bar{1},0)\rangle.$$

Thus, $a = \sqrt{1/2}$. Choosing the phase of the state basis $|(0,0,0)_2\rangle$ such that b is real positive, we have $b = \sqrt{2-1/2} = \sqrt{3/2}$. Applying $E_2F_2 = F_2E_2 + H_2$ to $|(0,0,0)_1\rangle$ we have

$$E_2F_2\,|(0,0,0)_1\rangle = c^2\,|(0,0,0)_1\rangle + cd\,|(0,0,0)_2\rangle$$
$$= (F_2E_2 + H_2)\,|(0,0,0)_1\rangle = (1/2+0)\,|(0,0,0)_1\rangle + \sqrt{3/4}\,|(0,0,0)_2\rangle.$$

Choosing the phase of the basis state $|(1,\bar{2},1)\rangle$ such that c is real positive, we have $c = \sqrt{1/2}$ and $d = \sqrt{3/2}$. The remaining basis states and the matrix entries of the lowering operators F_μ can be calculated similarly. The results are given in Fig. 10.1, where only those matrix entries of F_μ which are not equal to 1 are indicated.

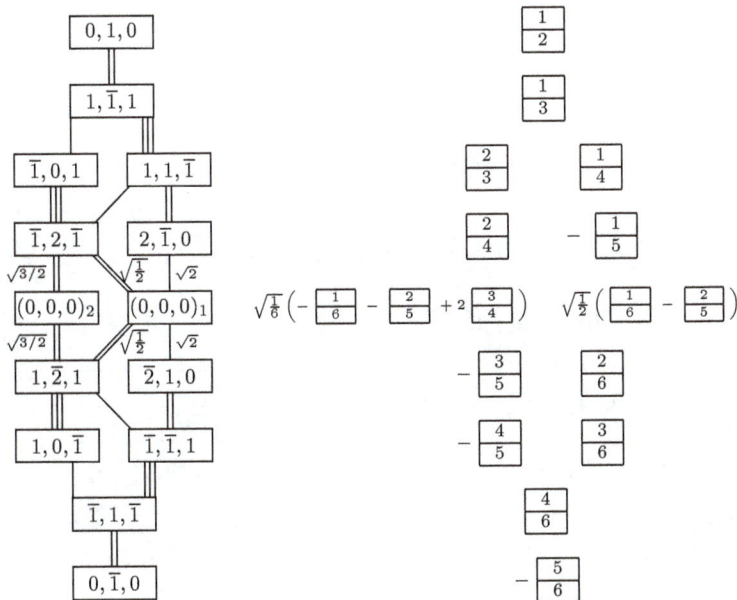

Fig. 10.1   The block weight diagram and the state bases for $[1,1,0]$ of Sp(6).

Now, we expand the orthonormal basis states in the standard tensor Young tableaux. The Young tableau $\mathcal{Y}^{[1,1,0]}$ is $\begin{array}{|c|}\hline 1\\\hline 2\\\hline\end{array}$. A standard tensor Young tableau is expanded as follows:

$$\begin{array}{|c|}\hline a\\\hline b\\\hline\end{array} = \mathcal{Y}^{[1,1,0]}\Phi_{ab} = \Phi_{ab} - \Phi_{ba}. \qquad a < b.$$

There is only one trace tensor in the tensor space of rank two projected by $\mathcal{Y}^{[1,1,0]}$. It belongs to the representation $[0,0,0]$ of Sp(6):

$$|(0,0,0),(0,0,0)\rangle = \begin{array}{|c|}\hline 1\\\hline 6\\\hline\end{array} + \begin{array}{|c|}\hline 2\\\hline 5\\\hline\end{array} + \begin{array}{|c|}\hline 3\\\hline 4\\\hline\end{array}.$$

Beginning with the highest weight state, we calculate the expansions of the basis states in terms of Eq. (10.27) and the block weight diagram. Due to the Wigner-Eckart theorem, the basis state belonging to the representation $[1,1,0]$ of Sp(6) is orthogonal to the basis state belonging to $[0,0,0]$, so it is a traceless tensor.

$$|(0,1,0)\rangle = \begin{array}{|c|}\hline 1\\\hline 2\\\hline\end{array},$$

$$|(1,\bar{1},1)\rangle = F_2|(0,1,0)\rangle = \begin{array}{|c|}\hline 1\\\hline 3\\\hline\end{array},$$

$$|(\bar{1},0,1)\rangle = F_1|(1,\bar{1},1)\rangle = \begin{array}{|c|}\hline 2\\\hline 3\\\hline\end{array},$$

$$|(1,1,\bar{1})\rangle = F_3|(1,\bar{1},1)\rangle = \begin{array}{|c|}\hline 1\\\hline 4\\\hline\end{array},$$

$$|(\bar{1},2,\bar{1})\rangle = F_3|(\bar{1},0,1)\rangle = \begin{array}{|c|}\hline 2\\\hline 4\\\hline\end{array},$$

$$|(2,\bar{1},0)\rangle = F_2|(1,1,\bar{1})\rangle = -\begin{array}{|c|}\hline 1\\\hline 5\\\hline\end{array},$$

$$|(0,0,0)_1\rangle = \sqrt{1/2}F_1|(2,\bar{1},0)\rangle = \sqrt{1/2}\left\{\begin{array}{|c|}\hline 1\\\hline 6\\\hline\end{array} - \begin{array}{|c|}\hline 2\\\hline 5\\\hline\end{array}\right\},$$

$$|(0,0,0)_2\rangle = \sqrt{2/3}\left\{F_2|(\bar{1},2,\bar{1})\rangle - \sqrt{1/2}|(0,0,0)_1\rangle\right\}$$
$$= \sqrt{1/6}\left\{-\begin{array}{|c|}\hline 1\\\hline 6\\\hline\end{array} - \begin{array}{|c|}\hline 2\\\hline 5\\\hline\end{array} + 2\begin{array}{|c|}\hline 3\\\hline 4\\\hline\end{array}\right\},$$

$$|(\bar{2},1,0)\rangle = \sqrt{1/2}F_1|(0,0,0)_1\rangle = \begin{array}{|c|}\hline 2\\\hline 6\\\hline\end{array},$$

$$|(1,\bar{2},1)\rangle = \sqrt{2}F_2|(0,0,0)_1\rangle = -\boxed{\begin{array}{c}3\\\hline5\end{array}}\,,$$

$$|(\bar{1},\bar{1},1)\rangle = F_1|(1,\bar{2},1)\rangle = \boxed{\begin{array}{c}3\\\hline6\end{array}}\,,$$

$$|(1,0,\bar{1})\rangle = F_3|(1,\bar{2},1)\rangle = -\boxed{\begin{array}{c}4\\\hline5\end{array}}\,,$$

$$|(\bar{1},1,\bar{1})\rangle = F_1|(1,0,\bar{1})\rangle = \boxed{\begin{array}{c}4\\\hline6\end{array}}\,,$$

$$|(0,\bar{1},0)\rangle = F_2|(\bar{1},1,\bar{1})\rangle = -\boxed{\begin{array}{c}5\\\hline6\end{array}}\,.$$

6. Calculate the orthonormal bases in the irreducible representation $[1,1,1]$ of the Sp(6) group by the method of the block weight diagram, and then, express the orthonormal bases by the standard tensor Young tableaux in the traceless tensor space of rank two for Sp(6).

Solution. The Lie algebra of Sp(6) is C_3. The relations between the simple roots $\mathbf{r}_\mu$ and the fundamental dominant weights $\mathbf{w}_\mu$ was given in Problem 5. The dimension of $[1,1,1]$ is

$$d_{[1,1,1]}(\text{Sp}(6)) = \frac{\boxed{\begin{array}{c}6\\\hline5\\\hline4\end{array}} + \boxed{\begin{array}{c}1\\\hline1\\\hline-2\end{array}}}{\boxed{\begin{array}{c}3\\\hline2\\\hline1\end{array}}} = \frac{\boxed{\begin{array}{c}7\\\hline6\\\hline2\end{array}}}{\boxed{\begin{array}{c}3\\\hline2\\\hline1\end{array}}} = 14.$$

The highest weight of the representation $[1,1,1]$ of Sp(6) is $\mathbf{M} = (0,0,1)$. The weights equivalent to the dominant weight $(0,0,1)$ are

$$(0,0,1),\quad (0,2,\bar{1}),\quad (2,\bar{2},1),\quad (\bar{2},0,1),$$
$$(2,0,\bar{1}),\quad (\bar{2},2,\bar{1}),\quad (0,\bar{2},1),\quad (0,0,\bar{1}).$$

From the highest weight state $(0,0,1)$ we have a doublet of $\mathcal{A}_3$ with $(0,2,\bar{1})$. From $(0,2,\bar{1})$ we have a triplet of $\mathcal{A}_2$ with $(1,0,0)$ and $(2,\bar{2},1)$. Both the matrix entries of F_2 are $\sqrt{2}$. The weight $(1,0,0)$ is a single dominant weight. The weights equivalent to the dominant weight $(1,0,0)$ are

$$(1,0,0),\quad (\bar{1},1,0),\quad (0,\bar{1},1),\quad (0,1,\bar{1}),\quad (1,\bar{1},0),\quad (\bar{1},0,0).$$

Therefore, all weights in the representation $[1, 1, 1]$ are single. It is easy to draw the block weight diagram for the representation $[1, 1, 1]$. In Fig. 10.2 we give the block weight diagram for the representation $[1, 1, 1]$ where only those matrix entries of F_μ which are not equal to 1 are indicated.

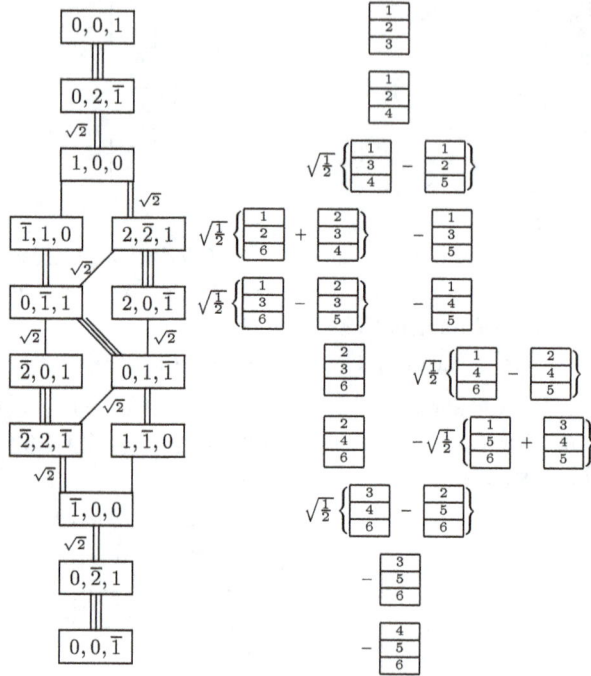

Fig. 10.2   The block weight diagram and the state bases for $[1, 1, 1]$ of Sp(6).

Now, we expand the orthonormal basis states in the standard tensor Young tableaux. The Young tableau $\mathcal{Y}^{[1,1,1]}$ is $\boxed{\begin{matrix}1\\2\\3\end{matrix}}$. The standard tensor Young tableau is

$$\boxed{\begin{matrix}a\\b\\c\end{matrix}} \;=\; \mathcal{Y}^{[1,1,1]}\Phi_{abc} = \Phi_{abc} + \Phi_{bca} + \Phi_{cab} - \Phi_{acb} - \Phi_{cba} - \Phi_{bac}.$$

The trace tensors in the tensor space of rank three projected by $\mathcal{Y}^{[1,1,1]}$ belong to the representation $[1, 0, 0]$ of Sp(6):

$$|(1,0,0),(1,0,0)\rangle = \begin{array}{|c|}\hline 1 \\\hline 2 \\\hline 5 \\\hline\end{array} + \begin{array}{|c|}\hline 1 \\\hline 3 \\\hline 4 \\\hline\end{array}\,,$$

$$|(1,0,0),(\bar{1},1,0)\rangle = -\begin{array}{|c|}\hline 1 \\\hline 2 \\\hline 6 \\\hline\end{array} + \begin{array}{|c|}\hline 2 \\\hline 3 \\\hline 4 \\\hline\end{array}\,,$$

$$|(1,0,0),(0,\bar{1},1)\rangle = -\begin{array}{|c|}\hline 1 \\\hline 3 \\\hline 6 \\\hline\end{array} - \begin{array}{|c|}\hline 2 \\\hline 3 \\\hline 5 \\\hline\end{array}\,,$$

$$|(1,0,0),(0,1,\bar{1})\rangle = -\begin{array}{|c|}\hline 1 \\\hline 4 \\\hline 6 \\\hline\end{array} - \begin{array}{|c|}\hline 2 \\\hline 4 \\\hline 5 \\\hline\end{array}\,,$$

$$|(1,0,0),(1,\bar{1},0)\rangle = \begin{array}{|c|}\hline 1 \\\hline 5 \\\hline 6 \\\hline\end{array} - \begin{array}{|c|}\hline 3 \\\hline 4 \\\hline 5 \\\hline\end{array}\,,$$

$$|(1,0,0),(\bar{1},0,0)\rangle = \begin{array}{|c|}\hline 2 \\\hline 5 \\\hline 6 \\\hline\end{array} + \begin{array}{|c|}\hline 3 \\\hline 4 \\\hline 6 \\\hline\end{array}\,.$$

Beginning with the highest weight state, we calculate the expansions of the basis states in terms of Eq. (10.27) and the block weight diagram. Due to the Wigner-Eckart theorem, the basis state belonging to the representation $[1,1,1]$ of $Sp(6)$ is orthogonal to the basis state belonging to $[1,0,0]$, so it is a traceless tensor.

$$|(0,0,1)\rangle = \begin{array}{|c|}\hline 1 \\\hline 2 \\\hline 3 \\\hline\end{array}\,,$$

$$|(0,2,\bar{1})\rangle = F_3|(0,0,1)\rangle = \begin{array}{|c|}\hline 1 \\\hline 2 \\\hline 4 \\\hline\end{array}\,,$$

$$|(1,0,0)\rangle = \sqrt{1/2}\,F_2|(0,2,\bar{1})\rangle = \sqrt{1/2}\left\{\begin{array}{|c|}\hline 1 \\\hline 3 \\\hline 4 \\\hline\end{array} - \begin{array}{|c|}\hline 1 \\\hline 2 \\\hline 5 \\\hline\end{array}\right\}\,,$$

$$|(2,\bar{2},1)\rangle = \sqrt{1/2}\,F_2|(1,0,0)\rangle = -\begin{array}{|c|}\hline 1 \\\hline 3 \\\hline 5 \\\hline\end{array}\,,$$

$$|(\bar{1},1,0)\rangle = F_1|(1,0,0)\rangle = \sqrt{1/2}\left\{\begin{array}{|c|}\hline 1 \\\hline 2 \\\hline 6 \\\hline\end{array} + \begin{array}{|c|}\hline 2 \\\hline 3 \\\hline 4 \\\hline\end{array}\right\}\,,$$

$$|(0,\bar{1},1)\rangle = \sqrt{1/2}\,F_1|(2,\bar{2},1)\rangle = \sqrt{1/2}\left\{\begin{array}{|c|}\hline 1 \\\hline 3 \\\hline 6 \\\hline\end{array} - \begin{array}{|c|}\hline 2 \\\hline 3 \\\hline 5 \\\hline\end{array}\right\}\,,$$

$$|(2,0,\bar{1})\rangle = F_3|(2,\bar{2},1)\rangle = -\begin{array}{|c|}\hline 1 \\\hline 4 \\\hline 5 \\\hline\end{array}\,,$$

$$|(\bar{2},0,1)\rangle = \sqrt{1/2}\,F_1|(0,\bar{1},1)\rangle = \begin{array}{|c|}\hline 2 \\\hline 3 \\\hline 6 \\\hline\end{array}\,,$$

$$|(0,1,\bar{1})\rangle = \sqrt{1/2}\,F_1|(2,0,\bar{1})\rangle = \sqrt{1/2}\left\{\begin{array}{|c|}\hline 1 \\\hline 4 \\\hline 6 \\\hline\end{array} - \begin{array}{|c|}\hline 2 \\\hline 4 \\\hline 5 \\\hline\end{array}\right\}\,,$$

$$|(\bar{2},2,\bar{1})\rangle = \sqrt{1/2}\,F_1|(0,1,\bar{1})\rangle = \begin{array}{|c|}\hline 2 \\\hline 4 \\\hline 6 \\\hline\end{array}\,,$$

$$|(1,\bar{1},0)\rangle = F_2|(0,1,\bar{1})\rangle = -\sqrt{1/2}\left\{\begin{array}{|c|}\hline 1 \\\hline 5 \\\hline 6 \\\hline\end{array} + \begin{array}{|c|}\hline 3 \\\hline 4 \\\hline 5 \\\hline\end{array}\right\}\,,$$

$$|(\bar{1},0,0)\rangle = F_1|(1,\bar{1},0)\rangle = \sqrt{1/2}\left\{\begin{array}{|c|}\hline 3 \\\hline 4 \\\hline 6 \\\hline\end{array} - \begin{array}{|c|}\hline 2 \\\hline 5 \\\hline 6 \\\hline\end{array}\right\}\,,$$

$$|(0,\bar{2},1)\rangle = \sqrt{1/2}\,F_2|(\bar{1},0,0)\rangle = -\begin{array}{|c|}\hline 3 \\\hline 5 \\\hline 6 \\\hline\end{array}\,,$$

$$|(0,0,\bar{1})\rangle = F_3|(0,\bar{2},1)\rangle = -\begin{array}{|c|}\hline 4 \\\hline 5 \\\hline 6 \\\hline\end{array}\,.$$

7. Calculate the Clebsch-Gordan series for the reduction of the direct product representation $[1,1,0] \times [1,1,0]$ of the Sp(6) group and the highest weight states for the representations contained in the series by the method of the standard tensor Young tableau.

Solution. There are a few methods to calculate the Clebsch-Gordan series of a reducible representation. The method of the standard tensor Young tableau is one of them. The main point in the calculation is to count the multiplicities of the dominant weights in the relevant representations. The difficulty in Sp(2ℓ) group, as well as SO(N) group, is how to remove the trace tensor subspace. In fact, the dimension of the tensor space projected by a Young operator $[\lambda]$ is nothing but the dimension $d_{[\lambda]}(\mathrm{SU}(2\ell))$ of SU(2ℓ) group. Subtracting the dimension $d_{[\lambda]}(\mathrm{Sp}(2\ell))$ of Sp(2ℓ) group, we obtain the dimension of the trace tensor subspace. Conversely, we can calculate the multiplicities of the dominant weights in the representation $[\lambda]$ of Sp(2ℓ) by counting the multiplicities both in the tensor space projected by the Young operator $\mathcal{Y}^{[\lambda]}$ and in the trace tensor subspace. As far as the highest weight state is concerned, the condition that each raising operator E_μ annihilates the highest weight state is useful to calculate the basis state.

First, we count the multiplicities of the dominant weights in the space of the direct product representation $[1,1,0] \times [1,1,0]$ of the Sp(6) group. Since the block weight diagram for the representation $[1,1,0]$ of Sp(6) was given in Fig. 10.1 of Problem 5, it is easy to count the multiplicities of the dominant weights. In the direct product space, there are one basis state with the dominant weight $(0,2,0)$, two basis states with the dominant weight $(1,0,1)$, four basis states with the dominant weight $(2,0,0)$, eight basis states with the dominant weight $(0,1,0)$, and 16 basis states with the dominant weight $(0,0,0)$.

Second, since there is one basis state with the dominant weight $(0,2,0)$ in the direct product space, one representation $[2,2,0]$ is contained in the Clebsch-Gordan series. Its highest weight state is

$$\begin{array}{|c|c|}\hline 1 & 1 \\\hline 2 & 2 \\\hline\end{array} = \begin{array}{|c|}\hline 1 \\\hline 2 \\\hline\end{array} \times \begin{array}{|c|}\hline 1 \\\hline 2 \\\hline\end{array}.$$

The decomposition of the tensor subspace projected by $\mathcal{Y}^{[2,2,0]}$ is

$$[2,2,0] \longrightarrow [2,2,0] \oplus [1,1,0] \oplus [0,0,0], \qquad 105 = 90 + 14 + 1.$$

The tensor subspace contains the basis states with the dominant weights as follows.

Weight   No. State bases

$(0,2,0):$ 1 $\begin{array}{|c|c|} \hline 1 & 1 \\ \hline 2 & 2 \\ \hline \end{array}$,

$(1,0,1):$ 1 $\begin{array}{|c|c|} \hline 1 & 1 \\ \hline 2 & 3 \\ \hline \end{array}$,

$(2,0,0):$ 2 $\begin{array}{|c|c|} \hline 1 & 1 \\ \hline 2 & 5 \\ \hline \end{array}$, $\begin{array}{|c|c|} \hline 1 & 1 \\ \hline 3 & 4 \\ \hline \end{array}$,

$(0,1,0):$ 4 $\begin{array}{|c|c|} \hline 1 & 1 \\ \hline 2 & 6 \\ \hline \end{array}$, $\begin{array}{|c|c|} \hline 1 & 2 \\ \hline 2 & 5 \\ \hline \end{array}$, $\begin{array}{|c|c|} \hline 1 & 2 \\ \hline 3 & 4 \\ \hline \end{array}$, $\begin{array}{|c|c|} \hline 1 & 3 \\ \hline 2 & 4 \\ \hline \end{array}$,

$(0,0,0):$ 9 $\begin{array}{|c|c|} \hline 1 & 2 \\ \hline 5 & 6 \\ \hline \end{array}$, $\begin{array}{|c|c|} \hline 1 & 5 \\ \hline 2 & 6 \\ \hline \end{array}$, $\begin{array}{|c|c|} \hline 1 & 3 \\ \hline 4 & 6 \\ \hline \end{array}$, $\begin{array}{|c|c|} \hline 1 & 4 \\ \hline 3 & 6 \\ \hline \end{array}$, $\begin{array}{|c|c|} \hline 2 & 3 \\ \hline 4 & 5 \\ \hline \end{array}$,

$\begin{array}{|c|c|} \hline 2 & 4 \\ \hline 3 & 5 \\ \hline \end{array}$, $\begin{array}{|c|c|} \hline 1 & 1 \\ \hline 6 & 6 \\ \hline \end{array}$, $\begin{array}{|c|c|} \hline 2 & 2 \\ \hline 5 & 5 \\ \hline \end{array}$, $\begin{array}{|c|c|} \hline 3 & 3 \\ \hline 4 & 4 \\ \hline \end{array}$.

We have to subtract the basis states in the representations $[1,1,0]$ and $[0,0,0]$: one state with $(0,1,0)$ and three states with $(0,0,0)$.

Third, since the difference between the multiplicities of the dominant weight $(1,0,1)$ in the direct product space and in the representation space $[2,2,0]$ is $2-1=1$, one representation $[2,1,1]$ is contained in the Clebsch-Gordan series. Its highest weight state is

$$\begin{array}{|c|c|} \hline 1 & 1 \\ \hline 2 & \\ \hline 3 & \\ \hline \end{array} = \sqrt{\frac{1}{2}}\left\{ \begin{array}{|c|} \hline 1 \\ \hline 2 \\ \hline \end{array} \times \begin{array}{|c|} \hline 1 \\ \hline 3 \\ \hline \end{array} - \begin{array}{|c|} \hline 1 \\ \hline 3 \\ \hline \end{array} \times \begin{array}{|c|} \hline 1 \\ \hline 2 \\ \hline \end{array} \right\}.$$

The coefficients are calculated by the condition that each raising operator E_μ annihilates the highest weight state and the normalization condition. The decomposition of the tensor subspace projected by $\mathcal{Y}^{[2,1,1]}$ is

$$[2,1,1] \longrightarrow [2,1,1] \oplus [2,0,0] \oplus [1,1,0], \qquad 105 = 70 + 21 + 14.$$

The tensor subspace contains the basis states with the dominant weights as follows.

Weight	No.	State bases
$(1,0,1)$:	1	$\begin{array}{cc}1&1\\2\\3\end{array}$,
$(2,0,0)$:	2	$\begin{array}{cc}1&1\\2\\5\end{array}$, $\begin{array}{cc}1&1\\3\\4\end{array}$,
$(0,1,0)$:	5	$\begin{array}{cc}1&1\\2\\6\end{array}$, $\begin{array}{cc}1&2\\2\\5\end{array}$, $\begin{array}{cc}1&2\\3\\4\end{array}$, $\begin{array}{cc}1&3\\2\\4\end{array}$, $\begin{array}{cc}1&4\\2\\3\end{array}$,
$(0,0,0)$:	9	$\begin{array}{cc}1&2\\5\\6\end{array}$, $\begin{array}{cc}1&5\\2\\6\end{array}$, $\begin{array}{cc}1&6\\2\\5\end{array}$, $\begin{array}{cc}1&3\\4\\6\end{array}$, $\begin{array}{cc}1&4\\3\\6\end{array}$,
		$\begin{array}{cc}1&6\\3\\4\end{array}$, $\begin{array}{cc}2&3\\4\\5\end{array}$, $\begin{array}{cc}2&4\\3\\5\end{array}$, $\begin{array}{cc}2&5\\3\\4\end{array}$.

We have to subtract the basis states in the representations $[2,0,0]$ and $[1,1,0]$: one state with $(2,0,0)$, two states with $(0,1,0)$ and five states with $(0,0,0)$. Note that the tensor space for $[2,0,0]$ is traceless and contains the basis states with the dominant weights as follows.

Weight	No.	State bases
$(2,0,0)$:	1	$\boxed{1\;1}$,
$(0,1,0)$:	1	$\boxed{1\;2}$,
$(0,0,0)$:	3	$\boxed{1\;6}$, $\boxed{2\;5}$, $\boxed{3\;4}$.

Fourth, since the difference between the multiplicities of the dominant weight $(2,0,0)$ in the direct product space and in the representation spaces $[2,2,0]$ and $[2,1,1]$ is $4 - 2 - (2 - 1) = 1$, one representation $[2,0,0]$ is contained in the Clebsch-Gordan series. Its highest weight state is

$$\boxed{1\,1} = \frac{1}{2}\left\{ \boxed{\tfrac{1}{2}} \times \left(-\boxed{\tfrac{1}{5}}\right) - \boxed{\tfrac{1}{3}} \times \boxed{\tfrac{1}{4}} + \boxed{\tfrac{1}{4}} \times \boxed{\tfrac{1}{3}} - \left(-\boxed{\tfrac{1}{5}}\right) \times \boxed{\tfrac{1}{2}} \right\}.$$

The coefficients are calculated by the condition that each raising operator E_μ annihilates the highest weight state and the normalization condition. The basis states with the dominant weights contained in the traceless tensor space of $[2,0,0]$ was given.

Fifth, since the difference between the multiplicities of the dominant weight $(0,1,0)$ in the direct product space and in the representation spaces $[2,2,0]$, $[2,1,1]$, and $[2,0,0]$ is $8-(4-1)-(5-2)-1 = 1$, one representation $[1,1,0]$ is contained in the Clebsch-Gordan series. Its highest weight state is

$$\boxed{\tfrac{1}{2}} = \frac{1}{4}\left\{ \sqrt{2}\,\boxed{\tfrac{1}{2}} \times \left[\sqrt{\tfrac{1}{6}}\left(-\boxed{\tfrac{1}{6}} - \boxed{\tfrac{2}{5}} + 2\,\boxed{\tfrac{3}{4}} \right) \right]\right.$$
$$- \sqrt{3}\,\boxed{\tfrac{1}{3}} \times \boxed{\tfrac{2}{4}} + \sqrt{3}\,\boxed{\tfrac{1}{4}} \times \boxed{\tfrac{2}{3}} + \sqrt{3}\,\boxed{\tfrac{2}{3}} \times \boxed{\tfrac{1}{4}}$$
$$\left. - \sqrt{3}\,\boxed{\tfrac{2}{4}} \times \boxed{\tfrac{1}{3}} + \sqrt{2}\left[\sqrt{\tfrac{1}{6}}\left(-\boxed{\tfrac{1}{6}} - \boxed{\tfrac{2}{5}} + 2\,\boxed{\tfrac{3}{4}} \right) \right] \times \boxed{\tfrac{1}{2}} \right\}.$$

The coefficients are calculated by the condition that each raising operator E_μ annihilates the highest weight state and the normalization condition. We have known from Fig. 10.1 of Problem 5 that the traceless tensor space for $[1,1,0]$ contains one basis state with the dominant weights $(0,1,0)$ and two basis states with the dominant weight $(0,0,0)$.

Finally, counting the multiplicities of the dominant weight $(0,0,0)$ in the direct product space and in the representation spaces $[2,2,0]$, $[2,1,1]$, $[2,0,0]$, and $[1,1,0]$, we have

$$16 - (9-3) - (9-5) - 3 - 2 = 1.$$

One representation $[0,0,0]$ is contained in the Clebsch-Gordan series. The expression for its highest weight state is quite long. We write it in the form of the basis states instead of the standard tensor Young tableaux. The latter form can be calculated from Fig. 10.1.

$$||(0,0,0),(0,0,0)\rangle = \frac{1}{14}\left\{|(0,1,0)\rangle|(0,\bar{1},0)\rangle - |(1,\bar{1},1)\rangle|(\bar{1},1,\bar{1})\rangle\right.$$
$$+ |(\bar{1},0,1)\rangle|(1,0,\bar{1})\rangle + |(1,1,\bar{1})\rangle|(\bar{1},\bar{1},1)\rangle - |(\bar{1},2,\bar{1})\rangle|(1,\bar{2},1)\rangle$$
$$- |(2,\bar{1},0)\rangle|(\bar{2},1,0)\rangle + |(0,0,0)_1\rangle|(0,0,0)_1\rangle + |(0,0,0)_2\rangle|(0,0,0)_2\rangle$$
$$- |(\bar{2},1,0)\rangle|(2,\bar{1},0)\rangle - |(1,\bar{2},1)\rangle|(\bar{1},2,\bar{1})\rangle + |(\bar{1},\bar{1},1)\rangle|(1,1,\bar{1})\rangle$$
$$+ |(1,0,\bar{1})\rangle|(\bar{1},0,1)\rangle - |(\bar{1},1,\bar{1})\rangle|(1,\bar{1},1)\rangle + |(0,\bar{1},0)\rangle|(0,1,0)\rangle\left.\right\} \ .$$

In summary, the dominant weight diagram for the direct product representation $[1,1,0] \times [1,1,0]$ is given in Fig. 10.3.

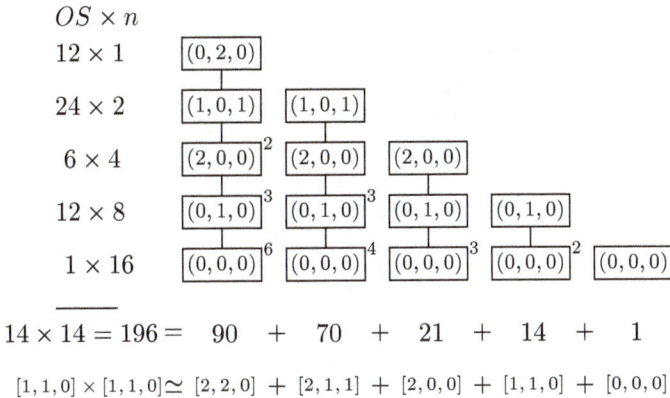

$OS \times n$

12×1	$(0,2,0)$			
24×2	$(1,0,1)$	$(1,0,1)$		
6×4	$(2,0,0)^2$	$(2,0,0)$	$(2,0,0)$	
12×8	$(0,1,0)^3$	$(0,1,0)^3$	$(0,1,0)$	$(0,1,0)$
1×16	$(0,0,0)^6$	$(0,0,0)^4$	$(0,0,0)^3$	$(0,0,0)^2$	$(0,0,0)$

$$14 \times 14 = 196 = \quad 90 \quad + \quad 70 \quad + \quad 21 \quad + \quad 14 \quad + \quad 1$$

$$[1,1,0] \times [1,1,0] \simeq [2,2,0] + [2,1,1] + [2,0,0] + [1,1,0] + [0,0,0]$$

Fig. 10.3 The dominant weight diagram for $[1,1,0] \times [1,1,0]$ of $Sp(6)$.

In the permutation of two factors in the direct product representation space $[1,1,0] \times [1,1,0]$, the basis states in the representations $[2,2,0]$, $[1,1,0]$ and $[0,0,0]$ are symmetric, and the representations $[2,1,1]$ and $[2,0,0]$ are antisymmetric.

Bibliography

Adams, B. G., Cizek, J. and Paldus, J. (1987). Lie Algebraic Methods and Their Applications to Simple Quantum Systems, *Advances in Quantum Chemistry*, **19** Academic Press, New York.

Berenson, R. and Birman, J. L.(1975). Clebsch-Gordan coefficients for crystal space group, *J. Math. Phys* **16**, p. 227.

Biedenharn, L. C. and Louck, J. D. (1981). Angular Momentum in Quantum Physics, Theory and Application, *Encyclopedia of Mathematics and its Application*, **8**, Ed. G. C. Rota, Addison-Wesley, Massachusetts.

Boerner, H. (1963). Representations of Groups, North-Holland, Amsterdam.

Bourbaki, N. (1989). Elements of Mathematics, Lie Groups and Lie Algebras, Springer-Verlag, New York.

Bradley, C. J. and Cracknell, A. P. (1972). The Mathematical Theory of Symmetry in Solids, Oxford: Clarendon Press.

Bremner, M. R., Moody, R. V. and Patera, J. (1985). Tables of Dominant Weight Multiplicities for Representations of Simple Lie Algebras, Pure and Applied Mathematics, A Series of Monographs and Textbooks 90, Marcel Dekker, New York.

Burns, G. and Glazer, A. M. (1978). Space Groups for Solid State Scientists, Academic Press, New York.

Chen, J. Q. (1989). Group Representation Theory for Physicists, World Scientific, Singapore. The second version: Chen, J.Q., Ping, J. L. and Wang, F. (2002). Group Representation Theory for Physicists, 2nd edition, World Scientific, Singapore.

Chen, J. Q. and Ping, J. L. (1997). Algebraic expressions for irreducible bases of icosahedral group, *J. Math. Phys* **38**, 387.

de Swart, J. J. (1963). The octet model and its Clebsch-Gordan coefficients, *Rev. Mod. Phys.* **35**, 916.

Deng Y. F. and Yang, C. N. (1992). Eigenvalues and eigenfunctions of the Hückel Hamiltonian for carbon-60, *Phys. Lett. A* **170**, 116.

Dirac, P. A. M. (1958). The Principle of Quantum Mechanics, Clarendon Press, Oxford.

Dong, S. H., Hou, X. W. and Ma, Z. Q. (1998). Irreducible bases and correlations of spin states for double point groups, *Inter. J. Theor. Phys.* **37**, 841.

Dong, S. H., Xie, M. and Ma, Z. Q. (1998). Irreducible bases in icosahedral group space, *Inter. J. Theor. Phys* **37**, 2135.

Dong, S. H., Hou, X. W. and Ma, Z. Q. (2001). Correlations of spin states for icosahedral double group, *Inter. J. Theor. Phys* **40**, 569.

Duan, B., Gu, X. Y. and Ma, Z. Q. (2001). Precise calculation for energy levels of a helium atom in P states, *Phys. Lett. A* **283**, 229.

Duan, B., Gu, X. Y., and Ma, Z. Q. (2002). Numerical calculation of energies of some excited states in a helium atom, *Eur. Phys. J. D* **19**, 9.

Eckart, C. (1934). The kinetic energy of polyatomic molecules, *Phys. Rev.* **46**, 383.

Edmonds, A. R. (1957). Augular Momentum in Quantum Mechanics, Princeton University Press, Princeton.

Fronsdal, C. (1963). Group theory and applications to particle physics, 1962, *Brandies Lectures*, **1**, 427. Ed. K. W. Ford, Benjamin, New York.

Gel'fand, I. M., Minlos, R. A. and Shapiro, Z. Ya. (1963). Representations of the Rotation and Lorentz Groups and Their Applications, translated from Russian by G. Cummins and T. Boddington, Pergamon Press, New York.

Gel'fand, I. M. and Zetlin, M. L. (1950). Matrix elements for the unitary groups, *Dokl. Akad. Nauk* **71**, 825.

Gell-Mann, M. and Ne'eman, Y. (1964). The Eightfold Way, Benjamin, New York.

Georgi, H. (1964). Lie Algebras in Particle Physics, Benjamin, New York.

Gilmore, R. (1974). Lie Groups, Lie Algebras and Some of Their Applications, Wiley, New York.

Gu, X. Y., Duan, B. and Ma, Z. Q. (2001). Conservation of angular momentum and separation of global rotation in a quantum N-body system, *Phys. Lett. A* **281**, 168.

Gu, X. Y., Duan, B. and Ma, Z. Q. (2001). Independent eigenstates of angular momentum in a quantum N-body system, *Phys. Rev. A* **64**, 042108(1-14).

Gu, X. Y., Duan, B. and Ma, Z. Q. (2001). Quantum three-body system in D dimensions, *J. Math. Phys.* **43**, 2895.

Gu, X. Y., Ma, Z. Q. and Duan, B. (2003). Interdimensional degeneracies for a quantum three-body system in D dimensions, *Phys. Lett. A* **307**, 55.

Gu, X. Y., Ma, Z. Q. and Sun, J. Q. (2003). Quantum four-body system in D dimensions, *J. Math. Phys.* **44**, 3763.

Gu, X. Y., Ma, Z. Q. and Sun, J. Q. (2003). Interdimensional degeneracies in a quantum isolated four-body system, *Phys. Lett. A* **314**, 156.

Gu, X. Y., Ma, Z. Q. and Sun, J. Q. (2003). Interdimensional degeneracies in a quantum N-body system, *Europhys. Lett.* **64**, 586.

Gu, X. Y., Ma, Z. Q. and Dong, S. H. (2002). Exact solutions to the Dirac equation for a Coulomb potential in $D+1$ dimensions, *Inter. J. Mod. Phys. E* **11**, 335.

Gu, X. Y., Ma, Z. Q. and Dong, S. H. (2003). The Levinson theorem for the Dirac equation in $D+1$ dimensions, *Phys. Rev. A* **67**, 062715(1-12).

Hamermesh, M. (1962). Group Theory and its Application to Physical Problems, Addison-Wesley, Massachusetts.

Heine, V. (1960). Group Theory in Quantum Mechanics, Pergamon Press, London.

Hirschfelder, J. O. and Wigner, E. P. (1935). Separation of rotational coordinates from the Schrödinger equation for N particles, *Proc. Natl. Acad. Sci. U.S.A.* **21**, 113.

Hou, X. W., Xie, M., Dong, S. H. and Ma, Z. Q. (1998). Overtone spectra and intensities of tetrahedral molecules in boson-realization models, *Ann. Phys. (N.Y.)* **263**, 340.

Itzykson, C. and Nauenberg, M. (1966). Unitary groups: Representations and decompositions, *Rev. Mod. Phys.* **38**, 95.

Joshi, A. W. (1977). Elements of Group Theory for Physicists, Wiley.

Koster, G. F. (1957). Space Groups and Their Representations in Solid State Physics, Eds. F. Seitz and D. Turnbull, Academic Press, New York, **5**, 174.

Kovalev, O. V. (1961). Irreducible Representations of Space Groups, translated from Russian by A. M. Gross, Gordon & Breach.

Lipkin, H. J. (1965). Lie Groups for Pedestrians, North-Holland, Amsterdam.

Littlewood, D. E. (1958). The Theory of Group Characters, Oxford University Press, Oxford.

Liu, F., Ping, J. L. and Chen, J. Q. (1990). Application of the eigenfunction method to the icosahedral group, *J. Math. Phys.* **31**, 1065.

Ma, Z. Q., Group Theory for Physicists, World Scientific, to be published.

Ma, Z. Q. (1993). Yang-Baxter Equation and Quantum Enveloping Algebras, World Scientific, Singapore.

Ma, Z. Q., Hou, X. W. and Xie, M. (1996). Boson-realization model for the vibrational spectra of tetrahedral molecules, *Phys. Rev. A* **53**, 2173.

Miller, W., Jr. (1972). Symmetry Groups and Their Applications, Academic Press, New York.

Racah, G. (1951). Group Theory and Spectroscopy, Lecture Notes in Princeton.

Roman, P. (1964). Theory of Elementary Particles, North-Holland, Amsterdam.

Rose, M. E. (1957). Elementary Theory of Angular Momentum, Wiley, New York.

Salam, A. (1963). The Formalism of Lie Groups, in *Theoretical Physics*, Director: A. Salam, International Atomic Energy Agency, Vienna, 173.

Schiff, L. I. (1968). Quantum Mechanics, Third Edition, McGraw-Hill, New York.

Serre, J. P. (1965). Lie Algebras and Lie Groups, Benjamin, New York.

Tinkham, M. (1964). Group Theory and Quantum Mechanics, McGraw-Hill, New York.

Tung, W. K. (1985). Group Theory in Physics, World Scientific, Singapore.

Weyl, H. (1931). The Theory of Groups and Quantum Mechanics, translated from German by H. P. Robertson, Dover Publications.

Weyl, H. (1946). The Classical Groups, Princeton University Press, Princeton.

Wigner, E. P. (1959). Group Theory and its Applications to the Quantum Mechanics of Atomic Spectra, Academic Press, New York.

Wybourne, B. G. (1974). Classical Groups for Physicists, Wiley, New York.

Yamanouchi, T. (1937). On the construction of unitary irreducible representation of the symmetric group, *Proc. Phys. Math. Soc. Jpn* **19**, 436.

Index